Companion Encyclopedia
of the
History and Philosophy
of the
Mathematical Sciences

Companion Encyclopedia
of the
History and Philosophy
of the
Mathematical Sciences

Volume 1

Edited by
I. GRATTAN-GUINNESS

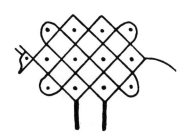

LONDON AND NEW YORK

First published in 1994
by Routledge
11 New Fetter Lane, London EC4P 4EE

Simultaneously published in the USA and Canada
by Routledge Inc.
29 West 35th Street, New York, NY 10001

Printed in Great Britain by
Clays Ltd, St Ives plc
Typeset in 10/12½ Times Compugraphic by
Mathematical Composition Setters Ltd, Salisbury, UK

Printed on acid-free paper

British Library Cataloguing in Publication Data

A catalogue record for this book is available from the British Library

Library of Congress Cataloging-in-Publication Data

Companion encyclopedia of the history and philosophy of the mathematical
sciences / edited by Ivor Grattan-Guinness.
    p.   cm.
Includes bibliographical references and index.
ISBN 0–415–03785–9
    1. Mathematics—History.   2. Mathematics—Philosophy.   I. Grattan-
Guinness, I.
QA21.E57   1992
510'.9—dc20   92–13707
                                                                    CIP

ISBN 0–415–03785–9 (set)
      0–415–09238–8 (Vol. 1)
      0–415–09239–6 (Vol. 2)

Illustration on title page:
An antelope drawn by the (Ts)Chokwe people of Angola in their tradition
of monolinear art; the main block of the pattern is composed of one
continuous line, and is so designed that a regular lattice of dots can be
inserted in the spaces (see P. Gerdes, *Lusona: Geometrical Recreations of
Africa*, 1991, Maputo, Mozambique: Eduardo Mondlare University Press, 15).
On this tradition, see §1.8 by C. Zaslavsky, Section 3.

Contents

Part 0
Introduction

Introduction

I. GRATTAN-GUINNESS

1 THE PLACE OF MATHEMATICS IN LIFE AND SOCIETY

After an unbroken tradition of many centuries, mathematics has ceased to be generally considered as an integral part of culture in our era of mass education. The isolation of research scientists, the pitiful scarcity of inspiring teachers, the host of dull and empty commercial textbooks and the general educational trend away from intellectual discipline have contributed to the anti-mathematical fashion in education. It is very much to the credit of the public that a strong interest in mathematics is none the less alive.

(Courant, foreword to Kline *1953*)

Mathematics occupies a peculiar position in cultural life today. 'Everybody knows' that it is one of the most basic, and also ancient, types of knowledge; yet it is not part of normal cultural discourse, and few people know much about its historical development, or even that it *has* a history. Even among the minority of people endowed with mathematical ability who train in the subject and function within the profession, questions such as 'Which century did Euler live in?' can be embarrassing; and to the rest the answer may well be 'Euler? Who?'.

It is not hard to find reasons why this situation has arisen. Mathematics quickly becomes quite a tricky subject, and beyond the natural capacities of many. Further, at school it comes over as arid and unmotivating as well as difficult, and also seemingly irrelevant to normal day-to-day needs. Thus, for example, innumeracy is normally regarded as a minor inability, though certainly a nuisance; but, unlike illiteracy, it is not usually seen as humiliating or incapacitating. At the same time, however, there remains the feeling that underneath this unpleasant and useless experience there must lurk a fascinating and important subject.

This is the spirit behind the opinion quoted above from Richard Courant, who was a major mathematician of the twentieth century (in Germany and then in the USA). His foreword was to a book which has gained much

3

favour for its presentation of the cultural aspects of mathematics: Morris Kline's *Mathematics in Western Culture*. My task here is somewhat similar: to furnish an introduction to this encyclopedia, which tries to recover our mathematical heritage by presenting an introduction to all the main branches, techniques, cultural traditions and applications of mathematics from ancient to modern times.

The scope of the encyclopedia, therefore, is even broader than that of Kline's book, for it takes in non-Western cultures and also covers a longer time-period and many more topics. For this purpose 133 authors have been assembled, of eighteen nationalities.

2 SCOPE AND CONTENT

2.1 Structure

The encyclopedia is divided into thirteen Parts. Each article is given a decimal number: for example, '§11.5' refers to the fifth article in the eleventh Part. Uniquely, this introduction is designated §0.

Part 1 deals with mathematics in various ancient and non-Western cultures from ancient up to medieval and Renaissance times. 'Non-Western' is used to indicate cultural traditions, of whatever period, that were not dependent on mathematical developments that have taken place over the last five centuries in the West; this latter phrase refers normally to the lands now occupied by Europe and the former Russian Empire, and during recent times also the USA. Western developments dominate the rest of this encyclopedia, for after the Middle Ages mathematics in non-Western countries declined greatly in status, reviving only within the last century or so, and then largely as clones of the mathematics of the West.

Part 2 treats developments in all the main areas of mathematics during the medieval and Renaissance periods, usually up to and including the early seventeenth century. (The question of defining the period and distinguishing the articles of this Part from those of the later ones arises here; it is considered somewhat further in §2.0, but it is best understood from the articles themselves.) After that the great increase in mathematical activity requires that the next eight Parts be divided into the main branches along which mathematics developed from the early seventeenth century onwards: the calculus and mathematical analysis, logic and foundations, algebras, geometries, mechanics and mathematical physics (including engineering aspects), and probability and statistics.

The final three Parts (11–13) review the history of mathematics from other points of view, in some cases worldwide. In turn are considered institutions of higher education and organization in various countries;

manifestations of mathematics in culture, art and society (ending with a review, §12.13, of the historical development of the history of mathematics itself); and general reference, consisting of a bibliography of principal works in (or for) the history of mathematics, a chronology of main events in the development of mathematics, brief information on the authors and on the main figures on whom they have written, and the index.

The details of this design are best examined in the table of contents and in the introduction to each Part; the index is most appropriate for specific matters. I shall now indicate some general aspects of the content.

2.2 The range of mathematics

One of the main lessons to learn from the encyclopedia is the huge variety of problems, theories and techniques of which mathematics is composed. It has encompassed analysing the flow of water in a canal, calculating the path of a comet from observations, introducing statistical tests of significance into medicine, recording the daily passages of the heavenly bodies on an animal horn with a pointed stone, finding properties of special functions in mathematical analysis, formulating the idea of recursion for use in computing, forming predator–prey models for the mathematics of animal behaviour, finding a complete axiom system for integral domains, examining the distribution of prime numbers, developing a theory of capillary flow in order to refine the accuracy of barometers, studying latent roots of matrices, searching for criteria to detect singular solutions to differential equations, interpreting the passage of light in terms of non-Euclidean geometry, and many, many more things.

This encyclopedia tackles or at least touches upon all principal branches and aspects of mathematics as it has developed over the centuries. This point deserves some emphasis, for most so-called general histories of the subject do not even begin to cover this range. Two vast areas of indefensible omission in that literature are worth stressing: the light treatment given to applications, especially from the eighteenth century onwards; and the common ignoring of most of probability and statistics. These points are taken up again in the closing article (§12.13).

2.3 Branches of mathematics

The division of Parts 3–10 of the encyclopedia by major branches of mathematics, and also of topics into articles within a Part, begs historical questions concerning the existence or identification of that branch or topic, as such, in the first place. This form of division of the encyclopedia was adopted partly in order to facilitate the task of each author but especially for the benefit of the reader, whose own understanding of mathematics is likely to follow divisions of this kind. That is, the encyclopedia has a structure determined by its chosen introductory level; the more advanced researcher would learn to ignore it when formulating more sophisticated questions. In their own research neither mathematicians nor historians bind themselves rigidly to these divisions, and indeed in the articles here the authors often refer to broader, or other, contexts. In addition, in appropriate contexts some articles describe divisions of mathematics into (some of) its main branches which were influential at the time in question.

A special word is needed on trigonometry. Today it is a staple but minor part of elementary mathematics; but its place in history is large, too much for a single article. It appears in many places in Part 2, and (with the trigonometric functions) frequently in several other Parts; §4.2 is a short 'overview' which binds these threads together.

2.4 The introductory nature of the articles

The quantity of mathematical literature that has been published over the centuries amazes even historians of mathematics, and so only main lines, or even merely examples of them, can be touched upon in the articles. There are also mounds of relevant manuscripts, not only technical ones, stored in archives (and family vaults!) around the world; while use is made of them in a few important contexts, no detailed discussion of their general role or significance is offered here.

An article attempts only to introduce the history of the topics treated. Sophisticated and important questions such as the various forms of influence of culture upon mathematics, and of mathematicians and branches of mathematics upon each other, are treated only lightly (although they may be treated in detail in the literature cited). Accordingly, footnotes, and the kind of detail they furnish, have been avoided.

2.5 The philosophies of mathematics

The role of the word 'philosophy' in the title of the encyclopedia needs clarifying. Given the range of mathematics surveyed, it would be a mistake to expect to look in this encyclopedia for the 'meaning' of mathematics, as if mathematics were a unified and coherent whole. Yes, there have been numerous interactions, but unity is something over and above that, something mathematics has yet to exhibit. However, it is fair (and in fact historically noteworthy) to record that at various periods some mathematicians have claimed to be able to express the 'whole' of mathematics (as known at the time) within certain principles; such ambitions have been evident especially in certain lines of research on foundations undertaken over the last hundred years and recorded in Part 5. But not only did the developments of mathematics at these various times outstrip the views proposed, in most cases much of the mathematics of the time was in fact not encompassed. Nevertheless, such enterprises were not worthless; fruitful methods and conceptions were produced, even if the main aim was not achieved.

Mathematics generates a wide spectrum of philosophical questions: types of proof and the relationship with logics; the place and purpose of axioms and definitions, and relationships between them; the supposed 'existence' of the objects which mathematics handles; the manner(s) of its application(s) to the physical world, and the supposed distinction between 'pure' and 'applied' mathematics; and so on. The foundational traditions just mentioned treated some of these questions in detail, as is duly recorded in several of the articles in Parts 1 and 5; but many other philosophical questions do not centre upon foundations in that sense, and they are treated in other articles.

2.6 Organization, bibliography and cross-referencing

In order to clarify its layout, each article is divided into numbered and titled sections. Since geometry and visualization have been a major component in mathematics, several articles contain diagrams or illustrations as well as text and formulas.

Each article is written in English, in some cases in translation from another language. Quotations from other languages have been translated into English unless some point is being made about the original phrase (as a name, for example, or as reflecting some intention by the historical figure). Titles of works, however, are often rendered in the original language.

Figures and tables are numbered from 1 in each article. Equations or mathematical expressions are numbered from (1), and if two or more occur on a line, subscripts indicate the one referred to: thus $(3)_2$ is the second expression in (3).

Each article has its own bibliography. It is meant to be read and not just used, for in many cases works are listed there which are not cited in the text. The method of citation in the text is as has already been used at the beginning of this Introduction, namely 'Kline *1953*'. Usually these items are historical books and papers, and occasionally a doctoral dissertation (which is often most easily obtainable from University Microfilms, London and Ann Arbor, Michigan). Space allows only a limited number of original sources to be referenced in full, although many are named in the text.

Some articles cite passages from Classical Greek authors such as

Aristotle; in these cases the so-called 'Bekker' system of citation is used (such as '249b17'), and no further reference is given, as this system is adopted in all standard editions of the works. Titles from Russian, Chinese, Arabic and other languages written in non-roman alphabets have been transliterated (from Chinese, according to the Pinyin system). Spellings used in titles of older works in foreign languages written in the Roman alphabet have been retained as originally printed.

No restriction on language has been placed on the citation of works; one of the beneficial side-effects of working in this field is the need to read several languages. Additional remarks, enclosed within square brackets, are appended to some references.

The number of items in a bibliography varies considerably. The choice is determined partly by the range and variety of topics covered in the article, and by the presence or lack of basic historical writings which contain extensive sets of references of their own.

In this way each article is self-contained. To complement this autonomy, cross-references to related articles are given by citing the designated number. (As the articles are quite short, particular sections are not normally indicated in the cross-references.) This number is carried in the running head at the top of each right-hand page, so that a search should be easily effected. Occasionally the cross-reference is prefaced by an imperative, such as 'compare', to stress that the same context or material is treated there but from a somewhat different point of view.

On the large scale, there is no definitive order in which the articles are to be read, although the order of Parts, and of articles within a Part, has been chosen with some care. I have provided each Part with its own introductory review.

3 LIMITS AND OMISSIONS

Having indicated some of the scope and ambitions of this encyclopedia, the equally important matter of limitations must be explained, over and above the caveat issued above concerning the introductory character of the articles. These limitations are of various different kinds.

3.1 Assumed knowledge

Probably the most important restriction is that this book is not intended to teach the mathematics whose history it records; familiarity with at least the main lines of each topic or theory is assumed. However, an important aspect needs to be stressed, especially for the mathematician reader. While this knowledge is probably known in some fairly modern form, it is to be applied here with historical sympathy. In recording the history of mathematics we try to remain faithful to the intellectual frameworks of past times, and not rethink them into the modern versions (which is how most mathematicians *mis*conceive of the history of mathematics). For example, matrix theory is often avoided because it became established only in the twentieth century (§6.7), and severe anachronism would arise from implanting it into earlier work.

The level of mathematical knowledge supposed varies quite considerably among the articles. The minimum would be upper school level, and applies largely to many articles in Parts 1, 11 and 12; the maximum should be that of courses in the last year of an undergraduate degree course in mathematics (although the topics discussed may themselves be more advanced). For all readers, no knowledge of the history itself is assumed; and particular technical terms or notations are explained, especially if they are not now current. However, there is no glossary of terms used in the book.

3.2 Time-period

The encyclopedia is *not* an introduction to modern mathematics: indeed, such a book (if properly conceived) would form a volume complementary to this one, starting where we usually stop. This point may be stressed in two ways.

First, the terminal period for an article is normally set at some convenient phase between the two world wars (if the topics covered were still current at that time); at most only very general remarks about recent trends are included, usually as part of the concluding section of an article. The only exceptions to this practice occur in articles which treat current topics of considerable philosophical interest (such as non-standard analysis in §3.3); in addition, §5.9 consists of a survey of current views in the philosophy of mathematics in the relatively narrow sense of that phrase described in Section 2.5 above.

Second, by and large the details furnished here about developments in the twentieth century are fewer than for earlier times. One reason is the need to taper off the details as the (normal) terminal period approaches; in addition (and to speak only a little crudely), the work in a given area has usually increased in quantity, specialization and detail in more recent times, thus rendering compact description all the more difficult.

3.3 Other sciences

Many developments in mathematics also involved other sciences (both natural and social): they often furnished the motivation, for example. Not all such sciences are treated here; furthermore, where they are, the articles are always centred on the mathematical aspects, even if these are not a major part of the whole story (for example, in §9.15 the mathematics involved in the origins of quantum mechanics). But the bibliographies of such articles usually cite historical (sometimes original) works where the broader story has been treated.

3.4 Details and aspects not covered

As with all history, especially of the summarizing kind presented here, the focus is on the highlights and the major steps: the essential working out of these ideas in detail does not receive the attention it deserves. Certain topics are too disparate to be successfully handled in articles of the type presented in this encyclopedia – for example notations, rigour, mathematical and scientific instruments (on these see Dyck *1892–3*). Neither are there articles on the deeper philosophical or sociological aspects, such as the influence of schools and groups of mathematicians, the possible occurrence of revolutions in mathematics, and the psychology of invention, although such issues do receive some coverage where appropriate.

In addition, the encyclopedia is not a dictionary in which all the (now) favoured theorems, methods, inequalities, constants, and so on would have an entry. While quite a few of these elements do appear in the book, I have not asked the authors to contrive their articles to be non-random walks around their subject-matter so that every pertinent element is named. One reason is that pressure of space would squeeze out other things, for the history of many of these results is often quite complicated (different forms, rediscoveries, the mathematician(s) after whom they are named not being the best choice, and so on), and they were not necessarily seen as important at the time of their arrival.

3.5 Biographies

There are no biographical articles in the encyclopedia; they would have extended its compass by too large a measure to be practical. Similarly, biographies of mathematicians are cited in the bibliography of an article only if they are of importance for the topics treated there. Forenames or initials are normally provided at the first appearance of the surname; where they are known in several languages (as with the Bernoulli family, for example), English versions are used. Some compensation for this partial information is provided in §13.2, which contains brief biographical details of major mathematicians, and of philosophers and historians of mathematics; years of birth and death are given there (and so do not normally appear in an article, where any dates are those of minor or non-mathematicians), and indications of any principal biographies or editions of works. The best general biographical source currently available is the *Dictionary of Scientific Biography* (Gillispie *1970–80*, and supplements recently published or in progress).

3.6 Notation

It has not been possible to copy or describe all original notations. Several modern conventions are followed, and many special symbols are explained in their context. Where choice is available, the partial derivative of the function z with respect to x is often denoted by z_x. Vectors are denoted by lower- or upper-case bold letters ($\mathbf{v}$ or $\mathbf{R}$, say), and matrices and tensors by capitals such as $\mathbf{M}$ and T, respectively.

4 HISTORY OF MATHEMATICS AND EDUCATION

The famous logician Kurt Gödel is credited with the following comment. When asked, 'What do you think is the main task of mathematicians?', he gave a brilliant answer: 'To see whether or not a theorem is interesting.' A mass of questions is raised by this answer, concerning not only mathematics as such but also its historical and pedagogical aspects.

The theme of education deserves special discussion from other points of view. As is discussed in more detail in §12.13, much of the recent increase of interest (at all educational levels) in the history of mathematics has been stimulated by reactions, usually negative ones, of mathematics students and professional mathematicians to the normal style of syllabus, lecture and examination. A workshop on the matter was held in Toronto in 1983, publishing its proceedings as Grattan-Guinness *1987*. There has also been formed an International Study Group on the Relations Between the History and Pedagogy of Mathematics, which issues a regular newsletter worldwide via a network of national representatives.

According to the view adopted by the advocates of a role for history, the greatest sin of the normal tradition of mathematics education is to forget completely that mathematics is thought up by people, and for a reason; yet it is still normal for textbooks to disgorge 'perfectly' formed definitions and theorems to the student reader with impressive cleverness, which in fact proceed from nowhere via nobody to arrive at nothing but more sophisticated versions of themselves: the answers may be complete, but the questions are non-existent. 'The traditional method of confronting the student not with the problem but with the finished solution,' opined Arthur Koestler, 'means depriving him of all excitement, to shut off creative impulse, to reduce the adventure of mankind to a dusty heap of theorems' (*1966*: 261).

But such views remain unorthodox among the communities in mathematical education. The normal view is that educators use the history of mathematics only to help the less talented pupils and students, recalling nostalgia for times long gone or telling little anecdotes (without, however, exploiting the power of story-telling). By contrast, we hope that this encyclopedia will prove of especial value in these contexts.

The account of the history of mathematical education in Part 11 (and in §2.11 for the medieval and Renaissance periods) is largely confined to the university level, or whatever equivalent was operating in the countries and times described. School-level education is treated very lightly, partly because most of its (enormous) history has been too poorly studied for reliable summaries to be given in the appropriate articles, and also because the majority of the mathematics described lies closer to the higher levels. Thus the concerns of histories such as Yeldham *1936* with the teaching of arithmetic, or the development of elementary methods of computation or of compilation of tables, or of book-keeping (Brown *1905*), are not given full place in this encyclopedia, notwithstanding the limitations discussed earlier. However, instances where educational issues have stimulated research – and there have been many significant occurrences of this process – are duly recorded where appropriate.

5 SOURCES OF INFORMATION

My review (§12.13) of the history of the history of mathematics contains some remarks on the available literature in the field. It is followed by a detailed bibliography (§13.1) of principal reference works: not only principal general histories of mathematics and journals in the field, but also large-scale handbooks and dictionaries of science and of knowledge in general. It is useful to give here some initial 'feel' of the literature and the bibliographical problems attending it.

For bibliographies of original ('primary') literature, J. D. Reuss's *Reportorium*, J. C. Poggendorff's *Handwörterbuch*, *The Royal Society Catalogue of Scientific Papers* and *The International Catalogue of Scientific Literature* are irreplaceable for the periods and types of item which they cover. Note also that §11.12 describes the development of journals in mathematics.

The historical ('secondary') and philosophical literature is much larger than one might guess (as with mathematics itself), but the collective coverage is far from exhaustive (especially in applied and engineering mathematics), and the quality of writing is extraordinarily variable. The bibliographies May *1973* and Dauben *1985* are essential sources, as is the *Dictionary of Scientific Biography*, already mentioned in Section 3. For past and current work, the best available sources are the abstracts department of the journal *Historia mathematica* and the appropriate sections of the review journals *Mathematical Reviews* and *Zentralblatt für Mathematik* (mainly Sections 00 and 01), and the annual 'critical bibliography' of the history of science journal *Isis* (including the cumulative bibliographies published in *Isis* since 1913, which are being reissued as accumulated volumes).

Among the secondary items themselves, some of the major encyclopedias of the past have contained very substantial and even authoritative articles on branches of mathematics: of special merit are the various editions of the *Encyclopaedia Britannica* up to the 13th edition of the 1920s, and the *Encyclopaedia Metropolitana* (which was completed in 1845). The famous *Encyclopédie* (1751–65) of Denis Diderot and Jean d'Alembert is of a different category; its historical importance lies in the positions adopted in its articles and their subsequent influence.

Above all I must mention the *Encyklopädie der mathematischen Wissenschaften*, a vast German project published between 1898 and 1935 in many thick volumes covering the six main areas into which mathematics was then divided. While many of the articles are reports for their time rather than histories, as the articles in this encyclopedia are intended to be, the mass of information provided is quite incredible, and the work as a whole will never be superseded. A revised and extended version of the *Encyklopädie* was begun by the French in the early 1900s, with the title *Encyclopédie des sciences mathématiques*; in several articles the additions were substantial, and a few new ones were commissioned. However, the project petered out during the First World War after the death of its general editor, J. Molk, with some articles published only in part; thus it never gained the publicity it deserved. Individual articles in those encyclopedias are cited in some articles here; I give a general recommendation to both of them for nineteenth-century developments in those branches of mathematics covered in Parts 3, 4 and 6–9. However, for Parts 5 and 10 (respectively on the foundations of mathematics and on probability and statistics), the treatments are rather light – which is an interesting point.

6 ACKNOWLEDGEMENTS

The idea for this encyclopedia came from Jonathan Price, then of the house of Routledge, and I am grateful to him and his colleagues for the courage of taking this unusual initiative. Of their staff, Alison Barr, Shân Millie, Emma Waghorn and Michelle Darraugh are especially to be thanked for their attention to the numerous details of administration. The intricate and burdensome task of copy-editing was accomplished with finesse by John Woodruff. I am also grateful to John Greenberg, Angela Hopkins, Albert Lewis, George L. Baurley and Pamela Marwood, who translated some articles from French or from German. To various authors and publishers thanks are extended for permission to reproduce copyright material (mainly figures). The vast load of secretarial work was shared by my wife.

Succour was received from the Commission for the History of Mathematics of the International Union of the History and Philosophy of Science, which readily gave its blessing to the project. The work of this Commission over the past two decades has done much to promote the field.

When the decision had been taken to go ahead, the first requirement of the benighted editor was to appoint an advisory editorial board. I am indebted to my colleagues on this board for advice on the design of the Parts and on the choice of authors, and for help in reviewing the articles. To the authors themselves I am appreciative not only of their articles but also of the advice they both sought and gave on their own and others' pieces, and the encyclopedia overall. In addition, for advice on the transliteration of Chinese characters the assistance of J. C. Martzloff was essential.

For myself and also for many of his other colleagues, there is great sadness that Eric Aiton did not live to see the completion of the encyclopedia. Until his death in February 1991, he carried out his duties as advisory board member and as author with all his normal promptitude and mastery. The later death, in March 1992, of Seymour Chapin deprived us of another leading historian of planetary physics.

This is the first book of its kind and scale to be devoted to the history of mathematics; maybe this has been the right time for such an enterprise to be undertaken, and for our mathematical heritage to be recovered to some extent. This mood of self-reflection is resumed in my concluding article, §12.13.

BIBLIOGRAPHY

I reiterate from Section 5 that this bibliography lists only items that have been cited in the text, and that §13.1 contains a survey of the historical literature.

Brown, J. (ed.) *1905, A History of Accounting and Accountants*, Edinburgh: Jack.
Dauben, J. W. (ed.) *1985, The History of Mathematics from Antiquity to the Present. A Selective Bibliography*, New York: Garland. [Excludes educational aspects.]
Dyck, W. (ed.) *1892–3, Katalog mathematischer und mathematisch-physikalischer Modelle, Apparate und Instrumente*, 2 vols, Munich: Wolf.

Gillispie, C. C. (ed.) *1970–80, Dictionary of Scientific Biography*, 16 vols, New York: Scribner's. [Succeeding volumes appearing and to continue.]

Grattan-Guinness, I. (ed.) *1987, History in Mathematics Education*, Proceedings of a Workshop held at the University of Toronto, Canada, July–August 1983, Paris: Belin. [Published under the auspices of the Société Française d'Histoire des Sciences et des Techniques.]

Kline, M. *1953, Mathematics in Western Culture*, New York: Oxford University Press. [Various reprints.]

Koestler, A. *1966, The Act of Creation*, London: Hutchinson.

May, K. O. *1973, Bibliography and Research Manual in the History of Mathematics*, Toronto: University of Toronto Press.

Yeldham, F. *1936, The Teaching of Arithmetic Through Four Hundred Years*, London: Harrap.

Part 1
Ancient and non-Western traditions

1.0

Introduction

Although the West has dominated the development of mathematics for around five centuries, the subject has far more ancient roots in some non-Western cultures (using this adjective as explained near the beginning of §0), and many of its basic theories owe their origins to these lines. This point is a principal concern of this Part, which concentrates largely on non-Western mathematics.

The order of the articles is partly determined by chronology, beginning with some of the most ancient cultures and proceeding to those such as the Mayan and the Jewish (§1.14–1.15), in which the principal achievements were during the Middle Ages (as Westerners conceived it) and the Renaissance. In modern geographical terms, the countries of the Middle East,

Greece and Africa are taken first, and then the Far East (Chinese characters are transcribed according to the Pinyin system). The two articles on number systems and fractions (§1.16–1.17) are placed at the end as they treat cross-national and cross-cultural topics; thus they review earlier articles to some extent. Some articles in Part 12 are also pertinent.

It has not been possible to treat every culture here. In some cases the history is too obscure to be summarized (as, for example, with Celtic methods; see Bain *1951*).

From the modern point of view, the mathematics itself is usually arithmetic and geometry, with early algebra and/or trigonometry and their uses in mechanics and astronomy. Astronomy was particularly important, with consequences for cosmology (often religiously conceived); no attempt has been made to treat these features in detail, as they were only secondarily mathematical. Within arithmetic, many notations were introduced; Cajori *1928* gives a good selection of them.

In recent decades there has been considerable growth in the study of non-Western science in general, and some specialist journals are devoted to it (for example, the *Journal of the History of Arabic Science*). In addition, *Archaeoastronomy* considers ancient mathematics of all cultures when appropriate.

Although note is taken of the arrival of Western mathematics in non-Western countries, no account is given of the more modern developments in those countries, and indeed in the modern period none of them became a major mathematical centre (the closest candidate would be Japan). Thus there are no successor articles in Part 11, which deals with institutional developments since the seventeenth century. However, contributions from individual mathematicians of these countries – at *any* period – are noted in the later Parts as they arise.

BIBLIOGRAPHY

Bain, G. *1951*, *Celtic Art: The Methods of Construction*, Glasgow: MacLellan. [Many reprints, most recently London: Constable.]

Cajori, F. *1929*, *A History of Mathematical Notations*, Vol. 1, Chap. 2, Chicago, IL: Open Court.

1.1

Babylonian mathematics

JENS HØYRUP

1 INTRODUCTION

Babylonian mathematics was, in its origin, an offspring of early state organization. It was transmitted by scribes and basically used for practical computation. Yet Babylonian mathematics was more than a set of practitioners' recipes. First, Babylonian calculators knew *what* they were doing and *why* they were doing it. Second, they produced a level of complex, 'pure' (i.e. not practically relevant) problems and pertinent techniques, especially in the field of algebra.

For lack of sources, only the mathematics of the Old Babylonian and the Seleucid periods were discussed in the literature until about 1970. Since then, however, a number of texts have been discovered which permit one to outline tentatively the development of Babylonian mathematics from the beginnings until the Late Babylonian and Seleucid periods.

2 PROTO-LITERATE AND SUMERIAN BEGINNINGS

From the eighth millennium BC, a system of arithmetical recording or accounting based on small clay tokens was used in the Near and Middle East. In the late fourth millennium, this system appears to have inspired the development both of writing and of numerical and metrological notations. Furthermore, a trend towards the harmonization of the various metrological systems emerged: the system of area measures (originally based on unconnected natural units) was keyed to the linear system, sub-unit metrologies were created, and so on.

Proto-literate mathematics was created for the purposes of practical administration in a 'redistributive economy' directed by a Temple institution; the replacement of 'natural' units by mathematically coherent metrologies corresponds to the needs of the planning and accounting official rather than to those of the immediate producer. But the complexity of the system appears to go beyond bureaucratic needs. The immediate cause of the reorganization of a bundle of arithmetical techniques as coherent mathematics seems to have been the teaching in the Temple school.

The early administration seems not to have distinguished bureaucratic from other priestly functions. Only from around the mid-third millennium is the term for 'scribe' found in the sources. At this time we also encounter non-bureaucratic uses of the professional tools of the scribes: literary texts and mathematical exercises beyond the context of daily administration, the exercises dealing with, for example, the division of extremely large round numbers by 'irregular' divisors like 7 and 33. Even though such problems would have played no significant role in practical administration, they were evidently a central concern to a scribal profession testing its own intellectual abilities.

The trend towards increasing regularization culminated under the 'third dynasty of Ur' ('Ur III', the twenty-first century BC). This regime made extensive use of systematic and extremely meticulous book-keeping. The sexagesimal place-value system was probably created for use in this context. Mathematical school exercises pointing beyond the administrative domain have not been found; parallels in other cultural domains suggest, furthermore, that the centralized state had drained away the resources for scribal intellectual autonomy and thus blocked further development of non-utilitarian mathematics.

The first survey of third-millennium mathematics was by Powell *1976*; recent discoveries concerning the earliest period are presented by Nissen *et al. 1991*; the mathematical texts from the peripheral Ebla area are discussed by Friberg *1986*.

3 MATURITY AND DECLINE

While Ur III mathematics appears to have been strictly utilitarian in orientation, non-utilitarian mathematics was central to Old Babylonian (OB) mathematics. This is the phase in the development of Babylonian mathematics which is best documented in the sources (1900 to 1600 BC, mainly the second part of this period). In this period, which was characterized by a highly individualized economy (compared to other Bronze Age cultures) and by an ideology emphasizing the individual as a private person, the scribal school developed a curriculum which stressed virtuosity beyond what was practically necessary; the triumphs of Babylonian 'pure' mathematics, not least the 'algebra', appear to be a product of precisely this OB scribal school and scribal culture.

Until Ur III, all mathematical texts had been in Sumerian; even in Semitic-speaking Ebla, Sumerian mathematics was taken over in the original language. On the contrary, OB mathematics was written in

Akkadian – supplementary evidence that it represents a new genre and a break with the (plausibly more purely utilitarian) Ur III tradition. Many texts, it is true, were written predominantly by means of word-signs of Sumerian descent; all but a handful of these word-signs, however, are simply elliptic representations of Akkadian words and sentences.

Many mathematical tablets from the OB period onwards are compilations, containing a variety of problems. Often, utilitarian and 'pure' problems are found together; but mathematical and non-mathematical matters are not treated in the same texts. Obviously OB mathematics was not divided into fully distinct disciplines yet mathematics as a whole was an autonomous concern, perhaps even – in the form of engineering, surveying or accounting, or as a teacher's speciality – a distinct vocation.

In 1600 BC, conquest by a warrior people put an end to the OB social order, to the age-old scribal school, to the characteristic OB scribal ideology – and at the same time to the characteristic form of OB mathematics. Scribal training was from now on based on apprenticeship within the scribal 'family'; to a certain degree, mathematics came to be mixed up with other subjects on the same tablets, having lost its disciplinary autonomy; and the 'mathematician' would from now on identify himself in the colophons of tablets as, for example, 'exorcist' (āšipu) or 'priest' (šangû).

For the first centuries after the conquest, mathematical texts are virtually non-existent; a few Late Babylonian mathematical tablets have been discovered recently. In the Seleucid era (311 BC onwards), the development of computational astronomy gave rise to a renaissance of numerical computation and, as a consequence, of some of the old interest in 'pure' problems.

4 NUMBERS AND PRACTICAL COMPUTATION

As has been said, Babylonian 'mathematics' means 'computation'. In intermediate calculations, it made use of the sexagesimal place-value system. The use of this system, and the conversion between metrological values and 'pure numbers', called for extensive use of mathematical, metrological and technical tables. The first group encompasses tables of multiplication and of reciprocals (m/n was calculated as $m \cdot 1/n$); tables of n^2 and $\sqrt{n}$, of n^3 and $n^3 + n^2$ and their roots; and tables of generalized composite interest a^n. The second group contains tabulated conversions of metrological values into sexagesimal multiples of the basic unit, rather like a list for pre-decimal British currency expressing shillings as fractions of a pound: 1s = £0.05, 2s = £0.1, 3s = £0.15, and so on. Finally, technical tables give constants to be used in technical computation (see also §8.15 on cartography).

Babylonian utilitarian mathematics was built on these tables. Multiplication tables, tables of reciprocals and metrological tables were aids for calculation, and the technical tables constituted the nexus between mathematical computation and administrative and engineering reality. Mathematics was taught in schools because scribes should be able to calculate the areas of fields and the volume of canals to be dug out and siege ramps to be built (and, not least, the manpower needed for that). The methods used were quite close to ours, with one important exception: the Babylonians had no concept of quantifiable angle and hence nothing similar to trigonometry. In practical mensuration, they would divide complicated fields into *practically right* triangles, *practically right* trapeziums and *practically rectangular* quadrangles (distinguishing, we might say, a 'right' from a 'wrong' angle); they would then calculate as we do, knowing that their results were not precise but apparently without any definite idea about the nature and size of the errors. Presumably, they would see no decisive difference between the imprecision of manpower calculations and those of area determinations.

With these qualifications, the Babylonians knew the area of a right triangle. Similarly, they would find correctly the area of a rectangle and of a trapezium considered 'right'. The area of an irregular quadrangle might be found by means of the 'surveyors' formula', average length times average width. In practical mensuration, this technique was probably used only for fairly regular quadrangles, for which it gives acceptable results. In school texts it was also used as a pretext for formulating algebraic problems in cases where it is extremely unrealistic. The circular area was found as one-twelfth times the square of the circumference (corresponding to $\pi = 3$), and the circumference as three times the diameter. (In spite of widespread assertions, $\pi = 3\frac{1}{8}$ was probably *not* used.)

Prismatic and cylindrical volumes were calculated as base times 'height' (i.e. a side approximately perpendicular to the base). The volume of a truncated cone was found as that of a cylinder with the average diameter, and that of a truncated pyramid in one text as height time average base (in another text perhaps correctly). When in doubt, once again, the Babylonians would opt for a (rather arbitrary) compromise instead of giving up in the face of theoretical difficulties.

5 RECIPE OR IMPROVISATION? PURE OR APPLIED? ALGEBRA OR GEOMETRY?

A specifically Babylonian type of geometric problem is the partition of areas. Initially, this may have been a practical problem. No later than the twenty-third century BC, however, it turns up as a 'pure' problem: what is the length of the transversal if a trapezium is bisected by a parallel transversal? In the OB period, complex problems of a similar kind are common, as are also a number of other more or less complex and more or less artificial division problems.

Many practical computations, of course, were concerned not with mensuration but with quantities of grain to be levied as dues, with commercial exchange, and so on. The techniques used can be illustrated by paraphrasing an illustrative problem: two fields I and II are given, from one of which 4 *gur* (1 *gur* = 300 *qa*, 1 *qa* ≈ 1 litre) of grain are to be levied per *bur* (1 *bur* = 1800 *sar*, 1 *sar* ≈ 36 m^2), while the other yields a rent of 3 *gur* per *bur*. The total yield and the difference between the two areas are given. First everything is converted into sexagesimal multiples of the fundamental units *sar* and *qa*, in part through calculation, in part by means of a metrological table. The yield of that part of field I which exceeds field II is found. The remainder of the yield must then be computed from the remaining area, which is composed of equal portions from field I and field II. The yield of one 'average *sar*' (half from each field) is found, this is divided into the remaining yield, giving the remaining area, and so on.

The idea behind the last step seems to be the 'single false position' also known from other Babylonian texts: had the remaining area been 1 *sar* it would consist of $\frac{1}{2}$ *sar* from each field, which permits the yield to be found as (say) *p qa*. In reality it is (say) *Q qa*, and therefore the remaining area must be *Q/p sar*.

The procedure gives an impression (confirmed by many other texts) of improvisation, built on concrete thought rather than on standardized techniques when it goes beyond the most basic methods (e.g. conversions). The same feature is also found in OB second-degree and higher 'algebra', perhaps the most astonishing accomplishment of the Babylonian mathematical tradition. 'Algebra' is placed in quotes because it is founded neither on symbols, like post-Renaissance algebra (§2.3), nor on words for unknown numbers, as in medieval Indian, Islamic and Italian algebra. Instead, it builds on 'naive' geometry: where modern algebra presents us with a problem $x^2 + x = A$ (which may be transformed into $x.(x + 1) = A$), the Babylonians would consider a geometric rectangle whose length is known to exceed its width by 1, and whose area is known to be A; where we transform the equation in order to isolate x the Babylonians would make corresponding cut-and-paste transformations of the rectangle. The way they did it would be intuitively obvious, and they would provide no formal proof that the procedure was correct (hence the term 'naive').

The basic transformations, such as the cutting up of rectangles, were made according to fixed schemes. But in the reduction of complex problems to simple ones, the Babylonians could call upon a repertoire of tricks, but no standard procedures – precisely as they did in arithmetical problems. Used with intelligence, OB 'algebra' is therefore highly flexible: as long as one sticks to problems in one or two variables and no higher than the second degree, it is almost as flexible as (and in its sequence of operations very similar to) modern symbolic algebra. Only in more complex cases (from which the Babylonians did *not* abstain) do the disadvantages of their techniques become manifest.

The geometry of OB 'algebra' dealt with concrete, measurable line segments and areas. But the technique was currently applied to non-geometric quantities. Where we would represent an unknown weight or an unknown price by a pure number x, the Babylonian would represent it by a line segment of unknown (but numerically knowable) length. Naive-geometric 'algebra' was an all-round way of finding unknown quantities involved in complex relations (only artificial relations, it is true: Babylonian scribal practice generated no real-life problems of the second or higher degree; these had to be and were constructed in order to allow the display of scribal virtuosity).

The character of OB 'algebra' and related techniques, in particular the philological and structural reasons for replacing the conventional arithmetical interpretation by a geometric reading, is discussed elsewhere (Høyrup *1990*).

A final important OB problem type is made up of numerical investigations. Some of these are connected to the computation of reciprocals, and hence to the needs of common computation. Others are inspired by the partition of the trapezium mentioned above, and lead to indeterminate problems for pairs or sets of numbers. The most famous of all such texts is the tablet Plimpton 322, which bears a table making use of Pythagorean triples (satisfying Pythagoras's theorem, as we call it). An overview of the various interpretations of this tablet has been given by Friberg *1981*.

6 A SPECIFIC MODE OF THOUGHT

Any mathematical corpus of knowledge is organized in a way which reflects its purposes, the modes of thought behind it and the underlying cognitive style. This is true of Babylonian mathematics also. A general characteristic is the dominance by methods, not problems. At the first, utilitarian level this betrays the fact that we know Babylonian mathematics from school texts which served to train future scribes in the methods of their profession. At the 'pure' level, however, we find the same primacy of methods, whereas (ideally speaking) Greek and modern pure mathematics take problems as their starting-point and develop the concepts and methods needed to surmount them. The difference can be found in the particular rationale of Babylonian 'pure' mathematics: not the acquisition of insight, but the display of professional virtuosity. This is also why it flourished in the OB era, but disappeared from the archaeological horizon with the death of the scribal school.

Even though Babylonian mathematics was governed by methods, it was taught through problems serving as paradigmatic examples. Only in two or three texts are these used as the basis for some sort of more general discussion of the method used (although oral teaching would probably have done so more often). Only in a couple of texts from Greek-ruled Uruk are rules formulated in the abstract.

It is in this feature that Babylonian mathematics is comparable to the make-up of Babylonian legal texts like the Codex Hammurapi. 'Hammurapi's law' is no law-book in the likeness of Roman law. It is a collection of legal decisions made by the king, but put together only because the royal decisions were supposed to serve as paradigms for the judges of the realm. One may also compare the listing of hundreds of separate cases in Babylonian 'omen science'.

One could say that Babylonian thought was more concrete and less inclined to abstraction than is the modern mind. But the systematization of the omen literature does suggest the presence of an underlying implicit abstraction, despite its origin in magic; and at least OB mathematics goes still further in this direction.

7 THE MATHEMATICS OF ASTRONOMY

Beginning after the fall of the Assyrian Empire and culminating during the Hellenistic era, Babylonian astrologer-priests created a highly sophisticated predictive planetary astronomy. We know nothing about possible physical models behind the techniques, but the mathematical methods which were used suggest that physical models and ideas were irrelevant.

Fundamental to the 'mathematics of astronomy' was the use of arithmetical schemes. On the basis of long sequences of observations (each of which need not have been particularly precise), and the assumption that the changing solar, lunar and planetary velocities could be adequately represented by 'step functions' (i.e. periodical, with stepwise constant values) or by 'linear zigzag functions' (equally periodical, and linearly ascending and descending), the Babylonian astronomers were able to predict planetary positions and eclipses with high accuracy. Since even the Sun's motion (which needed to be taken into account in every conversion of an astronomical to a calendar date) was known to vary − in one scheme is was assumed to be $30°$ per mean synodic month between longitudes $167°$

and 357° and 28;7,30° (in Neugebauer's sexagesimal place-value notation, that is $28 + 7/60 + 30/60^2$ degrees) for the rest of the year, in another system it was supposed to vary linearly between 28;10,39,40° and 30;1,59° – predictions of real phenomena would require several step or linear zigzag functions. Moreover, the time required for, say, the Sun to travel some distance along the ecliptic corresponding to a constant or a regularly increasing rate would not be an integer number of months. Combining the schemes for a planet or the Moon with that of the Sun in order to find the date and hour of a conjunction or a possible eclipse would thus require complex interpolation strategies, mostly linear but at times of the second or third degree.

The basic ideas behind these mathematical techniques can be traced back to OB mathematics, in particular to an interest in arithmetical progressions and to the solution of simple problems of proportionality (such as the 'grain problem' described in Section 5). However, while the simple numerical schemes of OB mathematics would require nothing more than the use of standard tables (if any at all), the astronomical calculations called for the tabulation of multi-place reciprocals, which were indeed developed in the same period, and which can have had no practical purpose other than astronomical computing (Neugebauer *1967, 1975*).

8 SOURCE COLLECTIONS AND GENERAL LITERATURE

The main source collections for OB and Seleucid mathematics have been published by Neugebauer *1935–7*, Thureau-Dangin *1938*, Neugebauer and Sachs *1945*, and Bruins and Rutten *1961*. Neugebauer *1955* also published the main source collection for mathematical astronomy.

The best overviews of Babylonian mathematics as known until 1975 are in Vogel *1959* and, especially, Vaiman *1961*. A more popular introduction is by van der Waerden (*1962*: 37–45, 62–81). Høyrup (*1985*: 7–17) discusses the interplay between the scribal profession, schooling and mathematical thought. A global overview including many of the recent discoveries has been published by Friberg *1990*, who has also written a recommendable selective bibliography (*1985*).

A concise but informative overview of the techniques of mathematical astronomy was Neugebauer's Henry Norris Russell lecture (*1967*).

BIBLIOGRAPHY

Bruins, E. M. and Rutten, M. *1961*, *Textes mathématiques de Suse* (Mémoires de la Mission Archéologique en Iran, Vol. 34), Paris: Paul Geuthner.
Friberg, J. *1981*, 'Methods and traditions of Babylonian mathematics. Plimpton

322, Pythagorean triples, and the Babylonian triangle parameter equations', *Historia mathematica*, **8**, 277–318.

—— *1985*, 'Babylonian mathematics', in J. W. Dauben (ed.), *The History of Mathematics from Antiquity to the Present. A Selective Bibliography*, New York and London: Garland, 37–51.

—— *1986*, 'The early roots of Babylonian mathematics: III. Three remarkable texts from ancient Ebla', *Vicino Oriente*, **6**, 3–25.

—— *1990*, 'Mathematik', in *Reallexikon der Assyriologie und vorderasiatischen Archäologie*, Berlin: de Gruyter, 531–85.

Høyrup, J. *1985*, 'Varieties of mathematical discourse in pre-modern socio-cultural contexts: Mesopotamia, Greece, and the Latin Middle Ages', *Science and Society*, **49**, 4–41.

—— *1990*, 'Algebra and naive geometry. An investigation of some basic aspects of Old Babylonian mathematical thought', *Altorientalische Forschungen*, **17**, 27–69, 277–369.

Neugebauer, O. *1935–7*, *Mathematische Keilschrift-Texte, I–III* (Quellen und Studien zur Geschichte der Mathematik, Astronomie und Physik, Abteilung A, Vol. 3), Berlin: Julius Springer. [Repr. 1973, Berlin: Springer.]

—— *1955*, *Astronomical Cuneiform Texts, I–III*, London: Lund Humphries.

—— *1967*, 'Problems and methods in Babylonian mathematical astronomy', *Astronomical Journal*, **72**, 964–72. [Repr. in his *Astronomy and History. Selected Essays*, New York: Springer, 255–63.]

—— *1975*, *History of Ancient Mathematical Astronomy*, Vol. 1, New York: Springer.

Neugebauer, O. and Sachs, A. *1945*, *Mathematical Cuneiform Texts* (American Oriental Series, Vol. 29), New Haven, CT: American Oriental Society.

Nissen, H. J., Damerow, P. and Englund, R. *1991*, *Frühe Schrift und Techniken der Wirtschaftsverwaltung im alten Vorderen Orient. Informationsspeicherung und -verarbeitung vor 5000 Jahren*, 2nd edn, Berlin: Franzbecker. [English edn in the press (University of Chicago Press).]

Powell, M. A. *1976*, 'The antecedents of Old Babylonian place notation and the early history of Babylonian mathematics', *Historia mathematica*, **3**, 417–39.

Thureau-Dangin, F. *1938*, *Textes mathématiques Babyloniens*, Ex Oriente Lux, Deel 1, Leiden: Brill.

Vaiman, A. A. *1961*, *Sumero-vavilonskaya matematika. III–I Tysyacheletiya do n. e.*, Moscow: Izdatel'stvo Vostochnoi Literatury.

van der Waerden, B. L. *1962*, *Science Awakening*, 2nd edn, Groningen: Noordhoff.

—— *1978*, 'Mathematics and astronomy in Mesopotamia', in *Dictionary of Scientific Biography*, Vol. 15, New York: Scribner's, 667–80.

Vogel, K. *1959*, *Vorgriechische Mathematik*, Vol. 2, *Die Mathematik der Babylonier* (Mathematische Studienhefte, · Vol. 2), Hanover: Schrödel, and Paderborn: Schöningh.

1.2

Egyptian mathematics

C. S. ROERO

Among the ancient civilizations, that of ancient Egypt is noted for the refinement of its art and the wealth of its technical and practical achievements connected with the construction of grand monuments, such as the pyramids, the obelisks and the colossi. Unfortunately few documents survive to attest to its mathematical developments. Most of them were written on papyrus, a material which was liable to deterioration over time. Very few of them are of a mathematical nature, and those that are often consist of series of calculations that may be interpreted in several ways.

1 EGYPTIAN WRITING AND MATHEMATICAL PAPYRI

With the discovery of the Rosetta Stone during the 1799 French expedition to Egypt, and its decipherment by J. F. Champollion in the 1820s, the problem of interpreting the Egyptian hieroglyphs and the cursive hieratic writing was solved. Egyptian writing had three different styles (hieroglyphic, hieratic and demotic), reflecting the usages and customs of different historical periods. Hieroglyphic writing was used for monumental inscriptions and tomb decorations, and consisted of single pictures or figures representing many different things: objects, plants, animals and human beings. Hieratic was the cursive style used by priests of the Ancient Kingdom, and demotic the writing used by people for everyday tasks from the seventh century BC onwards.

We can consider only those documents which preceded Greek civilization because of the great difficulty of separating out the Egyptian culture of the Hellenistic period from the influences of neighbouring countries. Egyptian mathematical sources, in order of their importance and with approximate dates, are listed below. They are all written in hieratic, though a hieroglyphic transcription is given by some of their editors.

1 The largest and best preserved is the Rhind Papyrus (1650 BC), now in the British Museum, London. Its name is derived from the Englishman

30

A. H. Rhind, who bought it at Luxor in 1858. The scribe of this papyrus, A'hmosé or Ahmes by name, declares that it is a copy of a papyrus from two centuries earlier. The first sentence reads: 'Accurate reckoning of entering into the knowledge of all existing things, and all mysteries and secrets.' The papyrus contains, on rectos, long lists of divisions, and on versos 87 mathematical problems of arithmetical, algebraic and geometrical content. The papyrus was first published in Eisenlohr *1877*, and then in Peet *1923b* and Chace *et al. 1927–9*.

2 The Moscow Papyrus (1850 BC) is now in the Pushkin Museum of Fine Arts, Moscow. It contains 25 problems, 11 on *pesus* of beer and bread (the word 'pesu' is explained in Section 4), and 6 on geometry. An edited version appeared in Struve *1930*, with a hieroglyphic transcription by B. A. Turajev.

3 The Reisner Papyrus (1880 BC), now in the Museum of Fine Arts in Boston, collects calculations of blocks of stone and problems on the distribution of salary to workmen for building a temple in the reign of Sesostris I. Simpson *1963–9* has published it with a translation.

4 The Berlin Papyrus (1850 BC) contains some defective problems on equations; it was published in Schack-Schackenburg *1900*.

5 The Kahun Papyrus (1850 BC) was found by W. M. F. Petrie at Kahun in 1889 and is now in the British Museum, London. It consists of some fragments relating to geometry and algebra, edited in Griffith *1898*. There is also a portion of the list of divisions contained in the Rhind Papyrus (recto).

6 The Leather Roll (1650 BC) remained unopened for 60 years because of its brittleness. It is now in the British Museum, and contains a table of the sums of 26 unit fractions, in duplicate. It was published in Glanville *1927*.

7 Two Thebes Wooden Tablets (2000 BC) are now in the Cairo Museum. These contain calculations relating measures of capacity; they were published by Daressy *1906*, and interpreted by Peet *1923a*.

2 NOTATION

The Egyptian system of numeration was decimal and additive. It used these hieroglyphic signs for the powers of ten from 10^0 to 10^6:

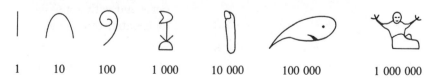

| 1 | 10 | 100 | 1 000 | 10 000 | 100 000 | 1 000 000 |

With these seven symbols of nine units each the ancient Egyptians were able to write the whole numbers from 1 to 9 999 999, which was enough for their everyday requirements.

We can find many specimens of Egyptian numerals written in hiero-glyphic notation on monuments. On Ptahotep's grave at Saqqara (1660 BC) there are represented the numbers 121 200 and 11 110 in the upper row and 121 022 and 111 200 in the lower row (Figure 1). This manner of writing and reading was sometimes modified between 2000 and 1600 BC. There are examples of a system of numeration which was multiplicative and additive at the same time. It occurs in the case of a number which is lower than another that precedes that higher one; for example, ℰ⋂⋂𝄾 symbolizes 120 000. The multiplicative and additive hieroglyphic numeration was probably connected with aesthetical needs, or perhaps was used for lack of space. Keeping in mind that this kind of writing was used for monumental inscriptions, it comes as no surprise to find the symbols sometimes placed one above the other. Only the context and the order of reading can decide the system of numeration. The hieratic numerals are cursive versions of hieroglyphic signs, drawn on papyri with brush and ink; their shape changed greatly over the centuries (Figures 2 and 3). Hieratic writing is always read from right to left.

Figure 1 Ptahotep's grave at Saqqara (from Guitel *1975*: 69)

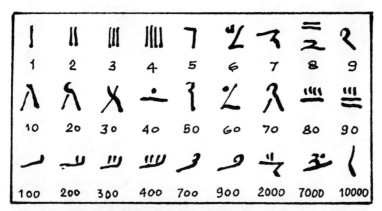

Figure 2 Hieratic Egyptian numerals in the Rhind Papyrus (from Roero *1987*: 128)

Figure 3 Hieratic Egyptian numerals in the Royal Papyrus (thirteenth century BC, Egyptian Museum, Turin) (from Roero *1987*: 128)

3 FRACTIONS

Egyptian fractions (see also §1.1 and §1.17) were unit fractions: only unity was allowed as a numerator, with the exception of $\frac{2}{3}$, written $\bigcirc$ as a hieroglyph and λ in hieratic. The numerator 1 of Egyptian fractions was never written. A number n became its reciprocal $1/n$ by putting the sign for a part above it. In general, the sign above the number was $\bigcirc$ in hieroglyphs, a dot in hieratics; 'r' is the phonetic transcription. The fraction $\frac{1}{2}$ was written with the sign of part reduced to a half ($\subset$ or $\supset$); $\frac{1}{4}$ was written as $\times$ to

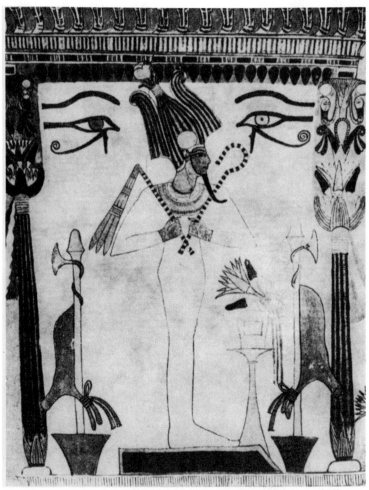

Figure 4 Osiris below Horus eyes, depicted on the grave of King Sennedjem (from Guitel *1975*: 79)

represent division by 4 in cursive hieratic (Figure 2). Other fractions were obtained as the sum of unit fractions, and there were tables for that purpose.

Some peculiar symbols were used by ancient Egyptians for fractions of *hekat*, the measure of capacity for grain. They are called Horus-eye fractions because of their connection with the famous myth of the eye of the

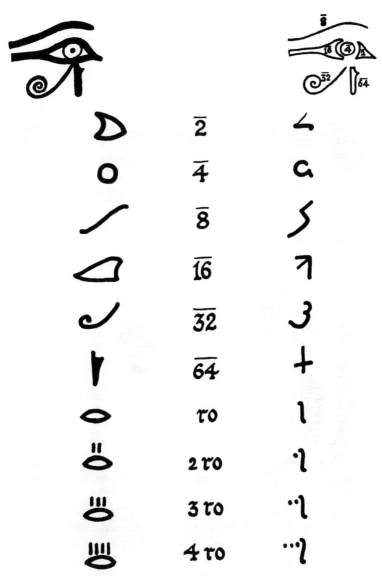

Figure 5 Horus-eye fractions (from Gillings *1972*: 211)

god Horus. The legend relates how the god Horus, the son of Osiris and Isis, had a fight with Seth to avenge the murder of his father Osiris and to obtain power over the Earth. During the struggle Seth tore away Horus's eye and broke it into pieces, which he dispersed everywhere. The broken parts of the eye were later found and restored by Thoth, the god of rules and calculations. The symbol of Horus eye was called *udjat*, 'whole', and was revered as a protection against evil (Figure 4). The mathematical meaning of this sign is connected with the fractions 1/2, 1/4, 1/8, 1/16, 1/32 and 1/64 symbolized by the parts of the Horus eye (Figure 5). According to the myth, the god Thoth, who realized that the sum of Horus eye fractions was 63/64, magically provided the missing fraction, 1/64, to complete the unit.

4 ARITHMETIC

The additive notation of Egyptian numbers made it possible to calculate at once the addition and subtraction of whole numbers. But the scribe must have known how to add units, tens, hundreds, and so on, even if they did not appear in the same column. In fact, the alignment of digits in their proper columns was not a feature of Egyptian calculations.

The technique of working out multiplication and division is very interesting; it is based on a process of repeated duplication or division by 2. To multiply 12 by 7, the scribe proceeded thus:

	1	7
	2	14
	\4	28
	\8	56
Totals	12	84

The first column contains the powers of 2 and the second one a factor of multiplication successively doubled. The scribe placed a check mark alongside the numbers in the first column that totalled 12, i.e. $4 + 8$; the sum of the corresponding numbers in the second column, i.e. $28 + 56$, then produced the result. This algorithm for multiplication was based on the arithmetical property that any integer can be expressed uniquely as the sum of certain terms of the geometrical progression $2^0, 2^1, 2^2, \ldots, 2^k, \ldots$.

The technique for division was the same as that for multiplication. For these operations the ancient Egyptians sometimes used not only 2 but other multipliers too, such as 10. For example, in the Rhind Papyrus (problem

69) $1120 \div 80$ is calculated, a division with an exact quotient:

	1	80
	10	800 /
	2	160
	4	320 /
Totals	14	1120

The scribe set out to obtain the dividend 1120 in the second column, so he multiplied the divisor 80 by 10 and then successively doubled it until a number was reached which, when added to 800, gave 1120. In this case the check mark was placed alongside the numbers in the second column that totalled 1120; the required quotient was then the sum of the corresponding numbers in the first column. This technique is simple enough when the two integers are divisible; when the quotient was not a whole number the scribe used fractions. This example is taken from the Rhind Papyrus (problem 24) for the division $19 \div 8$:

	1	8
	2	16 /
	$\frac{1}{2}$	4
	$\frac{1}{4}$	2 /
	$\frac{1}{8}$	1 /
Totals	$2 + \frac{1}{4} + \frac{1}{8}$	19

For the operations with fractions the ancient Egyptians made use of arranged tables, such as those on the rectos of the Rhind Papyrus and on the Leather Roll. In the first portion of the Rhind Papyrus there is a table that contains all fractions of the form

$$\frac{2}{2n+1}, \quad 1 \leqslant n \leqslant 50 \tag{1}$$

expressed as the sum of not more than four unit fractions: for example,

$$\tfrac{2}{29} = \tfrac{1}{24} + \tfrac{1}{58} + \tfrac{1}{174} + \tfrac{1}{232}. \tag{2}$$

Another short table shows the fractions $n/10$ for $n = 1, 2, 3, \ldots, 9$ written as the sum of unit fractions too. There are also tables for finding $\frac{2}{3}$ of seventeen different fractions in the Rhind Papyrus (problem 61). The Leather Roll contains a collection of twenty-six sums done in unit fractions (Figure 6). Egyptian scribes probably employed the tables in many ways, so as to take the greatest possible advantage of them. They were able to find prompt and efficient methods to make reductions. For example, using

Figure 6 The Leather Roll (from Guitel *1975*: 88–9)

Lignes	Fraction	Decomposition	Notes
1	$\frac{1}{8}$	$\frac{1}{40}\ \frac{1}{10}$	
2	$\frac{1}{4}$	$\frac{1}{20}\ \frac{1}{5}$	
3	$\frac{1}{3}$	$\frac{1}{12}\ \frac{1}{4}$	
4	$\frac{1}{5}$	$\frac{1}{10}\ \frac{1}{10}$	
5	$\frac{1}{3}$	$\frac{1}{6}\ \frac{1}{6}$	
6	$\frac{1}{2}$	$\frac{1}{6}\ \frac{1}{6}\ \frac{1}{6}$	
7	$\frac{2}{3}$	$\frac{1}{3}\ \frac{1}{3}$	
8	$\frac{1}{8}$	$\frac{1}{200}\ \frac{1}{25}\ \frac{1}{25}\ \frac{1}{25}$	un trait oublié transforme $\nearrow$ en $\wedge$
9	$\frac{1}{16}$	$\frac{1}{400}\ \frac{1}{150}\ \frac{1}{30}\ \frac{1}{50}$	
10	$\frac{1}{15}$	$\frac{1}{150}\ \frac{1}{50}\ \frac{1}{25}$	confusion avec la ligne suivante
11	$\frac{1}{6}$	$\frac{1}{18}\ \frac{1}{9}$	
12	$\frac{1}{4}$	$\frac{1}{28}\ \frac{1}{14}\ \frac{1}{7}$	
13	$\frac{1}{8}$	$\frac{1}{24}\ \frac{1}{12}$	
14	$\frac{1}{7}$	$\frac{1}{42}\ \frac{1}{21}\ \frac{1}{14}$	
15	$\frac{1}{9}$	$\frac{1}{54}\ \frac{1}{27}\ 18$	
16	$\frac{1}{11}$	$\frac{1}{66}\ \frac{1}{33}\ \frac{1}{22}$	un trait oublié transforme $\nearrow$ et $\wedge$
17	$\frac{1}{14}$	$\frac{1}{196}\ \frac{1}{98}\ \frac{1}{49}\ \frac{1}{28}$	un trait oublié transfor ne 4 en 3 $\frac{1}{98}$ a été omis
18	$\frac{1}{15}$	$\frac{1}{90}\ \frac{1}{45}\ \frac{1}{30}$	un trait oublié altère la valeur de 90

the 'red auxiliaries' (so-called because they were written in red), they were able to find the least common multiple.

Many theories about how and why the scribes determined the fractions given in those tables have been discussed and debated by historians of mathematics (Gillings *1974*; see also §1.17). Whether the values in the tables were obtained by trial and error, or whether they were derived from a particular rule, we must recognize that the ancient Egyptians built up a good knowledge of the first hundred integers, of the two times table and of certain arithmetical equalities. For example, in the Rhind Papyrus the scribe showed his ability to find $\frac{2}{3}$ of any number, integral or fractional. This rule, numbered 61B, states in modern terms that

$$\frac{2}{3}\frac{1}{2n+1} = \frac{1}{2(2n+1)} + \frac{1}{6(2n+1)}, \quad n > 0. \tag{3}$$

There were many practical problems with which the Egyptian officials had to deal, such as the distribution of wages or food among a number of workmen, and the reckoning of the amount of grain needed for the production of a given quantity of bread or beer. In the Rhind Papyrus we find numerous problems about the division of loaves between men in equal and unequal proportions (see, for instance, problems 3, 39, 40, 63 and 65). Twenty problems in the Rhind and Moscow Papyri called the '*pesu* calculations' are concerned with the determination of the amounts of grain needed for making bread or beer. The *pesu*, or 'strength', is the reciprocal of the grain density, which is the quotient of the number of loaves or jugs of beer divided by the quantity of grain.

However, there are also problems that do not concern specific objects. They are related instead to what we would now consider to be a kind of algebra, requiring the equivalent of solutions of linear equations of one unknown. A simple example of this kind of problem, called '*aha*-calculations' from the name of the unknown *aha*, or 'heap', is found in problem 24 of the Rhind Papyrus. Here it is required to find the value of a heap if the heap and a seventh of it make 19. The procedure followed by the scribe is what is now known as the method of false position. He starts with an arbitrary chosen number as the required quantity, in this case 7, because this makes the computation of the seventh part easy. So $7 + (\frac{1}{7} \times 7)$ gives 8, instead of the desired answer, which was 19. Hence, inasmuch as $8(2 + \frac{1}{4} + \frac{1}{8}) = 19$, one must multiply 7 by $2 + \frac{1}{4} + \frac{1}{8}$ to obtain the correct value for the heap. The scribe found the required result to be $16 + \frac{1}{2} + \frac{1}{8}$, and then verified the correctness of his answer.

This procedure is followed many times by the scribe A'hmosé, so that one can believe the Rhind Papyrus to be a handbook with exercises for young students. Besides practical problems there are some theoretical problems put in concrete form, and sometimes the scribe seems to have had puzzles or mathematical recreations in mind. Thus problems 28 and 29 of the Rhind Papyrus are examples of 'think of a number' problems. Another amusing example is problem 79 of the Rhind Papyrus (Figure 7). It reveals that the ancient Egyptians were able to calculate the sum of a finite number of terms of a geometric progression. The translation (from Chace *et al. 1927–9*) is the following:

A house inventory

\1	2 801	7	houses
\2	5 602	49	cats
\4	11 204	343	mice
Total	19 607	2 401	spelt
		16 807	*hekat*
		19 607	Total

The meaning was perhaps of the following nature (compare §7.13 on combinatorics); in each of 7 houses there are 7 cats, each of which eats 7 mice;

Figure 7 Problem 79 in the Rhind Papyrus (from Peet *1923b*: Plate W)

each mouse would have eaten 7 ears of spelt, each of which would have produced 7 *hekats* (i.e. 7 measures of grain). What is the total of all these?

The air of verisimilitude in this problem is apparent. Most probably it was an exercise for students in the field of numbers. The right-hand column shows, in fact, the sum of the first five terms of the geometric progression whose first term is 7 and whose common ratio is 7. The left-hand column presents the multiplication 2801 by 7. But unfortunately the scribe does not indicate whence came the number 2801, and which method was used. If it was an exercise for students, the aim, perhaps quite well known, was to explain the use of the tables. In this case, the scribe wished to illustrate the table for the powers of 7 and that for the sum of the terms of a geometric progression. This last one was based on the property, found empirically by the ancient Egyptians, that

$$S_{n+1} = (1 + S_n)7, \quad \text{where } S_n = 7 + 7^2 + 7^3 + \cdots + 7^n. \tag{4}$$

From this playful exercise were derived some similar rhymes, such as that from the *Liber abbaci* (1202) of Leonardo of Pisa (Boncompagni *1857*: Vol. 1, 311), and the familiar Mother Goose rhyme (§7.13):

As I was going to Saint Ives,
I met a man with seven wives;
Every wife had seven sacks,
Every sack had seven cats,
Every cat had seven kits.
Kits, cats, sacks and wives,
How many were going to Saint Ives?

5 GEOMETRY

The Greeks generally regarded the Egyptians as the inventors of geometry. Herodotus (fifth century BC), handing down to posterity the information that he had received from the Egyptian priests, relates in his *History* the following:

The king [Rameses II, known to the Greeks as Sesostris] moreover, so they say, divided the country among all the Egyptians by giving each an equal square parcel of land, and made this his source of revenue, appointing the payment of a yearly tax. And any man who was robbed by the river of a part of his land would come to Sesostris and declare what had befallen him; then the king would send men to look into it and measure the space by which the land was diminished, so that thereafter it should pay in proportion to the tax originally imposed. From this practice, to my thinking, the Greeks learned the art of geometry.

Proclus (fifth century AD) wrote similarly in his *Commentary* on Euclid, Book I:

> According to most accounts geometry was first discovered among the ancient Egyptians, taking its origin from the measurement of areas. For they found it necessary by reason of the rising of the Nile, which wiped out everybody's proper boundaries.

From the Egyptian papyri that have come down to us, we can only try to summarize which formulas were used by the scribes to calculate areas and volumes (see also §8.15 on cartography).

The working of the geometrical problems, as with the arithmetical ones, does not illustrate the use of methods and techniques. The ancient Egyptians were able to determine the areas of triangles, rectangles and trapezoids in the (to us) usual way. For instance, the area of an isosceles triangle was found by taking half the base and multiplying this by the height (Rhind Papyrus, problem 51). For determining the area of a circle (Rhind Papyrus, problem 50), the scribe stated the rule: 'Take away $\frac{1}{9}$ of the diameter and square the remainder': in modern terms,

$$A = (\tfrac{8}{9}d)^2 = \tfrac{256}{81} r^2. \tag{5}$$

This Egyptian rule corresponds to a very good approximation, $\pi = 3.1605$. The volumes of cylindrical granaries (Rhind Papyrus, problem 43; Kahun Papyrus, problem IV.3) were calculated by multiplying the area of the circular base, obtained by using the formula above, by the height.

The scribes were able to calculate the inclination of oblique planes, hence also the slope or inclination of the sides of a pyramid (Rhind Papyrus, problem 36), the volume of a pyramid and the volume of a truncated pyramid. The last of these, which has been called the gem of Egyptian geometry, is found in problem 14 of the Moscow Papyrus. The translation is as follows (Struve *1930*):

> Method of calculating a truncated pyramid. If it is said to thee, a truncated pyramid of 6 cubits in height, of 4 cubits at the base by 2 cubits at the top, reckon thou with this 4, its square is 16. Multiply this 4 by 2. Result 8. Reckon thou with this 2, its square is 4. Add together this 16, with this 8, and with this 4. Result 28. Calculate thou $\frac{1}{3}$ of 6. Result 2. Calculate thou with 28 twice. Result 56. It is 56. You have correctly found it.

If we denote the lower base of 4 cubits by a, the upper base of 2 cubits by b, and the height of 6 cubits by h, we find that the scribe was familiar with the modern formula for the volume of a frustum,

$$V = \tfrac{1}{3}h(a^2 + ab + b^2). \tag{6}$$

It is not known how these results were obtained by the Egyptian scribes. Many reconstructions have been suggested by Egyptologists and historians of mathematics (Gillings *1972*: 187–93).

6 THE NATURE OF EGYPTIAN MATHEMATICS AND ITS INFLUENCES

For historians, the mathematical attainments of the ancient Egyptians and the methods of calculation, presented in the papyri without any heuristic, pose two interesting questions. Was ancient Egyptian mathematics strictly practical, or was it a pure science with theoretical aims too? And did it, or did it not, influence early Greek mathematics? These historiographical problems have been discussed and debated for a long time. Here are two directly opposed judgements:

> A careful study of the Rhind Papyrus convinced me several years ago that this work is not a mere selection of practical problems especially useful to determine land values, and that the Egyptians were not a nation of shopkeepers, interested only in that which they could use. Rather I believe that they studied mathematics and other subjects for their own sakes . . . . Thus we can say that the Rhind Papyrus, while very useful to the Egyptian, was also an example of the cultivation of mathematics as a pure science, even in its first beginnings. (Chace *et al. 1927–9*: Vol. 1, 42–3).

> The Rhind and Moscow Papyri are handbooks for the scribe, giving model examples of how to do things which were a part of his everyday tasks . . . . The truth is that Egyptian mathematics remained at much too low a level to be able to contribute anything of value. The sheer difficulties of calculation with such a crude numeral system and primitive methods effectively prevented any advance or interest in developing the science for its own sake. It served the needs of everyday life . . . and that was enough. Its interest for us lies in its primitive character, and in what it reveals about the minds of its creators and users, rather than in its historical influence. (Toomer *1971*: 37, 45).

A complete answer to these questions is today very difficult because of the scarcity of mathematical documents in our possession and because of the poverty of concepts inherent in them. With these preliminary remarks it is also impossible to attempt a true reconstruction of early Egyptian mathematics or an exact evaluation of its influences on or interrelations with the mathematics of other ancient civilizations. That the Greeks took some elementary properties of arithmetic and geometry from the Egyptians is certain. The use of their multiplication and of their computations with unit fractions persisted in Greece for a long time (§1.17), and was prolonged well into the medieval period. The rules for the determination of areas and volumes were also derived from Egypt, according to what some Greek philosophers and mathematicians are said to have learned during their travels (Godel *1956*).

The ancient city of Naucratis, founded in the middle of the seventh century BC by the pharaoh Psammeticus on the Nile delta, became a free port, where a dozen Greek towns had their emporia (Miletus, Chios, Teo, Phocaea, Clazomenae, Mitilene, Cnidus, Alicarnasso, Phaselide, Lindus, Camirus and Iasilus). All these were Ionic towns, so perhaps it is not by chance that Greek science arose just from Ionia. Plutarch relates that Pythagoras lived a long time at Heliopolis (*De Iside*: 10), and Diodorus reports that Democritus spent five years in Egypt (*Bibliotheca Historica*: Vol. 1, 98). According to them, Plato and his friend Eudoxus of Cnidus lived in Egypt for a time. Our knowledge of the subject is too limited, however, to determine how these circumstances influenced Greek science. A sound appraisal of Egyptian mathematics will require the combined efforts of numerous experts in Egyptology, archaeology and the history of mathematics. Only a much broader and deeper understanding of the internal and external sources can lead to a solution of this important historiographic problem. One cannot pass over this passage in Aristotle's *Metaphysics* (981b, 20–25):

> Hence it was, after all, such practical arts were already established that those of the sciences which are not directed to the attainment of pleasure or the necessities of life were discovered; and this happened in the places where men had leisure. This is why the mathematical arts were first set up in Egypt, because there the priestly caste was allowed to enjoy leisure.

BIBLIOGRAPHY

Archibald, R. C. *1927–9*, *Bibliography of Egyptian Mathematics*, Oberlin, OH: Mathematical Association of America.

Boncompagni, B. (ed.) *1857*, *Leonardo Fibonacci Pisano Liber Abaci*, 2 vols, Rome: the author.

Chace, A. B., Bull, L. S. and Manning, H. P. *1927–9*, *The Rhind Mathematical Papyrus, British Museum 10057 and 10058*, 2 vols, Oberlin, OH: Mathematical Association of America.

Clagett, M., *1989*, *Ancient Egyptian Science*, Vol. 1, *Knowledge and Power*, Philadelphia: American Philosophical Society Memoirs, 184.

Daressy, G. *1906*, 'Calculs égyptiens du Moyen Empire', *Recueil de travaux relatifs à la philologie et à l'archéologie égyptiennes et assyriennes*, **28**, 62–72.

Eisenlohr, A. *1877*, *Ein mathematisches Handbuch der alten Ägypter (Papyrus Rhind des British Museum) übersetzt und erklärt*, 2 vols, Leipzig: Hinrichs. [Repr. 1972, Wiesbaden: Sandig.]

Gardiner, A. *1982*, *Egyptian Grammar*, 3rd edn, Oxford: Griffith Institute.

Gillings, R. J. *1972*, *Mathematics in the Time of the Pharaohs*, Cambridge, MA: MIT Press.

—— *1974*, 'The recto of the Rhind Mathematical Papyrus. How did the ancient Egyptian scribe prepare it?', *Archive for History of Exact Sciences*, **12**, 291–8.

Glanville, S. R. K. *1927*, 'The Mathematical Leather Roll in the British Museum', *Journal of Egyptian Archaeology*, **13**, 232–9.

Godel, R. *1956*, *Platon à Héliopolis d'Egypte*, Paris: Les Belles Lettres.

Griffith, F. L. *1898*, *The Petrie Papyri. Hieratic Papyri from Kahun and Gurob (Principally of the Middle Kingdom)*, 2 vols, London: Quaritch.

Guitel, G. *1975*, *Histoire comparée des numérations écrites*, Paris: Flammarion.

Knorr, W. *1982*, 'Techniques of fractions in Ancient Egypt and Greece', *Historia mathematica*, **9**, 133–71.

Peet, T. E. *1923a*, 'Arithmetic in the Middle Kingdom', *Journal of Egyptian Archaeology*, **9**, 91–5.

—— *1923b*, *The Rhind Mathematical Papyrus, British Museum 10057 and 10058. Introduction, Transcription and Commentary*, London: Hodder and Stoughton. [Repr. 1977, Leiden: Brill.]

Roero, C. S. *1987*, 'Numerazione e aritmetica nella matematica egizia', in W. Di Palma (ed.), *L'alba dei numeri*, Bari: Dedalo, 117–45.

Schack-Schackenburg, H. *1900*, 'Der Berliner Papyrus 6619', *Zeitschrift für ägyptische Sprache*, **38**, 135–40.

Simpson, W. K. *1963–9*, *The Papyrus Reisner*, 3 vols, Boston, MA: Museum of Fine Arts.

Struve, W. W. *1930*, Mathematischer Papyrus des Museums in Moskau', *Quellen und Studien der Geschichte der Mathematik*, Abteilung A, Vol. 1.

Toomer, G. J. *1971*, 'Mathematics and astronomy', in J. R. Harris (ed.), *The Legacy of Egypt*, 2nd edn, Oxford: Clarendon Press, 27–54.

Vogel, K. *1958*, *Vorgriechische Mathematik*, Vol. 1, *Vorgeschichte und Ägypten*, Hanover: Schrödel.

1.3

Greek mathematics to AD *300*

ALEXANDER JONES

1 THE SOURCES

The Classical and Hellenistic Greeks took various mathematical concepts and methods from the contemporary Mesopotamian and Egyptian cultures, but the mainstream of Greek mathematics was distinguished by an original deductive approach, most fully developed in geometry. Our sources for Greek mathematics consist primarily of two or three dozen original treatises by mathematicians, mostly dating from about 300 BC to about 200 BC, supplemented by mathematical and philosophical commentaries of late Antiquity (roughly AD 300–600; see §1.5). The texts have come down to us in medieval manuscripts that very seldom date from before the ninth century, and are not free from copying errors or even deliberate rewriting; such as they are, they represent only a small fraction of the mathematical literature that was still accessible to, say, Pappus in about AD 300. The majority of the available texts survive in Greek, but medieval Arabic translations exist for several writings that perished in the original language. Archeologically recovered documents such as papyri have so far added little to our knowledge of Greek mathematics, except at the most basic level.

Works on the history of mathematics did exist in Antiquity, beginning with the *History* of Eudemus (about 300 BC), and continued by Geminus (first century AD), but they survive only through fragmentary quotations in the works of later authors. In particular, we know very little of the lives of the Greek mathematicians. A sketchy biography of Archimedes is constructible, thanks to his ancient notoriety as an inventor and the informative prefaces to his surviving works; but more typical is the case of Euclid, for whom we have not even a reliable date or place of work (his traditional association with Alexandria has little basis in ancient documents).

At the worst extreme of obscurity are the sixth-century BC philosophers Thales and Pythagoras, to whom far later authors (e.g. Proclus) ascribe the foundations of Greek geometry and arithmetic, but whose real contributions are lost in a mire of legends (Burkert *1972*: 401–82). It is doubtful whether Pythagoras himself was the first author of any significant mathematical discovery – certainly not the theorem on right triangles that now bears his name, for it is already attested to in Babylonian texts of the second millennium BC (Neugebauer *1957*: 36; compare §6.10, Section 1). Even the historical importance of Pythagoras's later followers, the 'Pythagoreans', was probably exaggerated in late Antiquity, although the first serious advances in number theory and the discovery of incommensurable geometrical magnitudes are plausibly ascribed to the fifth-century Pythagoreans Archytas and Hippasus, respectively.

Throughout the period in question mathematicians were men of independent means, the same sort of 'intellectuals' as those who referred to themselves as philosophers (Eudoxus is a rare instance of a figure who made important contributions to both mathematics and philosophy). State patronage was probably much less important than is often assumed; and although the presence of such institutions as the Museum helped to make Alexandria the single most important centre of scientific activity after 300 BC, mathematicians of the first calibre lived in widely dispersed parts of the Greek-speaking world.

2 THE *ELEMENTS*

Euclid's *Elements*, a large treatise in 13 'books' (originally papyrus rolls), is the only surviving comprehensive treatment of the axiomatic foundations of Greek geometry and number theory (compare §2.1). Euclid probably composed, or compiled, the *Elements* in the first decades of the third century BC, but it underwent various 'editions' in later Antiquity. Two versions of the *Elements* have come down to us, one of them due to Theon of Alexandria (*circa* AD 370), the other believed to be closer to the original form of the work.

In structure the treatise falls into several sections directed broadly towards principal theorems. Thus the first two books concern the construction and properties of rectilinear figures, leading to the 'Pythagorean' Theorem (I: 47) and the quadrature of a given polygon (II: 14). Books III and IV treat the properties of circles and problems of inscribing and circumscribing polygons. Book V is devoted to a theory of proportion for geometrical magnitudes, which is applied in Book VI to the similarity of figures. The arithmetical Books VII to IX are concerned especially with the properties of composite and prime numbers, while the long Book X deals with incommensurability and an elaborate classification of incommensurable magnitudes. The final three books turn to solid geometry, and include demonstrations of the volumes of the cylinder and cone, and constructions of the five regular polyhedra.

Roughly two centuries of development culminated in the *Elements*, but the documentation for the formative period is so very poor that almost every detail is a point of controversy among historians (among the more successful reconstructions are those by van der Waerden *1954* and Knorr *1975*). Even the nature and quality of Euclid's own contribution have been disputed. It seems clear, at least, that the theory of numbers in Books VII to IX in large part goes back to the Pythagoreans, while much of the elementary geometry in Books I to IV was already known to Hippocrates of Chios in the mid-fifth century BC. Although the discovery of incommensurability is attributed fairly securely to the fifth-century Pythagoreans, how it occurred is open to conjecture; nor do historians agree about whether this discovery caused a 'foundational crisis' in early Greek mathematics. We do know that Euclid's elaborate classification of irrational magnitudes in Book X owes something to Theaetetus (early fourth century).

Eudoxus (first half of the fourth century) is credited with two innovations of considerable importance. The first of these is the ingenious theory of proportions that Euclid presents in Book V, based on the following definition of equal ratios: two pairs of magnitudes (A, B) and (C, D) are in the same proportion if for every pair of whole numbers m, n either ($mA > nB$ and $mC > nD$) or ($mA = nB$ and $mC = nD$) or ($mA < nB$ and $mC < nD$). It is likely that Eudoxus's ratio theory superseded a less powerful theory employing *anthyphairesis*, or repeated alternate subtraction, a technique mathematically equivalent to defining equal ratios as those leading to identical continued fractions A/B and C/D (Fowler *1987*; compare §6.3).

Eudoxus's second major contribution was the so-called 'method of exhaustion', applied to volumetric and stereometric theorems in Book XII. This method (for the planar case) depends on establishing an iterative procedure for inscribing or circumscribing polygons of controllable area in or about the perimeter of the area it is wished to determine, in such a way that the area between becomes arbitrarily small in a finite number of steps.

3 PROBLEM-SOLVING

The *Elements* is by no means a complete exposition of the scope of Greek mathematics at the beginning of the third century. In particular, it scarcely reflects the intense contemporary interest in problems of construction, and the expansion of the geometer's resources to tackle them (Knorr *1986*). The constructive postulates in the *Elements* are the 'compass-and-ruler' principles, that one can draw a line through two given points, or a circle with given centre and through a given point. Experience made it evident that these postulates were insufficient for the solution of many problems, including the classical problems of the quadrature of the circle, the trisection of a given angle, and the construction of two mean proportions B and C between given magnitudes A and D (i.e. $A:B = B:C = C:D$).

As early as Hippocrates, and at least as late as Archimedes, we find an additional elementary construction occasionally permitted – the *neusis* or 'verging' (Figure 1). This is the assumption that one can construct a line ABC passing through a given point A and with a segment BC of given length cut off between two given lines (straight or circular). *Neusis* was a powerful tool, although it had the drawback that the conditions under which a *neusis* is possible were not trivial to determine; among other applications, we have ancient solutions of the problems of angle trisection and the two mean proportionals involving *neuses*. During the fourth century BC, geometers began applying special curves to the solution of problems; and curves could be generated by intersections of simple curved and planar surfaces (Archytas) or by 'mechanical' constructions (Dinostratus).

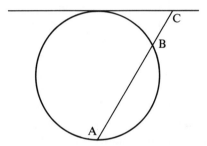

Figure 1 The principle of *neusis*

Another development of the fourth century was of the greatest importance for geometrical problem-solving: the method of 'analysis'. Greek geometrical analysis is described by Pappus (*Collection*: 7.1) as working step by step backwards from the thing sought, 'as if it has already been achieved', until one arrives at something that is known to be constructible. For this to have any demonstrative value, a special form of analytical argument had to follow, in which the quality of 'givenness' (i.e. constructibility) could be proved to adhere to various elements of the figure, including, ultimately, the thing sought. Euclid's *Data* (or 'things given') supplies the elementary theorems of givenness; his lost *Porisms* seems to have consisted of much more advanced propositions of the same analytic character, among which many have interesting affinities with problems in nineteenth-century projective geometry (§7.6).

The analysis of a problem thus consists of both an exploratory, heuristic stage (the 'working backwards') and a logical process to the thing sought, which serves as a thumbnail sketch of the final 'synthetic' construction and proof. The complete presentation of a problem generally included the analysis, the synthesis and the 'diorism', or deduction of conditions for one or more solutions to exist. Only synthetic proof seems to have been regarded as possessing full apodictic validity, but there is evidence that in Euclid's time analytic solutions of problems were sometimes published without syntheses.

The discovery of the conic sections is ascribed to Menaechmus (mid-fourth century), who used simple properties of the parabola and rectangular hyperbola to solve the problem of the two mean proportionals. (How Menaechmus proved that the curves were sections of a cone, if in fact he did, is not known.) Some fifty years later, a considerable body of theorems about the conic sections had accumulated, and Euclid is said by Pappus to have written an elementary treatise on conics. At this time the sections later known as the parabola, ellipse and hyperbola were called, respectively, the sections of a right-angled, an acute-angled and an obtuse-angled cone. Each conic section was associated with a single cone of the specified kind, such that the generating line of the cone through the section's vertex was perpendicular to the plane of the section. From this configuration a fundamental property or *symptoma* was deduced for each kind of conic section, which could be expressed analytically as a locus theorem within the plane of section. For example (Figure 2), the locus of a point, C, the square of whose perpendicular distance from a given line segment AB had a given ratio to the rectangle contained by AD and DB (i.e. $CD^2 : AD \times DB$ is constant), is a section of a given acute-angled cone – that is, a given ellipse.

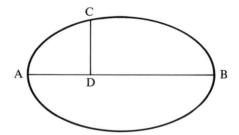

Figure 2 Fundamental property of the ellipse

The five lost books of *Solid Loci* by Aristaeus apparently consisted of locus theorems reduced analytically to the fundamental properties. It is probable that one of Aristaeus's more advanced theorems was the 'locus on three and four lines' mentioned by Apollonius, who informs us that Euclid achieved an incomplete synthesis of this locus. Pappus (*Collection*: 7.36) defined the four-line locus as the locus of a point whose (perpendicular) distances d_1, d_2, d_3, d_4 from four given straight lines are such that the ratio $d_1 \times d_2 : d_3 \times d_4$ equals a given ratio; for the three-line locus, one substitutes a constant line segment for d_4.

4 ARCHIMEDES

The mathematics of the mid-third century is dominated by Archimedes (Dijksterhuis *1956*). Two themes are especially prominent in the works of this masterly and versatile geometer: the measurement of areas and volumes of figures having curved boundaries, and the mathematical treatment of statics. Archimedes approaches the study of physical balance and stability via the application of axiomatic reasoning to idealized geometrical objects (both two- and three-dimensional) endowed with homogeneous density. Extensive use is made of the centre of gravity of figures, although this concept is never explicitly defined in the extant Archimedean corpus. The two books *On the Equilibrium of Plane Figures* deduce the law of the balance and the centres of gravity of triangles and parabolic segments. In *On Floating Bodies*, a masterpiece of mathematical physics, Archimedes reduces the problem of fluid stability to a single postulate separating vertical and horizontal components of pressure, and thereby establishes the laws of displacement and the conditions of stability of floating segments of a sphere and of a paraboloid of revolution. There is little reason to suppose that these problems were chosen for anything other than their intellectual interest.

Chief among Archimedes' treatises on planimetric and stereometric theorems is *On the Sphere and Cylinder*, in which he proves (among other things) that any sphere has a surface area four times that of its greatest circle, and a volume two-thirds that of the cylinder that contains it. This work, the *Quadrature of the Parabola* and *On Conoids and Spheroids* (concerning solids of revolution of conics) all use 'exhaustion' methods. The proofs, although excellent in their way, show the defects of the method: they are long, and one solution provides little clue towards finding the next.

In a remarkable treatise addressed to Eratosthenes, entitled *The Method of Mechanical Theorems*, Archimedes describes a heuristic application of his mathematical balances to the discovery of theorems on volumes, areas and centres of gravity. One example is illustrated in Figure 3. Line HAC is a 'balance' with fulcrum A, and we imagine a sphere with diameter AC (= HA), a right-angled cone with AC as axis, and the cylinder containing the cone. An arbitrary plane MN cuts the three solids in three circles with

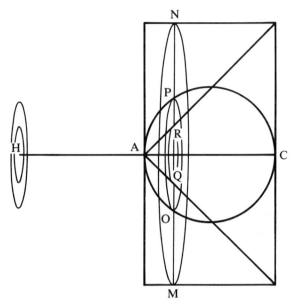

Figure 3 Archimedes' demonstration of the volumetric relationship
between sphere, cone and cylinder

respective diameters OP, QR, and MN. It is easily shown that the circle
MN, left where it is, balances circles OP and QR placed together at H.
Considering the solids as composed of such indivisible slices, Archimedes
concludes that the entire sphere and cone, placed with their centres of
gravity at H, balance the cylinder, from which it is easy to deduce the relationship between the volumes of the solids. In spite of its obvious affinities
with the methods of the precursors of the calculus, Archimedes' *Method*
had no influence on the development of modern mathematics, for it was lost
until J. L. Heiberg rediscovered it in 1906 in a palimpsest (a manuscript
whose leaves have been partially effaced and reused).

Among the mathematicians with whom Archimedes corresponded were
Conon and Eratosthenes, both active at Alexandria. Conon is reported by
Apollonius to have studied the intersection of conic sections in a work that
was subsequently attacked by one Nicoteles. Of Eratosthenes' mathematical
work, all that comes down to us is the 'sieve' for finding prime numbers
and a mechanical apparatus for finding the two mean proportionals;
Pappus knew, but unfortunately does not describe for us, Eratosthenes' lost
treatise *Loci on Means*. In the following generation, mathematicians derived inspiration from Archimedes' works. Thus Dionysodorus found the
volume of the torus, and Zenodorus attempted to prove that the circle and

sphere are the figures of maximum area for a given perimeter or surface area. The study of special curves, to which Archimedes' book *On Spirals* belongs, was furthered by Nicomedes (conchoids) and Diocles (the 'cissoid').

The focus of interest was still largely on problem-solving rather than the properties of the curves in their own right. Diocles' curve seems to have been a more or less *ad hoc* invention for solving the problem of the two mean proportionals; but the conchoid (the locus of points a given distance from a given straight line along a variable line passing through a given point) was applicable to the solution of a large class of problems involving *neusis* constructions. Archimedes' reduction of a problem in *On the Sphere and Cylinder* (II: 4) to a general construction equivalent to the solution of a cubic equation inspired solutions involving intersecting conics by Diocles and Dionysodorus; what may be Archimedes' own solution by conics is preserved by Eutocius.

5 APOLLONIUS

The *Conics* of Apollonius (in eight books, of which the last has not survived) was completed about 200 BC, when its author was about 40 years old. In his preface, Apollonius states that the first four books are 'in the manner of elements', a statement which led Pappus to the mistaken notion – often since repeated – that Apollonius lifted these parts bodily from Euclid. In fact, right from the start Apollonius undertakes a complete reform of the ancient study of the conic sections. He derives from the general section of any right or oblique cone a new set of *symptomata* for the three curves (by now called by their modern names). The old *symptoma* defined a relation between the length of ordinate lines (CD in Figure 2) dropped at right angles to the axis (AB) and the segment or segments of the axis cut off by the ordinates; Apollonius's general derivation of the conics leads to a *symptoma* in which the ordinates are drawn at *any* fixed angle to a *diameter* (as he calls it) of the section.

In Book I Apollonius proves that every conic has an infinite number of such diameters, including the axis of the old *symptoma*, and shows how to replace any diameter with any other. Book II is concerned with the properties of hyperbolas, with their conjugates and asymptotes. Book III consists of theorems useful for the synthesis of locus theorems, including the locus on three and four lines (the actual syntheses are not given). Book IV is a completion of the work on intersecting conics begun by Conon and Nicoteles, proving that two conics cannot intersect in more than four points. In Book V Apollonius undertakes a thorough investigation of the number and construction of minimal or maximal lines (i.e., as he proves, normal lines) from a given point to a conic. Book VI is concerned with equal and similar conics. Book VII presents several interesting theorems on the conjugate diameters of ellipses and hyperbolas. According to Apollonius's preface, these theorems were applied in the problems contained in the lost Book VIII.

Among Apollonius's other writings were six treatises related to problem-solving. Only one of these, *The Cutting Off of a Ratio*, survives (albeit only in Arabic translation), but the contents of the rest are described very accurately by Pappus (*Collection*: 7). In *The Cutting Off of a Ratio, The Cutting Off of an Area* and *The Determinate Section*, Apollonius solved a single geometrical problem or a set of related problems by analysis and synthesis. None of the problems is difficult, but he examines numerous cases of each, often with multiple solutions, and the treatises were accordingly very long indeed. We are told neither the motivation for the problems chosen (they may have arisen in the study of conics) nor why Apollonius felt the need to treat them in this exhaustive fashion. Perhaps they were meant as illustrations of what a truly complete solution of a problem could entail.

Apollonius's other three analytic treatises were apparently related to a classification of problems that is described by Pappus, but that probably came into use around Apollonius's time. The classification originates in the fact that many geometrical problems can be reduced by analysis to the determination of one or more loci on which a certain point must lie. Since the synthesis of a locus problem consisted of constructing the geometrical object (e.g. straight line, circle or cone) on which the variable point lies, it was natural to classify locus problems according to the nature of the object constructed. Circles and straight lines are 'planar' objects, cones and cylinders are 'solid', and various special curves that do not lie on the surfaces of elementary solids (e.g. Nicomedes' conchoid) are 'curvilinear'. This nomenclature is already apparent in Aristaeus's *Solid Loci*, which were conic sections; and Apollonius compiled a collection of loci on straight lines and circles, complementarily entitled *Plane Loci*.

The next logical step was to classify all problems according to the kind of locus to which they reduced. This leads to the realization that the three classes form a hierarchy: a problem that can be solved by 'planar means' (compass and straightedge) can generally also be solved using conics or special curves, but the reverse is not always true. For example, experience shows that planar methods do not suffice for the trisection of a given angle or the construction of two mean proportionals; but the problem of constructing a circle tangential to three given circles, which is easily reduced to the intersection of two hyperbolas, can also with more difficulty be solved by planar methods. This and a set of related tangency problems were so solved in Apollonius's *Tangencies*. Finally, there was no room for the *neusis* as an elementary construction in this scheme, so the challenge arose to reduce all varieties of *neuses* to locus problems. Of course, Nicomedes' conchoid already supplied a 'curvilinear' synthesis of most *neuses*, but solutions by means of conics or by planar methods must have seemed more economical. In his treatise entitled *Neuses* Apollonius reduced several *neuses* to planar constructions.

6 THE PERIOD OF DECLINE

The second century BC marked the beginning of a long decline in serious mathematics, except where it was applied in contexts such as astronomy (see §1.4). Various explanations have been suggested for this decadence, for example that an oral tradition of mathematical education, including heuristics, was broken by unknown causes, or that the strict rules for 'correct' solutions of problems that arose in Apollonius's time stifled invention, or that the mathematicians simply transferred their interest to other topics such as astronomy. There are nevertheless a few glimmers of original work among the swelling number of digests and commentaries. About 150 BC Hypsicles wrote a short work on the comparison of the dimensions of the icosahedron and dodecahedron (related to earlier researches by Aristaeus and Apollonius), which has come down to us as 'Book XIV' of Euclid's *Elements*. About the same time Hipparchus calculated the number of statements of a particular kind that could be composed from ten axioms according to the rules of Stoic logic; this is one of only a few traces that we can find of early efforts in combinatorics. An interesting development in the treatment of curves is revealed in a passage from Pappus (*Collection*: 4.51) which may derive from the otherwise unknown geometers Demetrius of Alexandria and Philo of Tyana (earlier than Menelaus, who lived about AD 100). It states that, from a curve on the plane or in space, it is possible to generate surfaces by letting a variable straight line travel along the curve while either maintaining a fixed direction (producing a 'cylindroid') or remaining perpendicular to a straight line (producing a 'plectoid'). The intersections of these surfaces with other surfaces are new curves with deducible properties.

Several compendia on planimetric and stereometric calculations are ascribed to Hero of Alexandria (*circa* AD 60). These are mere collections of formulas, without any attempt to prove their validity. Many of Hero's examples show striking affinities with Babylonian texts, but he also draws on the works of Archimedes and other Greek geometers. Book I of Diophantus's *Arithmetica* (probably *circa* AD 250), dealing with problems leading to linear and quadratic equations, has a similarly derivative

55

character. Of greater interest are the other Books of the *Arithmetica*, containing Diophantus's treatment of indeterminate equations. These are not 'Diophantine' equations in the modern sense (requiring integer solutions: see §6.10), but equations of the second degree and higher for which rational solutions are sought. The methods applied are rather haphazard, but generally involve the reduction of each problem to a single unknown. The character of Diophantus's approach may be illustrated by the following extract (Book I: 20):

> [Let it be required] to find two [rational] numbers such that the square of either of them added to the other makes a square.
>
> Let the first number be x, and the second $2x + 1$, so that the square of the first added to the second makes a square. It remains to have the square of the second added to the first make a square too. But the square of the second added to the first makes $4x^2 + 5x + 1$. This (has to be) equal to a square. I form the square from $2x - 2$. [The square] itself will then be $4x^2 - 8x + 4$. And x turns out to be 3/13 (by solving the equation $4x^2 + 5x + 1 = 4x^2 - 8x + 4$). The first number will be 3/13, and the second 19/13, and they solve the problem.

BIBLIOGRAPHY

[Bulmer-]Thomas, I. (transl.) *1939–41, Selections Illustrating the History of Greek Mathematics*, 2 vols, Cambridge, MA: Harvard University Press. [Anthology of reading , Greek and English, coordinated with T. L. Heath's history. Recent reprintings incorporate an extensive bibliography.]

Burkert, W. *1972, Lore and Science in Ancient Pythagoreanism*, Cambridge, MA: Harvard University Press. [Painstaking and trustworthy study of Pythagoras and his 'school'. Chapter 6 deals with the mathematics.]

Chasles, M. *1889, Aperçu historique des méthodes en géometrie*, 3rd edn, Paris: Gauthier-Villars. [Deals idiosyncratically but perceptively with some ancient topics, e.g. conics.]

Dijksterhuis, E. J. *1956, Archimedes*, Copenhagen: Munksgaard. [Brilliant analytic exposition of Archimedes' mathematics.]

Fowler, D. H. *1987, The Mathematics of Plato's Academy*, Oxford: Clarendon Press. [Interesting and speculative, with good discussion of sources.]

Heath, T. L. (transl.) *1910, Diophantus of Alexandria*, 2nd edn, Cambridge: Cambridge University Press.

—— *1922, A History of Greek Mathematics*, 2 vols, Oxford: Clarendon Press. [Thorough, competent survey of the documents. Out of date for the earliest periods.]

—— *1926, The Thirteen Books of Euclid's Elements*, 2nd edn, 3 vols, Cambridge: Cambridge University Press. [Excellent, wide-ranging commentary.]

Knorr, W. R. *1975, The Evolution of the Euclidean Elements*, Dordrecht: Reidel.

—— *1986, The Ancient Tradition of Geometric Problems*, Boston: Birkhäuser. [Insightful, often unconventional.]

Neugebauer, O. *1957, The Exact Sciences in Antiquity*, 2nd edn, Providence, RI: Brown University Press.

Tannery, P. *1912–50, Mémoires scientifiques*, 17 vols, Toulouse, private publication. [Vols 1–3 contain numerous important studies in Greek mathematics.]

van der Waerden, B. L. *1954, Science Awakening*, Groningen: Noordhoff. [Particularly strong for the period before Euclid and links with Babylonian and Egyptian mathematics.]

Ver Eecke, P. (transl.) *1923, Les coniques d'Apollonius de Perge*, Paris: Desclée, and Bruges: De Brouwer.

Zeuthen, H. G. *1886, Die Lehre von den Kegelschnitten im Altertum*, Copenhagen: Höst. [Profound study of ancient conics.]

1.4

Greek applied mathematics

ALEXANDER JONES

Mathematical concepts turn up in many contexts in Antiquity, including not only legitimate sciences, but also astrology and numerology, and even metaphysics (the Pythagorean tradition). The most interesting applications of mathematics, however, are to be found in the physical sciences. Mechanics, optics and astronomy are treated here; for music theory see §2.10, and for cartography see §8.15.

1 MECHANICS

Aristotelean physics distinguished between the 'natural' motion of bodies towards or away from the centre of the world and the 'forced' motion in any direction caused by contact with another moving body. In several passages of his works, Aristotle expresses rules of simple proportionality for both kinds of motion, for example the notorious assertion that the speed of a falling body is proportional to its weight, or that if a body A imparts a speed s to another body B, then it will impart a speed $2s$ to $\frac{1}{2}$ B. These rules are to some extent *ad hoc* elements in arguments on topics of immediate concern to Aristotle, such as whether void can exist; but they at least show that Aristotle was not wholly hostile to the mathematical treatment of movement, in spite of his well-known opinion that the objects of mathematics are independent of time and change. In the *Mechanics*, an early third-century BC work of the Peripatetics falsely ascribed to Aristotle, naive proportional relations between weight and speed are applied to static as well as kinetic situations (e.g. the lever), and here we also find the principle of the parallelogram of velocities.

Later work on mechanics seems to have continued the tendency already apparent in the pseudo-Aristotelean *Mechanics* of concentrating on problems related to machines. The *Mechanics* of Hero (*circa* AD 60), for example, contains numerical calculations of the power ratios of gearwork and pulleys; and treatises on military engines give the rule that the power of a catapult is proportional to the cube of the diameter of its spanner, an unusual instance of the discovery of a numerical formula from practical experimentation. At no stage, however, did the mathematical element in ancient mechanics reach a level of sophistication in any way comparable to that of the analysis of statics problems in Archimedes' *On the Equilibrium of Plane Figures* and *On Floating Bodies* (§1.3).

2 OPTICS

Mathematical optics seems to have developed around the beginning of the third century BC as an attempt to reconcile Aristotle's analysis of sensation with the physical theories of the Peripatetic philosophers, in which sight and illumination were supposed to occur through the emanation of thin, rod-like material rays from the eye or from the luminous body, respectively. Euclid's *Optica* represents the visual rays as a bundle of discrete lines directed from an eye point to various geometrically idealized objects, and demonstrates by geometrical reasoning that this hypothesis can account for the perception of Aristotle's 'common sensibles' (size, shape and move-ment), loss of resolution with distance, and various ambiguities and illusions of perception. Euclid's treatise is the oldest-known attempt to account for physical phenomena through axioms and proofs. A more inductive, experimental approach dominates Ptolemy's *Optica*. Some of Ptolemy's chief theoretical improvements upon Euclid in the field of direct vision are in his analysis of binocular vision and his rejection of the concept of discrete visual rays (Lejeune *1948*).

Catoptrics, the study of mirrors and reflecting surfaces, developed the consequences of the law of reflection at equal angles, which was known to the Peripatetics from about 300 BC (Knorr *1985*). The *Catoptrica* attributed to Euclid follows the interesting procedure of taking directly observable phenomena as axioms, and deducing geometrically the equal-angle law. This work already shows the division of catoptrics into two fields, one dealing with the location and properties of images seen in plane and curved mirrors, the other studying focal properties of mirrors reflecting solar rays.

Hero's *Catoptrica* (known only through a medieval Latin translation of an abridgement) anticipates Pierre de Fermat (§2.9) by inferring the equal-angle law from the assumption that the visual ray takes the shortest path from eye to object. Hero also displays the limits of mathematical treatment of images: although he begins with several geometrical theorems on the size and orientation of images in plane and circular mirrors (mostly plundered from the Euclidean treatise), he abandons all attempt at geometrical justification when he turns to more complex arrangements of illusion-producing mirrors. Diocles' *On Burning Mirrors*, a work recently dis-covered in an Arabic translation (Toomer *1976*), proves the focal property of a paraboloidal mirror, but makes heavy weather of other curved surfaces.

Refraction seems at first to have been treated only in a qualitative manner. From measurements of angles of refraction for the interfaces between various media, however, Ptolemy inferred (although he never explicitly stated) a second-order arithmetic progression relating angle of incidence and angle of refraction.

According to a passage in Vitruvius, perspective drawing originated in the fifth century BC (compare §12.6). The principles of linear perspective, however, can hardly have been known before geometrical optics had reached the stage represented by Euclid's *Optica*. The map projections that Ptolemy describes in his *Geography* are examples of qualitative perspective: the parallels and meridians on the globe are displayed as straight lines or circular arcs so as to give the illusion that one is looking at part of a sphere. Ptolemy also gives instructions for drawing a picture of a globe surrounded by rings representing the principal circles of the celestial sphere, which combine true linear perspective by central projection for the rings with qualitative perspective for the enclosed map. Pappus (*Collection*: 6) investigates problems connected with the perspective view of a circle.

3 ASTRONOMY

Greek mathematical astronomy was fundamentally geometrical, but the motions which had to be accounted for in astronomical models added a kinematic component to geometrical problems (Neugebauer *1975*). The first truly geometrical models in Greek astronomy were Eudoxus's homocentric spheres. Eudoxus hypothesized that all the apparent motions visible in the heavens — the daily revolutions of the fixed stars, and the slower and irregular movement relative to them of the Sun, Moon and planets — could be accounted for by combinations of revolving spheres concentric with the Earth. He proved that combining two spherical motions of equal but opposite speed and with slightly tilted axes results in a figure-of-eight path for a planet, a *hippopede*, which is the intersection of a sphere with an internally tangential cylinder. By adding a third revolution along the axis of the *hippopede*, Eudoxus achieved the desired alternation of direct and retrograde motions for the planets. How far Eudoxus was able to investigate the quantitative behaviour of his models is not known; in fact, the models cannot be made to reproduce even crudely the motions of Venus and Mars.

The epicyclic and eccentric models of later Greek astronomy, which combined more or less coplanar (but not concentric) circular motions, were more amenable to mathematical analysis. By the beginning of the second century BC it was known that the eccentric and epicyclic models were geometrically interchangeable; and Ptolemy (*Almagest*: XII) preserves for us an important theorem of Apollonius determining the points during a planet's revolution about an epicycle or eccentre where it will appear stationary to a terrestrial observer (Toomer *1984*).

Numerical data in astronomy were at first restricted almost entirely to estimates of Cosmic dimensions and periods of revolution. Two treatises from the first half of the third century BC, Aristarchus's measurement of the sizes and distances of the Sun and Moon, and Archimedes' *Sand Reckoner*, concern the calculation of numerical bounds to the dimensions of the *Cosmos*, but are more in the nature of mathematical exercises than serious astronomy. Exposure to Babylonian astronomy during the second century BC brought numerical schemes for predicting apparent positions of the heavenly bodies, together with sexagesimal arithmetic. Henceforth Greek astronomers usually carried out their calculations using sexagesimal notation, albeit only for the fractional parts of numbers. Ptolemy took it for granted that his readers knew how to manipulate sexagesimals; by the fourth century AD, the commentators Pappus and Theon found it necessary to explain at great length how to multiply, divide and take the square roots of sexagesimal numbers.

Hipparchus was probably the first to make quantitative measurements of the component circles of hypothetical eccentric or epicyclic models using selected observations. One of the prerequisites for this endeavour was trigonometry. Archimedes had already determined bounds for the ratio of a circle's perimeter to its diameter (subsequently improved upon by Apollonius), but it was left to Hipparchus to compile a table of the approximate lengths of the chords subtended by different angles in a standard circle. Hipparchus's chord table is lost (unless the Indian sine table is a descendant of it), but we do have Ptolemy's accurate chord table (*Almagest*: I), tabulated to three sexagesimal places and at intervals of half a degree. Since the chord of an angle is twice the sine of half the angle, the chord table can be used to solve most trigonometrical problems, but the lack of a tabulated function equivalent to the tangent was an obstacle in some calculations. Of particular importance for the solutions of astronomical problems involving trigonometry (e.g. the measurement of the radii of epicycles) was the chord equivalent of the sine theorem, which results directly from the elementary theorem that the angle at a circle's centre subtended by any chord is twice the angle subtended at the circle's periphery.

Problems connected with the risings and settings of stars on the celestial sphere motivated treatises on spherical geometry by Autolycus, Euclid and Theodosius, but a practicable spherical trigonometry only began with Menelaus's *Spherics* (*circa* AD 100), which first recognized the value of restricting attention to great circles, and gave the fundamental theorems for solving spherical triangles. Using these theorems, Ptolemy (*Almagest*: II) composed tables of the principal functions of spherical astronomy, which superseded approximate Babylonian-style tables based on arithmetical series.

Two other early methods of representing problems in spherical geometry on a plane are of mathematical interest. The so-called *analemma* was applied, especially in the theory of sundials, to determining angles on the celestial sphere by plane trigonometry. This technique, a sort of descriptive geometry, involved constructing a figure representing a plane section of the celestial sphere, and 'swinging' into this plane other circles occurring in the problem. Stereographic projection, a central projection of the celestial sphere from its south pole onto the equatorial plane that preserves all celestial circles as circles on the plane of projection, was the basis of the plane astrolabe.

Some astronomical problems gave rise to iterative procedures in which successive approximations converged on the desired result. For example, when Ptolemy (*Almagest*: XI) wishes to determine the eccentricities and the longitudes of the apogees of the planets Mars, Jupiter and Saturn according to his equant model, he begins by assuming a simpler model from which the apogees and eccentricities may be found by straightforward trigonometry. He then finds the correction necessitated by his real model by a subtle iteration process until he judges that the results are converging. In later treatises in the Ptolemaic tradition, the accurate moment of conjunction or opposition of the Sun or the Moon is found by using their calculated instantaneous positions and velocities to generate a succession of closer and closer approximations to the moment in questions.

No systematic approach to handling observational errors is to be found in ancient astronomy. Nevertheless, Hipparchus and Ptolemy did attempt to determine conditions for the choice of observations that would minimize the effect of errors on their subsequent calculations. On one occasion Ptolemy actually estimates from the characteristics of his instrument the possible error incurred in observing the dates of equinoxes and solstices (*Almagest*: III), but he does not discuss the rather large effects of such errors on his determination of the parameters of his solar model.

BIBLIOGRAPHY

Drachmann, A. G. *1963*, *The Mechanical Technology of Greek and Roman Antiquity*, Copenhagen: Munksgaard.

Knorr, W. *1985*, 'Archimedes and the pseudo-Euclidean *Catoptrics* [. . .]', *Archives Internationales d'Histoire des Sciences*, **35**, 28–105.

Lejeune, A. *1948*, *Euclide et Ptolémée, deux stades de l'optique géometrique grecque*, Louvain: Bibliothèque de l'Université.

—— (ed. and transl.) *1989*, *L'Optique de Claude Ptolémée*, 2nd edn, Leiden: Brill.

Neugebauer, O. *1957*, *The Exact Sciences in Antiquity*, 2nd edn, Providence, RI: Brown University Press.

—— *1975*, *A History of Ancient Mathematical Astronomy*, Berlin: Springer.

—— *1983*, *Astronomy and History, Selected Essays*, New York: Springer.

Simon, G. *1988*, *Le Regard, l'être et l'apparence dans l'optique de l'antiquité*, Paris: Seuil.

Toomer, G. J. (ed. and transl.) *1976*, *Diocles on Burning Mirrors*, Berlin: Springer.

—— (transl.) *1984*, *Ptolemy's Almagest*, London: Duckworth, and New York: Springer.

1.5

Later Greek and Byzantine mathematics

ALEXANDER JONES

1 GENERAL CHARACTER

With respect to the political and cultural history of the Greek-speaking civilization, the period called 'Antiquity' is conventionally said to end either with the founding of Constantinople in AD 324, or with Justinian's closing of the philosophical schools in AD 529, events that highlight the transition from the pagan Roman Empire with its Mediterranean cultural centres (e.g. Alexandria) to the new Byzantine Empire with its focus in the new capital, Constantinople. But for the history of mathematics, it is more convenient to make an earlier demarcation, about AD 300. Following a long interval for which we have only sparse evidence of mathematical activity, the fourth century brought a significant revival of writing on mathematics.

It has to be admitted that from this time on the Greek-speaking world made only a modest contribution to the growth of mathematics in terms of new concepts and methods. The final centuries of pagan Antiquity still witnessed a few attempts to add something new to the body of knowledge; but the subsequent history of Byzantine science up to the fall of Constantinople (AD 1452) is a pattern of declines and recoveries, in which the most fruitful interludes are marked by intelligent scholars seeking out, explicating and comparing old or foreign texts. Original work during this period is not so much feeble as non-existent. Byzantium was of crucial importance, however, as a channel for the survival and transmission of the mathematical sciences, passing on the works of Antiquity to its neighbours to the east (e.g. Syria, the Arabs) and to the west, and often absorbing from them new ideas in return (Vogel *1967*, Wilson *1983*).

The documentation for the later periods of Greek mathematics is in some respects better than it is for the Classical and Hellenistic periods: many works survive in manuscript, often at few removes from the original composition, and in some instances we even possess the autographs. But the subject has suffered from scholarly neglect, so that even serviceable editions are often lacking, especially for the later Byzantine authors.

2 THE SCHOOLS OF LATE ANTIQUITY

The various writings of Pappus (*circa* AD 320) that have come down to us under the title of *Collection* (Ver Eecke *1933*) are such a rich source for the exact sciences that, almost on the merits of this work alone, historians have described the fourth century AD as a 'Silver Age' of Greek mathematics. Although later authors sometimes refer to him as a 'philosopher', Pappus seems to have made his living primarily by teaching mathematics and astronomy at Alexandria, and many of his writings are commentaries or readers' aids to the classics. But Pappus read widely and had an eye for curious matter; his discussion of solutions of geometrical problems under restricted conditions (e.g. constructions using a compass of fixed opening) is just one example of the aspects of Greek mathematics that would be wholly unknown to us but for him. From time to time he claims credit for a proposition, for example a new way of constructing the two mean proportionals between given line segments, and a theorem on the volumes of solids of revolution. In both these instances he has merely added a veneer of Pappian originality to work substantially achieved by his predecessors. Pappus knew that he lived in a period of decline, and his reverence for the 'ancients' is matched by his disdain for his contemporaries.

The writings of Theon of Alexandria (*circa* 370) are even more didactic than Pappus's; we have much of his voluminous commentaries on Ptolemy's *Almagest* and *Handy Tables*. Theon published recensions of Euclid's *Elements* and *Data*; the principal aim of Theon's editing seems to have been to provide a reliable, mathematically correct text, although in the *Elements* he made some minor alterations. Modern scholarship has credited Theon with editions of several other works of Euclid and Ptolemy, though not always on strong evidence. The production of new editions of the classics was apparently a common activity in late Antiquity, with approaches ranging from conservatism to thorough rewriting. Theon's learned and ill-fated daughter Hypatia is supposed to have written commentaries on Diophantus, Apollonius's *Conics* and Ptolemy's *Handy Tables*, but the closest that we have to anything from her pen is the third book of Theon's commentary on the *Almagest*, which (according to the colophon) Hypatia proofread. (Incidentally, Hypatia is not the earliest known woman mathematician; Pappus had directed a polemic against a female teacher of mathematics named Pandrosion, and a certain Ptolemais is quoted in Porphyry's commentary on Ptolemy's *Harmonics*.)

The last independent treatises on mathematical topics that Antiquity has left us are the two short books by Serenus, *On the Section of a Cylinder* and *On the Section of a Cone*. In the former, Serenus is at pains to prove that the plane section of a cylinder is an ellipse; the poverty of the latter is apparent from the fact that it is concerned exclusively with plane sections of a cone *through its vertex*. From about this time come the 'lemma-books', disorganized farragos of propositions apparently compiled in the course of reading other works; some of these survive in Arabic, for example among the books that passed for Archimedes'.

The rigour of geometrical reasoning profoundly impressed the philosopher Proclus (mid-fifth century), who attempted to recast Aristotelean physics and Neoplatonist metaphysics in the pattern of axioms and propositions. Proclus wrote a historically very informative commentary on Book I of the *Elements* (Morrow *1970*), and a competent synopsis of Ptolemy's *Almagest*. He passed on his interest in the history of mathematics to his successors in the Neoplatonist schools of Athens (Simplicius) and Alexandria (Ammonius and Heliodorus). The Alexandrian school even produced one specialist in mathematical texts, Ammonius's disciple Eutocius of Ascalon (early sixth century). Eutocius had a flair for hunting out manuscripts, and his commentaries on works by Archimedes are valuable for their inclusion of related materials from sources afterwards lost. His annotated edition of the first four books of Apollonius's *Conics* (on which our text of these books entirely depends) is noteworthy for Eutocius's method of collating different recensions and reporting variant proofs.

3 BYZANTIUM

Eutocius's edition of the *Conics* was made at the behest of Anthemius of Tralles, the famous architect of the church of St Sophia in Constantinople. Perhaps the relations between these men provided the channel by which other manuscripts of Greek mathematics found their way to the imperial capital; from the sixth century on, Constantinople replaced Alexandria as the most important centre of scientific activity. Anthemius himself wrote two brief works on mirror optics, one an inept plundering of Hero's *Catoptrica*, but the other containing interesting constructions of paraboloidal and ellipsoidal mirrors. Anthemius's colleague Isidore of Miletus had a part in the ongoing recension of Archimedes' works and Eutocius's commentaries; he also wrote the unimpressive 'Book XV' of the *Elements*, dealing with inscribing regular solids inside one another.

From about the same time come the 'Bobbio mathematical fragments', in the oldest surviving manuscript of Greek mathematics (excluding papyri). This manuscript was either written in Italy or taken there before the end of the seventh century; it consists of several parchment leaves of works on burning mirrors, centres of gravity and sundial theory that were later erased and reused for a Latin text. In the early seventh century Stephanus of Alexandria, a pupil of the last pagan teacher of the Alexandrian school, Olympiodorus, came to Constantinople, where he wrote a long but elementary set of instructions to Ptolemy's *Handy Tables*; Stephanus's imperial patron Heraclius seems also to have had some pretensions to a knowledge of astronomy.

For the next two hundred years there is hardly a trace of anyone writing, copying or even reading mathematical texts. The revival of interest in the ninth century is usually associated with Leo the Mathematician, a scholar, teacher and sometime bishop whose proficiency in Euclidean geometry is said to have aroused the interest of the Caliph al-Ma'mūn. Leo certainly acquired manuscripts of important mathematical authors, including Archimedes and Apollonius, although it is not at all clear how well he understood the more technical aspects of ancient mathematics and astronomy. The ninth and tenth centuries saw the recopying of numerous ancient scientific authors (among them Euclid, Archimedes, Ptolemy, Pappus and Theon) in parchment codices, many of which have survived. These superb calligraphic books testify to the taste and wealth of their owners. Typical is the beautiful manuscript D'Orville 301 of Euclid's *Elements*, which the cleric and scholar Arethas commissioned in the year 888 for 14 gold pieces. What one misses is evidence that the texts were read and understood; trivial scribal errors in the mathematics were allowed to stand uncorrected, and the marginal annotations were usually copied from earlier manuscripts. Nor have we any significant original writings from this period.

4 THE MIDDLE AGES

Scholarly activity was vigorous during the eleventh and twelfth centuries – this was the time of Psellus, John Tzetzes and Eustathius – but the mathematical sciences played a small part in the learning of these men. Tzetzes, it is true, was fascinated with the personality of Archimedes, but the titles of the Archimedean writings that he claims to have read are mostly fictitious; and though he at one time had a rare copy of Pappus (possibly the present Vaticanus graecus 218), he consulted it only for general remarks on mechanics and for some names to drop. Very few extant mathematical manuscripts date from these centuries, while several important manuscripts became inaccessible, for example the Archimedes codex containing the *Method*, the pages of which were written over with a religious text. Scientific texts by, for example, Ptolemy and Euclid found their way to Sicily and southern Italy, where they were read and translated by western scholars. In the second half of the thirteenth century still more manuscripts were taken to Italy by the Flemish translator William of Moerbeke (see §2.1). Latin's gain was Greek's loss: what were probably the last two copies of Archimedes' works left the Byzantine world in this way.

On the other hand, mathematical astronomy started to show signs of renewed vigour in the eleventh century (Tihon *1983–*). Several texts show familiarity with Islamic astronomical treatises of the ninth and tenth centuries. Some of these Arabic works may have been known in Constantinople through the agency of Symeon Seth, who had sojourned in Egypt. The most impressive of the anonymous Byzantine adaptations of Arabic material is a handbook of astronomy composed between 1060 and 1072 that contains Al-Khwārizmī's method of predicting a solar eclipse and evinces familiarity with Islamic trigonometric tables (see §1.7). The author operates competently with sexagesimal numerals, although he shows a degree of self-consciousness in multiplications. It may be through the same circle of Arabophiles that knowledge of Arabic numerals first came to Byzantium; they appear in their eastern form in twelfth-century scholia to the *Elements*.

This movement seems to have died out around 1200. Most of the thirteenth century is a blank, no doubt partly because of the difficult conditions resulting from the fall of Constantinople in the Fourth Crusade (1204). Around the end of the century manuscripts began to be copied in large numbers (§2.1); these were written on paper and were the work of scholars, not calligraphers. Mathematics and astronomy had meanwhile recovered a respectable place in the libraries of the learned. Along with Euclid, Apollonius and Diophantus were again read, although it is difficult to guess how thoroughly the last two authors were understood. In about 1300, Maximus Planudes, a classical scholar of considerable talents, revised the texts of Diophantus and of Ptolemy's *Geography* (successfully reconstructing the map projections). Planudes also wrote an influential treatise on the use of the Arabic numerals; this was based on an earlier anonymous work of about 1252, but whereas that book used the western forms, Planudes chose the eastern.

Two anonymous compilations (Vogel *1968*, Hunger and Vogel *1963*) attest to an interest in practical arithmetical problems whose tradition was little affected by cultural boundaries; some of the problems would have looked familiar to an Egyptian scribe. The later of these collections dates from the very end of the Byzantine Empire, and its author has learned from the Turk al-Kāshī's decimal notation for fractions.

But the most significant developments of this period were in astronomy, involving on the one hand the complete mastery of Ptolemy's treatises by Theodore Metochites and his pupil Nicephorus Gregoras, and on the other the translation and explication of 'Persian' (in fact mostly Arabic) astronomical tables by Gregory (or George) Chioniades and others. The two traditions achieved a kind of synthesis in the *Tribiblos* of Theodore Meliteniotes in the middle of the fourteenth century. New translations and original treatises continued to be produced into the following century.

Even before the fall of Constantinople to the Turks in 1452, scholarly productions of all kinds had once more begun to decline. Manuscripts were passing in great numbers to Italy, together with such bibliophile Greeks as Bessarion and Isidore of Kiev. The works of the mathematicians (those who, unlike Archimedes and Pappus, had not already made the journey) shared in this migration, and entered a new and fertile milieu.

BIBLIOGRAPHY

Hunger, H. and Vogel, K. *1963, Ein byzantinisches Rechenbuch des 15. Jahrhunderts*, Vienna: Österreichische Akademie der Wissenschaften, Philosophisch-historische Klasse, Denkschriften 78.2.

Lemerle, P. *1971, Le Premier humanisme byzantin*, Paris: Presses Universitaires de France.

Morrow, G. R. (transl.) *1970, Proclus: A Commentary on the First Book of Euclid's Elements*, Princeton, NJ: Princeton University Press.

Tihon, A. *1981*, 'L'Astronomie byzantine (du Ve au XVe siècle)', *Byzantion*, **51**, 603–24.

—— (gen. ed.) *1983–, Corpus des astronomes byzantins*, Amsterdam: Gieben. [Ongoing series of editions and translations of Byzantine astronomical texts; 4 vols to date.]

Ver Eecke, P. (transl.) *1933, Pappus d'Alexandrie. La Collection mathématique*, 2 vols, Paris: Desclée, and Bruges: De Brouwer.

Vogel, K. *1967*, 'Byzantine science', in *Cambridge Medieval History*, Vol. 4, new edn, Cambridge: Cambridge University Press, 264–305, 452–70. [Compressed survey, but useful bibliography.]

—— *1968, Ein byzantinisches Rechenbuch des frühen 14. Jahrhunderts*, Vienna: Hermann Böhlaus Nachfolge (Wiener byzantinische Studien 6).

Wilson, N. *1983, Scholars of Byzantium*, London: Duckworth, and Baltimore, MD: Johns Hopkins University Press.

1.6

Pure mathematics in Islamic civilization

J. P. HOGENDIJK

1 INTRODUCTION

Mathematics in Islamic civilization between AD 700 and 1700 is called 'Islamic mathematics' by some and 'Arabic mathematics' by others, but neither term is completely appropriate. Most of the mathematicians whose work is considered in this article were Muslim, but there were also mathematicians with other religious backgrounds. There are a number of mathematical problems with a specific connection with Islam (see especially §1.7), but most of the mathematics in Islamic civilization was not directly related to religion. Reference to 'Arabic mathematics' and 'Arabic mathematicians' is therefore to be preferred, where the adjective 'Arabic' is to be understood in a linguistic sense. A large number of important 'Arabic' mathematicians were non-Arabs, who lived in Persia or other non-Arabic speaking areas, but Arabic was always the main language for mathematics (and science). In the tenth century and after, a few mathematical works were written in or translated into Persian; but their number is negligible compared with the works that were written in Arabic, and most of the technical mathematical terminology in these Persian works was borrowed from the Arabic.

Our main sources for Arabic mathematics are Arabic manuscripts copied between the tenth and nineteenth centuries by scribes who in most cases were not themselves mathematicians (Sezgin *1974*, Matvievskaya and Rozenfeld *1983*). The most important collections of these manuscripts are now in the Near East, Europe and India. Although the history of Arabic science has been investigated since the early nineteenth century, a large number of Arabic scientific manuscripts have not yet been studied, so our knowledge of Arabic mathematics is by no means complete. Information on the history of Arabic mathematics can also be obtained from the study of Arabic manuscripts on subjects outside the narrow limit of what is nowadays considered to be 'mathematics', such as astronomical texts and tables, and texts on optics, law, linguistics, and so on. Other sources for Arabic mathematics are instruments such as astrolabes and sundials, and Latin and Hebrew translations of Arabic mathematical treatises that are now lost.

Mathematics was studied in Islamic civilization for various reasons. Learning some elementary mathematics was often considered to be part of a general intellectual education. Elementary applications of arithmetic and geometry were made in commerce, law, state administration and land measurement. We know little about applications of geometry in Islamic architecture and the construction of mosaics (see §12.6). Applications of a very high level were made in astronomy, astrology and optics. Most people who took training in advanced mathematics did so in order to become astronomers (or astrologers). The most important creative contributions in Arabic mathematics were made by authors who studied mathematics for its own sake. A substantial part of the mathematics that was developed (for example, almost all of algebra) had no application whatsoever. Some medieval Arabic algebraists discussed 'applications' of algebra in the computation of inheritances, but this may be dismissed as propaganda for mathematics.

The applications of mathematics inspired by Islamic religion are noted in the next article, §1.7; see also §2.7 and §2.9 on astronomy and optics. The main concern here is with the 'pure' aspects of mathematics, where 'pure' is used in the sense of having little application in other sciences or daily life. Some subjects which have 'pure' and 'applied' aspects have been divided arbitrarily between this article and the next; thus trigonometry and stereographic projection are discussed in §1.6. Høyrup *1987* has discussed possible reasons why the connection between pure, applied and practical mathematics was more important in Islamic civilization than ever before.

2 THE TRANSMISSION OF MATHEMATICS INTO ARABIC

By 700 the Islamic empire included Persia and Syria, where remnants of ancient scientific traditions were still to be found. The history of Arabic science begins after the creation of the Abbasid Empire (in 760) and the foundation of the new capital, Baghdād. Towards the end of the eighth century Indian astronomers were received at the court of Baghdād, and some astronomical works were transmitted from Sanskrit into Arabic. In the early ninth century, the caliphs collected Greek scientific manuscripts and established at Baghdād the 'House of Wisdom', a kind of academy of science.

71

Many scientific treatises were translated from Greek into Arabic at that time, often more than once. For mathematics the transmission was by no means a trivial process, as is illustrated by the case of the *Conics* of Apollonius. In the ninth century the Banū Mūsā, three able geometers, wanted to translate the text, but they had only one defective manuscript of the first seven books, and they could not understand it. One of the brothers, Ḥasan, worked out for himself the theory of the sections of the cylinder, because he believed that this could serve as an introduction to the theory of the sections of the cone in the *Conics*. After his death, his brother Aḥmad found in Syria a manuscript of the first four books with the commentary by Eutocius. By means of the two manuscripts and Ḥasan's theory, Aḥmad and the third brother, Muḥammad, were able to make sense of Apollonius, and they inserted cross-references in the Greek text to make it more comprehensible. The Banū Mūsā then supervised the translation of the seven books by al-Ḥimṣi (I–IV) and Thābit ibn Qurra (V–VII) (Toomer *1990*: 620–29). The translation of scientific treatises into Arabic often required the creation of a new technical terminology. Foreign terms cannot easily be adapted to the structure of the Arabic language, so that a simple transcription is inelegant. Thus, for example, the Greek word *hyperbolè* (for hyperbola) was first transcribed as *ūbarbūlā* and later translated as *qaṭʿzāʾid* ('exceeding section'). After this term had been created, the word *ūbarbūlā* was no longer used.

The Arabic translators preserved several Greek mathematical works that are lost in the Greek original, such as the *Spherics* of Menelaus, *On Division* of Euclid, Books V–VII of the *Conics* of Apollonius and Books IV–VII of the *Arithmetica* of Diophantus. New traces of lost Greek mathematical works are still to be found in Arabic texts.

3 NUMBER SYSTEMS AND THE NUMBER CONCEPT

The Arabic mathematicians used different systems for writing numbers. The Hindu–Arabic numbers were transmitted from India into Arabic at the end of the eighth century as a system for writing positive integers. The use of these 'Indian numbers' is explained around 830 in a work by Muḥammad ibn Mūsā al-Khwārizmī (meaning from Khwārizm, now Khiwa, south of the Aral Sea), whose name was Latinized in the twelfth century as Algorismi (hence our word 'algorism' or, less correctly, 'algorithm'). In the tenth century, two varieties of symbols appeared: the *hindī* ('Indian') numbers, which are still in use in the Near East, and the *gubarī* ('dust') numbers, which were used in North Africa and Spain, and which are the immediate precursors of the modern number symbols 1, 2, 3, 4, 5, 6, 7, 8, 9 and 0.

The Arabic geometers and astronomers used another system, known as 'numbers of the astronomers', which they had taken from Ptolemy's *Almagest* and other Greek astronomical works. In this system integers were written as letters of the alphabet, and fractions were written sexagesimally, again using letters of the alphabet. The Arabic translators transcribed each Greek letter as the Arabic letter with the same numerical value, and they simply copied the Greek symbol for zero, a Greek omicron with a bar: ō. The sexagesimals in the fractional part of a number were called 'minutes' and 'seconds'. There was no separation symbol between integers and fractions. Decimal fractions (expressed in Hindu–Arabic numbers) were first used by al-Uqlīdisī (tenth century), and after him by al-Samaw'al (twelfth century, see Rashed *1984*: 122–8) and al-Kāshī (fourteenth century).

In the Hindu–Arabic system and the 'numbers of the astronomers' the symbol for zero was called *ṣifr* (Latinized as *cipherum* and *zephirum*, whence 'cipher' and 'zero'), meaning 'empty place' in the sexagesimal or decimal expansion of a number. Zero itself was not universally accepted as a number, and negative numbers were inconceivable in Arabic mathematics. The mathematicians worked with subtraction as an arithmetical operation, and 'minus 3' was accepted as a term in an equation or a polynomial. In classical Greek geometry, irrational segments, such as the diagonal of a square of side 1, were not considered to be numbers. The Arabic mathematicians were willing to consider the square root of 2 to be a number, and 'Umar al-Khayyāmī even assigned a numerical value to any ratio between arbitrary magnitudes. Hence, as soon as a unit segment is chosen, any line segment can be associated with a number, namely its length. What is happening here is not a conscious extension of the number concept; rather, the theory is being adapted to the practice of geometric and trigonometric computations, which had existed since the time of Ptolemy, in which the lengths of irrational segments were required to be computed. Complex numbers do not occur at all in Arabic mathematics.

4 ALGEBRA

By 'algebra' we mean here the art of solving equations and of manipulating equations and polynomials. It is unclear how and to what extent algebra was transmitted from pre-Islamic cultures into Arabic mathematics. A very successful early Arabic presentation of algebra is the book of 'restoration and confrontation' (*al-jabr wa'l-muqābala*) by al-Khwārizmī. He first presents the solution of six standard equations, in modern terms $bx = c$, $ax^2 = bx$, $ax^2 = c$, $ax^2 = bx + c$, $ax^2 + c = bx$ and $ax^2 + bx = c$, with $a, b, c > 0$. He then discusses the reduction of linear and quadratic equations to these standard forms, with many worked examples. Al-Khwārizmī did not use symbolism; instead of $ax^2 = bx + c$ he said 'properties are equal to roots plus numbers', and he expressed the modern equation $4x^2 - 3 = 2x$ as 'four properties except three dirhams are equal to two roots'. In order to solve this equation, one has to move the 'defect' of 3 dirhams to the other side of the equation. The technical term for this operation, *al-jabr* ('the restoration'), was Latinized to 'algebra'.

We now give some examples of the advances in Arabic algebra after al-Khwārizmī. Abū Kāmil studied quadratic equations with irrational coefficients and biquadratic equations that are trivially reducible (by a substitution) to quadratic equations. Al-Karajī and al-Samaw'al considered expressions in arbitrary (integer) powers of the unknown, but they lacked a convenient notation for the exponent. They expressed x^8 as 'property cube cube' (i.e. $x^2 \cdot x^3 \cdot x^3$) and $1/x^8$ as 'one part of property cube cube'. Al-Karajī observed that the square root of a polynomial $a_{n+k}x^{n+k} + \cdots + a_k x^k$ can be found by the then well-known algorithm for the extraction of the square root of a decimal number, $a_{n+k}10^{n+k} + \cdots + a_k 10^k$. Al-Karajī used this algorithm only in cases where the given polynomial is a square of another polynomial, so that the answer comes out nicely. Al-Samaw'al realized that the method for the division of decimal numbers can be generalized in the same way to the division of polynomials.

The Arabic mathematicians tried to solve the cubic equation algebraically, but they did not succeed. Their geometric solution of the cubic equation is discussed in Section 7 below. In the fifteenth century, the Western Arabic mathematician al-Qalṣādī used abbreviations of words in equations, in much the same way as Diophantus in his *Arithmetica* and the Renaissance algebraists in Europe (§2.4).

5 NUMBER THEORY

The number-theoretical parts of the *Elements* of Euclid and the works of Nicomachus and Diophantus on arithmetic were transmitted into Arabic and thoroughly studied by many Arabic mathematicians. A notable Arabic contribution is the formula for amicable numbers discovered and proved by Thābit ibn Qurra: if $p = 3(2^{n-1} - 1)$, $q = 3(2^n - 1)$ and $r = 9(2^{2n-1} - 1)$ are all prime, then the two numbers $a = 2^n pq$ and $b = 2^n r$ are amicable; that is to say, the sum of the proper divisors of a is equal to b, and the sum of the proper divisors of b is equal to a. Ibn al-Haytham discovered that any prime number p divides $(p - 1)! + 1$, and in the tenth century it was believed that $x^3 + y^3 = z^3$ had no integer solutions x, y, z (Rashed *1984*: 242, 220). It is unlikely that rigorous proofs of the two above-mentioned facts were given in the Middle Ages. The Arabic mathematicians solved some new Diophantine equations by skilful variations of Greek methods

(Sesiano *1977*), but no substantial methodological advances were made in number theory as a whole.

6 GEOMETRY

The Arabic geometers diligently studied, edited and commented on the Classical Greek geometrical works by Euclid, Apollonius, Archimedes and Menelaus, and Arabic geometry can be considered in part as a further development of these works. The *Elements* of Euclid inspired deep investigations of the foundations of geometry. Thus some mathematicians attempted to 'prove' Euclid's parallel postulate, while others, such as Ibn al-Haytham and Naṣīr al-Dīn al-Ṭūsī replaced it with another postulate which they believed to be simpler and therefore more appropriate (Jaouiche *1986*). Al-Māhānī replaced the definition of proportionality in Euclid's *Elements* by an alternative definition to the effect that $a:b=c:d$ if their continued-fraction expansions are the same.

In the tenth century, the Arabic geometers used the *Conics* of Apollonius in the solution of problems that cannot be solved by means of straightedge and compass, such as the construction of a regular heptagon (Hogendijk *1984*), the trisection of an angle, the construction of an equilateral pentagon in a square, and the solution of the following optical problem: given the positions of a concave or convex circular mirror, the eye and the object, to find the point of reflection. This problem was named the 'problem of Alhazen' after al-Ḥasan ibn al-Haytham, who solved it in his *Optics*. Archimedes' quadratures of the parabola had not been transmitted into Arabic, but the Arabic geometers knew from the preface of *On the Sphere and Cylinder* that Archimedes had worked on this problem. Thābit ibn Qurra then found the area of a segment of the parabola in his own way, and he also derived the volume of a solid of revolution of a parabola around its axis. Ibn al-Haytham determined the volume of the solid of revolution of a parabolic segment around an ordinate.

The Arabic mathematicians proved these theorems in the rigorous Archimedean way: they considered suitable inscribed and circumscribed figures, determined the areas or volumes of these figures, and then carried out an indirect passage to the limit by means of a *reductio ad absurdum*. In order to determine the volume of certain inscribed and circumscribed figures, Ibn al-Haytham solved a problem which is mathematically equivalent to expressing $1^4 + 2^4 + \cdots + n^4$ as a polynomial of degree 5 in n and proving that the expression is correct (Juschkewitch *1964*: 288–95). Al-Kūhī found the centres of gravity of several curvilinear areas and solids; he then conjectured (for mystical reasons) a theorem about the centre of gravity of a semicircle, from which he derived $\pi = 3\frac{1}{9}$, in conflict with the

Archimedean $3\frac{1}{7} > \pi > 3\frac{10}{71}$ (Berggren *1983*). Most Arabic mathematicians knew better than al-Kūhī, and in the fourteenth century the first 17 decimal places of π were correctly determined by al-Kāshī (Juschkewitsch *1964*, 312–19).

The above summary does not exhaust the non-trivial Arabic contributions to geometry. For example, several Arabic mathematicians gave new solutions to a problem posed by Apollonius: to construct (by means of straightedge and compass) a circle tangential to three given circles. There are also interesting Arabic treatises on the methodology and teaching of geometry; and see §2.9 on aspects of their optics.

7 ALGEBRA AND GEOMETRY

Algebra and geometry were considered to be two distinct fields, which, however, could be related in various ways. Geometry was often used as a foundation for algebra. Thābit ibn Qurra wrote a text in which he proved the correctness of al-Khwārizmī's rules for solving quadratic equations, on the basis of Book II of Euclid's *Elements*. A number of later Arabic algebraists such as al-Karajī, al-Samaw'al and Sharaf al-Dīn al-Ṭūsī attempted to put the science of algebra on a solid foundation, namely the theory of ratios of arbitrary magnitudes set out in Book V of Euclid's *Elements*. From the tenth century onwards, cubic equations were solved geometrically by means of two intersecting conic sections determined by the coefficients of the equation. 'Umar al-Khayyāmī's treatment is famous but incomplete, because he did not discuss the necessary and sufficient condition for the existence of a root in non-trivial cases. In the tenth and eleventh centuries, it was observed that the two conics in the geometrical construction of the roots of $x^3 + c = ax^2$ are tangent for $c = (4/27)a^3$, and hence the correct condition $c \leqslant (4/27)a^3$ was derived. In the twelfth century, Sharaf al-Dīn al-Ṭūsī found necessary and sufficient conditions for the existence of positive roots of all types of cubic equation $f(x) = c$ by skilful manipulation of expressions $f(y) - f(x)$ using the methodology of Book II of Euclid's *Elements* (Hogendijk *1989*).

All these are cases where geometry was applied in algebra. But algebra was also used in geometry, especially for trigonometric applications. Abū Kāmil derived a quadratic equation for the side x of a regular pentagon inscribed in a circle of radius 10, and then approximated x numerically. In the tenth century, the constructions of a regular heptagon and nonagon were shown to be equivalent to the solutions of certain cubic equations. Al-Kāshī showed that the trisection of any angle is equivalent to the solution of a cubic equation. He used this principle to obtain the value of $\sin 1°$, the

76

fundamental quantity in trigonometric tables, by an iteration process which is explained in the next section.

8 NUMERICAL MATHEMATICS

The Arabic mathematicians made important progress in the numerical solution of algebraic equations by means of what is now called the method of Ruffini–Horner (§6.1). The general idea is as follows (Luckey *1948*). Suppose that $P(x) = Q(x)$ is the equation to be solved (all terms are, of course, positive). (The description here is for finding x in the decimal system; to find x in the sexagesimal system, change 10 to 60.) One first determines the order of magnitude of x, that is to say k such that $10^k \leqslant x < 10^{k+1}$. One then finds (basically by trial and error) $x_0 = m \cdot 10^k$, m an integer, such that either $P(x_0) \leqslant Q(x_0)$, $P(x_0 + 10^k) > Q(x_0 + 10^k)$ or $P(x_0) \geqslant Q(x_0)$, $P(x_0 + 10^k) < Q(x_0 + 10^k)$. Clearly, m is the first digit in the decimal expression of x. If we put $x = x_0 + 10y$, then $P(x) = Q(x)$ is seen to be equivalent to a new identity $P_1(y) = Q_1(y)$ for certain polynomials P_1 and Q_1. We can now repeat the process and find the first digit in the decimal expression of y, which is the second digit in the decimal expression of x, and so on.

The method of Ruffini–Horner is basically an algorithm for obtaining P_1 and Q_1 from P, Q and x_0. This method was used in Antiquity (in sexagesimal form) to extract square roots, and in China and by Kūshyār ibn Labbān in the tenth century to extract cube roots. In the twelfth century or earlier, it was generalized to approximate the roots of arbitrary cubic equations and equations of higher degree (Rashed *1984*: 100). In the fourteenth century, al-Kāshī extracted the nth root for arbitrary n by this method (illustrated for $n = 5$ by Berggren *1986*: 53–62).

Iteration was also used for the numerical approximation of roots of equations. Al-Kāshī showed that $x = 60 \sin 1°$ is the root of the cubic equation $x^3 + c = bx$, with $c = 54\,000 \sin 3°$ and $b = 2700$; he then approximated this root by means of the iteration process $x_0 = c/b$, $x_n = (x_{n-1}^3 + c)/b$ for $n \geqslant 1$ (Juschkewitch *1964*: 321). Since the time of Ptolemy, iteration had also been used in astronomical and astrological problems. In such problems one is sometimes required to find, for given λ and a given function f, a quantity θ such that $\lambda = \theta - f(\theta)$. There are various instances of such problems being solved by the iteration $\theta_0 = \lambda$, $\theta_n = \lambda + f(\theta_{n-1})$, ... (Juschkewitch *1964*: 324 for the case $f(\theta) = m \sin \theta$). This process converges if $f(\xi)$ is small in comparison to ξ for ξ in a neighbourhood of θ. Note that al-Kāshī's iteration is the special case $\lambda = c/b$, $f(\theta) = \theta^3/b$.

9 MISCELLANEOUS

From the time of Abū Kāmil, the Arabic mathematicians were interested in computations involving complicated sums and differences of square roots. They found what amounts in modern terms to an expression for $(a + b)^n$ in terms of powers of a and b for at least $n \leqslant 12$. Al-Karajī gave, in a lost work quoted by al-Samaw'al, a table of binomial coefficients corresponding to the first 12 rows of what is known as Pascal's triangle (Rashed *1984*: 76–7). Around 1210, Ibn Mun'im discussed the binomial coefficients from a purely combinatorial point of view, without reference to $(a + b)^n$ (Djebbar *1983*). Al-Kāshī also gave binomial coefficients for $n = 5$ in connection with a simplified form of the method of Ruffini–Horner for the numerical approximation of the fifth root (see §10.1 on combinatorics).

10 THE INFLUENCE OF ISLAMIC MATHEMATICS

In the twelfth and thirteenth centuries many Arabic mathematical texts were translated into Latin in Spain and, to a lesser extent, in Sicily. These were Arabic translations of Greek mathematical works, but also works by Arabic authors, such as the treatises on Indian numbers and algebra by al-Khwārizmī and the *Optics* of Ibn al-Haytham. Thus the medieval European scholars came into contact with mathematics and astronomy of a much higher level than they had known before. Another type of transmission is exemplified by Leonardo of Pisa (Fibonacci), who travelled and studied in the Western Arabic world and wrote various books on mathematics after his return. It is important to bear in mind that the level of mathematics was much higher in the Eastern Islamic world (Iraq and Persia) than in the Western world, and that most of the mathematical discoveries were unknown in areas west of Egypt. Thus the transmission of Arabic mathematics into medieval Europe was very incomplete, and often the works that were transmitted (such as those of al-Khwārizmī) represent a stage that had long been surpassed in the East. The contributions of the most important Arabic mathematicians, such as Thābit ibn Qurra, Ibrāhīm ibn Sinān, al-Kūhī, al-Bīrūnī, 'Umar al-Khayyāmī, Sharaf al-Dīn al-Ṭūsī and al-Kāshī, have become known only as a result of modern historical research.

BIBLIOGRAPHY

Berggren, J. L. *1983*, 'The correspondence of Abū Sahl al-Kūhī and Abū Isḥāq al-Ṣābī: A translation with commentaries', *Journal for History of Arabic Science*, 7, 39–124.

—— 1986, *Episodes in the Mathematics of Medieval Islam*, New York: Springer. [Introductory.]

Djebbar, A. *1983*, *L'Analyse combinatoire dans l'enseignement d'Ibn Mun'im*, Orsay: Université de Paris-Sud.

Hogendijk, J. P. *1984*, 'Greek and Arabic constructions of the regular heptagon', *Archive for History of Exact Sciences*, 30, 197–330.

—— 1989, 'Sharaf al-Dīn al-Ṭūsī on the number of positive roots of cubic equations', *Historia mathematica*, 16, 69–85.

Høyrup, J. *1987*, 'The formation of "Islamic mathematics": Sources and conditions', *Science in Context*, 1, 281–329.

Jaouiche, K. *1986*, *La Théorie des parallèles en pays d'Islam*, Paris: Vrin.

Juschkewitsch, A. P. *1964*, *Geschichte der Mathematik im Mittelalter*, Leipzig: Teubner. [The best survey to date. Translated from the Russian *Istoria matematiki b srednie veka*, 1961, Moscow: Nauka. The section on Arabic mathematics was translated into French as: Youschkevitch, A. P. 1976, *Les Mathématiques Arabes (VIIᵉ–XVᵉ siècles)*, Paris: Vrin.]

Luckey, P. *1948*, 'Die Ausziehung der *n*-ten Wurzel und der binomische Lehrsatz in der islamischen Mathematik', *Mathematische Annalen*, 120, 217–74.

Mathematical Reviews. [The best source of bibliographical information on recent publications in the history of Arabic mathematics; subject code 01 A 30.]

Matvievskaya, G. P. and Rozenfeld, B. A. *1983*, *Matematiki i astronomy musul'-manskogo srednevekovya i ikh trudy (VIII–XVII vv.)*, Moscow: Nauka. [References to manuscripts of all authors up to 1700. Large bibliography. Knowledge of the Russian alphabet is all that is necessary to consult this work with profit.]

Rashed, R. *1984*, *Entre arithmétique et algèbre. Recherches sur l'histoire des mathématiques Arabes*. Paris: Les Belles Lettres. [Algebraical interpretation in the style of N. Bourbaki.]

Sesiano, J. *1977*, 'Le Traitement des équations indéterminées dans le *Badī' fīl-Ḥisab* d'Abū Bakr al-Karajī', *Archive for History of Exact Sciences*, 17, 297–379.

Sezgin, F. *1974*, *Geschichte des arabischen Schrifttums*, Vol. 5, *Mathematik bis ca. 430 H.*, Leiden: Brill. [References to all known Arabic manuscripts of authors before 1050, Hebrew and Latin translations of such texts, and Arabic translations of ancient texts.]

Toomer, G. L. *1990*, *Apollonius, Conics, Books V to VII. The Arab Translation of the Lost Greek Original*, New York: Springer.

1.7

Mathematics applied to aspects of religious ritual in Islam

DAVID A. KING

1 INTRODUCTION

In Islam, as in no other religion in the history of mankind, scientific procedures have been applied to assist the organization of various aspects of religious ritual. These are:

1 a calendar whose periods are based on the Moon,
2 five daily prayers whose times are based on the Sun, and
3 a sacred direction whose goal is a specific location.

In addition should be mentioned:

4 the distribution of inheritances, and
5 the geometry of Islamic decorative art.

The first three topics were treated not only by the scientists of medieval Islam, but also by the scholars of the religious law, albeit in quite different ways. The solutions of the former were based on the complicated techniques of mathematical astronomy (§2.7), those of the latter on the simple techniques of folk astronomy (see King *1990* for an overview); thus many parts of 'applied' mathematics were involved. Although the scholars of the sacred law and the scientists proposed different solutions to the same individual problem, there are few records of serious discord between the two groups in the medieval sources. The legal scholars criticized mathematical astronomy mainly in so far as it was used by some as the handmaiden of astrology, which was anathema to them; the scientists seldom spoke out against the simple procedures adopted by the legal scholars. The fourth topic is treated both in treatises on the sacred law and in treatises on arithmetic, the approach differing only in the complexity of the problems

80

discussed; although the fifth is rarely handled in texts, there is a wealth of actual examples in surviving artefacts and architecture.

2 THE REGULATION OF THE LUNAR CALENDAR

The Islamic calendar is strictly lunar, and the beginnings and ends of the lunar months, in particular of the holy month of Ramadan, as well as various festivals throughout the 12-month 'year', are regulated by the first appearance of the lunar crescent. Since 12 lunar months add up to about 354 days, the 12-month cycles of the Islamic calendar occur some 11 days earlier each tropical year, and the individual months move forward through the seasons (see Ilyas *1984* for a modern discussion).

For the scholars of the sacred law, the month began when the crescent Moon was actually sighted. The astronomers, on the other hand, knew that they could predict the possibility of sighting on a given day, given the positions of the Sun and the Moon relative to each other and to the local horizon. They found limiting conditions on such quantities as the apparent angular separation of the Sun and the Moon, or the difference in their setting times over the local horizon, or the altitude of the Moon at sunset, and also compiled tables for facilitating predictions (Kennedy *et al. 1983*, *Kennedy Festschrift 1987*, King *1993*). Their pronouncements for each month appeared in annual ephemerides or almanacs.

3 THE REGULATION OF THE FIVE DAILY PRAYERS

The times of the five daily prayers in Islam are defined in terms of astronomical phenomena, and depend on the position of the Sun in the sky. Thus they vary with terrestrial latitude and, unless measured with respect to a local meridian, also with terrestrial longitude (see Ilyas *1984, 1988* for a modern discussion). Each of the five prayers in the Islamic day may be performed during a specified interval of time, and the earlier during the interval the prayer is performed, the better. The day begins with the *maghrib*, the sunset prayer. The second prayer is the *isha*, the evening prayer, which begins at nightfall. The third is the *fajr*, the dawn prayer, which begins at daybreak. The fourth is the *zuhr*, the noon prayer, which begins shortly after astronomical midday when the Sun has crossed the meridian. The fifth is the *asr*, the afternoon prayer, which begins when the shadow of any object has increased beyond its midday minimum by an amount equal to the length of the object casting the shadow.

In the first few decades of Islam, the times of prayer were regulated by observing shadow lengths by day and twilight phenomena in the evening and early morning. The muezzins who performed the call to prayer from the minarets of the mosques knew how to regulate the times of the daylight prayers by shadow lengths and those of the night-time prayers by the stars of the lunar zodiac. On the other hand, the determination of the precise moments – expressed in hours and minutes, local time – when the prayers should begin, according to the standard definitions, could be accomplished only by applying complicated mathematical procedures of spherical astronomy. Accurate as well as approximate formulas for reckoning time of day or night from solar or stellar altitudes were available to Muslim scholars from Indian sources; these were improved and simplified by Muslim astronomers over the centuries (Kennedy *et al. 1983*, King *1986*, *1993*). Certain astronomers from the ninth century onwards compiled tables for facilitating the determination of the prayer-times. The tables would show for each day of the year the precise shadow lengths and solar altitudes for midday and afternoon prayers, or the lengths of the intervals between the prayer times, such as the duration of morning and evening twilight. More extensive tables, often with tens of thousands of entries, were compiled for timekeeping by the Sun and stars.

From the thirteenth century onwards timekeeping was in the hands of the *muwaqqits*, professional astronomers employed in the mosques. The tables available to the *muwaqqits* had to be used together with instruments which could establish when the given prayer-time had arrived. The most popular of these were the astrolabe and the quadrant. Another means of regulating the daytime prayers was available to the Muslims in the form of the sundial (Schoy *1988*, King *1987*).

4 THE DETERMINATION OF THE SACRED DIRECTION

In Islam, the Kaaba in Mecca is a physical pointer to the presence of God. Muslims all over the world perform their prayers and other ritual acts facing the Kaaba. This sacred direction is called *qibla* in Arabic and in all other languages of the Islamic commonwealth.

In the first two centuries of Islam, when mosques were being built from Andalusia to Central Asia, the Muslims used astronomical risings and settings for the *qibla*. These directions were advocated over the centuries by the legal scholars. Muslim astronomers from the eighth century onwards also concerned themselves with the determination of the *qibla* as a problem of mathematical geography. This activity required the measurement of geographical coordinates and the computation of the direction of one locality from another by procedures of geometry or trigonometry. The *qibla* at any locality was defined as the direction of Mecca along a great circle.

The Muslims inherited the Greek tradition of mathematical geography, together with Ptolemy's lists of localities and their latitudes and longitudes, and the need to determine the *qibla* in different localities inspired much of the activity of the Muslim geographers. Once the geographical data are available, a mathematical procedure is necessary to determine the *qibla* (Kennedy *1973*, *1983*). Muslim astronomers developed a series of approximate solutions and different exact solutions (mainly involving plane and spherical trigonometry). They also compiled a series of tables displaying the *qibla* for each degree of difference in latitude and longitude between a location and Mecca, based on both approximate and exact formulas, and developed cartographic grids for finding the *qibla* directly from a map with Mecca at its centre (King *1986*, *1993*).

5 THE ARITHMETIC OF INHERITANCE

The Koranic rules for the distribution of estates to various relatives are complicated, and their application calls for some skill in arithmetic and first-order algebraic equations. The Prophet Muḥammad is reported to have said that the laws of inheritance comprise one-half of all useful knowledge; this may explain why al-Khwārizmī devoted half of his book on algebra to problems of inheritance. Very few of the mathematical writings on this subject have ever been studied (important studies are by Gandz *1938* and Rebstock *1992*).

6 GEOMETRIC DESIGN

The Muslims developed geometric design for the decoration of both religious buildings and secular artefacts. Elegant combinations and juxtapositions of different varieties of polygons which could be extended infinitely in all directions feature widely. (The propriety of such ornamentation is discussed by various legal scholars, but their writings have yet to be properly studied.) Only two Muslim mathematicians are known to have included remarks on geometric design in their writings, a fact which confirms the suspicion that this was an art passed down among the craftsmen. Abu'l-Wafa al-Buzajani (*circa* 1000) wrote on the construction of regular polygons, and al-Kāshī (*circa* 1415) discussed the three-dimensional honeycomb lattice called *muqarnas* (Berggren *1986*).

BIBLIOGRAPHY

Berggren, J. L. *1986*, *Episodes in the Mathematics of Medieval Islam*, New York: Springer. [Deals with topics 2–5 listed at the beginning of Section 1.]

The Encyclopaedia of Islam, *1960*–, 2nd edn, Leiden: Brill. [Survey articles: see Fara'id (inheritance shares), 'Ilm al-hay'a (astronomy), Kibla (sacred direction), Makka as centre of the world (sacred geography) and 'Ilm al-mikat (astronomical timekeeping).]

Gandz, S. *1938*, 'The algebra of inheritance', *Osiris*, **5**, 319–91. [Outlines the basic rules of topic 5, and gives numerous examples.]

Ilyas, M. *1984, A Modern Guide to Astronomical Calculations of Islamic Calendar, Times and Qibla*, Kuala Lumpur: Berita Publishing. [A modern discussion of topic 1 by a Muslim astronomer.]

—— *1988, Astronomy of Islamic Times for the Twenty-first Century*, London and New York: Mansell. [A modern discussion of topic 2.]

Kennedy, E. S. *1973, A Commentary upon Biruni's Kitab Tahdid al-Amakin, An 11th Century Treatise on Mathematical Geography*, Beirut: American University of Beirut Press. [A translation of, and key to, the most important medieval treatise on mathematical geography.]

Kennedy, E. S., colleagues and former students, *1983, Studies in the Islamic Exact Sciences*, Beirut: American University of Beirut Press. [Reprints of 70 articles by the leading scholar in the field, some dealing with topics 1–3.]

Kennedy Festschrift, 1987, 'From deferent to equant: Studies in the history of science in the ancient and medieval Near East in honor of E. S. Kennedy' (ed. D. A. King and G. Saliba), *Annals of the New York Academy of Sciences*, **500**.

King, D. A. *1986, Islamic Mathematical Astronomy*, London: Variorum Reprints. [This, King *1987* and King *1993* contain reprints of various articles dealing with topics 1–3.]

—— *1987, Islamic Astronomical Instruments*, London: Variorum Reprints.

—— *1990*, 'Science in the service of religion', *Impact of Science on Society*, (159), 245–62. [Surveys Muslim activity in topics 1–3.]

—— *1993, Astronomy in the Service of Islam*, Aldershot: Variorum Reprints.

Rebstock, U. *1992, Rechnen im islamischen Orient – Eine Studie über die literarischen Spuren der praktischen Rechenkunst*, Darmstadt: Wissenschaftliche Buchgesellschaft. [The first survey of arithmetic in daily life in Islamic society.]

Schoy, C. *1988, Beiträge zur arabisch–islamischen Mathematik und Astronomie*, 2 vols, Frankfurt. [Reprints of numerous articles, including several on topics 2 and 3, published between 1911 and 1926.]

1.8

Mathematics in Africa: Explicit and implicit

C. ZASLAVSKY

1 MATHEMATICS IN ANCIENT AFRICA

Africa is the birthplace of humankind. It may also have been the birthplace of mathematical ideas. The oldest mathematical artefact yet discovered, according to Bogashi *et al. 1987*, is a small piece of baboon bone engraved with 29 notches, found in the mountains between South Africa and Swaziland, and estimated to be 37 000 years old. Said to resemble the present-day calendar-sticks of the San (popularly known as Bushmen), it antedates by perhaps 15 000 years the famous Ishango incised bone discovered in eastern Zaïre on the shore of Lake Edward. A microscopic examination of the groups of engraved marks on the more recent find suggests that they correlate with a six-month lunar calendar (Figure 1).

For thousands of years, Africa was in the mainstream of mathematics history. From ancient Egypt came a written numeration system, a computed calendar, practical applications of arithmetic operations and fractions, the beginnings of algebra and geometry, and a complete curriculum, including mathematics, for training members of the priesthood. Over 2000 years after the construction of the Great Pyramid at Gizeh, Pythagoras and other Greek scholars travelled to Egypt to pursue their higher education. The establishment in Egypt of the city of Alexandria as the intellectual centre of the Greek-speaking world marked the culmination of the scholarship of this period. Among the scholars who came from far and wide to use the great library and to enjoy the stimulation of other like minds were such notables as Archimedes of Syracuse (in Sicily), Eratosthenes of Cyrene (in Libya) and Apollonius of Perga (in Asia Minor) (Eves *1976*). It may have been here that Euclid composed his famous *Elements*.

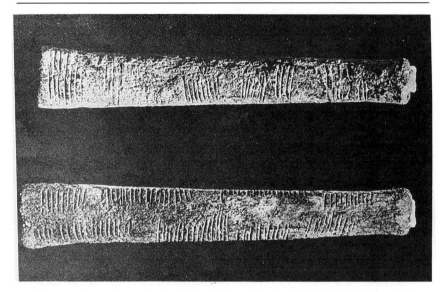

Figure 1 Two views of the Ishango bone, maybe 20 000–25 000 years old (from Zaslavsky *1979*: 19, on its dating see Marshack *1991*: 32). This carved bone indicates that a calendrical or numeration system was known to the fishing and hunting people of the region (Courtesy of Dr J. de Heinzelin)

2 AFRICAN MATHEMATICS IN THE ARABIC LANGUAGE

Northern Africa participated in the flowering of Arabic scholarship from the eighth to the fifteenth centuries. Many of the manuscripts of that period, written in the Arabic language by scholars of various ethnicities, have been analysed and translated only recently, and much work is still going on. Among those who made original contributions are the Egyptian Abū Kāmil (*circa* 850–930) with his work on algebra, and Ibn al-Haytham (Latinized as Alhazen), whose discoveries in optical geometry, carried out at the famous Cairo Academy of Science, influenced later European scientists (§1.6). Other scholars, working in Egypt and the Maghreb (north-west Africa), devised symbolism for fractions (§1.17), invented algebraic symbols for writing equations, and did work in the fields of combinatorial analysis and astronomy. The Hindu–Arabic numerals that we now use were introduced into Europe from North Africa. Toward the end of this historical period Muslim universities were established at Jenne and Timbuktu (in present-day Mali) with a curriculum that included mathematics. For further details about mathematics in northern Africa, see Djebbar *1981*, Lumpkin and Zitzler *1982*, Lumpkin *1983* and Pappademos *1983*.

Muḥammad ibn Muḥammad of Katsina (now northern Nigeria) was a later Muslim mathematician and astronomer. Part of his 1732 manuscript deals with the construction of magic squares in the Islamic tradition, some so large as to cover almost a whole page. Interest in mathematics remained strong among Muslim intellectuals. In 1826 the English traveller Hugh Clapperton paid a visit to Sultan Muḥammad Bello of northern Nigeria, bearing numerous gifts from the King of England. Most appreciated was a copy of Euclid's *Elements* in Arabic; the sultan's own copy had been destroyed in a fire (Zaslavsky *1979*).

3 MATHEMATICAL PRACTICES IN AFRICA

So far we have discussed written works that deal with mathematics as a formal discipline. Yet all societies have developed mathematical concepts and practices to serve their needs and interests, and one can apply ethno-mathematical principles to analyse them. Gerdes *1988b* speaks of 'hidden' or 'frozen' mathematics, meaning that when an artisan discovered a production technique, she or he was thinking mathematically. He goes on to analyse the geometric concepts implicit in such activities as house-building and basket-weaving in Mozambique. (See Denyer *1978* for a thorough discussion of the construction and decoration of buildings.) This information was passed down from one generation to the next by word of mouth and by example. Applying principles of graph theory, symmetry and number theory, Gerdes *1988a* and Ascher *1988* have done extensive analyses of the designs that the Chokwe (Tschokwe, Cokwe) of south-west Africa draw in the sand to illustrate the lore of their people and to instruct the young (see Figure 2, and the logo of this encyclopedia).

The level of a society's mathematical knowledge depends to a great extent on the cultural and technological level of that society. Although we cannot describe the mathematics they used, there is evidence: (a) that the Dogon people of Mali plotted the orbits of the Sirius star-system on the basis of knowledge acquired seven centuries ago (Adams *1983*, Griaule and Dieterlen *1965*); (b) for what seems to have been an astronomical observatory in north-west Kenya dating back over 2000 years (Lynch and Robbins *1978*); (c) for the practice of advanced metallurgy in central Africa at about the same time (Schmidt and Avery *1978*); (d) that Africans reached the New World long before the voyages of Columbus (Van Sertima *1977*); and (e) that eight centuries ago southern Africans constructed the massive complex of stone buildings called 'Great Zimbabwe' (Garlake *1973*).

All peoples count. Numeration systems may range from a few words to the extensive vocabulary of nations with a history of centuries of commerce. Among the ethnic groups of West Africa, numeration systems are usually

Figure 2 Chokwe (Angola) sand drawing illustrating the myth that describes how the world began (from Zaslavsky *1979*: 109)

based on grouping by twenties, with five and ten as subsidiary bases, while in southern and eastern Africa ten is the most common base. Frequently number-words are borrowed from neighbouring peoples or from Arabic and European languages. A characteristic of African counting is a standardized system of gestures to accompany, or even replace, the number-words.

One of the most interesting numeration systems is that of the Yoruba people of south-west Nigeria, a nation of urbanized traders and farmers. Not only is it based primarily on 20, with 10 as a subsidiary base, but it relies on subtraction to a great extent. For example, 45 is expressed as 'five from ten from three twenties'. Possibly the evolution of this complex system can be attributed to their method of counting cowrie shell currency, in use until the early twentieth century. The Yoruba word for 20 000 means 'bag', an allusion to the number of shells that a porter was required to carry on his head. Along the upper Niger River, this quantity was known by the French name *captif* ('slave'). Ironically, European traders (in slaves and other commodities) contributed to the extension of the numeration system by dumping quantities of these shells in the West African markets, thereby causing a devaluation of the currency and the need for larger numbers. European traders often commented upon the ability of their African counterparts to calculate mentally and to recall the details of transactions after many years. Women, too, came in for their share of praise on this score. See Zaslavsky *1979* for further discussion of numbers and their applications in Africa.

Mathematical ideas are evident in games of chance and games of strategy. Cowrie shells are asymmetrical, and how they will fall when tossed is not entirely predictable, yet winning had specific rewards based on concepts of probability. The universal African board game known by the generic name of 'mancala' (Arabic for 'transferring') is considered one of the world's ten best games (Figure 3). Called 'ayo', 'owari', 'bao' and dozens of other names, its rules vary from one area to another. Usually a given number of beans is placed in each cup (or in several of them), and the players take it in turns to move beans (anti)clockwise in a specified manner until some maximal number is found in a cup or until one player's assigned cups are empty. Play depends entirely on strategy, not at all on luck. For the rules of play and the cultural implications of the many versions of the game, see Russ *1984* and Zaslavsky *1979*; Béart *1955* gives a thorough discussion of games in francophone West Africa.

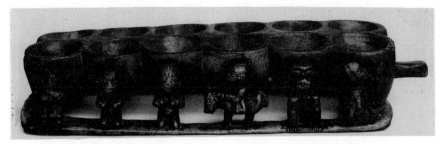

Figure 3 Board for the Yoruba game of ayo (Nigeria and Benin), British Museum (from Zaslavsky *1979*: 120)

Art is part of the moving continuum of daily and ritual existence in Africa. Everyday objects display the most intricate and beautiful motifs, often handed down from generation to generation and imbued with symbolic meaning. Among many peoples, the graphic arts play a role similar to that of writing in conveying ideas and recording history. Washburn and Crowe *1988* have applied the concepts of group theory to analyse the finite designs and repeated patterns found on textiles, wood carvings, basketry and other materials in diverse cultures, including African (Figure 4).

Current research in African mathematics covers many areas: analysing and translating medieval Arabic manuscripts, investigating the uses of numbers and the mathematics hidden in various local practices, and collecting children's games, to name but a few. Both Africans and expatriates participate in this research, the former playing a larger role as time goes on. Naturally, they are eager to reclaim their heritage from the ravages and disruption left by centuries of invasions, enslavement and colonial

Figure 4 Repeated geometric patterns in eighteenth-century embroidered
raffia cloth of the Kuba people (Zaïre), British Museum
(from Zaslavsky *1979*: 179)

exploitation, and to overcome the resulting denigration of their culture and mathematical underdevelopment (Rodney *1972*).

BIBLIOGRAPHY

The Newsletter of the African Mathematics Union Commission on the History of Mathematics in Africa (AMUCHMA), published semiannually since 1987, reports on current research in African mathematics. Contact Dr Paulus Gerdes, CP 915, Maputo, Mozambique.

Adams, H. H. III *1983*, 'African observers of the universe: The Sirius question', and 'New light on the Dogon and Sirius', in Van Sertima *1983*, 27–49.
Ascher, M. *1988*, 'Graphs in cultures (II): A study in ethnomathematics', *Archive for History of Exact Sciences*, **39**, 75–95.
Beart, C. *1955*, *Jeux et jouets de l'Ouest Africain*, 2 vols, Dakar: Institut Français d'Afrique Noire.
Bogashi, J., Naidoo, K. and Webb, J. *1987*, 'The oldest mathematical artefact', *Mathematical Gazette*, **71**, 294.
Denyer, S. *1978*, *African Traditional Architecture*, New York: Africana Publishing Company.
Djebbar, A. *1981*, *Enseignement et recherche mathématiques dans le Maghreb des XIIIe–XIVe siècles*, Paris: Publications Mathématiques d'Orsay, Vol. 81-02.
Eves, H. W. *1976*, *An Introduction to the History of Mathematics*, New York: Holt, Rinehart & Winston.
Garlake, P. *1973*, *Great Zimbabwe*, London: Thames & Hudson.
Gerdes, P. *1988a*, 'On possible uses of traditional Angolan sand drawings [...]', *Educational Studies in Mathematics*, **19**, 3–22.
—— *1988b*, 'On culture, geometric thinking [...]', *Education Studies in Mathematics*, **19**, 137–62.
Griaule, M. and Dieterlen, G. *1965*, *Le Renard pâle*, Paris: Institute d'Ethnologie, Musée de l'Homme.
Lumpkin, B. *1983*, 'Africa in the mainstream [...]', in Van Sertima *1983*, 100–109.
Lumpkin, B. and Zitzler, S. *1982*, 'Cairo – Science Academy [...]', *Journal of African Civilizations*, **4**, 25–38.
Lynch, B. M. and Robbins, L. H. *1978*, 'Namoratunga: The first archeo-astronomical evidence in sub-Saharan Africa', *Science*, **200**, 766–8.
Marshack, A. *1991*, *The Roots of Civilization*, rev. edn, Mount Kisco, NY: Mayer, Bell.
Pappademos, J. *1983*, 'An outline of Africa's role in the history of physics', in Van Sertima *1983*, 177–96.
Rodney, W. *1972*, *How Europe Underdeveloped Africa*, London: Bogle–L'Ouverture.
Russ, L. *1984*, *Mancala Games*, Algonac, MI: Reference Publications.
Schmidt, P. and Avery, D. H. *1978*, 'Complex iron smelting and prehistoric culture in Tanzania', *Science*, **201**, 1085–9.
Van Sertima, I. *1977*, *They Came Before Columbus*, New York: Random House.

—— (ed.) *1983, Blacks in Science, Ancient and Modern*, New Brunswick, NJ: Transaction Books.

Washburn, D. K. and Crowe, D. W. *1988, Symmetries of Culture: Theory and Practice of Plane Pattern Analysis*, Seattle, WA: University of Washington Press.

Zaslavsky, C. *1979, Africa Counts: Number and Pattern in African Culture*, Brooklyn, NY: Lawrence Hill Books.

1.9

Chinese mathematics

J. C. MARTZLOFF

'Chinese mathematics' may refer either to the traditional mathematics of Imperial China or to the mathematics of modern China, Republican or Communist. The latter is fully modernized, and does not differ in any essential way from the mathematics practised anywhere else in the world; by contrast, the former constitutes a typically Chinese traditional science relying almost wholly on algorithmic or algebraic rules rather than on logico-deductive reasoning. The language in which these mathematics was expressed – Classical Chinese – was understood everywhere in the sinicized world, and for this reason Korean, Japanese, Tibetan, Vietnamese and Mongol mathematics have been strongly influenced by it. But, unlike other Chinese traditional sciences such as medicine, Chinese mathematics is now studied solely as a historical subject (not including the special case of the abacus).

1 SOURCES

The most important primary sources for the study of Chinese traditional mathematics are printed books. The work of Ding Fubao (1874–1952) and Li Yan (1892–1963), among others, has set the bibliographical aspect of these mathematics on a sound basis (Ding Fubao *et al. 1957*, Li Yan *1954–5*: Vol. 4). But although Chinese woodblock printing goes back to the ninth and tenth centuries, the majority of extant Chinese mathematical books were printed as late as the eighteenth century.

Besides printed books, there also exist written inscriptions on wooden strips, bones, stone or bronze, measures, calculating instruments and manuscripts. Moreover, a hundred or so lost texts are known only by their titles: the most famous of them is the *Zhuishu* of Zu Chongzhi (*circa* 429–500), a book believed to contain a derivation of the approximation $\pi = 355/113$.

2 MATHEMATICS BEFORE THE IMPERIAL UNIFICATION OF CHINA (221 BC)

Contrary to the generally accepted opinion, Chinese mathematics does not go back to several thousand years BC. The oldest-known Chinese numerical symbols are none the less as ancient as Chinese writing itself.

As far back as the fourteenth century BC, as we know from oracle-bone inscriptions (*jiaguwen*), the Chinese were already in possession of a 'named' place-value decimal numeration system, rather similar to the present written system of contemporary China and even to the Western written system.

In the Warring States period (from the end of the fifth century BC to the Imperial unification in 221 BC), a very atypical sect of preaching friars, the Mohists (who were at the same time the most religious and the most logical of ancient Chinese thinkers), worked out geometrical definitions evoking those of Euclid. During the same period, astronomy became less and less vitiated by numerology and was eventually rationally mathematicized. There developed a cosmology based on an evaluation of the size of the Earth and of the distance between the Earth and the Sun, calculated by a method which used measurements of the shadows of gnomons, and which required knowledge of the 'Pythagorean' theorem, comparison of right-angled triangles, the extraction of square roots, and fractions. The earliest text into which this knowledge was systematically deployed is the *Zhoubi suanjing* ('Mathematical Canon of Zhou Dynasty Gnomons' AD 100?).

3 THE 'COMPUTATIONAL PRESCRIPTIONS IN NINE CHAPTERS' OF THE HAN (206 BC TO AD 220)

Under the Han, mathematical knowledge gradually accumulated during the Zhou and Qin periods, and began to appear in specialized mathematical works. Apart from a Han mathematical work on bamboo strips discovered in a tomb in 1983 and not yet wholly deciphered, the *Jiuzhang suanshu* ('Computational Prescriptions in Nine Chapters') is the oldest text accessible to us. (Available editions are derived from eighteenth-century reconstructions based on thirteenth- and fifteenth-centuries sources.) Many unsolved textual and chronological problems remain (Li Di *1982*). Its influence on the subsequent development of Chinese mathematics was considerable.

The *Jiuzhang suanshu* is an anonymous collection of 246 problems, always stated with their numerical answers and the corresponding algorithmic rules. The text is not intended for beginners since the basic arithmetical processes are supposed to be known. According to the most widely held view, computations were carried out with counting-rods (*chousuan*) on a counting-board in which blanks were left where we would put zeros. Numbers were expressed with their relevant decimal metrological units and, if need be, with fractions of the main unit. 'Abstract' decimal fractions (i.e. without units) were also allowed in special cases, especially in relation to procedures for extracting square and cube roots (see §1.17). Positive and negative numbers, in the form of black or red rods, were also used as fictive computational intermediaries (but never as answers to problems).

The subject-matter of the problems was extremely varied and strongly inspired by socio-economic life. Most of the problems, however, are not wholly realistic; one might be asked to determine the dimensions of a figure given its area or volume, or the total amount of money paid over a period of time, given a uniform rate of interest. Recreational problems also appear involving, for example, the filling of a reservoir, or the pursuit of a rabbit by a dog.

Mathematical rules are given in a general form (i.e. without numbers). The last three chapters of the *Jiuzhang suanshu* are each devoted to a unique family of related rules, all regrouped under the same heading. For example, chapter 7 deals exclusively with various aspects of the rule of double false position.

The historical interest of the *Jiuzhang suanshu* is not limited to its algorithms. From the third century onwards, numerous Chinese commentators, such as Liu Hui (*circa* AD 263), Zu Kengzhi (early sixth century), Li Chunfeng (602–670) and Yang Hui (*circa* 1270), subjected the text of the *Jiuzhang suanshu* not only to philological research but also to logical analysis.

Chinese proofs are very different from the proofs stemming from the Greek tradition. Lacking definitions, axioms and theorems proved deductively, they appear to be the result of complex interactions between various intellectual currents peculiar to the Chinese world, mainly Confucianism, Sophism and Taoism. Although it would be unwise to be too categorical about what should be attributed to each of these currents, it can safely be assumed that Confucianism was responsible for the overall pragmatic and didactic orientation of the demonstrations, that Sophism moulded the rhetoric, and that Taoism introduced a mistrust of the use of language as an exclusive means of access to reality. In these circumstances there was no sharp distinction between arithmetic and geometry, and calculations were viewed in a unified manner, a conception of proportionality pervading all the domains of mathematics (the rule of three and the comparison of triangles, for example, are dealt with in much the same way). More generally, all conceivable methods were freely used: empirical observations, generalizations from examples, recourse to intuitive heuristic principles such as that known in the West as Cavalieri's principle (the same as was used in calculus (§3.1), but only for volumes) or the principle of the invariance of areas or of volumes under dissections and recompositions (made manifest by manipulations of geometrical puzzles, real or idealized, composed of coloured components). The most remarkable examples of results obtained by such methods are the determination of the area of the 'circular field' (circle) and of the volumes of the 'ball' (sphere) and the pyramid.

There was also a tendency, apparently in connection with the linguistic structure of the Chinese language, to consider mathematical phenomena in pairs. The most striking example of such a conception appears in the *fangcheng* method (rectangular arrays of rod-numerals), a method which is based on the consideration of columns of numbers (or 'equations') not in isolation but in pairs, in a way which reminds one of the Gaussian elimination method for linear systems (see §6.6 on determinants).

4 SUI (518–617) AND TANG (618–907) MATHEMATICS

Under the Sui and Tang, mathematics was taught officially. According to the *Tang liu dian* ('Administrative Records of the Tang Dynasty'), students were trained for seven years and entered for civil mathematical examinations (*mingsuan*). The lectures were based on the *Suanjing shishu* ('The Ten Canons of Mathematical Computations'), a collection composed of recast ancient manuals (such as the *Zhoubi suanjing* and the *Jiuzhang suanshu*) and contemporary works. For the most part, the problems of the collection are more elementary than those of the *Jiuzhang suanshu* but there are some novelties, such as the problem of the 'hundred fowls', requiring indeterminate analysis for its solution: a cock is worth 5 *qian*, a hen 3 *qian* and 3 chicks 1 *qian*; with 100 *qian* we buy 100 of them; how many cocks, hens and chicks are there? It also contained the generalized techniques for extracting cube roots (third degree 'equations') of the *Qigu suanjing* ('Canon of the Continuation of Ancient Mathematics') by Wang Xiaotong (early seventh century).

During the same period, the diffusion of Indian Buddhism brought diverse elements into Chinese mathematics: some knowledge of Indian numerals (with a dot for zero, as explicitly stated in the *Kaiyuan zhanjing* ('Astrological and Calendrical Canon of the Kaiyuan Reign Period', 713–741), systems of notation for very large numbers and a table of sines, probably derived by the Indians from a table of chords by Hipparchus. One of the Indian texts transmitted was a part of the *Pañcasiddhāntikā* of Varāhamihira (*circa* 550), which is the basis of the *Jiuzhi* calendar. But the extant documentation suggests that these innovations had only a very limited influence on subsequent Chinese mathematics.

After the Tang, mathematics was still taught intermittently up to the twelfth century, but after that it played no further role in the Chinese educational system until the second half of the nineteenth century. Mathematics thus came to be practised by non-specialists, and for very varied motives. The social status of 'mathematicians' was generally somewhat low, but on occasion eminent literati such as Dai Zhen (1724–77) or Luo Shilin (died 1853) took its study seriously (Hummel *1943*).

5 LATE SONG AND EARLY YUAN MATHEMATICS (THIRTEENTH CENTURY)

The short period of unrest corresponding to the overthrow of the Song Dynasty and to the beginning of the Yuan (Mongol) Dynasty is generally considered to be the 'golden age' of Chinese mathematics. From 1247 to 1303 new computational tools became available:

1 A fully positional system of numeration, featuring decimal fractions and a written zero represented by a small circle (origin for China unknown).
2 The 'Pascal' triangle considered as a device for the computation of the powers of a binomial (compare §4.1), interpolation formulas computationally analogous to the later 'Newton–Stirling' formula (§4.3), summation formulas based on finite-difference techniques, and a family of methods similar to the 'Ruffini–Horner' procedure (*zeng–cheng kaifang fa*, or 'additive–multiplicative' root-extraction method) (Needham *1959*: 134; Libbrecht *1973*: 177; Martzloff *1987*: 222).
3 A fully elaborated method for solving systems of linear congruences, expressible in modern notation as

$$x \equiv r_i (\text{mod } m_i), \quad i = 1, \ldots, n \tag{1}$$

(the Chinese remainder 'theorem', or more strictly, 'algorithm'), applicable even when the moduli $\{m_i\}$ are not relatively prime in pairs. A particular case of the same problem appears in the *Sunzi suanjing* ('Mathematical Canon of Sunzi', late fifth century?), a manual included in the *Suanjing shishu* Tang collection (Libbrecht *1973*: 214–415).
4 A 'spherical trigonometry', based on approximation formulas for the computation of arcs, 'arrows' (i.e. versed sines) and chords.
5 A typically Chinese algebraic technique, the *tianyuan* (the 'celestial' origin, or unknown), whereby numbers are considered either as constants or as coefficients of powers of the unknown according to the particular

position they occupy on the counting-board, independently of any symbolism.

6 A generalized *tianyuan* technique, allowing up to four unknowns and methods for the elimination of the unknowns between systems of polynomial equations of high degree, up to the 14th.

Some of these methods (such as Horner's method and Pascal's triangle) are mentioned only in passing, leaving the impression that they were already known; but others are considered worthy of special study. For example, the algorithmic solution of the remainder problem is specifically introduced in the first chapter of the *Shushu jiuzhang* ('Mathematical Treatise in Nine Chapters', 1247) by Qin Jiushao (*circa* 1202–61); spherical trigonometry appears in the chapters of the *Yuan shi* ('Yuan Annals') devoted to Guo Shoujing's (1231–1316) *Shoushi* calendar; and the *tianyuan* algebra is developed at book length in many sources such as the *Ceyuan haijing* ('Sea Mirror of [Inscribed and Circumscribed] Circles Measurements', 1248) by Li Zhi (1192–1279) or in the *Siyuan yujian* ('Jade Mirror of the Four Unknowns', 1303) by Zhu Shijie. The latter text has been partly translated into French by J. Hoe using an ingenious method of translation based on a 'telegraphic' and semi-symbolic rendering of the terse Chinese text so as to avoid the kind of misleading verbal translations that often occur (Hoe *1977*).

The reasons for such an apparently sudden blossoming of Chinese mathematics are unknown, but the hypothesis of a purely autonomous development seems very unlikely since the Chinese world did not develop in isolation, and historians have endeavoured to detect foreign influences. It has been asserted, for example, that the Chinese remainder rule was derived from the more ancient Indian *kuṭṭaka* (§1.12). However, Libbrecht has convincingly argued that this hypothesis should be rejected since, unlike the Chinese rule, the *kuṭṭaka* proceed by substitutions (Libbrecht *1973*: 219ff).

The possibility of Islamic influences should also be mentioned. It has been known for a long time that exchanges between the Chinese and the Persian worlds were very well developed during the Mongol era, but the scientific consequences of such contacts are as yet poorly understood. The general opinion is that the Chinese were unreceptive to foreign ideas. Indeed, the conceptual bases of Chinese and Arab algebra (not to mention astronomy) are different: negative numbers are freely allowed only in Chinese algebra; the *tianyuan* algebra is 'instrumental' (i.e. it relies on counting-rods) and positional rather than written and symbolic. But note, for example, that around 1270, as we know from archaeological findings near Xi'an in 1956, Islamic magic squares of order 6, engraved on five iron tablets, were handed down to a Gengiskhnanid prince (the third son of

Khubilai, who died in 1280). Similar squares appeared around 1275 in a Chinese arithmetical manual, the *Yang Hui suanfa* ('Yang Hui's Arithmetic'), whereas before 1270 magic squares of order greater than three were unknown in Chinese mathematical works (see §12.2).

Last but not least, in 1477 a whole set of practical astronomical tables of Islamic origin, with their corresponding computational prescriptions (but with absolutely no indication on the rationale of the rules), was translated under the title *Qizheng tuibu* ('Computational Techniques of the Seven 'governors', i.e. the Sun, the Moon and the five classical planets) (Yabuuchi *1987*). Epicyclic astronomy really took root in China later on, during the seventeenth and eighteenth centuries, following the appearance of massive new Jesuit translations of astronomical texts (mainly Tychonic) of European origin.

6 THE MING (1368–1644) DECLINE AND CONTACTS WITH WESTERN MATHEMATICS UNDER THE QING (1644–1911)

Under the Ming, the major achievements of the Song and Yuan sank into oblivion. Mathematics had centred then on the abacus, an instrument whose origin is obscure but which is not known to have existed before the fourteenth century. At the same time, Zhu Shijie's *Suanxue qimeng* ('Introduction to Mathematical Studies'), written in 1299, later made its way into Korea where it was reprinted. Introduced into Japan via the Japanese military expeditions in Korea in the 1590s, it became a determining factor in the development of *Wasan* (Japanese mathematics of the Edo period, 1603–1867: see §1.10).

From late Ming to the end of the Qing (1600–1911), China came into contact with Western mathematics through the medium of two successive waves of translations initiated by Christian missionaries. Over this period about thirty mathematical works were adapted into Chinese (see the annotated bibliography in Martzloff *1987*: 334–9).

The first wave of translations was a consequence of Jesuit activities (1600–1773); the second a consequence of Protestant proselytism (1840–1911). In both cases the aim of the missionaries was the conversion of the Chinese to Christianity, but their respective policies were not based on the same strategy. The Jesuits thought that they should address themselves in the first place to the intellectual elite, whereas the Protestants imagined it would be best to attract the widest possible audience from the very beginning. Hence the orientations of the movements were different.

Towards 1600, the Chinese authorities were preoccupied with the inefficiency of their astronomy, and thus became more and more receptive to proposals of reform. The Jesuits offered their services and undertook a successful programme of translations of mathematical works containing knowledge they believed indispensable for astronomy: elementary arithmetic, calculating instruments such as Napier's rods and Galileo's proportional compass, geometry, plane and spherical trigonometry, and logarithms. The sources of the translations were manuals used in European Jesuit colleges. In particular, in 1607 the Italian Jesuit Matteo Ricci (1552–1610) and the Chinese high official Xu Guangqi (1562–1633) published the *Jihe yuanben* ('Elements of Geometry') based on Clavius's commentary on the first six books of Euclid's *Elements*.

By contrast, after 1830 Protestant missionaries translated books to meet the needs of the private schools they set up in Chinese cities. Thirty years later, they also played an active role in the translation offices established at Shanghai and Peking by Chinese reformers. The most important translators of this period were the Englishman Alexander Wylie (1815–87) and the American John Fryer (1839–1928). These translators did not work in isolation, but always in cooperation with Chinese whose social status was never very high and who made a living by translating large numbers of books in a variety of disciplines. The most important contributions of this period were the translation of the unfinished part of Euclid's *Elements* (1857), Augustus De Morgan's *Elements of Algebra* (1859), E. Loomis's *Elements of Analytical Geometry and of the Differential and Integral Calculus* (1859) and other now-forgotten manuals on conics and applied mathematics.

The influence of these translations was quite considerable. From the seventeenth century onwards the newly coined terminology was in general quickly adopted and, in scientific practice, computations were often performed by writing rather than with counting-rods. Logarithms and trigonometry were also well received and widely used.

Towards 1740, Miangat (died 1763?), an imperial astronomer of Mongol origin trained in European scientific methods under the personal guidance of the Emperor Kangxi, initiated a research project on infinite series. (Miangat is usually called Ming'antu from the sinicized form of his name.) Starting from a very limited knowledge of particular series for π, the sine and the versed sine, which the French Jesuit P. Jartoux (1669–1720) had communicated to him without proof, he obtained new results using special tabular methods of his own, involving a tremendous number of numerical computations, with no recourse to calculus. In the same spirit, authors such as Xiang Mingda (1789–1850), Dong Youcheng (1791–1823) and Dai Xu (1805–60) rectified the length of an arc of an ellipse and obtained various trigonometric and logarithmic expansions (Kawahara *1989*).

However, other important ingredients of Western mathematics were not accepted with the same enthusiasm. The deductive aspect of Euclidean geometry was considered obscure and worthless. From 1760, there grew a tendency to reject algebraic symbolism because the Chinese found superior their newly rediscovered ancient *tianyuan* algebra.

Nevertheless, after 1850, algebraic and analytic symbolism were imported, using adaptations either of European algebraic symbols or of Chinese characters. These innovations were only partly accepted, and at any rate did not wholly survive the 1911 revolution. At the same time, a powerful conservative current born of the rediscovery of Han, Song and Yuan mathematical sources became more and more influential. A typical representative of the contradictions of the epoch is the mathematician Li Shanlan (1811–82). As a professional translator associated with Wylie, he was the Chinese best acquainted with Western mathematics, but as a traditionalist he was also the most active supporter of native Chinese mathematics. We owe to him the very original summation formula

$$\sum_{0 \leqslant j \leqslant k} \binom{k}{j}^2 \binom{n+2k-j}{2k} = \binom{n+k}{k}^2, \qquad (2)$$

which he derived with no tools other than Zhu Shi Shijie's medieval *tianyuan* algebra.

7 THE WESTERNIZATION OF MATHEMATICS IN CHINA

After the 1911 revolution Chinese traditional mathematics was abandoned; China became progressively integrated into the international scientific community, not by means of new translations but through the enrolment of Chinese students in American, European and Japanese universities. Mathematicians invited to China – Knopp, Russell, Birkhoff, Wiener, Hadamard and others – also played a role in the process of modernization.

The Communist takeover of 1949 was followed by 10 years of close cooperation with the Soviet Union. In 1966, the Cultural Revolution inaugurated a period during which great mathematicians such as Hua Luokeng (1910–85) were associated by the authorities with the free-will aspects of operational research (Salaff *1972*; and compare §6.11). But instead of the expected rationalization of the economy, the attempt brought still more disorganization. Notwithstanding the hardships of a tormented period, Chinese mathematicians remained active (see Yuan Tong-li *1963*, as well as Springer-Verlag's recent editions of complete or selected works of Chinese mathematicians).

BIBLIOGRAPHY

The main periodicals in which papers on Chinese mathematics appear are *Archive for History of Exact Sciences* (Berlin), *Historia mathematica* (Toronto), *Sûgakushi Kenkyû* (Tokyo) and *Ziran Kexue Shi Yanjiu* (Beijing). Older books by Y. Mikami, A. P. Yushkevich and others are not listed here.

Ding Fubao and Zhou Yunqing (eds) *1957, Sibu conglu suanfa bian* ['General Catalogue of the Four Departments of Literature; Mathematics Section'], Shanghai: Shangwu yinshuguan.

Hoe, J. *1977, Les Systèmes d'équations polynômes dans le* Siyuan yujian *(1303)*, Paris: Collège de France, Institut des Hautes Etudes Chinoises.

Hummel, A. W. *1943, Eminent Chinese of the Ch'ing Period*, Washington, DC: United States Government Printing Office.

Kawahara, H. *1989*, 'Chûgoku no Mugenshô Kaiseki' ['The infinitesimal analysis of the Chinese'], in K. Yamada (ed.), *Chûgoku Kodai Kagaku Shiron* ['Historical Essays on Ancient Chinese Science'], Kyoto: Kyoto Daigaku Jinbun kagaku kenkyûjo, 223–316.

Lam Lay Yong *1977, A critical study of the* Yang Hui Suan Fa, *a Thirteenth Century Mathematical Treatise*, Singapore: Singapore University Press. [Review: J. Hoe 1982, *Annals of Acience*, **39**, 491–504.]

Li Di *1982*, 'Jiuzhang suanshu *zhengming wenti de gaishu*' ['An overview of controversial opinions (textual, chronological, etc.) on the *Jiuzhang suanshu*'], in Wu Wenjun (ed.), Jiuzhang suanshu *yu Liu Hui* ['Liu Hui and the *Jiuzhang suanshu*'], 5 vols, Beijing: Beijing Shifan Daxue chubanshe, 28–50.

Li Yan *1954–5, Zhong suan shi luncong* ['Collected essays on the history of Chinese mathematics'], 5 vols, Beijing: Kexue chubanshe.

Li Yan and Du Shiran *1987, Chinese Mathematics: A Concise History* (transl. J. N. Crossley and A. W. C. Lun), Oxford: Clarendon Press. [Based on a book written in 1963 by a famous Chinese historian of Chinese mathematics and his disciple. Representative of the historiography practised in the People's Republic of China.]

Libbrecht, U. *1973, Chinese Mathematics in the Thirteenth Century: The* Shu-shu Chiu-chang *of Ch'in Chiu-Shao*, Cambridge, MA: MIT Press. [Review: C. Diény 1979, *T'oung Pao*, **65**, (1–3), 81–93.]

Martzloff, J. C. *1985*, 'Aperçu sur l'histoire des mathématiques Chinoises telle qu'elle est pratiquée en République Populaire de Chine', *Historia scientiarum*, (28), 1–30. [Contains a quasi-exhaustive bibliography of 20 books and 241 papers from 63 Chinese periodicals published between 1949 and 1985.]

—— *1987, Histoire des mathématiques chinoises*, Paris: Masson. [English transl. to appear in *History of Chinese Mathematics*, New York: Springer.]

Needham, J. *1959, Science and Civilisation in China*, Vol. 3, *Mathematics and the Sciences of the Heavens and Earth*, Cambridge: Cambridge University Press. [Review: U. Libbrecht 1980, *Past and Present*, **37**, 30–39.]

Salaff, S. *1972*, 'A biography of Hua Luokeng', *Isis*, **63**, 143–83.

Tasaka, Kôdô *1957*, 'An aspect of Islam culture introduced into China', *Memoirs of the Research Department of the Tôyô Bunko*, **16**, 75–160. [Tentative identification of titles of Arabo-Persian mathematical and astronomical books imported into China at the beginning of the Yuan (Mongol) dynasty, around 1280, and known only through Chinese phonetical transliterations.]

Yabuuchi, K. *1987*, 'The influence of Islamic astronomy in China', in D. A. King and G. Saliba (eds), *Kennedy Festschrift*, *Annals of the New York Academy of Sciences*, **500**, 547–59.

Yuan Tong-li *1963*, *Bibliography of Chinese Mathematics 1918–1961*, Washington, DC: private publication.

1.10

Indigenous Japanese mathematics, Wasan

TAMOTSU MURATA

1 GENERAL SURVEY

Japan has acquired mathematical knowledge from abroad over three periods: Classical Chinese mathematics in the sixth to ninth centuries, contemporary Chinese mathematics in the fifteenth and sixteenth centuries, and European mathematics since the middle of the nineteenth century.

The mathematical learning which nowadays goes by the name of *Wasan* was developed on the basis of the second of these acquisitions. *Wa* means 'Japan' or 'Japanese', and *san* (*suan* in Chinese) means 'calculation' or 'arithmetic'. At the time, this learning was called by many names in Japanese, among them *san, sampô* (or *sanhô*) (*suanfa* in Chinese), *sangaku* (*suanxue*) and *sanjutsu* (*suanshu*), where *hô* (*fa*) means 'rule' or 'method', *gaku* (*xue*) 'learning' or 'science', and *jutsu* (*shu*) 'art' or 'technique'. The word 'mathematics' is derived from the Greek μάθημα, which originally meant something learned, or knowledge in general. The subject has a long and intricate history in the West, from the Pythagorean school up to the present, and has profoundly influenced the Western cultural tradition, including philosophy. However, although *san* has its own history accompanied by a different connotation, it has had little influence upon Japanese tradition in general. So, to what extent do *Wasan* and Western mathematics resemble each other, and how do they differ? It is questionable whether *san* may be translated as 'mathematics' without qualification, but it is beyond doubt that *Wasan* had many aspects thoroughly consistent with 'mathematics'.

Wasan, being independent of Western mathematics, was based upon a perfect understanding of Chinese mathematics. This was the most original contribution that Japan had ever made in the domain of science, or of culture in general. However, *Wasan* provides no trace of deductive reasoning in the Euclidean style, nor of the kind of philosophico-mathematical reflection upon the universe through which Europeans initiated modern natural philosophy backed up by mathematics. The only concern of *Wasanists* was to obtain elegant results, numerically or by construction, and for this purpose they made enormous calculations. Thus it could be said that many *Wasanists* were men of fine arts rather than men of mathematics in the European sense; in reality, only a few of them contributed even to calendrical science.

The principal *Wasanists* were Takakazu (or Kôwa) Seki (1640/42–1708), Katahiro Takebe (1664–1739), Yoshihiro Kurushima (died 1757) and some others cited below. It is said that almost all the basic ideas were laid down by these three men.

All the advancements of *Wasan* took place in the Edo era (1603–1867), during most of which time Japan kept her gates tightly closed to the West. In 1867 the Emperor was restored, and the new government adopted Western mathematics for the national educational system. The history of *Wasan* ended with this act. However, this decision was a wise one because, notwithstanding all its originality and excellence, *Wasan* would not have had the power to raise Japan to its present level of scientific achievement (Smith and Mikami *1914*, Murata *1975*, *1981*).

2 THE ORIGIN AND RISE OF *WASAN*

By the ninth century many Chinese texts in mathematics, especially the *Jiuzhang suanshu* (in its Japanese form, *Kyûshô Sanjutsu*, 'Computational Prescriptions in Nine Chapters', completed by the second or third century (§1.9)), were imported into Japan. But the Japanese at that time could not develop their own mathematics. It was left to a few scholars to preserve the study of the *Jiuzhang suanshu* for many centuries.

This situation had gradually changed by the seventeenth century through the importation via Korea of two mathematical textbooks (§1.11): Zhu Shijie's *Suanxue qimeng* (in Japanese, *Sangaku Keimô*, 'Introduction to Mathematical Studies', 1299) and Cheng Dawei's *Suanfa tongzong* (in Japanese, *Sampô Tôsô*, an 'Account of Arithmetic', 1593). The second of these is an introduction to elementary mathematics, including a manual for using the abacus. On this basis Mitsuyoshi Yosida compiled an excellent textbook, *Jingôki* ('Book of the Smallest and Largest Numbers', 1627), which not only laid the foundations for *Wasan* but also played a prominent role in promoting the diffusion of elementary mathematics among the people throughout the Edo era in its many revised editions.

What led directly to the rise of *Wasan* was the *Suanxue qimeng*, a textbook of higher instrumental algebra (called in Japanese *tengen jutsu*, and in Chinese *tianyuan shu*) making use of the abacus and calculating rods.

But by the time of its arrival in Japan it seems that nobody could understand its contents. It presented only various rules for solving numerical equations of higher degree and gave no explanations; a civil war in China had destroyed the scientific tradition in which the book had been written. The *Wasanists* were forced to decipher it, and by the time their task was completed, as is reflected in Kazuyuki Sawaguchi's *Kokon Sampôki* ('Mathematics, ancient and modern') of 1671, the level of *Wasan* had already surpassed that of its source. (Sawaguchi's dates are obscure.) Two other works representative of this achievement are Yoshinori Isomura's (died 1710) *Sampô ketsugi shô* (1661, revised 1684), and his disciple Yoshimasu Murase's (dates obscure) *Sampô futsutankai* (1673). Murase invented a method for the approximate solution of equations of third degree which had a considerable influence upon his successors, including Seki.

One piece of evidence for the independence of *Wasan* from Western mathematics is found in the *Wasanists'* studies of π. As is often the case with Oriental mathematical texts, the Chinese texts which they could refer to contained only approximations, $\sqrt{10}$ and 3.14, among them, with no indication of how the values were found. The *Wasanists'* first efforts were devoted to reproducing these results. In 1663 Shigekiyo Muramatsu (dates obscure) calculated the circumference of a regular polygon with 2^{15} sides, inscribed in a circle, and declared π to be 3.14, in spite of his more accurate result $3.14159264\ldots$. In 1672 Murase repeated the calculation for a polygon of 2^{17} sides, and gave π as 3.1415. These researches were continued by subsequent generations (see Section 3 below). However none of them, Seki and Takebe included, ever calculated the corresponding circumscribed polygons, as Archimedes did. So, strictly speaking, their values had no solid foundations. The one exception was Yoshikiyo Kamata (1678–1744) who, in 1722, estimated values bracketed by upper and lower limits for polygons up to 2^{44} sides, but his method was never taken up by anyone else. This shows a theoretical limitation in *Wasan*.

3 *WASAN*'S BRIEF FLOWERING

It was Seki (1640/42–1708) who put the finishing touches to the deciphering of the Chinese instrumental algebra, and invented a sort of notational algebra (*endan jutsu*) based upon it. He applied this to a theory of determinants, and then to his original theory of equations and elsewhere. The first theory, invented in his youth, was partly incomplete, but its publication preceded Leibniz's (§6.6). Although this algebra would substantially support the later advancement of *Wasan*, one must not overestimate its importance. Without a tradition of deductive mathematics, his algebra could never have reached the level of Cartesian algebra, but would at most have remained at the level of the Coss algebra of the sixteenth century (§2.3).

Seki contributed to many other fields of mathematics: number theory, including theories of Bernoulli numbers and of some Diophantine equations; numerical analysis, including Horner's method (§4.10) and Newton's interpolation formula (§4.3), a theory of equations of a sort; and calculations on conic sections and Archimedes' spiral. Concerning π, he also invented an ingenious extrapolation, applied it to a polygon with 2^{17} sides, and obtained a value of 'a little less than 3.14159265359'. His masterpieces, *Sambu shô* ('Short Account in Three Parts') and *Shichibu sho* ('Book of Seven Parts'), both compiled in 1683–5, were transmitted to his disciples. They have been translated into English by Hirayama *et al. 1974*, but the 'explanations' are not always penetrating.

Seki and his able disciple Katahiro Takebe (1664–1739) studied the calendar and astronomy, but even such a simple connection between *Wasan* and the real world would not be made by subsequent mathematicians. Takebe, together with his brother Kataaki(ra) (1661–1739), and under the editorship of Seki, made a compilation of all the mathematics of their forerunners in 20 volumes: *Taisei sankyô* ('Large Account of Mathematics', 1683–1710).

Takebe's work on π is really quite marvellous. By improving on Seki's method, he found a value to the 41st decimal place, from which he finally obtained a formula equivalent to Taylor's expansion of $\sin^{-1}\theta$, 15 years earlier than Euler. To do this, he made an enormous calculation and discovered a rule for determining the coefficients with an astonishing insight. But it has to be said that his method was not so general and powerful as those of Euler and others.

This accomplishment of Takebe's was published in his *Tetsujutsu Sankyô* (1722). *Tetsujutsu* literally means 'the art of composing', a title he took from the *Zhuishu* (in Japanese, *Tetsujutsu*), the greatest classic in Chinese mathematics, written by Zu Chongzhi (*circa* 429–500), an 'Archimedes in China'. However, this work was lost very early. All that remains are some results for π recorded in the *SuiShu* (636), the official history of the Sui Dynasty: $3.1415927 > \pi > 3.1415926$, and the two approximations $22/7$ and $355/113$. No doubt Takebe adopted this title of *Zhuishu* for his masterpiece, but he could not refer to Zu's book; so presumably the results he gave were his own findings (Murata *1980*).

Tetsujutsu Sankyô is perhaps the only book on the methodology and philosophy of mathematics in the history of *Wasan*. Takebe's 'philosophy of quality' is particularly interesting. First, he distinguished two qualities among mathematicians: an analytical quality, which characterizes those, like himself, who never hesitate to make enormous calculations; and an intuitive quality, which characterizes those who prefer to be an 'armchair detective', like his teacher Seki. Next, he classified three kinds of infinity in mathematics: infinity in number (infinite decimals), infinity in operation (e.g. square root, which can produce an infinity in number for a finite number) and infinity in quality, whose value will never be settled as a finite number by any finite operation. According to him, the reason he succeeded in calculating π, through making use of an infinite series, whereas Seki did not arrive at such a definitive result, is that the circle has a character of infinity in quality, and Seki's intuitive quality did not fit the quality of the problem. His conclusion is that, when the quality of a mathematician meets 'by chance or by a coincidence' with a problem of the same quality, the problem will be solvable. Otherwise it would be hard to resolve. This remark of his may be compared to Henri Poincaré's classification of types of mathematician: 'logicians *by analysis* versus intuitive ones *by geometry*' (Poincaré *1908*: 127). In Takebe's phrase 'by chance or by a coincidence', one can perceive the uniqueness of his philosophy, or rather of the Japanese spiritual atmosphere at that time (Murata *1987*).

Yoshihiro Kurushima (died 1757) also explored many subjects, revising Seki's theory of determinants and solving many optimization problems (through a sort of differentiation he invented himself) among other achievements. According to a simple criterion of priority, and neglecting the difference between their methods, some of his results were published earlier than Euler's. However, owing to his negligent character, most of his papers have been lost for ever.

Ryôhitsu Matsunaga (?1693–1744), a friend of Kurushima's, established many formulas equivalent to Taylor's expansion. He intended to compile a systematic textbook of *Wasan*, and asked Kurushima to cooperate with him. But Kurushima did not reply, and in the end the project came to nothing. Kurushima, eminent technician that he was, was probably hardly capable of such systematic thinking. This was perhaps the beginning of the decline.

4 DECLINE, AND FOREIGN CONTACTS

After Matsunaga and Kurushima the activity of *Wasan* gradually declined. Although investigations of infinite series were continued, the general trend of the time was to pursue studies largely out of curiosity. But there were one or two remarkable exceptions. In the late eighteenth century Naonobu Ajima (1739–98) established a theory of logarithms which he derived from studying an imported table of logarithms, and composed a table of definite integrals of a sort. Nei Wada (1787–1840), the eminent successor of Ajima, composed a further table of definite integrals, but he led a miserable life and died in the neediest of circumstances. His many contributions were scattered and lost, except for those published under his pupils' names – an episode symbolic of the fate of *Wasan*.

Even during Japan's self-imposed period of isolation (1639–1854), there were channels through which knowledge of Western civilization was acquired. Early contacts were with the Chinese and the Dutch, but after 1720 the importation of non-religious European books was allowed. So, by the end of that period, some Japanese had a fair knowledge of the world situation, as well as some aspects of Western culture. Indeed, there appeared abridged Japanese translations of books such as Aristotle's *Metaphysics* and Newton's *Principia*. Curiously, though, the *Wasanists* showed little interest in Western mathematics. It is said that when they saw a Chinese translation of Euclid's *Elements*, they could not see the importance of apodictic reasoning, dismissing the contents as too elementary.

5 INDEPENDENCE AND ORIGINALITY

Western technology and civilization were introduced mainly by scholars of Western learning. Mathematical activities in Japan today have no connection with *Wasan* at all. Influence that it has had (apart from any latent usefulness it may have had during the transitional period) may be negative, for in Japan, pure mathematics and other mathematical sciences have often been treated separately, and few attempts, technical or philosophical, have been made to bring them together. It is only very recently that this situation began to change.

The independence and originality of *Wasan* can be shown in three ways: (a) the period when *Wasan* progressed most rapidly was the period of international isolation (1639–1854); (b) one can trace through the sources the steps by which it grew, as is seen in calculations of π; and (c) some results were obtained in original ways, many of which were numerical. Some historians advocated that the infinitesimal calculus was invented by Seki, Takebe and others in the form of *enri* (the theory of circles), but this exaggerated claim has now been abandoned. And despite its originality, the sum of *Wasan*'s achievements is the group of fragmentary results outlined in this article. Even such fundamental concepts as angle, variable, function and differentiation had never appeared, to say nothing of the fundamental theorem of the calculus, the inverse relationship between integration and differentiation, nor of the mathematical natural philosophy developed by Descartes, Newton, Leibniz and others.

However, one must take the difference and isolation of Japan's cultural tradition into account. Most of *Wasan*'s results were accomplished in a world of cultural isolation from the Western world, especially from the traditions stemming from Greek mathematics. And the rapidity of its advancement is really exceptional. So, for all its weaknesses, *Wasan* should be appreciated at least as an unusual phenomenon in the world history of mathematics.

It must be noted, too, that all these points can be closely associated with the problem of the universality of mathematics as well as with the fundamental problem of comparative history. In fact, the independence of *Wasan* can be considered as plausible evidence of that universality, and a comparison between *Wasan* and Western mathematics raises some serious questions. On what basis should an impartial and objective comparison be made in such a case? Or rather, is it possible to establish such a basis? Just as the tradition of Western culture has its own bias, so too with the Japanese. This kind of comment, pedantic as it may appear, would produce some fruitful results not only for the comparative history of mathematics, or generally, that of human civilization, but also for the philosophy of mathematics.

BIBLIOGRAPHY

Mikami, Y. *1913*, *The Development of Mathematics in China and Japan*, Leipzig: Teubner. [Repr. 1961, New York: Chelsea.]

Hirayama, A., Shimodaira, K. and Hirose, H. (eds) *1974*, *Takakazu Seki's Collected Works, Edited with Explanations*, Ôsaka: Kyôiku Tosho. [Original texts with English translation.]

Murata, T. *1975*, 'Pour une interprétation de la destinée du Wasan – Aventure et mésaventure de ces mathématiques', in *Proceedings of the XIVth International Congress of the History of Science*, Vol. 2, Tokyo, 184–99.

—— *1980*, 'Wallis' *Arithmetica infinitorum* and Takebe's *Tetsujutsu Sankei* – What underlies their similarities and dissimilarities?', *Historia scientiarum*, (19), 77–100.

—— *1987*, 'Un Traité heuristique japonais. Contemporain de Wallis et de Newton', *Commentarii mathematici universitatis Sancti Pauli* (Tokyo), **36**, 235–53.

—— *1981*, *Nippon no Sûgaku, Seiyô no Sûgaku – Hikaku Sûgakushi no kokoromi* ['Japanese Mathematics and Western Mathematics – An Essay for a Comparative History of Mathematics'], Tokyo: Chûô-kôron sha.

Poincaré, H. *1908*, *Science et méthode*, Paris: Flammarion.

Smith, D. E. and Mikami, Y. *1914*, *A History of Japanese Mathematics*, Chicago, IL: Open Court.

1.11

Korean mathematics

YONG WOON KIM

1 CHARACTERISTICS OF KOREAN MATHEMATICS

In this examination of the history of Korean mathematics, emphasis is placed on the Choson Dynasty (AD 1392–1910) since a study of this period yields insight into the characteristics of modern Korean mathematics. The following are characteristic of traditional Korean mathematics: absence of philosophical reflection or observation, no fundamental change in methodology, no criticism of mathematical textbooks, and no basic change in form between the problem and the answer. From ancient times until the close of the Choson Dynasty, there was no change in this position. When Western mathematics (Euclidean geometry) was introduced into Korea in the early nineteenth century, it had no effect on traditional Korean mathematics, and both continued to exist side by side.

Unlike the mathematics of China and Japan, that of Korea enjoyed an uninterrupted tradition: the mathematical system first created and incorporated in the codes of laws and ethics in the Shilla Dynasty (682) continued to thrive until the end of the Choson Dynasty (1910).

During the Tang Dynasty, China first codified its mathematics in the *Tang liudian*, an educational programme designed to train professional mathematicians, and prescribed the number of mathematicians to be trained, their length of study, and curricula. The system had a great impact on Oriental mathematics and became the prototype for Korean and Japanese mathematics (Kim and Kim *1978, 1982*).

The system soon died out in China. It revived in the Song Dynasty, but failed to become an official system again. Meanwhile, Japan established its mathematics under the influence of the Paekche and Shilla dynasties in Korea. The *Taihorei* of 710 clearly defined its contents, but this too soon became extinct. In China and Japan official mathematics ceased to exist, but civilian mathematics prospered.

Korea, on the contrary, was fundamentally different from these two countries. However, its pragmatic character and mathematics became more

Table 1 Systems of mathematics education in Korea, China and Japan of around the seventh century AD

Country	Enrolment age	Admission	Subjects	Length of study
Korea (Shilla) AD 682	15–30	*Taesa* (grade 12 in a hierarchy of 17 grades) and those who had no official positions	Six-chapter arithmetics Nine-chapter arithmetics Three opening arithmetics Continuation technique[1]	9 years or longer[2]
China (Tang) AD 624	14–19	Children of grade 8, and lower-grade public officials[3]	Nine-chapter arithmetics Calculation of distance to island Sunzi's arithmetics Arithmetics for five bureaux Zhang Qiujian's arithmetics Calculation and gnomon Continuation technique Topics in number Three standard mathematics	7 years
Japan AD 710	13–15	Children of grade 8, and high-grade public officials and local officials	Nine-chapter arithmetics Calculation of distance to island Calculation and gnomon Sunzi's arithmetics Three opening arithmetics Continuation technique Arithmetics for six chapters	7 years

1 The arrangement of the Shilla curriculum here is hypothetical; originally only the names of the four subjects were given.
2 Nine years for classical studies; whether this applied to mathematics and other technical fields is yet to be explored.
3 In Tang, there were different qualifications for the *guozixue* (National Academy) and mathematics and other technical fields. But there was a difference in Korea and Japan between the fields.

idealized. Table 1 compares the educational systems in Korea, China and Japan of around the seventh century.

2 CHUNGIN MATHEMATICS OF THE CHOSON DYNASTY

The Choson Dynasty conserved and even strengthened the system of its predecessors, and *no* change was made to the paradigms (as often happened with mathematics in the West). The study of mathematics was encouraged to fulfil an administrative need and was incorporated into officialdom. During the reign of King Sejong (AD 1419–50), a Bureau of Mathematics and an Agency for Calendars were created, and mathematics was revived to match the level it had reached in the last days of the Koryo Dynasty (936–1392). The positions of *Sanhak Paksa* (Doctor of Mathematics), *Sanhak Kyosu* (Professor of Mathematics) and *Sansa* (mathematician) were created.

The Choson dynasty attached great importance to mathematics from its foundation, and the bureaucrats in charge of the technical civil service examination in ten fields began to play a greater role, ultimately forming the new social class of *chungin* ('middle men'). The *chungin* class may be unique in the history of the world. (The term began to be used officially during the reign of King Sukchong (1675–1720).) As technocrats, the *chungin* were recruited through a comparatively low-level civil service examination. In fact, however, most of them came from the *chungin* class. It was not that a son inherited his father's position; rather it seems that intermarriage among the *chungin* contributed to the preservation of the tradition. Among them, mathematicians showed an especially strong tendency toward heredity. In the roster of successful candidates in tests for mathematicians and the position of *Sanhak Sonsaeng* (mathematics teachers), there are listed a total of 1627 during 300 years from the fifteenth to the seventeenth century. Their fathers' occupations were as follows: 124 were herbalists, 75 translators and 6 astronomers; the rest were mathematicians (Kim and Kim *1978, 1982*).

The mathematicians in the *chungin* class lived in a very closed society. Moreover, as many of the mathematics books they wrote have been lost, it is difficult to evaluate their achievements, either as a group or as individuals. Only fragmentary information is available (Kim and Kim *1982*). The mathematics of the *chungin* is described by Hong Chong-ha, a professor of mathematics, in his eight-volume (plus a supplement) *Kuilchip* ('Nine Chapters on Arithmetic in One') which is still extant today. He was born in 1684 into a typical *chungin* family of mathematicians. His father, both his grandfathers, a great-grandfather and his wife's father were all mathematicians.

From his book, we find the following. First, mathematicians of the day were quite unfamiliar with the course of events in China, whereas the literati of the *yangban* class maintained indirect contact with Chinese culture and with European culture through China. The mathematicians were bound by the old system and continued to use only the traditional, handed-down manuals; they had no access to Chinese translations of Western mathematics books. Second, *tianyuan shu* (Horner's approximation theorem for an equation with real coefficients; §1.6) and calculation sticks thrived in Korea throughout the Choson Dynasty, long after they had fallen into disuse and were replaced by the abacus in China after the establishment of the Ming dynasty. Korean mathematicians were then isolated from the outside world and from the European mathematics which had already been introduced in China, but they preserved traditional mathematics using calculation sticks (Kim *1986*).

3 PHILOSOPHICAL MATHEMATICS IN THE CHOSON DYNASTY

If Hong Chong-ha was a typical mathematician of the class, Ch'oe Sok-chong was a mathematician of noble birth who was a fervent admirer of classical Chinese philosophy. As the author of *Kusuryak* ('Concise Nine Chapter Arithmetic'), similar in its style of description to that of early European monastic mathematics books, he was a Boethius (480–525) of the Orient. Boethius's mathematics was theological, metaphysical and number-theory centred (§2.11). Both gave a touch of mysticism to numbers, and Ch'oe studied magic squares; Figure 1 shows a typical one. The hexagon denotes 'water' in Chinese traditional philosophy, and Korean mathematicians attempted to indicate some philosophical meanings by means of mathematics.

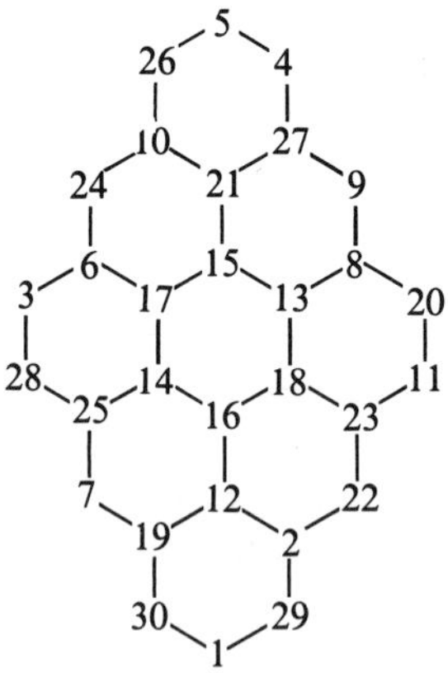

Figure 1 The six numbers forming each hexagon total 93; all the integers from 1 to 30 are arranged without duplication or omission

It was not by accident that Ch'oe, a Confucian scholar of the Choson Dynasty, invented magic squares, which had not existed in China for the purpose of divination; the dynasty as the standard-bearer of traditional Oriental mathematics made this possible. The study of this type of problem naturally led Ch'oe to the study of series and elementary number theory, in his book *Kusuryak*.

4 CULTURE AND MATHEMATICS IN KING SEJONG'S REIGN (1419–50)

King Sejong wished to see all the fields of Oriental learning – Confucianism, linguistics, music, astronomy, herbal medicine and agriculture – in the service of his country. Unlike the Greek system of learning that branched off into mathematics, natural sciences and metaphysics, the Oriental system tended to integrate all fields of learning into a whole; this was typified by Classical Chinese studies (compare the Western trivium and quadrivium, in §2.11). In, for example, the rejuvenation of music and the creation of the Korean alphabet, King Sejong remained true to the orthodox Oriental view of learning and was content with being the true inheritor of Oriental culture. His policy for the promotion of mathematics did not seek any new paradigm.

The achievements of some typical *Sirhak* scholars will now be noted. *Sirhak* is the Korean version of a neo-Confucian concept of 'practical learning', similar in nature to *jitsugaku* in Japan and *shixue* in China. They were active for about 300 years, from the mid-sixteenth to the mid-nineteenth centuries.

Yi Sugwang (1563–1628) wrote *Chibong yusol* ('Chibong's Miscellany') which treats astronomy, geography, bureaucracy, *belles-lettres*, human behaviour, technology, and even birds, animals and insects. The book is encyclopedic, following the trend in Chinese academic writing, a trend that *Sirhak* scholars continued to follow.

Typical of this encyclopedic coverage is *Ojuyon munjangjon san'go* ('An Oju's Multitude of Articles and Essays') by Yi Kyu-gyong (1788–?) in 60 volumes, which contained 1400 entries relating to the problems of all age-groups and countries. He had a very practical outlook, and in a commentary on the original text of a geometry book he regarded surveying as the purpose of the study of geometry. But he adhered to the traditional view of arithmetic, as did other *Sirhak* scholars.

Ch'oe Han-gi (1803–79) is said to have broken with the traditional position of neo-Confucianism and become an activist philosopher, adhering to a thoroughgoing empiricism. But he too remained orthodox in his attitude to arithmetic, as is evident in his comment, 'By the degree of knowledge one has acquired in arithmetic, we can judge his insight: we can see whether his attitude is reasonable or not by judging if his reasoning is arithmetical.'

In 1765 Hong Tae-yong (1731–83) visited a Catholic church in Beijing, China, and acquired first-hand knowledge of Western culture. He conversed with Hallerstein (the Chief Astronomer) and his deputy Gogeisl at a Chinese astronomical observatory, and thereby broadened his knowledge of astronomy. His work *Tamhonso* ('Tamhon's Writings') treats mathematics and astronomy in Volumes 4 to 6 of Book II. Hong consciously discussed the infinite; he was the first Korean to discuss an infinite Universe, and he also mentioned infinite decimals in discussing the value of π (compare §1.10 on Japan).

5 *YANGBAN–CHUNGIN* COLLABORATION

Nam Pyong-gil (1820–59) and Yi Sang-hyok (1810–?), perhaps the two greatest arithmeticians at the end of the Choson Dynasty, did not belong to the *Sirhak* school. None of the *Sirhak* scholars specialized in arithmetic. But Nam was born into a *yangban* family (that is, a family of literati) and held high government positions, and Yi was a professional arithmetican of *chungin* lineage. They joined hands in the study of arithmetic.

Included in Nam's works are many arithmetic books, but he made them look new by adding illustrated explanations to the names of the books. There is no trace of the metaphysical view, a dominant characteristic of the works of other *yangban* scholars.

Yi Sang-hyok was a typical *chungin* arithmetician. After passing the national test for arithmeticians, he was assigned to an astronomical observatory as a budget officer. Among his works are some astronomical books and the following books on arithmetic: *Iksan: Ch'agunbop monggu* ('Winged Mathematics: Hypothetical Method for the Theory of Roots') and *Sanhak Kwan-gyon* ('A Brief Survey of Arithmetic'). The first title probably implies 'Mathematics of two wings', one wing referring to traditional arithmetic and the other to modern arithmetic. *Ch'agunbop monggu* is an explanatory book on European algebraic equations, while *Sanhak Kwan-gyon* presents Yi's creative study of mathematics. The Japanese historian of mathematics Fujihara Matsusaburo lauded Yi Sang-hyok's work as 'breaking virgin soil in Korea' (Kim and Kim *1986*). Yi's single-minded devotion to higher mathematics, ignoring the traditional pattern of thought when classic arithmetic was reviving, leads us to assume that this type of mathematical research may have prevailed among *chungin* scholars.

6 CONCLUDING REMARKS

The mathematics of the Choson Dynasty set in motion by King Sejong gave rise to various schools. Among them were the *Sirhak* school which emerged in the latter part of the sixteenth century and prevailed for about 300 years thereafter, searching for new mathematics as seen in a collaboration between Yi Sang-hyok and Nam Pyong-gil. But their study of European mathematics was limited to algebraic equations and geometry at best, and they never went much beyond the traditional Choson dynasty mathematics even when they accepted European mathematics. The works of Hong Tae-yong, allegedly the most progressive of all *Sirhak* mathematicians, differ from other classical manuals only in that they deal with practical applications of old principles.

Yi Sang-hyok attempted an original study in mathematics, breaking with the old traditions. But he too was limited by the prevailing intellectual climate, which hindered the understanding and assimilation of modern mathematics. This rigidity in Choson dynasty mathematics was due to a lack of change in general cultural traditions.

Even if Korean mathematicians had managed to break free from traditional Confucian ideology, it would have faced insurmountable problems in gaining access to modern Western mathematics. One limiting factor was the hereditary nature of Choson Dynasty mathematics. Western mathematics was steeped in the tradition of intellectualism, but the bureaucratic mathematics of the Choson Dynasty was of a fundamentally different character (Kim *1986*).

BIBLIOGRAPHY

Kim, Y. W. *1973*, 'Introduction to Korean mathematics history', *Korea Journal*, **13**, (7), 16–23; (8), 26–32; (9), 35–9.
—— *1986*, 'Pan paradigm and Korean mathematics in Choson Dynasty', *Korea Journal*, **26**, (3), 24–46.
Kim, Y. W. and Kim, Y. G. *1978*, *Kankoku sugakusi* ['A History of Korean Mathematics'], Tokyo: Maki Shoten.
—— *1982*, *Han'guk suhak-sa* ['A History of Korean Mathematics'], Seoul: Yolhwa-dang.

1.12

Indian mathematics

TAKAO HAYASHI

The history of Indian mathematics can be divided roughly into three periods, namely, ancient (about 1200 BC to AD 500), medieval (about AD 500 to 1600) and modern (after about 1600). This article deals mainly with the first two periods, the last one being touched upon briefly in the last section, 'Foreign influences'. Chronological order has not been strictly maintained.

1 THE ANCIENT PERIOD

1.1 The *Śulbasūtras*

The *Śulbasūtras* – manuals for preparing Vedic sacrificial altars of various geometrical designs – are unique, rich sources of our knowledge about the mathematics that the Indo-Aryan people possessed in the first millennium BC. In them occur, among other things, the so-called Pythagorean theorem ('The diagonal rope of an oblong produces both [areas] which its side and length produce separately') as well as applications of Pythagorean triples $(3, 4, 5; 5, 12, 13; \ldots)$; a rule for computing the diagonal of a square (or $\sqrt{2}$),

$$d = a + \frac{a}{3} + \frac{a}{3} \cdot \frac{1}{4} - \frac{a}{3} \cdot \frac{1}{4} \cdot \frac{1}{34} \ (a = \text{side}, \ d = \text{diagonal}); \quad (1)$$

and several transformation rules, including ones for squaring a circle and circling a square without changing the area.

Major mathematical problems of the *Śulbasūtras* were related to a single theological requirement: namely, to construct altars having different shapes but the same area. The altars were built of five layers of brick; each layer contained 200 bricks of different shapes arranged so that the interstices between bricks in contiguous layers did not coincide. The geometers of the *Śulbasūtras* considered them from two different aspects: purely geometrical (without numbers) and numerical; the former aspect was to disappear

118

almost completely from Indian geometry after the time of Āryabhaṭa (born AD 476). For further details see Datta *1932*, Michaels *1978*, and Staal *1983* (especially Seidenberg *1983* therein).

1.2 Numbers and computations

Counting (*saṃkhyāna/gaṇanā*) was regarded as a basic subject for higher education by the Vedic Brāhmaṇas (in the *Chāndogyopaniṣad*), the Kṣatriyas (in the *Arthaśāstra*), the Jainas (in the *Kalpasūtra* and other works) and the Bauddhas (in the *Lalitavistara*). In the Vedic literature numbers up to 10^{12} (or 10^{13}) are named, while in the Jaina and Bauddha literatures names are given to much larger numbers (in the *Lalitavistara*, for example, *tallakṣaṇa* is the name given to 10^{53}). The Jainas even went on to speculate about different levels of infinity, but it was the Vedic names that became the core of the standard names of eighteen decimal places used in Hindu mathematical works from the eighth century AD onwards, to which core were added several terms of non-Vedic origin such as *lakṣa* (10^5) and *koṭi* (10^7) (Datta and Singh *1935–8*: Vol. 1).

The oldest extant work in India that used the decimal place-value system with a symbol (called a dot, *bindu*) for a vacant place is the *Yavanajātaka* (AD 269–70), a metrical version of a Sanskrit prose adaptation (*circa* AD 150) of a Greek (*yavana*) lost work (*circa* AD 100) on horoscope astrology (*jātaka*) (Pingree *1978b*; however, Shukla *1989*: 214 interprets this passage without a place-value system or a zero), while the oldest examples of the zero (*śūnya, kha, pūrṇa*, etc.) used as 'a number' which is subject to arithmetical operations occur in Varāhamihira's *Pañcasiddhāntikā*, a compendium of five astronomical systems, the epoch of the latest of which is AD 505 (Neugebauer and Pingree *1970–1*). By the seventh century Indian mathematicians became well versed in calculating negative and irrational numbers as well as zero, although they were indifferent to the logical proof of irrationality. The wide domain of their numbers made it possible for them to develop fully their computational skill and algebra (Datta *1926–31*).

They usually made computations on a board with chalk or on a board covered with dust. Brahmagupta called astronomical computations 'dust work' (*dhūlīkarman*). It has been pointed out, however, that some of the Bauddha philosophical works of the third to the fifth centuries AD mention a bead or a counter (*gulikā/vartikā*) which changes its value according to its place (Ruegg *1978*). This probably means that there existed in India a certain kind of abacus in the early centuries AD, although it perhaps became obsolete as calculations with numerical figures on a board became more and more popular. See Datta *1928* and Sarma *1985* on writing material, and Ifrah *1981* for numerical figures.

1.3 Jaina mathematics

While the mathematical knowledge of the Vedic Brāhmaṇas was embodied in the manuals for sacrificial altars, that of the Jainas, who denied Vedic rituals, found its way into their philosophy and cosmology. In the former field they developed, among other things, a theory of permutations and combinations, speculation about infinity, and laws of indices; and in the latter field mensuration rules such as those for a segment of a circle in order to measure various parts of the Universe. In his commentary (AD 629) on the *Āryabhaṭīya*, Bhāskara I quotes several mensuration rules written in a Prakrit, most probably of the Jainas. Jaina philosophy and cosmology are known to have influenced Āryabhaṭa's astronomy. Moreover, a Jaina canonical work ascribed to the third century BC lists ten mathematical topics. These examples, though fragmentary, seem to indicate that the Jainas played a very important role in the development of Indian mathematics in the centuries just before and after the birth of Christ (Datta *1929b*).

2 THE MEDIEVAL PERIOD

2.1 The astral science

Our main sources for Indian mathematics from about AD 500 onwards are astronomical as well as mathematical works. Traditionally, mathematics was included in 'the astral science' (*jyotiḥśāstra*), which consisted of 'three stems' (*triskandha*): namely, mathematical sciences (*gaṇita* or *tantra*), horoscope astrology (*horā* or *jātaka*) and divination (*saṃhitā*). The first 'stem' covers astronomy and calendrical science as well as mathematics proper (*gaṇita*). This classification dates at least from the sixth century AD, when Varāhamihira wrote several books on each subject. Some of the principal works on astronomy, called *siddhānta* (literally, 'the final conclusion'), contain a few chapters devoted solely to mathematics. As with other fields of Indian literature, in mathematics and astronomy the main texts were composed in metrical Sanskrit, and commentaries upon them mostly in prose. For the literature of the 'astral science' see Pingree *1981*.

2.2 The *Āryabhaṭīya* of Āryabhaṭa

The *Āryabhaṭīya* (AD 499/510) is the oldest datable Indian work on astronomy which contains a chapter on mathematics. It contains thirty-three verses: one verse (v.1) is the benediction, one (v.2) gives the names of ten decimal-places, three (vv.3–5) present algorithms for calculating the square, cube, square root and cube root in the decimal place-value system, seventeen (vv.6–22) are on 'field' (*kṣetra*), or geometry, and eleven (vv.23–33) are on 'quantity' (*rāśi*), or arithmetic and algebra. For modern translations see Shukla and Sarma *1976*, and Elfering *1975*.

The tenth verse gives 62 832 : 20 000 as an approximate ratio of the circumference of a circle to its diameter. This ratio remained the best available up to about 1400, when Mādhava made a breakthrough in trigonometry. The only ratio better than this was 355 : 113, which had been derived from the same ratio by the ninth century, but it was not known widely. Other values for π used in India include 3, 600/191, 22/7, $\sqrt{10}$ and 19/6 (Hayashi *et al. 1989*).

The last two verses give a general integral solution to the indeterminate system of equations:

$$n = ax + r = by + s, \quad 0 \leqslant r < a, \ 0 \leqslant s < b, \tag{2}$$

where n, x and y are unknown. Āryabhaṭa reduced the integral coefficients a and b by means of the so-called Euclidean algorithm until the equations could be easily solved by trial and error. Hence the name of the solution, 'pulverizer' (*kuṭṭaka*). The pulverizer was employed in calendrical computations. Mahāvīra in his *Gaṇitasārasaṃgraha* (*circa* AD 850) completed the solution by carrying out the mutual divisions of a and b until unity is obtained as a remainder (Ganguli *1931–2*).

2.3 The *Brāhmasphuṭasiddhānta* of Brahmagupta

Of the twenty-five chapters of the *Brāhmasphuṭasiddhānta* (AD 628), four are concerned with mathematics. Chapter 12 (entitled 'Mathematics', 66 verses) deals with twenty 'basic operations' (*parikarman*) and eight kinds of 'practical mathematics' (*vyavahāra*). The former category consists of the eight arithmetical operations ending with the cube root, five types of calculation with fractions, five rules for proportion, beginning with the rule of three and ending with the rule of eleven, the inverse rule of three, and barter; the latter consists of mixture, mathematical series, plane figures (including Brahmagupta's famous lemma for a cyclic quadrilateral), ditches, stacking bricks, sawing of timbers, piling of grain and the shadow.

Both the words *parikarman* and *vyavahāra* occur as the first two of the ten mathematical topics of the Jainas mentioned above, although their exact meanings are not known. According to Bhāskara I, a contemporary of Brahmagupta, there existed separate works on each of the 'eight kinds of practical mathematics' composed by Maskari, Pūraṇa, Mudgala, and so on, who were probably Jainas.

Chapter 18 ('Pulverizer', 103 verses) begins with the pulverizer (*kuṭṭaka*) itself, and gives rules for the six arithmetical operations involving negative and irrational numbers, zero and unknown numbers. As far as is known, this is the first systematic treatment of the subjects in India, and the rules given are correct except for '$0 \div 0 = 0$'.

Brahmagupta called the result of '$a \div 0$' ($a \neq 0$) a 'zero-divisor' (*khaccheda*), meaning '[a quantity] having zero as its divisor', but what value he assumed for it is not known. Bhāskara II in the twelfth century took it to be 'an infinite quantity' (*ananta-rāśi*), and compared it to the infinity of the Supreme God, Viṣṇu. In the sixteenth century Kṛṣṇa proved the infinity of the 'zero-divisor' by *reductio ad absurdum*.

The same chapter also deals with linear and quadratic equations including ones of the type $Nx^2 + t = y^2$, which was called 'the square nature' (*vargaprakṛti*). Brahmagupta showed, among other things, that this equation can be solved in integers for $t = 1$ (i.e. the so-called Pell's equation) if it is solved for $t = \pm 4$, ± 2 or -1. By the eleventh century Jayadeva improved the solution by giving a recursive method (called 'cyclic' (*cakravāla*)) which yielded a solution for $t = \pm 4$, ± 2 or ± 1 from a solution for any integer t (Selenius *1975*). For an English translation of Chapters 12 and 18, see Colebrooke *1817*.

Chapters 19 ('Knowledge about the gnomon and shadow', 20 verses) and 20 ('Enumeration of meters', 20 verses) may be regarded as supplements to Chapters 12 and 18. Chapter 20 has not been deciphered yet, but the technical terms in it suggest that it deals with permutations and combinations concerning Sanskrit meters, a topic which had already been treated to some extent in the *Chandaḥsūtra*, a work on Sanskrit prosody, the oldest stratum of which is assignable to the second century BC. One of its aphorism-like expressions (*sūtra*) was interpreted by the tenth-century commentator Halāyudha as describing the array now called Pascal's triangle (Bag *1966*; compare §4.1).

2.4 Proofs and verifications

Neither the *Āryabhaṭīya* nor the *Brāhmasphuṭasiddhānta* contains proofs of their mathematical rules, but this does not necessarily mean that their authors did not prove them. It was probably a matter of the style of exposition. In fact, later prose commentaries contain a number of demonstrations or derivations (*upapatti*), together with underlying principles (*yukti*, literally 'union', *nyāya*, 'reason', and *mārga*, 'way'). For further discussions about the methodological aspect of Indian mathematics, see Srinivas *1990*.

The recognition of the importance of proofs dates back at least to the time of Bhāskara I (around AD 600), who, in his commentary on the *Āryabhaṭīya*, rejected the Jaina value of π, √10, saying that it was only a tradition (*āgama*) and there was no derivation of it. He also emphasized the importance of verifications (*pratyayakaraṇa*, literally 'to make conviction') of solutions to mathematical problems. His typical style of exposition in his mathematical commentary was as follows:

Rule (*sūtra*) (in verse)   by Āryabhaṭa.
    Elucidation of the rule (derivations are rare) ⎫
    Example (*uddeśaka*) (usually in verse)   ⎪
    Setting (*nyāsa/sthāpanā*) of the numerical data ⎬ by Bhāskara I.
    Working (*karaṇa*) of the solution   ⎪
    Verification (*pratyayakaraṇa*) of the answer   ⎭

Verifications (called *ghaṭanā*, literally 'union') are still found in Siṃhatilakasūri's commentary (in the thirteenth century) on Śrīpati's *Gaṇitatilaka*, but occur very rarely thereafter, in contradistinction to the growing popularity of derivations.

A style close to Bhāskara I's is also employed in the Bakhshālī Manuscript. For this and several other reasons, I would assign the date of the work contained in the manuscript to the same century as Bhāskara I (Datta *1929a*: 38 ascribed it to 'the early centuries of the Christian era'; Kaye *1927–33* to 'the twelfth century'; the matter has remained controversial). Written in the early Śāradā script (which was used from the eighth to the twelfth centuries in Kashmir and its neighbouring districts) on birch bark, it is the oldest of all the extant Sanskrit mathematical manuscripts.

2.5 *Pāṭī* and *bījagaṇita*

We do not know why the two mathematical chapters, 12 and 18, of the *Brāhmasphuṭasiddhānta* are interrupted by five chapters on astronomy (it is possible that the chapters were rearranged in the course of transmission), but they were certainly precursors of the two major fields of medieval Indian mathematics, namely *pāṭī* and *bījagaṇita* (Datta *1929c*). Śrīdhara, in the eighth century, was probably the first to write separate works in each field: his *Triśatikā* and *Pāṭīgaṇita* on *pāṭī* are extant (Ramanujacharia and Kaye *1912–13*, Shukla *1959*), but his work on *bījagaṇita* referred to by Bhāskara II has been lost. Śrīpati's astronomical work, *Siddhāntaśekhara*, contains a chapter on each of the two fields (Sinha *1986*, *1988*). He also wrote a separate work on *pāṭī*, *Gaṇitatilaka* (Sinha *1982*). By 1150, Bhāskara II had completed the *Līlāvatī* on *pāṭī* and the *Bījagaṇita* on *bījagaṇita* (Colebrooke *1817*), both of which became very popular as textbooks at schools (*maṭha*) all over India. Two centuries later, Nārāyaṇa wrote the *Gaṇitakaumudī* on *pāṭī* and the *Bījagaṇitāvataṃsa* on *bījagaṇita*.

Pāṭī is believed to have been named after the board (*paṭṭa*) on which calculations were made. Just like Chapter 12 of the *Brāhmasphuṭasiddhānta*, a work on *pāṭī* consists mainly of basic operations *parikarman* and practical mathematics *vyavahāra*, to which Bhāskara II added chapters on permutations and combinations and on the 'pulverizer', and Nārāyaṇa on magic squares and on 'the square nature'. *Pāṭī* may be characterized as a collection of algorithms which can be applied mechanically to problems of specified types, and as a rule the domain of the numbers used in it is restricted to positive integers and fractions.

Bījagaṇita, on the other hand, literally means 'seed-mathematics', which could be interpreted either as 'mathematics by means of seeds' or as 'mathematics as a seed'. In support of the former is the fact that 'seed' stood either as a symbol for an unknown number (*avyakta*, literally 'invisible' hence *bījagaṇita* was also called *avyaktagaṇita* or 'mathematics with unknown numbers' as against *pāṭī* alias *vyaktagaṇita* or 'mathematics with known numbers'), or for an equation, since, just like seeds of plants, *bīja* had the 'potentiality to generate' solutions to mathematical problems. Thus, *bījagaṇita* itself was regarded by Bhāskara II as 'a seed' which generates algorithms to be included in a book on *pāṭī*.

According to Bhāskara I, there were 'four seeds' (*bījacatuṣṭaya*) which generated the eight kinds of practical mathematics (*vyavahāra*); he called them *yāvattāvat* (as much as), *vargāvarga* (?square), *ghanāghana* (?cube) and *viṣama* (odd). Although the exact meanings of these terms are not known, they are most probably the designations of four kinds of equation. Interestingly, four Prakrit terms similar to Bhāskara I's occur in the previously mentioned ten mathematical topics of the Jainas: *jāvaṃtāvati* (=*yāvattāvat*), *vagga* (=*varga*, a square), *ghana* (a cube) and *vaggavagga* (the square of a square).

A set of 'four seeds' is also mentioned in an anonymous commentary on Chapter 18 of the *Brāhmasphuṭasiddhānta*: equations with one colour (i.e. in one unknown) (*ekavarṇasamīkaraṇa*), elimination of the middle term (i.e. solution of quadratic equations) (*madhyamāharaṇa*), equations with more than one colour (*anekavarṇasamīkaraṇa*) and equations with 'the product' (i.e. of the type $ax + by + c = dxy$) (*bhāvitakasamīkaraṇa*). In fact, Brahmagupta's rules and examples for these topics occupy a fifth of that chapter; and the main part of Bhāskara II's *Bījagaṇita* has been modelled on the same scheme.

Āryabhaṭa used the words *gulikā* (literally, a bead or a ball) and *rūpaka* (a coin or a money measure) respectively for the unknown and the constants while giving a solution of the linear equation $ax + b = cx + d$. Brahmagupta called the unknown number *avyakta* (literally, 'invisible') and *varṇa* (a colour or a syllable), while Bhāskara I used the term *yāvattāvat* as a substitute for the *gulikā* of Āryabhaṭa (see §6.10, Section 1). In order to express equations (*samīkaraṇa*), Bhāskara II used the letters *yā, kā, nī, pī, lo*, and so on, which were the first letters of the word *yāvattāvat* and of the colour-names *kālaka* (black), *nīlaka* (blue), *pīta* (yellow), *lohita* (reddish brown), and so on. (It should be noted that *yā* and the other letters represent single characters, just like 'x', say, in the Roman alphabet, because the Indian alphabet was syllabic.) He ascribes the invention of these symbols to 'the venerable teachers'. Gaṇeśa, in his commentary (1545) on the *Līlāvatī*, even used these symbols for known (but indefinite) numbers while demonstrating an algorithm for squaring a number in the place-value system.

As mentioned above, the *gulikā* may also have been used as a counter, a type of abacus, around the time of Āryabhaṭa (or a little before). It may be conjectured that beads (or small balls) of different colours were used to express different unknown numbers before the written symbols, *yā, kā*, and so on, were introduced.

Thus Bhāskara II would write the equation

$$-5x^4 + 4x^3 + 3 = -2x^2 + 1 \tag{3}$$

as

yāvava $\overset{.}{5}$ *yāgha* 4 *yāva* 0 *rū* 3
yāvava 0 *yāgha* 0 *yāva* $\overset{.}{2}$ *rū* 1,

where *vava, gha, va* and *rū* stand respectively for *vargavarga* (square of square), *ghana* (cube), *varga* (square) and *rūpa* (either unity or a constant), and a dot above a numeral indicates that it is negative. Similarly, he would write

$$4x + 3y + 2 = xy \tag{4}$$

as

yā 4 *kā* 3 *rū* 2
yākābhā 1,

where *bhā* stands for *bhāvita*, the product (of different unknowns).

By means of these algebraic expressions and by making free use of the 'pulverizer' and the 'square nature', Bhāskara II solved not only linear and quadratic equations in one or more unknowns, but also special types of equation of the third and the fourth degrees. He often emphasized the importance of the intellect (*mati*) in *bījagaṇita*. For further information about Indian algebra, see Datta and Singh *1935–8*: Vol. 2.

2.6 Trigonometry

The first Indian sine table (a table of 24 half-chords in a quadrant) appeared in the astronomical works called *siddhānta* of the fifth century AD, and various trigonometrical formulas as well as sine tables, with or without interpolation rules, are found in a number of later works of the same category. Thus Indian trigonometry served its mathematical astronomy well, although a spherical trigonometry, like that of Ptolemy based on Menelaus's theorem, was never developed (Datta and Singh *1983*). Analemmas were also used.

Bhāskara II used a relationship equivalent to the differential $d(\sin\theta) = \cos\theta\,d\theta$ in his computations of the 'instantaneous motion' (*tātkālikī gati*) of planets; perhaps the relationship had already been recognized by Muñjāla in the tenth century. Bhāskara II also used a kind of integration, the summation of infinitesimal parts, to compute the volume and the surface of a sphere in his astronomical work *Golādhyāya* ('Chapter on the Sphere') (Sengupta *1932*).

The highlights of Indian trigonometry appeared in the fourteenth and subsequent centuries, when Mādhava and his successors (such as Nīlakaṇṭha and Śaṅkara) obtained a number of power series, either finite with a correction or infinite, for the circumference of a circle (or π) and for various trigonometric functions, including inverse ones. As only a few of Mādhava's own works are extant, it is difficult to differentiate his own discoveries from those of his successors, but the formulas undoubtedly assignable to him include

$$c(n) = \frac{4d}{1} - \frac{4d}{3} + \frac{4d}{5} - \cdots + (-1)^{n-1}\frac{4d}{2n-1} + (-1)^n \cdot \frac{4dn}{4n^2+1}, \quad (5)$$

where c and d are respectively the circumference and diameter of a circle, and the last term of the right-hand side is a correction factor. For further details see Sarasvati *1963* and *1979*, Rajagopal and Rangachari *1978*, and Datta and Singh *1980*.

3 FOREIGN INFLUENCES

It has been suggested that Greek and Vedic (Śulba) mathematics have a common, pre-Babylonian source (van der Waerden *1983*, Seidenberg *1988*). Naturally, the relationship of Indian mathematics to the Indus civilization is one of the most interesting problems to be investigated, but in effect we know nothing about the mathematics of the latter except for its highly developed system of weights and measures employed in building and construction (Chattopadhyaya *1986*).

In Sanskrit mathematical works we find the so-called 'typical problems', which occur throughout the ancient and medieval worlds (e.g. the 'hundred hens problem' and the 'cistern problem' in Śrīdhara's *Pāṭīgaṇita*), indicating that Indian mathematics was not isolated from that of other nations (Gupta *1989*). Moreover, no one can overlook Greek elements in Indian astronomy and horoscope astrology, such as epicycles, and a number of Sanskritizations of Greek astrological terms (Pingree *1978a*, *1981*). It might be natural, therefore, to hypothesize (with van der Waerden *1976*) a Greek influence on the early stages of Indian mathematics proper, but we have yet to find satisfactory testimonies to that effect except for trigonometry, which was mostly confined to astronomical contexts. Jagannātha's *Rekhāgaṇita* ('Mathematics of Lines', *circa* 1720) is the earliest extant Sanskrit translation (from the Arabic version by Naṣīr al-Dīn al-Ṭūsī) of Euclid's *Elements*, but its influence is evident in Kamalākara's earlier astronomical work, *Siddhāntatattvaviveka* (1658). These are examples of the increasing contacts and collaborations of Hindu and Muslim scholars under the Mughal Empire (Bag *1980*, Pingree *1987*).

Under the patronage of the East India Company, the Calcutta Madrasa (a college for Arabic and Persian studies) and the Benares Sanskrit College were established in 1781 and 1794, respectively. Mathematics was taught in both institutions in the respective languages. In 1857 the first three Indian universities, at Calcutta, Bombay and Madras, were founded on the British model in order to disseminate English/European knowledge, including mathematics (Kapur *1988*).

BIBLIOGRAPHY

For general histories see Datta and Singh *1935–8*, *1980*, *1983*, Juschkewitsch *1964*, Srinivasiengar *1967*, Sen *1971*, Sarma *1972*, Bag *1979*, Sarasvati *1979* and Gupta *1987*. For source materials see Pingree *1970–81*, *1981*. For the history of Indian astronomy, see Billard *1971*, Pingree *1978a*, and Sen and Shukla *1985*.

Bag, A. K. *1966*, 'Binomial theorem in ancient India', *Indian Journal of History of Science*, 1, 68–74. [Repr. in Kuppuram and Kumudamani *1990*: Vol. 4, 191–200.]
—— *1979*, *Mathematics in Ancient and Medieval India*, Varanasi: Chaukhambha Orientalia.
—— *1980*, 'Indian literature on mathematics during 1400–1800 AD', *Indian Journal of History of Science*, 15, 79–93. [Repr. in Kuppuram and Kumudamani *1990*: Vol. 2, 101–24.]
Billard, R. *1971*, *L'Astronomie indienne*, Paris: Ecole Française d'Extrême-Orient.
Chattopadhyaya, D. *1986*, *History of Science and Technology in Ancient India: The Beginnings*, Calcutta: Firma KLM.

Colebrooke, H. T. *1817, Algebra with Arithmetic and Mensuration from the Sanscrit of Brahmegupta and Bháscara*, London: Murray. [Repr. 1973, Wiesbaden: Sändig.]

Datta, N. *1926–31*, 'Early literary evidences of the use of the zero in India', *American Mathematical Monthly*, **33**, 449–54, **38**, 566–72.

—— *1928*, 'The science of calculation by the board', *American Mathematical Monthly*, **35**, 520–29.

—— *1929a*, 'The Bakhshālī mathematics', *Bulletin of the Calcutta Mathematical Society*, **21**, 1–60.

—— *1929b*, 'The Jaina school of mathematics', *Bulletin of the Calcutta Mathematical Society*, **21**, 115–45.

—— *1929c*, 'The scope and development of the Hindu gaṇita', *Indian Historical Quarterly*, **5**, 479–512.

—— *1932, The Science of the Śulba*, Calcutta: University of Calcutta Press.

Datta, B. and Singh, A. N. *1935–8, History of Hindu Mathematics*, 2 vols, Lahore: Motilal. [Repr. in one vol., 1962, Bombay: Asia Publishing House.]

—— *1980*, 'Hindu geometry' (revised by K. S. Shukla), *Indian Journal of History of Science*, **15**, 121–88.

—— *1983*, 'Hindu trigonometry' (revised by K. S. Shukla), *Indian Journal of History of Science*, **18**, 39–108.

Elfering, K. *1975, Die Mathematik des Āryabhaṭas*, Vol. 1, *Text, Übersetzung aus dem Sanskrit und Kommentar*, Munich: Fink.

Ganguli, S. *1931–2*, 'Indian contribution to the theory of indeterminate equations of the first degree', *Journal of the Indian Mathematical Society, Notes and Questions*, **19**, 110–20, 129–42, 153–68.

Gupta, R. C. *1987*, 'South Indian achievements in medieval mathematics', *Gaṇita Bhāratī*, **9**, 15–40.

—— *1989*, 'Sino-Indian interaction and I-Hsing', *Gaṇita Bhāratī*, **11**, 38–49.

Hayashi, T., Kusuba, T. and Yano, M. *1989*, 'Indian values for π derived from Āryabhaṭa's value', *Historia scientiarum*, (37), 1–16.

Ifrah, G. *1981, Histoire universelle des chiffres*, Paris: Seghers.

Juschkewitsch, A. P. *1964, Geschichte der Mathematik in Mittelalter* (transl. from Russian by V. Ziegler), Leipzig: Teubner.

Kapur, J. N. *1988*, 'A brief history of mathematics education in India', *Gaṇita Bhāratī*, **10**, 31–9.

Kaye, G. R. *1927–33, The Bakhshālī Manuscript: A Study in Medieval Mathematics* (Archaeological Survey of India, New Imperial Series, Vol. 43), 2 vols: Parts 1 and 2, Calcutta: Government of India, Central Publication Branch; Part 3, Delhi: Manager of Publications. [Repr. 1981, New Delhi: Cosmo Publications.]

Kuppuram, G. and Kumudamani, K. (eds) *1990, History of Science and Technology in India*, 12 vols, Delhi: Sundeep Prakashan.

Michaels, A. *1978, Beweisverfahren in der vedischen Sakralgeometrie*, Wiesbaden: Steiner.

Neugebauer, O. and Pingree, D. *1970–71*, (eds and transl.), *Pañcasiddhāntikā of Varāhamihira*, 2 vols, Copenhagen: Munksgaard.

Pingree, D. *1970–81*, *Census of the Exact Sciences in Sanskrit*, Series A, Vols 1–4, Philadelphia, PA: American Philosophical Society.

—— *1978a*, 'History of mathematical astronomy in India', in *Dictionary of Scientific Biography*, Vol. 15, New York: Scribner's, 533–633.

—— *1978b*, *The Yavanajātaka of Sphujidhvaja* (Harvard Oriental Series. Vol. 48), Cambridge, MA: Harvard University Press.

—— *1981*, *Jyotiḥśāstra: Astral and Mathematical Literature*. A History of Indian Literature (ed. J. Gonda), Vol. 6, Fasc. 4, Wiesbaden: Harrassowitz.

—— *1987*, 'Indian and Islamic astronomy at Jayasiṃha's court', in D. A. King and G. Saliba (eds), *Kennedy Festschrift*, *Annals of the New York Academy of Sciences*, **500**, 313–28.

Rajagopal, C. T. and Rangachari, M. S. *1978*, 'On an untapped source of medieval Keralese mathematics', *Archive for History of Exact Sciences*, **18**, 89–102.

Ramanujacharia and Kaye, G. R. *1912–13*, (transls) 'The *Triśatikā* of Śrīdharācārya', *Bibliotheca mathematica*, Series 3, **13**, 203–17.

Raṅgācārya, M. (ed. and transl.) *1912*, *Gaṇitasārasaṃgraha of Mahāvīra*, Madras: Government Press.

Ruegg, D. S. *1978*, 'Mathematical and linguistic models in Indian thought: The case of zero and *Śūnyatā*', *Wiener Zeitschrift für die Kunde südasiens und Archiv für indische Philosophie*, **22**, 171–81.

Sarasvati, T. A. *1963*, 'The development of mathematical series in India after Bhāskara II', *Bulletin of the National Institute of Sciences of India*, **21**, 320–43.

—— *1979*, *Geometry in Ancient and Medieval India*, Delhi: Motilal.

Sarma, K. V. *1972*, *A History of the Kerala School of Hindu Astronomy (in Perspective)*, Hoshiarpur: Vishveshvaranand Institute.

Sarma, S. R. *1985*, 'Writing material in ancient India', *Aligarh Journal of Oriental Studies*, **2**, 175–96.

Seidenberg, A. L. *1983*, 'The geometry of the Vedic rituals', in Staal *1983*: Vol. 2, 95–126.

—— *1988*, 'On the volume of a sphere', *Archive for History of Exact Sciences*, **39**, 97–119.

Selenius, C.-O. *1975*, 'Rationale of the *chakravāra* process of Jayadeva and Bhāskara II', *Historia mathematica*, **2**, 167–84.

Sen, S. N. *1971*, 'Mathematics', in D. M. Bose *et al.* (eds), *A Concise History of Science in India*, New Delhi: Indian National Science Academy, 136–212.

Sen, S. N. and Shukla, K. S. (eds) *1985*, *History of Astronomy in India*, New Delhi: Indian National Science Academy.

Sengupta, P. C. *1932*, 'Infinitesimal calculus in Indian mathematics – Its origin and development', *Journal of the Department of Letters, University of Calcutta*, **22**, 1–17.

Shukla, K. S. (ed. and transl.) *1959*, *Pāṭīgaṇita of Śrīdhara*, Lucknow: Lucknow University Press.

—— (ed.) *1976*, *Āryabhaṭīya of Āryabhaṭa with the Commentary of Bhāskara I and Someśvara*, New Delhi: Indian National Science Academy.

—— *1989*, 'The *Yuga* of the *Yavanajātaka*: David Pingree's text and translation reviewed', *Indian Journal of History of Science*, **24**, 211–23.

Shukla, K. S. and Sarma, K. V. (eds and transls) *1976*, *Āryabhaṭīya of Āryabhaṭa*, New Delhi: Indian National Science Academy.

Sinha, K. N. *1982*, 'Śrīpati's *Gaṇitatilaka*: English translation with introduction', *Gaṇita Bhāratī*, **4**, 112–33.

—— *1986*, 'Algebra of Śrīpati: An eleventh century Indian mathematician', *Gaṇita Bhāratī*, **8**, 27–34.

—— *1988*, 'Vyaktagaṇitādhyāya of Śrīpati's *Siddhāntaśekhara*', *Gaṇita Bhāratī*, **10**, 40–50.

Srinivas, M. D. *1990*, 'The methodology of Indian mathematics and its contemporary relevance', in Kuppuram and Kumudamani *1990*: Vol. 2, 29–86.

Srinivasiengar, C. N. *1967*, *The History of Ancient Indian Mathematics*, Calcutta: The World Press.

Staal, F. (ed.) *1983*, *Agni: The Vedic Ritual of the Fire Altar*, 2 vols, Berkeley, CA: Asian Humanities Press.

van der Waerden, B. L. *1976*, 'Pell's equation in Greek and Hindu mathematics', *Russian Mathematical Surveys*, **31**, 210–25.

—— *1983*, *Geometry and Algebra in Ancient Civilizations*, Berlin: Springer.

1.13

Tibetan astronomy and mathematics

GEORGE GHEVERGHESE JOSEPH

1 INTRODUCTION

Tibet has a living astronomical heritage which has been influenced by both China and India. A traditional calendar is still published in Lhasa for religious and ceremonial purposes. As a product of cultural mixing – a mixing of religions, of languages and of astronomical knowledge – Tibetan astronomy provides a useful illustration of how different traditions may join forces in the development of science. It is also valuable for the light it throws on the history of Indian astronomy before the appearance of the *Siddhāntas* in the fifth century AD. And it provides one of the few examples of Buddhism influencing the growth of Indian astronomy, for the bulk of evidence for the spread of Indian astronomy to Tibet comes from texts from a period that saw the decline of Buddhism as a religion in India (Petri *1968*).

Tibetan astronomy (or *rtsis*) may be broadly divided into two branches: *nag rtsis* ('black calculation') and *skar-rtsis* ('star calculation'), which may be loosely interpreted in modern terms as astrology and astronomy, respectively. A popular work on *nag rtsis* of the seventeenth century lists nine integral components of the subject. It is shown to be a blend of Chinese natural philosophy, such as the theory of the five basic elements and the *ying–yang* principle, and of Chinese astrological practices that include the association of each year of a twelve-year cycle and of each month in the year with a particular animal (rat, tiger, horse, sheep, and so on). There is also a description of the Chinese lunar zodiac which contains 28 stars, unlike the Indian lunar zodiac which has 27 or 28 *nakṣetras* (lunar mansions).

The origins of *skar rtsis* may be traced to an Indian Buddhist tantric text, the *Kālacakratantra*, and its commentary, the *Vimalaprabhā*. Little is known of the date and origin of the *Kālacakra* text, though it is likely that a version of it appeared in the eleventh century, since AD 1027 is given as the start of a 60-year cycle the duration of which is obtained as the product of the number of elements (5) and the number of animals in the 12-year cycle.

2 EARLY ASTRONOMY

During the later stages of Buddhism on Indian soil, there was great interest in cosmography and chronology. A system of teaching known as the *Kālacakra* (or the 'wheel of time') developed whose origin was traced to the revelations of the Buddha in Sambhala (a fabled country in the north of India). It was introduced into Tibet, where it has been preserved along with some associated texts. An abridged version of the original text (*Laghutantra*) is currently published in both Sanskrit and Tibetan.

The cosmography of the first chapter of this abridged text bears a characteristically Indian *Purānic* hallmark: the Earth is a flat disk; the stars orbit around Mount Meru, which is at the centre of the Universe. The mathematics is basic (merely a prelude to the astronomy), consisting of a brief description of a word-numeral system and the names of units of length and time. The circumference of a circle is given as three times its diameter. The discussion of astronomy that follows is cryptic to the point of obscurity, but a later commentary on the text, entitled *Kālacakrāvatāra*, provides a clearer explanation. It gives instructions for calculating how many days have lapsed since a given epoch (*ahargana*), and the longitudes of the planets (*grahaganita*). There are tables of corrections to the mean motions of the planets, corresponding to the two Indian epicycles which accounted for the displacements caused by the ellipticity of the planets' orbits and by the fact that we observe from the Earth. However, a surprising omission is the then widespread method from the *Siddhāntas* of 'halving the equation' in order to account for the influence of the two equations on each other (Schuh *1973*).

An encyclopedia (*Vaidūryadkarpo*, 'White Beryl'), compiled by the chief executive of the fifth Dalai Lama during the second half of the seventeenth century, contains a short description of Indian arithmetic, dealing mainly with the four fundamental operations of addition, subtraction, multiplication and division, and including a cryptic explanation of the determination of square roots along the lines of the chapter entitled 'Ganitadyāyah' in Brahmagupta's *Brāhmasphutasiddhānta* (AD 598). This work also contains lists of synonyms for numerical operations and for planets. Its astronomy is similar to that of the *Kālacakra* texts.

Noteworthy in all these books is the profound influence of Indian astronomy. This influence spread even to the sacred books of the Tibetans, which contain instructions on calendar construction and planetary calculations. A detailed list of the 27 stars (*naksetras* or, in Tibetan, *rgyu-skar*) in the Indian lunar zodiac in one of these books is ascribed to the legendary Indian sage-astronomer, Garga (Petri *1966*).

3 THE WORK OF BU-STON RIN-CHEN GRUB

The culmination of traditional Tibetan astronomy is found in the works of Bu-ston Rin-chen grub (1290–1344) and Rang-byung rdo-rje (1284–1339). Bu-ston wrote an astronomical text called *mKas-pa dga'byed* which summarizes Tibetan astronomy in seven chapters. Apart from the last chapter, which contains a discussion of word-numerals, the mathematics is seen very much as a tool for astronomical calculations. Space prevents a full discussion of Bu-ston's considerations, but they include the following.

3.1 The division of time

Three different kinds of day were distinguished. The civil day (*nyin zhag* or the Indian *sāvana* day) was the interval between two consecutive sunrises at a given place. Another kind of day, *tsheh zhag* (the Indian *tithi*), was taken to be 1/30 of a synodic month, so that a *tsheh zhag* was the time taken by the Moon to get $12°$ ahead of the Sun from the time of new Moon. To compute this day accurately would have required a knowledge of mean motions, velocities and the true longitudes of the Sun and Moon. What values the Tibetans had for these parameters, and what mathematical formulas they used, are not known.

Finally, there was the *khyim zhag* (the Indian *savra* or solar day), 1/360 of a solar year. Each day as well as each *nakṣetra* was divided into 60 *chu tshods*, each *chu tshod* into 60 *chu srangs*, and each *chu srang* into 6 *dbugs* (or *prāṇas* in Sanskrit, meaning 'breaths'). A simple relationship between the three kinds of day, known to Bu-ston, may be expressed in terms of the following formulas:

$$\text{length of 1 } \textit{tsheh zhag} = \text{length of 1 } \textit{nyin zhag} \times \frac{63 - 1/707}{64},$$

$$\text{length of 1 } \textit{khyim zhag} = \text{length of 1 } \textit{tsheh zhag} \times 67/65.$$

It can then be shown that one sidereal year, in terms of the *nyin zhag*, is equal to 365.2706 days. Incidentally, Bu-ston's value for the sidereal year is less accurate than the one given in the *Sūryasiddhānta*, 365.2588 days.

3.2 Equation of the centre of the Sun

For each of the 12 zodiacal signs the equation of the centre of the Sun was given, these signs being placed in four quadrants. By adding or subtracting the difference between the mean daily motion and the true daily motion of the Sun, the maximum equation (11 *chu tshod*, or $2°26'40''$) and the location of the solar apogee at what was known as the First Point of Cancer ($90°$ of celestial longitude), were obtained. Unlike many other aspects of Tibetan astronomy, these values are not found in any Indian astronomical text. The closest values are those in *Sūryasiddhānta* in which the maximum equation is $2°14'$ and the longitude of the solar apogee is $80°$.

3.3 Equation of the centre of the Moon

If one anomalistic month were taken to be approximately 28 *tsheh zhag*, a correction called *zla b'ai myur rkang* ('fast step of the Moon') was applied to the length of each *tsheh zhag*. Another correction to take account of the fact that 28 *tsheh zhags* is a slight overestimate of the length of an anomalistic month would give the maximum equation of the Moon as $5°23'20''$. The closest Indian value (from *Brāhmapaksa* was $5°10'28''$.

The mean motions of planets and their equations of the centre were estimated in a similar fashion. The values obtained were very similar to those found in Indian astronomy.

3.4 Epicycle corrections for the planets

To count the steps of an epicycle correction, the 'parameter of steps' was used. A period of 60 *chu tshods* was taken as one step. Now, 60 *chu tshods* also correspond to one *nakṣetra*, so 27 *nakṣetras* would give 1620 *chu tshods*, which described one cycle. If this cycle were divided into two halves, and each half had 14 steps, then the 14th step of the first half and the 1st step of the second half had only 30 *chu tshods*.

Two correction formulas were used. Let M be the true daily motion of the planet, D the daily motion of the planet as corrected by applying the equation of the centre of the planet, and m the 'fast step' correction factor. Then for the 1st step to the 13th, the equation used to estimate the true daily motion of the planets was

$$M = D + m/60. \tag{1}$$

For the 14th step only, the equation used was

$$M = D + m/30. \tag{2}$$

Table 1 presents the maximum epicycle corrections of Bu-ston and the nearest Indian values (that in the *Sūryasiddhānta* of the *Pāncasiddāntikā*). Venus is left out, since Bu-ston's description of the procedure for this planet is difficult to fathom. The closeness of these astronomical constants as found by Bu-ston to the Indian values is a clear indication of Tibetan astronomy's debt to India.

Table 1 Maximum epicycle corrections for four planets, as calculated by Bu-ston and the nearest Indian values

Planet	Bu-ston	Pañcasiddāntikā
Mercury	21°33'	21°30'
Mars	40°27'	40°30'
Jupiter	11°33'	11°30'
Saturn	6°13'	6°20'

4 NOTES ON SOURCES

Cosma de Koros *1834* was a pioneer in the study of the Tibetan language; his work includes a section on the chronological aspect of Tibetan astronomy. The basic reference in English on Tibetan astronomy is Petri *1968*, which contains a general discussion of early Tibetan astronomy and the debt it owed to other traditions. This is based on a more detailed exposition of the general features of Tibetan astronomy and astrology in Petri *1966*. A number of original texts on calendrical calculations and astrology have been reissued by Tibetan Buddhist missions based in Delhi or Dalhousie, of which the Tibetan Bonpo Monastic Centre and Damchoe Sangpo are the most important. Recent sources include Schuh *1973* and Yamaguchi *1973*, which contain a detailed examination of the chronological aspects of the Tibetan astronomy.

There is a journal for Tibetan studies, *Xiang-yanjiu* with articles in both Chinese and Tibetan, which has occasional articles on Tibetan sciences. The first issue in 1981 contains an article by Huang Mingxin and Chen Jiujin entitled 'Zangli yuanli yanjiu' ('A study of the Tibetan calendar'). The third issue includes a history of Tibetan astronomy by Tshul-khrims Chos-sbyor and bSod-nams Phun-tshos. In 1983, these two writers published a history of Tibetan astronomy in the Tibetan language. More recently, Yukio Ohashi of Japan has been studying the works of Bu-ston. This article owes a great debt to a mimeograph, privately circulated by Yukio Ôhashi, on Tibetan astronomy in general and Bu-ston's work in particular.

BIBLIOGRAPHY

de Koros, C. [=Csoma de Körösi, A.] *1834, A Grammar of the Tibetan Language,* Calcutta: Baptist Missionary Press.

Petri, W. *1966,* 'Indo-tibetische Astronomie', doctoral dissertation, Munich.

—— *1968,* 'Tibetan astronomy', *Vistas in Astronomy,* **9,** 159–64.

Schuh, D. *1973, Untersuchungen zur Geschichte der tibetischen Kalenderrechnung,* Wiesbaden: Steiner.

Yamaguchi, Z. *1973, Chibetto no Rekigaku* ['Calendrical Science of Tibet'], Tokyo.

1.14

Mathematics in medieval Hebrew literature

Y. T. LANGERMANN

Medieval Hebrew literature contains some noteworthy contributions to pure mathematics. However, medieval Hebrew mathematics is largely applied mathematics, most of which falls into two categories: astronomical theory and computations, and business arithmetic. Even the not infrequent appeals to mathematics that are found in the philosophical literature represent applications of mathematics to philosophy; little was ventured in the way of the philosophy of mathematics. Some basic Hellenistic texts such as the works of Euclid and Archimedes were translated (via Arabic); others, such as the works of Apollonius and Diophantus, were not. A number of more specialized Greek and Arabic books were also translated.

In this article the original Hebrew mathematical literature is surveyed first, followed by mathematics and philosophy, some interesting Hebrew translations and some miscellaneous items of interest. Some manuscripts are cited.

1 ORIGINAL HEBREW MATHEMATICS

The first work of note is the anonymous *Mishnat ha-Middot*, which presents algorithms for the solution of quadratic equations along with the geometrical analogues of these equations. This is also a feature of al-Khwārizmī's *Algebra*, and scholarly opinion as to which of the two is earlier has oscillated. Studies by G. Sarfatti indicate that the Hebrew *Mishnah* in fact depends upon al-Khwārizmī (Sarfati *1968*).

Perhaps the most important Hebrew treatise is Abraham bar Ḥiyya's *M^{e}shiḥah w^e-Tishboret*, written in 1145; its Latin version, *Liber embadorum*, played a major role in the diffusion of algebra through Christian Europe. This treatise contains much geometrical algebra, including solutions of some equations of higher degree. Bar Ḥiyya treats the

137

measurement of plane surfaces and some solids, including the truncated pyramid. He demonstrates the formula for the area of the circle by considering the circle to be fully covered by concentric rings of string, which are then snipped along a radius and folded down. The result is a triangle whose base is the circumference and whose height is the radius, and whose area, which equals the area of the circle, is $\frac{1}{2}bh = \frac{1}{2}(2\pi r)r = \pi r^2$. *Liber embadorum* contains some trigonometry, including a table of chords based on a radius of 14 units and a circumference of 88 units. The third chapter preserves remnants of Euclid's lost work on the division of figures. Bar Ḥiyya's encyclopedia *Migdal ha-Tᵉvunah* includes some algorithms and business arithmetic.

The contribution of the savant Abraham ibn Ezra is a lot harder to assess. Ibn Ezra's neo-Pythagoreanism manifests itself in scattered remarks throughout his works, including the monograph *Sefer ha-Eḥad* ('The Book of the One'). In the fourteenth century in particular quite a number of Jewish scholars strove to explain Ibn Ezra's mathematical 'hints'. He also wrote on arithmetic, and seems to have been instrumental in the diffusion of Indian mathematics.

Moses Maimonides (died 1204) applied his mathematical skill to the calculation of the visibility of the new Moon. According to him, the rounding errors that he introduced into his procedure all cancel out, such that simplification is achieved at no cost in accuracy. He also wrote, in Arabic, notes to Apollonius's *Conics* and to Ibn al-Haytham's *Completion of the Conics*, in which he filled in some steps in the proofs (Langermann *1984a*).

In the course of his extensive and original astronomical research, Levi ben Gerson (died 1344) displayed considerable mathematical ingenuity. For example, he devised a transversal scale in order to avoid the random errors that were introduced into observations when scientists tried to read off the minutes from the scale of an astrolabe calibrated only in degrees (Goldstein *1977*). His mathematical work, *Maʿaseh Ḥoshev*, contains interesting studies on combinations, permutations and induction. Another important fourteenth-century work is *Mᵉyashsher ʿAqov*, perhaps written by the apostate Abner of Burgos. Among other things, it discusses the construction of the conchoid and its asymptote.

Three little-studied Spanish treatises that date from the fourteenth or fifteenth century are interesting for their treatment of practical computations. Isaac of Orihuela's book in three parts instructs in the basic arithmetical operations, including the 'Hindu dust-board' method of extracting square roots. He cites the classical approximation $3\frac{10}{71} < \pi < 3\frac{1}{7}$, but notes that 'there exists no ratio between the straight and unstraight line, hence it is impossible to know the "quantity" of the circle exactly' (see Section 2 below for Maimonides' remarks about π). *ʿIr Siḥon*, by Yosef ben Moshe ha-Sarfati (?1422) is a somewhat more sophisticated work. Ha-Sarfati presents an interesting iterative procedure for finding $\sqrt{a}$, which is based on the following algorithm: take r, slightly larger than the true root, and find $r_1 = r - (r^2 - a)/2r$. He gives a clear formula for the sum of an arithmetic series, solves a variety of problems in business arithmetic and offers some interesting remarks of a philosophical character. The third treatise, by a silk-weaver named Aaron, survives in a unique manuscript which was, however, damaged in a fire (Turin A. V. 15). Aaron distinguishes between 'unlimited', 'immaterial' number, that cannot be expressed in word or thought, and number that has been made finite by our very thinking of it; the latter, of course, is the subject of his book. Unlike other works mentioned in this article so far, the terminology of this book is heavily influenced by Latin usage. There are extensive examples, potentially of great historical value, of computations of use to traders, money-changers, artisans (e.g. goldsmiths) and apothecaries.

Mordecai Finzi (Italy, mid-fifteenth century) had access to a wide range of scientific literature. He has left us an as-yet unstudied treatise on the mensuration of buckets and barrels (Langermann *1988*).

2 MATHEMATICS AND PHILOSOPHY

As noted above, philosophers occasionally resorted to mathematical arguments. Several remarks by Maimonides, who clearly enjoyed advanced mathematical training, are of great interest (Langermann *1984a*). In his commentary on the *Mishnah* (*'Eruvim* 1:5), he asserts unequivocally that π is irrational; our inability to express its value exactly is not, he says, due to any deficiency in our computation, 'as the ignorant claim, but rather the matter is by its nature inscrutable'. It seems to me that Maimonides is arguing here against the facile appeal, in the course of theological discussions, to such 'wonders' as the existence of irrational numbers.

An interesting episode developed out of Maimonides' remark in *Guide of the Perplexed* 1:73 that it is possible to demonstrate rigorously the existence of that which the imagination rejects. For example, Apollonius showed (*Conics* 2.14) that a straight line may continue to approach a curve without ever touching it. Because the *Conics* was never translated into Hebrew, students of the *Guide* produced several constructions of a curve and its asymptote; Lévy *1989a* has recently identified six distinct proofs. The same topic was discussed in some Arabic and Latin treatises, and the Hebrew material offers some important clues to the transmission of mathematics to Europe.

Medieval philosophers were deeply concerned with problems such as those of motion and divisibility, and discussions of these topics in the

Hebrew literature occasionally invoke mathematical reasoning. Of greatest promise are Levi ben Gerson's as-yet unstudied commentaries on Aristotle (as summarized and interpreted by ibn Rushd (Averroes)). Ḥasdai Crescas contradicted Aristotle outright and maintained that the infinite exists in actuality, not just *in potentia* (Rabinovich *1970*).

Whereas some Jews, following Ibn Ezra, made numerology an essential part of their philosophy (§12.5), most regarded mathematics as a preparatory science, necessary for metaphysics but not an end in itself. As Yosef Albo wrote (*'Iqqarim* 3:3):

> For what perfection can the soul attain by knowing that the angles of a triangle are equal to two right angles . . .? This knowledge gives perfection to the soul only as far as it is an introduction and a means to the understanding of the heavenly causes upon which the natural objects depend . . . Therefore the soul cannot acquire eternity from such knowledge.

3 TRANSLATIONS AND TRANSCRIPTIONS

The great bibliographer M. Steinschneider recorded detailed descriptions of the Hebrew translations known to him in his *Hebräische Übersetzungen* (*1893*). Only translations not covered in that monumental work are discussed here.

Clearly the most significant of these is the Hebrew version of a work on the regular polyhedra; it is closely related to a treatise in Arabic by Muḥyī al-Dīn on the same subject (Langermann and Hogendijk *1984*). Some of the same theorems are cited in an important, anonymous Hebrew mathematical encyclopedia, *Sefer Uqlidis*.

Abū Kāmil's *Algebra* was translated by Finzi; generally, though, the achievements of the Islamic algebraists were not rendered into Hebrew. The Hebrew translation of Eutocius' commentary on Archimedes' *On the Sphere and Cylinder* found in ms. Oxford–Bodley d.4 (the same codex that contains the book on polyhedra) differs significantly from the Greek original. This same manuscript also has unique copies of a solution to a criticism of Apollonius and a treatise on the triangle by one Abū Sa'adan. An interesting and lengthy commentary on Menelaus's *Spherics*, translated from the Arabic by Shmu'el ben Yehudah, exists at Milan (ms. Ambrosiana 97/1).

Jews living in Arabic-speaking countries did not translate Arabic works, but they did copy them into the Hebrew script; occasionally these transcriptions preserve important texts. Of the greatest historical interest from this body of material are some fragments of Simplicius's commentary on Euclid, written in a Yemenite hand and discovered by E. Wust.

4 MISCELLANEOUS

In the Latin manuscript tradition can be found some primitive attempts at graph representation (see the photograph in Pedersen and Pihl *1974*: 246). Specimens of graphs are found in the end-notes to ms. Oxford, Poc. 368, ff. 194b and 197b.

Although Jews showed little interest in geometrical optics, an interesting exception is Levi ben Gerson's detailed analysis of the halo, found in his *Meteorology* (e.g. ms. Vatican Heb. 342, 271a ff.)

Ms. Hamburg, Levi 113, is an interesting mathematical codex. The quality of the discussions is not consistent; however, the writer of this codex has drawn upon a wide range of materials, including Jābir bin Aflaḥ on Menelaus, al-Antākī, Euclid's *Elements*, *Data* and *Catoptrics*, and Ibn al-Haytham on Euclid.

Finally, Rabinovich *1973* argues that a tradition of probabilistic thinking can be traced both in medieval and ancient Jewish literature. The evidence he cites includes statistical inference in the Talmud, notions of relative frequency, and some treatment of combinations and permutations.

BIBLIOGRAPHY

Chemla, K. and Pahaut, S. *1992*, 'Remarques sur les ouvrages mathématiques de Gersonide', in G. Freudenthal (ed.), *Studies on Gersonides, a Fourteenth-Century Jewish Philosopher-Scientist*, Leiden: Brill, 149–91.

Freudenthal, Gad *1988*, 'Maimonides' "Guide of the Perplexed" and the transmission of the mathematical tract "On Two Asymptotic Lines" in the Arabic, Latin and Hebrew medieval Traditions', *Vivarium*, **26**, 113–40.

Gandz, S. *1970*, *Studies in Hebrew Astronomy and Mathematics* (ed. S. Sternberg), New York: Ktav.

Goldstein, B. R. *1974*, *The Astronomical Tables of Levi ben Gerson*, Hamden, CT: Connecticut Academy of Arts and Sciences.

—— *1977*, 'Levi ben Gerson: On instrumental errors and the transversal scale', *Journal for the History of Astronomy*, **8**, 102–16.

Guttmann, M. (ed.) *1913–14*, *Ḥibbur ha-Mᵉsiḥah wᵉ-ha-Tishboret*, Berlin: Mekize Nirdamim.

Langermann, Y. T. *1984a*, 'The mathematical writings of Maimonides', *Jewish Quarterly Review*, **75**, 57–65.

—— *1984b*, '"Sefer Uqlidis" by an anonymous author', *Kiryat Sefer*, **54**, 635. [In Hebrew.]

—— *1987*, *The Jews of Yemen and the Exact Sciences*, Jerusalem: Misgav Yerushalayim. [In Hebrew with English synopsis.]

—— *1988*, 'The scientific writings of Mordekhai Finzi', *Italia: Studi e ricerce sulla storia, la cultura e la letteratura degli ebrei d'Italia*, **7**, 7–44.

Langermann, Y. T. and Hogendijk, J. P. *1984*, 'A hitherto unknown Hellenistic treatise on the regular polyhedra', *Historia mathematica*, **11**, 325–6.

Lévy, T. *1989a*, 'L'Etude des sections coniques dans la tradition médiéval hébraïque: Ses relations avec les traditions arabe et latine', *Revue d'Histoire des Sciences*, **42**, 193–239.

—— *1989b*, 'Le Chapitre 1, 73 du *Guide des Egarés* et la tradition mathématique hébraïque au moyen âge: Un commentaire inédit de Salomon b. Isaac', *Revue des Etudes Juives*, **147**, 307–36.

Lorch, R. *1989*, 'The Arabic translation of Archimedes' *Sphere and cylinder* and Eutocius' *Commentary*', *Zeitschrift für Geschichte der arabisch–islamischen Wissenschaften*, **5**, 93–114.

Pedersen, O. and Pihl, M. *1974*, *Early Physics and Astronomy: An Historical Introduction*, London: MacDonald and James, and New York: Elsevier.

Rabinovich, N. L. *1970*, 'Rabbi Hasdai Crescas' (1340–1410) and numerical infinities', *Isis*, **61**, 222–30.

—— *1973*, *Probability and Statistical Inference in Ancient and Medieval Jewish Literature*, Toronto: University of Toronto Press.

Sarfatti, G. B. *1968*, *Mathematical Terminology in Hebrew Scientific Literature of the Middle Ages*, Jerusalem: Magnes Press.

Steinschneider, M. *1893*, *Die hebräische Übersetzungen des Mittelalters und die Juden als Dolmetscher*, Berlin: Kommissionverlag. [Repr. 1956, Graz: Akademische Verlagsanstalt.]

—— *1964*, *Mathematik bei den Juden*, Hildesheim: Olms. [Book-form repr. of articles published between 1893 and 1901.]

1.15

Maya mathematics

M. P. CLOSS

1 INTRODUCTION

The ancient Maya inhabited a region encompassing present-day Guatemala, Belize, the western parts of Honduras and El Salvador, and the lowlands of southern Mexico (the states of Yucatán, Campeche, Quintana Roo, most of Tabasco and the eastern part of Chiapas). Archaeologists have dated the earliest Maya village to 1000 BC. At the other extreme, the last independent Maya kingdom was not subdued until AD 1697. From around AD 300 to 900, the Maya civilization flourished during an era which archaeologists have termed the Classic Period.

The Maya are renowned for having developed and used a system of hieroglyphic writing which is unrelated to writing systems in the Old World. It consists of a large collection of intricate logographic and phonetic signs, commonly called glyphs, together with complex rules of orthography. Script elements in the mixed logographic–phonetic writing accurately reflected the spoken word, with fidelity to the grammar, syntax and sounds of ordinary speech. Scholars have made impressive progress in deciphering this once poorly understood script, and there is now wide agreement on the linguistic interpretation of many glyphs.

Students of Maya writing have at their disposal a large body of pre-Columbian texts which is being continually expanded through new archaeological discoveries. The current inventory includes four screen-fold books, called codices, thousands of carved stone monuments and thousands of ceramic vessels.

2 NUMERICAL NOTATION

The common numerals of the Maya were composed of 'bars', each having value 5, and 'dots', each having value 1. Numbers from 1 to 19 (the Maya number system was vigesimal) were represented by an economical combination of bars and dots yielding the desired value. Frequently, non-numerical crescents or other fillers were used to achieve a more aesthetic balance for the numeral. Examples of these numerals can be seen in Figure 1, from a Classic Period ceramic vessel, showing a classroom scene in which Pauahtun, a patron god of scribes, is giving a mathematics lecture to two apprentice scribes. The deity, wearing a net headdress, is shown leaning over a codex with a brush-pen in his left hand. A speech scroll issues from his mouth containing the numbers 11, 13, 12, 9, 8 and 7.

Figure 1 Detail from a Classic Maya vessel showing a classroom scene (drawing by M. P. Closs; after Robicsek and Hales *1981*: Vessel 56)

The Maya measured time intervals between calendar dates by a composite chronological count consisting of a vigesimal count of *tuns* (periods of 360 days) and distinct counts of *winals* (periods of 20 days) and *k'ins* (days). In some chronological contexts, bar and dot numerals were employed in a system of positional notation consisting of a vertical arrangement of numbers in which the lowest position was used to represent the number of *k'ins*, and successively higher positions to represent the number of each of the successively longer periods constituting the count. In this system, there were special symbols for zero.

The most extensive use of this positional notation is found in the astronomical tables of the Dresden Codex. The oldest securely dated Maya text employing it is found on Stela 1 at Pestac. This monument bears a chronological count, including a zero sign, which firmly dates it to AD 665.

The system of positional notation using bar and dot numerals is also found on several older monuments which predate those of undisputed Maya origin. The oldest of these goes back to 36 BC (Coe *1957*, Winfield Capitaine *1988*). None of these very early texts employs a zero sign.

On dynastic monuments, the Maya preferred to use a non-positional system in which time counts were expressed by prefixing numerals to chronological glyphs representing the *k'in*, *winal*, *tun*, *katun* (= 20 *tuns*) and *baktun* (= 400 *tuns*), and other longer time periods if needed. In this system the common usage of zero signs goes back further than the time of Stela 1 at Pestac. For example, both Stela 18 and Stela 19 from Uaxactun contain chronological counts, with zero glyphs, which date the monuments to AD 357. The oldest dated Maya monument with this type of notation, although without zero glyphs, is Stela 29 at Tikal which was dedicated in AD 292.

In addition to bar and dot numerals, the Maya sometimes used portrait heads and even entire anthropomorphic figures to represent numbers. These elaborate 'head variant' and full-figure numerals are singular in their beauty and are unrivalled in any other script. Their existence is not exceptional within the framework of Maya writing, which usually had several different glyphic signs for a given phonetic or logographic value.

The lengthiest, most detailed and most lavishly illustrated treatment of Maya numerical, calendrical and chronological notation is contained in Thompson *1971*. This work was originally published in 1950 and is badly out of date in its understanding of Maya writing and of the content of the monumental inscriptions. Closs *1986* summarizes and supplements the work of Thompson. It is sensitive to context, reflects modern glyphic research, and is well illustrated.

3 CALENDRICS AND CHRONOLOGY

The most sophisticated usage of Maya mathematics is found in problems arising in calendrics, chronology and astronomy. The most important Maya calendars were a sacred calendar of 260 days and an annual calendar of 365 days. These two were combined to yield a cycle of 18 980 paired dates known as the 'calendar round'. The most common problems were of two types: (1) to find the calendar-round date when a chronological count is added to or subtracted from a given date, and (2) to find the chronological interval separating two dates in the calendar round.

In addition to the calendar round, the Maya employed several other calendrical cycles. The most important of these were ritual cycles of 4 days, 9 days and 819 days. This led to other problems such as that of determining the closest 819-day station preceding a given calendar-round date.

The Maya had a system of absolute chronology whose zero date fell in 3114 BC. Scribes customarily anchored calendar-round dates in this absolute chronology by specifying the chronological interval separating it from the base date. They also performed calculations which ranged over thousands of years from contemporaneous dates to mythological dates preceding the zero point of their chronology. These mythological dates were anchored in the absolute chronology by specifying the negative chronological count separating them from the zero date. These negative counts are represented in a distinct notation which has prompted scholars to call them 'ring numbers'. Ring numbers are accompanied by companion numbers which bridge the gap between the mythological dates and the contemporaneous dates. They are characterized by being highly divisible into a product of relatively small prime numbers (Lounsbury *1976*). This phenomenon, together with other evidence including an error in a table of multiples (Closs *1977*: 92–3), a linguistic artefact of a computational strategy (Thompson *1971*: 248) and a few lists of residue classes in post-conquest Maya writing (*Ibid.*: 247), indicate that calendrical calculations were achieved by using residue arithmetic.

Discussions of calendars, ritual cycles, the absolute chronology, ring numbers and calendrical problems and their solutions are given by Thompson *1971*, Lounsbury *1978* and Closs *1986*.

4 ASTRONOMY

The greatest accomplishments in Maya mathematics are found in their arithmetically based astronomy. The Dresden Codex contains a table for Venus which enables its first appearances as 'morning star' and 'evening star' to be predicted. In creating this table, the mathematicians were able to resolve the problem of commensurating the natural periodicity exhibited by Venus with the regular periodicities of their calendars. Using an integer-based arithmetic, they devised mechanisms to maintain the astronomical integrity of the table over several hundred years (Thompson *1972*, Closs *1977*, Lounsbury *1983*).

Probably the supreme achievement of Maya mathematicians was the creation of the eclipse table in the Dresden Codex, which enabled the scribes to predict potential solar and lunar eclipses. This was achieved by commensurating a lunar calendar with the ritual calendar of 260 days in such a way as to highlight solar eclipse seasons (Thompson *1972*, Lounsbury *1978*, Aveni *1980*, Bricker and Bricker *1983*, Justeson *1989*, Closs *1989*). These eclipses might be visible anywhere on Earth, and so were not restricted to locally visible eclipses, the determination of which would be a much more complicated problem.

Astronomical calculations were also made in order to predict Moon ages in mythological times. From Classic Period texts containing such calculations it is possible to recover the Moon-age formulas that were used (Lounsbury *1978*: 175).

5 GEOMETRY

There is no information on Maya geometrical concepts in any of the native texts, but such knowledge is manifested in paintings, sculptures and site plans (Vinette *1986*). In a series of papers, Aveni, Hartung and others have discovered astronomical orientations and geometrical features in the site plans and buildings at a large number of Maya sites. There is space here to refer only to a sample of the work which has been done in this area (Aveni *1975*, Hartung *1975*, *1977*, Aveni and Hartung *1982*, *1986*).

As an example of this type of study, an analysis by Hartung *1977* of the site plan of Tikal, one of the largest ancient Maya cities in Guatemala, shows that geometrical notions are apparent, and gives indications that they were not coincidental. For instance, there is an isosceles right triangle, with the right angle at Temple III formed by an east–west baseline from Temple I to Temple III and a south baseline from Temple III to Structure 5D-90 on the south side of the Plaza of the Seven Temples. Temple I and Temple III were erected around AD 700 and 810, respectively. Equinoctial observation could have been made only before the construction of Temple III.

As a second example, there is a right angle ($89°57'$) at Temple I formed by the lines from the doorways of Temple IV to Temple I and from Temple I to Temple V. The deviation of the directions of both lines by about $14°$ from the cardinal points is not frequently found in the orientations of the buildings at Tikal. Temples I and V were constructed around AD 700, and Temple IV around AD 750.

6 MATHEMATICAL SPECIALISTS

Diego de Landa, the third bishop of Yucatán, wrote a book in about 1566 containing a wealth of information on the ancient Maya (Tozzer *1941*). In it we learn that the Maya scribes were members of the priestly and noble class. In particular, Landa mentions that 'the computation of the *katuns*...was the science to which they gave the most credit, and that which they valued most and not all the priests knew how to describe it'. This reference indicates that mathematics was a highly regarded specialization within scribal studies, that it carried its own prestige, and that not all scribes mastered the subject.

Mathematics, as a discipline, had sufficient presence and concreteness in Maya thought for it to show up in Classic Period iconography. This can be seen in a number of ceramic vessels which carry scenes portraying mathematical specialists at work. These specialists are rendered as scribal figures, distinguished by a vegetal scroll containing numbers emanating from the armpit. An example of such a scribal figure, illustrated in Figure 2, shows a scribe, in seated position, holding a brush-pen and writing in an opened codex. A large vegetal scroll emerging from his armpit contains the single number 13. Above the head of the scribe a panel of rectangular cartouches, known as a 'skyband', bears astronomical symbols, the first of which is a sign for the planet Venus. The scene portrays a mathematician at work on an astronomical topic.

Figure 2 Detail from a Classic Maya vessel showing a mathematician at work (drawing by M. P. Closs; after Robicsek and Hales *1981*: Vessel 61)

BIBLIOGRAPHY

Aveni, A. F. *1975*, 'Possible astronomical orientations in ancient Mesoamerica', in A. F. Aveni (ed.), *Archaeoastronomy in Pre-Columbian America*, Austin, TX: University of Texas Press, 163–90.
—— *1980*, *Skywatchers of Ancient Mexico*, Austin, TX: University of Texas Press.
Aveni, A. F. and Hartung, H. *1982*, 'Precision in the layout of Maya architecture', *Annals of the New York Academy of Sciences*, **385**, 63–80.
—— *1986*, 'Maya city planning and the calendar', *Transactions of the American Philosophical Society*, **76**, (7).

Bricker, H. M. and Bricker, V. R. *1983*, 'Classic Maya prediction of solar eclipses', *Current Anthropology*, **24**, 1–23.

Closs, M. P. *1977*, 'The date-reaching mechanism in the Venus table of the Dresden Codex', in A. F. Aveni (ed.), *Native American Astronomy*, Austin, TX: University of Texas Press, 89–99.

—— *1986*, 'The mathematical notation of the ancient Maya', in M. P. Closs (ed.), *Native American Mathematics*, Austin, TX: University of Texas Press, 291–369.

—— *1989*, 'Cognitive aspects of ancient Maya eclipse theory', in A. F. Aveni (ed.), *World Archaeoastronomy*, Cambridge: Cambridge University Press, 389–415.

Coe, M. D. *1957*, 'Cycle 7 monuments in Middle America: A reconsideration', *American Anthropologist*, **59**, 597–611.

Hartung, H. *1975*, 'A scheme of probable astronomical projections in Mesoamerican architecture', in A. F. Aveni (ed.), *Archaeoastronomy in Pre-Columbian America*, Austin, TX: University of Texas Press, 191–204.

—— *1977*, 'Ancient Maya architecture and planning: Possibilities and limitations for astronomical studies', in A. F. Aveni (ed.), *Native American Astronomy*, Austin, TX: University of Texas Press, 111–29.

Justeson, J. S. *1989*, 'Ancient Maya ethnoastronomy: An overview of hieroglyphic sources', in A. F. Aveni (ed.), *World Archaeoastronomy*, Cambridge: Cambridge University Press, 76–129.

Lounsbury, F. G. *1976*, 'A rationale for the initial date of the Temple of the Cross at Palenque', in M. Greene Robertson (ed.), *The Art, Iconography and Dynastic History of Palenque*, Part 3, Pebble Beach, CA: The Robert Louis Stevenson School, 211–24.

—— *1978*, 'Maya numeration, computation, and calendrical astronomy', in *Dictionary of Scientific Biography*, Vol. 15, New York: Scribner's, 759–818.

—— *1983*, 'The base of the Venus table of the Dresden Codex, and its significance for the calendar-correlation problem', in A. F. Aveni and G. Brotherson (eds), *Calendars in Mesoamerica and Peru: Native American Computations of Time*, Oxford: British Archaeological Reports International Series, No. 174, 1–26. [Proceedings of the 44th International Congress of Americanists.]

Robicsek, F. and Hales, D. M. *1981*, *The Maya Book of the Dead: The Ceramic Codex*, Charlottesville, VA: University of Virginia Art Museum.

Thompson, J. E. S. *1971*, *Maya Hieroglyphic Writing: An Introduction*, Norman, OK: University of Oklahoma Press.

Thompson, J. E. S. *1972*, 'A commentary on the Dresden Codex', *Memoirs of the American Philosophical Society*, **93**, 156 pp.

Tozzer, A. M. *1941*, *Landa's Relacion de las Cosas de Yucatan*, Papers of the Peabody Museum, No. 18, Cambridge, MA: Harvard University. [A translation, edited with notes.]

Vinette, F. *1986*, 'In search of Mesoamerican geometry', in M. P. Closs (ed.), *Native American Mathematics*, Austin, TX: University of Texas Press, 387–407.

Winfield Capitaine, F. *1988*, *La estela 1 de la Mojarra, Veracruz, Mexico* (Research Reports on Ancient Maya Writings, No. 16), Washington, DC: Center for Maya Research.

1.16

The beginnings of counting and number

JOHN N. CROSSLEY

1 THE DEVELOPMENT OF COUNTING

The development of mathematics in all known societies has always begun with the treatment of numbers and counting. Depending on the needs of a particular society, such number and counting systems have developed to a greater or lesser degree. In such developments we can distinguish many stages. These include enumerating specific objects, counting small numbers of objects, grouping objects for counting, abstracting numbers, constructing sufficiently large number systems, constructing larger and larger number systems, developing the idea of an unending sequence of numbers, logically codifying a number system, and developing arithmetic operations and relations. These stages occur, very roughly speaking, in the order given, but there are overlaps. It is important to distinguish between the practice of counting or of using numbers from the abstract principles which govern such a practice. As everyone knows, it is much easier to learn to perform a task like counting than it is to describe the principles and rules involved. Thus a young child easily learns to count up by ones from thirty, forty, fifty, having learnt how to count twenty, twenty-one, twenty-two, up to twenty-nine. On the other hand, everyone eventually runs out of new number-words for very large numbers beyond, say, billion, trillion or quadrillion (though some words have been invented, see e.g. Johnstone *1975*) and the principles here are quite complicated.

2 PRIMITIVE COUNTING

Accounts of number systems were collected mainly in the nineteenth century, especially as a result of the discovery of new peoples and languages by explorers and missionaries. The most compendious collection is by Conant *1896*. Examples may also be found in Tylor *1871*, Lévy-Bruhl *1912*,

Codrington *1885*, Seidenberg *1960*, Dawson *1881* (for Australia), Kluge *1937–41*, Lean *1985–* (for Papua New Guinea), Closs *1986* (for America; compare his §1.15), Zaslavsky *1973* (for Africa; compare her §1.8), Menninger *1969* and Hurford *1975*. These books usually describe the number systems and counting practices.

In what we may regard as very primitive counting it is not clear that there is any idea of a sequence (such as $1, 2, 3, \ldots$); it appears rather that certain configurations or groups impress themselves on the human mind. Indeed, it appears that this happens with animals too. Birds notice if, say, one egg is missing from their nest. On the other hand, it is by no means clear that any animal can actually count (Thorpe *1960*, Katz *1953*).

Languages which do not have extensive number systems may have words for small groups, say those of up to three or four members. These are the languages of societies which found no necessity to use numbers. A member of the Pitjatjantjara aboriginal tribe in Australia, when asked how many people are in a group, will answer not with a number but with a listing of those in the group by name (personal communication from Miss J. Bain). Again in Walbiri, a central Australian aboriginal language, there are determiners – *tjinta, tjirama, wirkadu, panu* – which may be rendered as *singular, dual, a few, many*, but these do not constitute a system of number-words, and in fact they are not used in counting (Hale *1975*). If Walbiri speakers wish to count, they use slightly modified forms of the usual English words which have been imported through contact with English speakers.

There are languages in which the word used for a certain number of objects depends on the nature of the objects themselves. This happens among the Tauade in New Guinea where there is no commonly used word for any number larger than two, but there are different words for talking about pairs of males, pairs of females and a mixed pair (Hallpike *1979*: 243).

3 COUNTING BY GROUPING

The basic grouping is two. Languages in various parts of the world have small number systems where counting goes $1, 2, 2 + 1, 2 + 2, 2 + 2 + 1$ and perhaps a little further; Table 1 gives three examples. Such a system is said to use 'two-counting'.

For slightly larger groups the usual practice is tallying; that is, correlating objects with standard procedures. Counting on one's fingers is a version of such a practice. However, for large numbers marks or notches may be made on, say, a piece of wood – as was done in the British Treasury even into the late eighteenth century (Flegg *1984*: 44), or a fern frond may be used

151

Table 1 Three examples of two-count number systems (Seidenberg *1960*: 216)

Gumulgal (Australia)	Bakairi (South America)	Bushman (South Africa)
1 *urapon*	*tokale*	*xa*
2 *ukasar*	*ahage*	*t'oa*
3 *ukasar–urapon*	*ahage tokale* (or *ahewao*)	*'quo*
4 *ukasar–ukasar*	*ahage ahage*	*t'oa-t'oa*
5 *ukasar–ukasar–urapon*	*ahage ahage tokale*	*t'oa-t'oa-t'a*
6 *ukasar–ukasar–ukasar*	*ahage ahage ahage*	*t'oa-t'oa-t'oa*

to count by leaving attached the appropriate number of leaflets. Such tallying procedures are very old indeed. What are thought to be examples have been found on palaeolithic bones, dating from 20 000–70 000 BC (see §1.8). We know, from microscopic examination, that the marks were sometimes made with different instruments and presumably at different times. So this may be regarded as the earliest known counting (Marshack *1972*).

The most common grouping after two is five. Here again, some languages do not go beyond five, even though finger counting may be used. Thus the Zuñi of North America (Conant *1896*: 48) count

1	*Töpinte*	'taken to start with'
2	*kwilli*	'put down together with'
3	*ha'i*	'the equally dividing finger'
4	*awite*	'all the fingers all but done with'
5	*öpte*	'the notched off'.

Again, in a number of Malay-Polynesian languages the word for 'hand' is the same as the word for 'five', e.g. *lima* (Codrington *1885*: 235; Lean *1985*–).

Some languages use quite different correlates for numbers up to five. Dobritzhoffer (*1784*: Part 2, 172; *1822*: Vol. 2, 169) records that the Abipones of South America use the word *geyenkňatè*, 'emu's [in fact rhea's] toes', for 'four' and the word *neenhalek*, which denotes a beautiful skin with five colours, for 'five'. In the Marquesas in the South Pacific, *pona*, meaning 'knot', is used for 'four' and refers to the way of tying breadfruit in knots of four (Conant *1896*: 93).

Once such a group has been established, it is possible to proceed further. Thus the Inuit (Eskimo) people count using one hand for 'five', two for 'ten', then go to feet, while 'twenty' is expressed by a phrase meaning 'a man is completed' (Cassirer *1953*: Vol. 1, 230). A similar thing happens with the Galibi of Brazil (Conant *1896*: 138), for whom *poupou patoret oupoume*, meaning 'feet and hands', is 'twenty'.

Other configurations give rise to groups of different sizes. There are many examples of such groups in the languages of Papua New Guinea (Lean 1985–). In particular, the Kewa use a group of 47 (Wolfers 1972: 218–19). Such a number is obtained by counting round the body: after counting on the fingers and thumb, the count moves to the heel of the thumb, then the palm, and so on *via* the arms, neck, eye to between the eyes, and then back down the other side of the body.

The locations do not have to be unique; it is the whole sequence of actions which correlates with the numbering. (There may be no one-to-one correspondence between locations and numbers.) Thus Roth (*1908*: 80) records a system of counting on one hand in which 'two' and 'eight' are counted at the same place (see Crossley *1987*: 15). Lévy-Bruhl (*1926*: 202) and Hallpike (*1979*: 239, 243) both conclude that this form of counting on the body may exist as a series of actions without the corresponding number-words.

Such counting in groups is a very concrete process, and in some languages it is clearly tied to the objects being counted and is not an abstract process, as counting is for school-educated Westerners. In many languages it is necessary to use classifiers along with counting words, these classifiers varying with the type of object being counted. In his study of Chontal, a Mayan language spoken in southern Mexico, Keller records 78 different classifiers, for example -*tek* for plants and standing trees, and -*kuc* for loads of things carried with a headstrap (*1955*: 260).

The view that a further abstraction process is necessary is supported by Gell's account (*1975*: 162) of Umeda children in Papua New Guinea learning to count numbers. He says:

> As it happens, I was present during one delirious afternoon when the children finally did catch on to the basic principles of number – the fact that with numbers you can count *anything*. Released from the schoolhouse, the excited children ran hither and thither in little groups, applying their new-found insight: they counted the posts of the houses, the dogs, the trees, fingers and toes, each other – and the numbers worked, every time.

The most familiar grouping is, of course, tens, and English number-words give an example of 'ten-counting'.

Groups of two, five, ten and twenty are the most commonly used. Others can be found especially in Papua New Guinea, but there are vestiges in European languages. Thus English possesses 'dozen', 'gross' = twelve dozen and 'score', together with specialized words used particularly in trade, for example 'ream' means 20 quires = 20×24 sheets of paper. In French the system of counting in groups of ten has a hiatus from 70 to 90 where, instead of the expected *septante, octante, nonante* (which can still be found in some dialects) we find *soixante-dix, quatre-vingts, quatre-vingt-dix*, thus exemplifying a mixed system of ten- and twenty-counting (Menninger *1969*: 55).

4 LIMITS OF COUNTING

The next stage is the counting of groups. In English this means counting: (one) ten, two tens, three tens, ... or, as we now write: ten, twenty, thirty, ... Counting in this way with groups of ten (or counting in or with base ten, as it is also called) allows us to get as far as ten tens (i.e. 100). In a number system with base twenty we can get to twenty twenties (i.e. 400). Thus in Chol, a Mayan language, a new number-word is introduced for 400 (Merrifield *1968*: 98).

This process of regarding a group as one (new) unit can be repeated. The Chol language has special words for 20, $400 = 20 \times 20$ and $8000 = 20 \times 20 \times 20$ (but no word for $160\,000 = 20 \times 20 \times 20 \times 20$). Similarly, English has special words for 10, 10×10 and $10 \times 10 \times 10$, but only the esoteric word 'myriad' for $10 \times 10 \times 10 \times 10$ and no special word for $10 \times 10 \times 10 \times 10 \times 10$.

The idea of being able to continue this process indefinitely is quite separate from carrying it out a small number of times. Thus Mimica (*1988*: 31, 36) records how the Iqwaye of Papua New Guinea know how to count, using twenty-counting, up to 400; but his informant then added:

'This [i.e. 400] they [the other Iqwaye] would call *hyepu, hyepu* [that all, that much].' By this the informant meant that other people would merely refer to their own digits and specify their numerical plurality in an indefinite sense rather than having a clear grasp of the notion that 20 digits represent 20 men and therefore 400 (20×20).

Such large numbers are of little or no use in everyday life so it is not surprising there are no words for them even if the rules are clear. Other examples may be found in America (Closs *1986*: 13); an amusing example comes from Houton La Billardière's account of collecting number-words in Tonga at the end of the nineteenth century while his boat was stranded. The Tongan informants did keep on producing words in response to the request for words for numbers of one billion or more, but closer examination shows they were making fun of the enquirer and simply swearing at him (Crossley *1987*: 26).

In general, at the limits of useful counting in a particular society, instead of a continuing number system we find either a vagueness or a repetition indicating imprecision. Thus Dawson (*1881*: xcviii), after listing numbers below one hundred in the Chaap Wuurong language of Australia, concludes with one hundred (*larbargirrar*):

Larbargirrar, which concludes expressed numbers; anything beyond one hundred is *larbargirrar larbargirrar*, signifying a crowd beyond counting, and is always accompanied by repeated opening and shutting the hands.

Although the principle of extending the counting words by identifying groups with units, e.g. 20 as 1, so one can talk of 'two twenties', 'three twenties', and so on, this process inevitably leads to pure repetition. As Mimica (*1988*: 41) puts it when talking about extension of the Iqwaye twenty-counting:

> And the next cycle would be from 8000 to 160 000 (20^4), and again 'one person'! But these are merely formal possibilities, which in indigenous lingual expression would lose their conceptual clarity. The symbolic concreteness of the system and its lingual articulation impose insurmountable limitations upon its generativity.

In English, when it was not common to talk of billions or trillions (in the UK sense), one spoke of million million million..., thus exemplifying the inevitable repetition. The only way this is avoided is by having a place-notation system such as that using figures $0, 1, 2, \ldots, 9$. In this case each extra figure indicates another step of the grouping-together process. Thus $100 = 10$ tens, $1000 = 10$ hundreds, and so on. The first such notation systems were in written form; besides the familiar one, which is of Hindu–Arabic origin, an example may also be seen in the rod-numerals of ancient Chinese mathematics.

With the advent of such notation the number system became potentially infinite, capable of continuing for ever without the repetition associated with verbal systems, for example the 'million million million...' noted above.

5 CONSTRUCTING NUMBER-TERMS

In counting we have referred to grouping units together to make larger units, but this is not the only way that number-phrases are formed. Many languages use something that looks like addition. For example, even in two-counting saying 'two two one' for 'five' makes it look as though one is adding 2, 2 and 1. Likewise, 'twenty' $= 2 \times 10$ and similar words suggest multiplication. Finally, a grouping in which we take, say, one (new) unit = 20 (old) units, so that we go through 20, 20×20, $20 \times 20 \times 20$ (or in ordinary mathematical notation 20, 20^2, 20^3), would appear to be based on exponentiation. In reality these are isolated facts which are only to be seen as examples of addition, multiplication and exponentiation when these operations have been developed and understood.

Piaget's analyses of how children's knowledge of number and measure develop clearly show that the understanding of logical relationships such as $2 \times 10 = 4 \times 5$ comes a long time after counting abilities have been acquired (see e.g. Piaget *1941*). It is therefore reasonable to suppose that basic counting ability precedes our understanding of multiplication. For example, the Kpelle of Liberia can do addition and subtraction (at least for numbers below about forty), but seem to have no operation properly corresponding to multiplication (Gay and Cole *1967*: 50). Similarly, Australian Aborigines with little contact with Europeans appear to have understood addition but not multiplication (de Lacey *1974*: 361). The same reservations apply to notions of equality or relative size (fewer than, more than). The idea of equality is not inherent in the counting system, but requires additional insight (Hallpike *1979*: 246–50).

Again, in constructing number-words a few languages use what we would call 'subtraction' in some instances. For example, in Latin we have *undeviginti* = 1 from 20 for 'nineteen', and in the Ainu language from Japan 'eight' is *tu-pesan*, meaning 'two steps down' (from 10) and 'nine' is *shine-pesan*, meaning 'one step down'. A more complicated example comes from the Mayan of Yucatán in Mexico, where thirty-five is *holhu-cakal*, which looks like 'fifteen two twenty', and presumably means '15 on the way to 40' (= two twenty) (Flegg *1984*: 30 or Menninger *1969*: 61; compare the German *halb zwölf*, literally 'half twelve', meaning 11.30, i.e. half an hour on the way to twelve). In the Pomo language of California, tens between 50 and 90 are counted below the next 20 group. Thus 50 is *hadagal-e-xomka-xai*, meaning '10 below 3 sticks', that is, 10 below 3 twenties (Closs *1986*: 36).

6 ZERO

It is hard to say anything about the origins of zero in spoken language. Of course 'zero' is usually not necessary in verbal or aural communication (and many synonyms can be used), but it is essential in most written number systems employing place-value notation (for example, decimal notation).

The use of a symbol for zero originated in India no later than the end of the third century AD. Claims have been made that the Babylonians had a symbolic zero, and that the zero symbol 'ō' occurs in the Greek notation for sexagesimal numbers at least as early as the second century BC (§1.6). Menninger *1969* says nothing about a spoken 'zero', but shows how confusing the symbolic use of 0 was when it was imported from India, first into China in the sixth century AD (Needham *1959*), and later into the West. Negative numbers do not seem to have been used in Greece or India, and not by the Romans.

7 NOTATION

With the exception of zero, the main concern here has been with number concepts and the concept of counting. Of course, notation gets deeply involved (Menninger *1969*: *passim*).

Notation follows the groupings described in Section 3 above. For example, Roman numerals have a special symbol, X, for ten, another, C, for a hundred, and likewise for higher named numbers. Multiples of a group are represented by repetitions of the symbol for the group. Thus 23 is represented by XXIII: two tens and three ones. (V, for 5, is an exception to these rules.) Varieties of this style of notation occur in Mayan and ancient Greek as well as Roman numerals.

The other method in common use is to have a place-value notation, as, for example, in our present system. The more digits a digit in a numeral has to its right, the bigger the value this digit represents. Thus the 1 in 1924 means one thousand, whereas in 9412 it means one ten. This system appears to have begun in China more than two thousand years ago. Other systems exist but are not so widely used. Thus scientists use floating-point notation, in which for example 9124 is represented as 9.124×10^3.

8 MYSTICAL NUMBERS

Finally, besides the logical relationships between numbers, other associations have at various times been considered important (§12.5). Thus in ancient Greece, 2 was identified with 'feminine' and 3 with 'masculine' (Aristotle, *Metaphysics* 1078b23; Burkert *1972*: 33–4); the Konso of East Africa make the same identification (Hallpike *1979*: 244). Presumably these identifications come from the configurations of female breasts and male genitalia (see also Jung *1964*: 385–6), illustrating how the powerful forces of the human mind can enter such a seemingly abstract notion as number.

BIBLIOGRAPHY

Burkert, W. *1972*, *Lore and Science in Ancient Pythagoreanism* (transl. E. L. Minar), Cambridge, MA: Harvard University Press.

Cassirer, E. *1953*, *The Philosophy of Symbolic Forms* (English transl.), 3 vols, New Haven, CT: Yale University Press.

Closs, M. P. *1986*, 'Native American number systems', in M. P. Closs (ed.), *American Mathematics*, Austin, TX: University of Texas Press, 3–43.

Codrington, R. H. *1885*, *The Melanesian Languages*, Oxford: Clarendon Press.

Conant, L. L. *1896*, *The Number Concept. Its Origin and Development*, New York: Macmillan.

Crossley, J. N. *1987, The Emergence of Number*, 2nd edn, Singapore: World Scientific Publishing.

Dawson, J. *1881, Australian Aborigines*, Melbourne: George Robertson. [Facsimile edn, 1981, Canberra: Australian Institute of Aboriginal Studies, and Atlantic Highlands, NJ: Humanities Press.]

Dobritzhoffer, M. *1784, Historia de Abiponibus* [...], 3 vols, Vienna: Kurzbek.

—— *1822, An Account of the Abipones* [...] (transl. S. Coleridge), 3 vols, London: Murray.

Flegg, G. *1984, Numbers, Their History and Meaning*, Harmondsworth: Penguin.

Gay, J. and Cole, M. *1967, The New Mathematics and an Old Culture*, New York: Holt, Rinehart & Winston.

Gell, E. F. *1975, Metamorphosis of the Cassowaries: Umeda Society, Language and Ritual* (London School of Economics Monographs in Social Anthropology, No. 51), London: Athlone Press.

Hale, K. *1975*, 'Gaps in grammar and culture', in M. Dale Kinkade, K. L. Hale and O. Werner (eds), *Linguistics and Anthropology. In Honour of C. F. Vogelin*, Lisse: de Ridder, 295–315.

Hallpike, C. R. *1979, The Foundations of Primitive Thought*, Oxford: Clarendon Press.

Hurford, J. R. *1975, The Linguistic Theory of Numerals*, Cambridge: Cambridge University Press.

Johnstone, W. D. *1975, For Good Measure*, New York: Holt, Rinehart & Winston.

Jung, C. G. *et al. 1964, Man and his Symbols*, London: Aldus Books. [Repr. 1968, New York: Doubleday.]

Katz, D. *1953, Animals and Men*, Penguin: Harmondsworth. [Transl. by H. Steinberg and A. Summerfield from the German edn of 1937.]

Keller, K. C. *1955*, 'The Chontal (Mayan) numeral system', *International Journal of American Linguistics*, **21**, 258–75.

Kluge, T. K. *1937–41, Ein Beitrag zur Geistesgeschichte der Menschen*, (1937), *Ein zweiter Beitrag* [...] (1938), *Ein dritter Beitrag* [...] (1940), *Ein vierter Beitrag* [...] (1941), all vols Berlin: the author. [Partly typescript.]

de Lacey, P. R. *1974*, 'A cross-cultural study of classificatory ability in Australia', in J. W. Berry and P. R. Dasen (eds), *Culture and Cognition: Readings in Cross-cultural Psychology*, London: Methuen, 353–66.

Lean, G. A. *1985–, Counting Systems of Papua New Guinea*, 12 vols to date plus research bibliography, draft edn Lae, Papua New Guinea: PNG Institute of Technology.

Lévy-Bruhl, L. *1926, How Natives Think*, London: Allen & Unwin. [Transl. by L. A. Clare from French original of 1912.]

Marshack, A. *1972, The Roots of Civilization*, London: Weidenfeld & Nicolson. [2nd edn 1991.]

Menninger, K. *1969, Number Words and Number Symbols* (transl. P. Broneer), Cambridge, MA: MIT Press.

Merrifield, W. *1968*, 'Number names in four languages in Mexico', in H. B. Corstius (ed.), *Grammars for Number Names* (Foundations of Language Supplementary Series, Vol. 7), Dordrecht: Reidel, 91–102.

Mimica, J. *1988, Intimations of Infinity*, Oxford: Berg.

Needham, J. *1959, Science and Civilisation in China*, Vol. 3, Part 1, Cambridge: Cambridge University Press.

Piaget, J. *1941, La Genèse du nombre chez l'enfant*, Neuchâtel: Delachaux & Niestlé. [English transl. by C. Gattegno and F. M. Hodgson as *The Child's Conception of Number*, 1952, London: Routledge & Kegan Paul.]

Roth, W. E. *1908*, 'Counting and enumeration', *Records of the Australian Museum, Sydney*, **7**, 79–82.

Seidenberg, A. *1960*, 'The diffusion of counting practices', *University of California Publications in Mathematics*, **3**, 215–300.

Thorpe, W. H. *1960, Learning and Instinct in Animals*, London: Methuen.

Tylor, E. B. *1871, Primitive Culture*, 2 vols, London: Murray. [Partly repr. as *The Origin of Culture*, 1958, New York: Harper.]

Wolfers, E. P. *1972*, 'Counting and numbers', in *Encyclopaedia of Papua and New Guinea*, Vol. 1, Melbourne: Melbourne University Press, 216–19.

Zaslavsky, C. *1973, Africa Counts*, Boston, MA: Prindle, Weber & Schmidt.

1.17

Some ancient solutions to the problem of fractioning numbers

KARINE CHEMLA

Of the four basic arithmetical computations on integers, division has played a specific role in that it has led mathematicians to deal with residual quantities which are different from the original terms of an operation. Division has been approached in different ways in the past, which has led to different ways of elaborating the concept of a fraction. Three such approaches, recorded in Babylonian, Egyptian and Chinese texts (§1.1, 1.2, 1.9), are considered in this article, which examines how the result of a division was given in each tradition, and how mathematicians reintroduced these results into computations.

1 BABYLONIAN CLAY TABLETS

Mathematical tablets dating from 1800 to 1600 BC attest to the use of a sexagesimal place-value system. This may be considered as a floating-point system: the significant digits were arranged and operated on without giving any indication of the order of magnitude of the whole number, which had to be stated at the end of the computation for the result. Hence arithmetic was only dealing with integers. Division of one number by another, m/n, was performed by multiplying m by the inverse of n, which was to be looked up in an appropriate table. Such tables contained only 'invertible' numbers – those whose inverses, in base 60, may be written with a finite number of digits. This set of numbers depends on the factors of 60; each of its elements is of the form $2^p 3^q 5^r$, where p, q, r are integers. Available evidence suggests that mathematicians avoided other forms of number or, when facing them, approximated the result by reading a multiplication table in the reverse way. Babylonian mathematics seems not to have needed fractions, but turned the related difficulties into problems concerning integers and approximations (§1.1).

2 EGYPTIAN PAPYRI

Egyptian papyri written at about the same time as the Babylonian tablets discussed above demonstrate a radically different arithmetical practice and, specifically, a different approach to fractions. The result of a division of two integers is expressed as an integer (if the dividend is greater than the divisor) plus the sum of a sequence of unit fractions (we shall denote by 'n''', the unit fraction associated with n). Such a sequence had several characteristics, for instance it cannot contain two identical unit fractions. The integral part of the quotient having been found, tables would be used to obtain the result of dividing the remainder by the divisor. Many tables of this kind have been preserved, including one at the beginning of the Rhind Papyrus, where divisions of 2 by the odd numbers up to 101 are performed. On the basis of the setting-up of such divisions and of the structure of the table, many hypotheses have been put forward about how the table was devised (Knorr *1982*).

Let us consider one of these divisions, the division of 2 by 13 in the Rhind Papyrus, which proceeded as follows:

	1	13
	1/2	6 1/2
	4'	3 4'
\	8'	1 1/2 8'
\	52'	4'
\	104'	8'

The leading term of the left-hand column is 1; the right-hand column has as its leading term the divisor, 13. Repeated halvings are carried out until the number in the right-hand column is less than the dividend, 2. Fractions are then entered in the right-hand column to make this number up to 2. These fractions are divided by 13 and the results written alongside them in the left-hand column. Back-slashes indicate the lines to be added together to give the result: $8'52'104'$. The whole procedure can be understood as the generation of a sequence of unit fractions m', n', p', ... such that $13(m' + n' + p' + \cdots) = 2$.

Such an algorithm makes use of unit fractions. First they must be multiplied; tables for such operations are still extant. Then the scribe must complete a set of fractions in such a way that they add up to the next higher integer. In the example considered above, one needs to know that $1/2 + 4' + 8' + 8' = 1$. Again, there were tables listing such identities, in the

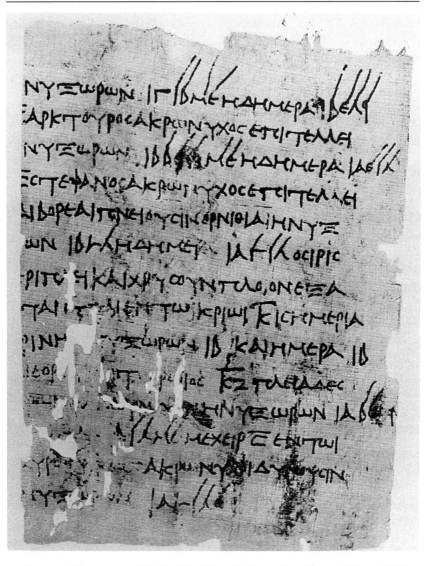

Figure 1 The papyrus Hibeh i 27 (300 BC) (from Grenfell and Hunt *1906*: Plate VIII). The long strokes over some letters indicate unit fractions. In the first line, 13 4/45 is written according to the Greek numeration system, and in the Egyptian way as 13 1/12 1/45 (a scribal error has probably changed 1/15 into 1/12)

'Leather Roll' for instance (§1.2). In the above example a succession of halvings led to the result. The succession of parts calculated depends on the actual numbers involved in the division. The choice takes into account their arithmetical properties. Numbers that are produced by division are subsequently used in all sorts of computations (addition, subtraction, multiplication, . . .).

Interestingly enough, this tradition of computation seems to have been continued in Greek writings up to the seventh century AD at least. A fragment of Greek papyrus from 300 BC containing astronomical data uses this Egyptian representation of numbers as integers plus a sum of unit fractions, reproduced in Figure 1 (Fowler and Turner *1983* comment on it). Other Greek papyri from the second to the seventh or eighth century AD present tabulated values for a set of divisions as sums of unit fractions. They are comparable in their structure and in their results to the Egyptian ones, and computations they contain are similar to those found in Egyptian documents (Knorr *1982*). Thus the Egyptian arithmetical process was kept alive for more than 2000 years. One is tempted to conclude that, whereas Babylonian ways of calculating were used in Greek astronomical texts from the second century BC, they did not replace Egyptian ways of calculating in all spheres. Besides the practical aspect of it, similarities may be noticed between number-theoretic problems posed by such computations and the treatment of the ratio of integers in Book VII of Euclid's *Elements*.

3 CHINESE BOOKS

The earliest Chinese mathematical book available, the *Jiuzhang suanshu* ('Computational Prescriptions in Nine Chapters'), is a compilation, made in the first century AD or perhaps earlier (§1.9), of knowledge previously acquired. A complete arithmetic of fractional quantities is presented in its first chapter. The algorithm for division is taken for granted in the book, yet various hints suggest that division was being done with numbers represented in a decimal place-value numeration system in much the same way as ours (Chemla, in Benoit *et al. 1992*). The result is given as the integral part (quotient) plus a fractional quantity, the numerator of which is the remainder, and the denominator the divisor — simplified if necessary. What was still a problem for the Egyptian mathematician (m/n) is here expressed as a fraction, characterized by a pair of numbers (numerator and denominator), giving a value which is always less than one. In the 'Nine Chapters', computations with fractions are carried out according to algorithms dealing with their numerators and denominators, such as the following one for addition:

> Multiply by the denominators the numerators that do not correspond to them; add up; take this as the dividend.
> The denominators being multiplied by one another, make the divisor. Divide.
> What does not fill the divisor, with the divisor name it.
> If the denominators are equal, only add them [i.e. the numerators] to one another.

What we would express as the formula $b/a + b'/a' = (ab' + ba')/aa'$ is here described as an algorithm yielding a result, produced by division, in the form 'integer + fraction'. Note that the number of fractions to be added is not specified, and that the procedure is applicable whether the denominators are equal or different. Similar algorithms are provided to simplify fractions, to subtract or multiply them, to compare them or to compute their mean value. To divide fractions, though, another kind of algorithm is recommended which makes use of what we would now call reduction to a common denominator. Unlike the Egyptian way of dividing, the Chinese algorithm does not depend on the numbers to be divided, nor does the arithmetic of fractions depend at all on specific properties of the numbers involved.

In his commentary on the 'Nine Chapters', Liu Hui (third century AD) gives proofs for these algorithms. First, he notices that a quantity may be represented by different numerical fractions, and relates this to the actual parts that compose a fractional quantity:

> When quantities are expressed in a simplified way, their parts are coarser; when they are expressed in a complex way, their parts are finer. Although the degrees of fineness differ, the quantities that are expressed are the same.

Then, in the case of the addition of fractions, for instance, he shows how these properties help in understanding the algorithm: parts of the various fractions should be refined so that they may be added. He explains the role of steps:

> Every time denominators multiply a numerator that does not correspond to them, we call this to *homogenize*. To multiply with one another the set of denominators, we call this to *equalize*. If one equalizes, it is in order to make the parts communicate as they correspond to one another; the equalized is the same denominator, common to all of them; if one homogenizes, numerators and denominators are homogeneous, so the procedure cannot have lost the original quantities.

Hence in order to prove the algorithm, Liu Hui makes explicit the meaning of its steps, for he explains the results in terms of the fractional quantities themselves; he also emphasizes the overall form of the algorithm, that it proceeds by equalizing some quantities and homogenizing others.

On this basis, Liu Hui goes on to establish analogies between algorithms for dealing with fractions and others such as the algorithm for solving systems of simultaneous linear equations presented in his Chapter 8. Given the equations

$$ax + by = c \quad \text{and} \quad a'x + b'y = c', \tag{1}$$

this algorithm amounts to transforming them into

$$aa'x + ba'y = ca' \quad \text{and} \quad aa'x + ab'y = ac' \tag{2}$$

so as to obtain y from the equation

$$(ab' - a'b)y = ac' - a'c. \tag{3}$$

Liu Hui recognizes here that the procedure amounts to equalizing the coefficients of x and homogenizing the other terms of the equations. Even though the operations of equalizing and homogenizing have different meanings in each context (reduction to a common denominator on the one hand, elimination on the other), they are identical from a formal point of view, and this is what Liu Hui stresses. In his commentary, fractions thus play a key role since they introduce general operations which are then used to draw analogies between different algorithms. Not only was the arithmetic of fractions fully developed in China, but it also gave rise to theoretical activity.

4 CONCLUSION

Three approaches to dealing with fractions, each with its own specific features, have been developed in different traditions. Egyptian mathematicians used procedures based on the number-theoretic properties of integers, whereas Chinese ones gave uniform algorithms, using a decimal place-value system. Arithmetic based on such a numeration system is met with in Indian mathematical texts (the oldest extant, the $\bar{A}ryabha\underline{t}\bar{\imath}ya$, dates from the late fifth or early sixth century AD) and in Arabic texts (from the ninth century onwards). It attests to a way of dealing with fractions that presents the same features present in Chinese texts. All these traditions found their way to the Arabic texts, where we find a synthesis of these various arithmetics. Indeed, the Egyptian way of dealing with fractional quantities presents problems that are still unsolved and generate activity in the field of number theory today.

BIBLIOGRAPHY

Benoit, P., Chemla, K. and Ritter, J. (eds) *1992, Histoire de fractions: Fractions d'histoire*, Basel: Birkhäuser. [Many bibliographical items on this topic.]

Fowler, D. H. and Turner, E. G. *1983*, 'Hibeh Papyrus i 27: An early example of Greek arithmetical notation', *Historia mathematica*, **10**, 344–59.

Grenfell, B. P. and Hunt, A. S. *1906, The Hibeh Papyri*, Vol. 1, London: Egypt Exploration Fund.

Knorr, W. *1982*, 'Techniques of fractions in ancient Egypt and Greece', *Historia mathematica*, **9**, 133–71.

Ritter, J. *1989*, 'Chacun sa vérité: Les mathématiques en Egypte et en Mésopotamie', in M. Serres (ed.), *Eléments d'histoire des sciences*, Paris: Bordas, 38–61.

Part 2
The Western Middle Ages and the Renaissance

2.0

Introduction

The concern with Western mathematical traditions begins to come to the fore in this Part, in which the spectrum of medieval and Renaissance mathematics is surveyed. Issues of space and time need to be clarified at once. As described in §0, 'Europe' is normally taken to refer to the area now occupied by the countries of eastern and western Europe, and to the pre-Revolutionary Russian Empire. The normal period covered is up to and including the early seventeenth century, although the 'end-points' vary according to the subject-matter of the articles. A few of them refer to, and even briefly look forward to, developments during later periods; conversely, several articles in later Parts start off within this period.

The articles here describe branches of mathematics which showed sufficient individuality of their own to be distinguished from their descendents (for example, the algebras of §2.3–2.4 from those of Part 6, or the mechanics of §2.6–2.7 from that of Part 8), or in which the principal

THE
SVRVEYOR
in
Foure bookes
by
AARON RATHBORNE

LONDON
Printed by W. Stansby for W. Burre.
1616.

advances were achieved during this time (such as logarithms, §2.5). Among other relevant articles, examples from this period arise especially in §5.10 on algorithmic thinking, in the overview §4.2 of trigonometry and its functions, in §6.9 on the algebra of François Viète, in §8.15 on cartography, and in several articles in Part 12 on mathematics and culture.

The articles are ordered so that pure mathematics is treated first, then applied mathematics; but such a strong division did not apply at the time, when applications often drove the subject. The order has been determined by the fact that the 'purish' branches such as geometry and algebra are more prerequisite to the reader's understanding of the mechanics and technology than vice versa. Further, in general there is more continuity in the history of mathematics than the 'isolation' of this Part suggests; nevertheless, the period possesses sufficient characteristics of its own to justify the assignment to it of a Part.

The final two articles cross branches in different ways: §2.11 sketches out the slowly growing status of universities, and §2.12 surveys the philosophical context of mathematical (and scientific) research, including the concept of research itself and our historical understanding of questions such as the diffusion of knowledge at that time.

During this period Latin was the principal written language; but some texts were written in living languages, and the names of some authors were known in Latin and at least one other form. No strict convention has been imposed here, although a well-known form is always used.

There was a growing distinction between the professional and amateur

---

Figure 1 Title page of an influential text of the early seventeeth century on surveying. Rathborne, a surveyor and mathematical writer, advocated the use of angle measurement and triangulation to replace the traditional reliance on linear measure (see §2.2). He lauded the approach of the geometrical craftsman ('Artifex') such as himself above the practices of the unlearned surveyors, and showed Artifex trampling these supposed swindlers and charlatans down from his own level to that of the amateur's 'busy laziness' (*inertia strenua*, a tag from Horace).

From attitudes such as these came the Western practice, maintained from the seventeenth until well into the nineteenth century, of calling the professional mathematician a 'geometer', to be distinguished from a mere practising 'mathematician'. However, as mathematics became fully professionalized during the nineteenth century, the distinction gradually disappeared, and all mathematicians called themselves 'mathematicians'.

Thanks are due to Dr J. Bennett (University of Cambridge) for drawing this title page to my attention.

(Photograph © Whipple Museum, Cambridge)

mathematician towards the end of this period. Figure 1 provides an amusing and instructive example of the class difference.

The period has excited much distinguished scholarship, with substantial revisions of interpretation proposed in recent decades. Grant *1974* is a major source-book for the sciences in general; Cajori *1928* contains a wide survey of the notations of the time, pertaining especially to arithmetic and algebra.

BIBLIOGRAPHY

Cajori, F. *1928*, *A History of Mathematical Notations*, Vol. 1, Chicago, IL: Open Court.

Grant, E. (ed.) *1974*, *A Source Book in Medieval Science*, Cambridge, MA: Harvard University Press.

Herzog August Bibliothek (Wolfenbüttel) *1989*, *Mass, Zahl und Gewicht. Mathematik als Schlüssel zu Weltverständnis und Weltbeherrschung*, Weinheim: VCH, Acta humaniora. [Exhibition catalogue, containing substantial articles, by M. Folkerts, E. Knobloch and K. Reich.]

Juschkewitsch, A. P. *1964*, *Geschichte der Mathematik im Mittelalter*, Leipzig: Teubner. [Russian original 1961.]

Lindberg, D. C. (ed.) *1978*, *Science in the Middle Ages*, Chicago and London: University of Chicago Press.

2.1

Euclidean and Archimedean traditions in the Middle Ages and the Renaissance

JÜRGEN G. SCHÖNBECK

1 EUCLID AND ARCHIMEDES: TWO 'ALEXANDRINE' MATHEMATICIANS

Scientific life in Classical Greece had, up to the fourth century BC, been determined essentially by philosophical debate and discussion. This dominance of philosophy also characterized the subsequent third century; it was, however, modified by an increasing shift of interest towards the formulation of more particular, subject-specific questions. The outstanding intellectual achievements of scholars such as Herophilus, Erasistratus, Eratosthenes, Apollonius, Ctesibius and Philo in their various fields were comparable to those of the former two most significant philosophers of the Western world, Plato and Aristotle. This group of prominent specialist scholars also includes Euclid, the geometrician, and Archimedes, the mathematician, physicist and engineer.

For Girolamo Cardano, the Italian mathematician, doctor and natural scientist of the Renaissance, Euclid and Archimedes were emphatically among the ten most important scientists in history; without doubt they are among the most influential and significant mathematicians of ancient Greece. Both, tradition has it, lived at least part of their lives (around 300 and 250 BC, respectively) in Alexandria, the scientific capital of the old world, and possibly Archimedes was a young pupil of Euclid. Apart from anecdotes, however, almost no details of their lives are known, and there have been great differences in their importance and in the influence of their scientific work.

The writings of both Euclid and Archimedes are today widely known through Greek, Arabic and Latin version – albeit not in all detail, not complete and not in absolutely original form, but probably with full content.

The history of how these writings were handed down as part of the history of mathematics is the subject of this article.

2 EUCLIDEAN LEGACIES

2.1 Sources

Euclid's work includes writings on mathematics and mathematical physics (Table 1), although Euclid's authorship is not certain in every case (Bulmer-Thomas *1971*). Of prime importance in the development of mathematics and its teaching was his main work, the *Elements*, a systematic, axiomatically deductive presentation of 'Euclidean' elementary geometry and elementary arithmetic, which leads in 13 books (i.e. chapters) to a classification of quadratic irrationalities and the description and complete definition of all regular *polyeds* (Platonic bodies). Only a few fragments of the text of the *Elements* (on shards or as papyrus fragments) have survived from Greek Antiquity. Our present knowledge of the *Elements* rests mainly on Greek manuscripts (Table 2) dating from the period between the ninth and twelfth centuries and which, with one exception, go back to a recension of Euclid's work by the fourth-century Greek mathematician Theon of Alexandria. The oldest of these manuscripts, the Codex Bodleian (manuscript B), dates from the year 888 and was produced by the Byzantine calligrapher Stephanus. The aforementioned exception is the famous manuscript with the 'pre-Theonic' content, brought to light only in 1809 by François Peyrard (manuscript P); it dates from the tenth century and is the

Table 1 Euclid's mathematical works

1	*Elements*	6	*Surface Loci*
2	*Book of Fallacies*	7	*Conics*
3	*Data*	8	*Phenomena*
4	*On Division of Figures*	9	*Optics*
5	*Porisms*	10	*Elements of Music*

Table 2 Greek manuscripts of the *Elements*

B	Oxford	Bodleian Library	ninth century
P	Rome	Biblioteca Vaticana	tenth century
F	Florence	Biblioteca Laurenziana	tenth century
b	Bologna	Biblioteca Comunale	eleventh century
V	Vienna	Nationalbibliothek	twelfth century (?)
p	Paris	Bibliothèque Nationale	twelfth century

most important basis of all modern editions of Euclid. This pre-Theonic edition was already known to medieval Arab scientists.

2.2 The 'Arabic' Euclid

The *Elements* had already become a standard work in ancient times (§1.3), widely published and much copied, edited and commented upon (Murdoch *1971*, Schönbeck *1984*). Examples are a Greek commentary to Book I of the *Elements*, which was written by the neo-Platonic Proclus, and Euclidean sections in the Latin writings of the Roman statesman and philosopher Boethius. Nevertheless, the real, comprehensible history of Euclid's legacy begins only with the work of Arab and Persian scholars (Sezgin *1974*). This was somewhat late – over a thousand years after Euclid's death!

How the *Elements* made its way into the Islamic countries still remains a mystery. (A connection with a Syrian edition of the *Elements* has been suggested; see §1.6.) Reports about the origins of the Arabic Euclid, however, mention two complete translations of the *Elements* as early as the eight and ninth centuries, both in two different editions, by al-Ḥajjāj and Isḥāq ibn Ḥunayn. They go back to different master-copies – 'pre- Theonic' and 'Theonic' – and in their open-minded endeavour they represent an early phase in the Arabic transmission of Euclid's legacy. In a second phase, there appeared numerous extracts, summaries, explanations and annotations of Euclid's work based on these translations. They were in no way interpretations of Euclid, but distinct didactic initiatives to grasp questions and develop them, to suggest criticisms and offer solutions and results.

Wide circulation was enjoyed by the detailed and extensive excerpt undertaken by Ibn Sīnā (Avicenna), one of the most important philosophers of his time (around 1000), in his multi-volume encyclopedia *Kitāb as-Sifā*. This work sought to unify the ancient and oriental body of thought. After being translated into Latin (by Gerard of Cremona), it exercised great influence on the Christian West, for example on Albertus Magnus, the 'doctor universalis', and on Thomas Aquinas, the 'princeps philosophorum' of scholasticism. More ambitious mathematically and more significant in the history of mathematics was a commentary on Euclid by the mathematician and astronomer an-Nayrīzī (Anaritius). It contained excerpts from lost Greek commentaries (by Hero and Simplicius) and later had a strong influence on Euclid's reception in Europe. Finally, in a third phase, there appeared revised editions and recensions of single books from the *Elements*, or even all 13.

Outstanding, and of similar significance to the Theonic edition in Antiquity, was the complete revised edition of the *Elements* by the universal scholar and mathematician Naṣīr al-Dīn al-Ṭūsī, who lived in the thirteenth century. He deliberately aimed at restoring Euclid's text, often handed down only in distorted versions, and to this end he applied not only mathematical but also philological and historical criteria – a procedure that shows him to have been completely modern in his approach to research (Folkerts *1980*). His edition of Euclid is undoubtedly a high point in Arabic mathematics.

In all, there are known to have been more than two hundred Arabic works handing on the *Elements* in this period, although not all have survived. They are evidence of an interest not only in historical literature, but also, markedly, in mathematics. In particular the 'Euclidean' problem of parallels (in connection with Book I), the 'Eudoxean' theory of proportions for common sizes (in connection with Book V) and the 'Theaetetan' theory of irrationals (in connection with Book X) prompted far-reaching, independent critical research. Mathematicians like Ibn al-Haytham (Alhazen), al-Ḥaiyam and Naṣīr al-Dīn al-Ṭūsī discovered 'theorems' of non-Euclidian geometry (the 'Saccheri' rectangle) and succeeded in producing a well-founded extension of the concept of numbers ('real numbers'). They also established the topics which have attracted the interest of Western mathematicians right up to the present day – sometimes with reference to their Arabian predecessors. Euclid's *Optics* was handed on in a similar way: the *Kitāb al-Manazīr* ('Book of Optics') by Ibn al-Haytham, based on Euclid's work (and that of Ptolemy), constitutes the beginning of optics as a physical discipline, one to which both Johannes Kepler and Isaac Newton felt themselves committed (§2.9).

2.3 The 'Latin' Euclid

Throughout several centuries mathematics in the Latin countries of the West referred to the work of Boethius with his Euclidean fragments. (Such a 'Boethius–Euclid' was used by Gerbert d'Aurillac, the leading Western scholar of his time and later Pope Sylvester II, in his mathematical studies in the famous monastery library of San Colombano.) Only through the intermediacy of Arab scholars did the more comprehensive and fundamental achievements of Greek mathematics become really available to the Latin scientists of the Middle Ages. This happened in the twelfth and thirteenth centuries, first in Spain, at the western limit of the Islamic world. A great number of translations from Arabic into Latin appeared here, both from original Arab works and those previously translated into Arabic from Hellenistic Greek literature. Among the classics, which had been translated and edited several times and from different versions, appear the works of Euclid (Murdoch *1971*).

The tradition of the *Elements* remains double-tracked, as with the Arabs. In the first place an Isḥāq master-copy was used by Gerard of Cremona for his translation. He was the leading personality in a translation school in Toledo, and passed on not only Euclid's *Elements* but also his *Data* and *Optics*, in addition to an-Nayrīzī's Euclid commentary, of great importance in the history of mathematics. His work, which included translations of Archimedes (*On the Measurement of the Circle*), Menelaus (*Spheres*) and Ptolemy (*The Almagest*), collected by al-Khwārizmī, Thābit ibn Qurra and Ibn al-Haytham, is testimony to the desire to open up Arab and Greek mathematics for the Christian West.

Even greater influence was gained by the older translations, many of which were done by the much-travelled English scholar, philosopher, mathematician and translator Adelard of Bath. There are several versions of his *Elements*, based on both the al-Ḥajjāj master-copies. Greater importance was achieved by an evidently drastically abridged but mathematically all the more content-rich version (known as *Adelard II*): from the thirteenth century to the fifteenth it served as model and master for a whole series of Euclid recensions and translations.

To this 'Adelard legacy' belongs an annotated edition of the *Elements* produced in the mid-thirteenth century by the north Italian mathematician and astronomer Campanus of Novara. With its numerous additional systems and methods, it clearly treated the *Elements* as a potential schoolbook and was soon to serve as the basis for the teaching of mathematics in the faculties of Arts. For a long time this 'Adelard–Campanus' remained the standard Latin translation of the *Elements* from the Arabic. In 1482 it became one of the very first printed mathematics textbooks, and of lasting importance especially for the numerous Renaissance editions of Euclid.

Such translations were not without influence on the thought of medieval scholars. The Augustinian Hugh of Saint Victor placed mathematics as a pure theoretical discipline next to physics and theology in his epistemology *Didascalion* (§2.2). The Franciscan Roger Bacon, who was familiar with both Euclid's *Elements* and his *Optics*, attempted to construct a comprehensive scheme of all worldly sciences, arithmetic and geometry included. The Dominican Albertus Magnus, chief representative of Christian Aristotelianism in the Middle Ages, produced a commentary on the *Elements*, influenced by an-Nayrīzī. And in support of Campanus, Thomas Bradwardine, later to become Archbishop of Canterbury, wrote a synopsis of the *Elements* for philosophers and theologians under the title of *De geometria speculativa*; it was considerably compressed but original, with references to Aristotle and Boethius. In his work Bradwardine also treated the Archimedean calculation of circles, the isoperimetric problem and the question of the complete filling of space by regular bodies.

However, despite these examples the level of mathematics in the time of Scholasticism remained poor. Medieval monastery pupils and university students knew little more than Book I of the *Elements*, and scientific argument mainly concerned comparison between neo-Platonism and Aristotelianism; mathematicians of the standing of Leonardo of Pisa (Fibonacci) or Jordanus de Nemore were the exceptions. (The latter is thought to be Jordanus Saxony, but see Høyrup *1988*.) Basic research directly connected with Euclid, comparable to that of the Arabs, is hardly known in this period. In the mathematical respect the work of Euclid handed down in Latin was certainly less varied and extensive, less lively and original, less mathematically creative, than that in Arabic.

2.4 The 'modern European' Euclid

Despite individual outstanding personalities, mathematics in the Middle Ages moved only within very narrow bounds. Neither did the literary Humanist movement produce much in the way of new contributions; its philological studies, however, did ensure that the writings of the great scientists of Antiquity became generally available, not only those of Euclid and Archimedes, but also those of Apollonius, Pappus, Ptolemy and Diophantus. The Italian Humanist Vittorino da Feltre insisted that the pupils in his boarding-school, *casa giocosa*, should read Euclid in the original. Nevertheless, the historical dimension of Greek mathematics was not recognized until the Renaissance period.

Three outstanding events in the history of Euclid's legacy occurred early in the modern European period: in 1482 the first printing of a Latin Euclid text (by Campanus); in 1505 a new Greek–Latin Euclid translation by Bartolomeo Zamberti; and in 1533 the first printing ('editio princeps'), by Simon Grynaeus, of a Greek Euclid text together with the commentary by Proclus to Book I. These first printings were followed by an abundance of reprints and recensions. From the sixteenth century alone more than a hundred printed editions of the *Elements* are known (a new Euclid every year!), some of which included a few of Euclid's minor works such as *Phenomena*, *Optics* or *Data*. Particularly influential was a 'synoptical' edition of the *Elements* by the founder of the French Humanist school, Jacques Lefèbre d'Etaples (Jakob Faber Stapulensis). One of his aims was to reconcile the various divergent sources and traditions by comparing the Campanus text (from Arabic sources) with the Zamberti text (from Greek sources); another was to overcome the medieval bias towards Aristotle, which had been reflected even in Scholastic editions of Euclid.

The Zamberti and Campanus legacies were superseded in the second half of the sixteenth century by a new Latin translation prepared by Federico

Commandino from the Greek. He was not only a distinguished mathematician (famous for his centre-of-gravity determinations after Archimedes), but also a tireless translator, his work including Hero's *Pneumatics*, Apollonius's *Conic sections*, Pappus's *Collectio* and almost the entire opus of Archimedes. His translation of Euclid, distinguished by its mathematical and philological excellence, became the most influential work in the history of the legacy up till then. Its effect is noticeable in the exhaustively annotated edition by the Jesuit Christoph Clavius, and again in the famous Greek–Latin edition of 1703 by David Gregory, the most significant and textually critical edition of the *Elements* before the discovery of the pre-Theonic Manuscript P (Table 2). It was Commandino, moreover, who drew attention to a long-standing biographical error: that Euclid the geometrician from Alexandria, and Euclid the philosopher from Megara, were thought to be one and the same person. This error explains the inscription appearing in earlier printed editions ('Euclidis megarensis mathematici clarissimi Elementorum geometricorum libri': 'The *Elements*, a book of geometry by Euclid from Megara, most illustrious mathematician').

The translations that appeared towards the end of the sixteenth century and the works of Greek (and Arab) mathematicians printed earlier prepared the ground for the further development of mathematics (Schönbeck *1988*). Through them, among other things, the direct influence of Euclid's *Elements* becomes clear in three areas. First, in the history of non-Euclidean geometry: starting with the question of parallels in Book I and picking up from Arab predecessors, Girolamo Saccheri and Johann Heinrich Lambert arrived at theorems of hyperbolic and elliptical geometry (§7.4). Second, in the development of analytical geometry: on the basis of the 'geometrical algebra' in Book II and preparatory work by François Viète, Pierre de Fermat and René Descartes laid the foundations of coordinate geometry (§7.1). Third, in the process of developing a new concept of number: in connection with the classification of irrationalities in Book X, Michael Stifel attempted to give a comprehensive arithmetical theory of all rational and irrational numbers (§2.3). 'Pure' elementary mathematics thus reached a degree of complexity which not only surpassed that of every earlier stage of development, but at the same time strengthened interest in advanced, higher mathematics, including the mathematical work of Archimedes, Euclid's most famous student.

3 ARCHIMEDEAN LEGACIES

3.1 Sources

Archimedes was, after Euclid, the most influential mathematician of the

Alexandrine school and undoubtedly its most important representative, well known even in ancient times (Clagett *1964–84*). He pursued a lively exchange of letters with the widely travelled Conon and with Eratosthenes, was friends with Dositheus, and was quoted by the later mathematicians Hero, Pappus and Theon. Nevertheless, his contemporaries were more inclined to admire him for his astonishing technical and engineering inventions (including winches, block-and-tackle, levers, water-pumps, planetaria and battle-machines) than to study him for his outstanding and unique mathematical achievements. But he founded no school of his own, in contrast to Euclid and the circle with which he may have been associated. He left extensive written work (Table 3), known thanks primarily to the Byzantine and Arabic mathematicians of the Middle Ages (Schneider *1979*). In addition, there are further works ascribed to Archimedes by Arabic writers.

Table 3 Archimedes' mathematical works

1	*Elements of Mechanics*	7	*On the Equilibrium of Planes*
2	*On the Quadrature of the Parabola*	8	*On the Method of Mechanical Theorems*
3	*On the Sphere and the Cylinder*	9	*On Floating Bodies*
4	*On the Measurement of the Circle*	10	*The Sand-reckoner*
5	*On Spirals*	11	*The Cattle Problem*
6	*On Conoids and Spheroids*	12	*Stomachion*

3.2 The 'Arabic' Archimedes

As in the case of Euclid, mathematicians in Islamic countries played an important part in handing down the legacy of Archimedes' works. But the 'Arabic Archimedes' is younger than the 'Arabic Euclid': Archimedes' works were probably not translated into Arabic until the middle of the ninth century. And his are less complete: while most of Euclid's writings, even the lesser ones, were translated by the Arabs, only a few of Archimedes' were known to them.

Considerable influence over the development of Arab mathematics started with the book *On the Sphere and the Cylinder*. It was translated at least twice, and was still being studied in the thirteenth century, by Naṣīr al-Dīn al-Ṭūsī. Back in the ninth century the Banū Mūsā, three sons of the Mūsā of Baghdād, knew the work and completed and expanded it through their own research – fresh evidence for the belief that Arabic mathematicians did rather more than simply pass on ancient mathematics (§1.6). A discourse on the measurement of flat and spherical figures contained formulas for calculating the contents of circles and spheres, and dealt with problems such as the trisection of an angle and duplicating the cube.

Translated into Latin early on (by Gerard of Cremona) and generally known as *Verba filiorum (Moysi filii Sekir)*, it introduced to the West the rudiments of higher mathematics. At the same time it had a lasting effect on the adoption of Archimedes in medieval Latin countries, noticeable in the work of Leonardo of Pisa and Jordanus de Nemore, right down to Regiomontanus and Nicolaus Copernicus.

3.3 The 'Byzantine' Archimedes

The history of the texts of Archimedes' works can, however, be traced even further back. It starts in the sixth century with the didactic and historical interest demonstrated by scholars like Isidorus of Miletus, architect of the Hagia Sophia in Constantinople, and his pupil Eutocius of Askalon: they brought about a new evaluation of mathematics after its decline in Roman Antiquity.

To start with, there was a now vanished collected edition, edited by Isidorus, of the best-known writings of Archimedes in his day, including *On the Sphere and the Cylinder*, *On the Measurement of the Circle* and *On the Equilibrium of Planes*. The accompanying commentaries by Eutocius, still extant, had an effect on the history of mathematics. They contain not only cross-references to Greek geometry, but also exhaustive comments on Archimedes' calculation of circles. It was these that opened an early door to Archimedes' work and then, centuries later, influenced the work of Western scholars such as Nicolaus of Cusa, Leonardo da Vinci, the Abbot Francesco Maurolico and Niccolò Tartaglia.

The Byzantine contribution to the handing-down of the Archimedes legacy reached its high point in the ninth century, with the appearance of two separate Greek manuscript collections (known as Codex A and Codex B, both now lost). Ultimately both probably go back to the time of Isidorus and Eutocius. Codex A was ordered by the mathematician Leon, regarded as the new founder of the University in Constantinople and wielding great influence over the intellectual life of the Byzantine Empire. His exemplary collection, far more extensive than that of Isidorus and even more extensive than that of the (later) Arabic Archimedes, contained almost all Archimedes' writings known today and laid the ground for the Archimedes renaissance in Latin countries in the Middle Ages. Codex B contained the 'mechanical' writings, including *On the Quadrature of the Parabola* and *On Floating Bodies*.

There is a third Greek manuscript (Codex C), extraordinarily important to the history of mathematics and apparently dating from the tenth century. It includes – unlike Codices A and B – the treatise *On the Method of Mechanical Theorems*, which is of fundamental importance in evaluating the development of Archimedes and his work. This manuscript was long believed to be lost. It was probably unknown in either the Arabic or the Latin Middle Ages, and only in 1906 was it rediscovered as a palimpsest by J. L. Heiberg, who produced editions of both Archimedes and Euclid. Parts of this manuscript may however have already been transmitted, perhaps only indirectly, at the beginning of the modern era, and could have led to the genesis of indivisible calculus in the Galileo school (Schneider *1979*).

3.4 The 'Latin' Archimedes

The Archimedes tradition of the Latin Middle Ages drew upon the Byzantine and Arabic traditions. It began with translations from Arabic sources. *The Measurement of the Circle* was passed on by Gerard of Cremona, and results from *On the Sphere and the Cylinder* were known through the *Verba filiorum*. A little later, there appeared Latin treatises taken from Greek, such as *De curvis superficiebus Archimenidis*. Such texts evince more than a strictly mathematical interest in Archimedes' work. For example, the measurement of curved planes and the associated question of the comparability of 'curved' and 'flat' led to dialectic arguments that touched on general questions of philosophy, physics and cosmology.

Of notable importance for the inheritance of Archimedes' work were the Greek–Latin translations by the Dominican William of Moerbeke, a friend of Thomas Aquinas and later Bishop of Corinth, at about the same time as Campanus was working on his Euclid text. The sources for this Latin Archimedes were the two Byzantine Codices A and B. The main effect of this complete edition of his works was to make Archimedes the physicist well known, and it exerted lasting influence on medieval concepts of the specific gravity of a body, equilibrium and the centre of gravity, the development of statics, and discussions about the law of falling bodies. Galileo himself referred expressly to the 'superhuman Archimedes', his 'master'. Parts of Galilean mechanics derive directly from Archimedean hydrostatics, as do corresponding works by Simon Stevin and Blaise Pascal. On the other hand, Archimedes the mathematician and the methods he evolved were understood by only a few medieval scholars, his influence on the curricula in schools and high schools was slight, and until well into the sixteenth century 'studying geometry' meant primarily 'studying Euclid'.

3.5 The 'modern European' Archimedes

Although single parts of the Moerbeke legacy were printed early on, for example *The Measurement of the Circle*, *The Quadrature of the Parabola* and *On Floating Bodies*, ultimately by Niccolò Tartaglia, it was not until the late sixteenth century, then increasingly in the seventeenth, that Archimedes' mathematical work began to have a formative influence on the development of mathematics.

Of exceptional significance for the beginnings of this modern European inheritance and handing-on of the Archimedes legacy were the first edition ('editio princeps'), in 1544, of an almost complete Greek and Latin Archimedes text based on Codex A, with Latin text by Jacob Cremona; and a new Latin edition by Federico Commandino in 1558, then expanded in 1565 with *On Floating Bodies*. Commandino here represents the intense sixteenth-century concern not only to understand Archimedes' conclusions in advanced mathematics (*The Quadrature of the Parabola*), but also to elaborate and consolidate them. This is also true of the Abbot Francesco Maurolico, one of the sixteenth century's best geometricians, who had already mastered Archimedes' entire work.

Such efforts were crowned with lasting success in the seventeenth century, after the advances in Euclidean elementary mathematics and with the new methods developed primarily by Viète, Fermat and Descartes. In addition, Fermat solved quadrature problems, Kepler carried out the determination of volumes, Bonaventura Cavalieri developed his method of indivisibles, Christiaan Huygens conducted centre-of-gravity research and Gottfried Wilhelm Leibniz founded his new method for 'the greatest and the least'. Thus the ground was laid for the higher level of abstraction of modern mathematics as compared with ancient geometry (§3.1). This came about essentially, as the above works testify, through the legacy of Archimedes, the greatest mathematician of the ancient world.

4 EUCLIDEAN AND ARCHIMEDEAN METHODS OF DISCOVERY AND VERIFICATION

The lines down which the works of Euclid and Archimedes have been inherited and handed on are certainly not known in full detail, and are still far from being fully recorded (Murdoch *1971*, Knorr *1989*, Mueller *1991*). But more important than further and more detailed lines of development are the methodological traditions rooted in the work of the two 'Alexandrine' mathematicians which have influenced the history of mathematics and science ever since: the axiomatic and deductive method, and the method of analysis and synthesis.

As Pappus reported in the fourth century AD, it was principally Euclid who, in his writings (*Data* and *Elements* – to mention them yet again), created a two-pronged method of mathematical research: the 'analytical', working backwards from unknown to known, and the 'synthetic', working forwards from known to unknown. Greek mathematics already made a sharp distinction between discovery and verification of truth – a good two thousand years before Descartes! This was also endorsed by Archimedes in his textbook on method (and by the classic methods of exhaustion and *reductio ad absurdum*). Galileo, with reference to his ancient predecessors, even transferred this heuristic process (analysis–synthesis–procedure) to the natural sciences as well (Dijksterhuis *1956*). He made a distinction between the *metodo risolutivo* (the 'conjectural tracking down' of the causes for appearances) and the *metodo compositivo* (the 'well-founded explanation' of appearances on the grounds of those causes). Again, it was Euclid and Archimedes who, fully in the tradition of Plato and Aristotle, established mathematics and the natural sciences as axiomatic–deductive theories: axiomatic in their fundamental principles, and deductive in their logical procedures. Their work has so far been the model for Thomas Hobbes (*Politics*), Baruch de Spinoza (*Ethics*), Leibniz (*Jurisprudence*), William Whiston (*Cosmology*) and many more, and has thus become the unique paradigm for strict science in the Western sense: as science '*more geometrico*'.

BIBLIOGRAPHY

Bulmer-Thomas, I. *1971*, 'Euclid. Life and works', in *Dictionary of Scientific Biography*, Vol. 4, New York: Scribner's, 414–37.

Clagett, M. *1964–84*, *Archimedes in the Middle Ages*, 5 vols, Philadelphia, PA: American Philosophical Society.

—— *1970*, 'Archimedes', in *Dictionary of Scientific Biography*, Vol. 1, New York: Scribner's, 213–31.

Dijksterhuis, E. J. *1956*, *Die Mechanisierung des Weltbildes*, Berlin: Springer. [English transl. 1961, *The Mechanization of the World Picture*, Oxford: Clarendon Press.]

Folkerts, M. *1980*, 'Probleme der Euklidinterpretation und ihre Bedeutung für die Entwicklung der Mathematik', *Centaurus*, **23**, 185–212.

Høyrup, J. *1988*, 'Jordanus de Nemore', *Archive for History of Exact Sciences*, **38**, 307–63.

Knorr, W. *1989*, *Textual Studies in Ancient and Medieval Geometry*, Basel: Birkhäuser.

Mueller, I. (ed.) *1991*, 'Peri Tōn Mathēmatōn', *Apeiron*, **24**, 1–251.

Murdoch, J. *1971*, 'Euclid. Transmission of the *Elements*', in *Dictionary of Scientific Biography*, Vol. 4, New York: Scribner's, 437–59.

Schneider, I. *1979*, *Archimedes*, Darmstadt: Wissenschaftliche Buchgesellschaft.

Schönbeck, J. *1984*, 'Euklid durch die Jahrhunderte', *Jahrbuch Überblicke Mathematik*, **17**, 81–104.

—— *1988*, 'Euklid und die "Elemente" der Geometrie', *Der mathematische und naturwissenschaftliche Unterricht*, **41**, 204–10.

Schreiber, P. *1987*, *Euklid*, Leipzig: Teubner.

Sezgin, F. *1974*, *Geschichte des arabischen Schrifttums*, Vol. 5, Leiden: Brill.

Szabó, A. *1969*, *Anfänge der griechischen Mathematik*, Munich and Vienna: Oldenbourg.

2.2

Practical geometry in the Middle Ages and the Renaissance

H. L'HUILLIER

The tradition of practical geometry was very firmly initiated in the *Practica geometriae*, a short work about geometry written by Hugh of Saint Victor in Paris around 1130 (§2.1). It was innovative not so much in its content as in the title deliberately chosen by the author, making him the first to establish a clear distinction between 'practical' and 'theoretical' geometry. Subsequent writers on geometry throughout the Middle Ages and the Renaissance kept strictly to the practical field, judging by the limited number of 'speculative' geometries. This seems at first paradoxical for a science which is considered to be the very essence of abstraction. In fact, the Roman and medieval epochs contributed nothing to the history of geometry; not until the mid-sixteenth century was it revived with François Viète, Johannes Kepler and Bonaventura Cavalieri. With the numerous texts produced during this long dark period, however, practical geometry must have answered a need, otherwise the tradition would not have endured.

1 THE ORIGINS OF THE DISTINCTION BETWEEN PRACTICAL AND THEORETICAL GEOMETRY

The distinction established by Hugh of Saint Victor between practical and theoretical geometry, however innovative it may appear, actually bears the mark of a triple encounter between the Latin culture of agrimensure (agricultural mensuration), the discovery of fuller versions of Euclid's *Elements*, and the epistemological thought expressed in the attempts at classification within the various sciences.

The initial distinctions between theoretical and practical geometry arise at the beginning of the twelfth century. They appear clearly in the commentaries on the *Elements* long attributed to Adelard of Bath, of which Hugh of Saint Victor could have been aware (Victor *1979*: Chap. 1). Theoretical geometry is the privilege of the *demonstrator*, who exercises it at his desk using propositions, theorems, examples and conclusions; the second, by contrast, is practised in the field with the appropriate tools.

For several decades, in Spain, the notions of *theorica* and *practica* were the subject of an attempt at definition under the influence of Al-Fārabī, an attempt which applied equally well to various sciences in particular as to science in general. In the twelfth century this movement found a Latin exponent in the person of Gundissalvi, the Archdeacon of Segovia. According to this train of thought, theoretical geometry is characterized by the rational consideration of lines, surfaces and volumes, in a research or teaching context, on the basis of demonstrations. In comparison, practical geometry is defined by the measurement of real dimensions, in the process of making something, using instruments and tools. This leads to two professional streams, agrimensure and the *artes mechanicae*.

It was in this setting, which had become highly intellectual, that the tradition of practical geometry developed, based on Latin texts consisting mostly of surviving fragments of Roman agrimensure. Basically, the content of geometry treatises has changed less than the consideration given to geometry itself, although this content has been progressively expanded from its original nucleus.

2 THE CONTENT OF PRACTICAL GEOMETRY

Hugh of Saint Victor never himself claimed that his work was original; he had merely gathered together fragments of earlier works. Predating him, in fact, the *Geometria Gerberti* and the *Geometria incerti auctoris* may be regarded as belonging to the tradition of practical geometry (Bubnov *1899*).

This tradition is fundamentally characterized by a tendency to tackle geometric space in a utilitarian manner, organizing it around the three immediately visible concepts of lengths, surfaces and solids; alternatively length, which determines straight lines, width, which combines with length to determine surfaces, and height, which combines with the other two to determine solids. The measurement of lengths, surfaces and solids is the most concrete possible domain where practical geometry can be applied. It does not seek to define notable points inside or outside the figures, nor the relations between the angles, nor the particular shapes of lines. In this respect it departs from Euclid's *Elements*.

Within its strict domain it operates with just a few rules, which are founded on proportions, notably on the geometry of triangles. For this reason the works generally expound at length on triangles. They sometimes also contain a fairly substantial chapter on fractions or proportions. In theory, practical geometry is exclusively for operators in the field, who will encounter the problem of units of measurement, which were very diverse in the Middle Ages and rich in multiples and sub-multiples.

In all, practical geometry is a geometry of procedures and of instruments. It has been dubbed 'artificialist', a term that satisfactorily renders account of its principle, although not totally covering the preoccupations of the medieval writers. It does not set out to demonstrate anything, but invents processes, sometimes not fully applicable in the field, the main tools for which are the quadrant, or the back of the astrolabe, and the geometric square. The quest for instruments to measure casks proceeds basically from the same principle.

3 THE TRADITION OF PRACTICAL GEOMETRY FROM THE TWELFTH TO THE SIXTEENTH CENTURY

On these bases the tradition of practical geometry flourished, essentially in university milieux, until the sixteenth century. The course of its development may be regarded as having consisted of two main streams. The first, structured around an instrument, the astrolabe, lies in the direct line of agrimensure and is concerned essentially with measurement of terrain; the second, less instrument-based and at first glance less immediately practical, is found in works which today would tend to be classed as treatises on applied geometry.

The first stream is by far the most abundant. The teaching of practical geometry as defined by Hugh of Saint Victor in his *Practica geometriae* is an integral part of the quadrivium (§2.11), which thus served it as a vehicle throughout Western Europe. The works dealing with this are fairly limited in number, but have been abundantly copied, existing in numerous versions and contributing to the crystallization of knowledge. The most important of them date from the late twelfth and the thirteenth centuries; they are the treatises known by their opening words, *Geometriae duae sunt partes* and *Artis cujuslibet consummatio*, and the *Tractatus quadrantis*. These works were the subject of thorough studies between 1970 and 1985 (Victor *1979*). It is known that they were still used in the fourteenth century. They are also sometimes found to figure in certain bodies of work consisting of various treatises sharing a knowledge of the astrolabe (Beaujouan *1975*).

In the fourteenth and early fifteenth centuries the knowledge of practical geometry provided material for more exhaustive works. The most famous is the *Practica geometriae* composed in Paris around 1350 by an Italian of whom nothing is known but his name, Dominicus de Clavasio (Busard *1965*). Also produced in northern France was Jean Fusoris's *Geometrie pratique*. From Italy the most complete works are those of Leonardo Cremonensis (*Ars metrica*), Prosdocimo de Beldomandi (*Trattato sull'astrolabio*) and Francisco di Giorgio Martini (*La praticha di gieometria*).

Some works of this sort appeared very early in the vernacular, but until the fifteenth century they were very few. Known works include notably a French geometry in the Picardy dialect dating from the early thirteenth century, and an English treatise on the measurement of heights and distances written in the mid-fourteenth century.

The development of Renaissance learning did not stop this stream continuing to flow into the sixteenth century; its vigour is demonstrated by some very fine works, abundantly illustrated. In Italy there were Georg Peurbach's *Quadratum geometricum* and the works of Cosimo Bartoli (partially adapted from a prolific French mathematician, Oronce Fine), and in England the treatises of Thomas Digges.

From the start, a number of treatises devoted several paragraphs to the problem of measuring casks, generally likened to two truncated cones joined base to base. In the course of the fifteenth and sixteenth centuries this problem formed the subject of specialized works devoted to the fabrication of gauges for measuring the capacity of casks and, notably in Italy, the compilation of calculation tables giving their content when only partially filled.

The other tendency is made up of works less concerned with the measurement of terrain and therefore less centred on the use of the astrolabe or the quadrant. These are in fact 'applied' geometries, usable for measurements on the ground, but mainly for all types of measurement. The tradition goes back to Hero of Alexandria (§1.3). There is room to suppose that it was brought to the West by Arab intermediaries (a point that seems to merit deeper studies) while becoming richer with time. In particular, there are similarities between certain passages in Western works and Arabic tradition known as *misāha*. But further similarities may be revealed between the works of Abu'-l-Wafa and the major treatises of this stream, notably in research into the tracing of figures with ruler and compass.

With this stream are associated some complete major works, the most elaborate in medieval geometry. The principal ones are the *Practica geometriae* by Leonardo of Pisa (*circa* 1220), the *Ars mensurandi* by Jean de Murs (*circa* 1340), the treatise on practical geometry by Nicolas Chuquet (1484), the *Trattato d'abaco* by Piero della Francesca (*circa* 1480) and the geometric scheme in the *Summa* by Luca Pacioli (1494). In general they include a small section on practical geometry such as conceived by those favouring artificial representations; a detailed account of the basic figures; various problems, sometimes resolved with algebra; and finally some problems on tracing figures. For this last part they join the treatises on speculative geometry, which emphasizes the fact that these works contained a part devoted to research.

In all there are many works on practical geometry, far more numerous than the treatises on speculative geometry. The lineage is evident, but at the same time the evolution is real and various: the family likeness between Oronce Fine and Piero della Francesca is not immediately evident. In fact, from their content it is not clear why their authors gave them titles which placed them in the family of practical geometry. They respond to a concern persisting throughout the Middle Ages which we must try to comprehend.

4 THE PROBLEM OF INTERPRETING THIS TRADITION

It is now accepted that the works on practical geometry were not aimed at operators, surveyors or 'site managers'. The problems they contain are not sufficiently precise; until the later period their language is Latin, and the texts are frequently bound in heavy volumes along with various other treatises, to be discreetly shelved in the libraries of monks or canons. They are intended primarily for the education of the children of persons of note, initiating them into the concrete problems involved in the measurement of spaces and certain objects. The problems were chosen from the fields that were expanding in the Middle Ages, at the same time as practical geometry was emerging from the restrictive framework of the quadrivium: surveying and measurement of land, and architecture, but they remain general or purely hypothetical constructs. This did not prevent certain works preferring more utilitarian problems which might be useful in the practice of certain professions, for instance the measurement of casks, or the cutting of stones along the individual outlines of figures.

This is why historians of science have justifiably sought to start by examining the problems posed by practical geometry, and by considering how medieval figures saw the connection between theory and practice. The result of their investigations is inadequate, and only reinforces what has been said above. On the one hand exists the written material which, although its level seem to us weak, embodies a pedagogic (albeit scarcely scientific) concern; on the other hand are a few objects or pieces of evidence revealing approximative practices. Nevertheless, even from the works themselves, we can discern an interplay between the theorem or the theoretical formula and observation and the practical problem. Thus knowledge of certain rules leads to the invention of processes that are unworkable because they are impractical or too imprecise, such as the case of measurements made with a mirror placed on ground where its horizontality could not be established. Conversely, the existence of a poorly resolved concrete problem would prompt theoretical research in order to achieve a better result; thus Jean de Murs was induced to liken a cask to an ellipsoid of revolution cut off at the extremes. In the end the perception of a concrete problem can lead to the discovery of a practial solution, as happened with the rules for measuring casks. Generally speaking, the nearer to the modern era, the greater the concern to resolve concrete problems.

Technical problems associated with the real 'engineers' of the fifteenth century may spring to mind, but they arose as much from the development of economic activity and a certain amount of interest in scientific research (§2.8). The measurement of objects is equally necessary to experimental science, to trade and to the implementation of technical processes, which is what the tradition of practical geometry sought to address. The calculation of the diameter of a clock wheel which does a given number of turns while another does more turns or fewer may seem a childish geometry problem, but its does reveal a concern for precision and a desire to develop new ways of working that may be encountered among 'mechanicians' and traders who have no wish to cheat their customers. The practical geometer must know how to liken an object to a figure or combination of figures, and be able to calibrate a measuring instrument in full knowledge of its limits, the best example of which is the proposal for measuring the volume of any metallic object by calculating the volume of water it displaces when immersed.

5 CONCLUSION

The tradition of practical geometry in the Middle Ages and the Renaissance perhaps leaves the historian of mathematics unsatisfied when he finds no new theorems and a general backwardness in comparison with works from the Greek world. However, one can discern an approach to reality (still relevant today) through instruments and measuring processes, proceeding to the definition of rules equally used and understood by men in their dealings. A similar spirit is encountered in the physics of the late Middle Ages, and the sixteenth century saw the appearance or perfecting of numerous measuring instruments in several fields. This whole tradition is therefore not devoid of scientific spirit, but consistently indicates that geometry is both an experimental and an exact science.

BIBLIOGRAPHY

The bibliography lists editions of geometrical works from the Middle Ages (those from the sixteenth century are available in print). Recent publications are generally accompanied by commentaries. In addition, the reader may seek out the other works by the various authors mentioned, notably those of Arrighi, Curtze, Shelby and Victor.

Arrighi, G. (ed.) *1966, La pratica di geometria de Gherardo de Dino*, Pisa: Domus Galilaeana.

—— (ed.) *1970, La pratica di geometria de Francesco di Giorgio Martini*, Pisa: Domus Galilaeana.

Baron, R. (ed.) *1966, Practica geometriae de Hugues de Saint-Victor*, South Bend, IN: University of Notre Dame Press.

Beaujouan, G. *1975*, 'Réflexions sur les rapports entre théorie et pratique au Moyen Age', in Murdoch and Sylla *1975*: 437–84.

Boncompagni, B. (ed.) *1862, Practica geometriae de Leonardo Fibonacci*, Rome: the author.

Bubnov, N. *1899, Gerberti postea Silvestri II papae opera mathematica (972–1003)*. Berlin: Friedländer. [Repr. 1963, Hildesheim: Olms.]

Busard, H. L. L. (ed.) *1965*, 'Practica Geometriae de Dominicus de Clavasio', *Archive for History of Exact Sciences*, **2**, 520–75.

—— (ed.) *1968*, 'Le Livre sur la mensuration des figures d'Abu Bakr traduit par Gérard de Crémone', *Journal des Savants*, 65–124.

Crombie, A. G. *1952, From Augustine to Galileo*, London: Heinemann.

Curtze, M. *1896*, 'Über die im Mittelalter zur Feldmessung benutzen Instrumente', *Bibliotheca mathematica*, Series 3, **10**, 65–72.

Geldner, F. (ed.) *1965, Geometria deutsch de Mätthaus Roriczer*, Wiesbaden: Steiner.

Herzog August Bibliothek (Wolfenbüttel) *1989, Mass, Zahl und Gewicht. Mathematik als Schlüssel zu Weltverständnis und Weltbeherrschung*, Wienheim: VCH, Acta humaniora. [Exhibition catalogue, by M. Folkerts, K. Reich and E. Knobloch, with substantial commentaries. Chaps 6 and 7 treat practical geometry, including pertinent instruments.]

l'Huillier, H. (ed.) *1979, La Géométrie de Nicolas Chuquet*, Paris: Vrin.

Murdoch, J. E. and Sylla, E. D. (eds.) *1975, The Cultural Context of Medieval Learning*, Dordrecht and Boston: Reidel.

Shelby, L. R. *1972*, 'The geometrical knowledge of medieval master masons', *Speculum*, **47**, 395–421. [Several other articles by this author since 1961 on this topic.]

Victor, S. *1979*, 'Practical geometry in the High Middle Ages', *Memoirs of the American Philosophical Society*, **134**. [Edn of 'Artis cuiuslibet consummatio' and 'La Pratike de geometrie'.]

2.3

The 'Coss' tradition in algebra

KARIN REICH

In the Islamic world, Abū Jaʿfar Muḥammad ibn Mūsā al-Khwārizmī (*circa* 780–850) was the first to write a treatise on the Hindu–Arabic numerals and the basic arithmetical operations. With al-Khwārizmī began a tradition of similar texts (§1.6). He was also the author of *Ḥisāb al-jabr wa'l-muqābala*, the work that gave algebra its name (*al-jabr* means 'restoration'). The unknown quantity and its powers were expressed not by means of symbols but in words: x was *shay'* ('thing' or 'something') or *jidr* ('root'); x^2 was *māl* ('wealth', 'property'); and x^3 was *kaʿb* ('cube'). Muslim algebra became known in the Western world only in the twelfth century, when Robert of Chester translated al-Khwārizmī. The Arabic terminology was carried over into Latin as 'res' (or 'radix'), 'census' and 'cubus'.

1 ITALY

In 1202 Leonardo of Pisa wrote a detailed compendium on mathematics, entitled *Liber abbaci* ('Book of the Abbacus'). This 'abbacus' has nothing in common with the abacus calculator (§5.11); Leonardo used the term for the new kind of mathematics based on the Hindu–Arabic numerals and the new methods of counting and calculating they made possible. He drew heavily on Arabic sources. For the unknown quantity Leonardo used 'res' and 'radix'; a third unknown was introduced as 'pars'. For x^2 he employed 'quadratus', 'census' and 'avere' (wealth); for x^3 'cubus'; for x^4 'census de censu' and 'censuum census'; and for x^6 'cubus cubi'. The development of modern algebra began with Leonardo of Pisa.

In Italy in the thirteenth century an economic revolution took place (§2.4). The universities flourished, with Latin as the basic language, but there also emerged another type of school, the abbacus school. The students who attended abbacus schools did not know Latin; their aim was to get a solid grounding in practical mathematics – arithmetic and geometry and sometimes also algebra. Most of these students intended to pursue a career as a merchant; others were would-be cartographers, artists, architects. The heads of these schools were the *maestri d'abbaco*. They composed textbooks especially for their students, the *Trattati d'abbaco* or similar. (For more discussion of the tradition and its commercial context, see §2.4.) According to the particular need, these works were written in Italian or Latin. The oldest vernacular treatise extant today is the *Livero del abbecho*, written in about 1290 (Van Egmond *1980*: 156*f*), which included no algebra. Among the oldest Italian texts to present arithmetic and algebra were Paolo Gerardi's *Libro di ragioni* (1328) and Magister Dardi's *Aliaabra-Argibra* (1344). The topics covered by this kind of book varied, but typical contents were calculation with roots, calculation with algebraic monomials and polynomials, rules for solving algebraic equations, and problem-solving by means of algebra (Franci and Toti Rigatelli *1988*: 12). The unknown quantity was called 'cosa', a translation of the Latin 'res'. Antonio de Mazzinghi da Peretola, a pupil of the famous Paolo Dagomari, was perhaps the first to use two unknowns, one of them designated a 'cosa' and the other a 'quantità'.

In the fourteenth century Florence became a noted centre of abbacus schools. The corresponding *libri* were written in the Tuscan vernacular. Algebra, referred to at first, following the Arabic, as *regola dell'algebra ammucabala*, was soon transformed into *regola della cosa*, 'rule of the thing'. Although there was no word for equation, the *maestri* solved what in modern terms are linear and quadratic equations, and sometimes special equations of the third and fourth degree, or even higher. It became quite common to abbreviate the unknown 'cosa' by '*co*', its second power, 'censo', by '*ce*', and its third power, 'cubo', by '*cu*'. The fourth power was denoted by 'censo di censo' or 'quadro'; higher powers were not normally used in the fourteenth century, but appeared only in the fifteenth. There were three ways of denoting the fifth power: 'duplici cubo', 'chubo di censi' or 'censo di chubo'. This last was also used by several writers for x^6, which was also denoted as 'cubo di cubo' (Franci and Toti Rigatelli *1988*: 15). In Italy, mathematical texts dealing with arithmetic and algebra were very common in this period, and the number of surviving manuscripts is about three hundred (Van Egmond *1988*: 129; *1980*).

In the fifteenth century mathematics came to have a close relationship with painting (§12.6). The famous painter Piero della Francesca, who wrote a treatise on perspective, was also the author of a *Trattato d'abaco*. Although Luca Pacioli contributed few original ideas himself, his *Summa de arithmetica, geometria, proportioni e proportionalità* (1494) was the first printed compendium which included algebra. Pacioli was a friend of Leone

Battista Alberti and Leonardo da Vinci. His *Summa* was well recognized, and a second edition followed in 1523. Pacioli designated the sixth power, and not the fifth, as 'cubus de censo', so he had to adopt a special terminology for prime powers, calling x^5 'primo relato', x^7 'secondo relato', and so on. In the sixteenth century these names were quite common. In 1490, Raffaele Canacci introduced special symbols, which were also applied by Francesco Ghaligai in his *Summa de arithmetica* (1521, 1548, 1552) (Franci and Toti Rigatelli *1985*: 56*f*, 68):

Numero	n°	chubo di censo	▱ ▱
chosa	c°	Promicho	⊞
censo	▱	censo di censo di censo	▱ ▱ ▱
chubo	▱▱	chubi di chubi	▱▱ ▱▱
censo di censo	▱ ▱	relato di censo	⊟ ▱
relato	⊟		

The development reached its culmination with the solution of the general cubic equation by means of radicals. Influenced by the former *maestri d'abbaco*, Scipione del Ferro solved the equation $x^3 + ax = b$. Niccolò Tartaglia, *maestro d'abbaco* himself, added the solution of $x^3 = ax + b$. Girolamo Cardano was the first to publish these solutions, in his *Ars magna* (1545), causing a dispute with Tartaglia, who published his ideas in *Trattato di numeri e misure* (1556–60). In the chapter 'Segni per numeri', he introduced the common abbreviations for the unknown and its powers, but he also added the corresponding number of the exponents:

n	0	*cecu*	6
co	1	*2° rel.*	7
ce	2	*cecece*	8
cu	3	⋮	⋮
cece	4		
pri.rel.	5	*9° rel.*	29

2 FRANCE

The economic and social conditions in pre-sixteenth-century France were different from those in Italy. Abbacus texts were not very common; only about 25 manuscripts survive from this time (Van Egmond *1988*, 129).

In 1343, John de Murs finished his *Quadripartitum numerorum*, which also contained algebra; this text was more or less copied by Rolland of Lisbon (1430). There was a tradition of abbacus texts in the Provençal language. Of the six French abbacus texts that appeared between 1475 and 1485, only one was also devoted to algebra: Nicolas Chuquet's *Triparty en la science des nombres* (1484) (Flegg *et al. 1985*). Chuquet denoted the unknown and its powers as 'premiers', 'champs', 'cubicz' and 'champs des champs'. He spoke of *règle des premiers* when he meant algebra; this is the French translation of *regola della cosa*. There was only one mathematician in France who was influenced by Chuquet, Estienne de la Roche. His work *Larismethique*, published in Lyon in 1520 and in 1538, was closely linked to the *Triparty*.

3 GERMANY

Developments in Germany took their lead from Italy, beginning in the fifteenth century. Abbacus schools, so-called *Rechenmeisterschulen*, were founded mainly in those towns which traded with Italy, one of the first being in Nuremberg. It was not until the sixteenth century that such schools became common. In contrast to the Italian schools, the *Rechenmeisterschulen* were quite often a cross between an elementary writing school and an abbacus school. In Nuremberg, Johann Neudörffer maintained a *Rechenmeisterschule* together with a school for calligraphy. He is known as the founder of German calligraphy, and was in contact with his fellow-townsman Albrecht Dürer.

The foundation of abbacus schools had an economic background, and has to be seen as part of a sudden rapid development. With these schools the Hindu–Arabic numerals began to spread; they were first used in sales ledgers. But Roman numerals continued to be used, and even in some printed textbooks the Roman numerals in combination with counters were preferred. In several treatises the methods of reckoning by placing counters on lines (*Linienrechnen*) were the only ones that were presented, for example in the *Trienter Algorismus* (1475), in the first textbook by Adam Ries, *Rechnung auff der linihen* (1518, 1525, 1527) and in Jakob Köbel's *Ain new geordnet Rechenbiechlin* (1514).

At about the same time, the Hindu–Arabic numerals and the basic operations with them were propagated, for example in the *Algorismus Ratisbonensis*, which exists in several manuscripts, the two main ones written in 1449–50 (Clm 14783) and 1457–59 (Clm 14908). In this *Algorismus* the operations with integers and fractions were presented together with a large collection of problems. The operations were described by means of abbreviations:

Clm 14783: ⱬ ⱬ ⱡ (ligature of 'et'), ⱹⱳ ⱬ ⱦ. ⱬⱳ (ligature of 'minus')
Clm 14908   ⱬ ⱬ   ⱬⱳ

The Codex Dresden C 80 contains several treatises on algebra. On fol.288r the minus sign was at first written like m, but later Johannes Widman erased it and replaced it with the now common sign $-$. On the same folio Widman added marginal notes which contain three different forms of the plus sign:

$$\text{ʒ}, +, \times.$$

Later in the codex, however, the now common signs $+$ and $-$ were preferred. In the so-called *Deutsche Algebra* (Dresden Codex C 80, fols. 368r–378v), the unknown quantity and its powers were denoted and abbreviated as follows:

x^0:   czall ϕ

x^1:   dingk $\mathcal{X}$ (ligature of re(s))

x^2:   czensi $\mathcal{z}$

x^3:   chubi $\mathcal{C}$

x^4:   wurcell von der worcell $\mathcal{ff}$

In the Codex Leipzig 1470, the series of symbols was continued up to x^9:
x^5, $\mathcal{Xff}$; x^6, $\mathcal{fff}$; x^7, $\mathcal{Cff}$; x^8, $\mathcal{ffff}$; x^9, $\mathcal{Xffff}$ (Tropfke *1980*: 281–4).

Widman, who was professor of mathematics at the University of Leipzig, had also used the operational signs $+$ and $-$ in a lecture on algebra which he delivered in 1486 (Glaisher *1921*). This is the first lecture on algebra in Germany to have been preserved: there are two copies written in Latin, one in the Dresden Codex C 80 (fols. 350r–364v), and one in the Leipzig Codex 1470 (fols. 479r–493v). Widman was also the author of an elementary textbook for merchants, *Behende vnnd hubsche Rechnung auff allen Kauffmanschafft*, first printed in 1489 and reprinted in 1508, 1519 and 1526. This book guaranteed the dissemination of the operational signs $+$ and $-$.

In this period algebra changed its form. Previously expressed in words, algebraic quantities came more and more to be replaced by symbols as the sixteenth century progressed. In German, Coss became synonymous with algebra, the word coming from the Italian *cosa* ('thing'). The cossic symbols were still in use, though there were also other kinds of representation. For example, Henricus Grammateus published a textbook for merchants, *Ein new kunstlich behend Rechenbüchlin uff alle Kauffmanschafft* (1518); in the chapter on algebra he introduced the unknowns in the following forms: 'prima quantitas', abbreviated as *pri*; 'secunda quantitas', abbreviated as *se*; then *ter, quart, quint* and *sex*. He would write an equation as '2 *se* + 18 *N sein gleich* 15 *pri*' ($2x^2 + 18 = 15x$), but he still described the classical forms of the quadratic equation which were in use at that time by means of words; he called them the *regel Cosse*.

A dominant role was played by Christoph Rudolff's *Coss (Behend und hubsch Rechnung durch die kunstreichen regeln Algebre so gemeinicklich die Coss genennt werden*), published in 1525. Michael Stifel re-edited it in a revised, extended form as *Die Coss Christoffs Rudolffs* (1553, 2nd edn 1615). Rudolff and Stifel used the usual cossic symbols for $x^1, \ldots, x^4$ and beyond that

$$x^5: \mathcal{B} \quad \text{(sursolidum)} \qquad x^{10}: \mathcal{Z}\mathcal{B}$$
$$x^6: \mathcal{Z}\mathcal{C} \quad \text{(zensicubus)} \qquad x^{11}: c\mathcal{B}$$
$$x^7: b\mathcal{B} \quad \text{(bisursolidum)} \qquad x^{12}: \mathcal{Z}\mathcal{Z}\mathcal{C}$$
$$x^8: \mathcal{Z}\mathcal{Z}\mathcal{Z} \qquad x^{13}: d\mathcal{B}$$
$$x^9: \mathcal{C}\mathcal{C}$$

and so on.

They described in detail the operations of addition, subtraction, multiplication and division of binomials $ax \pm b$:

$$(a_1 x \pm b_1) \pm (a_2 x \pm b_2), \qquad (a_1 x \pm b_1) \cdot : (a_2 x \pm b_2).$$

These results were used for solving quadratic equations. Rudolff had given the solution of the following eight types, illustrated by many examples (original language retained):

1 Wann zwo quantitetn natürlicher ordnung einander gleych werden: $ax^{n+1} = bx^n$,

e.g. $\qquad 32x = 6, \qquad 4x^2 = 8x, \qquad 5x^3 = 10x^2, \qquad 6x^4 = 12x^3.$

2 Wann zwo quantitetn einander gleych werden, zwischen welchen eine natürlicher ordnung geschwigen ist: $ax^{n+2} = bx^n$,

e.g. $\qquad 2x^2 = 8, \qquad 3x^3 = 12x, \qquad 4x^4 = 16x^2, \qquad 5x^5 = 20x^3.$

3 Wann zwo quantitetn einander gleych werden zwischen welchen zwo andere natürlicher ordnung geschwigen sind: $ax^{n+3} = bx^n$,

e.g. $\qquad 2x^3 = 16, \qquad 3x^4 = 24x, \qquad 4x^5 = 32x^2, \qquad 5x^6 = 40x^3.$

4 Wann zwo quantitetn einander gleych werden zwischen welchen drey andere natürlicher ordnung geschwigen sind: $ax^{n+4} = bx^n$,

e.g. $\qquad 2x^4 = 32, \qquad 3x^5 = 48x, \qquad 4x^6 = 64x^2, \qquad 5x^7 = 80x^3.$

5 Werden einander vergleycht drey quantitetn natürlicher ordnung, also das die grössern zwo werden gleich gesprochen der kleinern: $ax^{n+2} + bx^{n+1} = cx^n$,

e.g. $\qquad 3x^2 + 4x = 20, \qquad 5x^3 + 6x^2 = 32x, \qquad 7x^4 + 8x^3 = 44x^2.$

6 Werden einander vergleycht drey quantitetn natürlicher ordnung also das die kleyner und grösser samptlich werden gleich gesprochen der mitteln: $ax^{n+2} + cx^n = bx^{n+1}$,

e.g. $\qquad 4x^2 + 8 = 12x, \qquad 5x^3 + 9x = 14x^2, \qquad 6x^4 + 10x^2 = 17x^3.$

7 Werden einander vergleycht drey quantitetn natürlicher ordnung also das die kleynern zwo werden gleych gesprochen der grössern: $ax^{n+2} = bx^{n+1} + cx^n$,

e.g. $\quad 4x + 12 = 5x^2, \quad 5x^2 + 14x = 6x^3, \quad 6x^3 + 16x^2 = 7x^4.$

8 Wann einander vergleycht werden drey quantitetn also das ye zwischen zweyen eine zwo oder drey quantitet ausgelassen sind: Procedir nach laut der fünfften, sechsten oder sibenden equation:

$$ax^{n+2p} + bx^{n+p} = cx^n \quad \text{durch die fünfte,}$$

$$ax^{n+2p} + cx^n = bx^{n+p} \quad \text{durch die sechste,}$$

$$ax^{n+2p} = bx^{n+p} + cx^n \quad \text{durch die siebente,}$$

e.g. $\quad 2x^4 + 5x^2 = 53, \quad 3x^5 + 6x^3 = 72x, \quad 4x^6 + 7x^4 = 92x^2;$

$\qquad\quad 2x^4 + 12 = 11x^2, \quad 3x^5 + 16x = 16x^3, \quad 4x^6 + 20x^2 = 21x^4;$

$\qquad\quad 2x^4 + 40 = 18x^2, \quad 3x^5 + 64x = 28x^3, \quad 4x^6 + 96x^2 = 40x^4.$

In his *Deutsche Arithmetica* (1545) Stifel denoted the unknown quantity x by *sum* (abbreviation of 'summa'), x^2 by *sum sum*, x^{10} by *sum sum sum sum sum sum sum sum sum sum*, and so on; he explicitly wanted to avoid the cossic symbols so as to make it easier for the reader. In his *Arithmetica integra* (1544), however, he did use the cossic symbols. This work was a source for Jacques Peletier (1517–1582) (Van Egmond *1988*: 141), who also employed the cossic symbols which were not usual at that time in France.

Teachers in abbacus schools had to be familiar with the coss; this is proved by a manuscript from the year 1646 in which the training and the necessary knowledge of a future *Rechenmeister* was described (Folkerts and Reich *1989*: 213–15). But in the sixteenth century the textbooks used by the *Rechenmeister* often did not include algebra (Coss). Adam Ries, for example, the most famous of the German *Rechenmeisters*, wrote many textbooks, but his *Coss* was not one of them (Ries *1992*).

The tradition of the *Rechenmeister* survived until the beginning of the eighteenth century. But the time of Coss was over when François Viète published his short but highly influential book *In artem analyticem isagoge* (1591). He represented unknowns by upper-case vowels and knowns by consonants, and raised the status of algebra by associating it with analysis in the traditional sense of Greek philosophy, to which he referred (§1.3); the title of his book is translated as 'Introduction to the Analytic Art'. This connection was to endure for centuries as algebra grew in importance (§6.9).

BIBLIOGRAPHY

Bartolozzi, M. and Franci, R. *1990*, 'La teoria delle proporzioni nella matematica dell'abaco da Leonardo Pisano a Luca Pacioli', *Bolletino di Storia delle Scienze Matematiche*, **10**, 3–28.

Flegg, G., Hay, C. and Moss, B. (eds and transls) *1985*, *Nicholas Chuquet, Renaissance Mathematician*, Dordrecht: Reidel.

Folkerts, M. and Reich, K. *1989*, 'Rechenmeister', in *Mass, Zahl und Gewicht, Mathematik als Schlüssel zu Weltverständnis und Weltbeherrschung*, exhibition catalogue, Herzog August Bibliothek (Wolfenbüttel), Weinheim: VCH, Acta humaniora, 188–215.

Franci, R. and Toti Rigatelli, L. *1985*, 'Towards a history of algebra from Leonardo of Pisa to Luca Pacioli', *Janus*, **72**, 17–82.

—— *1988*, 'Fourteenth-century Italian algebra', in Hay *1988*: 11–29.

—— *1989*, 'La matematica nella tradizione dell'abaco nel XIV e XV secolo', in *Storia delle scienze in Italia*, Bramante: Busto Arsizion, 68–94.

Glaisher, J. W. L. *1921*, 'On the early history of the signs + and −, and on the early German arithmeticians', *Messenger of Mathematics*, **51**, 1–144.

Hay, C. (ed.) *1988*, *Mathematics from Manuscript to Print: 1300–1600*, Oxford: Clarendon Press.

Kline, J. *1968*, *Greek Mathematical Thought and the Origin of Algebra*, Cambridge, MA: MIT Press. [Includes translation of Viète's *Isagoge*.]

Ries, A. *1992*, *Adam Ries. Coss*, Stuttgart: Teubner. [Facsimile edn, with commentary by W. Kaunzner and H. Wussing.]

Tropfke, J. *1980*, *Geschichte der Elementarmathematik*, Vol. 1, *Arithmetik und Algebra*, 4th edn (ed. K. Vogel *et al.*), Berlin and New York: de Gruyter.

Van Egmond, W. *1980*, *Practical Mathematics in the Italian Renaissance: A Catalog of Italian Abbacus Manuscripts and Printed Books to 1600*, Florence: Istituto e Museo di Storia della Scienza.

—— *1983*, 'How algebra came to France', in Hay *1988*: 127–44.

2.4

Abbacus arithmetic

W. VAN EGMOND

1 THE ABBACUS TRADITION

Abbacus arithmetic constituted a distinct tradition of mathematics in the later Middle Ages and the Renaissance, one that played an important role in shaping the form and character of Western arithmetic and algebra as it emerged during the sixteenth and seventeenth centuries. It is represented in the historical record by a distinct genre of mathematical books, known in Italian as the *libri d'abbaco*.

Contrary to the impression given by the name, abbacus arithmetic had nothing to do with the operation of the abacus, the classical reckoning device in which numbers are represented by counters or pebbles on a ruled board, and calculations are performed by moving the counters from line to line (§5.11); nor with the more familiar Chinese abacus in which the counters are strung on wires. There are two b's in 'abbacus' to distinguish the abbacus tradition from that of the abacus device, a convention which derives from modern Italian. The abacus was first used by the Greeks and Romans and remained in use in Northern Europe until the sixteenth century, but was largely abandoned in Italy during the later Middle Ages (Pullan *1969*). Italian mathematicians instead used the Hindu–Arabic numeral forms and methods of calculation that are in common use today. Indeed, they are largely responsible for popularizing the use of these numerals and creating these methods.

The Hindu–Arabic numeral system was first introduced into Europe in the twelfth century via a number of Latin translations of an Arabic work on the Hindu numeral system that had been written by Muḥammad ibn Mūsā al-Khwārizmī in the ninth century AD, of which the best known are the *Carmen de algorismo* by Alexander de Villa Dei (Halliwell *1839–41*: 73–83) and the *Algorismus vulgaris* by Johannes de Sacrobosco (Halliwell *1839–41*: 1–26; Grant *1974*: 9–101). These works, which were commonly called 'algorisms' from a corruption of al-Khwārizmī's name, are however easily distinguished from the abbacus books. Although both works use the ten Hindu–Arabic numerals (including zero) in place-value notation, the algorisms teach the Arabic 'dust-board' method of calculation which requires the repeated erasure and shifting of numerals as the calculation is performed (§1.6). They are also comparatively short treatises, confined to outlining the basic methods with few or no practical examples. The abbacus books, in contrast, use more modern methods of calculation in which all the digits are retained and laid out in a distinctive pattern, and also provide hundreds of practical examples and problems which illustrate these methods. Indeed, it is the large collection of problems that is the most distinctive feature of the abbacus books. Some even dispense with instruction in the numeral forms and methods of calculation altogether. Thus, it is the use of modern methods of calculation combined with a large collection of practical problems that makes abbacus arithmetic a distinctive form of mathematics.

The immediate origin of this tradition lies unquestionably in the Latin *Liber abbaci* ('Book of the Abbacus') written in 1202 by Leonardo of Pisa – more familiar to modern readers by his nickname, Fibonacci (Boncompagni *1857*). In the preface to this book Leonardo tells us that his father was the head of the Pisan customs house in Bugea, a major Arabic trading port on the coast of North Africa, in modern Algeria. There, as a boy, he acquired a knowledge of the Arabic methods from Arab teachers. In order to bring this knowledge to readers of Latin, he compiled a massive encyclopedia of all the basic types of problem and methods of problem-solving then current in the Arab world, including a long section on the method of algebra. The origins of Leonardo's work thus lie unquestionably in the medieval Arab world, but the precise sources of his work are in most cases still unknown and remain a subject of current research.

2 THE COMMERCIAL REVOLUTION

Leonardo's work became widely known in Italy during the thirteenth century, and was a major factor in turning the Italian merchants towards the use of the Hindu–Arabic system. It was during this period that Italy was blossoming into the economic centre of Europe, serving as an entrepôt between the countries of Western Europe and those of the Near East. The agricultural economy of feudal Europe, based on manor farms and local fairs, was gradually giving way to a commercial economy based on money and international trade. The Italian merchants had taken the lead in this development, using their strategic position in the central Mediterranean to trade the raw materials of France and Western Europe for the precious silks and spices of the Near East. But unlike earlier merchants, who had travelled with their goods one load at a time and operated primarily on a barter basis, the new merchants established international companies with permanent representatives in all the major trading cities of Europe and the East. A company representative might, for example, buy a raw material such as wool in Britain and France, ship it to Florence where other company officials farmed it out to workshops that manufactured it into high-quality cloth, and then shipped the finished goods on the Levant, where yet another agent would trade them for silks and spices to be sent back to Europe to pay for more raw materials. The merchants themselves became sedentary businessmen, operating their far-flung empires through letters and written orders, and enjoying the fine life that comes from great wealth.

In order to manage this continuing flow of money and goods, the Italian merchants developed many important new business tools, among the most important of which were continuous accounting with double-entry bookkeeping. Modern banking and insurance companies also had their origins at this time, inventing bills of exchange, letters of credit and the principles of underwriting to help smooth the flow of goods and money. Italian merchants thus learned how to settle accounts, balance books, compute interest on loans, distribute the profits of a partnership, and all the other necessary techniques of modern banking and business. In all these developments, a good working knowledge of mathematics was essential.

Since Renaissance children were expected to follow their father's trade, the new merchants of the Italian cities had a strong interest in seeing to it that their children had a thorough knowledge of the mathematical techniques essential to the functioning of their new businesses. This led to the development of a series of special schools which taught basic mathematics to children between the ages of 8 and 10. Such schools became a regular part of the standard educational programme in Renaissance Italy. Generally known as the *scuole* or *botteghe* (shops) *d'abbaco*, they were run by a professional class of teachers known as the *maestri d'abbaco*, or 'abbacists' (Goldthwaite *1972*; Van Egmond *1976*: Chap. 3). It was these teachers and schools that were largely responsible for the composition of the many surviving abbacus books. Teachers used them to record problems and exercises that might be useful in their classes, and students collected problems that interested them, while wealthier parents commissioned some of the books to help their children's education. The abbacus books preserve the mathematical skills that were essential to the operation of this new commercial economy.

Abbacus arithmetic is thus no more than the basic elementary mathematics of calculating and problem-solving that is necessary for the operation of any commercial society, utilizing the Hindu–Arabic numerals and adapted to the special circumstances of Renaissance Italy.

3 METHODS OF CALCULATION

While the basic numeral forms came from the Arabs, the actual methods of calculation, particulary the methods of multiplication and division used in the abbacus books, did not. As noted earlier, the Arabic treatises of reckoning and their Latin translations used a method of calculation that required the repeated erasure and rewriting of the digits as the calculation proceeded, something that was easy to do on the dust-board (a board covered with a fine layer of sand that was commonly used for calculations in the East), but impossible to perform with pen and paper, which is precisely why it is impossible to print an example here. All that can be done is to show how the operation appears at various stages of the calculation (Figure 1).

2326	428326	496486	497764
214	214	214	214

| (a) | (b) | (c) | (d) |

Figure 1 Successive stages in the dust-board or algorism method of multiplication: (a) initial arrangement, (b) after the first digit has been multiplied, (c) before the last digit is multiplied and (d) after the multiplication has been completed

Instead of using this method, the abbacists developed other forms of calculation that were better suited to execution with pen and ink, the medium most favoured among the new merchant classes. Leonardo himself used a compact method called multiplication *in croce*, 'in cross', so called because the digits of each number were multiplied in pairs according to a specific pattern which included a cross. Intermediate calculations were carried mentally or 'on the fingers', and only the digits of the final answer were written down. For example, in Figure 2(a), to multiply 37 by 49 one first multiplies the last two digits, $7 \times 9 = 63$, writes the 3 as the last digit of the product, and carries the 6. Next, one multiplies all four digits crosswise (hence the name), $(3 \times 9) + (4 \times 7) = 55$, adds the carry to obtain 61, writes the 1 and carries the 6. Finally, one multiplies the first two digits, 3×4, and adds 6 to obtain 18, which is written down to complete the product, 1813. The drawback of this system is that every combination of multipliers and multiplicands requires its own pattern. Figure 2(b) shows the steps involved in multiplying two 3-digit numbers; Figure 2(c) shows the pattern used in multiplying two 4-digit numbers.

In order to make it easier to remember the patterns, some later abbacists wrote down the intermediate calculations in a particular order so as to form

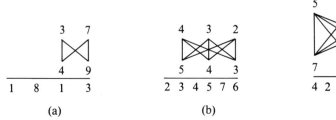

Figure 2 Multiplication *in croce*: cross-multiplication patterns for (a) two 2-digit numbers, (b) two 3-digit numbers and (c) two 4-digit numbers. The stages for (b) are:

$2 \times 3 = 6$; write 6, no carry
$(2 \times 4) + (3 \times 3) = 17$; write 7, carry 1
$1 + (2 \times 5) + (3 \times 4) + (4 \times 3) = 35$; write 5, carry 3
$3 + (3 \times 5) + (4 \times 4) = 34$; write 4, carry 3
$3 + (4 \times 5) = 23$; write 23

distinctive figures like those shown in Figure 3. In Figure 3(a), one starts with the cross-products of the first and last digits (5×5 and 7×7) to give the first two rows, 25 and 49, then proceeds to multiply the last two digits of the top number by the first two digits of the bottom number, 2×7 and 7×2, to make the third row, and vice versa, 3×5 and 5×8, to give the fourth. The figure is completed by taking the last three digits of each with the first three digits of the other, and finally each digit with the one above or below it. This procedure was called multiplication *per canpana*, 'in [the shape of] a bell'. In Figure 3(b), 432 and 543 are multiplied by first multiplying the digits directly above and below each other to give 20, 12 and 6 in the first row, then multiplying the last two digits of the top number, 3×5 and 2×4, to make the second row, and the last digit of the top with the first digit of the bottom, 2×5, to give the third row. Taking the last digit of the bottom number in the fourth row and its last two digits in the fifth produces the final fgure. This procedure was called *per coppa*, 'in a cup'. Figures 3(c) and 3(d) show how different orders could produce patterns that were outlined with a diamond and a circle, respectively.

The origins of our modern method are seen in a method the Renaissance abbacists called multiplication 'in square', in which the multiplier and multiplicand are written at the top and right sides of a square (or rectangle, as need be) which is subdivided into cells. Each digit of the multiplier is then multiplied by the multiplicand and the product written in the cells of its row, as in Figure 4(a). The final product is found by adding the numbers in the cells diagonally to give the product along the left and bottom sides. Another way to do this, which avoids the need to carry any numbers, is to

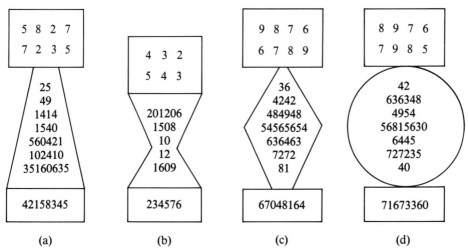

5 8 2 7		9 8 7 6	8 9 7 6
7 2 3 5	4 3 2	6 7 8 9	7 9 8 5
	5 4 3		
25		36	42
49		4242	636348
1414	201206	484948	4954
1540	1508	54565654	56815630
560421	10	636463	6445
102410	12	7272	727235
35160635	1609	81	40
42158345	234576	67048164	71673360
(a)	(b)	(c)	(d)

Figure 3 Multiplication (a) *per campana*, (b) *per coppa*, (c) in the pattern of a diamond and (d) in the pattern of a circle

subdivide the cells with diagonal lines and write the product of each pair of numbers in the corresponding cell, as in Figure 4(b). A natural evolution of the first method led to the modern method of partial products, as the lines of the square were abandoned and the rows were shifted to maintain the proper place position. There was a similar initial diversity and long process of development in the operation of division that led eventually to the current method of long division.

The many different methods of calculation found in the abbacus arithmetics help to show that these methods were not taken from outside sources but were developed independently by the abbacists themselves, with much trial and error. After a long period of experimentation, by the sixteenth century the abbacists seem to have settled on the methods of multiplication

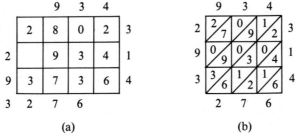

Figure 4 Multiplication 'in square': (a) the basic form and (b) subdivision of cells to avoid the need to carry

and division that are used exclusively in Western countries today. These methods, then, are not a contribution of the Hindus or Arabs, as is commonly assumed, but a unique and significant contribution of Italian mathematicians to the development of modern mathematics.

4 PROBLEM TYPES AND METHODS OF SOLUTION

In addition to teaching new methods of multiplication and division, the other significant feature of the abbacus books is the large number of problems they contain. These can be divided into three general classes. The first and largest class consists of business problems which teach such basic but essential tasks as calculating the price of an item, fair exchange in barter, interest and discount on a loan, and dividing the profits of a partnership. These are obviously practical problems that would be of use in the profession for which the majority of students were being trained. But there is a second and equally large class of problems which might be called 'recreational' or 'fictitious', whose only possible function could have been to provide the enjoyment of solving a challenging puzzle, a motivation seen down through the ages and still present in popular books of 'mathematical recreations'. In this class, we find such problems as the hare and the hound, the travellers, the cistern and the duplication of the chessboard. Finally, one often finds in the abbacus books a sizeable number of practical geometric problems, involving tasks like computing areas and volumes and finding the height of a tower or the breadth of a river.

There is nothing particularly innovative about these problems. They are the same types found in mathematical documents that survive from the earliest civilizations of Egypt and Mesopotamia, and are still found in elementary arithmetic books today. Even the methods of solution used in the treatises are quite traditional. The most common method used is the 'rule of three' or 'golden rule', in which three known terms are used to find a fourth from the proportion $a:b::c:d$. This is quite satisfactory for simple pricing and exchange problems. Another popular method is the 'rule of false position', in which an initial guess is made at the value of the solution and the error that it produces is used in a proportion to find the correct value. A somewhat more complicated procedure is the 'rule of double false position', in which two guesses are made at the solution and the ratio between their false answers is used to find the correct solution. All these methods were well known to the Arabs, and can in some cases be traced back to ancient Egyptian and Babylonian mathematical texts.

5 ALGEBRA

One of the newer and more interesting methods the abbacus books used to solve problems is algebra. It too comes from the Arabs (§1.6) – the fundamental treatise was written by the same al-Khwārizmī who wrote the first treatise on the Hindu–Arabic numerals (Rosen *1831*) – and it too was translated into Latin during the twelfth century by Gerard of Cremona (Libri *1838*: 253–97) and Robert of Chester (Karpinski *1915*), and was included by Leonardo of Pisa in his *Liber abbaci*. None of these works, however, bears much similarity to the algebra familiar to students today as elementary 'school algebra' or the theory of equations. (There is of course no similarity at all to what is now called 'modern algebra'.) For one thing, al-Khwārizmī's algebra, and all the books based on it, was 'rhetorical' rather than symbolic, meaning that all equations and operations were written out in words and complete sentences. Even the unknowns and their powers were given words, for example *cosa* and *censo* for x and x^2. The range of algebra was also restricted to the solution of the five simple linear and quadratic equations, each of which was treated as a special case with a rule of solution unique to it.

Despite its awkwardness and restricted use, the abbacists found in this simple algebra a powerful tool for the solution of many problems that were of special interest to them, particularly the calculation of compound interest, which leads naturally to quadratic and higher-order equations. As a consequence, a treatment of algebra is a regular feature of abbacus books from the earliest times to the sixteenth century, when Niccolò Tartaglia (himself an abbacus teacher) and Girolamo Cardano discovered the general solution to third- and fourth-degree equations and initiated the series of changes that led to the development of modern algebra (§6.1). Thus, during the 400-year period from 1150 to 1550, it was the abbacists who preserved and fostered the study of algebra in Western Europe.

A few abbacists even made some notable contributions to the field. For example, all the abbacists were challenged by the goal of extending al-Khwārizmī's rules for solving quadratic equations to higher degrees. Many fruitless attempts were made to produce such rules, but one abbacist, known to us only by the name of Maestro Dardi, succeeded in finding the correct solution to several special cases in a work written in 1344 (Van Egmond *1983*). Another fourteenth-century mathematician, Antonio de' Mazzinghi of Florence, was the first to use two unknowns in the solution of a problem (Arrighi *1967*: 28). And throughout the period there was a continuing movement towards the development of algebraic symbolism, replacing the wordy rhetorical expressions of al-Khwārizmī with more concise abbreviations. But it was not until the sixteenth century that algebraists abandoned abbreviations for arbitrary and abstract symbols. Even such basic symbols as + and − did not appear until the late fifteenth century, and did not come into general use until the seventeenth century.

6 THE SPREAD OF ABBACUS ARITHMETIC

Since Italy was a leading country in Europe during the Renaissance, the influence of the abbacus system gradually spread to the other countries of Europe. In France a number of teachers and writers began to translate and write abbacus books, most notably Nicolas Chuquet and his student Estienne de la Roche (Marre *1880–81*). In Germany the teachers took the name of *Rechenmeister*, and set up schools in the centres of German trade (§2.3). In this way, the methods of arithmetic and algebra conveyed from the Arabs by Leonardo of Pisa and modified to meet the needs of Italian merchants spread to become the dominant system of doing mathematics throughout Western Europe. What we consider today to be elementary arithmetic and algebra is in fact the system of reckoning and problem-solving that was developed and adapted by the abbacus teachers of Renaissance Italy during the fourteenth and fifteenth centuries.

BIBLIOGRAPHY

Arrighi, G. (ed.) *1967, Antonio de' Mazzinghi, trattato di fioretti*, Pisa: Domus Galilaeana.

Bautier, R.-H. *1971, The Economic Development of Medieval Europe*, London: Thames & Hudson.

Boncompagni, B. (ed.) *1857, Scritti di Leonardo Pisano*, Vol. I, Rome: the author.

Cajori, F. *1928, A History of Mathematical Notations*, Vol. 1, La Salle, IL: Open Court. [Wide range of notations recorded in Chap. 3.]

Flegg, G., Hay, C. and Moss, B. (eds) *1985, Nicolas Chuquet, Renaissance Mathematician*, Dordrecht: Reidel.

Franci, R. and Toti Rigatelli, L. *1985*, 'Towards a history of algebra from Leonardo of Pisa to Luca Pacioli', *Janus*, **72**, 17–82.

Goldthwaite, R. *1972*, 'Schools and teachers of commercial arithmetic in Renaissance Florence', *Journal of European Economic History*, **1**, 418–33.

Grant, E. (ed.) *1974, A Source Book in Medieval Science*, Cambridge, MA: Harvard University Press.

Halliwell, J. O. (ed.) *1839–41, Rara arithmetica*, 2 vols, London: Parker. [Repr. 1977, Hildesheim: Olms.]

Karpinski, L. C. (ed.) *1915, Robert of Chester's Latin Translation of the Algebra of Al-Khowārizmī*, New York: Macmillan. [Repr. as *Contributions to the History of Science*, 1930, Ann Arbor, MI: University of Michigan Humanistic Series, Vol. XI.]

Libri, G. *1838, Histoire des sciences mathématiques en Italie*, vol. 1, Paris: Renouard.

Lopez, R. S. *1971, The Commercial Revolution of the Middle Ages, 950–1350*, Englewood Cliffs, NJ: Prentice-Hall.

Marre, A. (ed.) *1880–81*, 'Nicolas Chuquet, "Triparty en la science des nombres"', *Bulletino di Bibliografia e di Storia delle Scienze Matematiche e Fisiche*, **13**, 555–659, 693–814; **14**, 417–60.

Miskimin, H. A. *1969, The Economy of Early Renaissance Europe, 1300–1460*, Englewood Cliffs, NJ: Prentice-Hall.

Pullan, J. M. *1969, The History of the Abacus*, New York: Praeger.

Rosen, F. *1831, The Algebra of Mohammed ben Musa*, London: Oriental Translation Fund.

Tropfke, J. *1980, Geschichte der Elementarmathematik*, 4th edn, Berlin: de Gruyter.

van der Waerden, B. L. *1985, A History of Algebra from al-Khwarizmi to Emmy Noether*, Berlin: Springer.

Van Egmond, W. *1976*, 'The commercial revolution and the beginnings of Western mathematics in Renaissance Florence, 1300–1600', PhD Dissertation, Indiana University.

—— *1980, Practical Mathematics in the Italian Renaissance: A Catalog of Italian Abbacus Manuscripts and Printed Books to 1600*, Florence: Istituto e Museo di Storia della Scienza.

—— *1983*, 'The algebra of Master Dardi of Pisa', *Historia mathematica*, **10**, 399–421.

2.5

Logarithms

WOLFGANG KAUNZNER

1 PREHISTORY OF THE LOGARITHM

From the time of Regiomontanus in the second half of the fifteenth century, trigonometric tables with finely interpolated values of arguments have been constructed. These were effected by a method, developed by Johannes Werner between 1505 and 1513, of reducing the multiplication and division of large numbers to addition and subtraction (Tropfke *1980*: 297). This method was called *prosthaphaeresis*, a word meaning 'setting down and taking away', which can be rendered as 'the method of addition and subtraction' (Cantor *1900*: 454). It proved especially valuable in astronomy where, for example, Tycho Brahe, shortly before mathematicians and reckoners worked with logarithms, applied the formulas

$$\sin a \sin b = \tfrac{1}{2}(\cos(a - b) - \cos(a + b)) \tag{1}$$

and

$$\cos a \cos b = \tfrac{1}{2}(\cos(a - b) + \cos(a + b)). \tag{2}$$

There was a long period of development in the theories of proportions and series, as well as in calculations with powers, which prepared the way for the construction which characterizes logarithmic calculation. The regularities of correspondence between the terms of arithmetic and geometric sequences had been known at least since the time of Archimedes. Was it suspected that there was not only one inverse involved in calculations with powers? In any case, the search for the calculation rules governing these regularities, and how to represent them, lasted until the Renaissance. The seventeenth century saw the formulation of several logarithmic principles and the calculation of tables of values based on them which enabled reckoning by logarithms, by various means, to find a place as a 'quick calculating technique' in the calculating stock-in-trade of the time, despite it having no definitive formal notation. Leonhard Euler offered this definition: '$a^z = y$; iste autem valor ipsius z, quatenus tanquam functio

210

ipsius *y* spectatur, vocari solet Logarithmus ipsius *y*' (Euler *1770*: 106); that is, the logarithm is the exponent of a power. With this the logarithm became firmly established in modern analysis.

The development of logarithms can be traced back in different ways: calculations with powers and roots, exponential problems, the theory of proportion, and the theory of arithmetic and geometric series.

In the fifteenth century the natural numbers, the positive rational and irrational numbers, and zero were available for use. The transition from Roman to Hindu–Arabic numerals was completed in Western Europe by about 1480 (§2.3–2.4). The symbols $+$, $-$, / for division and the root sign were applied in contemporary arithmetic and algebra. The exponent did not yet have a representation. The idea that calculations should follow basic, established methods had yet to emerge. A calculation was verified by performing the inverse operation; this fostered the advance into the domain of negative numbers and strengthened the suspicion that extracting the root was not the only analytical operation corresponding to the arithmetical operation of raising to a power.

2 MATHEMATICAL MANUSCRIPTS

Calculations with powers or roots were used, for example, in fourteenth-century formulas for compound interest

$$K_n = K_0(1 + Z/K_0)^n \qquad (3)$$

connecting K_n, K_0 and Z (Tropfke *1980*: 542–3). Calculations with powers were required for the intensive work being done in algebra from the second half of the fifteenth century; they eventually took their place as a discipline in their own right. The symbols for powers of the algebraic unknown, i.e. for our present-day x^0, x, x^2, x^3, and so on, were formed from the initial letters of the corresponding technical terms 'dragma' or 'numerus', 'res' or 'radix', 'census', 'cubus', and so on. Numerical values $(1, 2, 4, 8, 16)$ were used for greater clarity and ease of understanding since the concepts of base and exponent were not yet known.

Exponential problems with 'fractional' exponents can be found, among other places, in the appendix to *Le Triparty en la science des nombres* (1484) by Nicolas Chuquet, where we find the following (Marre *1881*: 438*f*): how does a capital of 10 units of money grow at a rate of 100% for $3\frac{3}{4}$ years? In other examples the 'exponent' is sought: in $(\frac{9}{10})^x = \frac{1}{2}$ it is discovered that $6 < x < 7$ by trial and error; the method for correctly determining intermediate values had not yet appeared (Marre *1881*: 439).

The Pythagorean theory of line segments, through the theory of proportions, had a further fundamental influence on logarithms. The expression 'to add (or subtract) two relations' on a monochord meant their multiplication (or division). Thus the major and minor thirds are 'added' to form a pure fifth, but $\frac{5}{4} \cdot \frac{6}{5} = \frac{3}{2}$. Likewise $(m:n)^3$ is the 'threefold relation' of $m:n$. In the fourteenth century Thomas Bradwardine and, in particular, Nicole Oresme added significantly to the theory of proportions, thereby benefiting the theory of fractional powers by the special notation they developed. These developments were propagated further by their use in algorithmic calculations. From around 1500, in the manuscript *Algorismus de proportionibus*, i.e. instructions in methods of calculating proportions (Dresden Codex C 80^m, fol. 41^r), we find:

> To add is to multiply. To subtract is to divide. To double or to multiply is to take in itself. To determine the mean or to divide is to extract the root.

Lining up corresponding terms of the arithmetic and geometric series of the natural numbers, including zero, one under the other, led in 1480 to the derivation of the characteristic notational rules (Dresden Codex C 80, fol. 288^v):

> It is a universal truth that, whenever some sign is taken into another, by as much distance that the first is in order from the first sign, that is to say from ϕ [x^0], so is the distance of the second in order from the other sign towards the right, from this comes the length.

In practice points were placed above the constant ϕ [x^0], and above x^m and x^n, in order to determine x^{m+n} from the distances between these points. This was conceptually more advanced than Archimedes who, in the *Sandreckoner*, had the two corresponding series begin with 1 (Archimedes *1913*: 241); in modern form:

$$1 \quad 2 \quad 3 \quad 4 \ldots n$$
$$1 \quad a \quad a^2 \quad a^3 \ldots a^{n-1}$$

In the *Triparty*, Chuquet wrote $5^{2\overline{m}}$, $5^{1\overline{m}}$, 5^0, 5^1, 5^2, ... for $5x^{-2}$, $5x^{-1}$, 5, $5x$, $5x^2$, ..., and $R^2 3^1$, $R^3 3^2$, $R^4 3^3$, ... for $\sqrt{3x}$, $\sqrt[3]{3x^2}$, $\sqrt[4]{3x^3}$, ... (Marre *1880*: 737–42). He placed the terms of arithmetical and geometrical series in correspondence thus:

Nombres	1	2	4	8	16	...	1 048 576
Denominacion	0	1	2	3	4	...	20

In the Dresden Codex C 80^m, there is a 'table of division' (Figure 1) next to a 'table of multiplication'.

Figure 1 'Table of division' from the Dresden Codex C 80^m, fol. 35^v.
The first row gives the dividends in the forms x^0, x, x^2, x^3 and x^4;
the first column gives the divisors in the same forms.
The boxes corresponding to quotients such as x^0/x,
which would give negative powers of x, are left empty

3 MATHEMATICAL PRINTED BOOKS

These techniques made their way into printed books. Heinrich Schreyber in
Ayn new kunstlich Buech (1521, fol. G 1^{r}f.) wrote:

When now such a number is to be written after another according to a proportion, then write each such quantity with the number of its order, so that in
the case of double proportion the number 1 is placed over 2, 2 over 4, ...

That is,

$$N \quad 1 \quad 2 \quad 3 \quad 4 \quad 5 \quad 6 \quad 7 \quad ... \quad 16$$
$$1 \quad 2 \quad 4 \quad 8 \quad 16 \quad 32 \quad 64 \quad 128 \quad ... \quad 65\,536$$

Schreyber set out four calculating principles in the terminology of his time
(fol. G 1^v to G 3^r):

1 To extract the root from a 'quadratic number':

$$8 \qquad 8:2 = 4$$
$$256 \qquad \sqrt{256} = 16.$$

2 To construct a 'quadrangular number' from a 'proportioned' number:

$$7 \qquad 3 + 4$$
$$128 \qquad 8 \times 16.$$

213

3 To extract a cube root:

$$6 \qquad 6:3=2$$
$$64 \qquad \sqrt[3]{64}=4.$$

4 To decompose a 'corporeal' number:

$$10 \qquad 2+3+5$$
$$1\,024 \qquad 4\times 8\times 32.$$

We can see the beginnings of the transition to the use of negative superscripts in a division table of Schreyber for powers of unknowns, shown in Figure 2.

Die quantitet welch man taylt

	1a	2a	3a	4a	5a	6a	7a	8a
1a	N	1a	2a	3a	4a	5a	6a	7a
2a	$\frac{1a}{2a}$	N	1a	2a	3a	4a	5a	6a
3a	$\frac{1a}{3a}$	$\frac{2a}{3a}$	N	1a	2a	3a	4a	5a
4a	$\frac{1a}{4a}$	$\frac{2a}{4a}$	$\frac{3a}{4a}$	N	1a	2a	3a	4a
5a	$\frac{1a}{5a}$	$\frac{2a}{5a}$	$\frac{3a}{5a}$	$\frac{4a}{5a}$	N	1a	2a	3a
6a	$\frac{1a}{6a}$	$\frac{2a}{6a}$	$\frac{3a}{6a}$	$\frac{4a}{6a}$	$\frac{5a}{6a}$	N	1a	2a
7a	$\frac{1a}{7a}$	$\frac{2a}{7a}$	$\frac{3a}{7a}$	$\frac{4a}{7a}$	$\frac{5a}{7a}$	$\frac{6a}{7a}$	N	1a
8a	$\frac{1a}{8a}$	$\frac{2a}{8a}$	$\frac{3a}{8a}$	$\frac{4a}{8a}$	$\frac{5a}{8a}$	$\frac{6a}{8a}$	$\frac{7a}{8a}$	N

Die quantitet welche taylt.

Was auß solcher taylung kumbt.

Figure 2 Division table from Schreyber's *Ayn new kunstlich Buech* (1521), fol. G VIII^{v}. The first row gives dividends corresponding to $x, x^2,\ldots, x^8$; the first column gives the same as divisors. For example, x^2/x^5 is represented as 2a/5a

It is an open question just when non-integral steps were introduced into an arithmetic series or when the terms of a geometric series were produced in a non-integral fashion. In Christoff Rudolff's *Behend vnnd Hubsch Rechnung durch die kunstreichen regeln Algebre* (1525, fol. D III^v) we find, in modern notation:

$$x^0 \quad x \quad x^2 \quad x^3 \quad \ldots \quad x^9$$
$$1 \quad \tfrac{2}{3} \quad \tfrac{4}{9} \quad \tfrac{8}{27} \quad \ldots \quad \tfrac{512}{19\,683}$$

Michael Stifel, in his *Arithmetica integra* (1544, fol. 249^v), completed the shift to negative terms which had only been hinted at before:

-3	-2	-1	0	1	2	3	4	5	6
$\tfrac{1}{8}$	$\tfrac{1}{4}$	$\tfrac{1}{2}$	1	2	4	8	16	32	64

The upper numbers are the 'exponents' of the lower. On fol. 35^{r}f. he presents more clearly than had his predecessors the rules governing logarithmic calculations:

1 Addition in arithmetic progressions corresponds to multiplication in geometric progressions. As, for example, in this arithmetic progression, 3, 7, 11, 15, the two extreme terms added make the quantity of the middle terms added together, both are indeed valued 18. Thus in this geometric progression, 3, 6, 12, 24, the two extreme terms multiplied together make the quantity of the middle terms multiplied together, both are indeed valued 72, and so forth for infinitely many examples.

2 Subtraction in arithmetic progressions corresponds to division in geometric progressions...

3 Simple multiplication, (i.e. numbers into numbers) as is done in arithmetic progressions corresponds to self-multiplication as done in geometric progressions...

4 Division in arithmetic progressions corresponds to extraction of roots in geometric progressions...

In this way the ground was prepared for the breakthrough to logarithms. The word 'logarithm' was used by Stifel's student Caspar Peucer in his *Commentarivs de praecipvis generibvs divinationvm* (1572, fol. 227^r–228^r), but in a different sense from that of today (Vogel *1978*: 59, 65); the concept of the logarithm comes from John Napier. Peucer and Napier had in common only their method of looking at two series of elements that were ordered in a one-to-one fashion. Napier chose, without explanation, the word logarithm as 'number of the ratio', that is λόγον ἀριθμός (Vogel *1978*: 59).

Further progress required a new notation and a means of incorporating arbitrarily many intermediate values in the number series. In place of decimal fractions, which had not come into use in the second half of the sixteenth century, fractions (or large numbers implicitly referred to correspondingly large denominators) were used. Analogously, trigonometric tables used a large radius, usually 6×10^n or, later, 10^n.

4 THE FIRST TABLES

The first logarithmic tables originated with Jost Bürgi and John Napier. Both based their tables on the short but thorough directions that were formulated by Stifel, who juxtaposed corresponding members of arithmetic and geometric series. The tables were produced in response to the need of the practical mathematicians and astronomers of the time to reduce the laborious procedures met with in higher calculations, principally those involving functions of angles. Consequently it was primarily sine and tangent tables that were provided, with logarithms in supplementary columns, and the whole subject was treated in a 'Trigonometria' chapter rather than as part of the fundamentals of calculation. In his textbook on analysis, Leonhard Euler was the first to give logarithms their appropriate place in the structure of arithmetic (Euler *1748*: 75–82).

5 JOST BÜRGI

According to the writings of Raimarus Ursus, Johannes Kepler and Benjamin Bramer, it was probably between 1588 and 1611 that Bürgi drew up the tables which first appeared in Prague in 1620 as *Aritmetische vnd Geometrische Progress Tabulen, sambt gründlichem vnterricht, wie solche nützlich in allerley Rechnungen zugebrauchen, vnd verstanden werden sol* (Gieswald *1856*: 13*f*, 21*f*; Voellmy *1948*: 12*f*, 17*f*). The 'gründliche Unterricht', or 'basic lesson', is known only in manuscript (Gieswald *1856*: 26–36); in the relevant passage, where Simon Jacob and Moritius Zons are mentioned, Bürgi writes (Gieswald *1856*: 26):

> Considering the properties and correspondences of the two series, namely the arithmetical and the geometrical: multiplication in the latter is addition in the former, division in the latter is subtraction in the former, extraction of roots in the latter is halving in the former . . . . I found it most useful to extend these tables so that all numbers in question are discovered there, independently of which basic consideration these tables come from. Not only can we circumvent the difficulties inherent in multiplication, division and extraction of roots, however advantageous and useful this is for algebra or coss, but, much more importantly, we can also place as many geometric means between two given numbers as we wish. How difficult it is to calculate without these tables, anyone who has done a little work in this area knows.

Subdivision in the number series, which had not been explicit before, is thus addressed.

Bürgi did not speak of logarithms. He only applied the corresponding rules of calculating with powers and adjoined the 'red numbers' in an arithmetical series with double entry to the 'black numbers' in a geometrical series. From the method of construction this was later called an 'antilogarithm table', since the 'logarithms' (red numbers) are equally spaced whereas the Numeri (the black numbers) are variably spaced.

In Bürgi's numeric–logarithmic progression table (Figure 3), the first two black-number entries in the first column, $100\,000\,000$ (meaning 1) and $[1000]\,10\,000$, give $q = 1.0001$ as the ratio of its geometric series. The ten-thousandth part of each Numerus (black number) is added to the Numerus to give the new Numerus. Thus $100\,501\,227 +$ $10\,050\,|\,1227 = 100\,511\,277\,|\,1227$. In the margin and above are placed the red numbers at equal intervals – in this example, from 500 to 510 – and these are to be the exponents or logarithms.

	0	500	1000
0	100 000 000	100 501 227	101 004 966
10	. . . . 10 000	. . . . 11 277	. . . . 15 067
20	. . . . 20 001	. . . . 21 328	. . . . 25 168
30	. . . . 30 003	. . . . 31 380	. . . . 35 271
40	. . . . 40 006	. . . . 41 433	. . . . 45 374

Figure 3 Part of the logarithmic–numeric progression table produced by Bürgi (1620)

In the table Bürgi proceeds so that the Numerus $N = 10^8 \times q^n$ and the logarithm $l = 10n$. For example, from column 3, row 5:

$$N = 101\,045\,374 = 10^8 \times 1.0001^n,$$

$$n = \log_{1.0001} \frac{101\,045\,374}{10^8} = \log_{1.0001} 1.010\,453\,74$$

$$= \frac{\ln 1.010\,453\,74}{\ln 1.0001} = 104,$$

$$l = 1040.$$

If we required that the red numbers corresponding to the black be 'densely' laid out – that is, that Bürgi placed the series

0	0.0001	0.0002	0.0003	...	$0.0001 \times n$
1	1.0001	1.0001^2	1.0001^3	...	1.0001^n

one above the other (Voellmy *1948*: 17) – then $0.0001 \times n = 1$; $n = 10^4$ and the resulting hypothetical base of the fundamental system would be

$$(1 + 1/10^4)^{10^4} = 2.718\,145\,9, \tag{4}$$

which is close to e (Mautz *1919*: 29, 31). The title page to Bürgi's tables shows this as well, for the 'whole' red number 230270.022 is coordinated with the whole black number 1000000000. Thus

$$N = 1\,000\,000\,000 = 10^8 \times 1.0001^n,$$

$$n = \log_{1.0001} \frac{1\,000\,000\,000}{10^8} = \log_{1.0001} 10$$

$$= \frac{\ln 10}{\ln 1.0001} = 23\,027.002\,21,$$

$$l = 10n = 230\,270.0221 \approx 230\,270.022.$$

6 JOHN NAPIER

John Napier published the first logarithmic work, namely the *Mirifici logarithmorum canonis descriptio* (Napier *1614*; on his life and work, see Knott *1915*). He probably worked on the subject in 1592 or even earlier (Brahe *1924*: 335). Edward Wright translated Napier's work as *A*

Gr.   18 $+ | -$

18								
min	*Sinus.*		*Logarithmi*	*Differentia*	*logarithmi*		*Sinus*	
30	3173047		11478926	10948332	530594		9483237	30
31	3175805		11470237	10938669	531568		9482314	29
32	3178563		11451556	10929013	532543		9481390	28
33	3181321		11452883	10919364	533519		9480465	27
34	3184079		11444219	10909723	534496		9479539	26
35	3186837		11435563	10900090	535473		9478612	25
36	3189594		11426915	10890464	536451		9477685	24
37	3192351		11418275	10880845	537430		9476757	23
38	3195108		11409644	10871234	538410		9475828	22
39	3197864		11401021	10861630	539391		9474898	21
40	3200620		11392406	10852033	540373		9473967	20
41	3203375		11383800	10842444	541356		9473035	19
42	3206130		11375202	10832862	542340		9472103	18
43	3208885		11366612	10823287	543325		9471170	17
44	3211640		11358030	10813719	544311		9470236	16

Figure 4 Part of a page from Napier *1614*

Description of the Admirable Table of Logarithmes (1616). Here too it is a matter of shortening calculations:

> Since there is nothing (most beloved mathematical practitioners) that is so distressing in the practice of mathematics, or that so hinders and retards calculators, than the multiplication, division, and square and cubic extractions of large numbers, which, besides the tedious expense of time, is also subject in large part to many slippery errors. (Napier *1614*: fol. A 3^r)

Napier calculated the logarithms of values of the sine function, not of natural numbers. He presented a trigonometric–logarithmic table in which the angles are given in intervals of a minute of arc. For the 'sinus totus' (sin 90°, or radius) he chose 10^7 and thus his values for sines lay between 0 and 10^7. The seven-place values of sine, up to seven places, are his 'Numeri', whose logarithms he gives to eight places. The table runs from 0° to 45° and, complementarily, from 45° to 90°. The middle columns denote 'Differentiae' (Figure 4):

$$\log\left(\frac{\sin a}{\sin(90° - a)}\right) = \log \sin a - \log \sin(90° - a) = \log \tan a. \qquad (5)$$

A point 'traverses' the line segment $\overline{\alpha\omega} = 1$ (Figure 5) at a speed which is proportional in each time unit to the mth part of the remaining distance:

$$\overline{\alpha\gamma} = \frac{1}{m} \cdot 1 = \frac{1}{m}, \qquad \overline{\gamma\omega} = 1 - \overline{\alpha\gamma} = 1 - \frac{1}{m} = \frac{m-1}{m},$$

$$\overline{\gamma\delta} = \frac{1}{m} \cdot \overline{\gamma\omega} = \frac{1}{m} \cdot \frac{m-1}{m},$$

$$\overline{\delta\omega} = 1 - \overline{\alpha\delta} = 1 - \left(\frac{1}{m} + \frac{1}{m} \cdot \frac{m-1}{m}\right) = \left(\frac{m-1}{m}\right)^2,$$

$$\overline{\delta\varepsilon} = \frac{1}{m} \cdot \overline{\delta\omega} = \frac{1}{m} \cdot \left(\frac{m-1}{m}\right)^2, \qquad \overline{\varepsilon\omega} = 1 - \overline{\alpha\varepsilon} = \left(\frac{m-1}{m}\right)^3 \quad \ldots .$$

This non-uniform decreasing motion is coordinated with a uniform one proceeding from A (Figure 6) at speed d, which is the mean speed in the segment $\overline{\alpha\gamma}$ (Matzka *1860*: 341). Thus after m time units the point A has travelled the course $\overline{AB}_m = m \times d$.

Figure 5 Non-uniform decreasing motion of a point along a line segment

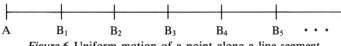

Figure 6 Uniform motion of a point along a line segment

Napier's definition reads: 'Logarithms of proportional numbers or quantities are equally differing' (Napier *1614*: 5), and indeed this is so for the logarithm of the distance the point α has still to traverse (Matzka *1860*: 342). As stated in the *Mirifici logarithmorvm canonis constrvctio* (first published posthumously in 1619): 'Thus a geometrically moving point approaching a fixed one has its velocities proportional to its distances from the fixed one' (Napier *1620*: 13, article 25). This method of coordination suggests that logarithms may be seen as an introduction to the concept of continuity in number theory (Matzka *1860*: 342; Mautz *1919*: 3).

Taking $\overline{\alpha\omega} = 10^7$, the remainders

$$\overline{\alpha\omega}, \quad \overline{\gamma\omega} = \left(\frac{m-1}{m}\right) 10^7, \quad \overline{\delta\omega} = \left(\frac{m-1}{m}\right)^2 10^7, \quad \overline{\varepsilon\omega} = \left(\frac{m-1}{m}\right)^3 10^7, \quad \ldots$$

can be regarded as terms of a sine series. The technique referred to by Napier is expounded in Napier *1620* after a preliminary introduction to the estimation of errors:

1 In a first series, series I, m is put equal to 10^7; with

$$1 - 1/m = 1 - 1/10^7, \qquad q_\mathrm{I} = \frac{9\,999\,999}{10\,000\,000},$$

it follows that

$$a_{n+1} = a_n - \frac{a_n}{10\,000\,000},$$

and

1) $10\,000\,000$
2) $9\,999\,999$
3) $9\,999\,998.000\,000\,1$
⋮
101) $9\,999\,900.000\,495\,0$.

2 In a series II, starting again with 10^7, the second term is taken as the last term of series I rounded off; with

$$q_\mathrm{II} = \frac{9\,999\,900}{10\,000\,000} = 1 - \frac{1}{10^5}$$

220

the terms are:

1) 10 000 000
2) 9 999 900
3) 9 999 800.001
$\vdots$
51) 9 995 001.224 804.

Napier miscalculated the last term as 9 995 001.222 927; consequently the last places of subsequent terms are inexact.

3 Series III begins with $m = 10^7$ and has as its second term the last term of the previous series rounded off; this gives, with

$$q_{III} = \frac{9\,950\,000}{10\,000\,000} = 1 - \frac{1}{2000},$$

1) 10 000 000
2) 9 995 000
$\vdots$
21) 9 900 473.578 08.

4 Correspondingly, in series IV with

$$q_{IV} = \frac{9\,900\,000}{10\,000\,000} = 1 - \frac{1}{100},$$

the range up to 4 998 609.403 4 is covered in 69 columns.

This attains approximately half the starting value of 10^7. All desired intermediate numerical values could be represented in series I in terms of the initial term 10^7 and q_I as

$$10^7 \times \left(1 - \frac{1}{10^7}\right)^n = 10^7 \times \left(\frac{9\,999\,999}{10\,000\,000}\right)^n.$$

In principle, Napier arranged the series as shown in Figure 7. This finely

Geometric series	Arithmetic series
10 000 000	0
9 999 999	1.000 000 05
9 999 998	2.000 000 10
9 999 997	3.000 000 15
$\vdots$	$\vdots$
9 999 900	100.000 005
$\vdots$	$\vdots$
5 000 000	6 931 469.22

Figure 7 Napier's arrangement of series

221

meshed grid of $69 \times 20 \times 50 \times 100 + 1 = 6\,900\,001$ interpolations makes it easy to cover the range with greater accuracy. Divided into minutes, for a range between $\sin 90° = 10\,000\,000$ and $\sin 30° = 5\,000\,000$, there are $60 \times 60 + 1 = 3601$ entries and thus the grid is $3601/6\,900\,001$, or $0.000\,521\,884$. Angles of less than $30°$ are taken care of by the formula

$$\sin 2a = 2 \sin a \cos a = 2 \sin a \sin(90° - a) \qquad (6)$$

(Tropfke *1980*: 307) or are found via their complements.

Napier, in two of the 60 articles in the *Constructio*, addresses the limits of logarithms, especially in the case of small differences of angles (Napier *1620*: 14, 19). In Article 28 he writes:

> Hence also it follows that the logarithm of any given sine is greater than the difference between radius and the given sine, and less than the difference between radius and the quantity which exceeds it in the ratio of radius to the given sine. And these differences are therefore called the limits of the logarithm.

Thus

$$\sin 90° - \sin a < \log \sin a < (\sin 90° - \sin a) \frac{\sin 90°}{\sin a}, \qquad (7)$$

and consequently

$$\log \sin a \approx \frac{1}{2} (\sin 90° - \sin a)\left(1 + \frac{\sin 90°}{\sin a}\right). \qquad (8)$$

For $a = 88°$ and $\log \sin 88° = 6094$, it follows that $6091.73 < 6094 < 6095.44$.

Again, in Article 39:

> The difference of the logarithms of two sines lies between two limits; the greater limit being to the radius as the difference of the sines to the less sine, and the less limit being to the radius as the difference of the sines to the greater.

Thus

$$\frac{\sin b - \sin a}{\sin b} \sin 90° < \log \sin a - \log \sin b < \frac{\sin b - \sin a}{\sin a} \sin 90°. \qquad (9)$$

For $a = 88°$, $b = 89°$ and $\log \sin 89° = 1523$, we have $4569.38 < 4571 < 4571.47$.

In the case of numbers the corresponding sine values are looked up in the geometric series and their logarithms obtained from the arithmetic series. For example, $\sin 18°41' = 0.320\,337\,444\,7$; in the geometric series

$0.999\,999\,9^x = 0.320\,337\,444\,7$; $\quad x \ln 0.999\,999\,9 = \ln 0.320\,337\,444\,7$; $\quad x =$ 11 383 803.24; and in the arithmetic series $11\,383\,803.24 \times 1.000\,000\,05 =$ 11 383 803.81. Napier had the values 3 203 375 and 11 383 800.

From the 'kinematic' definition of Napierian logarithms, their hypothetical base can be calculated as

$$(1 - 10^{-7})^{10^7/(1 + (1/2) \times 10^{-7})},$$

or approximately $1/e$ (Mautz *1919*: 18).

Among other authors of the time, Benjamin Ursinus in his *Cursus mathematici practici* (1618) printed eight-place Napierian logarithms to six places in steps of $1'$. Also, he calculated them himself to nine places at $10''$ intervals in *Trigonometria cum magno logarithmor. Canone* (1625). John Speidell published six-place tables patterned after Napier in the *New Logarithmes* (1619) with a conversion factor of $454/2909$.

Johannes Kepler provided more accurate values for the Napier series with the aid of successive proportions between two given terms. In the *Tabulae Rudolphinae* (1627), he was the first to divide a table of logarithms into numerical and trigonometric parts.

Peter Crüger in his *Praxis trigonometriae logarithmicae* (1634) published, following Napier, a logarithmic table of the numbers from 1 to 10 000, and a table for the first quadrant (0–90°) in minutes, in both cases up to six places.

7 THE BASE-10 LOGARITHMS

In an appendix to the *Constructio*, Napier (*1620*: 36) is probably the first to suggest, in addition to $\log 1 = 0$, a base for a logarithmic system, namely 10 or $1/10$:

> Among the various improvements of logarithms the more important is that which adopts zero as the logarithm of unity and 10 000 000 000 as the logarithm of either one-tenth of unity or ten times unity. Then this being once fixed, the logarithms of all other numbers necessarily follow. But the methods of finding them are various, of which the first is as follows.

Henry Briggs in his *Logarithmorum chilias prima* (Briggs *1617*) published without explanation the base-10 logarithms of the first thousand numbers to 14 places (Figure 8). With the *Arithmetica logarithmica* (Briggs *1624*), he presented a collection of 14-place tables for the numbers from 1 to 20 000 and 90 000 to 100 000, along with the corresponding differences. In the tables each group of logarithms of a hundred numbers appears on one page. Augmenting this, a ten-place table from 1 to 100 000, *Arithmetica logarithmica: una cum canone triangulorum*, by Adriaan Vlacq, was published in 1628 in Gouda. The appended *Canon triangulorum, sive tabula*

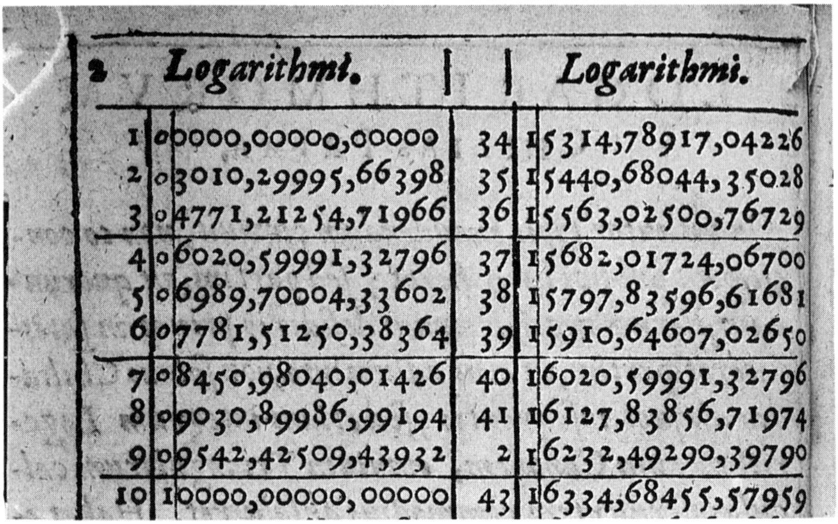

	Logarithmi.			*Logarithmi.*
1	0,0000,00000,00000		34	1,5314,78917,04226
2	0,3010,29999,66398		35	1,5440,68044,35028
3	0,4771,21254,71966		36	1,5563,02500,76729
4	0,6020,59991,32796		37	1,5682,01724,06700
5	0,6989,70004,33602		38	1,5797,83596,61681
6	0,7781,51250,38364		39	1,5910,64607,02650
7	0,8450,98040,01426		40	1,6020,59991,32796
8	0,9030,89986,99194		41	1,6127,83856,71974
9	0,9542,42509,43932		42	1,6232,49290,39790
10	1,0000,00000,00000		43	1,6334,68455,57959

Figure 8 Part of a page from Briggs (*1617*: 2)

artificialium sinuum, tangentium et secantium goes to ten places in 1′ intervals, and is based on a radius of 10^{11}. The *Logarithmicall Arithmetike* by Briggs (1631) is constructed in exactly the same way.

In 1633 Vlacq published in Gouda a *Trigonometria artificialis* which contained the base-10 logarithms of the natural numbers from 1 to 20 000, with some eighty appended corrections. It also contained what are today still-usable logarithms of the angular functions in 10″ intervals calculated to ten decimal places and with a radius of 10^{10}. A ten- and fourteen-place trigonometric base-10 table of logarithms running to hundredths of a degree, *Trigonometria Britannica*, was almost completed by Briggs and published from his manuscripts by Henry Gellibrand in Gouda in 1633. In the first part of this book there was a compilation of the methods of calculating plane and spherical triangles which took into account logarithmic formulas.

8 CALCULATION METHODS

The progression tables of Bürgi exerted little noticeable influence since they lacked the basic information which we now know from copies in Danzig and Graz. The Danzig text has been published (Gieswald *1856*); on the copy in Graz University Library see Seidel (*1985*: i, 15). Napier's tables were initially enthusiastically adopted; but they were soon overtaken by the base-10 logarithms of Briggs whose works, with Vlacq's augmentation, were to become the basis for later tabular computations – a field in which the British and Dutch excelled (Bierens de Haan *1875*: 6–11). Around 1630 trigonometry was approaching a state of completion, thanks to the increased use of logarithmic methods (§4.2).

These methods became more and more refined. Bürgi let the 'red numbers', the logarithms, run at equal intervals and added to the corresponding number, the 'black number', the ten-thousandth part, in order to obtain 1.0001 times it as the new number. Napier provided the angle in steps of 1' and next under them their sines which are, in effect, the numbers he obtained starting from $\sin 90° = 10\,000\,000$ by repeated multiplication by 0.999 999 9. For the logarithms of this sine series he chose an increasing arithmetic series beginning with 0 and 1.000 000 05.

Base-10 logarithms can be determined by mean proportions; for example $\log 4$ comes from $1 : \sqrt{10} = \sqrt{10} : 10$; $\log 1 = 0$; $\log 10 = 1$:

$$\log\sqrt{10} = \log 3.162\,277\,66 = \log 10^{1/2} = \tfrac{1}{2}\log 10 = \tfrac{1}{2}, \qquad 3.162\,277\,66 < 4;$$

$$\log\sqrt{3.162\,277\,66 \times 10} = \log 5.623\,413\,25 = \log(10^{1/4} \times 10^{1/2}) = \tfrac{3}{4}\log 10 = \tfrac{3}{4},$$
$$5.623\,413\,25 > 4;$$

$$\log\sqrt{3.162\,277\,66 \times 5.623\,413\,25} = \log 4.216\,965\,03 = \log(10^{1/4} \times 10^{3/8}) = \tfrac{5}{8},$$
$$4.216\,965\,03 > 4;$$

$$\log\sqrt{3.162\,277\,66 \times 4.216\,965\,03} = \log 3.651\,741\,27 = \log(10^{1/4} \times 10^{5/16}) = \tfrac{9}{16},$$
$$3.651\,741\,27 < 4.$$

Log 4 can be approached arbitrarily closely. Tabular calculations were very time-consuming; Briggs dedicated his life to logarithmic reckoning and its improvement.

9 THE FURTHER DEVELOPMENT OF LOGARITHMS

Further developments included a reasonable coordination of the intervals of the numbers with the number of places of the logarithms. The early tendency in the production of logarithmic tables was to calculate many places: nine for Bürgi, eight for Napier, fourteen for Briggs (*1617*, *1624*) and nine for Ursinus (1625). This greatly hastened the introduction of decimal notation. Kepler specified how terminal digits are rounded off. The intervals between tabular numbers were whole numbers or, in the case of angles in the trigonometric tables, ran mainly from minute to minute, and because of this the accuracy of calculations using interpolations is degraded after just a few places. Consequently, for normal use the number of calculated digits was reduced and more angular values given, every 10″ in the case of Ursinus (1625) or Vlacq (1633). The compilers of new tables aimed usually at excluding the errors of earlier tables. Bürgi improved the readability by making divisions into groups of three rows, Vlacq (1628) into groups of five. Bürgi replaced the repeated initial digits of numbers with dots; the failure to do this made the tables of Ursinus (1618 and 1625) somewhat more difficult to read. New tables were made noticeably quicker and clearer to read: Charles Babbage in his *Table of Logarithms of the Natural Numbers, from 1 to 108000* (1827) used a smaller size of type in his eight-place work at the point in a row where the fifth digit changed.

Questions concerning the history of the logarithmic function and its graph should be left open since it was certainly investigated before Christiaan Huygens, who gave it its name; perhaps Evangelista Torricelli was the first to do so.

The theory of series opened up new possibilities for logarithmic calculations. For example,

$$\ln(1 + x) = x - \frac{x^2}{2} + \frac{x^3}{3} - \frac{x^4}{4} + \cdots, \quad |x| < 1 \tag{10}$$

can be read into Nicolaus Mercator's *Logarithmo-technia* (1668) (Maseres *1791–1807*: Vol. 1, 222). The series

$$\ln\left(\frac{1+x}{1-x}\right) = 2\left(x + \frac{x^3}{3} + \frac{x^5}{5} + \cdots\right) \tag{11}$$

converges faster. Putting $x = (a-1)/(a+1)$,

$$\ln a = 2\left[\frac{a-1}{a+1} + \frac{1}{3}\left(\frac{a-1}{a+1}\right)^3 + \frac{1}{5}\left(\frac{a-1}{a+1}\right)^5 + \cdots\right], \tag{12}$$

which converges very quickly. Thus the last digits of a logarithm of many places can be determined correctly without great effort.

For determining new values, the functional equation

$$\log(ab) = \log a + \log b \tag{13}$$

can be employed, and for transformation to another base the formula is

$$\log_b a = \frac{\log_c a}{\log_c b}. \tag{14}$$

Greater accuracy of the tabular values required higher and higher orders of differences, analogously to the way they seem to have been used by Raimarus Ursus in the *Fundamentum astronomicum* (1588), fol. 9^r, mentioned above in connection with Bürgi.

Charles Babbage posited twelve conditions which an adequate work ought to satisfy, among them: '5th. Those figures which are first sought on entering a table ought to be so distinguished, either by position or by magnitude, as to strike the eye readily', and '12th. Coloured paper is more favourable to distinctness than white'.

10 NOTES ON THE HISTORICAL LITERATURE

Finally, some remarks on historical works. Francis Maseres produced a general edition of logarithmic works (Maseres *1791–1807*). Bierens de Haan *1875* provided a compilation of the tables that appeared up to 1874. For a comprehensive historical account, Naux *1966–71* can be recommended. On Gaspard Riche de Prony's project to produce a collection of tables larger than anything previously known, see Grattan-Guinness *1990*. D'Ocagne *1911* and Knott *1915* are recommended for their handling of many detailed questions.

BIBLIOGRAPHY

Archimedes *1913*, *Opera omnia* (ed. J. L. Heiberg), 2nd edn, Vol. 2, Leipzig: Teubner. [Repr. 1972, Stuttgart: Teubner.]

Bierens de Haan, D. *1875*, 'Tweede Ontwerp eener Naamlijst van Logarithmentafels [...]', *Verhandelingen der Koninklijke Akademie van Wetenschappen, Amsterdam*, **15**, 1–35.

Brahe, T. *1924*, *Opera omnia* (ed. J. L. E. Dreyer), Vol. 7, Copenhagen: Gyldendal.

Briggs, H. *1617*, *Logarithmorum chilias prima*, London.

—— *1624*, *Arithmetica logarithmica*, London: Jones. [Repr. 1952, Cambridge: Cambridge University Press.]

Cantor, M. *1900*, *Vorlesungen über Geschichte der Mathematik*, Vol. 2, 2nd edn, Leipzig: Teubner.

Euler, L. *1748*, *Introductio in analysin infinitorum*, Vol. 1, Lausanne: Bousquet. [Also *Opera omnia*, Ser. 1, Vol. 8; cited here.]

—— *1770*, *Vollständige Anleitung zur Algebra*, St Petersburg Academy. [Also *Opera omnia*, Ser. 1, Vol. 1; cited here.]

Gieswald *1856*, 'Justus Byrg als Mathematiker und dessen Einleitung in seine Logarithmen', *Bericht über die St. Johannis-Schule Danzig*, 1–36.

Goldstine, H. H. *1977*, *A History of Numerical Analysis from the 16th through the 19th Century*. New York: Springer. [See especially Chap. 1.]

Grattan-Guinness, I. *1990*, 'Work for the hairdressers: The production of de Prony's logarithmic and trigonometric tables', *Annals of the History of Computing*, **12**, 177–85.

Knott, C. G. (ed.) *1915*, *Napier Tercentenary Memorial Volume*, Edinburgh: Royal Society of Edinburgh.

Marre, A. *1880*, 'Notice sur Nicolas Chuquet et son *Triparty en la science des nombres*', *Bullettino di Bibliografia e di Storia delle Scienze Matematiche e Fisiche*, **13**, 555–659, 693–814.

—— *1881*, 'Appendice au *Triparty en la science des nombres* de Nicolas Chuquet Parisien', *Bullettino di Bibliografia e di Storia delle Scienze Matematiche e Fisiche*, **14**, 417–60.

Maseres, F. (ed.) *1791–1807*, *Scriptores logarithmici*, 6 vols, London: White. [Includes reprints of many basic texts.]

Matzka, W. *1860*, 'Ein kritischer Nachtrag zur Geschichte der Erfindung der Logarithmen', *Archiv der Mathematik und Physik*, **34**, 341–54.

Mautz, O. *1919*, 'Zur Basisbestimmung der Napierschen und Bürgischen Logarithmen', *Jahresbericht des Gymnasiums Basel*, 1–49.

Napier, J. *1614*, *Mirifici logarithmorum canonis descriptio, ejusque usus, in utraque trigonometria; et etiam in omni logistica mathematica, amplissimi, facillimi, et expeditissimi explicatio*, Edinburgh: Hart.

—— *1620*, *Mirifici logarithmorvm canonis constrvctio; et eorvm ad natvrales ipsorum numeros habitudines*, Lyons: Vincentius.

Naux, C. *1966–71*, *Histoire des logarithmes de Néper à Euler*, 2 vols, Paris: Blanchard.

d'Ocagne, M. *1911*, 'Calculs numériques', in *Encyclopédie des sciences mathématiques*, Tome 1, Vol. 4, 196–452 (article I 23).

Seidel, E. *1985*, 'Bibliotheca mathematica von Euclid bis Gauss', in *Katalog Ausstellung Graz 16.–21. September*, Graz University Library.

Tropfke, J. *1980*, *Geschichte der Elementarmathematik*, Vol. 1, 4th edn, Berlin and New York: de Gruyter.

Voellmy, E. *1948*, 'Jost Bürgi und die Logarithmen', in *Beihefte zur Zeitschrift 'Elemente der Mathematik'*, No. 5, Basel: Birkhäuser, 1–24.

Vogel, K. *1978*, 'Bemerkungen zur Vorgeschichte des Logarithmus', in *Beiträge zur Geschichte der Arithmetik*, Forschungsinstitut des Deutschen Museums, München: Minerva, 54–66.

2.6

Medieval and Renaissance mechanics

A. G. MOLLAND

1 STATICS

In the pseudo-Aristotelian *Mechanica*, a work unknown (or virtually unknown) in the Latin Middle Ages, we are told (847a15–19) that, 'When . . . we have to do something contrary to nature, the difficulty of it causes us perplexity and art has to be called to our aid. The kind of art which helps us in such perplexities we call mechanical skill.' The assignment of mechanics to the realm of art (in the traditional division between art and nature) may help to provide a linkage for various medieval activities which are often lumped together anachronistically under the blanket term 'mechanics', although this is not to say that we can neglect nature. The author of the *Mechanica* informs us that mechanical problems 'are not quite identical nor yet entirely unconnected with physical [or natural] problems. They have something in common both with mathematical and physical speculations; for the fact is known by mathematics, but the circumstances by physics.'

The distinction between the roles of mathematics and natural philosophy is often obscure in Aristotelian writings, and Aristotle himself characterized mechanics as one of what came to be known as the 'middle sciences', belonging half to mathematics and half to natural philosophy. Here it is the mathematical side (to which Aristotle sometimes assigned the more important causal role) that principally concerns us.

Despite some high-flown reference that it makes to the circle as the principal cause of mechanical phenomena, especially as regards the lever, the *Mechanica* is not especially mathematical, although Archytas of Tarentum (whose work on the subject had been lost by the Middle Ages) may earlier have gone further in that direction. The most famous mathematical treatment of the lever in Greek Antiquity was by Archimedes. This was translated into Latin in the thirteenth century by William of Moerbeke, but

229

exercised little influence before the sixteenth century. Archimedes' approach was highly mathematical, and as the title of his treatise *On the Equilibrium of Planes* indicates, statics was almost completely reduced to geometry. The work is formulated in strict axiomatic form, with postulates which can appear self-evident, and make a strong appeal to intuitions of symmetry; for example, 'We postulate that equal weights at equal distances are in equilibrium and that equal weights at unequal distances are not in equilibrium, but incline towards the weight which is at the greater distance.' There has often been the suspicion among writers who emphasize the strong dependence of science upon sensory experience, such as Ernst Mach, that here something was being illegitimately pulled out of a hat: this may depend upon the almost surreptitious ascription of certain properties to centres of gravity (Dijksterhuis *1956*: 286–304). Archimedes also wrote a less clearly axiomatized work, *On Floating Bodies*, which included some very sophisticated considerations of stability. For a long period it was only known from William of Moerbeke's Latin translation, but (except for some sixteenth-century treatments of falling bodies) it had little substantive influence during the Middle Ages and Renaissance.

The main sources for medieval 'statics' were treatises associated with the name of the enigmatic Jordanus de Nemore, which themselves depended on earlier Greek and Arabic writings. It has been conventional to call this tradition Aristotelian, as opposed to Archimedean (§2.1), and to see it as appealing to something like the later principle of virtual velocities (compare §8.1). The label 'Aristotelian' is disputable if taken to impute a substantial influence to the *Mechanica*, but more importantly the reference to virtual velocities easily encourages one to read history backwards, when it might be more illuminating to see the tradition's guiding ideas as arising from common experiences of their time. For example, the first three postulates of Jordanus de Nemore's *Elementa* read: 'The motion of every weighty body to be towards the middle. What is heavier to descend more swiftly. To be heavier in position when in the same position the descent is less oblique' (Moody and Clagett *1960*: 128). There is vagueness in the formulation (as also occurs with some of Euclid's definitions in the *Elements*), but the approach is geometrical, and consciously so, as is indicated both by the postulational form and by what is probably a Latin comment added to the so-called *Liber karastonis*: 'The causes of the karaston are derived from geometrical figures' (Moody and Clagett *1960*: 88).

The concepts associated with weight are more heavily nuanced than we might expect, and usually refer somewhat imprecisely to a power of doing something. From this arises the concept of what has been called 'positional gravity', outlined in the third postulate of the *Elementa* (quoted above), which is closely linked to the situation of a body in a particular mechanical set-up. Mathematical reasoning was extensively employed, but if not rigorously controlled it could lead to dubious conclusions. One such is that a weightless lever carrying equal weights at equal distances from the fulcrum would return to the horizontal if displaced from it (Moody and Clagett *1960*: 130–32, 176–8).

The Latin Middle Ages saw few significant mathematical developments in the tradition, but more of attempts to integrate the doctrines within an Aristotelian context, and this makes one wonder how originally 'Latin' the treatises were in the first place. In the Renaissance there were more diversified developments. This was partly a result of a closer union of theory and practice, which served to bring 'statics' more firmly back into the realms of art.

A paradigmatic example is the work of Leonardo da Vinci, on whom Pierre Duhem *1906–13* centred three large volumes in the course of his monumental re-evaluation of medieval mechanics. Nevertheless, Leonardo himself occupies relatively few of Duhem's pages, and in any case it is hard to portray him as a crucial furtherer of mathematical statics. Leaving aside the question of how much influence he actually wielded, he acted mainly from within the Jordanus tradition, and his principal divergences were to attack what he saw as excessive mathematical idealizations:

> I have found that the ancients were in error in their reckoning of weights, and that this error has arisen because in a considerable part of their science they have made use of poles which had substance and in a considerable part of mathematical poles, that is such as exist in the mind or without substance. (Leonardo *1939*: 503–4; compare §2.8)

Leonardo does not figure prominently in *Mechanics in Sixteenth-Century Italy* (Drake and Drabkin *1969*), which concentrates on Niccolò Tartaglia, Girolamo Cardano, Giovanni Battista Benedetti, Federico Commandino, Guido Ubaldo, Bernardino Baldi and Galileo Galilei, whose writings probably offer a more representative picture of Italian statical developments at the time. In them one finds quite frequent reference to practical situations (balances, levers, wheels, pulleys, whirlpools, screws, etc.), but also a strong concern with theoretical form, often abetted by the example of Archimedes' extreme mathematizing tendencies. In northern Europe something very similar can be said of Simon Stevin and others, but in general, Western medieval and Renaissance statics may be of less importance in itself than as a spur to other sciences, notably that of motion.

2 KINEMATICS

This is another anachronistic title, but one which is not too misleading, since it refers to what in medieval terms would have been called motion considered with respect to its effects. Aristotle, in a manner that did not satisfy all his medieval commentators, divided simple natural motions into two kinds, the straight and the circular. Circular motions were appropriate to celestial bodies, from the sphere of the Moon upwards, and by late Antiquity had become the subject of an empirically and mathematically very sophisticated 'middle science'. This sophistication was conveyed with greater and lesser degrees of perfection to the Middle Ages, and by the sixteenth century a renewed process of reform was under way which was soon to displace the Aristotelian and Ptolemaic foundations of astronomy. In the sublunary (non-astronomical) world, where rectilinear motion was the norm, things were far more chaotic, and no adequate middle science for it was developed during Antiquity. Part of the difficulty arose from the fact that Aristotle's approach was holistic, the primary object of attention being the whole motion of the whole body from its beginning to its end. This meant that providing quantitative descriptions became a process of prodding, and of taking different sightings from different angles. Notions such as the instantaneous velocity of mass points, which we would regard as fundamental, were reached derivatively as the result of a process of analysis — or not at all.

Further, because of the absence of an established 'middle science', together with Aristotle's immense stature in the medieval universities, it was his writings that provided the starting point for the 'mathematical physics' that has assumed such prominence in twentieth-century scholarship. Aristotle was not nearly as anti-mathematical as is sometimes supposed: in his *Posterior Analytics* geometry he provided the model for a proper demonstrative science; in his scheme of knowledge he allowed an important if ill-defined role for the mathematical 'middle sciences'; and in his writings on sublunary motions he hinted at the possibility of further mathematization, hints which were extensively developed by some of his medieval successors. This had to do with the measure of motions. Aristotle had spoken somewhat vaguely of the divisibility of motion (a concept leading to its measure) both 'in virtue of the time that it occupies' and 'according to the motions of the several parts of that which is in motion' (*Physics* VI.4, 234b21–23).

In the fourteenth century in the hands of the famous French schoolman and court educator Nicole Oresme, this developed into something like a conception of motions or speeds as five-dimensional objects, with three spatial dimensions (those of the mobile), one temporal, and one of intensity of speed. The mathematical tone is prominent, but Oresme's dialectical proce-dures (in common with many of his contemporaries) mean that the edges are often ragged, as when a uniform speed that lasts for three days is equated with a speed three times as intense that lasts for one day. There is a temptation to translate into modern terms, and render only Oresme's 'intensity of speed' as (instantaneous) speed in the modern sense; but even here there are quantificational difficulties, for Oresme's measure of intensity of speed depends intimately on how the motion is described. (For further details see Molland *1982*; Souffrin and Segonds *1988*: 13–30.) Earlier there had been a treatise *De motu* by one Gerard of Brussels (re-edited by Clagett *1964–84*: Vol. 5, 1–142). This also gave a quantitative treatment of whole motions, but had more the form of a 'middle science' than does Oresme's account; Sylla (*1970*: B1) has described it as finding 'average or mean velocities'. *De motu* is often found in the same manuscript codices as works by Jordanus de Nemore, and, as with Jordanus, we may doubt the genuine Latin paternity of many of its leading ideas, even when no earlier potential source is extant.

This doubt is reinforced by the very limited influence that Gerard had on his medieval successors. Thomas Bradwardine (*1955*: 148–50) mentioned one of his early conclusions (together with preceding postulates, but without proof) to the effect that the speed of a rotating straight line should be equated with that of its mid-point, only to rebut it in favour of the fastest-moving point, and other references in the later Middle Ages are also mathematically unprofound. But it is towards Bradwardine and his fol-lowers, a group of mathematically minded natural philosophers often referred to as the Merton School or the Oxford Calculators, that we should more profitably look for precursors of Oresme. Here we meet with the clas-sification of speeds into uniform, uniformly difform, difformly difform (roughly equivalent to our uniformly and non-uniformly accelerated or retarded), and so on, and a similar strategy is adopted towards qualities, accompanied in both cases with strings of complicated arguments, both mathematical and disputational.

In all this the intensities of qualities and speeds were at least implicitly represented by segments of straight lines. In Oresme's work this representa-tion was fully explicit, and in a remarkable piece of scientific vision he imagined the intensities of qualities and speeds as 'graphed' across the several parts of the subjects that they informed, and for motions across the times also. The particular 'configurations' thus achieved were then held to explain the properties of bodies and their diverse actions (including those of musical sounds) in a way similar to ancient atomism. In this we are nearer to nature than to art, something that may be held appropriate to the theoretical strains in medieval thought.

3 DYNAMICS

Medieval kinematics had strong affinities with nature; but the medieval dynamics (to introduce a further anachronism) of the sublunary realm had also to look much towards art, for it sought to quantify the forces that produced motions, and this usually meant interference in the natural order of things; in Aristotle's case this could be a bullock drawing a cart, or men hauling ships (*Physics* VII.5, 250a17–19). Aristotle himself paid only infrequent and fragmentary attention to quantitative relations between forces, resistances and speeds, as may be gleaned from the following snippet:

> If A the movent have moved B a distance C in time D, then in the same time the same force A will move 1/2 B twice the distance C, and in 1/2 D it will move 1/2 B the whole distance C: for thus the rules of proportion will be observed. (*Physics* VII.5, 249b30–250a4)

Nevertheless, the mention of proportion, here as elsewhere in the corpus, encouraged the search for something that looked more like a general law, and in his now-famous *Tractatus de proportionibus* of 1328, Bradwardine, after rejecting what he saw as four erroneous positions, plumped for the rule that 'The ratio of speeds in motions follows the ratio of the power of the motor to the power of the thing moved' (Bradwardine *1955*: 110).

This statement is not so innocent as may be thought. For a long time Bradwardine's 'law' was thought to be a mere tidying up of Aristotle, but in the 1940s historians began to realize that it can be translated into the modern form, $v = k \log F/R$ (with v the speed, F the force and R the resistance), and writers have often spoken of an exponential relationship. But this obscures the essential simplicity of Bradwardine's own formulation, and demands explication. This depends intimately upon the meaning and syntatic relations of the term 'ratio' (*proportio*). In both the ancient and the medieval traditions the result of compounding two ratios $A : B$ and $B : C$, with A greater than B and B greater than C, was the ratio $A : C$. This is relatively unproblematic, and we may be tempted to assimilate the operation to multiplication of fractions, a step which was taken by some medieval writers. But ratios were not identical to fractions, and there was another possibility as is clearly revealed by the musical tradition.

Theoretical music is a good candidate for recognition as the first mathematical 'middle science' (§2.10), and Bradwardine himself appealed to its example to justify the possibility of a mathematical science of dynamics: 'If there were not ratios between powers because they are not quantities, for the same reason there would not be between sounds, and then the modulation of the whole of music would perish' (Bradwardine *1955*: 106). Music made much use of ratios, where they provided the measure of musical intervals: in, for example, the fundamental treatise by Boethius, it is clear that composition of ratios was regarded as addition, not multiplication. This provides one of the leading ideas behind the later emergence of logarithms (§2.5), and gives warrant for the 'exponential' translation of Bradwardine.

With this point made, it is also important to note that in one sense Bradwardine's 'law' was not a law at all, for it had a very imprecise empirical reference. This also applied to Oresme's doctrine of configurations, as he himself (almost tacitly) noted (Oresme *1968*: 404), and served to give medieval mechanics far more of the flavour of the disputation hall than of the laboratory. And Bradwardine's 'law' provided ample new resources for extremely complicated debates on quantitative issues in natural philosophy, as is evidenced to the highest degree by the so-called *Liber calculationum* of another fellow of Merton College, Richard Swineshead, who for Robert Burton 'well nigh exceeded the bounds of human genius'.

4 IMPETUS AND FREE FALL

Two medieval topics remain to be considered, less for their own medieval mathematical relevance than for their alleged influence on later work. These are impetus and free fall. Both relate to the Aristotelian doctrine of causality, in which events were explained in terms of four kinds of cause – material, formal, efficient and final. Ideally, all four causes would come into operation, but sometimes one or another tended to be eclipsed. There was also the division of motions into natural and violent. In natural motion a body was fulfilling its own nature, as when a heavy body, left to its own devices, moves towards its proper place at the centre of the World. However, if it is moving upwards, this is contrary to nature, and it is necessary to seek out the efficient cause that is doing violence to it. If it is being moved upwards in my hand, then I, as a voluntary agent, am the mover; but if I throw it upwards then there is no obvious efficient cause after it has left my hand, since it is only in contact with the air.

Aristotle, with impeccable logic, decided that the air itself was the mover, to which I had communicated a certain power of continuing the motion. But this was too much for the common sense of many commentators, Greek, Arabic and Latin alike, and the usual solution was to say that, rather than having communicated a power to the air, I had given to the body itself an immaterial moving force. To the fourteenth-century Parisian schoolman Jean Buridan and many of his followers, this force was known as 'impetus', but other names also appeared. Impetus had little direct impact on mathematical physics, but the use of an internal force has been seen as an important step on the way to the principle of inertia, and has been compared to Newton's *vis insita* (also called *vis inertiae* by him), 'by which every body, as much as in it lies, perseveres in its state either of resting or of moving uniformly in a straight line' (*Principia* (1687): Book I, Def. 3).

The 'free fall' of heavy bodies also made difficulties for Aristotelian causality. These motions were definitely in the realm of nature, but the suspicion that they should have an efficient as well as a final cause contributed also a whiff of art. Aristotle himself gave scant attention to the nature of the efficient cause, but the problem considerably exercised the tradition. For instance, the Arabic commentator Ibn Rushd (Averroes), who was very influential in the Latin West, appealed to the air, while Buridan and other impetus theorists looked towards an internal moving principle.

Swineshead gave a different twist to the problem. He considered a falling body, part of which had already passed the centre of the World. Did that part act as a separate moving principle, and resist the motion of the whole body? He decided, by means of an argument that was impeccable in his own terms, that an affirmative answer was impossible, and retreated to a position with strong emphasis on final causality in which each part of the body acted in such a way as to fulfil the nature of the whole. Swineshead's argument was highly mathematical, but more routine discussions of the acceleration of falling bodies had only a vaguely quantitative tinge, and we must regard even the assertion of the sixteenth-century schoolman Domingo de Soto that 'uniformly difform' motion was proper to falling bodies (Clagett *1959*: 555; compare Wallace *1981*: Chap. 6) as a little more than a whimsy. The problem could not become really significant for mathematical mechanics until there were more radical developments.

5 THE BEGINNINGS OF A NEW MECHANICS

In the sixteenth century, several thinkers considered that Aristotelian mechanics was by no means satisfactory, but a more precise diagnosis, together with the possibility of a remedy, was not easily made. Galileo was not atypical in this respect, and because of his later achievements it is convenient to take him as an example. After an unsatisfactory stab at medicine at the University of Pisa, Galileo studied mathematics extramurally under Ostilio Ricci from 1583. He became immensely impressed by the work of 'the superhuman Archimedes, whose name I never mention without a feeling of awe', and not only set out to emulate him in statics, but tried to extend his ideas to the dynamics of falling bodies. This involved him in a thrust towards a more rigidly quantitative approach (in which he was arguably less successful than Bradwardine), and also made use of hints from Archimedean hydrostatics. It followed that the relation between the force of the mobile and the resistance of the medium should be thought of in terms of the reduction in weight effected on the former by the latter, and that this weight should be thought of in specific terms, and not increased as the mobile became simply larger in bulk. Despite the Archimedean influence, there was still a strong Aristotelian focus on the causal agency behind the motion, and the associated interaction between force and resistance. It was similar with projectile motion. Galileo (*1960*: 76) affirmed that 'Aristotle, as in practically everything that he wrote about locomotion, wrote the opposite of the truth on this question too'; but he himself still looked for the cause, and, not surprisingly, came down in favour of an internal impressed force.

Soon Galileo came to realize that a more radical approach was desirable. In particular, he moved from asking why motions happened to asking how they would happen in certain ideal circumstances (perhaps unrealizable in practice). His revaluation involved him in both abstract thought and experiment, and there is still much controversy as to how much weight should be assigned to each, although the former can seem to dominate in his published work. There Galileo, by means of an ingenious series of real and thought experiments, established that a body moving on a perfectly smooth horizontal surface, with all other impediments removed, would continue moving indefinitely at uniform speed. A complication which Galileo, like Archimedes, took seriously was that, strictly speaking, a horizontal surface was spherical and centred on the centre of the Earth, but again like Archimedes, he was usually prepared to approximate this by a plane.

Galileo also reasoned that all bodies would fall at equal speed in a vacuum. For the acceleration of fall he looked for a simple mathematical law, and concluded that the only viable one was that the speed increased uniformly with time. To test this he described a famous experiment in which balls were rolled down an inclined plane. The way was now open for combining the two 'natural motions' and producing the famous parabolic trajectory for projectiles (which, however, was of little use to gunners, since they never met such ideal conditions). Thus, although the relevant work is a dialogue set in the Venetian Arsenal, the role of art is necessarily limited.

It is often asked how much Galileo owed to his medieval predecessors, in particular to fourteenth-century practitioners of 'mathematical physics'. There is no definitive answer, but it seems more profitable to look not at particular results but at general leading ideas, such as the representation of speeds (or their intensities) by segments of straight lines – obvious to us, but probably not always so. It is also possible to see another, perhaps more important, current leading in the opposite direction in the switch from medieval holistic attitudes to modern 'atomistic' ones, in which objects and motions were conceived as being made up of indivisible elements from which any mathematical picture should be constructed (Molland *1982*).

This new current may also have affected Galileo's contemporary Johannes Kepler, but Kepler's concern was with celestial rather than sublunary mechanics. Nevertheless, his discussion of the causes of heavenly motions helped pave the way towards a unification of the two domains, and his use of forces that varied in magnitude according to some function of the distance from a centre may be seen as important background to Isaac Newton's universal gravitation, and to his mathematical concern with laws of force in Book I of the *Principia* (1687). Another important influence on Newton (sometimes in a negative direction) was René Descartes, who often used to be dismissed from the history of mechanics as merely an impractical philosopher who propounded some hideously incorrect laws of impact. But Descartes was intent on constructing a more radical world-picture than Galileo's, and in the course of this he produced a very near equivalent to Newton's first law of motion; also, his concern with conservation principles influenced many later men of science. Descartes claimed that his physics was nothing but geometry, but (save for a few specialized areas) it contained surprisingly little mathematics. The real paradigm for the new mechanics was set by Newton's *Principia*, the *Mathematical Principles of Natural Philosophy*, whose title, besides emphasizing the role of mathematics, may perhaps be seen as indicating a greater concern with nature than with art (§2.12). In the preface, Newton divided mechanics into rational and practical branches, and chose the former as his principal concern, establishing the subject under the name that has become standard.

BIBLIOGRAPHY

Bradwardine, T. *1955, Tractatus de proportionibus* (ed. H. L. Crosby), Madison, WI: University of Wisconsin Press.

Caroti, S. (ed.) *1989, Studies in Medieval Natural Philosophy*, Florence: Olschki.

Clagett, M. *1959, The Science of Mechanics in the Middle Ages*, Madison, WI: University of Wisconsin Press.

—— *1964–84, Archimedes in the Middle Ages*, Vol. 1, Madison, WI: University of Wisconsin Press, Vols 2–5, Philadelphia, PA: American Philosophical Society.

Dijksterhuis, E. J. *1956, Archimedes*, Copenhagen: Munksgaard.

Drake, S. and Drabkin, I. E. *1969, Mechanics in Sixteenth-Century Italy: Selections from Tartaglia, Benedetti, Guido Ubaldo, and Galileo*, Madison, WI: University of Wisconsin Press.

Duhem, P. *1906–13, Etudes sur Léonard de Vinci*, 3 vols, Paris: Hermann.

Galileo Galilei, *1960, On Motion and On Mechanics* (transl. I. E. Drabkin and S. Drake), Madison, WI: University of Wisconsin Press.

Grant, E. and Murdoch, J. E. (eds) *1987, Mathematics and its Applications to Science and Natural Philosophy in the Middle Ages: Essays in Honor of Marshall Clagett*, Cambridge: Cambridge University Press.

Koyré, A. *1968, Metaphysics and Measurement: Essays in Scientific Revolution*, London: Chapman & Hall.
—— *1978, Galileo Studies*, Hassocks: Harvester.
Leonardo da Vinci *1939, The Notebooks of Leonardo da Vinci* (transl. E. MacCurdy), New York: Reynal & Hitchcock.
Lindberg, D. C. (ed.) *1978, Science in the Middle Ages*, Chicago, IL: University of Chicago Press.
Maier, A. *1982, On the Threshold of Exact Science: Selected Writings of Anneliese Maier on Late Medieval Natural Philosophy* (ed. and transl. S. D. Sargent), Philadelphia, PA: University of Pennsylvania Press.
Molland, A. G. *1968*, 'The geometrical background to the "Merton School": An exploration into the application of mathematics to natural philosphy in the fourteenth century', *British Journal for the History of Science*, **4**, 108–25.
—— *1982*, 'The atomisation of motion: A facet of the scientific revolution', *Studies in History and Philosophy of Science*, **13**, 31–54.
Moody, E. A. and Clagett, M. *1960, The Medieval Science of Weights*, Madison, WI: University of Wisconsin Press.
Oresme, N. *1968, Nicole Oresme and the Medieval Geometry of Qualities and Motions: A Treatise on the Uniformity and Difformity of Intensities Known as Tractatus de configurationibus qualitatum et motuum* (ed. and transl. M. Clagett), Madison, WI: University of Wisconsin Press.
Souffrin, P. and Segonds, A. P. (eds) *1988, Nicolas Oresme: Tradition et innovation chez un intellectuel du XIVe siècle*, Padua: Programma e 1 + 1, and Paris: Les Belles Lettres.
Sylla, E. D. *1970*, 'The Oxford Calculators and the mathematics of motion, 1320–1350. Physics and measurement by latitudes', PhD thesis, Harvard University.
Wallace, W. A. *1981, Prelude to Galileo: Essays on Medieval and Seventeenth-Century Sources of Galileo's Thought*, Dordrecht: Reidel.

2.7

Astronomy

K. P. MOESGAARD

1 NEWTON, WHY? AND COPERNICUS, HOW?

In his *Mathematical Principles of Natural Philosophy* (1687), Isaac Newton presented a dynamical basis for the age-old doctrine of positional astronomy. For the first time in history, general laws of motion in combination with the assumption of an attractional force of gravitation, active everywhere in the Universe, made possible the derivation of particular positions in specific orbits of the celestial bodies. Expressed in the new mathematics of infinitesimal changes – the calculus – the laws of physics proved capable of governing celestial phenomena. Theoretical astronomy had become dynamical, and it was possible to answer the 'why?' of the motions in the sky.

In particular, it is trivial to form, from Newton's laws, the concept of the rather small terrestrial globe in orbital motion around the Sun. But the idea that the Earth moved had had a long history, although it had never gained general acceptance. That the Earth was in motion was suggested in Antiquity by Aristarchus, and from time to time it was the subject of academic discussions among natural philosophers in medieval universities. In his *De revolutionibus orbium coelestium* (1543), Nicolaus Copernicus made the moving Earth a constituent of his new architecture of the Solar System, in essence the modern heliocentric world-picture. But Copernicus, like his predecessors as far back as Hellenistic Greece, worked within the purely geometrical kinematics of theoretical astronomy, concerned with 'saving the phenomena' – with answering the 'how?' of the heavenly motions. He could advance nothing but kinematics in his defence of what was basically a dynamical feature of the Solar System.

So that is why astronomers, natural philosophers and theologians spent two centuries discussing whether Copernicus was right or wrong, and why twentieth-century historians of astronomy still debate whether he ought to have been right. Anyway, Copernicus's work triggered great efforts on the part of his followers to deal both with the 'how?' and the 'why?' of

240

positional astronomy, which on that account went into the melting-pot time and again until the final Newtonian synthesis.

2 THE PTOLEMAIC–ARISTOTELIAN BASIS

The framework for astronomical research, at least until Copernicus, originated in Aristotle's cosmology from the fourth century BC and in the geometrical doctrine of astronomy subsequently shaped by Hipparchus and Ptolemy (§1.7).

Hellenistic astronomy was empirically founded on observational data, partly handed down from the Babylonians as the outcome of unaided naked-eye observations, and partly gathered by Greek astronomers using newly invented sighting instruments, such as the parallactic ruler and the zodiacal armillary sphere, which remained important instruments of observation until the time of Tycho Brahe.

Hellenistic astronomy, the first branch of applied mathematics, reached a high degree of perfection in Ptolemy who, by his *Syntaxis*, later known as the *Almagest*, founded a tradition of astronomical literature which was to last, again, until Tycho Brahe. Here spherical geometry yielded a realistic picture of all events, like rising, culmination and setting, connected with the assumed daily rotation of the starry heavens and of the Moon, Sun and planets. In addition each of the planets, 'wandering stars', moves at its own period of revolution against the background of the 'fixed' stars. Plane geometrical models, consisting of circular components, proved capable of imitating the actual complex of motions observed in the sky. These featured eccentric circles and extra circles, epicycles, to generate variable rates of motion and, for the planets, the retrograde loops of their motions. With the radii, eccentricities and periods of revolution of all the circles calibrated by means of observational data, this entire geometrical structure allowed the derivation of tables which, in their turn, by comparatively simple procedures of calculation, enabled the positions of all the heavenly bodies to be calculated for any future or past date. The geometrical kinematics of astronomy as thus sketched served as an adequate foundation for practical applications such as time-reckoning, navigation and astrology. So, because of both its utility and its lofty, almost divine subject, mathematical astronomy was always held in high esteem.

From its very beginning in Antiquity, this empirical and theoretical astronomy was connected to Aristotelian 'physics', a natural philosophy of a rather speculative nature. But there was no cause-and-effect relation: one could not deduce the Ptolemaic models of planetary motion from the Aristotelian cosmology. In this cosmology the Earth is globular and rests at the midst of the Universe, the outermost confines of which are also spherical. Below the Moon is the changing, sublunary world of the four elements with their rectilinear natural motions, down towards the centre – of the Earth and of the Cosmos – for the heavy elements, earth and water, and upwards away from the centre for the light elements, air and fire. In the celestial world everything is perpetually unchangeable and made out of the fifth element, aether. All motions occurring here must be uniform rotations along circles around the centre, without beginning or end, and thus eternal. Within positional astronomy proper, the principle of uniform circular partial motions could serve as an abstract mathematical axiom, but in natural philosophy it was thought of as reflecting the rotations of actual spheres formed out of the celestial material (Pedersen and Pihl *1974*).

From sound observational data, Hellenistic astronomy placed the Moon at the right distance, about 60 times the Earth's radius. No other real dimensions could be determined for the Solar System, not to speak of the world of the fixed stars. So, any of the kinematic planetary models could be scaled up or down to any degree. A couple of principles, however, drawn from natural philosophy, established some limits. It was assumed first that no unused empty space can exist anywhere in the Universe, and second that a planet occupies its own domain, never to be entered by neighbouring planets. Now, the eccentricity and the epicycle radius of a planetary model determine the thickness, or the depth of space, occupied by the domain of the planet in question. So once the Moon has been been placed in a spherical shell of well-determined dimensions, the next planet, Mercury, must have the innermost confines of its domain coinciding with the outermost boundary of the lunar sphere; and so forth for Venus, the Sun, Mars, Jupiter and Saturn, and finally the rather thin sphere of the fixed stars.

This argument completes the picture of a perfectly round heavenly space, divided into spherical shells, and everywhere stuffed with aether. The Sun was placed at a distance of about 1200 times the Earth's radius (too close by a factor of 20), and the outermost boundary of the entire visible Universe lay at a distance of roughly 20 000 Earth radii (the actual distance of the nearest 'fixed' star is over 300 000 times greater).

Anyway, for fifteen hundred years astronomers worked on the kinematics of the circle within this tiny sphere of the Cosmos, outside which there was nothing, not even empty space; space had to contain something, space containing nothing was inconceivable.

3 THE HOUSE OF WISDOM
IN NINTH-CENTURY BAGHDĀD

After the disintegration of the Roman Empire, a learned tradition of science survived in the eastern Byzantine part, from where time and again during the later Middle Ages Greek scholars and manuscripts found their way to the Latin West, providing new inspiration (§1.5). A fifteenth-century example of particular importance in the history of astronomy is the joint project by Cardinal Bessarion and Georg Peurbach to produce a new free translation of the *Almagest*; it was completed by Regiomontanus and used constantly by Copernicus.

The transmission of classical astronomy and cosmology began with a move towards the east – to Syria, Persia and, eventually, India. From about AD 500 Indian astronomers produced several astronomical textbooks, called *Siddhāntas* (§1.11–1.12). They were based on Greek geometrical astronomy, with elements of earlier Babylonian influence, and adapted to specific Indian methods of calculation, including the use of their sine tables instead of Ptolemy's table of chords. This Indian tradition became in turn the first source of inspiration for the rise of Islamic astronomy around 800, and consequently its influence may be traced in the development of astronomy in Latin Europe in universities from about 1100 (Billard *1971*, Pingree *1978*).

Within a century of the death of Muḥammad, the Islamic Empire stretched from India to Spain. In their new capital of Baghdād, the ʿAbbāsid caliphs established a centre of sciences and letters, *Bayt al-Ḥikma*, the 'House of Wisdom'. It became the model for several similar well-organized institutions in the Islamic world, and may be compared to the Royal Academies founded in western Europe in the seventeenth century. Here was a large library with a staff busily engaged in translating scientific and philosophical works, and giving high-level training in all branches of learning.

Attached to the House of Wisdom were large departments of observational and theoretical astronomy, as well as geodetic surveying. Here, during the last few years of the reign of Caliph al-Ma'mūn (813–33), an intensive period of activity made Islamic astronomy take off, and in less than a century the classical picture was revised. A new basis was established for astronomical research and teaching that would last until the time of Copernicus and Tycho Brahe (§1.7).

Both al-Khwārizmī and Ḥabash al-Ḥāsib ('the Calculator') were leading members of the academy's staff, writing extensively on mathematics and astronomy. Their early *zījes*, astronomical tables with instructions for use, were in the Indian tradition with some model parameters adjusted in the light of contemporary check observations. In particular, al-Khwārizmī's *Zīj al-Sinhind* became highly influential through a Latin version, produced in the early twelfth century by Adelard of Bath, and also through the Toledan Tables, which for a century around 1200 dominated European astronomy. In or around 820, direct translations into Arabic of the *Almagest* became available at Baghdād. Al-Farghānī then prepared a well-organized descriptive compendium of Ptolemaic astronomy, which in its Latin version, known as *Rudimenta astronomica* or *Liber 30 differentiarum*, served as a standard introduction to astronomy until it was replaced in the thirteenth century by Johannes de Sacrobosco's *De sphaera mundi*.

The arrival of the *Almagest* posed the problem of having to choose between the Greek and the Indian doctrines of astronomy. There were differences in numerical parameters, as well as in the geometrical structures of the planetary models. Fresh observations were necessary before a correct choice could be made; repeated solstice observations from Baghdād proved mutually incompatible. So a large-scale programme for observing the heavenly bodies (in particular the Sun and the Moon) every day for a whole year was set up at a new observatory at Mount Qāsīyūn in Damascus. To ensure high precision the observatory was equipped with large instruments, including a marble mural quadrant with a radius of 5 metres. Thus a tradition was founded for building observatories to carry out programmes of observation ideally covering 30 years to take in a complete revolution of slowly orbiting Saturn. Most famous among the numerous later observatories are the thirteenth-century Marāgha institution founded by Naṣīr al-Dīn al-Ṭūsī in the reign of Hulāgu Khan, and Ulugh Beg's fifteenth-century observatory at Samarkand. In Latin Europe, nothing comparable appeared until the late sixteenth century when Tycho Brahe established his unique centre of astronomical research, generously sponsored by the Danish king, Frederick II.

Immediate results of the activity under al-Ma'mūn were a geographical survey of the Islamic Empire and new, 'tested' tables on Ptolemaic principles, possibly produced by a team of authors under the supervision of Ḥabash al-Ḥāsīb. Ptolemy had saved selected phenomena spread over some centuries in Antiquity. When extrapolated to the ninth century his tables were no longer in step with contemporary phenomena, but it was hoped that this could be remedied by improving the estimated rates of circular rotations by means of check observations.

However, trying to reconcile the data in the *Almagest* with the evidence of fresh observations created long-term problems of an unexpected complexity at the very root of astronomical science. The Arabic astronomers could not confirm the *Almagest*'s values of basic parameters, naturally thought of as constants of nature. These included the duration of the (tropical) solar year, the obliquity of the ecliptic (its inclination to the celestial equator) and the rate of precession (the slow displacement of the vernal equinox among the fixed stars). To save phenomena that spanned a period of more than a millennium proved to be anything but trivial. Either one had to trust the old data, and accordingly establish a theory for the secular variation of the basic parameters, or one had to be satisfied with producing, on the basis of contemporary data, tables of an admittedly ephemeral validity.

Thābit ibn Qurra, in his works 'On the Solar Year' and 'On the Motion of the Eighth Sphere' (i.e. of the shell containing the fixed stars), made the first comprehensive attempt at a solution for the first of these two options. First, behind the seemingly secular variations of the tropical year (i.e. of the seasonal period of revolution of the Sun as measured from equinox to equinox), he found a virtually constant value of the sidereal year, which measures the solar motion against the background of the fixed stars. Here was the foundation for a 'sidereal' doctrine of astronomy with the sphere of the fixed stars as basic reference against which steady motions and constant parameters could be deduced.

But this called for a non-linear precession of the equinoxes in relation to the fixed stars. Now, the equinoxes are the points where the plane of the celestial equator, the projection onto the celestial sphere of the terrestrial equator, intersects the ecliptic. The path of the ecliptic among the stars was believed to be fixed, because no changes of stellar positions had ever been noticed. So either the Earth with its equator or the eighth sphere of the stars with the ecliptic had to perform a non-linear motion. With the Earth fixed *a priori*, Thābit proposed his 'theory of trepidation' in which the eighth sphere oscillates, with the effect that both the obliquity of the ecliptic and the equinoxes slowly oscillate around mean values. Later astronomers advanced modified theories for the motion of the eighth sphere. In cosmology this demanded first a ninth and then a tenth sphere, completely empty of any visible objects, to explain the precessional components of motion.

Almost seven centuries after Thābit, Copernicus was advancing arguments in the same tradition, but he had a longer series of observational data at his disposal. He succeeded in saving a variety of coupled long-term anomalies. But with more far-reaching consequences, he was the first to take seriously the idea of 'sidereal' astronomy. He stopped the fixed stars and thus rendered superfluous all speculation of a ninth and a tenth sphere. In return he had to let the Earth's axis, and hence its equator, perform the slow motion of precession superimposed by oscillations.

Another main result of the early Islamic period is the *Zīj al Ṣābī*, or *Opus astronomicum*, by Muḥammad ibn Jābir al-Battānī (Albategnius), the first original Arabic work comparable in size and scope to the *Almagest*. It was based on a 40-year programme of observation at Raqqa where al-Battānī worked with rather small, but nevertheless very accurate, instruments. He did not accept Thābit's 'theory of trepidation', but used an improved average rate of precession. So both in his practical activity and his theoretical outlook, al-Battānī resembles Tycho Brahe rather than Copernicus.

During the remaining centuries of the Middle Ages, Islamic science and philosophy contributed in many ways to the development of astronomy, producing improved instruments and new sets of tables, elaborating a systematic trigonometry, and debating thoroughly the physical and cosmological problems associated with the sophisticated mathematical models used by the astronomers. But here we leave Islamic astronomy, with the observation that, as early as AD 900, it had reached a level of institutional organization, in observational practice and in theoretical outlook, that would not be equalled in Europe until the sixteenth century (King *1986*, *1987*; Sayili *1960*).

4 THE HOUSE OF LEARNING IN THIRTEENTH-CENTURY PARIS

In Latin Europe astronomy progressed slowly throughout the Middle Ages. But both in the early cathedral and monastic schools, and from about 1100 in the universities, astronomy was part of the scientific quadrivium which, together with the humanistic trivium, made up the seven liberal arts to be mastered by any student before he could enter the higher faculties of law, medicine or theology (§2.11). So astronomy belonged to the general education of the entire world of learning.

In the (Western) early Middle Ages the instruction was based on encyclopedic works containing only the most scanty terminology and results handed down from Antiquity. The most notable original contribution is Bede's admirable eighth-century work on *Computus*, dealing with rules for calculating the calender. From the twelfth century a steady flow of translations of Arabic and Greek works constantly added to the curricula of the growing number of universities (§2.1, 2.11). The university teachers commented upon the translated material, and eventually produced their own treatises for educational purposes (Crombie *1964*).

Around the middle of the thirteenth century, at the University of Paris, a curriculum of astronomy was established which, with later additions, became normative all over Europe during the remaining period of the Middle Ages. Obviously the goal was to teach astronomy to the level necessary for applications within time-reckoning and astrology; every cleric had to master computing the calendar. Teaching at Paris was based on three treatises by Sacrobosco: *Algorismus*, about elementary calculation with Hindu–Arabic numerals; *De sphaera mundi*, with a description of spherical cosmology and of the phenomena caused by the supposed daily rotation of the heavens (Thorndike *1949*); and *Compotus*, with detailed rules for computing the calendar. There were also lunar tables, one or more calendars, and tracts on small and simple instruments and their use for observation, calculation or demonstration (Lindberg *1978*).

Late in the thirteenth century the corpus grew with the addition of the anonymous treatise *Theorica planetarum*. It contained a clear and compendious description of the terminology and the geometrical principles of the Ptolemaic planetary theories with their steady circular components of motion. With the final addition of a set of planetary tables – after 1300, ordinarily the Alphonsine Tables – and rules (*canones*) for their use, the curriculum provided the necessary basis for calculating planetary positions corresponding to any date. In the late Middle Ages codices often included also a manual of astrological rules of interpretation to be used in meteorology, in medicine or for the purpose of casting horoscopes (Pedersen *1975*). Basic astronomical research was rare; the fifteenth-century works of Peurbach and Regiomontanus mark the beginning of the western astronomical reform and paved the way for a real understanding of the *Almagest* as the necessary platform for new research.

5 EXTRA-MURAL REFORM OF ASTRONOMY

In the few decades before and after 1600 a handful of astronomers, mostly active outside university circles, laid the foundation of modern 'physical' astronomy. Copernicus, struggling in the tradition founded by Thābit ibn Qurra to find a theory which would reconcile ancient and modern observations, was led to accept that there were secular oscillations of the Earth's axis. From this he progressed to the idea of the Earth spinning daily on its axis, and completing every year its orbit around a central and stationary 'mean' Sun. Copernicus spent the rest of his career transforming in great detail all the Ptolemaic models to conform with his new cosmological outlook. He calibrated the resulting models with two sets of parameters: one, valid for Antiquity, derived from ancient observational data, and another drawn from contemporary observations. Comparing the two sets of models led him to suggest that there were secular variations in, for example, the eccentricity of the terrestrial orbit. Close examination of Copernican calculations reveals his efforts, towards the end of his career, to accomplish the full shift from a 'mean-Sun-centric' to a genuine heliocentric system with the true physical body of the Sun as the immovable centre of the planetary system.

But this went far beyond the drawing of reasonable conclusions from observational data accurate to the traditional level of ancient standards. Copernicus refused to publish what in his heart he knew to be a torso; the publication of *De revolutionibus* in the year of his death, 1543, was mainly due to the enthusiasm of Georg Joachim Rheticus, Copernicus's only pupil.

After Copernicus the science of astronomy, for the first time in history, was presented with two radically different doctrines to choose between. The young and well-educated Tycho had a keen eye for this situation, and he also had the faculty – and royal sponsorship – for providing the right means to build a totally new foundation for astronomy. This happened during the last quarter of the sixteenth century at his research institution on the small island of Hveen between Scania and Zealand (Dreyer *1890*, Thoren *1990*).

Tycho's observational results vastly surpassed anything obtained before in their scope, comprehesiveness and accuracy. He recorded his copious observations in diaries for later use in some appropriate theoretical context. Tycho built some thirty astronomical instruments. Having used a particular instrument so much that he knew its flaws and merits, Tycho would take it as a model for a new instrument of the same type, but perhaps scaled up or down, or made of other materials. He developed new and more accurate sights, with parallel slits instead of holes, with which he reached an accuracy of about one minute of arc, close to the limit of the optical resolving power of the naked eye. But perhaps more importantly, Tycho would perform the same observation, say of the angular distance between two stars, many times and with different instruments. And his demand for agreement between the different observations, or for reproducibility as a condition for the acceptance of a result, introduced the ideal of objectivity into empirical science.

But Tycho also knew how to utilize observations. By an enormous amount of calculation he reduced the observations of a thousand stars to catalogue entries of their positions and magnitudes. To simplify the numerical calculations, Tycho, together with Paul Wittich, developed the method of *prosthaphairesis*, using trigonometrical tables just as logarithms would later be used. Not since the time of Hipparchus had an astronomer found it necessary to begin a revision of the theories of the motions of the 'wandering stars' (i.e. the planets) by establishing improved positions for the fixed ones.

Tycho completed an improved solar theory, and his preliminary lunar theory contained no fewer than four new components of motion. He demonstrated that comets moved among the planets, at distances far beyond the Moon. So, just as the 'new star' of 1572 was no atmospheric phenomenon, so neither were the comets. This was a blow for the Aristotelian assumption of the existence of hard impenetrable spheres in the sky.

Yet Tycho was not prepared to accept the full Copernican cosmology, so he formulated his own compromise: around the immovable and central Earth, the Moon and the Sun rotate, and around the Sun, as a secondary movable centre, the remaining five planets perform their orbital revolutions. As for the geometry of planetary models, Tycho adhered faithfully to the traditional principle that any motion in the sky should be composed of uniform and circular components.

Within less than a decade of Tycho's death, Galileo turned a telescope towards the sky, and Johannes Kepler removed the circles from planetary models and replaced them with 'true' elliptical planetary orbits in space. Only after the development of telescopic sights and micrometers some half a century later would telescopes outperform Tycho's sighting devices. But the magnifying power alone of the telescope helped Galileo to discoveries that demolished stone by stone the credibility of traditional cosmology, revealing mountains on the surface of the Moon, hosts of previously unseen stars, sunspots, and four satellites encircling Jupiter.

Kepler built on Tycho's stock of empirical data. He was the indefatigable calculator *par excellence*, using traditional geometry as well as summations of series, laboriously tabulated for want of the methods of calculus to be created half a century later. But Kepler's most original contribution to astronomy came about through his insistence on carrying the analysis beyond geometrical kinematics and all the way through to unveil 'true' physical causes.

Kepler himself saw the physics behind the phenomena as 'true', because it had to mirror the actual design on the part of the Creator of His creation. With hindsight, we know that all of Kepler's particular ideas about physical explanations were wrong, be they concerned with solar light, planetary minds, regular polyhedra, magnetic fibres or musical harmonies. Yet he produced sound theoretical astronomy, including the celebrated laws of planetary motion, guided by his firm belief that the heavenly motions were the consequence of forces acting on real bodies. Kepler's physics was not philosophical, based on axioms and logic, but an empirical physics raised upon the firm foundation of knowledge built by the Greek, Arab and Latin students of mathematical astronomy (Stephenson *1987*: 205).

René Descartes formulated the first all-embracing and overall successful alternative to Aristotelian cosmology. Tied to his hierachy of vortices, the Copernican world-system conquered learned Europe (Aiton *1972*). So the final synthesis of Newtonian mathematical physics, based on the law of universal gravitation and entailing the concept of action at a distance, had to fight for decades against Cartesianism before gaining general acceptance. For the next stages, see §8.1 and 8.8.

BIBLIOGRAPHY

Aiton, E. J. *1972, The Vortex Theory of Planetary Motions*, London: MacDonald, and New York: American Elsevier.

Billard, R. *1971, L'Astronomie indienne*, Paris: Publications de l'Ecole Française d'Extrême-Orient, Vol. 83.

Crombie, A. C. *1964, Augustine to Galileo*, 2 vols, London: Mercury Books.

Dreyer, J. L. E. *1890, Tycho Brahe. A Picture of Scientific Life and Work in the Sixteenth Century*, Edinburgh: Edinburgh University Press. [Repr. 1963, New York: Dover.]

King, D. A. *1986, Islamic Mathematical Astronomy*, London: Variorum Reprints.

—— *1987, Islamic Astronomical Instruments*, London: Variorum Reprints.

Lindberg, D. C. *1978, Science in the Middle Ages*, Chicago, IL: University of Chicago Press.

Pedersen, O. *1975*, 'The *Corpus astronomicum* and the traditions of mediaeval Latin astronomy', *Studia Copernicana*, **13** [= *Colloquia Copernicana*, Vol. 3], 57–96.

Pedersen, O. and Pihl, M. *1974, Early Physics and Astronomy*, London: MacDonald, and New York: American Elsevier.

Pingree, D. *1978*, 'History of mathematical astronomy in India', in *Dictionary of Scientific Biography*, Vol. 15, New York: Scribner's, 533–633.

Sayili, A. *1960, The Observatory in Islam*, Ankara: Publications of the Turkish Historical Society, Series 7, No. 38.

Stephenson, B. *1987, Kepler's Physical Astronomy* (Studies in the History of Mathematics and Physical Sciences, Vol. 13), New York: Springer.

Swerdlow, N. M. and Neugebauer, O. *1984, Mathematical Astronomy in Copernicus's* De revolutionibus, Parts 1 and 2 (Studies in the History of Mathematics and Physical Sciences, Vol. 10), New York: Springer.

Thoren, V. E. *1990, The Lord of Uraniborg. A Biography of Tycho Brahe*, Cambridge: Cambridge University Press.

Thorndike, L. *1949*, The Sphere *of Sacrobosco and its Commentators*, Chicago, IL: University of Chicago Press.

Yoder, J. G. *1988, Unrolling Time. Christiaan Huygens and the Mathematization of Nature*, Cambridge: Cambridge University Press.

2.8

Mathematical methods in medieval and Renaissance technology, and machines

EBERHARD KNOBLOCH

1 MASONRY AND ARCHITECTURE

The book *De architectura* ('On Architecture') by the Roman architect and engineer Vitruvius (first century AD) became known in the twelfth century; nearly all the works of Archimedes (third century BC) were translated by William of Moerbeke in 1269 from Greek into Latin; and the peripatetic *Mechanica* ('Mechanical Problems') were translated into Latin in the early thirteenth century. But these writings had no influence on the practice of medieval craftsmen who did not understand Latin. Moreover, even Vitruvius did not know mathematical statics, only empirical rules. Arches or vaults did not play a real role in his thinking. Utility had traditionally not been a consideration in science, only in the arts and crafts, while technology had got along quite successfully without any assistance from science (Drake *1976*). The physics of Aristotle was designed to explain the causes of things, not to be of use to the engineer, the architect or the builder.

Medieval builders undoubtedly used techniques of design and construction, knowledge of which was lost with the phasing out of Gothic building at the end of the Middle Ages (Shelby *1976*). Their design technique of deriving the elevation from the ground-plan was a characteristic use of what might be called the 'constructive geometry' of the medieval mason. It was almost entirely non-mathematical. It permitted a great many variations to be played on a basic set of geometrical forms.

The architecture of the Middle Ages was founded on empirical knowledge which was transmitted within the building lodges (*Bauhütten* in Germany). The apprentice had to learn from his master the step-by-step procedures for every design and construction problem. There were no technical secrets, or 'secrets of the craft'. The medieval Gothic cathedrals were built by means of empirical rules, rules of thumb; there were no engineering calculations based on mathematically formulated laws of nature, nor did medieval architects formulate any new fundamental law.

The Middle Ages produced no literature describing the construction of the foundations underlying its cathedrals. It is improbable that the tradition of the craft included written instructions for this kind of work, with a few exceptions (Prager *1968*). While arches were pointed, piers and pillars were more graceful and slender than in the Romanesque or Norman style, the weight of stone vaulted ceilings being supported by exterior flying buttresses as well as by the walls and pillars. The great central octagon of Ely Cathedral in England, for example, is one of the most beautiful and original designs to be found in the whole of Gothic architecture. The machines that were used in building such structures were mainly for lifting or shifting material; in the fourteenth century gearing was developed to levels of great complexity (White *1962*).

2 RENAISSANCE ARCHITECTURE

Builders of the late Middle Ages had available common devices such as simple winches, pulleys, wheels and cog-wheels. Technical refinements in hoists appeared in the machines of Filippo Brunelleschi (1377–1446) for the construction of the cupola of the Florentine cathedral Santa Maria del Fiore (1420–36). These machines were recorded in architectural sketch-books of the late fifteenth century (Scaglia *1966*). Brunelleschi was the first artist and engineer in whom we see art and technology combined with a search for a scientific basis. He was the main architect of the early Renaissance, one of the great developers of Gothic building and an important early figure in the evolution of modern structural design and analysis (Prager and Scaglia *1970*). Debates about the construction of cathedrals dealt mainly with the reinforcements required by the large vaults, and their aesthetic effects. While Gothic buildings prior to Brunelleschi had risen to great heights, none had ever covered a space of such breadth as did the Florentine cupola.

Little was known about the statics of cupolas; larger ones of every type had collapsed. Vaults were one of the most important supporting systems of construction engineering, but the first theoretical investigations into their operation date only from the end of the seventeenth century. Architects writing after Brunelleschi, like Leon Battista Alberti (1404–72), François Blondel (1618–86) and Carlo Fontana (1634–1714), only recorded empirical knowledge; there was no theoretical approach to statics (Straub *1964*).

Brunelleschi's technical work was characterized by the connection with mathematics and mechanics (Klemm *1965*). He probably knew the work of

Blasius of Parma and earlier mathematicians and physicists, and was personally acquainted with mathematicians of his time. He developed the painter's perspective, thus reproducing reality by means of mathematical instruments (§12.6). His solution of the cupola construction for the cathedral was based on theoretical considerations. He used the ellipse in his solution, enabling him to omit the conventional but expensive centring armatures (that is, trusswork of timber), as well as Gothic buttressing. Vaulting without armatures, that is the freehand vaulting method, which may be seen as a development of the Gothic 'permanent centrings' method, became his most famous invention. He proposed to construct a large vault as a double shell reinforced by modified Gothic buttresses and hollow ribs, combined with tie-rings, which were modified Roman features. Alberti praised him in *Della pittura* (1434), saying:

> Who is so dull or jealous that he would not admire Filippo the architect, in the face of this gigantic building, rising above the vaults of heaven, wide enough to receive in its shade all the people of Tuscany, and built without the aid of any trusswork or mass of timber.

Brunelleschi was forced to look for a completely new solution, because no one then could build a rigid wooden armature as required by the known methods, even if unlimited sums of money were spent on it (as Brunelleschi's biographer, Giorgio Vasari, remarked in 1568). The Florentine vault spanned over 40 metres.

The work of completing the cathedral during Brunelleschi's tenure was something of a proving-ground for engineering inventions. It is difficult to decide whether his inventions were founded on empirical or theoretical knowledge, or on both; his passion for experimentation in mechanics through practice was certainly not without foundation in theoretical studies. He probably promoted the rediscovery of worm-gear drive. The possible source of its development may have been the writings of Archimedes. The rotary crane on a mast, or hoist-crane, is his invention from the period 1417–19. Although Gian Poggio discovered a manuscript of Vitruvius's *De architectura* in 1417, it came too late to influence Brunelleschi's planning of mechanical devices. The worm-gear with screw and slider was conceived in Florence as early as 1436. At all events, Brunelleschi's achievements made the first half of the fifteenth century one of the great periods in technological experimentation and progress, preparing the ground for the contributions of Leonardo da Vinci (Scaglia *1966*). But it is not clear how long Brunelleschi's machines remained in use; building techniques may have taken one step back as a result of the Renaissance of Vitruvian mechanics.

The first printed work on construction engineering was Alberti's book

De re aedificatoria (1485), composed between 1443 and 1452. Written in Latin, it was of little use to the practitioner, but would have held some appeal for the humanistically minded patron. It continued and even tried to surpass Vitruvius, whose book appeared in print for the first time two years after Alberti's work. Alberti discussed new subjects like a new theory of cupola construction and rules for vaults. The numerical proportions he gives for stone bridges are of course not based on statics: they were derived from mathematically formulated empirical rules. Alberti himself was at the same time a scientist and a practitioner. In his Italian text *Ludi matematici* (*circa* 1450) he deals with problems of practical geometry, drawing on ancient, medieval and contemporary writings as well as on his own experience.

At around the same time, Antonio Francesco Averlino (*circa* 1416–70) wrote in Italian a treatise on architecture, which was never printed during his lifetime. The only sources he cited were Vitruvius and Alberti.

Only from the sixteenth century did Vitruvius's book enjoy a great vogue. Daniel Barbaro, for example, published his commentary on Vitruvius in 1556 (it was reprinted in 1567 and 1584). Of special interest was a water-screw or water-conveyor, described by him as a movable screw turning in a fixed casing. When Guiseppe Ceredi (*circa* 1520–70) published his *Tre discorsi sopra il modo d'alzar acque da' luoghi bassi* ('Three Discourses on the Method of Lifting Water from Low Places') in 1567, he found that no general rules on the optimum construction of water-screws were available. The device was by no means new – its design and construction were discussed by Vitruvius – but he soon realized that the formula given by Vitruvius (length to central shaft diameter, 16 : 1) was both pointless and impracticable. Ceredi improved on this ratio, and applied for a patent on the Archimedean water-screw, which was not known in western Europe in the Middle Ages; the patent was granted in 1566. He may have been the first to advocate in print the building of models to different scales and testing them as simulations as a means of optimizing practical efficiency. The history of technology abounds in examples of this practice, most notably with Leonardo da Vinci (Drake *1976*).

3 STEPS TOWARDS SCIENTIFIC TECHNOLOGY AND TECHNICAL MECHANICS

New elements of the technological development are already to be found in Konrad Kyeser's (1366 – after 1405) *Bellifortis* (1405). He provided the first secure evidence of the crank, the most important single mechanical device next to the wheel: it is the chief means of transforming continuous rotary motion into reciprocating motion, and vice versa. The Venetian physicist

Giovanni Fontana (1395–1455) described a compound crank in his notebook, in about 1420. The essential step in exploring the kinetic possibility of crank and connecting-rod was taken in Italy (White *1962*). The earliest evidence of such a combined device is due to the 'Archimedes of Siena', Mariano Daniello di Jacopo, called Taccola (1381–1453/1458), who produced the series of four books *De ingeneis* between 1427 and the end of his life, and the work *De rebus militaribus* in 1449. He put forward many ideas for the construction of pumps, pump systems, mills and hoists (Prager and Scaglia *1972*), but he was unaware of many of the more refined devices developed in Antiquity. The Archimedean spiral, for example, was misconceived by him, though Vitruvius has described it correctly. Indeed, the amount of technical knowledge accumulated and distributed in books and available in Taccola's time was small. He did not mention Euclid, Archimedes or Vitruvius, especially the famous machines of Vitruvius's tenth book; nor did he apparently know the peripatetic 'Mechanical Problems'. But he did contribute to the gradual re-emergence of rational teaching and accumulation of knowledge.

The fifteenth century saw the elaboration of theories of crank, connecting-rod and governor. The technicians strove to achieve continuous rotary motion, and showed enthusiasm for the flywheel–crank combination. But progress in the development of machinery came about as a result of studies in scientific technology, such as the mathematical and experimental analysis of gear-tooth profiles, lubricated bearings, and load-bearing posts or beams, as exemplified in particular by Leonardo (Section 4 below). This new direction is characterized by the greater attention paid to mathematical considerations. Brunelleschi, Fontana and Alberti occupied themselves with science, mathematics and physics. Bonaccorso Ghiberti (1451–1516) and his contemporary Bartolomeo Neroni studied problems of pure geometry, in the case of Ghiberti for example, before discussing traditional military engineering. Ghiberti wrote his *Zibaldone* between 1472 and 1483.

Francesco di Giorgio Martini (1439–1501), who was well acquainted with Taccola's drawings, tried to find the correct relation between the length, diameter and thickness of a gun-barrel, and between powder charge and the weight of a cannonball. He recognized that a water-pipe which is narrowed at its end produces a stronger jet of water. However, in his *Trattato di architettura ingegneria e arte militare* (*circa* 1480) he gives no general law; his simple rules always concern particular examples. The machines are the truly new element among his drawings. He studied them systematically because he wanted to rehearse all possible solutions and to combine all known mechanisms. This applies especially to the fifty or so corn-mills that he designed in his treatise.

4 THE ACHIEVEMENTS OF LEONARDO DA VINCI

Leonardo was personally acquainted with Francesco di Giorgio, and knew Taccola's drawings at least to a certain extent. He made more effort than any of his predecessors to experiment, to carry out calculations and to look for general rules whenever he studied technical devices. Leonardo was artist, engineer, scientist and philosopher. He divined the methodological importance of mathematics for natural sciences and technology (though only the Baroque period created mathematical physics which made modern technology possible); for him, therefore, mathematics was the model of all genuine knowledge (Fleckenstein *1965*). His respect for mathematics grew when he applied it to solve mechanical problems: 'There is no certainty in science when one of the mathematical sciences cannot be applied', he noted. Without modern scientific principles to help him, he drew upon those guiding rules that he could grasp (Reti and Dibner *1969*). Nevertheless, he appreciated the importance of experiment: 'Before making this case a general rule, test it by experiments two or three times and see if experiments produce the same effect.' He assigned numbers to things measurable: they were added, subtracted and multiplied to tally with observed results. He recognized the universal validity of the law of action and reaction, known today as Newton's third law. He also gained an amazing knowledge of the relation between the volume, temperature and pressure of vapours and gases.

The two Madrid Codices I and II (which were rediscovered at Madrid in 1965) are a pattern-book of mechanical engineering. Madrid I belongs to the period 1492–7; Madrid II was written between 1491 and 1505. They contain some of Leonardo's most brilliant machine assemblies and engineering designs. He aimed at a systematic analysis of conditions and constructive details that led towards the rational assembly of useful machines. He was always anxious to integrate theory with application. He laid the foundations for a completely novel general theory of the construction of machines (Maschat *1989*).

He had clear ideas about the virtues and limitations of water power. In the Codex Atlanticus, which contains a lifetime's notes, he wrote: 'Falling water will raise as much weight as its own, adding the weight of its percussion . . . . But you have to deduce from the power of instrument what is lost by friction in the bearings.' This remark can be interpreted as the first formulation of the basic definition of potential energy. Advice is offered to the engineer for correcting the theoretical efficiency, by taking into account energy losses caused by friction.

Leonardo studied intensively problems related to friction in gearwheel trains and in bearings. He recognized the faults of the devices of his day, proposing solutions that have been adopted only in recent times. This applies to worm-gears, screw-jacks, ball-bearings and disc bearings. He suggested, for example, anti-friction devices based on bodies rolling between the working surfaces of a bearing. He studied the transmission of power and movement in belt drives, chain drives and sprocket-wheels.

Experiments with perpetual-motion machines were certainly one of the reasons for the great and even increasing interest in friction and in methods of reducing it. Leonardo's attitude towards all these attempts was critical. He said: 'O speculator about perpetual motion, how many vain chimeras have you created in the like quest? Go and take your place with the seekers after gold.'

He was aware of the relation of the speed of work to its magnitude, as can be concluded from his power-analysis diagram. In it, the power derived from a hydroturbine is associated schematically with each multiplication of mechanical advantage as the power is transmitted from a large turbine to a small worm-gear acting on a large gear. The faster-revolving worm created in the slower-revolving gear an increment in power in reverse proportion to its speed: 'The movement of the pinion and the surface of the wheel is as much more rapid than that of its axis as the circumference of the pinion is contained in the circumference of the wheel.' Many of his drawings show for the first time a reaction turbine based on an inverted Archimedean screw, thus preceding by a hundred years Giovanni Branca, to whom the invention of those driving spirals is generally attributed.

Leonardo also analysed the efficiency of a treadmill. Simon Stevin, the first scholar after Leonardo to analyse mathematically this efficiency, arrived at the same conclusion. In a similar way he made an intensive study of elevating mechanisms like pulley blocks, toothed wheels, springs, screws and nuts, rope-and-chain gearings and wedges, and deduced lasting results in technical mechanics and mechanical engineering.

5 CONCLUDING REMARKS

The engineers of the sixteenth century continued the mathematization of technical problems, thus preparing the way for the scientifically applied technology of the eighteenth century. In general they were the first to try to develop their work scientifically. Niccolò Tartaglia began to mathematize ballistics (*La nova scientia*, 1537: see §8.11). Guidobaldo del Monte (1545–1607), mathematician, physicist and chief of staff of the Toscanian fortifications, revived the idea of a new mechanics (Keller *1975*). Finally, Domenico Fontana (1543–1607) became famous in setting up a Roman obelisk. He was not provided with statical calculations, but calculated the obelisk's weight before estimating how many turns of the mechanism would raise it to the vertical by means of pulley blocks.

BIBLIOGRAPHY

Drake, S. *1976*, 'An agricultural economist of the late Renaissance', in Hall and West *1976*: 53–73.

Fleckenstein, J. D. *1965*, 'Die Einheit von Technik. Forschung und Philosophie im Wissenschaftsideal des Barock', *Technikgeschichte*, **32**, 19–30.

Gille, B. *1964*, *Les Ingénieurs de la Renaissance*, Paris: Hermann. [English edn: *The Renaissance Engineers*, 1966, Cambridge, MA, and London: MIT Press. German edn: *Ingenieure der Renaissance*, 1968, Vienna and Düsseldorf.]

Hall, B. S. and West, D. C. (eds) *1976*, *On Pre-modern Technology and Science. A Volume of Studies in Honor of Lynn White, Jr.*, Malibu, CA: Undena Publications.

Keller, A. G. *1975*, 'Mathematicians, mechanics and experimental machines in Northern Italy in the sixteenth century', in M. P. Crosland (ed.), *The Emergence of Science in Western Europe*. London and Basingstoke: Macmillan, 15–34.

Klemm, F. *1965*, 'Die Rolle der Technik in der italienischen Renaissance', *Technikgeschichte*, **32**, 221–43.

Maschat, H. *1989*, *Leonardo da Vinci und die Technik der Renaissance*, Munich: Profil.

Prager, F. D. *1968*, 'A manuscript of Taccola, quoting Brunelleschi, on problems of inventors and builders', *Proceedings of the American Philosophical Society*, **112**, 131–49.

Prager, F. D. and Scaglia, G. *1970*, *Brunelleschi. Studies of his Technology and Inventions*, Cambridge, MA, and London: MIT Press.

—— *1972*, *Mariano Taccola and his Book* De ingeneis, Cambridge, MA, and London: MIT Press.

Reti, L. and Dibner, B. *1969*, *Leonardo da Vinci, Technologist. Three Essays on Some Designs and Projects of the Florentine Master in Adapting Machinery and Technology to the Problems in Art, Industry and War*, Norwalk, CT: Burndy Library.

Scaglia, G. *1966*, 'Drawings of machines for architecture from the early Quattrocento in Italy', *Journal of the Society of Architectural Historians*, **25**, 90–114.

Shelby, L. R. *1976*, 'The "secret" of the medieval masons', in Hall and West *1976*: 201–19.

Straub, H. *1964*, *Die Geschichte der Bauingenieurkunst. Ein Überblick von der Antike bis in die Neuzeit*, 2nd edn, Basel and Stuttgart: Birkhäuser.

Taccola, M. *1984*, *De rebus militaribus (De machinis, 1449)* (ed. E. Knobloch), Baden-Baden: Koerner. [Includes a manuscript.]

White, L. Jr 1962, *Medieval Technology and Social Change*, Oxford: Clarendon Press. [German edition: 1966, *Die mittelalterliche Technik und der Wandel der Gesellschaft* (transl. G. Quarg), Munich: Moos.]

2.9

Mathematical optics from Antiquity to the seventeenth century

A. MARK SMITH

1 THE EVOLUTION OF RAY THEORY

The history of mathematical optics begins effectively with Euclid's articulation of the visual-ray theory (*circa* 300 BC). According to this theory, the eye emits discrete lines of visual flux (subtle optical 'fire') in rectilinear bundles to form a cone whose vertex defines the centre of sight and whose base is the visual field. Whatever the flux touches within the field is thereby seen, the resulting visual information being conveyed back through the flux-lines to the centre of sight. Elaborated on over the next four and a half centuries, this theory found mature expression in Ptolemy's *Optics*, by which time the structure of optical analysis had taken canonical form according to the tripartite division into *optics* (unimpeded radiation), *catoptrics* (fully broken, or reflected radiation), and *dioptrics* (partially broken, or refracted radiation) (§1.4).

Despite apparent similarities between ancient and modern ray theory, there is one crucial difference: the former was intended to explain sight, not light. The Euclidean–Ptolemaic ray was thus presumed to establish both a physical sense-link between viewpoint and external objects, and a simple, mathematically determined spatial relationship between viewpoint and point viewed. Accordingly, visual perception of such properties as shape and size could be explained in terms of angles and ray lengths. The Euclidean–Ptolemaic ray thus represented a line of sight rather than a path of light; in fact, visual-ray theorists all but ignored light, regarding it as a mere precondition for, not an actual object of, sight (Lejeune *1948*, Simon *1988*).

Perhaps the clearest systematic flaw with the visual-ray theory is its redundancy: why, after all, posit both an outward physical reach of visual flux and an inward perceptual reach when the two can be united in a single inward reach of visual information, somehow physically radiated from external objects to the eye? This unification is precisely what the Arab thinker, Ibn al-Haytham, achieved some 850 years after Ptolemy. At the heart of his account in the *Kitāb al-Manāẓir* ('Book of Optics') is the notion that every point of light or luminous colour on a visible surface replicates itself continuously and omnidirectionally as a formal effect through transparent media. The result is a sphere of propagation, with each radius a sort of trajectory for point-forms of light and illuminated colour. On reaching the eye, such point-forms make physical impressions on the surface of the crystalline lens, which then senses them visually. But only those point-forms striking orthogonally make an effective impression, so there is a perfect point-to-point correspondence between a visible surface and its physical/visual impression on the lens. In effect, then, Ibn al-Haytham created a cone of visibility mathematically identical, but physically opposite, to the Euclidean–Ptolemaic visual cone (Lindberg *1976*).

Ironically, the destiny of Ibn al-Haytham's ray theory lay not in the Islamic East but in the Christian West. Translated into Latin (*circa* 1200) under the title *De aspectibus* (and ascribed to 'Alhazen'), the *Kitāb al-Manāẓir* was soon recognized as authoritative within scholastic circles. As such, it inspired a number of derivative works, such as Roger Bacon's *Perspectiva* (*circa* 1265), Witelo's *Perspectiva* (*circa* 1275) and John Pecham's *Perspectiva communis* (*circa* 1280), which formed the core of the so-called Perspectivist optical tradition. While the Perspectivists did little to alter the basic structure of Ibn al-Haytham's theory, they did provide a fuller causal account of the physics of light radiation. Particularly significant were their efforts to explain the action of light in quasi-mechanistic terms, according to which light radiation was treated virtually, albeit not literally, in terms of physical projection through space.

Ibn al-Haytham and his Perspectivist followers went a long way towards transforming the ray from a line of sight to a trajectory for light, but the full transformation had to await the seventeenth century. The crucial figures in this process were Johannes Kepler, René Descartes, Pierre de Fermat and Christiaan Huygens, all of whom sought to recast the account of light radiation and vision in mechanistic terms. Thus, in his *Ad Vitellionem paralipomena* ('Additions to Witelo', 1604), Kepler reduced the eye to a mere 'camera' within which incoming colour images are radially projected via the crystalline lens onto the retinal screen (Kepler also provided the definitive account of the projection of pinhole images in the *camera obscura*). Subsequently Descartes outlined a physical theory of light in terms of mechanical impulses transmitted rectilinearly and instantaneously through a continuous, unyielding aethereal medium. Fermat went even further in supposing that light consists literally of minute particles hurtling radially through space at enormous speed. And, finally, Huygens reformulated the Cartesian account of light radiation in terms of longitudinal wavefronts passing swiftly through a continuous, but highly elastic, aethereal medium. Thus, by the second half of the seventeenth century, not only had light been completely (and literally) transformed from a formal to a material effect, but also the focus of mathematical optics had been irrevocably shifted from sight to light (Lindberg *1976*, Smith *1987*).

2 CATOPTRICS

2.1 The general study of reflection

The equal-angles law of reflection was known by Euclid's time at the latest. Also known was that point images in plane mirrors are located at the intersection of the line joining the eye to the point of reflection and the perpendicular dropped from the object-point to the mirror's surface (what we now call the cathetus of incidence). This knowledge underpins the somewhat unsophisticated treatment of mirrors in both the *Catoptrics* ascribed (perhaps falsely) to Euclid and the later *Catoptrics* of Hero of Alexandria (? mid-first century AD) (§1.4). What makes the latter work particularly important is Hero's attempt at the beginning to demonstrate the necessity of the equal-angles law by proving that the shortest possible ray-couple linking centre of sight to point of reflection and point of reflection to object-point is the one that subtends equal angles with the mirror's surface, or the tangent to its surface, at the point of reflection. Reflection, he concludes, must therefore follow the minimum path defined by this ray-couple, because Nature does nothing in vain. Despite its ultimate failure (with regard to concave mirrors), Hero's least-lines proof none the less served as an influential model of argument whose trace can be seen in Fermat's least-time proof of the sine law of refraction (see Section 3 below).

At a practical level, Ptolemy's account in books III and IV of the *Optics* raises the study of reflection to a far higher level than that reached by Euclid or Hero. For instance, Ptolemy offers an experimental verification of the law of equal angles for plane as well as cylindrical concave and convex mirrors. He then undertakes a systematic investigation of image location in all three kinds of mirror, attempting to explain precisely how the size and location of the image vary with distance between object and mirror surface, how images are distorted in curved mirrors, how to determine multiple image locations in concave mirrors, an so forth. Granted the extraordinary difficulty of some of these problems, it is small wonder that Ptolemy's solutions to them are sometimes less than satisfactory (Lejeune *1957*).

Whatever its shortcomings, though, particularly in regard to concave mirrors, Ptolemy's account provided the blueprint for subsequent studies of reflection. Thus, Ibn al-Haytham followed Ptolemy's general lead in books IV–VI of the *Kitāb al-Manāẓir*, where we see much the same pattern of analysis according to type of mirror, ranging in complexity from plane, through convex (subdivided into spherical, cylindrical and conical) to concave (equivalently subdivided). The greater complexity and specificity of Ibn al-Haytham's treatment is, of course, matched by the greater sophistication of his mathematical approach. Book V of the *Kitāb al-Manāẓir* provides a good example of what Fermat later dubbed 'Alhazen's problem': given a centre of sight and a point on a visible object, to find the point of reflection on a convex or concave circular mirror. Ibn al-Haytham's solution demands a far higher level of geometrical knowledge (including a command of conic sections) than is evinced in Ptolemy's *Optica*. Accordingly, the overall study of reflection in the *Kitāb al-Manāẓir* remained unsurpassed until the application of algebraic techniques in the seventeenth century.

2.2 Burning mirrors

For all its fascination to the ancients, the study of burning mirrors has left little documentary trace before the Middle Ages, perhaps because, having nothing to do with vision, it fell outside the mainstream of optical analysis. Possibly the earliest such trace is found in the final theorem of the *Catoptrica* attributed to Euclid, where it is maintained that a spherical concave mirror will focus enough rays to the centre to cause burning. The first known demonstration that a parabolic mirror focuses all parallel rays to a single point – a fact curiously ignored by Apollonius of Perga – occurs in Diocles' *On Burning Mirrors* (*circa* 200 BC). Although Diocles' analysis seems to have had little long-term influence in the Greek world (for instance, we see no evidence of it in Anthemius of Tralles' clumsy treatment in *On Paradoxical Devices* (*circa* AD 500) (compare §1.6)), it apparently found a readier reception among medieval Arabic thinkers. Prominent among these was Ibn al-Haytham, whose 'On Burning Mirrors' represented the most sophisticated treatment of its subject to that time. Rendered into Latin around 1200, this brief treatise was a crucial vehicle for the knowledge of parabolic sections throughout the European Middle Ages and Renaissance (Toomer *1976*, Clagett *1980*).

3 DIOPTRICS

As with reflection, so evidently with refraction, the first truly systematic study was undertaken by Ptolemy, in the fifth book of his *Optica*. The

262

centrepiece of this study is an experimentally based effort to determine the index of refraction from air to water (as well as from air to glass, and water to glass). Ptolemy's apparatus consisted of a hollow glass semicylinder and a circular plaque divided into quarters by two diameter-lines and marked off in $1°$ subdivisions along its circumference. Having filled the semicylinder with water, Ptolemy placed the plaque upright in it so that one of its diameter-lines coincided with the water's surface, while the other cut it along the normal. After affixing a small marker to the plaque's centre, Ptolemy attached another $10°$ from the normal along the section of the circumference that lay above the water. Sighting along the line connecting the two markers, he adjusted another marker on the circumference below the water until all three markers lined up, the arc between the normal and the marker below the water's surface representing the angle of refraction. Repeating this process at ten-degree intervals, Ptolemy found that, as the angle of incidence, i, was increased from $10°$ to $80°$ in $10°$ steps, the corresponding values for the angle of refraction, r, were $8°$, $15.5°$, $22.5°$, $29°$, $35°$, $40.5°$, $45.5°$ and $50°$. The pattern of those results indicates that Ptolemy was systematically adjusting his raw data according to constant 'second differences' of $0.5°$. Thus, when $i = 20°$, $r = 8° + (8° - 0.5° = 7.5°) = 15.5°$, when $i = 30°$, $r = 8° + (8° - 0.5° = 7.5°) + (7.5° - 0.5° = 7°) = 22.5°$, and so forth (for more discussion, see Smith *1982*).

However accurate these results may appear by comparison to modern values, it is clear from their actual derivation by constant second differences that Ptolemy was badly off track in his search for the law of refraction. Still, his was the track that was followed for the next millennium and a half, with one signal exception – the tenth-century Arab mathematician Ibn Sahl, who somehow managed to adduce the correct relationship of sines (i.e. $\sin i : \sin r = \text{constant}$) and, on that basis, to prove that the surface of refraction that will focus parallel rays to a single point (the anaclastic surface) is hyperbolic (Rashed *1990*). Unfortunately, Ibn Sahl's discovery seems to have gone unnoticed (the problem of the anaclastic had to be re-solved by Kepler and Descartes in the early seventeenth century), so it was primarily through Ptolemy's misguided approach, subsequently taken up by Ibn al-Haytham and his Perspectivist followers, that refraction was understood throughout the Middle Ages and Renaissance.

For a variety of reasons, not the least of which was an acute interest in lenses (particularly after the invention of the telescope in 1608 (§9.3)), the study of refraction was pursued with renewed intensity at the turn of the sixteenth and seventeenth centuries. Among those involved, three in particular stand out. Kepler, although he ultimately failed to find the correct law, did much to undermine the traditional understanding of refraction.

Wilibrord Snel and Descartes, on the other hand, succeeded where Kepler failed. Yet, while Snel and Descartes (and perhaps Thomas Harriot) share the credit for 'discovering' the sine law, Descartes was the first to publish this discovery, along with a theoretical derivation. Formulated in the *Dioptrique* (1637), this derivation is based on two key suppositions: that the speed of light in any transparent medium is directly proportional to the medium's optical density, and that when light passes obliquely through the interface between media of different optical densities, its motion along the horizontal vector is conserved. Given this suppositional basis, it is little wonder that the resulting 'proof' of the sine law aroused such immediate critical reaction. Indeed, it was in just such reaction that Fermat adduced his own counter-proof (perfected by 1662) on the assumption that, *pace* Descartes, the speed of passage for light through any given medium is inversely, not directly, proportional to the medium's optical density. Assuming, further, that any light-particle passing through the interface between media of different optical densities will follow the path that is most temporally, rather than spatially, economical, Fermat proved that such a path is the one dictated by the sine law (for further details, see Sabra *1967*, Smith *1987*). The next stages of physical optics and optical instruments are traced in §9.1–9.3; geometrical optics is noted in §7.7 on line geometry.

BIBLIOGRAPHY

Boyer, C. B. *1959*, *The Rainbow, from Myth to Mathematics*, New York: Yoseloff. [Repr. 1987, Princeton, NJ: Princeton University Press.]

Clagett, M. *1980*, *Archimedes in the Middle Ages*, Vol. 4, Philadelphia, PA: American Philosophical Society.

Lejeune, A. *1948*, *Euclide et Ptolémée: Deux stades de l'optique géometrique grecque*, Louvain: Bibliothèque de l'Université.

—— *1957*, 'Recherches sur la catoptrique grecque', *Mémoires de l'Académie Royale de Belgique*, 2ᵉ série, Classe des Lettres, **52**, No. 2.

Lindberg, D. C. *1976*, *Theories of Vision from Al-Kindi to Kepler*, Chicago, IL: University of Chicago Press.

Rashed, R. *1990*, 'A pioneer in anaclastics: Ibn Sahl on burning mirrors and lenses', *Isis*, **81**, 464–91.

Sabra, A. I. *1967*, *Theories of Light from Descartes to Newton*, London: Oldbourne. [Repr. 1981, Cambridge: Cambridge University Press.]

Simon, G. *1988*, *Le Regard, l'être et l'apparence dans l'optique de l'Antiquité*, Paris: Seuil.

Smith, A. M. *1987*, *Descartes's Theory of Light and Refraction: A Discourse on Method*, Philadelphia, PA: American Philosophical Society.

—— *1982*, 'Ptolemy's search for a law of refraction: A case-study in the classical methodology of "Saving the Appearances" and its limitations', *Archive for History of Exact Sciences*, **26**, 221–40.

Toomer, G. J. *1976, Diocles on Burning Mirrors*, Berlin: Springer.

Van Helden, A. *1977, The Invention of the Telescope*, Philadelphia, PA: American Philosophical Society.

2.10

Musical intervals

H. FLORIS COHEN

1 MUSIC BEYOND AESTHETICS

Systematic, theoretical thought about music has always transcended the feelings of pleasure that music may arouse in us. Not only is there a discipline of musicology in the sense of a set of notions about the aesthetic appeal of a piece of music as perceived by the human ear; besides the theory of music as an art-form there has also existed from Pythagorean times onwards a science of music that, in its turn, was strongly associated with certain traditions in natural philosophy. The principal concept linking these varied aspects of musical theory together was that of harmony. In its widest sense, music was thought to express, or to mirror, both the supreme harmony that pervades the Universe (macrocosm) and man as the active creator as well as passive recipient of musical harmonies. From Boethius's summing-up of ancient doctrine on the subject until far into the seventeenth century, it was customary to distinguish between *musica instrumentalis* (studying melody, singing, the mathematical basis of musical theory), *musica humana* (how the incorporeal soul mingles with the physical world) and *musica mundana* (the music of the spheres, the harmony of the elements, the cycle of the seasons).

In the sense here indicated, music formed part of a very broad conceptual framework that, in the Western tradition to which it belongs, retained its vitality at least until the Enlightenment. Inside this huge thought-complex a more specialized branch which may be called 'quantitative musical theory' can be discerned; it equally takes its origins in Pythagorean ideas about harmony. Out of some of the ideas that belong to this domain there arose two distinct scientific problem-areas of a broadly mathematical nature, which have remained under scrutiny up to the present. These problem-areas emerge when musical intervals are connected with number, as was done for the first time in the Pythagorean Brotherhood.

266

2 THE TWO PROBLEM-AREAS IN MUSICAL ARITHMETIC

The connection discussed here between music and number is not the well-known one of how number may be directly utilized in musical composition, for example J. S. Bach writing 365 notes in his chorale *Das alte Jahr vergangen ist* (BWV 614) (see §12.5). Rather, the connection is established by observing that dividing a string by consecutive integers yields consonant intervals (up to a point). The discovery is traditionally attributed to Pythagoras (Figure 1). In the huge continuum of intervals established by making two arbitrary musical notes sound together, only a few strike our ear as harmonious and pleasing (consonant) rather than harsh and jangling (dissonant). Pythagoras's discovery came down to the observation that these rare consonances are produced by sounding together notes from strings whose lengths are in ratios given by the first few integers (other properties of the strings being equal). Thus, unison is given by the ratio of string lengths 1 : 1; the octave by 1 : 2; the fifth by 2 : 3, and the fourth by 3 : 4. Here, for the Pythagoreans, the range of consonant intervals came to an end. Meanwhile, sufficient material has been generated by means of these

Figure 1 How Pythagoras discovered the ratios of the consonances, according to Franchinus Gafurius, *Theorica musice* (1492)

267

successive divisions to build a tonal scale. This is achieved primarily by 'subtracting' the fourth 3/4 from the fifth 2/3, which is done arithmetically by division, yielding 8/9 for the whole tone. Thus having 1/2 for the octave, and $(8/9)^5$ for five consecutive whole tones, the two remaining semitones in the Pythagorean scale are given by 243/256 each.

We can now define the two problems that have vexed many able mathematical minds over the centuries: the problem of consonance, and the division of the octave. The problem of consonance is, simply, to ask 'why': why is it that those few consonant intervals are the very ones given by the ratios of the first few integers? A closely related question is why the demarcation between consonant and dissonant intervals occurs where it does – in the Pythagorean tradition, after the interval of the fourth or, in terms of ratios of string lengths, after the number 4 (justified with reference to the tetraktys; see §12.5).

The most questionable aspect of the division of the octave accomplished by the Pythagoreans is that piling up seven octaves $(1/2)^7$ from a given C upwards yields a C that is very close, but not identical, to the high C (really B sharp) reached by piling up twelve fifths $(2/3)^{12}$. The difference equals 524 288/531 441, or roughly 73/74, which ratio is known as the 'Pythagorean comma'. If nothing is done to eliminate the gap, it would make itself heard as the so-called 'wolf-fifth' – a pseudo-fifth that exceeds the normal magnitude of 2/3 by the comma, and makes itself heard in the howling that gives it its name. How best to eliminate the gap, as well as some others that were to emerge later, is the core issue in the division of the octave.

3 EARLY COMPLICATIONS

The Pythagorean scale is straightforward only as long as it remains a simple, diatonic scale; but as soon as chromatic alterations are introduced, many complications arise. Such complications occupied those ancient musical theorists who followed the mathematical approach originated by the Pythagoreans, among them Euclid and Ptolemy. Besides these men, there has always been a school of musical thought (customarily named after Aristoxenos) which preferred to stick to the purely empirical evidence provided by hearing alone. In this approach the validity of the problem of consonance is denied because the connection between musical sound and number is assumed to provide no insight into otherwise hidden properties of the real world.

Even though Greek thought produced two distinct theories of the production of sound (one based on vibrational motion, one based on the emission of sound corpuscles), no systematic connection was made with the problem of consonance. The ratios of the consonance were abstracted at once from the vibrating string that brought them forth. In other words, the science of music was treated as a branch of applied arithmetic rather than as quantifiable acoustics (Barker *1989*). This remained true in the Middle Ages, when *musica* became one of the four academic disciplines that made up the quadrivium. The mathematical content of the domain remained largely confined to operations with numerical ratios deriving from a variety of possible divisions of the octave.

Meanwhile, in actual music-making drastic innovations took place which altered some fundamental parameters of the mathematical domain. Chief among them were the rise of polyphony, the acceptance of the major third as consonant and the emancipation of instrumental music.

After the introduction of the consonant major third in musical practice, it was recognized by theorists to be given by the ratio of string lengths 4/5. It carries in its wake three more consonances: the minor third 5/6, the major sixth 3/5 and the minor sixth 5/8. Hence, the problem of consonance appeared in a partly novel guise: how to explain that with the range of consonances now accepted the series comes to an end? The authoritative answer given by G. Zarlino (1558) centred around the *senario* – the range of the first six integers, with the 5/8 ratio for the minor sixth being worked into the explanation on an *ad hoc* basis.

In Pythagorean tuning the major third is the dissonant interval 64/81. The difference with the pure major third 4/5 is 80/81, known as the 'syntonic comma'. Scales in so-called 'just intonation' (i.e. based as far as possible on both pure fifths and pure thirds) are inherently unstable, as was demonstrated by G. B. Benedetti (1563, 1583) and others. That is to say, perfectly normal progressions, if made in none but pure intervals, result in clearly audible gains or losses in pitch. The question of how singers resolve the dilemma gave rise to heated debates in the sixteenth and seventeenth centuries, notably between Zarlino and his erstwhile disciple, Vincenzo Galilei (the father of Galileo) (Palisca *1961*, Walker *1978*).

For keyboard music, too, just intonation raises tough problems. Whereas for singers two different whole tones exist (8/9 and 9/10, together producing the pure major third, 4/5), keyboard players must either extend the range of keys available to their fingers and feet, or make concessions to the purity of either the major third or other consonant intervals. The sixteenth to eighteenth centuries saw a proliferation of solutions advanced in order to circumvent, by means of one compromise or another, the inescapable consequences of the existence of the Pythagorean comma, the syntonic comma and the 'lesser and greater dieses' (all these micro-intervals expressing incompatibilities between certain consonances). Among such solutions, the principle of 'temperament' emerged as the most practicable. It takes advantage of the empirical fact that our hearing is willing to put up with small deviations from purity in most musical intervals. Hence, temperament comes down to making concessions, at appropriate points, to the purity of some consonant intervals, with the integrity of the octave 1/2 always being taken as inviolable (Barbour *1951*, Lindley *1984*).

4 TEMPERAMENT

A widespread division of the octave that was practicable for keyboard players of the time is known as 'mean-tone temperament'. It preserves the purity of most of its major thirds (which implies a slightly enlarged fifth), and equalizes its whole tone by splitting the difference between the 'large' whole tone 8/9 and the 'small' whole tone 9/10 produced in just intonation, thus yielding $2/\sqrt{5}$ for the whole tone. The chief drawback from a present-day point of view is that relatively little chromatic alteration within a piece is possible, because all 'black keys' are uniquely determined (e.g. E flat cannot serve as a D sharp at the same time). As long as music-making required little modulation or transposition this did not matter a great deal. With the introduction in the early decades of the seventeenth century of new stylistic devices in musical composition, however, more chromatic alteration became gradually necessary than mean-tone temperament allows.

One mathematically simple solution is equal temperament. This cuts through the Gordian knot of commas and dieses by radically abolishing the difference in size between the diatonic and chromatic semitones. Eleven mean proportionals between 1 and 2 are made to define the ratios of consecutive intervals within the range of the octave. Equal temperament is explained (though by no means advocated) together with many alternative temperaments in the works of Zarlino and F. Salinas (second half of the sixteenth century). An early computation of the numerical values involved is often ascribed to Simon Stevin – an attribution which can stand only if one realizes that he did not so much advocate a temperament as compute what he believed to be the one and only true, or 'natural', division of the octave. Marin Mersenne, in *Harmonie universelle* (1636–7), presented calculations for a great variety of temperaments without unambiguously favouring any of them (Cohen *1984*). After all, there were good reasons to reject equal temperament, since its adoption requires a major musical sacrifice in terms of subtle distinctions between intervals.

During the centuries over which the battle of the temperaments raged, proposals for suitable divisions proliferated. Whereas most solutions sought by practising musicians (for example, Andreas Werckmeister) were mathematically uninteresting, those advanced by the 'scientific party' were often as elegant mathematically as they were irrelevant musically. To what extent one side may have benefited from the work of the other is a historical question that still awaits its student.

Most mathematical scientists who involved themselves in the issue came out in favour of multiple divisions. One example is Huygens' division of the octave into 31 equal parts. For him the chief benefits were the remarkable closeness to his preferred temperament – the mean-tone variety – and the fact that his division yielded a closed 'harmonic cycle'. Another interesting feature is that Huygens was among the first to use logarithms in calculations of the division of the octave (Huygens *1691*, Cohen *1984*).

5 THE PROBLEM OF CONSONANCE TRANSFORMED

During the early decades of the seventeenth century the problem of consonance underwent a decisive transformation (Cohen *1984*). As mentioned, its treatment in the quadrivium had been of an exclusively numerical nature, and this had been true of Zarlino's *senario* as well. Early attempts to break through the numerical boundaries customarily imposed on the problem were made by Benedetti, Vincenzo Galilei and Stevin. Benedetti linked the vibrational nature of the production of musical sound to the quantitative properties of the consonant intervals, even though he did not pursue this novel line of approach beyond trivial, numerical conclusions. Vincenzo Galilei extended the range of numbers to be associated with the consonances by considering string tension as a variable that is as relevant to the problem as string length. This yields squared ratios for the intervals (e.g. 4/9 for the fifth). He also investigated experimentally unisons produced by strings possessing diverse properties. Stevin radically denied (in line with the Aristoxenian tradition) that number and consonant sound are related in any meaningful sense. Johannes Kepler, on the contrary, fully endorsed this fundamental Pythagorean view, save that he shifted the meaning of these ratios from the numerical to the geometrical realm. In *Harmonice mundi* (1619) he erected an intricate structure of astounding beauty upon the idea that the geometrical derivation of the consonant intervals through divisions of circular arcs by successive regular polygons yields the clue to the mathematical regularities (*logoi kosmopoietikoi*) God had employed in creating the world (Dickreiter *1973*, Walker *1985*).

Kepler's contemporaries Galileo Galilei, Isaac Beeckman and Mersenne meanwhile explored a physical interpretation of the consonances. Taking the Pythagorean ratios as expressions of vibrational frequencies rather than abstract numbers, they established a theory which explained consonance through the relative coincidence of vibrations. Beeckman and René Descartes, in particular, made attempts to extend the theory to the physiological domain. It is from theorizing along these lines that much early acoustics (§9.8) took its origin, particularly in Mersenne's work (Dostrovsky *1974*).

One virtue of the coincidence theory was that it yielded a neat table of degrees of consonance. The table starts with how often vibrations unite in the case of the unison (once every $1 \times 1 = 1$ vibration), and ends with the minor sixth (once every $5 \times 8 = 40$ vibrations). Obviously, it can now be asked why the boundary separating the consonant intervals from the dissonances is situated at 40 rather than anywhere else. In other words, the boundary between consonance and dissonance appears to have become rather fluid. Another tricky matter is the existence of septimal ratios like $4/7$. Its musical counterpart – the augmented sixth – was customarily taken as dissonant, whereas in the scale of relative consonance it would be assigned a place before even the minor third $5/6$. Despite these and other difficulties, the coincidence theory held sway as the solution to the problem of musical consonance until it was replaced by the overtone theory of harmony advanced by Jean Philippe Rameau and elaborated by Jean le Rond d'Alembert in the first half of the eighteenth century (§3.15).

The turn to the physical domain taken by Galileo and Beeckman, in particular, was not followed by all subsequent investigators of the quantitative aspects of music. Gottfried Wilhelm Leibniz and Leonhard Euler, for example, while accepting the ratios of the consonant intervals as expressive of ratios of vibrational frequencies, went on to operate with these numbers as entities in their own right, abstracting largely from their roots in physical and physiological realities. In Euler's case this went together with ingenious contributions to the issue of the division of the octave, involving extensions of the triad that were meant to create a novel foundation for musical harmony.

A solution to the problem of consonance where mathematics, physics, physiology and harmonic theory came together in an all-encompassing synthesis was proposed by Hermann von Helmholtz in his *Die Lehre von den Tonempfindungen* ('Sensations of Tone', 1863). He founded his theory upon the phenomenon of beats as the ultimate causal agent of dissonance. The complicated sound waves of which every real musical tone is composed (blending the fundamental tone with upper partials and combination tones) are subjected, inside the human ear, to a Fourier analysis; the degree of consonance perceived depends inversely on the amount of beating the analysis detects. 'Ultimately, then, the reason of the rational numerical relations of Pythagoras is to be found in the theorem of Fourier, and in one sense this theorem may be considered as the prime source of the theory of harmony' (Helmholtz *1863*: 227).

6 A FINAL CONSIDERATION

The discovery made some 25 centuries ago in the Pythagorean circle has continued to intrigue mathematical scientists ever since. Whence comes this connection between, on the one hand, an objective, numerical regularity and, on the other, the subjective experience of musical pleasure that takes place inside us humans? Time and again the validity of the problem has been denied, often for very interesting reasons. Indeed, any attempt to explain the full pleasure of music through the mathematical regularity behind the production of the consonances seems doomed to failure – much more would be needed for such an explanation to be valid, if it is attainable at all. Yet denying the importance of the problem altogether, such as was done by the Aristoxenians, and by later followers of their line of thought like Stevin and Descartes, seems sterile. Somehow the regularity appears to serve as a clue to some aspect or other of reality. A part – albeit perhaps a very small part – of the riddle of musical pleasure appears to be reducible to a property that can be quantitatively expressed. Musical intervals are subject to mathematical law in a more direct and tangible manner than is true of any other art form. However large or small the ultimate significance of this state of affairs may turn out to be, the intriguing task of finding out how far the connection discovered by the Pythagoreans can get us in understanding the pleasure of musical sound has not yet been exhausted.

BIBLIOGRAPHY

Barbour, J. M. *1951, Tuning and Temperament. A Historical Survey*. East Lansing, MI: Michigan State College Press. [Biased in favour of equal temperament, and antiquated in many respects; none the less, still a useful survey.]

Barker, A. (ed.) *1989, Greek Musical Writings*, Vol. 2, *Harmonic and Acoustic Theory*, Cambridge: Cambridge University Press. [A collection of the most important pieces of ancient Greek musical theory to survive.]

Boethius, Anicius Manlius Severinus *c. 505*, De institutione musica. [English transl.: *Fundamentals of music* (ed. C. V. Palisca, transl. C. M. Bower), 1989, New Haven, CT: Yale University Press.]

Burnett, C., Fend, M. and Gouk, P. (eds) *1991, The Second Sense: Studies in Hearing and Musical Judgement from Antiquity to the Seventeenth Century*, London: The Warburg Institute, University of London.

Cohen, H. F. *1984, Quantifying Music. The Science of Music at the First Stage of the Scientific Revolution, 1580–1650*, Dordrecht: Reidel.

Dickreiter, M. *1973, Der Musiktheoretiker Johannes Kepler*, Berne and Munich: Francke.

Dostrovsky, S. *1974*, 'Early vibration theory: Physics and music in the seventeenth century', *Archive for History of Exact Sciences*, **14**, 169–218.

Gouk, P. *1988*, 'The harmonic roots of Newtonian science', in J. Fauvel *et al.* (eds), *Let Newton Be!*, Oxford: Oxford University Press, 101–25.

Gozza, P. (ed.) *1989*, *La musica nella rivoluzione scientifica del seicento*, Bologna: Il Mulino. [An anthology with an excellent introduction to the entire field and an extensive, up-to-date bibliography.]

Helmholtz, H. von *1863*, *Die Lehre von den Tonempfindungen*, Braunschweig: Vieweg. [Edns to 4th, 1877. English transl.: *On the Sensations of Tone*, 1st edn 1875, London: Longmans, Green; 2nd edn 1885, cited here.]

Huygens, C. *1691*, 'Le Cycle harmonique', in *Oeuvres complètes*, Vol. 10, 169–74, with related pieces in Vol. 20, 1–173. [A more recent edn by R. Rasch, with introduction and transl. in English: 1986, Utrecht: Diapason Press.]

Lindley, M. *1984*, *Lutes, Viols and Temperaments*, Cambridge: Cambridge University Press.

Palisca, C. V. *1961*, 'Scientific empiricism in musical thought', in H. H. Rhys (ed.), *Seventeenth Century Science and the Arts*, Princeton, NJ: Princeton University Press, 91–137.

Walker, D. P. *1978*, *Studies in Musical Science in the Late Renaissance*, London and Leiden: Brill.

—— *1985*, in P. Gouk (ed.), *Music, Spirit and Language in the Renaissance*, London: Variorum Reprints.

2.11

The teaching of mathematics in the Middle Ages and the Renaissance

G. R. EVANS

The teaching of mathematical disciplines throughout the medieval period and into the Renaissance was shaped in principle by Boethius's sixth-century division of the four subjects of the quadrivium: arithmetic (number considered *per se*), music (number considered 'relatively'), geometry (magnitude considered *per se*) and astronomy (magnitude considered 'relatively', or as 'mobile'). In practice, coverage of the quadrivium subjects always lagged far behind that of the grammar, logic and rhetoric of the trivium, although it is clear from the manuscripts and their glosses that Beothius's *Arithmetica* was studied more fully from the eleventh or twelfth century (Chadwick *1980*).

1 THE TEXTBOOKS

The reason for this comparative neglect of mathematics lay partly in the patchy coverage of available classical textbooks, and partly in the uneasy separation of the theoretical from the practical sciences. There was extensive work on the use of the abacus from late Carolingian times to beyond the twelfth century, when Arabic numerals came more commonly into use in the West (on the later development of the abacus tradition, see §2.4). Leonardo of Pisa (Fibonacci) composed one of the first arithmetics with Arabic numbers in about 1202; Johannes de Sacrobosco's *Algorismus vulgaris* followed in about 1230. Treatises on surveying survived from Roman times, so there was no shortage of *agrimensores* (authorities on surveying) to guide the student of practical geometry (§2.2). Musicians of the eleventh century already knew that one might become proficient in *cantus* (singing) without studying Boethius's *Musica*. For astronomy at the practical level, the great need was met by Bede in his work of the 730s on the computation of dates.

On the theoretical side, Boethius's *Arithmetica* was available (heavily indebted to Nichomachus of Gerasa's *Introduction to Arithmetic*), as was his *Musica* (derived from several ancient sources); both could be found without difficulty in early medieval libraries. For geometry there were at first few books available. It is clear from the *Geometria* by Gerbert of Aurillac (Pope Sylvester II, 999–1003) that he himself had seen Euclid's *Elements*, but it was not until the twelfth century that translations appeared (particularly notable are the multiple versions, ascribed to Adelard of Bath, which would make the work more generally known in the West; see §2.2). It was also during the twelfth century that some knowledge of Arabic astronomy came into the West, with texts to meet the need for the theoretical study of that branch of the quadrivium. By the thirteenth century Ptolemy's *Almagest* was important, and there was a popular modern textbook, Sacrobosco's *De sphaera*, first published in 1220. This work survives with commentaries by Robertus Anglicus, Michael Scot, Cecco d'Ascoli and a number of anonymous authors; and rival works exist by John Peckham and Robert Grosseteste.

2 PHILOSOPHY OF MATHEMATICS AND MATHEMATICAL METHOD IN TEACHING

In the early universities, the formal academic study of the theoretical textbooks of mathematics suffered in the twelfth century and thereafter from the continuing disadvantage of being overshadowed in the Arts course by the study of the trivium. In fourteenth-century Oxford, as elsewhere, space was allowed in the timetabling of lectures to cover the main textbooks, but the lectures were always unavoidably hurried and cursory because so little time was allowed for each book (although the lectures themselves lasted about an hour). The commentaries which survive indicate that interest was focused chiefly on those aspects of the mathematical disciplines which had a bearing on philosophical or methodological questions. Adelard of Bath's 'On the Same and the Different' (*De eodem et diverso*) – although it draws also on other influences – clearly reflects this sort of preoccupation. In the case of Boethius's *Arithmetica* there was a tendency during the Middle Ages to concentrate on the discussion of the nature of number and quantity, concepts of the odd and even, and so on, with which Boethius deals in his opening remarks. For geometry there was the particular interest of its claims to be the paradigmatic science, having an elegance which no other could match, its proofs carrying demonstrative force convincing to all rational minds.

Later medieval logicians were much interested in its demonstrative method, and attempts were made to apply it outside logic and mathematics. A series of authors from the late twelfth century reflect on the possibilities of applying the method to other sciences: for example, John of Salisbury in some bafflement in the *Metalogicon*, Grosseteste commenting on the *Posterier Analytics* of Aristotle; Dante in the *Monarchia*, and Marsilius of Padua in the *Defensor pacis*. The use of demonstrative method became controversial among theologians. Alan of Lille and Nicholas of Amiens experimented at the end of the twelfth century and the beginning of the thirteenth with the possibility of 'demonstrating' the truth of all articles of faith. But by 1277 the assertion that 'one should not hold anything unless it is self-evident or can be manifested from self-evident principles' had been formally condemned, and Duns Scotus went on to argue that demonstrative method is not enough on its own to prove all truths.

Among the foremost teachers to take up the philosophical possibilities of the mathematical sciences in the medieval universities were the so-called Oxford Calculators of the second quarter of the fourteenth century (see §2.6 on mechanics). Thomas Bradwardine wrote *De proportionibus velocitatum* (1328), William Heytesbury the *Regulae solvendi sophismata* (1335) and Richard Swineshead the *Liber calculationum* in the middle of the century. Around these may be grouped works by logicians and natural philosophers touching on related matters: Walter Burley on 'instants' and on 'forms', Roger Swineshead on 'natural motions', for instance. The intellectual focus of this work was logical rather than strictly mathematical; but there was also a substantial element of natural philosophy and natural science in the distinction drawn between dynamics and kinematics, and in the development of a notion of instantaneous velocity (Clagett *1968*).

3 MATHEMATICS TEACHING IN DIFFERENT PARTS OF EUROPE

While nominally at least the mathematical disciplines had a standard place in the Arts course, there was some variation in the extent to which they were lectured on from country to country during the Middle Ages. There is evidence that Liège was an important centre during the eleventh century, and England produced many treatises and some leading figures in the late eleventh and twelfth centuries, chief among them perhaps Adelard of Bath. England retained its prominence: Sacrobosco was almost certainly an Englishman, and he attracted a number of English commentators.

But we find French versions of Latin texts, and other vernacular texts on mathematical subjects. These indicate not only that innovative work was being done all over Europe, but also that there was a demand for mathematical textbooks outside the universities. A striking case is the *Arithmetic* written in about 1475 by Jehan Adam, secretary to the auditor of Louis XI of France. It describes practical calculation, using an abbacus system with counters. He divided arithmetic into numeration, addition, subtraction, halving, duplication, multiplication, division, progression and extraction of roots. He knew enough of the scholarly background to cite Aristotle, Plato, Pythagoras, Isidore, Boethius, Albertus Magnus, Alexander of Villa Dei, Bartholomew de Roumanis, Sacrobosco, Johannes de Lineriis, Jean de Meun and Jehan Loquemeren. But his chief purpose was to help meet a practical need. He and his near-contemporary Nicolas Chuquet, whose *Triparty* was composed in 1484, seem to have been the first authors to speak of trillions (10^{12}), which Adam called 'trimillions'; Chuquet also mentioned billions (10^9) (Flegg *et al.* 1985).

4 THE EARLY PRINTED BOOKS

The early printed books on mathematical subjects are principally concerned with arithmetic. Luca Pacioli's *Summa de arithmetica* (1494) draws heavily on Leonardo's *Liber abbaci*. Among the early French vernacular arithmetics is one by Estienne de la Roche (1520). Gemma Frisius, Girolamo Cardano and Michael Stifel all published 'Arithmetics' in the 1530s, Robert Recorde his *Grounde of Artes* in 1542, Cuthbert Tunstall his *Arts supputandi* in 1552. The emphasis was practical, but not without a sense that the subject had dignity. Recorde, John Dee, and Leonard and Thomas Digges attempted in mid-sixteenth century England to apply mathematcs to the needs of the instrument-maker, and to meet the growing demand for better techniques of navigation, fortification, surveying and cartography. Recorde stressed that mathematics is both intellectually reliable ('certain'), and 'useful'.

There were also advances in sixteenth-century arithmetic. For example, the scale of notation was extended to include the place value, was well as actual numbers, in fractions of the unit. Leonardo included some algebra, but Recorde took the subject further in his *Whetstone of Witte* (1577) (Yeldham *1936*).

Music began to drop out of the list of mathematical disciplines as such, and went its own way in the Renaissance (§2.10), although Boethius's *Musica* retained a place. Among the early printed geometry books were editions of Archimedes, Apollonius's *Conics*, the *Spherics* of Theodosius of Tripoli and Bradwardine's *Geometria speculativa*. Only Euclid was printed before 1500 (with the commentary by Campanus of Novara). Geometry, like arithmetic, proved important in the sixteenth century to the makers of instruments and to those engaged in the construction of buildings and fortifications, and in surveying, ballistics and optics.

5 INFORMAL STUDIES AND NEW THINKING

For most of the sixteenth century progress seems to have depended upon the informal exchange of ideas and on the borrowing and copying of instruments. Designs were circulated, or even stolen and pirated. All this went on mostly outside the universities. In 1597 Gresham College was founded in London, with chairs of geometry and astronomy – the first mathematical professorships in England. Lectures were to be in English as well as in Latin; there was to be practical as well as theoretical geometry on the syllabus, navigation as well as astronomy. In short, there was a deliberate move away from the traditional stress on theoretical study in the universities and from the limited range of established textbooks.

It was in astronomy that the mathematical disciplines made the most significant strides during the Renaissance. Nicolaus Copernicus taught mathematics and astronomy at Rome from 1500. In the course of his work he began to see that the hypothesis that the Sun is the centre around which the Earth and the planets revolve solved many of the difficulties presented by conventional accounts. He also suggested that the Earth might rotate on its axis, and that the stars were at an immense distance. These theories he put into the *De revolutionibus orbium coelestium*, which was eventually published in 1543 (§2.7).

Digges and Recorde both took up the theory. Recorde's *Castle of Knowledge* (1556) was the first comprehensive and original treatise on astronomy in English, and it includes a brief account of the Copernican system. Digges's *Perfit Description of the Caelestiall Orbes* (1576) rendered Book I of the *De revolutionibus* into English and gave a diagram of the heliocentric universe.

The most notable proponent of the Copernican system was Galileo; his work on the phases of Mercury, Venus and Mars supported Copernicus's theory (§2.7). But he was to depend on support from outside the universities to continue his work. In 1589 he was made professor of mathematics at the University of Pisa, and in 1592 he was appointed professor of mathematics at Padua, where he remained for 18 years. But it was the patronage of Cosmo II, Grand Duke of Tuscany, who appointed him his grand-ducal mathematician and philosopher in 1610, that made it possible for him to work at leisure. His university background encouraged him to draw on Classical and some medieval prototypes in constructing his diagrams, and there were respects in which its geocentric presuppositions proved a hindrance to his thinking at first.

The old pattern of mathematics teaching of the late classical and medieval world was of diminishing usefulness in the Renaissance, and it was on the whole those individuals who broke free of it and went their own way who made discoveries. A student at Cambridge about 1635, John Wallis later recalled that in his youth there the mathematical disciplines had lost their status as higher disciplines of the mind: 'Mathematics, (at that time, with us) were scarce looked upon as Academical studies, but rather Mechanical; as the business of Traders, Merchants, Seaman' (Hearne *1725*: Vol. 1, 147–8).

BIBLIOGRAPHY

Chadwick, H. *1980, Boethius*, Oxford: Oxford University Press.

Clagett, M. *1968, Nicole Oresme and the Medieval Geometry of Qualities and Motions*, Madison, WI: University of Wisconsin Press.

Flegg, G., Hay, C. and Moss, B. (eds) *1985, Nicholas Chuquet. Renaissance Mathematician*, Dordrecht: Reidel.

Hay, C. (ed.) *1988, Mathematics from Manuscript to Print 1300–1660*, Oxford: Clarendon Press.

Hearne, T. (ed.) *1725, Peter Langtoft's Chronicle*, 2 vols, Oxford: Theatre. [Various later edns.]

Lindberg, D. *1978, Science in the Middle Ages*, Chicago, IL: University of Chicago Press.

Sullivan, J. W. N. *1924, The History of Mathematics in Europe* [...], Oxford: Oxford University Press.

Thorndike, L. *1929, Science and Thought in the Fifteenth Century*, New York: Columbia University Press.

Yeldham, F. *1936, The Teaching of Arithmetic Through Four Hundred Years*, London: Harrap.

2.12

The philosophical context of medieval and Renaissance mathematics

A. G. MOLLAND

1 INTRODUCTION

Mathematics does not flourish in a vacuum, although at times of relative stability, and when afforded strong institutional protection, it can seem to be carried along solely by its own momentum. But in general, besides an appropriate institutional environment (discussed in §2.11), it has needed theoretical and/or practical motivation. Practical motivation (be it calculatory, calendrical or architectural) is alluded to in several other articles; the concern here is with the interactions of mathematics with philosophical doctrines and attitudes.

In the earlier Middle Ages these were largely of a 'Platonic' kind, strongly reinforced by Boethius's highly influential treatises on arithmetic and music from the early sixth century. Mathematics not only led the mind towards the contemplation of higher things, but also allowed greater insight into the structure of the world, which Boethius compared to a musical harmony, and music of course had been one of the earliest mathematical sciences. Both the mathematical and the philosophical literature available to the Latins in the early part of the Middle Ages was limited; but from the twelfth century onwards things changed, and far more Greek works were made accessible in Latin translation (§2.1), although it must be admitted that on the mathematical side only a very limited number of people read them very deeply.

2 ARISTOTELIANISM

The chief philosophical influence became Aristotle, who, when compared with Plato, is often regarded as very unmathematical, if not anti-mathematical. This is at best a half-truth. Certainly Aristotle did not accord to mathematical objects the same elevated status in the realm of Being as had Plato, nor did he impute a mathematical structure to the very heart of physical reality. Yet mathematics was for him a subject of great scientific importance. To start with, it was itself a science (or several sciences), and in his discussion of the logic of science Aristotle used geometry as a prime example of how a properly organized demonstrative science should look. By his own time, other sciences were being formed on a similar model, notably astronomy, music, optics and statics. These had to take rather more account of the vagaries of nature than had arithmetic and geometry. Aristotle referred to them as 'the more physical of the branches of mathematics', and by the later Middle Ages they were often called 'middle sciences'. Their very existence produced an urge to form other sciences of the same kind, as for example Thomas Bradwardine's attempt to provide a mathematical account of sublunary (i.e. non-astronomical) motions (§2.6), and Nicole Oresme's doctrine of configurations, in which many properties of qualities and motions were to be explained in terms of the geometrical shapes formed by graphing their intensities across the bodies that they informed.

For Plato, mathematical objects (usually) occupied a realm of their own, midway between the world of sensible objects and that of the eternal Forms or Ideas. Aristotle was more down-to-earth, and insisted that they were rooted in sensible bodies, but considered in a different way – in abstraction from their material constituents. Thus the mind had a significant role in determining their nature, which also came into play during geometrical reasoning, when constructions were performed; for, as was generally admitted, the geometer was speaking not about his diagrams but about what was understood through them. This mental dimension allowed some medieval thinkers (but by no means all) to minimize the role of mathematics in a scientific understanding of nature. Albertus Magnus, for instance, held that 'Many of the geometer's figures are in no way found in natural bodies, and many natural figures, and particularly those of animals and plants, are not determinable by the art of geometry.' The drawing of examples from biology is significant, for biology often went hand in hand with holistic habits of thought, in which in some sense wholes did not add up to the sum of their parts. This often complicated, if not precluded, mathematical science, although paradoxically Aristotelian holism was well in accord with geometrical orthodoxy in discussions of the nature of continuity. For further details see J. Murdoch in Kretzmann et al. 1982: 564–91; and A. G. Molland in Caroti 1989: 227–35.

3 NEW DEVELOPMENTS

During the Middle Ages there was little intellectual urge towards mathematical progress. To be sure, from the twelfth century onwards much effort was put into the recovery of Greek and Arabic texts and their dissemination in Latin form (§2.1); but thereafter the focus was on assimilation and analysis rather than on emulation, and even such a monument of medieval mathematics as Campanus of Novara's version of Euclid's *Elements* was deeply imbued with what William Whewell called the 'commentatorial spirit' of the Middle Ages.

The aim in general was to see how mathematical arguments worked and how they fitted into wider schemes of knowledge, rather than to produce new mathematics in the spirit of the ancient Greeks. Paradoxically, much of the motivation for straying into unfamiliar mathematical fields came from attempts to reconstruct ancient mathematical works which had become lost and were now known only from tantalizing reports giving some idea of their content, such as those by Pappus of Alexandria. This movement, particularly characteristic of the sixteenth and seventeenth centuries, became enveloped in a more general urge towards intellectual progress which was manifested in a variety of scientific and philosophical fields. It was also entwined with the search for a correct procedure for seeking out new knowledge, for many held that Archimedes and other ancient mathematicians had developed new mathematics by such a procedure, which they had jealously kept hidden from posterity. These beliefs were important features in the background to the formation of analytical geometry (§7.1) and of the infinitesimal calculus (§3.1).

This was not the only way in which, from the Renaissance onwards, a new and more positive valuation was placed on mathematics, and in particular mathematical activity as opposed to contemplation, a development often rather simplistically associated with a resurgence of Platonism at the expense of Aristotelianism. In the context of scholastic Aristotelian debate, it was often asked whether mathematics provided the most potent form of demonstration, which – whatever the answer – shows that the issue was much in the air. In more self-consciously Platonic circles there was renewed interest in numerology and related studies, such as the significance of magic squares. These are fields which can now seem to be mere superstition, but this was not obviously so at a time when it could be thought that the world had been created a finite time ago on arithmetical principles.

On surer ground, we may note how in the seventeenth century many undoubted pioneers of the scientific revolution strongly propagandized for the use of mathematics as an essential tool in the advancement of physical science. Johannes Kepler asserted that geometry provided the archetype for the Creation; Galileo Galilei held that the book of the universe was written in the language of mathematics, whose characters were 'triangles, circles and other geometrical figures, without which it is humanly impossible to understand a single word of it' (Drake and O'Malley *1960*: 184); René Descartes went so far as to assert that his physics was nothing but geometry; and in 1687 Isaac Newton was to produce *Mathematical Principles of Natural Philosophy*, a title whose first two words are highly significant.

At the same time as physics (and especially that branch of it which we call mechanics) was thus being increasingly subjected to mathematical conceptions and procedures, mechanics was playing an increasing role within mathematics itself. Although from Greek Antiquity onwards there had been frequent references in geometrical contexts to motion and to the use of instruments, these had usually been regarded with suspicion as endangering the integrity of pure mathematics. In the sixteenth and seventeenth centuries they became more thoroughly legitimized. The correlated motions of two points played a key role in John Napier's development of his concept of logarithms (§2.5), and Descartes's imagination of articulated instruments was crucial in his demarcation of what were to be regarded as properly geometrical curves (Molland *1976*). In this tradition much effort was devoted later in the seventeenth century to providing 'organic' (i.e. instrumental) solutions of geometrical problems, even in contexts that we would regard as more appropriately dominated by algebra (§6.9). In all these ways mathematics was intimately involved in one of the major revolutions in Western thought: a move away from Aristotelianism that is often referred to as the 'mechanization of the world picture' (Dijksterhuis *1961*).

BIBLIOGRAPHY

Bochner, S. *1966*, *The Role of Mathematics in the Rise of Science*, Princeton, NJ: Princeton University Press.

Brunschvicg, L. *1972*, *Les Etapes de la philosophie mathématique*, Paris: Blanchard.

Burtt, E. A. *1932*, *The Metaphysical Foundations of Modern Physical Science*, 2nd edn, London: Routledge & Kegan Paul.

Caroti, S. (ed.) *1989*, *Studies in Medieval Natural Philosophy*, Florence: Olschki.

Dijksterhuis, E. J. *1961*, *The Mechanization of the World Picture*, Oxford: Clarendon Press.

Drake, S. and O'Malley, C. D. (transls) *1960*, *The Controversy on the Comets of 1618*, Philadelphia, PA: University of Pennsylvania Press.

Heath, T. L. *1949*, *Mathematics in Aristotle*, Oxford: Clarendon Press.

Kretzmann, N., Kenny, A. and Pinborg, J. (eds) *1982*, *The Cambridge History of Later Medieval Philosophy*, Cambridge: University Press.

Molland, A. G. *1976*, 'Shifting the foundations: Descartes's transformation of ancient geometry', *Historia mathematica*, **3**, 21–49.

—— *1978*, 'An examination of Bradwardine's geometry', *Archive for History of Exact Sciences*, **19**, 113–75.

Rashed, R. (ed.) *1991*, *Mathématiques et philosophie de l'antiquité a l'age classique*, Paris: Editions du Centre National de la Recherche Scientifique.

Strong, E. W. *1936*, *Procedures and Metaphysics: A Study in the Philosophy of Mathematical–Physical Science in the Sixteenth and Seventeenth Centuries*, Berkeley, CA: University of California Press.

Vuillemin, J. *1960*, *Mathématiques et metaphysique chez Descartes*, Paris: Presses Universitaires de France.

Part 3
Calculus and Mathematical Analysis

3.0

Introduction

For most of the rest of this encyclopedia, the tale is dominated by Western mathematics since around the mid-seventeenth century. The calculus and related subjects are placed first, as they came to form the most prominent branch. Indeed, there is so much to cover that two Parts have been assigned: this one concentrates largely on general theories and foundational questions, while Part 4 treats techniques and methods. There are, of course, numerous interactions, which the cross-references help to identify.

The order of the articles is roughly chronological as far as §3.8: the history of the fundamental notions of tangent and area is presented as a more or less continuous thread running from the seventeenth to the early twentieth century (Grattan-Guinness *1980*). The background to fractal theory (which arouses so much interest today) is placed as §3.8 as it draws mostly on ideas described in preceding articles.

The next three articles deal chronologically with topics which have developed largely in parallel in the twentieth century. Up to §3.11, this Part is concerned almost entirely with real-variable analysis; complex analysis enters in §3.12–3.13, and both the real and complex kinds feature thereafter. The branches of analysis described in the rest of the Part also ran chronologically very much in parallel from the seventeenth century onwards. Among related articles are §7.9 on early modern algebraic geometry and §9.4 on heat diffusion.

A special word is in order about §3.14–3.15, where differential equations are treated. They became the chief source for problems and results during the later eighteenth century, and in the nineteenth they grew into an enormous industry. Often motivated by problems in applied mathematics, they also raised 'pure' questions of their own, such as the generality and even the existence of a solution. Some of the work was of very high quality, but a measure of rather useless 'erudition' is also discernible. (Forsyth *1890–1906* is a good and quite comprehensive survey of the field at that time, with some historical remarks.) The authors of these articles have restricted themselves to the principal forms and solutions (a few turn up also in §4.2), but differential equations are treated in many other articles, especially in this Part and also in Parts 4, 8 and 9. In addition, Volume 3 of the *Encyklopädie der mathematischen Wissenschaften*, praised in general in §0, is enormously helpful in bringing out the status of differential equations, and indeed of analysis as covered in this and the next Parts; further attention is paid to applications in its Volumes 4–6.

Various general histories of the calculus and mathematical analysis have been produced. Boyer *1939* is a well-known classic source, although many of its findings have been challenged or superseded; Birkhoff *1973* is a source-book of several primary texts, although some of the original authors would not recognize the notations into which their works have been cast.

BIBLIOGRAPHY

Birkhoff, G. (ed.) *1973*, *A Source Book in Classical Analysis*, Cambridge, MA: Harvard University Press.

Boyer, C. B. *1939*, *The Concepts of the Calculus*, New York: Hafner. [Repr. as *A History of the Calculus and its Conceptual Development*, 1949, New York: Columbia University Press; and 1959, New York: Dover.]

Edwards, C. H. *1979*, *The Historical Development of the Calculus*, New York: Springer.

Forsyth, A. R. *1890–1906, Theory of Differential Equations*, 6 vols, Cambridge: Cambridge University Press.

Grattan-Guinness, I. (ed.) *1980, From the Calculus to Set Theory, 1630–1910: An Introductory History*, London: Duckworth.

3.1

Precalculus, 1635–1665

KIRSTI ANDERSEN

1 INTRODUCTION

The years from 1635 to 1665 – the period between the publication of Bonaventura Cavalieri's method of indivisibles and the birth of Isaac Newton's ideas on a method of fluxions – witnessed the emergence of numerous methods for solving problems which today belong to calculus, and which for convenience are referred to here as calculus problems. A majority of the various types of problem were inherited from Greek mathematics: problems of quadratures and cubatures – the equivalent of finding areas and volumes, questions concerning centres of gravity, problems of determining tangents, and problems about extreme values (§1.3). Apart from the last, all these problems concerned curves, and the study of curves was the main analytical activity during this period. The development of the concept of a function as a fundamental mathematical tool came after the creation of the calculus.

The Greeks would have liked to construct a line segment equal to the length of the circumference of a circle, but there is no evidence that they generally searched for the ratios between segments of curves and lines. This theme – rectification – was taken up in the 1640s; in the late 1650s it became a central topic after the English architect Christopher Wren had shown that the arc length of a cycloid is eight times the radius of the generating circle, and Hendrick van Heurat had suggested a general procedure for transforming the problem of rectification of an algebraic curve to a problem of quadrature. At about the same time Christiaan Huygens introduced another new topic into mathematics which was inspired by his work on the pendulum clock: he wanted to control the curve along which the pendulum is swinging, and was thereby led to the study of evolutes and evolvents (see §8.13).

The list of seventeenth-century calculus problems is longer, but the problems already mentioned were the most influential in the process that led to the emergence of the calculus. It should be added, however, that as early as the late 1630s, several mathematicians were working on problems corresponding to differential equations, formulated as questions of determining curves whose tangents have given properties.

To find methods for solving the various types of calculus problems was the chief aim of mathematical research in the period 1635–65, and all the leading mathematicians displayed much ingenuity in this work. The investigations started in Italy, and were first guided by two of the mathematicians close to Galileo, Cavalieri and Evangelista Torricelli; later Pietro Mengoli and Stefano degli Angeli made some interesting quadratures and cubatures. Knowing that their Italian colleagues were working on calculus problems (but, at the beginning, not knowing how), French mathematicians also engaged themselves in such problems. Important results were obtained by René Descartes, Pierre de Fermat, Florimond Debeaune, Gilles Personne de Roberval and Blaise Pascal. From France, inspiration spread to Belgium and the Netherlands, where in particular Grégoire de Saint-Vincent, René François de Sluse, Frans van Schooten, Johannes Hudde, van Heurat and Huygens contributed new insights. Around 1650, calculus problems also became fashionable in Britain, and outstanding contributions to their solutions were made by John Wallis, Isaac Barrow, James Gregory and Nicolaus Mercator (who came from Holstein, but worked in England).

The following sections deal with some of the precalculus methods that were used to solve calculus problems. The aim is to give an impression of the basic ideas underlying the various methods, concentrating in particular on ideas which had some influence on Newton's and Leibniz's later work.

2 ALGEBRAIC METHODS OF NORMALS AND TANGENTS

The two leading French mathematicians in the 1630s, Descartes and Fermat, each created a method for determining normals or tangents. Their methods were based on different ideas, but both yielded algebraic procedures for calculating the lengths of subnormals or subtangents to algebraic curves.

Descartes treated the problem of determining the normal to a given algebraic curve at a given point, C, say (Figure 1). He considered circles having their centres on the axis of the curve and passing through C. The circle having its centre at the point of intersection of the normal and the axis touches the curve, whereas each of the other circles cuts it and has, Descartes said, at least two points in common with the curve. From this observation he concluded that C is a double meeting-point between the curve and the touching circle. Translating this into algebra, he found that the subnormal, $v - y$, is determined from the condition that the

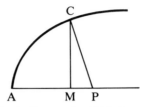

Figure 1 Descartes's method of normals. CP is the normal to the curve at C,
AM = y, MC = x, AP = v and the subnormal MP = $v - y$

equation for either the y-coordinate or the x-coordinate of the point of intersection of the circle and the curve must have a double root. As an example, Descartes showed how to calculate the subnormal to an ellipse. If C is the given point (y, x), P is the centre of the circle and CP = s, then the circle has the equation

$$x^2 + (v - y)^2 = s^2. \tag{1}$$

Together with this, Descartes used the following equation of the ellipse (where r and q are constants):

$$x^2 = ry - (r/q)y^2. \tag{2}$$

Elimination of x gives a second-degree equation in y which has a double root when

$$v - y = (r/2) - (r/q)y. \tag{3}$$

This determines the subnormal, and knowing this one can easily construct the normal – and thus the tangent – at the point C.

It was with some pride that Descartes presented his solution to the problem of determining normals in *La Géométrie* (1637): he claimed that 'I dare say that this is not only the most useful and most general problem in geometry that I know, but even that I ever desired to know' (Descartes *1637*: 95). He had indeed obtained a method which in principle can be applied to all algebraic curves – the only ones he allowed in geometry. However, the actual calculations often turn out to be rather laborious. According to his description, v should be determined in the following way: the coefficients in the (polynomial) equation of, let us say, the y-coordinate of the point of intersection of the curve and the circle should be equated to the coefficients in a polynomial which can be written as a product of $(y - e)^2$ times another polynomial, e being the double root.

This determination is complicated when the curve does not have a simple equation. Nevertheless, Descartes's method was inspiring for several mathematicians. Debeaune adopted the idea of a double intersection and applied it directly to the tangent. Thus he determined the subtangent, t, to a point (y_0, x_0) by requiring that the equation of one of the coordinates of the point of intersection of the curve and the line through (y_0, x_0) having the declination $x_0 : t$ should have a double root. By exchanging the circle with a straight line, Debeaune simplified the calculations (Giusti *1986*).

Debeaune's result was published as a comment to Descartes's method in the first Latin edition of *La Géométrie*, which was issued in 1649. The edition published 10 years later contains a rule by Hudde for finding a double root which leads to less cumbersome calculations than those originating from the comparison of coefficients. Hudde's rule consists of transforming the problem of finding a double root in the polynomial $p(y)$ into that of finding a root in a polynomial, which in modern notation can be written as

$$ap(y) + byp'(y), \qquad (4)$$

where a and b are constants (Descartes *1659*: 507). These constants can be chosen so that an unpleasant term in $p(y)$ vanishes in the expression (4).

In the autumn of 1664, Newton worked on the problem of determining normals, using a combination of Descartes's method and Hudde's rule. After having calculated a few examples he recognized a pattern and gave a general formula for the subnormal to algebraic curves (Newton *1967*: 236). Keeping the order of the coordinates from Descartes's example, Newton's formula can be described as determining the subnormal to an algebraic curve with equation $f(y, x) = 0$ from

$$v - y = -xf_y/f_x. \qquad (5)$$

It is, however, based on algebraic rules and does not contain the idea of a derivative.

At about the same time as Descartes worked on his method of normals, Fermat invented a method of determining maxima and minima which apparently was also purely algebraic. He did not publish his method, but described it in a number of memoirs which were circulated from around the late 1630s among French mathematicians. In his first memoir, Fermat – using Viète's notation (§2.3) – introduced an algorithm for determining the A for which an expression corresponding to an algebraic function $f(A)$ has an extreme value. It consisted of the following steps (Fermat *1891*: 133–4):

1 A is replaced by $A + E$,
2 $f(A)$ and $f(A + E)$ are treated as if they were in some way equal, i.e.

$$f(A + E) = f(A), \qquad (6)$$

3 common terms in the relation (6) are removed,
4 all terms are divided by E,
5 terms still containing E are ignored,
6 A is determined from the obtained equation.

By and large this corresponds to determining A from the relation

$$\left.\frac{f(A+E)-f(A)}{E}\right|_{E=0} = 0. \tag{7}$$

This method is based on the observation that, locally, the equation $f(A) = B$ has two solutions when B is not an extreme value, but only one when B is a maximum or minimum. Having worked out an algorithm for determining A, Fermat realized that it, or parts of it, could be used for solving problems other than those concerning maxima and minima. He applied the procedure to the problems of determining tangents to curves, points of inflection on a curve and centres of gravity, and furthermore to derive the sine law for refraction. As early as in his first memoir, Fermat illustrated the application to tangent problems, and later he clarified the ideas behind it. (In setting out these ideas below, modern notation is used, in particular x instead of Fermat's A.) To determine the tangent at the point (x, y) on the algebraic curve with equation $f(x, y) = 0$, Fermat searched for a method which would give him an expression for the subtangent t (Figure 2). A neighbouring point on the curve has the coordinates $(x + e, y + d)$ and lies on the same vertical as the point $(x + e, y + ye/t)$ on the tangent. Instead of applying the equation of the curve to the former point, Fermat applied it to the latter, setting

$$f(x + e, y + ye/t) = 0. \tag{8}$$

From this relation he obtained an expression for t by performing steps 3–6 in the above procedure for determining maxima and minima; for each

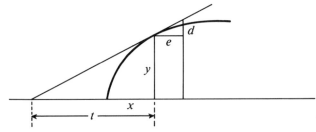

Figure 2 Fermat's method of tangents

case he obtained a result corresponding to

$$t = -yf_y/f_x. \tag{9}$$

Fermat himself claimed that his determination of tangents was related to the method of maxima and minima; perhaps he saw it as an inverse application of the method. It is known that at the point (x, y) the distance between the curve and the tangent is minimal, so the subtangent t is determined by being the expression that provides a minimum; this knowledge used in the method of extreme values should give t. His procedure is, however, not a direct translation of this idea. It is a challenge to grasp exactly what Fermat meant and why his method worked. Attempts to clarify these points were made by a number of mathematicians, among them Huygens. He found it worrying that Fermat divided through by E and then took a step which corresponds to setting E equal to 0; wanting to avoid this, Huygens worked out another method which combined Fermat's idea of two roots becoming one in a point of extreme value with Descartes's idea of determining double roots. However, Huygens had to admit that Fermat's method was easier to use than his own (manuscript from 1652, printed in Huygens *1910*: 60–68).

The problem which Huygens tried to solve remained a problem in the calculus up to the nineteenth century; in particular we find it in Newton's proof of the formula for determining the ratio between the fluxions of x and y when an algebraic relation between x and y is given. In Newton's set-up, Fermat's E has become an infinitely small moment or increment, often denoted by o, and this was set equal to zero after having served as a divisor. It is natural to interpret Fermat's E also as a very small quantity, but Fermat himself never expressed this idea in his writings, not even when he introduced some *ad hoc* assumptions to make the method applicable to transcendental curves like the cycloid.

Other mathematicians developed algorithms similar to Fermat's for determining subtangents to algebraic curves. Around 1655 Sluse, probably as the first, discerned a pattern in the calculations and formulated a rule which directly gives the result (corresponding to (9), yet without the use of derivatives) for algebraic curves (Sluse *1673*).

3 THE KINEMATIC METHOD OF TANGENTS

While Descartes and Fermat solved the problem of tangents analytically, Torricelli and Roberval approached it, independently and about the same time, kinematically. An example of Torricelli's method appeared in his *De motu* of *1644*, whereas Roberval's method was published only in *1693*. Presumably inspired by the fact that most curves were defined by motion, Roberval and Torricelli started from the concept of a curve as the path of a moving point. To this they added the idea that the tangent is the instantaneous direction of the motion. Before the invention of the calculus there was no general way of finding this direction; nevertheless, Roberval managed to apply the method to a dozen various curves, while Torricelli concentrated his investigations on the parabola and the 'higher-order parabolas', the curves given by $y = kx^n$.

Roberval proceeded by first decomposing the generating motion into two motions whose directions and relative speeds were known; sometimes – as for instance with the Archimedean spiral – the definition of the curve immediately provides the two motions. The next step was to recompound the two motions by the parallelogram rule; thus the required tangent was obtained as the diagonal in the parallelogram formed by the representatives of the two motions. The method has the advantage that it is directly applicable to transcendental curves. Without much work, Roberval was able to confirm Archimedes' result concerning the tangent to the spiral, and he supplied a simple and elegant determination of the tangent to the cycloid.

The application of the method is, however, complicated by the fact that not all generating motions, for instance not the ones generating the quadratrix, can be compounded according to the parallelogram rule; such motions are often called 'dependent'. Roberval realized that the quadratrix and the cissoid were problematic and succeeded in determining their tangents correctly; his procedure was to compensate for the dependence of the generating motions by using the parallelogram rule several times. In 1665–6, before Roberval's method was published, Newton struggled with the tangent to the quadratrix. After a wrong determination, he found it in exactly the same manner as Roberval had done about a quarter of a century earlier (Newton *1967*: 418).

Newton too applied the kinematic method of tangents – combined with his formula for determining the ratio between fluxions of algebraically related quantities – to derive a general formula for subtangents to algebraic curves. Roberval, who did not have at his disposal a method for determining the ratio between the velocities of generating motions along axes, had less success with algebraic curves. Dealing with the conic sections, he considered the motions away from the foci, or for the parabola the motions away from the focus and away from the directrix. Contrary to the case of the quadratrix, he overlooked the problem of dependence in connection with these generating motions of the conics and composed them according to the parallelogram rule. One of the reasons why Roberval did not become aware of his mistake is presumably that, because of the special properties of the conics, he got the right answers. However, when Gaspard Monge later generalized Roberval's special procedure for the conic sections, he got wrong results.

Applying Galileo's result on motions, Torricelli managed to find the ratio between the velocities of motions along the axes generating a parabola; these motions he then easily compounded, and found correctly the tangent to the parabola (Torricelli *1919*: Vol. 2, 122–4). He also generalized his method to higher-order parabolas (Torricelli *1919*: Vol. 1, Part 2, 311).

It is interesting to note that, in his *Lectiones geometricae* (1670), Barrow had a general theorem relating motions and subtangents. In transcription the theorem states that if $(x(t), y(t))$ is a point on a curve, and the sub-tangent on the y-axis is $s(t)$, then

(velocity in direction y) : (velocity in direction x) $= s(t) : x(t)$.   (10)

He does not seem to have made any important applications of this theorem; nevertheless his kinematic thinking has some historical relevance because it probably inspired Newton. The kinematic approach was fruitful in other connections, for example for determinations of arc lengths and of centres of curvature (as instantaneous centres of rotation).

4 METHODS OF QUADRATURE

4.1 General remarks

In presenting quadratures, the phrase 'quadrature of a curve' – which was common in the seventeenth century – is used to mean the quadrature of the area under the curve, and 'squaring' a curve or an area is used to mean quadratures of areas.

Whereas only a very few pages were published on differential methods during the 1630s, there appeared in 1635 a book of almost 700 pages on an integration method. This was written by Cavalieri and contains a fascinating and very special attempt to represent an intuitive understanding of a plane figure, F, as composed of infinitesimal rectangles. Using moving planes, Cavalieri introduced the concept of all the lines – *omnes lineae* – of F taken with respect to a given direction. By and large, all the lines taken together, denoted here by $O_F(l)$, correspond to all the chords in F having the given direction. Cavalieri's idea was that $O_F(l)$ as a collection could be treated as a magnitude having the same properties as classical magnitudes; in particular this meant that the theory of proportion applied to collections of lines. His basic assumption – occurring as a 'proved' theorem – was that the ratio between two figures F and G is equal to the ratio between their collections of lines:

$$F : G = O_F(l) : O_G(l).$$ (11)

Building upon a series of *ad hoc* concepts, Cavalieri obtained results which may be characterized as geometrical equivalents of integrating x and x^2. Later he also squared higher-order curves, and, seeing a pattern, he came to the conclusion that the result we would now express as

$$\int_0^a x^n \, dx = \frac{a^{n+1}}{n+1} \tag{12}$$

must be true when n is a natural number.

In 1644 Torricelli presented 21 different ways of squaring the parabola, among them some in which he used Cavalieri's concept of all the lines; he maintained that they constituted the figure, in other words that

$$F = O_F(l). \tag{13}$$

Cavalieri himself had avoided drawing conclusions about the composition of the continuum and had therefore not assumed the relation (13) (which would make (11) trivial), but in the title of his book and in a few other places he had called all the lines of a figure its 'indivisibles'. This fact, combined with Torricelli's claim (13), gave rise to the idea that Cavalieri's method of integration was founded on the assumption that an area is composed of indivisibles which are line segments; moreover, the composition was understood as an addition, so it became common to talk about Cavalieri's method of indivisibles as a method whereby a plane figure F is determined by the relation

$$F = \sum_F l. \tag{14}$$

Some mathematicians saw a dimension problem in (14), and required that the line segments l should be replaced by infinitesimal rectangles. However, since quadratures were about determining ratios, and since all the infinitesimal sides of the rectangles could often be chosen to be equal, let us say of length Δ, there was no practical difference between the method called Cavalieri's method of indivisibles and the method of infinitesimals, because

$$F:G = \left(\sum_F l\Delta\right) : \left(\sum_G l\Delta\right) = \left(\sum_F l\right) : \left(\sum_G l\right). \tag{15}$$

All in all, the seventeenth-century mathematicians did not make a sharp distinction between the various methods of quadrature emerging before the calculus, tending to call all of them 'methods of indivisibles'. Moreover, the expressions *omnes lineae* and 'the sum of the lines' were used synonymously; an example of this is found in one of the important steps which Leibniz took in 1675 on his way to the invention of the calculus, namely replacing his abbreviation *omn* of *omnes lineae* by the first letter, ∫, of 'summa' (§3.2).

In the rest of this section are outlined the methods used to obtain results which correspond to integrating x^n (n being an integer) and some transcendental expressions.

4.2 Results concerning x^n

It has already been mentioned that Cavalieri, in his special geometrical calculation with *omnes lineae*, found the quadratures of x^n for $n > 0$. Most of his successors applied a method corresponding to the relation (15), where F is now an area defined by the curve $y = x^n$ ($n > 0$) and straight lines, and G is the area of the circumscribed parallelogram. If the area is squared from $x = 0$ to $x = a$, and x is set equal to $(ia)/k$, then

$$F : G = \lim_{k \to \infty} \left(\sum_{i=1}^{k} \frac{(ia)^n}{k^n} : \sum_{i=1}^{k} a^n \right) = \lim_{k \to \infty} \left(\sum_{i=1}^{k} i^n : \sum_{i=1}^{k} k^n \right), \qquad (16)$$

which means that quadrature became an arithmetical exercise. The group of mathematicians using this method included Roberval, Pascal and Wallis; in the late 1630s Roberval found the results for specific values of n, and Wallis and Pascal found the general result. Around 1654 Pascal carefully determined the sums $\Sigma(a + id)^n$ and indicated how quadratures could be obtained from these (*Potestatum numericarum summa*, published in Pascal *1665*). Wallis, on the other hand, was rather daring, using a deduction by analogy. By this he came to the conclusion that the limit at the right-hand side of (16) is $1/(n + 1)$ for any rational number different from -1, and he did not doubt that this is also the case for other numbers, for instance $\sqrt{3}$.

Fermat made a more rigorous quadrature of the 'higher-order hyperbolas', $y = kx^{-n}$, around 1658 (published in 1679), finding a relation corresponding to

$$\int_a^\infty x^{-n} \, dx = \frac{a^{-n+1}}{n-1}. \qquad (17)$$

His result was a surprise for several of his contemporaries, who firmly believed that a quadrature involving an infinite base could never result in something finite. In dividing the interval $[a, \infty)$, Fermat did not use an equidistant subdivision, but one in which the intervals formed an increasing geometrical progression with ratio $(a + \Delta_1) : a$ (the notation is not original). Thus he had, in Figure 3,

$$a : (a + \Delta_1) = \Delta_1 : \Delta_2 = \Delta_2 : \Delta_3 = \cdots, \qquad (18)$$

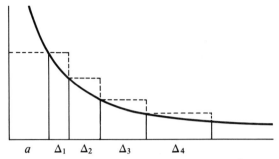

Figure 3 Fermat's squaring of $y = kx^{-n}$

which is equivalent to

$$a:(a+\Delta_1) = (a+\Delta_1):(a+\Delta_1+\Delta_2)$$
$$= (a+\Delta_1+\Delta_2):(a+\Delta_1+\Delta_2+\Delta_3) = \cdots . \qquad (19)$$

This subdivision has the advantage that the areas R_i of the rectangles circumscribed around the strips of the curves defined by having sides Δ_i are in a geometrical progression with ratio less than 1, and that this ratio can be calculated without the use of binomial coefficients.

Fermat carried out the calculations for $n = 2$, but claimed that they could be generalized. In the general case (n a natural number larger than 1),

$$R_{i+1}:R_i = \Delta_{i+1}(a+\Delta_1+\cdots+\Delta_i)^{-n}:\Delta_i(a+\Delta_1+\cdots+\Delta_{i-1})^{-n}$$
$$= \Delta_{i+1}(a+\Delta_1+\cdots+\Delta_{i-1})^n:\Delta_i(a+\Delta_1+\cdots+\Delta_i)^n; \qquad (20)$$

the relation (18) implies that

$$\Delta_{i+1}:\Delta_i = (a+\Delta_1):a, \qquad (21)$$

and (19) implies that

$$R_{i+1}:R_i = ((a+\Delta_1):a)\times(a+\Delta_1+\cdots+\Delta_{i-1})^n:(a+\Delta_1+\cdots+\Delta_i)^n$$
$$= (a+\Delta_1):a\times a^2:(a+\Delta_1)^2\times(a+\Delta_1):(a+\Delta_1+\Delta_2)$$
$$\times(a+\Delta_1+\Delta_2):(a+\Delta_1+\Delta_2+\Delta_3)\times$$
$$\vdots$$
$$= a:(a+\Delta_1+\cdots+\Delta_{n-1}). \qquad (22)$$

Summing the R_i, where $R_1 = a^{-n}\Delta_1$, Fermat found the following relation for the sum S:

$$(\Delta_1+\cdots+\Delta_{n-1}):\Delta_1 = a^{-n+1}:(S-a^{-n}\Delta_1). \qquad (23)$$

To make S approximate to the area under the curve, Fermat assumed that

302

Δ_1 is small, which according to (18) implies that the Δ_i are almost equal. From this he concluded that the left-hand side of (23) is equal to $n - 1$, and by ignoring the term $a^{-n}\Delta_1$ he found that $S = a^{-n+1}/(n - 1)$, the equivalent of (17). Fermat himself stressed that this result could easily be confirmed by a lengthy proof carried out in the manner of Archimedes; he also pointed out that it is not valid for $n = -1$, because then the circumscribed rectangles become equal (see (22)) (Fermat *1891*: 255–60).

The quadrature of the hyperbola $y = kx^{-1}$ was published by Grégoire de Saint Vincent in 1647; he saw (although he did not state it explicitly) that this quadrature has logarithmic properties, and soon after it became a standard procedure to reduce results involving logarithms to the quadrature of the hyperbola (Whiteside *1960*: 221).

The results for x^n led to general quadratures of polynomial curves; moreover, the subject of squaring more general algebraic curves was dealt with, but not systematically.

4.3 Quadratures of transcendental curves

Along with algebraic curves, various transcendental curves were squared; the cycloid was, for instance, squared by several mathematicians, particularly elegantly by Roberval. However, in contrast to quadratures of polynomial curves, no pattern emerged; each transcendental curve had to be treated separately by *ad hoc* procedures.

One could expect that one of the main concerns was to find quadratures corresponding to integrating the elementary transcendental functions: the trigonometric functions, the logarithm and the exponential function. However, although these functions played an important role in calculations, their graphs were not among the most studied curves. The logarithmic curve was the first to get special attention, and since one coordinate was not preferred to another, this also served as the exponential curve. That squaring a logarithmic curve can be reduced to the quadrature of a hyperbola was realized around 1670.

Earlier, Pascal had made some investigations of the cycloid by which he was led to the problem of finding 'sums' of sines. He solved this problem very elegantly and published the result in *Traité des sinus du quart du cercle* (1659). Besides giving the solution to a concrete problem, this tract contains some basic ideas which would inspire Leibniz in the 1670s, among them the idea of using the triangle later to be called the characteristic triangle (§3.2). As indicated by the title of his work, Pascal considered the sines as line segments in the quarter of a circle; in Figure 4 the sines are the segments DI. What Pascal needed was the sum corresponding to $\int \sin v \, dv$ which, in his geometrical interpretation, is the sum of the sines taken with respect to the

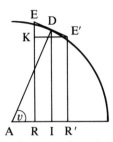

Figure 4 Pascal's determination of the sum of sines.
DI $= r \sin v$, AI $= x = r \cos v$, RR$' = $ KE$' = \Delta x$ and EE$' = \Delta(rv)$

circumference of the circle. For a circle of radius r, the sum Pascal wanted to determine can be described as (his own formulation is completely verbal)

$$\sum r \sin v \, \Delta(rv). \tag{24}$$

Applying the characteristic triangle, he transformed the problem of determining this sum to the problem of taking a sum with respect to the axis.

In modern notation, we may set $x = r \cos v$, the similarity of the characteristic triangle EE$'$K (Figure 4) and triangle ADI implies that

$$r \sin v : r = \Delta x : (\Delta(rv)) \tag{25}$$

or

$$r \sin v \, \Delta(rv) = r \, \Delta x. \tag{26}$$

Pascal could calculate the sum of the right-hand side of (26); when the summation was made over the interval from rv_0 to rv_1, he obtained

$$\sum r \sin v \, \Delta(rv) = \sum r \, \Delta x = r \sum \Delta x = r^2 (\cos v_1 - \cos v_0). \tag{27}$$

Similarly, he proved for $n = 2$, 3 and 4 – and claimed that it was generally true – that

$$\sum (r \sin v)^n \, \Delta(rv) = r \sum (r \sin v)^{n-1} \Delta x, \tag{28}$$

which corresponds to

$$\int (r \sin v)^n \, d(rv) = r \int (r \sin v)^{n-1} \, d(r \cos v). \tag{29}$$

4.4 Transformations of quadratures

In dealing with more complicated quadratures, mathematicians used geometrical considerations to obtain results corresponding to transformations of integrals. A result such as

$$\int_0^a y \, dx = ab - \int_0^b x \, dy, \tag{30}$$

where (x, y) is a point on a monotonic curve through $(0, 0)$ and (a, b), was for them an immediate geometrical consequence. This view was generalized to solids formed *ad hoc* and led to several new results. One of them is the geometrical formulation of the result that, if $y = f(x)$ and $g(x)$ are monotonic functions, and $f(a) = 0$, $f(0) = b$ and $g(0) = 0$, then

$$\int_0^a f(x)g(x) \, dx = \int_0^b \int_0^{f^{-1}(y)} g(t) \, dt \, dy. \tag{31}$$

Similarly, in working with centres of gravity an intuitive application of the rule for the equilibrium of a lever guided mathematicians to other transformations.

In 1658 Pascal systematized the insights gained and made a list of which sums were needed for which kinds of problem (*Traité des triangles rectangles*, published in Pascal *1659*). The big problem remaining was then to determine these sums.

Shortly after 1665 a new tool was introduced into analysis, namely infinite series. They gave a new and useful method of quadrature which consisted in a termwise squaring (§4.3).

5 A SURVEY OF THE FOUNDATION AND THE IMPORTANCE OF THE VARIOUS METHODS

All the methods described here were created to solve infinitesimal problems. Infinitesimals as such did not, however, play an important role in the methods; in two of them they were actually hidden. Thus Descartes obtained his method (3) by taking a double intersection between a curve and a circle as an intuitively understandable characteristic of the normal. Similarly, the kinematic method of tangents built upon an intuitive understanding, namely of the concept of velocity, and thus avoided the use of infinitesimals. Fermat's method of extreme values (7) contains a term, E, which we naturally interpret as an infinitely small increment, but Fermat himself – seeing his method as completely algebraic – refrained from commenting upon the size of E.

In the methods of integration the infinitesimals were more difficult to avoid. However, mathematicians were not keen to discuss what it meant to use them. Cavalieri made an attempt to avoid them by creating the concept of *omnes lineae* and a series of *ad hoc* concepts. His method as such had very little influence, but the fact that there seemed to be a method of indivisibles encouraged other mathematicians to work out similar methods. In the beginning some concern about the absent foundation was shown, and doubters were invited to provide a classical proof of the results obtained. The fact that the new methods provided new results in a much easier way than the classical method made them gradually accepted. Thus many of the second generation of mathematicians working on quadratures were much less worried than the first generation; this applied for instance to Wallis and Pascal. In a letter to Carcavi, Pascal stated that he had used 'the doctrine of indivisibles which cannot be rejected by those who demand to be ranged among the geometers' (Pascal *1659*).

Neither Newton nor Leibniz solved the problem of foundations, and, like several of their predecessors, they were not too worried about it. A significant contribution from the period 1635–65 was that some mathematicians were daring enough to abandon the standard of Greek mathematics which they had learnt and which they admired so much. They overcame, so to speak, a *horror infinitesimalis*. This was far from the only contribution. Newton's and Leibniz's work made most of the previous methods superfluous, but it is unthinkable that they had achieved their results without being able to draw on the insights of their predecessors. Each of the three methods described in Sections 2–4 above contributed to Newton's and Leibniz's experience. The algebraic methods taught them that some methods do not only give procedures but lead to formulas from which a required magnitude can be calculated. The kinematic approach was very inspiring for Newton, and he actually based his method of fluxions on this (§3.2). In this method the intuitive understanding of a fluxion or a velocity survived, since Newton did not define fluxions, but only showed how one can calculate with them. Quadratures and, particularly, the intuitive transformations of results were very stimulating for Leibniz. For some time he held the view that, by coming to a deeper understanding of the transformations, he would be able to create a calculus which could solve all problems of quadrature. It turned out that his aspirations were much too high, but this does not alter the fact that one of his starting-points was the study of relations between various quadratures.

Among the things Newton and Leibniz did not inherit was the idea that finding tangents (or velocities) and quadratures are inverse processes. This very important insight exists in disguise in several of the results obtained by their predecessors – but only in disguise. It was also left to Newton and

Leibniz to develop a notation in which the new kind of calculus could conveniently be described, as we shall now see.

BIBLIOGRAPHY

Andersen, K. *1986*, 'The method of indivisibles: Changing understandings', *Studia Leibnitiana*, Sonderheft 14, 14–25.

Baron, M. E. *1969*, *The Origins of the Infinitesimal Calculus*, Oxford: Pergamon Press.

Cavalieri, B. *1635*, *Geometria indivisibilibus* [...], Bologna: Ferronij.

Descartes, R. *1637*, *La Géométrie*, appendix to *Discours de la methode* [...], Leiden. [*The Geometry of René Descartes* (ed. D. E. Smith and M. L. Latham), 1925, Chicago, IL: Open Court; repr. 1954, New York: Dover.]

—— *1659*, *Geometria a Renato des Cartes* [...] *cum notis* [...], Amsterdam: Elsevier.

Fermat, P. *1891*, *Oeuvres*, Vol. 1, Paris: Gauthier-Villars.

Giusti, E. *1986*, 'Le Problème des tangentes de Descartes à Leibniz', *Studia Leibnitiana*, Sonderheft 14, 26–37.

Huygens, C. *1910*, *Oeuvres complètes*, Vol. 12, The Hague: Martinus Nijhoff.

Mahoney, M. S. *1973*, *The Mathematical Career of Pierre de Fermat*, Princeton, NJ: Princeton University Press.

Newton, I. *1967*, *The Mathematical Papers* (ed. D. T. Whiteside), Vol. 1, Cambridge: Cambridge University Press.

Pascal, B. *1659*, *Lettres de A. Dettonville* [...], Paris: Desprez.

—— *1665*, *Traité du triangle arithmétique, avec quelques autres petits traités* [...], Paris: Desprez.

Pedersen [later Andersen], K. M. *1980*, 'Techniques of the calculus', in I. Grattan-Guinness (ed.), *From the Calculus to Set Theory, 1630–1910: An Introductory History*, London: Duckworth, Chap. 1.

Roberval, G. P. de *1693*, *Divers ouvrages* [...] *par Messieurs de l'Académie Royale des Sciences*, Paris: Academy of Sciences.

Sluse, R. F. de *1673*, 'Methodum [...]', *Philosophical Transactions of the Royal Society*, 7, 5143–7; 8, 6059.

Torricelli, E. *1644*, *Opera geometrica*, Florence: Masse and de Laudis.

—— *1919*, *Opere*, 3 vols (ed. G. Loria and G. Vassura), Faenza: Montanari.

Wallis, J. *1656*, 'Arithmetica infinitorum', in *Operum mathematicorum*, Oxford: Sheldonian Theatre. [2nd edn, 1695, 364–478 used; repr. 1972, Hildesheim: Olms.]

Whiteside, D. T. *1960*, 'Patterns of mathematical thought in the later seventeenth century', *Archive for History of Exact Sciences*, 1, 179–388.

3.2

Three traditions in the calculus: Newton, Leibniz and Lagrange

NICCOLÒ GUICCIARDINI

A student of mathematics nowadays knows very well what the calculus is about, and what the fundamental concepts and methods are in this important area of pure and applied mathematics. However, during a very long and fruitful period, beginning with Isaac Newton and Gottfried Wilhelm Leibniz and continuing at least as far as Augustin Louis Cauchy and Karl Weierstrass, the calculus was approached and developed in several different ways, and there was debate among mathematicians about its nature. We can identify several different traditions before the time of Cauchy; one approach is to concentrate on three 'schools': the Newtonian, the Leibnizian and the Lagrangian.

1 NEWTON'S APPROACH

Isaac Newton, with his researches carried out over the period 1664–71, has the credit of being the 'first' inventor of the calculus. Even though several techniques already existed in the works of Bonaventura Cavalieri, Evangelista Torricelli, Blaise Pascal, John Wallis and others (§3.1), Newton was the first to recognize that in order to study the properties of a 'function' (as we would call it nowadays), one has to introduce two new functions, equivalent to the modern derivative and indefinite integral. He also understood the centrality of the inversion theorem, which in modern terms can be expressed as

$$\frac{\mathrm{d}}{\mathrm{d}x} \int_a^x f(t)\,\mathrm{d}t = f(x), \quad \int_a^x F'(t)\,\mathrm{d}t = F(x) - F(a). \tag{1}$$

All the known problems of geometry were reduced by Newton into two

308

broad classes: direct problems of differentiation and inverse problems of integration.

Newton was also one of the first to understand the importance of power-series expansions. Among other results on series, he proved the binomial theorem for fractional powers (§4.1). The simple law

$$\int_0^x t^n \, dt = \frac{1}{n+1} x^{n+1}, \ n \neq 1 \tag{2}$$

together with the binomial theorem and other techniques of series expansion (§4.3), offered a means of studying the properties of many curves which were not defined by algebraic equations.

Newton's algorithm for differentiation was equally straightforward. The essential characteristic of the algorithm employed in his early writings can be understood from a simple example. Let $y = x^2$, and let a be an 'infinitely small' increment of the variable x, and b an 'infinitely small' increment of the variable y; then

$$(x + a)^2 = y + b, \qquad x^2 + 2xa + a^2 = y + b, \qquad b/a = 2x + a = 2x. \tag{3}$$

This procedure is based on a way of cancelling higher-order infinitesimals according to which, if x is finite and a is infinitely small, then $x + a = x$. However, Newton became dissatisfied with this justification of his algorithm. The status of infinitesimals (or *moments*, as he called them), quantities greater than zero but smaller than any finite quantity, was unclear. Furthermore, in dealing with problems of dynamics in which accelerations occurred, and with problems of geometry such as the calculation of radii of curvature, he was led to introduce second- and higher-order infinitesimals, which seemed to him even more mysterious.

Then, around 1670, Newton introduced two new concepts: fluent and fluxion. A fluent (denoted by, for example, x) was to be understood as any quantity which varies, or 'flows', in time, while the fluxion (denoted by $\dot{x}$) was its velocity, or rate of change. Subsequently, in his masterpiece, the *Philosophiae naturalis principia mathematica* (1687), he proposed basing geometric proofs on a new method which he called the method of 'prime and ultimate ratios'. He conceived of the geometrical quantities as being generated in time: the motion of a point generates a line, the motion of a line generates a plane, and so on. The direct problem was then that of calculating the rate of change, given the motion; the inverse problem was that of calculating the motion, given the rate of change. Newton (*1687*: Book I, Section 1, Lemma 1) stated: 'Quantities and the ratios of quantities; which in any time converge continually to equality, and before the end of that time approach nearer to each other than by any given difference, become ultimately equal.'

In his first published work on the calculus, the *De quadratura* published as *1704* as an appendix to the *Opticks*, Newton defined fluxions in terms of the limiting value of the ratio of finite increments. For instance, if $y = x^p$ and Δt is a *finite* time interval, then $\dot{y}\Delta t$ $(= dy/dt\,\Delta t)$ is the finite linear increment of the 'fluent' y, which in time Δt reaches the value $y + \Delta y$. Meanwhile, $\dot{x}\Delta t$ is the finite linear increment of the 'fluent' x; if we assume that x flows with uniform velocity, we have that $\dot{x}\Delta t = \Delta x$; then

$$y + \Delta y = (x + \dot{x}\Delta t)^p = (x + \Delta x)^p \tag{4}$$

$$= x^p + px^{p-1}\Delta x + \frac{p(p-1)}{2!}\,x^{p-2}(\Delta x)^2 + \cdots. \tag{5}$$

Taking into consideration that $y = x^p$, and dividing both sides by Δx,

$$\frac{\Delta y}{\Delta x} = px^{p-1} + \frac{p(p-1)}{2!}\,x^{p-2}\Delta x + \cdots. \tag{6}$$

When Δt 'vanishes', Δx and Δy 'vanish' simultaneously; therefore the 'limiting ratio' to which $\Delta y/\Delta x$ approaches is

$$\lim(\Delta y/\Delta x) = px^{p-1}. \tag{7}$$

To the modern reader this procedure is reminiscent of Cauchy (§3.3); however, the context in which Newton expressed his theory of limits is completely different. In fact, Newton's 'limits of prime and ultimate ratios' spring from a geometrical context, and should perhaps be compared to the limiting processes employed in the so-called 'method of exhaustion' of Eudoxus and Archimedes (§1.3). As a matter of fact, Newton lacked a theory of convergence as we would understand it nowadays. This is clear from the way he dealt with the problem of the existence and uniqueness of a limit $\Delta y/\Delta x$ when Δy and Δx 'vanish' simultaneously. Newton appealed to our geometrical and kinematical intuition. He wrote:

> Perhaps it may be objected, that there is no ultimate proportion of evanescent quantities; because the proportion, before the quantities have vanished, is not the ultimate, and when they are vanished is none. But by the same argument it may be alleged that a body arriving at a certain place, and there stopping, has no ultimate velocity; because the velocity, before the body comes to the place, is not its ultimate velocity; when it has arrived, there is none. But the answer is easy; for by the ultimate velocity is meant that with which the body is moved, neither before it arrives at its last place and the motion ceases, nor after, but at the very instant it arrives; that is, that velocity with which the body arrives at its last place, and with which the motion ceases. And in like manner, by the ultimate ratio of evanescent quantities is to be understood the ratio of the quantities not before they vanish, nor afterwards, but with which they vanish. (Newton *1687*: Book I, Section 1, Scholium)

2 LEIBNIZ'S APPROACH

Around 1675 Leibniz discovered, independently of Newton, the fundamental rules of the differential and integral calculus (Hofmann *1949*). Leibniz approached the calculus with motivations which differ profoundly from those of Newton. He was a philosopher interested in the development of a *characteristica universalis*, a universal language in which one could express any reasoning. Therefore, when he began to study mathematics he focused on methods rather than on particular problems: he sought a general method of mathematical reasoning. Furthermore, Leibniz had studied combinatorics (compare §6.6 on determinants), considering in particular successions of differences such as

$$b_1 = a_1 - a_2, \ b_2 = a_2 - a_3, \ b_3 = a_3 - a_4, \ \ldots . \tag{8}$$

He noted that it was possible to obtain the sum $b_1 + b_2 + \cdots + b_n$ as a difference, $a_1 - a_{n+1}$. This simple law extrapolated to the infinite he employed in two apparently unrelated problems: the summation of infinite series and the quadrature of curves.

For instance, in order to find the sum of the series

$$\sum_{n=1}^{\infty} \frac{2}{n(n+1)} = \sum_{n=1}^{\infty} b_n, \tag{9}$$

Leibniz noted that the terms of this series may be thought of as differences:

$$b_n = \frac{2}{n} - \frac{2}{n+1} = a_n - a_{n+1} . \tag{10}$$

Therefore

$$\sum_{n=1}^{s} b_n = a_1 - a_{s+1} = 2 - \frac{2}{(s+1)} . \tag{11}$$

So, if we 'sum' all the terms, we obtain 2.

Leibniz found that dealing with successions of differences was useful also in the geometric problem of 'squaring' curves; that is, in the problem of calculating areas, arc lengths, and so on. Let us take a curve C in a Cartesian coordinate system, and let us subdivide the x-axis into a sequence of infinitesimal intervals $x_2 - x_1, \ x_3 - x_2, \ x_4 - x_3, \ldots .$ The sequence of ordinates $y_1, y_2, y_3, \ldots$ generates another succession of infinitesimal differences. Leibniz stated that the area below the curve C is equal to the infinite summation of the rectangles $y \, dx$, where dx (the *differential* of x) is the difference between two successive values of the variable x, while the slope of the tangent is given by $dy/dx \ (= dy \div dx)$.

Note that dx has the same geometrical dimension as x: if x is a line, dx

is a line segment of infinitesimal length; if x is an angle, dx is an infinitesimal angle; and so on. The fact that geometrical dimension is preserved is important for two reasons. In the first place, once we have generated first-order differentials, we can repeat the process of taking differences and obtain an endless succession of higher-order differentials which are all of the same dimension. In the second place, we are allowed to extrapolate to the world of infinitesimals the rules of geometry: we can deal, for instance, with triangles with infinitesimal sides. These two characteristics were crucial in the applications of the calculus to geometry and mechanics because they allowed a geometrical representation of the relations among infinitesimals (Bos *1974*).

The inverse relationship between taking differences and taking sums, already employed in finite sequences and series, took the form

$$\int dx = d \int x. \tag{12}$$

Higher-order differentials and the rules of cancellation for higher-order differentials (such as $dx + ddx = dx$) played a fundamental role in the Leibnizian calculus. The first textbook on the differential calculus, the Marquis de l'Hôpital's *Analyse des infiniment petits* (*1696*), began with the postulate:

Grant that two quantities, whose difference is an infinitely small quantity, may be used indifferently for each other: or (which is the same thing) that a quantity, which is increased or decreased only by an infinitely smaller quantity, may be considered as remaining the same.

The use of this postulate can be illustrated by a simple example. Let us suppose that we wish to find the differential of a product of two quantities, x and y. Leibnizians proceeded as follows:

$$d(xy) = (x + dx)(y + dy) - xy = x\,dy + y\,dx + dx\,dy. \tag{13}$$

But $dx\,dy$ is infinitely smaller than $y\,dx$ or $x\,dy$, and therefore we can write

$$d(xy) = x\,dy + y\,dx. \tag{14}$$

3 THE PROBLEM OF FOUNDATIONS: BERKELEY AND MACLAURIN

The Leibnizian approach to the calculus had an enormous impact on eighteenth-century mathematics. The great majority of eighteenth-century mathematicians either adopted the Leibnizian approach as such, or used it in applications, while referring to the Newtonian approach as a more logical but more cumbersome presentation. In fact, Leibnizian arguments could be retranslated in terms of Newtonian limits, but very few mathematicians adhered coherently to Newton's method of limits.

The power of the Leibizian differentialist approach is evident especially in applications to the mechanics of extended bodies. It is in fact extremely convenient, in order to obtain the system of differential equations that describes a certain physical situation, to subdivide the solid body (or the fluid) into infinitesimal components (§8.5). One can then seek physical laws (hopefully linear ones) which relate the differential components, and thus obtain an equation which was appropriately called a 'differential equation'. (This terminology is still used, even though a differential equation is no longer an equation among differentials, but an equation in which there occur an unknown function and its derivatives.)

The Newtonian approach to the calculus has the *De quadratura* (*1704*) as its starting point. Eighteenth-century mathematicians such as Colin Maclaurin, Benjamin Robins and Jean d'Alembert maintained that the calculus dealt with finite quantities and their rates of change. Furthermore, the method of proof consisted in calculating limits of ratios and sums. Infinitesimalist techniques, which even Newton had employed in his early writings, were regarded as ungrounded.

The fluxionist approach was followed, mainly in Britain, by loyal Newtonians such as Brook Taylor, James Stirling, Maclaurin and Abraham De Moivre; but even in Britain limiting ratios were generally mentioned only in the introduction or preface of a mathematical work, then abandoned when it came to applications and actual calculations (Guicciardini *1989*).

The Leibnizians (mainly Continentals) and the Newtonians (mainly British) agreed on results – their algorithms were in fact equivalent – but differed over methodological questions. In some cases this confrontation was influenced by chauvinistic feelings, and a quarrel between Newton and Leibniz and their followers, over the priority in the invention of the calculus, soured the relationships between the two schools.

Mathematicians interested in establishing which were the true foundations of the calculus had to confront themselves with one of the most acute philosophers of the Enlightenment, George Berkeley. In a short pamphlet entitled *The Analyst* (*1734*), he maintained that Newtonian and Leibnizian mathematicians were guilty of ontological misunderstandings and logical fallacies. According to Berkeley, there was no ontological justification for attributing existence to limits or infinitesimals. A limit of a ratio is either a limit of two finite quantities, and therefore not the 'ultimate' ratio, or it is an undetermined 0/0. An infinitesimal, a quantity which is not zero and not finite, has a contradictory definition, and therefore nothing corresponds to it. From a logical point of view mathematicians were guilty of a *fallacia suppositionis*: that is, from an assumption A ('the increments of the variables are finite'), they derived several propositions and then, in the middle of the proof, used 'not A' to reach the conclusion (see equations (3), (4)–(7), (13) and (14)).

The most authoritative answer to Berkeley came from Maclaurin, one of the most talented of Newton's disciples. In his *Treatise of fluxions* (*1742*), he tried to show that infinitesimals were used by Newton only to abbreviate proofs; these proofs could be re-expanded, following the style of the *De quadratura* (*1704*), in terms of limits. Furthermore, Maclaurin elaborated the idea that Newtonian proofs in terms of limits were equivalent to the venerated method of exhaustion of the ancient Greek mathematicians. Newton's 'method of prime and ultimate ratios' was just the direct version of the indirect (*ad absurdum*) method of Archimedes.

4 THE EMERGENCE OF THE CONCEPT OF FUNCTION

The progress of the calculus was not hindered by foundational problems. During the eighteenth century, mathematicians such as John, James and Daniel Bernoulli, Leonhard Euler, Alexis Clairaut, Joseph Louis Lagrange and Pierre Simon Laplace developed new techniques and introduced new concepts in the calculus, including the calculus of variations and the study of partial differential equations. These developments changed the calculus profoundly. The original Newtonian and Leibnizian calculus had dealt mainly with the study of geometrical objects (typically, curves); it investigated the relationships between variable quantities, x, y and z, which represented geometrical lines, areas and angles. The eighteenth-century calculus did not have an immediate geometrical interpretation: mathematicians began to think mainly about equations (e.g. differential or partial differential equations), how to classify them, how to solve them. Meanwhile, from mechanics emerged the problem of minimizing certain 'functionals' (to use modern terminology), and this prompted the study of the calculus of variations (§3.5). The first systematic treatment of this new discipline is found in Euler's *Methodus inveniendi lineas curvas* (1744). In this new context emerged the importance of the concept of function: the calculus about variable quantities and their differentials (or fluxions) became a calculus about functions and their derivatives (Yushkevich *1976*).

Perhaps the man who made the most important contribution to shaping this new calculus of functions was Euler. In his *Introductio in analysin infinitorum*, he adopted John Bernoulli's definition: 'a function of a variable quantity is an analytical expression composed in whatever way of that variable and of numbers and constant quantities' (Euler *1748*: Vol. 1, para. 4).

Even though he gave prominence to the concept of function, Euler did not abandon the concept of differential: he saw the calculus as very closely related to mechanics, and in this context, as we have already noted, a differentialist concept is very useful. However, it was Euler who regularized the practice of choosing a variable x for which $d\,dx = 0$; and this is equivalent, as Bos *1974* has shown, to specifying the independent variable in terms of which the other variables are functionally related. Therefore the idea of functionality was present in his 'differential' calculus.

5 LAGRANGE'S ALTERNATIVE APPROACH

Lagrange chose to abandon differentials, fluxions and limits in favour of the concept of function. In a paper published in 1772, and especially in his *Théorie des fonctions analytiques* (*1797*), he tried to prove algebraically that any function $f(x)$ can be expanded in a Taylor series

$$f(x) = A + B(x - a) + C(x - a)^2 + D(x - a)^3 + \cdots. \qquad (15)$$

The derivatives could then be defined in terms of the coefficients of the Taylor series:

$$A = f(a), \ B = f'(a), \ 2!C = f''(a), \ 3!D = f'''(a), \ldots. \qquad (16)$$

One of the advantages of this proof is that the calculus would have been freed from the foundational uncertainties which had embarrassed eighteenth-century mathematicians. Berkeley was not alone in his scepticism about the calculus. In the second half of the eighteenth century, mathematicians often had to face a number of questions. What is a differential, a quantity which is different from zero and less than any finite quantity? What is the limit of the ratio of two 'vanishing' quantities, since it is not 0/0 and it is not the ratio of these quantities before they have 'vanished'? Is it legitimate to put $x + dx = x$?

Lagrange's plan to deduce the calculus employing only algebraical tools was certainly motivated by foundational concerns: it would make the calculus as safely grounded as algebra. Lagrange's programme also had the merit of focusing only on the concept of a function and its 'derivatives'. The derivative of $f(x)$ is another function, obtained by an (algebraic) operation performed on $f(x)$: the supposed algebraic proof of Taylor's theorem was the way to pass from a function to its derivatives. Even though Lagrange's proof of Taylor's theorem was faulty, his new approach to the calculus as a mathematical theory of functions and operations on functions marked an important step ahead.

Another motivation for Lagrange *1797* was related to educational problems. He was teaching at the Ecole Polytechnique, a newly founded military institution where, for the first time, advanced mathematics needed to be taught in a systematic way (§11.1). Textbooks were required, and in order to write a good textbook on the calculus one had to think about a logical and systematic presentation of the basic concepts and propositions. The choice of Lagrange as author led to the calculus being based on the familiar rules of algebra.

Sylvestre Lacroix, Lagrange's successor at the Ecole Polytechnique, wrote a treatise in three volumes entitled *Traité du calcul différentiel et du calcul integral* (1797–1800). To facilitate the students' study of the mathematical literature, he presented side by side the Leibnizian, Newtonian and Lagrangian approaches and notations. This treatise is very important for the historian: it is an encyclopedic work on the eighteenth-century calculus which gives the reader a clear idea of the richness of alternatives which coexisted two centuries ago in the world of mathematics.

BIBLIOGRAPHY

Berkeley, G. *1734*, *The Analyst*, London: Tonson. [Also in *Works* (ed. A. A. Luce and T. E. Jessop), Vol. 4, 53–102.]

Bos, H. J. M. *1974*, 'Differentials, higher-order differentials and the derivative in the Leibnizian calculus', *Archive for History of Exact Sciences*, **14**, 1–90.

Boyer, C. *1939*, *The Concepts of the Calculus*, New York: Columbia University Press. [Repr. as *The History of the Calculus and its Conceptual Development*, 1949, New York: Dover.]

Edwards, C. H. *1979*, *The Historical Development of the Calculus*, New York: Springer.

Euler, L. *1748*, *Introductio in analysin infinitorum*, 2 vols, Lausanne: Bousquet. [Also *Opera omnia*, Series 1, Vols 8–9.]

Grattan-Guinness, I. (ed.) *1980*, *From the Calculus to Set Theory, 1630–1910: An Introductory History*, London: Duckworth.

—— *1986*, 'French *calcul* and English fluxions around 1800: Some comparisons and contrasts', *Jahrbuch Überblicke Mathematik*, 167–78.

Guicciardini, N. *1989*, *The Development of Newtonian Calculus in Britain, 1700–1800*, Cambridge: Cambridge University Press.

Hess, H.-J. and Nagel, F. (eds) *1989*, *Der Ausbau des Calculus durch Leibniz und die Brüder Bernoulli*, Stuttgart: Steiner (*Studia Leibnitiana*, Vol. 17).

Hofmann, J. E. *1949*, *Die Entwicklungsgeschichte der Leibnizschen Mathematik während des Aufenthalts in Paris (1672–1676)*, Munich: R. Oldenbourg. [English transl.: *Leibniz in Paris 1672–1676*, 1974, Cambridge: Cambridge University Press.]

Kitcher, P. *1973*, 'Fluxions, limits, and infinite littlenesse: A study of Newton's presentation of the calculus', *Isis*, **64**, 33–49.

Lagrange, J. L. *1797, Théorie des fonctions analytiques* [. . .], Paris: Imprimerie de la République.

Leibniz, G. W. *1849–63, Leibnizens mathematische Schriften* (ed. C. F. Gerhardt), 7 vols, Berlin: Asher, and Halle: Schmidt. [Repr. 1962, Hildesheim: Olms.]

—— *1920, The Early Mathematical Manuscripts of Leibniz* (ed. J. M. Child), Chicago, IL, and London: Open Court.

L'Hôpital, G. F. A. *1696, Analyse des infiniment petits* [. . .], Paris: Imprimerie Royale.

Maclaurin, C. *1742, Treatise of fluxions*, 2 vols, Edinburgh: Ruddimans.

Newton, I. *1687, Philosophiae naturalis principia mathematica*, London: Streater. [English transl. of the 3rd edn by A. Motte, revised by F. Cajori, 1934, Berkeley: University of California Press.]

—— *1704, Tractatus de quadratura curvarum*, in Newton *1964–7*. [Preparatory manuscripts in Newton *1967–81*, Vol. 8, 92–167.]

—— *1964–7, The Mathematical Works of Isaac Newton*, 2 vols (ed. D. T. Whiteside), New York and London: Johnson.

—— *1967–81, The Mathematical Papers of Isaac Newton*, 8 vols (ed. D. T. Whiteside), Cambridge: Cambridge University Press.

Scriba, C. J. *1964*, 'The inverse method of tangents. A dialogue between Leibniz and Newton (1675–1677)', *Archive for History of Exact Sciences*, **2**, 113–37.

Yushkevich, A. P. *1976*, 'The concept of function up to the middle of the 19th century', *Archive for History of Exact Sciences*, **16**, 37–85.

3.3

Real-variable analysis from Cauchy to non-standard analysis

DETLEF LAUGWITZ

1 INTRODUCTION

Two periods of major innovations in real-variable analysis in the early nineteenth century were influenced by the requirements of university teaching, by the accumulation of paradoxical results during the preceding decades, and by the needs of physics. In the 1820s, Augustin Louis Cauchy, then in his thirties and a professor at the Ecole Polytechnique in Paris, published his textbooks; during the second half of the nineteenth century numerous mathematicians spread the gospel of epsilontic rigour which had been preached by Karl Weierstrass at Berlin.

Both Cauchy and Weierstrass sought to clarify foundations, one of their main objectives being a systematic reorganization of the material they found at hand; moreover, they developed and propagated new techniques. Their methods were readily accepted, even by those who were not primarily attracted by aspects of foundations. One incentive for reconsidering foundations was the strange behaviour of trigonometrical series which had been appearing in mathematical physics since the middle of the eighteenth century. Fourier analysis can be seen as a linking thread in the developments to be considered here, through to generalized functions (delta and other distributions) and to the hyper-real numbers of the twentieth century (§3.11). The latter cast new light on the early use of infinitesimals, and on the work of the *polytechniciens*. Also, the invention of set theory by Georg Cantor had been inspired by trigonometrical series (§3.6).

The best source for nineteenth-century analysis is Bottazzini *1986*, in spite of the deplorable shortcomings of the American translation. For special topics Grattan-Guinness *1970*, Lützen *1982* and Laugwitz *1986* will help. Related chapters of Kline *1972* and Edwards *1979* are confined to the spirit of the Weierstrassians. Excellent information on Cauchy and his time is given in Grattan-Guinness *1990*, especially Chapters 9 and 10.

2 CONTINUITY AND CONVERGENCE

2.1 Infinitesimals with Cauchy

Before Cauchy, the calculus had dealt with expressions. A description of continuity by Leonhard Euler had been popular: the law or expression of $f(x)$ continues to hold when x varies. This became unsatisfactory when it was observed that an expression as a trigonometrical series could represent a function with different laws for various values of x. Treating infinite series like finite sums had led to inconsistencies, and Cauchy banned divergent series. He said 'only God is infinite'; and Niels Abel wrote in a letter that divergent series were 'an invention of the devil'.

The subjects of Cauchy's new analysis in 1821 are not expressions but variable quantities. Dependent variables are restricted to one-valued functions. An infinitesimal i is a variable having 0 as its limit; this is abbreviated here as '$i \approx 0$'. A variable quantity has L as its limit if its difference from L can be made smaller than any assignable quantity. Analytical conclusions are drawn not from expressions but from concepts, or from properties.

The fundamental property which a function can have is to be continuous on some interval, or domain: $f(x + i) - f(x) \approx 0$ whenever $i \approx 0$. For infinite series Σu_n, or sequences

$$s_n = u_0 + u_1 + \cdots + u_n, \tag{1}$$

the fundamental property is that of convergence: there exists some s, and $s - s_n \approx 0$ for all infinitely large n. Cauchy's analysis deals with series of functions, series of numbers appearing as a particular case.

The basic concepts, convergence and continuity, are compatible with each other by the following theorem, Theorem 1 from Cauchy *1821*: if a series (1) of continuous functions u_n converges on some interval, then its sum s is, again, a continuous function on that interval.

Every student should doubt that theorem, and Abel soon did. Obvious counter-examples are the Fourier series of jump functions which Cauchy himself used in his research papers. We shall return to Theorem I in Section 5.

2.2 The emergence of arithmetization

In taking a continuous variable as an undefined fundamental concept, Cauchy was following in a long tradition of physics and philosophy. Later, and possibly beginning with Abel and J. P. G. Lejeune Dirichlet in the 1820s, variables gradually lost their fundamental role and became replaced by collections of numbers. This process, usually called (after Felix Klein) 'the arithmetization of analysis', led eventually to the subsuming of variables under 'sets' of numbers or points. It was a consideration of exceptional points of Fourier series that led Cantor to invent set theory, around 1870 (§3.6). New concepts of convergence were needed in conceptual frameworks different from that of Cauchy.

Four years before Cauchy's *Cours d'analyse* was to appear in 1821, Bernard Bolzano of Prague saw that, in order to prove the intermediate-value theorem for continuous functions, definitions of convergence and continuity were needed. These he gave and used in a booklet of 1817, avoiding infinitesimals. He also attempted a theory of real numbers including their completeness property. His theory of number expressions, which even included infinitely small and large quantities, was developed at some time after 1830 but came to light only in the 1960s. Bolzano never contributed to technical progress in analysis, and had no influence on his contemporaries. The Weierstrassians were to acknowledge his statements: $f(x)$ is continuous on an interval if at any x of that interval $|f(x + \omega) - f(x)|$ can be made as small as one wishes by taking ω sufficiently small; and, a sequence $F_1(x)$, $F_2(x)$, ... converges to a fixed quantity if and only if we can make $|F_{n+r}(x) - F_n(x)|$ smaller than any given quantity, no matter how large r is.

Up to the middle of the nineteenth century everybody, with the exception of Bolzano, was satisfied with the undefined concept of a continuous variable which could accept real values, represented by decimal numbers, and possibly augmented by infinitesimals. But what did 'continuity' of a variable or a line mean? The answer given by Richard Dedekind in 1872 was one of the first occasions on which the concept of a set was mentioned explicitly. Every 'cut' or partition of the rational numbers into two sets, each of the members of the first set being smaller than each member of the second one, defines one and only one 'real number'. It follows easily that these real numbers form a complete ordered field. (If you still feel that Dedekind did not fill in all of the gaps, wait for Section 5 and the resurrection of infinitesimals.)

At about the same time, Cantor and Heinrich Eduard Heine defined the real numbers as equivalence classes of rational fundamental sequences, (a_n) being equivalent to (b_n) if $\lim(a_n - b_n) = 0$. This more constructive approach could be generalized to sets of functions, and to metric spaces.

Uniformity of convergence and of continuity may have been implicit in the definitions put forward by Cauchy and Bolzano. Although E. G. Björling, P. L. von Seidel and George Stokes took some steps in that direction during the 1840s, it was Weierstrass who stated those concepts explicitly, stressing their significance. (A list of events in this field was given by Hardy *1918*.) Theorem I of Cauchy is an easy exercise if uniform convergence is assumed.

2.3 Epsilons with Weierstrass

Cauchy and others had used epsilons and deltas as occasional tools; Weierstrass made them the essence of rigorous definitions and deductions, exactly in the way we learn about them today. In his lectures in Berlin in 1861, Weierstrass offered this definition:

> If it is possible to determine a bound δ for h such that, for all $|h| < \delta$, $|f(x+h) - f(x)|$ will be smaller than a given arbitrarily small quantity ε, then we say that infinitely small changes of the variable correspond to infinitely small changes of the argument.

If this is true for each fixed x, then $f(x)$ is pointwise continuous. And it is uniformly continuous if ε does not depend on x. For manuscripts of Weierstrass's lectures, see Dugac *1973* and Weierstrass *1988*.

Weierstrass and his followers proved lemmas that had been taken for granted by Cauchy, and even by Dirichlet and Bernhard Riemann in the 1850s: on a closed interval, a continuous function is uniformly continuous; it is bounded and assumes its extreme values. Weierstrass shared with Cauchy a preference for continuous functions. It was with great satisfaction that he announced in 1886, near the end of his long teaching career, his famous approximation theorem: a continuous function is, on a closed interval, equal to a uniformly convergent series of polynomials. In a precise sense the analytical expressions of Euler seemed vindicated, with a revival of his polynomials of infinite degree.

3 THE LIMITATIONS OF THE BASIC CONCEPTS

Why were mathematicians no longer content with (uniform) convergence and (piecewise) continuity? There were more snares in these simple concepts than had been expected. Warnings came, once more, from trigonometrical series used in applications.

After futile efforts by Carl Friedrich Gauss, Cauchy, Siméon-Denis Poisson, Dirichlet and others to show that the Fourier series of a continuous function is equal to that function, Paul Du Bois-Reymond gave counter-examples in 1876. A different property, that of bounded variation, was more satisfactory, covering in addition many useful discontinuous functions. Although each continuous function had a continuous primitive function or integral, Bolzano, Riemann and Weierstrass – again by using a Fourier series – displayed continuous functions having (almost) nowhere a derivative. On the other hand, Gaston Darboux observed in 1875 that the intermediate-value property of continuous functions did not characterize this class of functions.

Uniform convergence turned out to be unsuitable for trigonometrical series and other systems of functions defined by differential equations of mathematical physics. Here, convergence in the quadratic mean,

$$\int [s_n(x) - s(x)]^2 \, dx \to 0 \quad \text{as } n \to \infty, \tag{2}$$

was a more suitable concept.

Despite Cauchy, divergent series frequently made sense in applications, even when numerical values were asked for (and even in Cauchy's own research work). In the 1880s some of the techniques of divergent series were put on the level of Weierstrassian rigour.

Emile Borel, then in his twenties, began a fairly general vindication of divergent series in 1896. He started from the observation that, for a power series $f(x)$ with ρ as its radius of convergence,

$$f(x) = a_0 + a_1 x + a_2 x^2 + \cdots = \int_0^\infty e^{-t} F(tx) \, dt, \quad \text{with } |x| < \rho \tag{3}$$

and

$$F(u) = a_0 + a_1 u + a_2 \frac{u^2}{2!} + a_3 \frac{u^3}{3!} + \cdots. \tag{4}$$

Now, the integral on the right-hand side of (3), or its analytic continuation, may have a value for some $|x| \geqslant \rho$, even if $\rho = 0$, and this value may be taken as a Borel sum of the divergent series. Under certain conditions Borel sums can be manipulated as convergent series. If $f(x)$ is a formal power-series solution of a differential equation, then its Borel sum will be a proper solution whose formal properties may be more easily obtained from the divergent series $f(x)$. (See Kline 1972: Chap. 47 for this and the following; also Tucciarone 1973.)

Another type of divergent series, called 'semi-convergent' or 'asymptotic', had been successfully used since the eighteenth century to find numerical values of functions. If, for large x,

$$g(x) = b_0 + b_1 x^{-1} + b_2 x^{-2} + \cdots + b_n x^{-n} + x^{-n} o_n(x), \tag{5}$$

where $\lim o_n(x) = 0$, for each n and $x \to +\infty$, the series $\Sigma \, b_n x^{-n}$, though possibly divergent, is called an asymptotic expansion of $g(x)$. Experience had shown that, for any given large x, the series gives a good approximation of $g(x)$ when cut off at the nth term where $|b_n x^{-n}|$ is the smallest term of the infinite sum. In 1886 Henri Poincaré and T. J. Stieltjes were the first to justify these empirical results, and to pave the way for further applications. Others, including Georg Frobenius, E. Cesàro, O. Hölder and L. Fejér reinvented summation methods which had already been known in the infinitesimal analysis of Cauchy and his contemporaries (see the end of Section 5).

Although the restrictive concepts of convergence and continuity as used by Cauchy and Weierstrass were known to be too narrow even in their times, they still govern present-day courses in analysis.

4 REFINEMENTS OF TECHNIQUES AND DEVELOPMENTS OF CONCEPTS

4.1 Sequences, series, elementary functions

These precalculus topics were treated systematically by Cauchy *1821* in his algebraic analysis. From his famous convergence criterion, Cauchy derived the majorant test, the ratio test of Jean d'Alembert and the root test for infinite series. He obtained the correct formula for the radius ρ of convergence for (real and complex) power series, given by

$$\rho^{-1} = \limsup \sqrt[n]{|a_n|}, \qquad (6)$$

explaining lim sup as 'the greatest of the values' to which a sub-sequence can converge. The series for the exponential and logarithmic functions, and the binomial series for real exponents, were obtained by ingenious and correct applications of his Theorem I. He introduced products of series and double series, and investigated their convergence properties.

These techniques became widely known through the texts of others, among them the textbooks by J. C. F. Sturm, Joseph Liouville and J. M. C. Duhamel in France, by Augustus De Morgan in England, and by O. Schlömilch in Germany. However, the conceptual framework developed by Cauchy *1821*, *1823* and the *Cours d'analyse* itself were apparently soon forgotten. In 1888 Hadamard had to rediscover formula (6). In the 6th edition of 1880 of the famous *Cours d'analyse* by Sturm (1st edition 1857), out of more than 1100 pages no more than nine lines on p. 309 are dedicated to the definition of the integral, which is treated in the eighteenth-century style of an antiderivative. By the end of the nineteenth century, the editors of Cauchy's collected works had difficulty in tracing a copy of Cauchy *1821* in France!

Conceptual innovations were accepted with rather more reluctance, and then only by a few. Abel from Norway was 24 years old, and Dirichlet from

Germany 21, when they inhaled the Paris air of rigour in 1826. Abel continued the work on series by demonstrating the general binomial theorem, and by his result that $f(x)$ is continuous on $0 \leqslant x \leqslant 1$ if $f(x)$ is a power series having $\rho = 1$ which still converges for $x = 1$. This was a result for which Cauchy had been wrong in using his Theorem I. In 1829 Dirichlet published his famous proof that a Fourier series of a continuous function converged to that function if there were only finitely many maxima and minima in each finite interval. Cauchy underlined the significance of remainder terms and, in the 1830s, developed strict rules for error bounds in his 'calcul des limites'. In these papers, epsilontics appears as an occasional trick and not as a general technique.

Weierstrass was almost 50 when he became a full professor at the University of Berlin in 1864. Virtually self-taught, he had achieved his most significant results while a schoolteacher at remote places in East Prussia. Now his fundamental lectures on real and complex analysis attracted hundreds of students and scholars, among them Cantor and K. H. A. Schwarz. A product of a rich experience gained in both teaching and research, Weierstrass's epsilons clarified the significance of uniform as opposed to pointwise concepts, and led to safe conditions for reversing orders of limiting processes.

4.2 Derivatives and integrals

In Cauchy *1823* the derived function $f'(x)$, if it exists, is defined on an interval, by the condition that

$$f'(x) - \frac{1}{i}\, [f(x+i) - f(x)] \approx 0 \quad \text{if } i \approx 0. \tag{7}$$

The differential $dy = f'(x)\,dx$ is a function of its two variables, x and dx, and, in contrast to earlier conceptions, is not restricted to $dx \approx 0$. Again, (7) is not a pointwise definition – implicitly, it contains the continuity of $f'(x)$. Others were to return to the pointwise definition of $f'(x_0)$ as a limit of slopes of chords.

Cauchy made much use of inequalities. In particular, $f'(x)$ has, on an interval, the same bounds as the ratio of differences. Cauchy used the mean-value theorem and Taylor's theorem with a remainder term. He recalled the Leibnizian definition of the integral as a sum, proved by his infinitesimal reasoning the integrability of continuous functions, and re-established the fundamental theorem $(\int^x f(x)\,dx)' = f(x)$ as a provable statement. Following Joseph Fourier, he introduced the notations of integrals and their bounds as they became generally accepted (§3.7). He also established the main properties and notations in the calculus of several variables. A few of his deductions lack clarity, among them that of the statement that $f_{xy} = f_{yx}$; correct conditions were given by Schwarz in 1873.

Cauchy's texts laid a firm foundation for the study of ordinary and partial differential equations, which had become more and more important in physics. On the whole, the technical framework of real analysis built up by Cauchy would remain unchanged for the remainder of the century.

4.3 Existence and approximation of solutions

There had been isolated proofs of existence for solutions of particular problems before the 1820s. D'Alembert, Euler, Gauss and Jean Argand had recognized that the fundamental theorem of algebra needed an (analytical) proof (§6.1). That problem had also been the starting-point for Bolzano's attempt at the intermediate-value theorem for continuous functions in 1817.

To Cauchy and Weierstrass, rigour also meant that the existence of a solution of some analytical problem should not be assumed on the basis of geometrical or physical reasons. Cauchy proved existence theorems for differential equations, using power series, a method which was further developed by Weierstrass and his student Sonya Kovalevskaya (§3.15). In addition to rigour, there are two more technical requirements of existence proofs: that the mathematical model should actually fit, and meet the expectations of physicists; and that they should supply approximations to the solutions, possibly with error bounds. Only gradually did these insights come to be shared by more than a few mathematicians.

Until the 1850s Gauss, Dirichlet and Riemann had taken for granted that the following variational problem had a solution. Consider the integral of $u_x^2 + u_y^2$ over some region of the x, y-plane when the values of $u(x, y)$ on the boundary are prescribed. It follows that

$$\Delta u := u_{xx} + u_{yy} = 0 \tag{8}$$

inside the region if u minimizes the integral. The existence of a minimizing u was simply stated as 'Dirichlet's principle'. In 1870, Weierstrass criticized that principle, showing that the integral over $(xu_x)^2 + (yu_y)^2$, though bounded from below, had no minimizing function u (see also Weierstrass *1988*; and §3.17 on potential theory). That criticism eventually led to the invention of direct methods for the solution of variational and other extremum problems, in around 1900. The problem was vital since Laplace's equation $\Delta u = 0$ appears almost everywhere in mathematical physics; its solutions are called 'harmonic functions', and they appear also as the real and imaginary parts of complex analytic functions (§3.11–3.12).

4.4 Spaces of functions

Technical progress on existence theorems in the twentieth century came mainly from fixed-point theorems in spaces of functions. More generally, several of the techniques of real-variable analysis had begun to merge by the end of the nineteenth century, eventually to be called 'functional analysis' (§3.9).

Consider the set (or 'space') $C[a, b]$ of real functions continuous on $a \leqslant x \leqslant b$, with the distance of two functions measured by the 'norm'

$$\| f - g \| = \max_x | f(x) - g(x) |. \tag{9}$$

This space is complete in the sense that every fundamental sequence has a limit inside $C[a, b]$. Uniform convergence of $f_n(x)$ to $f(x)$ means that $\| f_n - f \| \to 0$. The Weierstrass approximation theorem now becomes: 'the set of polynomials is dense in $C[a, b]$'.

This norm is not suitable for trigonometric series, for which the orthogonality of functions is a useful property. Introducing the 'inner product'

$$(f, g) := \int f(x) g(x) \, \mathrm{d}x \tag{10}$$

and a norm $\| f \|^2 = (f, f)$ makes it possible to express orthogonality simply by $(f, g) = 0$, and convergence in the quadratic mean by $\| f_n - f \| \to 0$.

5 THE RESURRECTION OF INFINITESIMALS

Once the concept of a variable quantity had been replaced by that of sets of numbers, infinitesimals lost their position of importance. Nevertheless they survived, mainly in the rather imprecise languages of physics and differential geometry. The occasional attempts that were made to build rigorous foundations implicitly relied on a principle dating back to Leibniz: the rules of the finite remain valid for the infinite; in other words, infinitesimals behave like very small real numbers.

In some of his texts Cauchy took i to be a 'base' of infinitesimals; $f(i)$ was said to represent an infinitesimal whenever $f(x)$ was continuous near $x = 0$, and $f(0) = 0$. Later, Stolz reconsidered the 'moments' of Newton (§3.2) in a similar manner; and Du Bois-Reymond used the behaviour of functions $f(x)$ for $x \to +\infty$ to obtain different grades of the infinitely large and small. In this *Fondamenti di geometria* (1891), Giuseppe Veronese extended the real numbers by adjoining a symbol ∞_1 and all rational expressions formed with it; moreover, he considered expressions such as $\infty_1{}^{\infty_1}$ and infinite sums of expressions already formed. The symbol ∞_1 was to stand for an infinitely large number, and $1/\infty_1$ for an infinitesimal. His student Tullio Levi-Cività gave descriptions of these quantities in terms of what would later become 'modern algebra'.

These attempts, and others made during the first half of the twentieth century, succeeded in defining number systems but failed to show how even the most basic theorems of classical ('standard') analysis could be obtained. Moreover, it was unclear how a standard function like $y = \sin x$ could be extended from the standard numbers to some greater number system. Although non-standard 'delta functions' such as the rational function

$$\delta(x) = \frac{\omega}{\pi(\omega^2 + x^2)}, \quad 0 < \omega \approx 0, \quad \delta(x) \approx 0 \text{ for } x \neq 0 \qquad (11)$$

made sense in such number systems, how could one prove analytical properties like

$$\int \delta(x)\,dx = 1, \qquad \int \delta(x - t)f(t)\,dt \approx f(x), \qquad (12)$$

used by Paul Dirac and other physicists since 1925?

Around 1960 several mathematicians began to consider variations on the theme of Leibniz's principle mentioned above. A first account of that development can be found in the historical remarks at the end of Robinson *1966*.

Though rudimentary, the following approach will suffice to throw some light on the basic idea, and on the history of real variable analysis including the work of Cauchy. Assume that an ideal element ν is adjoined to the real numbers, corresponding to the ∞_1 of Veronese or the i^{-1} of Cauchy. A precise version of Leibniz's principle is then the following. Suppose that some formula $S(n)$, built up from the symbols of 'standard' mathematics, is valid for all sufficiently large 'standard' natural numbers n; then, by definition, $S(\nu)$ is valid in 'non-standard analysis'. Since each of the formulas

$$S(n): n > 1; \quad S(n): n > 2; \quad S(n): n > 3; \quad \ldots \qquad (13)$$

is valid for all sufficiently large n, it follows that all of

$$\nu > 1; \quad \nu > 2; \quad \nu > 3; \quad \ldots \qquad (14)$$

are valid, and that ν is infinitely large. Similarly, $\omega = \nu^{-1}$ is a positive infinitesimal. (Take care in advance that your standard mathematics does not use either of the symbols ν and ω.)

If $x(n) = \hat{x}(n)$ for all sufficiently large n, then $x(\nu) = \hat{x}(\nu)$, which is an equality of two non-standard objects $\xi = x(\nu)$, $\hat{\xi} = \hat{x}(\nu)$. If $x(n)$, $\hat{x}(n)$ are natural or real numbers, we say that ξ, $\hat{\xi}$ are 'hypernatural' or 'hyper-real' numbers. Now, the equation $\eta = \sin \xi$ is well defined for all hyper-real numbers, and $|\sin \xi| \leqslant 1$ and

$$[\sin(\xi + \varepsilon) - \sin \xi]/\varepsilon \approx \cos \xi \qquad (15)$$

are easily verified for all hyper-real ξ and $\varepsilon \approx 0$. Cauchy's definitions of continuity, of the derived function and of the integral as a sum, make sense, and so do (11) and (12).

Since, for real numbers, the formula

$$S(n): \ x(n) > \hat{x}(n) \ \text{ or } \ x(n) = \hat{x}(n) \ \text{ or } \ x(n) < \hat{x}(n) \qquad (16)$$

is valid for all n, we formally obtain

$$\xi > \hat{\xi} \ \text{ or } \ \xi = \hat{\xi} \ \text{ or } \ \xi < \hat{\xi} \qquad (17)$$

as a valid statement for any pair of hyper-real numbers ξ, $\hat{\xi}$. The properties of an ordered field follow in the same way. To our stock of infinitesimals belong $\sqrt{\omega}$, ω^2, ω^ν, $e^{-\nu}$.

A few examples will show how Cauchy's proofs can be retraced here. Suppose that $f(\xi, \eta)$ is a function which is continuous in each of its variables separately. Consider

$$f(\xi + \delta, \eta + \varepsilon) - f(\xi, \eta) = [f(\xi + \delta, \hat{\eta}) - f(\xi, \hat{\eta})] + [f(\xi, \eta + \varepsilon) - f(\xi, \eta)],$$
$$(18)$$

where $\hat{\eta} = \eta + \varepsilon$ and $\delta \approx 0 \approx \varepsilon$. It follows that the function is continuous in both of its variables together. This is correct here (but not in standard analysis) since we have assumed a property (that of continuity in one variable) to be valid not only for standard real values, but also for hyper-real numbers ξ, η, $\hat{\eta}$. This shows how uniformity is circumvented – or rather, how it is smuggled in through infinitely small gaps between the real numbers!

Similarly, the controversial Theorem I of Cauchy *1821* discussed in Section 2 is now valid, even with Cauchy's own formal proof. Suppose that $s_n(\xi)$ is, for each finite n, a continuous function on $a \leqslant \xi \leqslant b$, and that $s_n(\xi)$ converges to $s(\xi)$ for each ξ of that interval. Clearly,

$$|s(\xi + \delta) - s(\xi)| \leqslant |s(\xi + \delta) - s_n(\xi + \delta)|$$
$$+ |s_n(\xi + \delta) - s_n(\xi)| + |s_n(\xi) - s(\xi)|. \qquad (19)$$

For $\delta \approx 0$ and n finite, the second term on the right-hand side is, by the continuity of s_n, infinitesimal. For any standard $\varepsilon > 0$, take n sufficiently large but finite to render both the first and third terms smaller than $\varepsilon/3$. This is possible because of convergence. It follows that $|s(\xi + \delta) - s(\xi)| < \varepsilon$ for each standard $\varepsilon > 0$. In other words, $s(\xi + \delta) - s(\xi) \approx 0$, and $s(\xi)$ is continuous.

In these examples, standard theorems are proved using infinitesimals. Cauchy and his contemporaries went beyond that and used hyper-real functions like delta functions which have no 'standard' graph. Fourier's argument in 1822 was based on a 'smooth' unit jump function (defined for some infinitely large p),

$$\eta(x) = \frac{1}{\pi} \int_0^p \frac{\sin xt}{t} \, dt \tag{20}$$

with

$$\eta(x) \approx \begin{cases} \frac{1}{2}, & x \geq p^{-1/2} \\ -\frac{1}{2}, & x \leq -p^{-1/2}. \end{cases} \tag{21}$$

In terms of

$$\delta(x) = \eta'(x) = \frac{1}{\pi} \frac{\sin px}{x} = \frac{1}{2\pi} \int_{-p}^p \cos kx \, dk, \tag{22}$$

the famous Fourier integral theorem is

$$\int_a^b \delta(x-t) f(t) \, dt \approx f(x), \quad a < x < b \tag{23}$$

for well-behaved $f(t)$. This can be obtained through integration by parts from $\delta = \eta'$. Dirichlet and others translated the argument into the more clumsy language of limits and epsilons.

The graph of the delta function (22) oscillates rapidly, and it is not true that $\delta(\xi) \approx 0$ for non-infinitesimal ξ. Not satisfied with that misbehaviour, Poisson and Cauchy introduced auxiliary factors $h(k)$ under the integral sign, with $h(k) \approx 1$ for all k not infinitely large, the most popular example being $h(k) = e^{-|\omega k|}$ ($\omega \approx 0$). Then,

$$\delta(x) = \frac{1}{2\pi} \int_{-\infty}^{\infty} h(k) \cos kx \, dk \approx 0, \quad \text{for } x \neq 0, \tag{24}$$

but still $\int \delta \approx 1$, and (23) is valid again. They applied the same procedure to Fourier series and showed that, up to an infinitesimal, any continuous periodic function was equal to its Fourier series (modified by factors $h(k)$ which did not sensibly differ from 1). They stated as a general method for divergent series $\Sigma u_k(x)$ of continuous functions that, if, for some continuous function $s(x)$,

$$s(x) \approx \sum h(k) u_k(x) \tag{25}$$

on an interval, then $s(x)$ did not depend on the particular choice of $h(k) \approx 1$, and could be taken as the sum of the divergent series. These

329

summability methods were soon forgotten after the decline of infinitesimals, and had to be reinvented in epsilontic disguise after 1880.

For further information on non-standard methods and their historical roots, see Laugwitz *1986*; for generalized functions, see Lützen *1982*; and for Cauchy and summability problems, see Laugwitz *1989*.

BIBLIOGRAPHY

Bottazzini, U. *1986*, *The Higher Calculus: A History of Real and Complex Analysis from Euler to Weierstrass*, New York: Springer.

Cauchy, A. L. *1821*, *Cours d'analyse: Analyse algébrique*, Paris: De Bure. [Also in *Oeuvres complètes*, Series 2, Vol. 3.]

—— *1823*, *Résumé des leçons données à l'Ecole Royale Polytéchnique sur le calcul infinitésimal*, Paris: De Bure. [Also in *Oeuvres complètes*, Series 2, Vol. 4, 5–261.]

Dugac, P. *1973*, 'Eléments d'analyse de Karl Weierstrass', *Archive for History of Exact Sciences*, **10**, 41–176.

Edwards, C. H. *1979*, *The Historical Development of the Calculus*, New York: Springer.

Grattan-Guinness, I. *1970*, *The Development of the Foundations of Mathematical Analysis from Euler to Riemann*, Cambridge, MA: MIT Press.

—— *1990*, *Convolutions in French Mathematics, 1800–1840*, 3 vols, Basel: Birkhäuser.

Hardy, G. H. *1918*, 'Sir George Stokes and the concept of uniform convergence', *Proceedings of the Cambridge Philosophical Society*, **19**, 148–56. [Also in *Collected Papers*, Vol. 7, 505–13.]

Kline, M. *1972*, *Mathematical Thought from Ancient to Modern Times*, New York: Oxford University Press.

Laugwitz, D. *1986*, *Zahlen und Kontinuum*, Mannheim: Bibliographisches Institut.

—— *1989*, 'Definite values of infinite sums: Aspects of the foundations of infinitesimal analysis around 1820', *Archive for History of Exact Sciences*, **39**, 195–245.

Lützen, J. *1982*, *The Prehistory of the Theory of Distributions*, New York: Springer.

Robinson, A. *1966*, *Non-Standard Analysis*, Amsterdam: North-Holland.

Tucciarone, J. *1973*, 'The development of [...] summable divergent series from 1880 to 1925', *Archive for History of Exact Science*, **10**, 1–40.

Weierstrass, K. *1988*, *Ausgewählte Kapitel aus der Funktionenlehre* (ed. R. Siegmund-Schultze), Leipzig: Teubner. [In addition to lecture notes of 1886, there are reprints of fundamental papers, commented upon by the editor.]

3.4

Differential geometry

KARIN REICH

1 PREHISTORY

The Greeks did not have a general theory of curves, although they did distinguish between several different kinds of curve: straight lines, circles, conic sections, epicycles, the Archimedean spiral, the quadratrix, the conchoid of Nicomedes, the cissoid of Diocles, and so on (§1.3). All these curves were located in the plane; there was no general way of describing them.

In his *Elements*, Euclid gave a first definition of a tangent to the circle (Book III, def. 2): 'A straight line is said to contact the circle when its elongation does not intersect the circle.' This definition shows the lack of any ideas about infinitesimals, but it was to remain valuable until the seventeenth century. The Greeks also lacked a surface theory, but they knew several kinds of surfaces such as the sphere, the cylinder, and conoids and spheroids related to the conic sections (Gericke *1984*: 120–24).

The first treatment of curvature was made in medieval times by Nicole Oresme. In his *Treatise on the Configuration of Qualities and Motions*, he distinguished between 'uniform' and 'difform' qualities (§2.6). He applied this idea to curvature: a circular curvature is uniform, the intensity of the uniform curvature is to be measured by the quantity of the radius. Every other curvature is difform; an Archimedean helix, however, has a 'difformly difform' curvature (Gericke *1990*: 150–53).

The seventeenth century saw the development of analytical geometry (§7.1), which provided a much better way of describing curves – by means of analytical functions. The newly invented method of determining maxima and minima (§3.1) made it possible to think of tangents in a new, more general way.

When Gottfried Wilhelm Leibniz and Isaac Newton laid down the fundamental principles of the calculus (§3.2), they provided the main method of differential geometry. This is the crucial event, and it is only from this time on that one can really talk of differential geometry.

2 CLASSICAL DIFFERENTIAL GEOMETRY

2.1 The use of calculus methods

Both Leibniz and Newton applied their new ideas to geometry. In 1684 Leibniz published a paper entitled 'Meditatio nova de natura anguli con-tactus et osculi', in which he tried to give a first analytical idea of curvature. In the same way that a straight line is suitable for determining direction, because its direction is the same everywhere, the circle is suitable for determining curvature, because its curvature is the same everywhere. Therefore Leibniz considered a given curve to be 'osculated' by a circle at every point. The complete theory of osculations was later developed by James I Bernoulli (Haas *1881*).

Prior to Leibniz, Newton found an analytical expression for the 'quantity of curvature', but these results were published only much later, in his *Method of fluxions* (1736). The circle of curvature is the circle with the property that it is not possible to draw another circle within the angle of contingence between the curve and this circle. The curve is said to have the same curvature as this circle in the point of contact. Newton calculated its radius by examining the normals of the curve in three points which approach each other (Stiegler *1968*):

$$r = \frac{(1 + z^2)^{3/2}}{\dot{z}}, \quad \text{with } z = \dot{y}/\dot{x}. \tag{1}$$

Newton had thought only of plane curves. The first extensive investigation of space curves was presented by Alexis Clairaut in his book *Recherches sur les courbes à double courbure* (1731). He used spatial coordinates based on the three planes Oxy, Oyz and Oxz, onto which the space curve is projected. The result is three plane curves, with the equations $y = f(x)$, $z = g(y)$ and $x = h(z)$, each of which has its own curvature. Two of these equations are sufficient to determine the space curve; this justified the term 'curve of double curvature'. A space curve also allowed the interpretation as the intersecting curve of two surfaces, which Clairaut proved for the case of conic sections. He went on to determine tangents and subtangents, normals and subnormals, rectifications and the volume of solids produced by rotating curves.

Surface theory was initiated by Leonhard Euler. In his *Introductio in analysin infinitorum* (1748) he dealt mainly with curve theory, expanding functions into series, in which linear terms represented the connections with the tangents. In his paper 'Recherches sur la courbure des surfaces' (1767) he remarked at the beginning that until then it had only been possible to give an expression of curvature for spheres. For general surfaces, Euler proposed normal sections. For the curves produced by the surface and the normal section, it was possible to examine their curvatures; indeed, he found that there were special normal sections where the curvatures of the curve had a maximum and a minimum, their radii given by f and g. It was possible to prove that these two normal sections (which Euler called 'principal sections') were normal to each other. As a consequence, he determined the radius of curvature of an arbitrary normal section as

$$r = \frac{2fg}{f + g - (f - g)\cos\phi}, \tag{2}$$

with ϕ denoting the angle between the normal section and one of the principal sections.

In 1776, Jean-Baptiste Meusnier gave a supplement to Euler's formula, a result of considering skew sections also. For a plane σ which intersects the surface in a curve with radius of curvature r_σ and the normal of the surface with an intersecting angle α, Meusnier found that

$$r_\sigma = r \cos\alpha. \tag{3}$$

In 1813, Charles Dupin converted Euler's formula into the more elegant form

$$\frac{1}{r} = \frac{\cos^2\phi}{f} + \frac{\sin^2\phi}{g} = \frac{\cos^2\phi}{R_1} + \frac{\sin^2\phi}{R_2}. \tag{4}$$

In 1770, Euler had used the line element of a surface which he represented in parametric form as

$$dx = l\,dT + \lambda\,dU, \qquad dy = m\,dT + \mu\,dU, \qquad dz = n\,dT + \nu\,dU, \tag{5}$$

with T and U as parameters, and proved the important theorem that all surfaces which are developable into the plane without distortion or bending are tangential surfaces of a space curve – 'tangential developables'. Therefore they can be conical surfaces, cylindrical surfaces or tangent surfaces.

Gaspard Monge made many contributions to differential geometry, especially in those fields which were related to differential equations, concerning families of surfaces, enveloping surfaces and their characteristics, and minimal surfaces. The basic partial differential equation is due to Joseph Louis Lagrange, who had found it by means of the calculus of variations. Monge summarized his own results together with others available at the time in the first textbook on differential geometry: *Application de l'analyse à la géométrie* (1795, 5th edn 1850). Monge has to be regarded as the founder of the French school in differential geometry: Meusnier and Dupin, for example, were his pupils (Taton *1951*). This tradition was continued by others and still persists; the current terminology is to a great extent of

French origin. Among those who promoted differential geometry at the beginning of the nineteenth century was Augustin Louis Cauchy. In 1826 he published his *Leçons sur l'application du calcul infinitésimal à la géométrie*, a textbook based on lectures he gave at the Ecole Polytechnique (§3.3).

2.2 The use of invariant methods

With Carl Friedrich Gauss's 'General investigations of curved surfaces' (*1828*) a new epoch began; Gauss is sometimes called the 'father of differential geometry'. He was the first to introduce the idea of an invariant. Using the line element in terms of the parametric variables p and q,

$$ds^2 = E\,dp + 2F\,dp\,dq + G\,dq^2, \tag{6}$$

he was able to show that his newly defined measure of curvature, k, in

$$4(EG - FF)^2 k = E\left(\frac{dE}{dq}\frac{dG}{dq} - 2\frac{dF}{dp}\frac{dG}{dq} + \left(\frac{dG}{dp}\right)^2\right)$$

$$+ F\left(\frac{dE}{dp}\frac{dG}{dq} - \frac{dE}{dq}\frac{dG}{dp} - 2\frac{dE}{dq}\frac{dF}{dq} + 4\frac{dF}{dp}\frac{dF}{dq} - 2\frac{dF}{dp}\frac{dG}{dp}\right)$$

$$+ G\left(\frac{dE}{dp}\frac{dG}{dp} - 2\frac{dE}{dp}\frac{dF}{dq} + \left(\frac{dE}{dq}\right)^2\right)$$

$$- 2(EG - FF)\left(\frac{d^2E}{dq^2} - 2\frac{d^2F}{dpdq} + \frac{d^2G}{dp^2}\right), \tag{7}$$

was invariant against distortion and bending of the surface. The main theorem is the so-called 'remarkable theorem' (*theorema egregium*): if a surface is developed upon any other surface whatever, the measure of curvature in each point remains unchanged. The original Latin version has *invariata* for 'unchanged'.

Gauss also defined surfaces from a new point of view, as what we would in modern terms describe as two-dimensional manifolds:

> When a surface is regarded, not as the boundary of a solid, but as a flexible, though not extensible solid, one dimension of which is supposed to vanish, then the properties of the surface depend in part upon the form . . . and in part are absolute and remain invariable. To these latter properties . . . belong the measure of curvature and the integral curvature . . . also the theory of shortest lines. (Gauss *1828*)

Here he again used the expressions 'invariant' and 'absolute'. Yet in some notes written between 1822 and 1825 he had referred to the geodesic curvature as 'absolute curvature'.

In a paper published in 1864, Eugenio Beltrami used Gauss's definition of a surface but distinguished between absolute and 'relative' properties, the latter depending on the special form of the surface. In the same paper he succeeded in determining new absolute operators, the differential parameters later named after him.

Ideas about invariants also proved fruitful in curve theory. The most important formulas, the so-called Frenet–Serret formulas, relate the tri-hedral vectors $\mathbf{t}$, $\mathbf{n}$ and $\mathbf{b}$ (tangent, normal and binormal) to the arc length s:

$$\frac{d\mathbf{t}}{ds} = \varkappa\mathbf{n}, \qquad \frac{d\mathbf{n}}{ds} = -\varkappa\mathbf{t} + \tau\mathbf{b}, \qquad \frac{d\mathbf{b}}{ds} = -\tau\mathbf{h}, \tag{8}$$

where $\varkappa(s)$ is curvature and $\tau(s)$ is torsion. Equations $(8)_{1,3}$ are in Cauchy's *Leçons* (1826). The Estonian mathematician Karl Eduard Senff was the first to produce all three equations, in 1831. Without recognizing their importance, Frédéric Jean Frenet presented these formulas in his thesis (1847, 1852); his countryman Joseph Alfred Serret underlined their fundamental significance in 1851.

The fundamental equations in surface theory play a role equivalent to those of the Frenet–Serret formulas in curve theory. Gauss had used integrability conditions without any discussion. The problem was recognized by Karl Peterson, who was born in Riga and had studied in Dorpat. In 1853, he finished his thesis 'On the bending of surfaces', which was written in German and not published until later. Peterson set up all the possible integrability conditions:

$$\mathbf{x}_{iik} = \mathbf{x}_{iki}, \qquad \mathbf{x}_{kik} = \mathbf{x}_{kki}, \qquad \mathbf{n}_{ik} = \mathbf{n}_{ki} \tag{9}$$

(where the subscripts indicate partial differentiation of the surface and normal vectors $\mathbf{x}$ and $\mathbf{n}$). Equation $(9)_3$ was first published in 1862 by Julius Weingarten. Equations $(9)_{1,2}$ were first published in 1857 by Gaspare Mainardi, and independently in 1867–73 and 1883 by Delfino Codazzi; they are called the Mainardi–Codazzi or sometimes the Peterson–Codazzi equations. On the basis of these integrability conditions, Peterson (1853) and Ossian Bonnet (1867) propounded and proved the fundamental theorem of surface theory: that the coefficients of the first and second fundamental form g_{ik} and $L_{ik} = \mathbf{n}\mathbf{x}_{ik}$ determine a surface, apart from its position in space (Reich *1973*: 301–4; Phillips *1979*).

3 RIEMANNIAN GEOMETRY

Gauss had considered surfaces as two-dimensional manifolds. The concept of an n-dimensional manifold was introduced into differential geometry by Bernhard Riemann, who quoted the philosopher Johann Friedrich Herbart, by whom he was strongly influenced (Scholz *1980*: Chap. 2; Scholz *1982*; Portnoy *1982*). In 1854, Riemann delivered his doctoral thesis 'On the hypotheses which lie at the foundations of geometry', where he introduced the new concept of an n-dimensional manifold (Nowak *1989*):

> In a concept whose instances form a continuous manifold, if one passes from one instance to another in a well-determined way, the instances through which one has passed form a simply extended manifold, whose essential characteristic is, that from any point in it a continuous movement is possible in only two directions, forwards and backwards. If one now imagines that this manifold passes to another, completely different one, and once again in a well-determined way, that is, so that every point passes to a well-determined point of the other, then the instances form, similarly, a double extended manifold. In a similar way, one obtains a triply extended manifold when one imagines that a doubly extended one passes in a well-determined way to a completely different one, and it is easy to see how one can continue this construction. If one considers the process as one in which the objects vary, instead of regarding the concept as fixed, then this construction can be characterized as a synthesis of a variability of $n + 1$ dimensions from a variability of n dimensions and a variability of one dimension.

In 1861 Riemann submitted a paper to the Académie des Sciences in Paris, as an entry for a prize competition. In the second part of this paper he considered the transformation of the expression $\Sigma_{i,i'}\, b_{i,i'}\, \mathrm{d}s_i\, \mathrm{d}s_{i'}$ into the given form $\Sigma_{i,i'}\, a_{i,i'}\, \mathrm{d}x_i\, \mathrm{d}x_{i'}$ (Farwell and Knee *1990*). By using the Christoffel symbols (as they are now called), he was able to give the analytical expression for Riemannian curvature:

$$R^i_{jkl} = \frac{\partial \Gamma^i_{jl}}{\partial u^k} - \frac{\partial \Gamma^i_{jk}}{\partial u^l} + \Gamma^p_{jl}\Gamma^i_{pk} - \Gamma^p_{jk}\Gamma^i_{pl} \tag{10}$$

(in modern notation). Riemann did not win the prize; the paper was published only in 1876, in his collected works. But with the publication of his thesis in 1868, this work found acceptance and acknowledgement.

Its influence was twofold. First, it led to the further development of Riemannian geometry. Beltrami immediately picked up Riemann's n-dimensional concept and used it to investigate manifolds with constant curvature (1868). These kinds of manifold became a very interesting field, to which many mathematicians contributed.

Differential geometry at the end of the nineteenth century included classical differential geometry and normally also the whole spectrum of Riemannian geometry. But it made almost no use of vector analysis or the absolute differential calculus − that is, tensor analysis. This field was not thought to belong to differential geometry. The main textbooks of this time reflect this situation: in particular, Luigi Bianchi's *Lezioni di geometria differenziale* (1886, 1902–9, 1922–3) was widely read, and even gave its name to the whole discipline. By contrast, Gaston Darboux's *Leçons sur la théorie générale des surfaces et les applications géométriques du calcul infinitésimal* (1887–96, 1914–15) did not include Riemannian geometry, as is clear from the title (Reich *1989*).

4 THE ABSOLUTE DIFFERENTIAL CALCULUS

The other direction from Riemann led to the development of the absolute differential calculus. This was intimately connected with invariant theory, at that time a well-established discipline within algebra (§6.8) in which J. J. Sylvester introduced the terms 'covariant', 'contravariant' and 'invariant'. Elwin Bruno Christoffel referred to algebraic invariant theory when in 1869 he introduced invariants and covariants into the theory of binary differential forms. He determined the R_{klji}, which he called the coefficients of a quadrilinear form, as a result of integrability conditions; there was no mention of curvature. He defined what would come to be called the covariant derivation as

$$a_{i_1 \ldots i_r, i} = \frac{\partial a_{i_1 \ldots i_r}}{\partial u^i} - \sum_{m=1}^{r} a_{i_1 \ldots i_{m-1} h \, i_{m+1} \ldots i_r} \Gamma^h_{i_m i}, \qquad (11)$$

an operation which allowed a series of differential covariants to be created. He called the $a_{i_1 \ldots i_r}$ a 'system of transformation relations of the rth degree'.

Gregorio Ricci-Curbastro developed Christoffel's ideas in creating a calculus for the systems of transformation relations, which were soon to become known simply as 'systems'. In a series of papers from 1884 onwards, Ricci began by referring especially to Christoffel and Beltrami, that is to Christoffel's differential parameters and their significance as absolute operators. In 1887, he named the operation (11) 'covariant derivation', and derived the very important theorem that the second covariant derivation enjoys the property of commutativity only for plane (i.e. Euclidean) manifolds. This implies a connection between the covariant derivation and the R_{klji}. A year later Ricci derived this property also for contravariant systems; he used only covariant or only contravariant systems, never both together. He applied his new calculus to the theory of surfaces, the theory of elasticity and the propagation of heat, and in 1893 named it 'absolute differential calculus'. In 1896 he introduced the contraction R_{hk} to stand for R^j_{hkj}, a magnitude which became important later in the theory of relativity.

A summary of these results was first published in French, in 1893; an extensive presentation followed in 1901 as 'Méthodes du calcul différentiel absolu et leurs applications', which Ricci published jointly with his former student Tullio Levi-Cività. They treated the addition, the multiplication and the contraction and transvection of covariant and contravariant systems as calculus operations (Struik *1933*).

Ricci's new calculus was eagerly discussed among mathematicians who worked on differential invariants. In this field he soon found recognition, not only in Italy but also internationally. Like invariant theory, the theory of differential invariants was also regarded at that time as belonging to algebra and not to differential geometry.

5 EINSTEIN'S THEORY OF RELATIVITY

In 1905 Albert Einstein published his paper on special relativity, 'On the electrodynamics of moving bodies'. He managed to present his theory without using 'higher mathematics' – that is, without the application of vectors and tensors. But the relativity principle was to demonstrate their usefulness (§9.13).

At this time, unfortunately, vector analysis was in a rather chaotic state; there were several kinds, differing to some degree or another in notation and interpretation (§6.2). Similarly, tensor analysis existed as three disciplines:

1 The absolute differential calculus.
2 Calculus with vector functions. Josiah Willard Gibbs had in 1884 introduced linear vector functions, which he called 'dyadics'; their components were transformed in the way that tensor components were. The multiplication of dyadics was defined analogously to the multiplication of matrices.
3 Tensor calculus. In 1898, the crystallographer Woldemar Voigt introduced tensors (as he called them) as magnitudes which were at first related to stress and strain (§9.17). These tensors were offsprings of vector calculus. In his early works Voigt only used symmetric tensors of the second order, located in three-dimensional Euclidean space, but later he introduced symmetrical tensors of the nth order.

Voigt's tensors were soon widely adopted. His colleague in Göttingen, Max Abraham, applied them in electrodynamics (§9.10). Abraham's *Einführung in die Maxwellsche Theorie der Elektrizität* (1904–5), a new edition of August Föppl's work of 1894, became a well-known textbook.

The interpreters of Einstein's special relativity were physicists, familiar with Voigt's interpretation. In 1908 Hermann Minkowski introduced four-dimensional vectors (spacetime vectors). He applied two different kinds: vectors with four components, ρ_i, and vectors with six components f_{ik} and $f_{ik} = -f_{ki}$, where $i, k = 1, 2, 3, 4$. These vectors of the second kind were rooted in matrix calculus (§6.7). Arnold Sommerfeld knew Voigt personally and appreciated his tensor calculus, and in 1910 he transformed Voigt's three-dimensional tensors into four-dimensional ones and used them in special relativity. Abraham (1910), Max von Laue (1911) and others pursued similar aims.

In 1911, when Einstein was searching for a more efficient mathematical calculus, his colleague in Prague, Georg Pick, pointed him in the direction of the absolute differential calculus. This was a crucial step. Einstein and his friend Marcel Grossman studied Ricci and Levi-Cività, and interpreted the absolute differential calculus as a generalized vector analysis, something Ricci had not originally envisaged. It was Einstein who recognized that Ricci's calculus included vectors and Voigt's tensors.

In 1913 and 1914 Einstein and Grossman developed their own notation and terminology (§9.13): Ricci's 'systems' were now called 'tensors'. These were at first covariant tensors $T_{i_1 \ldots i_k}$ and contravariant tensors $\Theta_{i_1 \ldots i_r}$. But Einstein and Grossman soon introduced mixed tensors, $T_{i_1 \ldots i_\mu}^{k_1 \ldots k_\nu}$, which made contraction easier. The fundamental tensor appeared, in the form still usual today, as g_{ik}, the 'g' recalling Gauss and gravitation. In the mathematics of general relativity, Einstein used the old Christoffel symbols $\{^{\mu\nu}_\tau\}$ and $[^{\mu\nu}_\tau]$; in the physics he introduced $\Gamma^\tau_{\mu\nu} = -\{^{\mu\nu}_\tau\}$, underlining with this notation the important role that the Christoffel symbols played in gravitational theory.

General relativity was intimately connected with Riemannian geometry. This view was emphasized in particular by Hermann Weyl, who gave tensor calculus a more geometrical interpretation in his textbook *Time, Space, Matter* (1918). For example, he called R^i_{hkj} the 'curvature tensor' (Einstein had called it the 'Riemannian differential tensor'). This geometric stand-point received support in 1917 when Levi-Cività interpreted the covariant derivation in terms of the parallel displacement of a vector.

6 THE PARALLEL DISPLACEMENT

Levi-Cività first showed that Riemann's 'hypotheses' were only an embryo, because they lacked the concept of parallelism. He defined parallelism with a system of n equations:

$$d\xi^{(i)} + \sum_{j,l=1}^{n} \Gamma^i_{jl} dx_i \xi^{(l)} = 0, \quad i = 1, \ldots, n, \tag{12}$$

and pointed out these main properties.

First, a direction through an arbitrary point P being parallel to another direction determined through P_0 is generally dependent on the path from P_0 to P. Only in Euclidean manifolds is parallelism independent of the path. Second, as far as a geodesic is concerned, the directions of its tangents are parallel. This means that the geodesics in a Riemannian manifold play the same role as straight lines in Euclidean space. Third, the angle produced by two different directions through P is the same after their parallel displacement to P_0. Gerhard Hessenberg and Jan Arnoldus Schouten developed similar ideas, which were published shortly after Levi-Città's paper (Reich *1992*).

This definition of parallelism showed the tremendous importance of the Christoffel symbols Γ. Weyl *1918* created the 'affine connection', the components of which were the Γs. He emphasized the fact that otherwise Levi-Città's parallelism would be 'distant' (i.e. directions at separate points A and B any distance apart could be compared without recourse to parallel displacement from A to B). He explained the new meaning of lines and geodesics in an affine-connected space: 'The line, i.e. geodesic, is created, if one displaces a vector paralleledly, as the curve of the initial point of this vector.' In his famous textbook *Ricci Calculus* (1924, 1954), Schouten suggested that there are as many differential geometries as there are different kinds of connection.

Luther Pfahler Eisenhart called this new field 'new differential geometry' or 'non-Riemannian geometry', which is also the title of one of his books (1927). This new differential geometry was propagated by Wilhelm Blaschke's textbooks, which appeared between 1921 and 1960, as well as by the books of the Dutch mathematicians Schouten (1924) and Dirk Jan Struik (1935, 1938). Thereafter both differential geometry and tensor calculus were widely taught and used.

BIBLIOGRAPHY

Berwald, L. *1927*, 'Differentialinvarianten in der Geometrie. Riemannsche Mannigfaltigkeiten und ihre Verallgemeinerung', in *Encyklopädie der mathematischen Wissenschaften*, Vol. 3, Part 3, 73–181 (article III E 2).

Farwell, R. and Knee, C. *1990*, 'The missing link: Riemann's "Commentatio", differential geometry and tensor analysis', *Historia mathematica*, 17, 223–55.

Gauss, C. F. *1828*, 'Disquisitiones generales circa superficies curvas', *Commentarii Societas Göttingensis*, 6, 99–146. [Also in *Werke*, Vol. 4, 217–58.]

Gericke, H. *1984*, *Mathematik in Antike und Orient*, Berlin: Springer.

—— *1990*, *Mathematik im Abendland. Von den römischen Feldmessern bis zu Descartes*, Berlin: Springer.

Haas, A. *1881*, *Versuch einer Darstellung der Geschichte des Krümmungsmasses*, Tübingen: Fues.

Herbert, D. *1992*, *Die Entstehung des Tensorkalküls* [. . .], Stuttgart: Steiner.

Nowak, G. *1989*, 'Riemann's "Habilitationsvortrag" and the synthetic *a priori* status of geometry', in D. Rowe and J. McCleary (eds), *The History of Modern Mathematics*, Vol. 1, Boston, MA: Academic Press, 17–46.

Phillips, E. *1979*, 'Karl M. Peterson: The earliest derivation of the Mainardi–Codazzi equations and the fundamental theorem of surface theory', *Historia mathematica*, **6**, 137–63.

Portnoy, E. *1982*, 'Riemann's contribution to differential geometry', *Historia mathematica*, **9**, 1–18.

Reich, K. *1973*, 'Die Geschichte der Differentialgeometrie von Gauss bis Riemann (1828–1868)', *Archive for History of Exact Sciences*, **11**, 273–382.

—— *1989*, 'Das Eindringen des Vektorkalküls in die Differentialgeometrie', *Archive for History of Exact Sciences*, **40**, 275–303.

—— *1992*, 'Levi-Civitasche Parallelverschiebung, affiner Zusammenhang, Übertragungsprinzip: 1916/17–1922/23', *Archive for History of Exact Sciences*, **42**, 77–105.

—— *1993*, *Die Entwicklung des Tensorkalküls. Vom absolutem Differentialkalkul zur Relativitätstheorie*, Basel: Birkhäuser.

Rowe, D. *1989*, 'Interview with Dirk Jan Struik', *The Mathematical Intelligencer*, **11**, (1), 14–26.

Scholz, E. *1980*, *Geschichte des Mannigfaltigkeitsbegriffs von Riemann bis Poincaré*, Basel and Boston: Birkhäuser.

—— *1982*, 'Herbart's influence on Bernhard Riemann', *Historia mathematica*, **9**, 413–40.

Stiegler, K. *1968*, 'Newtons Lösung des Problems der Bestimmung einer ebenen Kurve mit Hilfe seiner Fluxionsrechnung', in *Rechenpfennige*, Munich: Deutsches Museum, 167–74.

Struik, D. *1926*, 'Über die Entwicklung der Differentialgeometrie', *Jahresbericht der Deutschen Mathematiker-Vereinigung*, **34**, 14–25.

—— *1933*, *1934*, 'Outline of a history of differential geometry', *Isis*, **19**, 92–120; **20**, 161–91.

—— *1989*, 'Schouten, Levi-Civita, and the emergence of tensor calculus', in D. Rowe and J. McCleary (eds), *The History of Modern Mathematics*, Vol. 2, Boston, MA: Academic Press, 99–105.

Taton, R. *1951*, *L'Oeuvre scientifique de Monge*, Paris: Presses Universitaires de France.

Weitzenböck, R. *1922*, 'Neuere Arbeiten der algebraischen Invariantentheorie. Differentialinvarianten', in *Encyklopädie der mathematischen Wissenschaften*, Vol. 3, Part 3, 1–71 (article III E 1).

Weyl, H. *1918*, 'Reine Infinitesimalgeometrie', *Mathematische Zeitschrift*, **2**, 384–411.

3.5

Calculus of variations

CRAIG G. FRASER

1 OPTIMIZATION PROBLEMS

A basic problem of the differential calculus is to find the point at which a given curve is a maximum or minimum. An immediate generalization is posed by problems in which it is required to find a curve from among a class of curves that renders a certain integral quantity an extremum. For example, it might be required to find the curve joining two points that makes the distance between them a minimum (a straight line), or to find the curve for which the solid formed by revolving that curve about an axis yields a surface of minimum area (a catenary). In the problem of the brachistochrone, one considers a smooth curve joining two points in a vertical plane and seeks the shape that minimizes the time it takes a body to slide down the curve (a cycloid). In each of these examples the quantity to be extremalized is expressed as a definite integral of the form $\int_a^b f(x, y, y') \, dx$, where $y = y(x)$ is a function of x, and $y' = dy/dx$ is its derivative. The fundamental problem of the calculus of variations is to investigate necessary and sufficient conditions for $y(x)$ to make this integral an extremum.

The fundamental problem may be modified by demanding that the extremalizing function satisfy an auxiliary condition or constraint. One thereby obtains the so-called 'isoperimetrical problems' which figured prominently in the early history of the subject. Thus it may be necessary to find the curve with a fixed perimeter that encloses a maximum area (the circle), or to find the shape of a hanging chain of given length whose centre of gravity is lowest (the catenary).

The theory may be further modified or generalized in a number of ways: by introducing into the integrand function f a variable s of the form $s = \int_a^x g(x, y, y') \, dx$; by permitting higher-order derivatives y'', y''', and so on in the integrand; by allowing the end-points of the curve to vary; by considering more than one independent variable, so that the quantity to be extremalized becomes a multiple integral; or by formulating the problem in parametric form with a parameter and several dependent variables in the integrand.

2 SURVEY OF THE HISTORY

The calculus of variations began as a disparate set of problems and techniques in the early eighteenth century. It was synthesized into a branch of mathematics by Leonhard Euler in the 1740s and was radically reformulated by Joseph Louis Lagrange using his δ-algorithm in the next decade. Both men also used variational ideas extensively in mechanics (§8.1). In his famous calculus textbooks published at the end of the century, Lagrange returned to the subject and developed it along formal analytical principles (§3.2).

In the nineteenth century several traditions of research emerged. Siméon-Denis Poisson, Mikhail Ostrogradsky, Charles Delaunay, Frédéric Sarrus and Augustin Louis Cauchy investigated variational problems involving multiple integrals, a subject that was closely connected to problems in potential theory and mathematical physics. William Rowan Hamilton, Carl Jacobi and Joseph Liouville concentrated on the formal study of analytical dynamics by means of variational principles. Jacobi continued Legendre's analysis of the second variation along more strictly mathematical lines, and achieved highly significant results concerning necessary conditions for an extremum. His students and successors at Königsberg, Ludwig Hesse, Alfred Clebsch and Christian Mayer, extended his theory in various directions.

Throughout the nineteenth century many writers understood variational mathematics in terms of the concepts and methods of operator and formal calculus. It was Karl Weierstrass's achievement in the 1870s to systematically ground the subject in the theory of a function of a real variable (§3.3). Weierstrass also investigated the basic problems of the calculus of variations in parametric form. His work was presented in his lectures at the University of Berlin and was disseminated by his students Paul Du Bois-Reymond, K. H. A. Schwarz and, especially, Oscar Bolza.

Significant research into the early twentieth century was carried out by David Hilbert, Adolf Kneser, Hans Hahn, Constantin Carathéodory, William Osgood and Gilbert Bliss. The modern 'classical' formulation of the calculus of variations embodying the results of the German 'school' was presented in textbooks by Kneser (*1900*) and Bolza (*1904, 1909*). Significant contributions since then (which will not be explored here) include, in the 1930s, Marston Morse's calculus of variations in the large, and the development in recent decades of optimal control theory. Classical variational mathematics continues to play an important role in engineering mechanics and physics.

In addition to the interest inherent in its development as a technical branch of mathematics, the calculus of variations is historically noteworthy for two reasons. The close relationship that existed between it and mechanics, at least until the late nineteenth century, illuminates the historical links that have connected mathematics and physics. Originating as a generalization of the ordinary calculus, variational mathematics is of foundational interest in illustrating changing conceptions of analysis. Its development by Lagrange serves as an example of the algebraic understanding of calculus that was dominant in the later eighteenth century; in the nineteenth century the different traditions that emerged indicate the diversity of research during the period and help to define the context, character and historical significance of the Weierstrassian programme of real analysis that has informed the modern organization of the subject.

3 THE EIGHTEENTH CENTURY

Early work in what later became known as the calculus of variations was part of the new infinitesimal calculus pioneered by Isaac Newton and Gottfried Wilhelm Leibniz. In the *Principia* (1687), Newton investigated the problem of the shape of a solid of revolution that yields the least resistance as it moves through a fluid in a direction parallel to its axis. A substantial British programme of variational research continued in the writings of Brook Taylor, Colin Maclaurin and Thomas Simpson into the middle of the eighteenth century. Nevertheless, it was Continental mathematicians following Leibnizian principles who contributed most decisively to the subsequent development of the subject, and it is this tradition that forms the subject of the following account.

The early Leibnizian calculus consisted of a sort of geometrical analysis in which differential algebra was employed in the study of 'fine' geometry (§3.4). The curve was analysed in the infinitesimal neighbourhood of a point, and related by means of an equation to its overall shape and behaviour. An important curve that was the solution of several variational problems was the cycloid, the path traced by a point on the perimeter of a circle as it rolls without slipping along a straight line. The cycloid could be simply described in terms of the infinitesimal calculus. Let the generating circle roll along the y-axis, and let the vertical distance be measured downward from the origin along the x-axis (Figure 1). An elementary geometrical

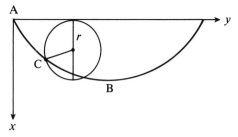

Figure 1 The cycloid

argument revealed that the equation of the cycloid is

$$\left(\frac{ds}{dy}\right)^2 = \frac{2r}{x},\tag{1}$$

where $ds = (dx^2 + dy^2)^{1/2}$ is the differential element of path length.

The cycloid was most notably the solution to the brachistochrone problem. Following John Bernoulli's public challenge in 1696, solutions to this problem were proposed by his elder brother James, by John himself, and by Newton, Leibniz and the Marquis de l'Hôpital. All these people showed that the condition that the time of descent be least for the given curve leads to equation (1), and hence to the conclusion that the curve is a cycloid. John Bernoulli's solution was based on an optical analogy which, although interesting, did not lead anywhere within the subject. The analysis by his brother James was illustrative of the ideas that would develop into the calculus of variations. He considered three arbitrary, infinitesimally close points C, G and D on the given curve, and constructed a second neighbouring curve identical to the first except that the arc CGD was replaced by CLD (Figure 2). Because the curve minimizes the time of descent, it was clear that the time taken to traverse CGD is equal to the time taken to traverse CLD. Using this condition and the dynamical relation $ds/dt \propto \sqrt{x}$, Bernoulli was able to derive equation (1).

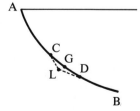

Figure 2 John Bernoulli's treatment of the brachistochrone problem

James Bernoulli also investigated isoperimetrical problems in which the extremalizing curve satisfies an auxiliary integral condition. His idea was to vary the curve at two successive ordinates, thereby obtaining an additional degree of freedom, and use the side constraint to obtain a differential equation. Following James's death in 1705, his brother John, then 38 years old, adopted his approach and developed it extensively in a paper that appeared in 1718 (Bernoulli *1991*).

In two memoirs published in the St Petersburg Academy of Sciences in 1738 and 1741, Euler extracted from the various solutions of John and Daniel Bernoulli, as well as the researches of Taylor, a unified approach to integral variational problems. These investigations were further developed and made the subject of his classic treatise *Methodus inveniendi lineas curvas maximi minimive proprietate gaudentes* ('Method of Finding Curved Lines that Show some Property of Maximum or Minimum') (*1744*). Published when he was 37 years old, this book was a remarkable synthesis in which he virtually created the calculus of variations as a branch of mathematics. He realized that the different integrals in the earlier problems were all instances of the single form $\int_a^b f(x, y, y')\,\mathrm{d}x$. He derived a canonical differential equation, known today as the Euler or Euler–Lagrange equation, as the fundamental necessary condition of the variational problem:

$$\frac{\partial f}{\partial y} - \frac{\mathrm{d}}{\mathrm{d}x}\left(\frac{\partial f}{\partial y'}\right) = 0. \tag{2}$$

This equation was obtained by investigating the expression

$$\mathrm{d}f = M\,\mathrm{d}x + N\,\mathrm{d}y + P\,\mathrm{d}y', \tag{3}$$

introduced in a slightly different form by Taylor in 1715, when the curve was altered at a single ordinate.

Of general interest was Chapter 4 of Euler's treatise. He noted that, although the derivation of the variational equations had been carried out with respect to an orthogonal coordinate system, the derivation and the equations themselves remained valid when alternate coordinate variables were employed. He proceeded to investigate problems using various coordinate descriptions. Euler's observation here is of considerable historical significance: it is the explicit recognition in a definite mathematical context that the essential nature of calculus is defined by its analytical character, and not by any geometrical interpretation conferred upon the formalism.

This theoretical insight notwithstanding, the analytical scope of Euler's variational calculus was restricted in practice by the detailed process involved in each individual derivation. Euler himself called for further research on algorithms that would facilitate the procedures of the subject. The call was answered in 1760 when the 24-year-old Lagrange published his δ-algorithm in the memoirs of the Turin Society. Formalistic in spirit, his method was based on an entirely new idea involving the simultaneous variation of the range of values of the extremalizing curve. In this formulation the variational problem

$$\delta \int_a^b f(x, y, y') \, dx = 0 \tag{4}$$

was reduced by means of an integration by parts to

$$\int_a^b \left[\frac{\partial f}{\partial y} - \frac{d}{dx} \left(\frac{\partial f}{\partial y'} \right) \right] \delta y \, dx = 0. \tag{5}$$

From this equation he inferred the basic equation (2).

Lagrange showed that Euler's theory could be efficiently recovered using his new method, and demonstrated its superiority in handling problems involving variable end-points. Euler, in turn, adopted the δ-algorithm and coined the name 'calculus of variations' for the new mathematics. In the years that followed, both men published memoirs exploring the analytical possibilities of the subject. Considerably later, in the second edition of his *Leçons sur le calcul des fonctions* (1806), the elderly Lagrange attempted to develop the calculus of variations algebraically as part of his larger project of basing analysis on Taylor's theorem.

In a series of researches that began in 1760 and culminated with his *Méchanique analitique* (1788), Lagrange employed variational techniques in mechanics. He provided variational formulations for the principle of least action, the principle of virtual work and d'Alembert's principle, and derived the 'Lagrangian' differential equations of motion (§8.1). His emphasis throughout was on the formal mathematicization of mechanics as opposed to conceptual development or experimental verification.

With Lagrange's theory, the calculus of variations had completed the formative phase of its historical development (Fraser *1985*). Although later researchers would introduce new concepts, deepen the theory and extend its geometrical applications, the basic structural character of the subject was in place.

4 THE NINETEENTH CENTURY

Throughout the nineteenth century a growing number of researchers worked on the calculus of variations. In Lecat's bibliography *1913*, the number of papers increases from 65 in the period 1800–24 to 130 in 1825–50. This trend continued in subsequent decades as the subject became established as a major branch of real analysis. Although research was carried out in France, Italy, Britain and the United States, German-speaking Europe came to dominate the field, particularly in the closing years of the century.

In the early decades considerable effort was devoted to extending the established theory for single-variable integration to multiple integrals. The integration by parts in Lagrange's original derivation of equation (2) required in this case results that correspond to the later Green's theorem and the divergence theorem (§3.17). In the simplest case, where $z = z(x, y)$ and the boundary R is unvaried in the variational process, the condition

$$\delta \iint_R f\left(x, y, z, \frac{\partial z}{\partial x}, \frac{\partial z}{\partial y}\right) dx\,dy = 0 \tag{6}$$

leads to the partial differential equation

$$\frac{\partial f}{\partial z} - \frac{\partial f_p}{\partial x} - \frac{\partial f_q}{\partial y} = 0, \quad p = \frac{\partial z}{\partial x}, \quad q = \frac{\partial z}{\partial y}. \tag{7}$$

In the derivation of equation (7), a line integral appears that is zero because the variation is taken to be zero on the boundary. The problem was to find expressions for this integral when the boundary is itself varied. This question, as well as the corresponding one for triple integrals, occupied the attention of Poisson and Ostrogradsky in long memoirs published in the 1830s. The most successful assault on the problem was by Frédéric Sarrus, in a paper published in 1846. His approach was adopted and extended by Cauchy and by François Moigno.

Although there is little mention of Sarrus's theory in the modern literature, it was highly regarded by his contemporaries. It was awarded a prize by the Paris Academy in 1842, when Sarrus was 44 years old, and was judged by Isaac Todhunter *1861* to be 'the most important original contribution which has been made to the Calculus of Variations in this century'. The theory's historical context is the French interest in differential geometry of surfaces, and it forms part of the background to Gaston Darboux's own researches in this subject later in the century.

In his seminal paper of 1836, Carl Jacobi, then 32 years old and professor of mathematics at Königsberg, investigated the question of when solutions to the single-integral variational problem lead to an actual minimum. He began with a result proved by Legendre in 1786 concerning the second variation. Legendre had shown that, in addition to the Euler–Lagrange equation (2), a necessary condition that a given function $y = y(x)$ on $[a, b]$ be a minimum is that $\partial^2 f/\partial y'^2 > 0$ on $[a, b]$. He assumed in his demonstration that a certain differential equation has a finite integral on $[a, b]$. Jacobi investigated this equation further and showed that its integrals were given by differentiating the family of solutions $y = y(x, \alpha, \beta)$ of equation (2) with respect to α and β. He introduced a certain function $\Delta(x, a)$, defined in terms of these solutions, and considered values k of x for which $\Delta(k, a) = 0$. Such values are known as conjugate values with respect to the one-parameter family of solutions to equation (2) passing through $(a, y(a))$. He showed that, if there is a conjugate value k such that $k \leqslant b$, then the given arc is not a minimum; if $k > b$, he inferred, the arc is a minimum.

Jacobi provided a simple geometric interpretation for his theory in terms of the envelope to the family of extremals of the variational problem. Over the next twenty years a series of papers appeared that elaborated and extended his results.

Mathematical writers early in this century regarded Weierstrass's Berlin lectures of 1872–81 as a turning-point in the history of the subject. Bliss (*1925*: 177) wrote that 'The memoirs and treatises on the calculus of variations up to the latter part of the nineteenth century frequently leave one in doubt as to the validity and the precise character of the results which they contain.' He credited Weierstrass with bringing to the field a sense of rigour and an awareness of the need for sharp formulation of fundamental concepts (§3.3). This critical spirit was evident in the writings of Weierstrass's student Du Bois-Reymond, who in 1879 provided a rigorous demonstration of the fundamental lemma, the result that allows one to infer the validity of the Euler–Lagrange equations from the vanishing of the first variation.

Among Weierstrass's many contributions was the parametric theory, in which x and y are regarded as functions of a parameter t and the extremalizing integral is written as

$$\int_{t_0}^{t_1} f(x, y, x', y') \, dt, \tag{8}$$

where $x' = dx/dt$ and $y' = dy/dt$. The class of comparison arcs here includes curves whose derivatives differ by finite amounts from the given curve. Kneser later introduced the term 'strong' variation to distinguish this case from the usual one where the variations are 'weak'. Weierstrass developed the parametric theory of the first and second variations, discovered a new necessary condition and completed important researches on sufficiency.

The strength of the parametric approach is its ability to handle geometric applications. The problem of finding a geodesic or shortest line on a surface was explored by Darboux, Ernst Zermelo and Kneser in papers published in the 1880s and 1890s. Then, in a related series of analytical researches, David Hilbert began a very general investigation of existence results and sufficiency conditions. His work brought the classical phase of the subject to a close. The modern research that led to Morse's theory, carried out to a considerable extent by Hilbert's American students, evolved by bringing his methods and 'the tools of topology to bear on the classical calculus of variations in order to develop a macro-analysis' (Goldstine *1981*: 371).

BIBLIOGRAPHY

Bernoulli, James and John *1991*, *Die Streitschriften von Jacob und Johann Bernoulli. Variationsrechnung* (ed. H. Goldstine and P. Radelet-de Grave), Basel: Birkhäuser.

Bliss, G. A. *1925*, *Calculus of Variations*, Chicago, IL: Mathematical Association of America and Open Court.

Bolza, O. *1904*, *Lectures on the Calculus of Variations*, Chicago, IL: University of Chicago Press.

—— *1909*, *Vorlesungen über Variationsrechnung*, Leipzig: Teubner.

Euler, L. *1744*, *Methodus inveniendi* [...], Geneva: Bousquet. [Also *Opera omnia*, Series 1, Vol. 24.]

Fraser, C. G. *1985*, 'J. L. Lagrange's changing approach to the calculus of variations', *Archive for History of Exact Sciences*, **32**, 151–91.

Goldstine, H. H. *1980*, *A History of the Calculus of Variations from the 17th Through the 19th Century*, New York: Springer.

Kneser, A. *1900*, *Lehrbuch der Variationsrechnung*, Braunschweig: Vieweg. [2nd edn, 1925.]

Lecat, M. M. *1913*, *Bibliographie du calcul des variations 1850–1913*, Ghent and Paris: the author.

—— *1916*, *Bibliographie du calcul des variations depuis les origines jusqu'à 1850*, Ghent and Paris: the author.

Stäckel, P. *1894*, *Abhandlungen über Variations-Rechnung*, Leipzig: Engelsmann (Ostwald's Klassiker, Nos. 46 and 47). [Contains German translations of works by John I Bernoulli, James I Bernoulli, Euler, Lagrange, Legendre and Jacobi.]

Todhunter, I. *1861*, *A History of the Progress of the Calculus of Variations During the Nineteenth Century*, Cambridge: Macmillan. [Repr. 1961, New York: Dover.]

Woodhouse, R. *1810*, *A Treatise on Isoperimetrical Problems and the Calculus of Variations*, Cambridge: Deighton. [Repr. as *A History of the Calculus of Variations in the Eighteenth Century*, (no date), New York: Chelsea.]

3.6

Set theory and
point set topology

JOSEPH W. DAUBEN

1 INTRODUCTION

The creation of modern set theory is usually ascribed to the German mathematician Georg Cantor. But ideas that may be easily regarded as set-theoretic in one form or another can be traced back to Antiquity: the Greeks considered infinite collections of prime numbers, for example, and constructed theorems dealing with point sets, such as the locus of points describing a curve such that each point is a given distance r from a given point P (in this case the curve is a circle of radius r about P in the plane, or a sphere or radius r about P in three dimensions). Also, problems of continuity and infinity have always been closely linked; some historians have even ascribed an awareness of non-denumerably infinite sets to certain medieval authors (Ashworth *1977*). Set theory in the Jaina school of Indian mathematics has also been investigated (Jain *1973*).

Similar hints of set theory and interest in questions of continuity and infinity can be found in the works of many mathematicians of the seventeenth and eighteenth centuries, but the first person to study sets in themselves as collections of objects was Bernard Bolzano, whose *Paradoxien des Unendlichen* (1851) contains many interesting results that are today considered basic parts of elementary set theory and topology. The important property of 1–1 correspondence was used by Bolzano in much the same way as it would later be used (to much greater advantage) by Cantor and Richard Dedekind. Bolzano, although understanding that an infinite set S was characterized by a 1–1 correspondence between S and a proper subset $S' \subset S$, did not see that this also meant that they necessarily contained the same number of elements (an essential part of Cantor's investigations of equivalent, equipotent infinite sets). Dedekind actually used this property as a means of defining infinite sets, as did the American philosopher-mathematician Charles Sanders Peirce (Dauben *1979*). Instead, Bolzano

351

seems to have had only a vague idea of what constituted the equality of two infinite sets, saying only that they were equal if they were defined 'on the same basis'.

Increasingly, however, following Riemann's *Habilitationsschrift* (1854) on trigonometric series, in which he introduced his generalization of the concept of the integral (§3.7), mathematicians became increasingly interested in various properties of point sets in a function's domain of definition, especially when connected with unusual or pathological functions. Among those to advance such studies were Hermann Hankel, who reformulated Riemann's criteria for integrability in connection with his study of oscillating and discontinuous functions (1870), and introduced an important method of the 'condensation of singularities'. Although he confused sets of zero content with nowhere-dense sets, he was nevertheless a forerunner of those who advanced modern and increasingly comprehensive theories of integration. Hankel's work was also noteworthy for including an example of a continuous function that was non-differentiable at an infinite number of points (Hawkins *1970*).

2 RICHARD DEDEKIND AND GEORG CANTOR ON NUMBERS AND SETS

In 1872, in order to give a rigorous definition of real numbers, Dedekind introduced his now famous Dedekind cut. This was formulated in lectures on the calculus he had given at Zürich in 1858, in which he sought to give a purely arithmetic definition of the continuum, especially of irrational numbers. Dedekind's other major contribution to set theory was a book of 1887 on the nature of number, in which he took the definition of an infinite set I to be the 1–1 correspondence that was always possible between I and one of its proper subsets.

Cantor's earliest interest in set theory was closely linked to problems involving the representation (and uniqueness) of functions by trigonometric series. This led him, in 1872, not only to formulate a rigorous definition (different from Dedekind's) of real numbers, but also to contemplate sets of 'exceptional points' for which the function might be discontinuous and yet for which the representation was still possible and unique. Soon thereafter Cantor published one of his most important discoveries, that the set of real numbers is non-denumerably infinite, and he spent the next decade working out the earliest version of his theory of point sets and transfinite numbers, which were finally summarized in 1883 in an important monograph that was both technical and philosophical. In 1891 he introduced his famous method of diagonalization to prove that, given any infinite set, the set of all its subsets is of a greater power or cardinality. In 1895–7 he summarized his life's work in two essential articles, which also introduced his famous transfinite alephs.

3 REACTIONS TO CANTOR

It was not long before mathematicians – despite the opposition of some like Leopold Kronecker (§5.6) – began to take up Cantor's new set-theoretic ideas, which were found to have important applications especially in analysis. Among the first to appreciate this were Italian mathematicians like Giulio Ascoli (1877), Salvatore Pincherle (1880) and, especially, G. Vivanti (1889, 1891, 1899), although papers presented by Jacques Hadamard and A. Hurwitz, among others, at the International Congress of Mathematicians in 1897 also did much to stimulate interest in Cantor's ideas.

Among early works written to introduce, explain and evaluate Cantor's set theory and transfinite numbers were monographs and articles by E. V. Huntington (1905) and P. E. B. Jourdain (1906–13). Of these, the most influential text was Felix Hausdorff's *Grundzüge der Mengenlehre* (1914), which introduced such set theoretic/topological notions as Hausdorff dimension and Hausdorff measure. Earlier, topological concepts had already been introduced by Maurice Fréchet (among others), but it was Felix Hausdorff who succeeded in creating a comprehensive theory of topological spaces which may be taken to mark the beginning of the study of both topological and metric spaces (Dierkesmann *et al. 1967*). As for his contributions to set theory, Hausdorff is also known for the Hausdorff maximal principle, as well as for work on partially ordered sets. He also studied the cardinality of Borel sets (1916) and introduced what are now called Hausdorff operations (1927).

One of the most basic theorems of elementary set theory was proved by Felix Bernstein in one of Cantor's seminars at Halle University. He was able to show that, if two sets A and B are each equivalent to a subset of the other, then they must be equivalent to each other. This theorem is essential to the theory of transfinite cardinality and the comparability of transfinite numbers. Bernstein went on to write his dissertation *1901* on set theory at Göttingen, followed a few years later by an article on transfinite ordinal numbers (*1905*).

Arthur Schönflies made important contributions to the promotion of set theory by writing an influential article on the subject for the *Encyklopädie der mathematischen Wissenschaften* in 1899. He also produced extensive reports for the *Deutsche Mathematiker-Vereinigung* which were published in 1900 and 1908, and (a measure of its influence) revised as Schönflies and Hahn *1913*. Until the appearance by Hausdorff's *Grundzüge der*

Mengenlehre (1914), these reports by Schönflies were considered to be the standard introduction to set theory (Bieberbach *1923*).

The Polish mathematician Wacław Sierpiński was the first to teach a systematic university course on transfinite set theory, beginning in about 1908. Sierpiński investigated set theory, point set topology and the theory of functions of a real variable (Fryde *1964*). Among his textbooks on set theory are Sierpiński *1934* and *1958*. Set theory and topology were both of fundamental importance in the development of the theory of functions and the birth of functional analysis, to which early contributions were made by Vito Volterra, Erik Ivar Fredholm and David Hilbert, among others. Functional analysis was pioneered in a famous paper by Stefan Banach *1922*, best known for his work on function spaces (§3.9). Banach's study of linear spaces in the 1920s relied heavily upon set theory and transfinite constructions. He also did important work on locally meagre sets (§7.10), and his approach to measure theory was closely related to that taken in axiomatic set theory. Later developments in function spaces are closely related to René Baire's theorem on complete metric spaces and descriptive set theory (Steinhaus *1963*).

4 THE PARADOXES OF SET THEORY, AND SOME REACTIONS

Although Cantor was the first to appreciate the 'paradox' of considering any largest transfinite ordinal or cardinal number, he did not see this as problematic. Others, like Cesare Burali-Forti, who also considered the question of the existence of a 'largest ordinal', simply concluded that comparability of very large transfinite numbers was impossible. The paradoxes are discussed in §5.3.

The paradoxes produced a number of reactions. Henri Poincaré noted that they were generated by impredicative definitions; limiting mathematics to only predicative definitions was one solution (but a number of important impredicative concepts would then be inadmissible, including the definition of the upper bound of an infinite set). Another response was L. E. J. Brouwer's, who became interested in set theory by reading Schönflies's reports on the subject. Brouwer's dissertation of 1910 considered indecomposable continua. In 1914 he reviewed the Schönflies–Hahn report *1913* on the development of set theory, and in the years that followed began to develop a comprehensive, constructive foundation for set theory which sought to reject, among other principles, that of the excluded middle (§5.6). He wrote a number of texts devoted to providing constructive theories of measure, functions and related mathematical concepts (Heyting *1956*).

354

5 SET THEORY AND THE AXIOM OF CHOICE

Cantor had long hoped to prove the theorem that every set could be well ordered, which was closely related to his most famous conjecture, the continuum hypothesis, which states (in one of its many variant forms) that the cardinality of the non-denumerable set of real numbers follows immediately from the cardinality of the denumerable set of natural numbers. If $\aleph_0$ (read 'aleph-null') is the cardinal number of denumerably infinite sets (like the natural, rational or algebraic numbers), since the cardinality of the non-denumerably infinite set of real numbers may be written as $2^{\aleph_0}$, the continuum hypothesis is equivalent to the conjecture that $2^{\aleph_0} = \aleph_1$, where $\aleph_1$ is taken to be the second transfinite cardinal number, and the one immediately following $\aleph_0$.

Julius König launched an attack on Cantor's set theory in 1904 by announcing a proof that the continuum hypothesis was false. But Ernst Zermelo found an error in König's argument, from which he went on to produce his famous well-ordering theorem. König spent the last part of his life working on his own approach to set theory and foundations of mathematics, and this work was published posthumously as König *1914*.

It was in 1904 that Zermelo proved that every set could be well-ordered, using a controversial idea that soon came to be known as the axiom of choice (see also §5.4). Zermelo was teaching at the University of Göttingen, where he lectured on set theory. His proof of 1904 was soon criticized for its use of infinite subsets of infinite sets arising from the axiom of choice. One major criticism focused on the lack of any clear means of carrying out the infinitely many 'choices' that the axiom assumed. In 1908 Zermelo attempted to meet the objections of his critics with a 'new proof'. Among the most important of his critics was Henri Poincaré, who objected especially to the impredicative nature of the axiom of choice.

6 SET THEORY AND THE THEORY OF FUNCTIONS

One of the first mathematicians to accept Cantor's new transfinite set theory was Ernst Schröder, whose *Algebra der Logik* also helped to develop algebraic logic (§5.1). Schröder was also interested in the foundations of mathematics, as well as the theory of functions of a real variable (Lüroth *1903*). Erhard Schmidt was among those to help simplify and extend Hilbert's work on the theory of integral equations, formalizing these into the idea of the Hilbert space (§3.9).

Poincaré and others began to use Cantor's new work in applications to the theory of functions. In the early 1880s Poincaré considered perfect, non-dense sets in his work on automorphic functions and differential equations. But when, at the turn of the century, the paradoxes of set theory and other foundational questions arose in the context of transfinite set theory, he became wary and adopted a position with respect to set theory that was an early form of intuitionism (§5.6). While he accepted the potential infinite, he came to repudiate use of actual infinite sets in mathematics, preferring to eliminate impredicative definitions altogether. He also considered axiomatics to be entirely sterile, and was unsympathetic to Hilbert's attempts to formalize mathematics.

Among the most important applications of set theory, however, was the theory of integration advanced by Henri Lebesgue (§3.7). Camille Jordan had already taken a measure-theoretic approach to the Riemann integral in 1893. Emile Borel, in his work on the theory of complex functions, also introduced ideas of measure, but it was Lebesgue, drawing on the results of both Jordan and Borel, who fashioned a powerful general theory of measure. (W. H. Young and Giuseppe Vitali also generalized Jordan's theory of measure in similar ways.) Lebesgue was especially successful in applying his theory of measure and the Lebesgue integral to the theory of trigonometric series (May *1966*).

Borel, early in the 1890s, once said he was 'extremely seduced' by Cantor, and in his thesis he took up the theory of functions of a real variable, applying a theory of measure to divergent series, non-analytic continuation, denumerable probabilities, Diophantine approximations, and the metrical distribution theory of values of analytic functions. All these were related in one way or another to basic ideas of Cantor's new transfinite set theory, especially the idea of a dense set. Among the best known results is the Heine–Borel covering theorem, and the proof that a denumerable set is of measure zero. Despite his success in applying the new set theory, Borel always remained suspicious of the actual infinite, and was always uncomfortable in using transfinite numbers in proofs, although he was much less strict than other of his French colleagues – notably Poincaré – in allowing their usefulness as a method of discovering new theorems. Others in France, including Baire, Lebesgue and Fréchet, succeeded in pushing the ideas of set and measure much further, laying the foundations of twentieth-century abstract analysis (Fréchet *1965*).

Baire, for example, devoted his thesis to a theory of functions of a real variable based upon set theory. In particular, he was interested in the problem of pointwise discontinuities on perfect sets. He developed the idea of semicontinuity, and took another important step in defining functions over compact sets. Using Cantor's ideas of derived sets of the first and second species, he realized that the derived sets were what underlay the legitimacy of the transfinite numbers (whatever one might think of them; see Dauben *1979*).

Other important applications were made by Charles Jean De La Vallée-Poussin, whose *Cours d'analyse* was transformed in its second edition of the introduction of set theory, measure, the Lebesgue integral, trigonometric series and the famous Jordan curve theorem (Burkill *1964*). In discussing the Lebesgue integral, Vallée-Poussin was among the first to discuss analytic sets (later developed further by the Russians M. Y. Suslin and N. N. Luzin).

Although best known for his postulates for the natural numbers (1898) and his well-known space-filling curve (§7.11), Giuseppe Peano also made important applications of set theory to analysis, especially in the courses he began to teach at the University of Turin in 1885. In his textbook *Applicazioni geometriche del calcolo infinitesimale* (1887) he discussed the measure of point sets and the additive functions of sets (Kennedy *1974*).

7 THE RUSSIAN AND POLISH SCHOOLS

Luzin's Moscow thesis (1915) contained impressive results on the structure of measurable sets and functions in a study of convergence and the representation of functions by trigonometric series. These studies led him to establish his C-property: every measurable function almost-everywhere finite over a given segment can be made continuous over the segment by varying its value on a set of arbitrarily small measure. In 1915 Luzin studied the power of B sets, and later M. Y. Suslin went even further, examining an extensive class of analytical (A) sets, and in 1925 developed a theory of projective sets (from B sets) by successively performing projections and taking complements.

In Poland, along with Sierpiński and Stefan Mazurkiewicz, Zygmunt Janiszewski was a founding member of a school of mathematics which relied heavily on the new methods of set theory and mathematical logic introduced at the beginning of the century (§5.7). Among its important contributions was the founding of the journal *Fundamenta mathematicae*, devoted primarily to set theory and related subjects. Janiszewski had taken his doctorate at the Sorbonne in Paris in 1911, working on a topic proposed by Lebesgue, which led to important applications of both set theory and topology. Among the important subjects he studied were continua and the topology of the plane.

8 SET THEORY AND FOUNDATIONS

The intuitionist Brouwer approached set theory and logic with important studies that, among other characteristic features, questioned the law of the excluded middle (§5.6). Others did so as well, including the Russian Samuel

O. Shatunovsky. His master's thesis argued the logical inadmissibility of the law of the excluded middle (but he was not as radical in his views as Brouwer and the intuitionists). In 1906 he taught an introductory course on functions and introduced descriptive set theory and rigorous definitions of irrational and complex numbers.

The most influential group of mathematicians to take up set theory as a foundation for mathematics, however, is the French Bourbaki. In the 1930s a group of mathematicians calling themselves 'Bourbaki' began with set theory and sought to build upon it the rest of mathematics, from abstract algebra and general topology to a general theory of functions and integrals (Halmos *1957*).

9 SET THEORY AND LOGIC

The first to provide an axiomatic system for set theory was Zermelo, in 1908. He used seven axioms and two primitive terms, set and inclusion, hoping to exclude the famous 'paradoxes' that had raised serious questions about set theory in theorems proposed by Burali-Forti and Bertrand Russell, among many others (including Cantor). Zermelo limited the formation of sets with a 'condition of definiteness' restricting the creation of subsets (a problem the axiom of choice had raised), in the hope of eliminating the paradoxes from set theory (§5.4).

Axiomatic set theory was given a revised formulation in 1921 by Abraham Fraenkel, who was interested in showing the independence of the axiom of choice, and found certain problems with Zermelo's system. The 'condition of definiteness' was too vague to use in proofs of independence (or consistency). Zermelo's system also failed to be categorical, and the axiom of infinity was too weak. Fraenkel added a more powerful 'axiom of replacement' (Moore *1982*) which enabled him to prove the independence of the axiom of choice, but only by reference to an infinite set of objects that were not themselves sets. Later, Thoralf Skolem interpreted a definite property as a property expressible in first-order logic. Fraenkel wrote one of the early influential textbooks on set theory and foundations, *Einleitung in die Mengenlehre* (1919), and later *Abstract Set Theory* (1953) and *Foundations of Set Theory* (1958).

The axiom of choice was shown by Kurt Gödel (1938) to be independent of the Zermelo–Fraenkel axioms of set theory. The independence of the continuum hypothesis was proved by Paul Cohen in 1963 (§5.4). Another important axiomatization was given for set theory by John von Neumann in 1925; his system was later adopted by Gödel in his work on set theory and logic (§5.5).

BIBLIOGRAPHY

Ashworth, E. J. *1977*, 'An early 15th century discussion of infinite sets', *Notre Dame Journal of Formal Logic*, **18**, 232–4.

Banach, S. *1922*, 'Sur les opérations dans les ensembles abstraites et leur application aux équations intégrales', *Fundamenta mathematicae*, **3**, 133–81.

Bernstein, F. *1901*, 'Untersuchungen aus der Mengenlehre', Dissertation, Halle.

—— *1905*, 'Über die Reihe der transfiniten Ordnungszahlen', *Mathematische Annalen*, **60**, 187–93.

Bieberbach, L. *1923*, 'Arthur Schoenflies', *Jahresbericht der Deutschen Mathematiker-Vereinigung*, **32**, 1–6.

Burkill, J. C. *1964*, 'Charles Jean De La Vallée-Poussin (1866–1962)', *Journal of the London Mathematical Society*, **39**, 165–75.

Dauben, J. *1979*, *Georg Cantor: His Mathematics and Philosophy of the Infinite*, Cambridge, MA: Harvard University Press. [Repr. 1990, Princeton, NJ: Princeton University Press.]

Dierkesmann, M., Lorentz, G. G., Bergmann, G. and Bonnet, H. *1967*, 'Felix Hausdorff zum Gedächtnis', *Jahresbericht der Deutschen Mathematiker-Vereinigung*, **69**, 51–76.

Fréchet, M. *1965*, 'La Vie et l'oeuvre d'Emile Borel', *L'Enseignement mathématique*, **11**, 1–95.

Fryde, M. *1964*, 'Wacław Sierpiński – Mathematician', *Scripta mathematica*, **27**, 105–11.

Halmos, P. *1957*, 'Nicolas Bourbaki', *Scientific American*, **196**, 88–99.

Hawkins, T. *1970*, *Lebesgue's Theory of Integration: Its Origins and Development*, Madison, WI: University of Wisconsin Press. [Repr. 1975, New York: Chelsea.]

Heyting, A. *1956*, *Intuitionism. An Introduction*, Amsterdam: North-Holland.

Jain, L. C. *1973*, 'Set theory in Jaina school of mathematics', *Indian Journal of History of Science*, **8**, (1/2), 1–27.

Kennedy, H. C. *1974*, *Life and Works of Giuseppe Peano*, Dordrecht: Reidel.

König, J. *1914*, *Neue Grundlagen der Logik, Arithmetik und Mengenlehre*, Leipzig: Veit.

Lüroth, J. *1903*, 'Nekrolog auf Ernst Schröder', *Jahresbericht der Deutschen Mathematiker-Vereinigung*, **12**, 249–65.

May, K. O. *1966*, 'Biographical sketch of Henri Lebesgue', in *Lebesgue Measure and the Integral*, San Francisco, CA: Holden-Day.

Moore, G. *1982*, *Zermelo's Axiom of Choice: Its Origins, Development, and Influence*, New York: Springer.

Nagell, T. *1963*, 'Thoralf Skolem in memoriam', *Acta mathematica*, **110**, i–xi, 303.

Schönflies, A. with Hahn, H. *1913*, *Entwicklung der Mengenlehre und ihrer Anwendungen*, Leipzig: Teubner.

Sierpiński, W. *1934*, *Hypothèse du continu*, Warsaw and Lwów: Z Subwencji Funduszu Kultury Narodowej. [Repr., with some papers: 1956, New York: Chelsea.]

—— *1958*, *Cardinal and Ordinal Numbers*, 1st edn, Warsaw: Państwowe Wydawnictwo Naukowa. [2nd edn 1965.]

Steinhaus, H. *1963*, 'Stefan Banach', *Studia mathematica*, **1**, 7–15.

3.7

Integral, content and measure

I. GRATTAN-GUINNESS

1 INTRODUCTION

One of the consequences of the emergence of real-variable mathematical analysis and the (later) growth of point set topology was that the concept of the integral took on a life of its own, in contrast to being merely the inverse of the derivative in earlier treatments of the calculus. This article outlines developments in this direction from the 1820s onwards.

Two issues provided stimulus: extending current definitions of the integral; and extending the sufficient conditions which validated the processes of analysis (differentiating under the integral sign, integrating an infinite series term by term, changing the order of component integrals in a multiple integral, and so on). The first of these issues receives most attention here, as it was the prime mover; but this fact is surprising, for the second issue seemed to raise the more pressing mathematical problems. This finding is one lesson to learn from its history, which has been well studied: Hawkins *1970* and Medvedev *1974* are the best sources available.

2 THE INTEGRAL WITH CAUCHY AND RIEMANN

One of the principal innovations made by Augustin Louis Cauchy in his teaching of analysis at the Ecole Polytechnique (§3.3) was to define the integral independently of the derivative; thus the fundamental theorem of the calculus,

$$d/dx\left(\int f(x)\,dx\right) = \int f'(x)\,dx = f(x), \tag{1}$$

became at last a genuine theorem, requiring conditions (especially continuity conditions) to be imposed on $f(x)$ to guarantee its truth.

In Figure 1, consider the partition $\{x_r\}$ of the interval $[x_0, x_n]$, and the sum of the areas of the rectangles that have $y_0, y_1, \ldots$ as their vertical sides. Cauchy defined the integral as the limiting value S of this sum ($\Sigma_r y_r \Delta x_r$) as

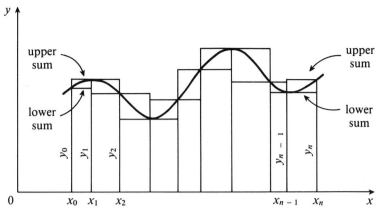

Figure 1 The integral according to Cauchy and Riemann

the partition becomes ever finer, *if* such a limit existed; he also showed that existence is assured if $f(x)$ is continuous (in his sense of the term), and that S is independent both of the choice of the point within each sub-interval from which a rectangle is defined, and of the sequence of partitions through which S is attained. Then he extended his definition to infinite-valued functions, infinite intervals and multiple integrals, and found conditions under which the processes of analysis are valid.

Among early users of this approach was the young German mathematician J. P. G. Lejeune Dirichlet, who used it in 1829 to show that if $f(x)$ possesses only a finite number of discontinuities, maxima and minima over an interval, then its Fourier series converges, and to that function (§3.11). This result forged a strong link between integration and Fourier series, which was to recur ever after.

Now consider the upper and lower sums of Figure 1 defined relative to the partition. This was the concern of Bernhard Riemann, in a doctoral thesis written in 1854 but published only posthumously in 1867. His subject was trigonometric series, which use integrals in their definition of the (Fourier) coefficients (§3.6); so he devoted part of the thesis to reworking the definition of the integral in terms of Archimedes' idea of sandwiching the area from above and below as the partition became finer. His conclusion was that integrability is guaranteed if the collection of sub-intervals over which the $f(x)$ ranges in value by more than some given magnitude (its 'oscillation', as we now say) becomes vanishingly small in overall length when the partition becomes sufficiently fine.

Riemann's formulation and proof of this result was cryptic, for he was inventing parts of set and measure theory in the process without fully realizing it; but it seemed clear from his examples that his conception was more powerful than Cauchy's. In addition, he found series for various functions which went beyond Dirichlet's conditions in possessing infinite sets of points of discontinuity or optimal value; and he also posed the question of whether or not a trigonometric series is a Fourier series. All these matters excited great interest among mathematicians when Riemann's paper was published. In particular, in 1875 Gaston Darboux gave a version of the integral which is usually known today as the 'Riemann integral': instead of working with oscillation, he formed the analytical expressions for the upper and lower sums and formulated the condition of integrability in terms of the difference between them decreasing to zero as the partition becomes finer.

3 SET TOPOLOGY AND CONTENT

By the late 1870s, Georg Cantor's theory of sets (which, as shown in §3.6, grew out of Riemann's thoughts on Fourier series) was gaining some currency. One application was to treat an area as a set of points in the plane (or space), define the length, area or volume of ordinary objects such as straight lines and spheres in the usual way, and then deploy the new techniques to define the 'content' (to use the word that came in) of point sets in general. The usual properties would have to be satisfied by such definitions: two congruent sets had equal content, the subset of a set had a smaller content, and especially the 'additivity' property that the content of the union of disjoint sets was the sum of the individual contents.

It turned out that Riemann's definition could not only be clarified but actually extended. However, there were some problems: a major one was that several of the first definitions (including Cantor's) gave the same value to the content of a set P as to its closure (that is, P together with the set of its limit points). But if P were, say, the rational numbers within $[5, 6]$, then the closure was $[5, 6]$ itself and so naturally had content 1; but P could be shown to have content 0. Another blunder was the assumption that the properties 'denumerable', 'with content zero' and 'nowhere dense' of a set were equivalent; but then nowhere-dense sets were found which possessed a positive content.

In order to avoid such mistakes, in 1892 Camille Jordan imitated but extended Darboux's use of upper and lower sums. Following some work by Giuseppe Peano, he defined the 'outer content' of a bounded set P in the plane as the greatest lower bound of all polygonal regions (which had 'orthodox' contents) containing P, and its 'inner content' as the least upper bound of regions contained in P, and he offered the equality of these two values as the criterion that P had a content. He then reformulated Riemann's upper and lower sums in these terms to give a definition of integrability of a function. In the course of extending the domain of functions, he stressed those of 'bounded variation' (a notion closely linked with oscillation). However, his definition allowed the additivity property to be satisfied only by a finite number of sets, so that term-by-term integration of an infinite series was beyond its purview.

4 MEASURE THEORIES IN THE TWENTIETH CENTURY

The contributions of Darboux and Jordan heralded a great rise of interest among French mathematicians in this German subject of set theory; a school developed, largely around Emile Borel, which applied it to many areas of analysis (Medvedev *1976*). Borel himself made very specific all the properties that the 'measure' – the new current word – of a set should have (he found later that they could be adapted to define probability). The task was then to give a definition of the measure of a set, and the integral of a function, which satisfied them all.

In the early 1900s three mathematicians succeeded: Borel's colleague Henri Lebesgue, the English mathematician W. H. Young (then working in Switzerland) and the Italian Giuseppe Vitali. Their formulations were mathematically different, but logically equivalent. Lebesgue's is outlined here; it was the first to appear (in 1902) and soon became dominant.

First, Lebesgue imitated Jordan's definition of the outer content of the bounded set P to specify the 'outer measure' of a set, to define inner measure as the outer measure of the complement of P relative to some bounding 'orthodox' region, and to state measurability in terms of the equality of these two measures (as $m(P)$). However, he involved infinitely many covering intervals, and so handled infinite additivity.

Second, to define the integral Lebesgue focused on values of $f(x)$, on the sensible ground that the discontinuities of $f(x)$ occurred there. Figure 2 shows the same function (which is much better behaved than those of main concern to him) as in Figure 1; but this time the Oy-axis was partitioned, into $q + 1$ values $\{y_r\}$ between its extreme values y_0 and y_q, and intervals of values of x were formed for which $f(x)$ lay between y_r and y_{r+1}. For each r these points formed a set (P_r, say), along the Ox-axis, whose five components (in the case illustrated) are marked by the thicker lines. If all sets P_r were measurable in this sense, then (in the formulation given here) the integral would be defined as

$$\int f(x)\,\mathrm{d}x := \lim_{r \to \infty} \sum_r y_r m(P_r) + (x_n - x_0)y_0, \quad x_0 \leqslant x \leqslant x_n, \quad (2)$$

when the partition of the Oy-axis became ever finer; again, *if* such a limit exists.

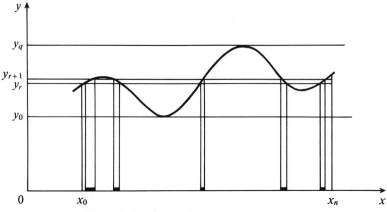

Figure 2 The integral according to Lebesgue

To develop a full theory of integration, Lebesgue now defined infinite and multiple integrals and sought conditions to validate the processes of analysis. His extensions were especially welcome, for the sufficient conditions were more general than those of uniform convergence and continuity which had become routine requirements of the Weierstrassian tradition of analysis (of which the Riemann integral was a part). Among applications elsewhere in mathematics, trigonometric series continued to be prominent: for example, Lebesgue himself followed up a book-length presentation of his integral in his *Leçons sur l'intégration* (1904) with *Leçons sur les séries trigonométriques* (1906).

However, some altercations attended these developments. In particular, the axioms of choice made their controversial entry into set theory exactly at this time (§5.4), and Vitali quickly used them to define a set which could not have a measure because it did not satisfy the basic properties. Was choice or measure at fault? On the one hand, Lebesgue was not much concerned by the axioms and used them rather casually; but Borel felt that their non-constructive character rendered them unacceptable in mathematics and wanted to restrict sets (and thereby measure theory) only to those constructible by the operations of set combination.

Despite these doubts, the mathematical community took to measure theory with enthusiasm and several generalizations and extensions were introduced to 'integrals' over topological spaces of various kinds. When Lebesgue issued the second edition of his 1904 *Leçons* in 1928, he had much more to report.

5 TWO EXTENSIONS, AND AN APPLICATION

5.1 The Stieltjes integral

In the 1890s, in connection with continued fractions, T. J. Stieltjes extended Cauchy's idea of the integral as the limit of a partition sum so as to accommodate a function $g(x)$ as the base variable instead of x. That is, he took the limit (if it existed) of this sum:

$$\int f(x)\,dg(x) := \lim_{r \to \infty} \sum_r f(t_r)\,[g(x_r) - g(x_{r-1})], \quad x_{r-1} \leqslant t_r \leqslant x_r. \quad (3)$$

For his purposes it was sufficient to let f be continuous; later work on equation (3) would weaken this condition, and extend the definition into the measure-theoretic domain (Hawkins *1970*: epilogue).

5.2 Multiple, surface and line integrals

The double or multiple integral was usually handled as the limit of a double or multiple sum, in the manner of Cauchy or Lebesgue. There was less attention to detail than with the single integral, but – especially from the Weierstrassian era on – these integrals were distinguished from their related repeated integrals, and care was taken over the interchange of limits, especially when (as in Fourier integral theory, for example) infinite ranges of values of variables were involved.

In contrast, the treatment of surface and line integrals was much more casual. They are, of course, more 'difficult', requiring notions from differential geometry (§3.4), an area from which they did indeed emerge in the late eighteenth century. Applications of (proto-)potential theory (§3.17) to mechanics and mathematical physics were a prime source, but only with the emergence of divergence and related theorems did they start to gain proper respect, and even then not quickly. Their history is similarly neglected.

5.3 Minkowski and the 'geometry of numbers'

A striking application of measure theory was made in the 1900s by Hermann Minkowski, in the course of his study of quadratic forms in number theory (§6.10). They were represented in the plane by ellipses, and his concerns had led him to consider whether particular sets of ellipses intersected or not; the answers involved their areas. He realized that the reasoning could be extended to arbitrary symmetric convex curves, and he now drew upon measure theory to express their areas. He called his topic 'the geometry of numbers', and its study greatly enhanced the understanding of convexity (Hancock *1939*).

BIBLIOGRAPHY

Grattan-Guinness, I. (ed.) *1980, From the Calculus to Set Theory, 1630–1910: An Introductory History*, London: Duckworth, and New York: Columbia University Press. [Hawkins summarizes his *1970* here; chapters by J. W. Dauben and the editor are also pertinent.]

Hancock, H. *1939, Development of the Minkowski Geometry of Numbers*, New York: Macmillan.

Hawkins, T. W. *1970, Lebesgue's Theory of Integration. Its Origins and Development*, Madison, WI: University of Wisconsin Press. [Repr. 1975, New York: Chelsea.]

Medvedev, F. A. *1974, Razvitie ponyatiya integrala*, Moscow: Nauka.

—— *1976, Frantsuzskaya shkola teorii funktsii i mnozhestv na rudezhe XIX–XX bb.*, Moscow: Nauka.

Monna, A. F. *1972*, 'The integral from Riemann to Bourbaki', in D. van Dalen and A. F. Monna (eds), *Sets and Integration. An Outline of their Development*, Groningen: Wolters–Noordhoff, 77–154.

Pesin, I. N. *1966, Razvitie ponyatiya integrala*, Moscow: Nauka. [English transl.: *Classical and Modern Integration Theories*, 1970, New York: Academic Press.]

Saks, S. *1937, Theory of the Integral*, 2nd edn (transl. L. C. Young), New York: Stechert. [Reprinted ?1950, New York: Hafner; and 1964, New York: Dover. Valuable presentation and bibliography of later developments.]

3.8

The early history of fractals, 1870–1920

J.-L. CHABERT

Fractals, familiar in modern times from the work of Benoît Mandelbrot in the 1970s, were first met with a century earlier, during the 1870s, when continuous functions without derivatives were discovered. Current work on the notion of dimension and on holomorphic dynamics stems from studies carried out before 1920.

1 CONTINUOUS CURVES WITHOUT TANGENTS

For a long time, it appeared natural that a continuous function always has a derivative, except, perhaps, at a finite number of points. Indeed, a demonstration of this, inspired by A. M. Ampère, can be found in textbooks on infinitesimal calculus from the second half of the nineteenth century, at the very time when counter-examples were first being developed.

1.1 The analytic way: Riemann and Weierstrass

By exhibiting the sum of the series

$$\sum_{p=1}^{+\infty} \sin(p^2 x)/p^2,\tag{1}$$

which has no derivatives at points that are irrational multiples of π, Bernhard Riemann was apparently the first to put forward a contrary opinion. In 1872 Karl Weierstrass gave as a better example the sum of the series

$$\sum_{n=0}^{+\infty} b^n \cos(a^n \pi x),\tag{2}$$

in which b is an odd number, $0 < a < 1$ and $ab > 1 + (3\pi/2)$; this has no derivative for any value of x.

These first examples of objects we today refer to as fractals generated interest, but also rejection: 'I turn away in fright and horror from that awful affliction of functions without derivatives', said Charles Hermite.

1.2 The geometric way: Koch and Bolzano

In 1904 Helge von Koch proposed 'a continuous curve without a tangent at any point obtained by an elementary geometric construction'. Since there is no tangent at a point, he successively introduces an infinity of points (Figure 1). Let us specify the algorithm, a prototype for constructing fractal curves (Figure 2). At each step, we replace each segment by four smaller segments in the same way as when passing from step 1 to step 2. The corner points which appear at each step are preserved. The Koch curve is the limit curve, formed from the set of corner points thus obtained and of their limit points. This continuous curve has no tangent at any point.

In fact, in a paper written in the 1830s but published only at a later date, Bernard Bolzano had already obtained an example of a continuous function without derivative by an algorithm similar to Koch's.

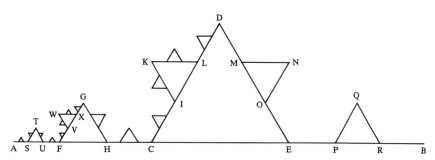

Figure 1 Koch's curve: continuous, but with nowhere a tangent
(from Koch *1904*: 685)

step 1   step 2   step 3

Figure 2 First steps in the generation of Koch's curve

368

1.3 The experimental way: Brownian motion

The fast and irregular motion of small particles suspended in a liquid discovered by the botanist Robert Brown was interpreted at the end of the nineteenth century as resulting from the thermal agitation of molecules. Figure 3, taken from physicist Jean Perrin's book *Les Atomes* (1913), illustrates this Brownian motion. The position of a particle is recorded at regular intervals of time. But if the positions were recorded twice as often, each segment would be replaced by two new segments, the sum of whose lengths would be greater than that of the initial segment, and so on. Therefore the length of the curve cannot be finite, and no tangent can exist at any point. Perrin had written:

> Functions with derivatives are the exception, or, to use geometric language, curves with no tangent at any point become the rule and it is the very regular curves we have been taught to imagine that are seen to be peculiar: no doubt highly interesting cases, but very particular all the same. (Perrin *1906*)

The trajectory-curves of Brownian motion have a random, Markovian character in contrast to extreme regularity of the singularities of the Koch curve.

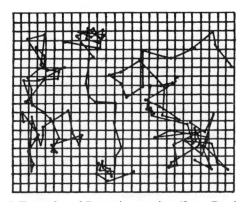

Figure 3 Examples of Brownian motion (from Perrin *1913*)

2 NOTIONS OF DIMENSION

To formalize the intuitive idea of 'number of dimensions', let us introduce a Cartesian coordinate system, and imagine that the dimension of an object is the minimum number of parameters necessary to determine the coordinates of the points of the object. Thus, a curve would be a set of points whose coordinates depend on a single parameter.

2.1 Cantor and Peano

Having demonstrated the impossibility of a 1–1 correspondence between the set of whole numbers and the set of real numbers, Georg Cantor in 1877 showed a correspondence between the points of a line segment and those of a square. He thought this called into question the notion of dimension, but Richard Dedekind doubted this because of the discontinuity of the correspondence, which he emphasized (§7.11).

In 1890 Giuseppe Peano defined a space-filling curve in the plane: a continuous, surjective mapping of the unit interval [0, 1] onto the unit square. David Hilbert *1891* explained a similar example, but using a geometric construction, and E. H. Moore *1900* described the algorithm which allows us to imagine Peano's curve (Figure 4). Obtained with the help of successive approximations, the mapping is defined at points of the form $t = n/9^k$ (analogues of the decimal numbers in base 9). Extended by continuity, the mapping becomes surjective, but is not injective.

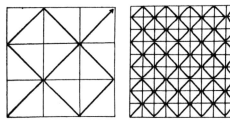

Figure 4 First steps in the generation of Peano's curve
(from Moore *1900*: 77)

2.2 Topological dimension and Hausdorff dimension

To exclude Cantor's correspondence and Peano's curve, let us impose both bijection and continuity. Indeed, there can be no bicontinuous bijection between a line segment and a square. The more general question of the existence of a bicontinuous bijection between two Euclidean spaces of different dimensions was resolved negatively by L. E. J. Brouwer (§7.11). The dimension of a Euclidean space is therefore an invariant for bicontinuous bijective transformations. We see the current notion of a homeomorphism which preserves the topological dimension making an appearance, a notion which corresponds to our intuitive understanding of the simple cases.

In 1919, Felix Hausdorff proposed another notion, making it possible to associate with these pathological examples a number which takes into account the greater or lesser filling of the spaces. The principle is as follows.

To measure the volume of an object, we associate with any union of spheres containing that object the sum of the volumes of these spheres. Then we consider the lower bound of these sums. For an area, we replace the volume of a sphere $((4/3)\pi r^3)$ by the area of a circle (πr^2) and, for a length, we replace it with the length of the line segment $(2r)$. Koch's curve, however, has both an infinite length and an area equal to zero; this gave Hausdorff the idea of making the calculations with expressions of the form r^d, where d is any positive number. The result is called the content of the object with respect to d. For $d = 1$, 2 or 3, we come back essentially to length, area or volume.

The Hausdorff dimension of an object is then the intermediate value d_0 such that, for $d < d_0$, the content is infinite, and for $d > d_0$, the content is zero. This new dimension is no longer a whole number -- it is a metric notion linked, however, to the topological dimension by

$$0 \leqslant \text{topological dimension} \leqslant \text{Hausdorff dimension} \leqslant 3. \qquad (3)$$

Koch's curve, for example, has a topological dimension of 1, whereas its Hausdorff dimension is $\log 4/\log 3 = 1.26\ldots$.

2.3 The Cantor ternary set

In his studies on sets, Cantor considered the notion of a set derived from a set (i.e. the set of its accumulation points) and, in particular, the notion of a perfect set (i.e. a set identical to its derived set, such as a closed interval). In 1883 he gave an example of a perfect set which is not dense in any interval, no matter how small: the set of real numbers in the interval $[0, 1]$ having the form

$$z = c_1/3 + c_2/3^2 + \cdots + c_k/3^k + \cdots, \qquad (4)$$

in which the c_k are equal to 0 or 2.

This set, today referred to as 'the Cantor ternary set', has a length, a Lebesgue measure and a topological dimension all equal to zero, but its Hausdorff dimension is $\log 2/\log 3$, or $0.63\ldots$. It has the following remarkable property: if two trigonometric series of period 1 have the same sum except, perhaps, on this set, then they have the same sum everywhere.

3 THE ITERATION OF RATIONAL FRACTIONS

An entirely different point of view, the iteration of rational fractions (i.e. the study of the behaviour of the sequence of iterates $z_n = R^n(z_0)$, in which z_0 is a complex number and R a rational fraction), leads to the discovery of other fractal objects. The fundamental results were obtained by Pierre Fatou and Gaston Julia around 1917–19.

When a satisfies $R(a) = a$ and $|R'(a)| < 1$, a is called an attractive fixed point of the fraction R, or, to put it another way, the sequence z_n tends towards a as soon as z_0 is chosen close enough to a. But here we are interested in the entire domain of attraction of a, that is to say the set of all complex numbers z_0 such that the sequence z_n tends to a. In his study of Newton's method and the domains of attraction of different roots, Arthur Cayley *1879* had already seen things in these terms.

The domains of attraction of attractive fixed points and also of attractive periodic points of a fraction R are in general non-connected domains of the complex plane; but they have a single, common boundary, today called the Julia set of the fraction R. This set, which is generally fractal, therefore comes into being as a limit: not a limit of curves, as in Koch's curve, but instead as a boundary separating different domains. It is either connected or, on the contrary, of Cantorian type. The theoretical work by Julia and Fatou was not properly followed up until the 1980s when computer simulations enabled a better understanding of the phenomena.

4 THE FRACTALS OF BENOÎT MANDELBROT

In introducing the term 'fractal', Mandelbrot *1975* brought back into fashion all the old work on curves and strange objects. He used them systematically in the most varied disciplines: in physics, geography, astronomy and biology, in phenomena such as turbulence. Here, fractals serve as models, and the Hausdorff dimension associates with them a number which can be interpreted.

In a later book, Mandelbrot *1982* put the computer to good use, pursuing Julia's and Fatou's work in the more general context of holomorphic dynamics. The mysterious aesthetics of the pictures obtained has since generated a considerable volume of work. The Mandelbrot set, the set of the complex numbers c such that the Julia set of the polynomial $R_c(z) = z^2 - c$ is connected, corresponds to the distinction made above (Figure 5).

Thus, in a world of great diversity, the term 'fractal' provides a degree of unity. Having the singular property that an enlarged part is similar to the whole, fractal objects can be either extremely regular in their singular features, as in Koch's curve, or of chaotic appearance although they are deterministic when obtained by the iteration of fractions, or yet again truly the result of chance, as in Brownian motion.

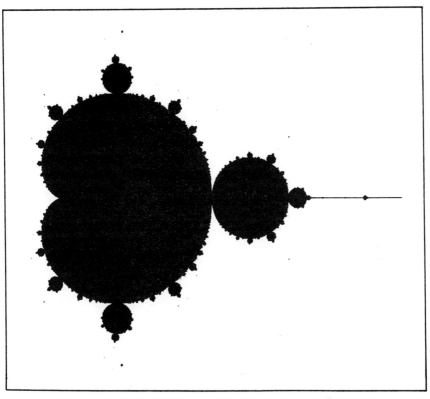

Figure 5 The Mandelbrot set (a version kindly made available
by the author)

BIBLIOGRAPHY

Cayley, A. *1879*, 'The Newton–Fourier imaginary problem', *American Journal of Mathematics*, **2**, 97. [Also in *Mathematical Papers*, Vol. 3, 405–6.]

Chabert, J.-L. *1990*, 'Un demi-siècle de fractales: 1870–1920', *Historia mathematica*, **17**, 339–65.

Hilbert, D. *1891*, 'Über die stetige Abbildung einer Linie auf ein Flächenstück', *Mathematische Annalen*, **38**, 459–60. [Also in *Gesammelte Abhandlungen*, Vol. 3, 1–2.]

Koch, H. von *1904*, 'Sur une courbe continue sans tangente obtenus par une construction géométrique élémentaire', *Arkiv for Mathematik, Astronomi och Fysich*, **1**, 681–704.

Mandelbrot, B. *1975*, *Les Objets fractals: Forme, hasard et dimension*, Paris: Flammarion. [3rd edn, 1989.]

—— *1982*, *The Fractal Geometry of Nature*, San Francisco, CA: Freeman.

Moore, E. *1900*, 'On certain crinkly curves', *Transactions of the American Mathematical Society*, **1**, 72–90.

Peitgen, H.-O. and Richter, P. *1986*, *The Beauty of Fractals*, Berlin: Springer.

Perrin, J. *1906*, 'La Discontinuité de la matière', *Revue de Mois*, **1**, 323–43.

—— *1913*, *Les Atomes*, Paris: Alcan. [English transl.: *Atoms*, 1916, London: Constable.]

3.9

Functional analysis

REINHARD SIEGMUND-SCHULTZE

1 SOME BASIC NOTIONS

The essence of the development of functional analysis was the transfer of a number of concepts from n-dimensional Euclidean space $\mathbb{R}^n$ and the functions defined on it to infinite-dimensional 'function spaces' of various types and their 'operators' – concepts such as compactness, boundedness, convergence, distance, continuity, completeness, dimension, scalar product and linearity. To bring this about, a way was needed to pass from the finite to the infinite; but the form of this passage was the object of great concern and even strife among the early functional analysts. Often it was only through generalizing – through the increasingly axiomatic definition of the new spaces, where $\mathbb{R}^n$ was subordinated as a special case – that the relations of the original concepts, and their partial logical dependence or independence, became recognizable. Concepts such as that of convergence became diversified, while equivalent properties such as boundedness and compactness separated from each other. In addition, new fundamental principles and concepts appeared that made no sense in the finite realm (e.g. the Hahn–Banach extension theorem, category theory and separability) and could be introduced only with the help of Georg Cantor's set theory (§3.6).

Important bridges for the 'passage from the finite to the infinite' were generalized geometric intuition and ways of speaking (e.g. with the help of the Schmidt orthogonalization process), and conceptual analogies with linear algebra and the theory of real functions, as well as the adoption of approximation and iteration principles (e.g. Neumann's series). Concrete applied problems on the one hand, and, on the other, that search for a unifying, generalizing point of view which is inherent in mathematics, were of equal historical importance in stimulating the development of functional analysis.

2 THE ROOTS IN THE THEORY OF SYSTEMS OF LINEAR EQUATIONS AND INTEGRAL EQUATIONS

In the search for a solution to the partial differential equation

$$\Delta v := v_{xx} + v_{yy} = 0 \tag{1}$$

(the potential equation) in the 1800s, Joseph Fourier came up against the infinite system of linear equations

$$\sum_{q=1}^{\infty} a_{pq} x_q = c_p, \quad p = 1, 2, \ldots . \tag{2}$$

He found the solution $(x_1, x_2, \ldots)$ of this system by putting $x_k = 0$ and $c_k = 0$ for $k > n$, calculating the solution for the resulting finite system, consisting of the first n equations for fixed n, and then letting n increase to infinity. This method is not rigorous, and Fourier's success relied on the fact that the coefficients a_{pq} and c_p had 'convenient' values.

This was convincingly demonstrated in the following example by the Austrian Eduard Helly in 1921 (Monna *1973*: 10):

$$\begin{aligned}
x_1 + x_2 + x_3 + \cdots &= 1 \\
x_2 + x_3 + \cdots &= 1 \\
x_3 + \cdots &= 1 \\
\vdots
\end{aligned} \tag{3}$$

The system of the first n equations has the solution $x_i = 0$ $(i \neq n)$, $x_n = 1$ for each n; but the system (3) does not have the solution $x_i = 0$ $(i = 1, 2, \ldots)$, and in fact has no solution.

Fourier's fundamental idea nevertheless proved fruitful. It was used by the American George William Hill in his 1877 work on the motion of the Moon (Heuser *1986*: 612) and by Henri Poincaré (1886) and Helge von Koch (1890) in a rigorously worked-out theory of infinite determinants (Dieudonné *1981*: 77*ff*; Bernkopf *1968*).

Integral equations (the term comes from Paul Du Bois-Reymond: §3.10) became significant in potential theory (J. P. G. Lejeune Dirichlet's boundary-value problem) through the works of the physicists A. Beer (1856) and Carl Neumann (1877). Beer already knew that the integral equation of the second type (David Hilbert's term),

$$x(s) + \int_a^b K(s, t) x(t) \, dt = f(s), \tag{4}$$

where x, f and K are continuous, and x is unknown, could be handled with the help of the method of successive approximation (Neumann's series), known since Augustin Louis Cauchy (§3.14).

Vito Volterra was the first, in 1896, to indicate the possibility of treating integral equations as boundary cases of infinite systems of linear equations. He spoke of a 'passage from the discontinuous to the continuous' (Heuser *1986*: 616).

The Swede Erik Ivar Fredholm systematically carried out this conceptual analogy between 1900 and 1903. He conceived of equation (4) as a boundary case of the discrete version of the problem

$$x_p + \sum_{q=1}^{n} \frac{1}{n} K_{pq} x_q = f_p, \quad p = 1, 2, \ldots, \tag{5}$$

where x_p, f_p and K_{pq} are values of the continuous functions x, f and K at intermediate points in the n sub-intervals into which the interval $[a, b]$ is divided, or, in the case of K, the n^2 sub-intervals of $[a, b] \times [a, b]$. Fredholm, with the aid of infinite determinants, provided solutions for equation (4) and, under the application of orthogonality relations, formulated the famous 'Fredholm alternative'. In Fredholm we can see the beginnings of a theory of operators in function spaces.

3 THE ROOTS OF THE CALCULUS OF VARIATIONS AND OF THE ITALIAN *CALCOLO FUNZIONALE*

This last idea was more familiar in the older, richly traditional calculus of variations, the second principal source of functional analysis. In the calculus of variations, concrete functionals (the term *fonctionnelle* comes from Jacques Hadamard), functions that are defined on sets of functions, had been investigated for a century (§3.5).

In 1887, Volterra also took a general approach, and sought to analyse a so-called 'line function' (*funzione di linee*) $y = y \mid [\phi(x)] \mid$, where y and ϕ are given general properties rather than particular ones (as in the calculus of variations). Volterra defined 'derivatives' of his line functions, in effect differentiating with respect to a parameter as in the classical calculus of variations, and tried to build up a line-function theory analogous to the (complex) theory of Riemann functions (Siegmund-Schultze *1982*: 40*ff*). These efforts had little significance for applications, unlike the results achieved in the theory of integral equations. Hadamard had this to say about Volterra's motivation: 'Why was the great Italian geometer led to operate on functions as infinitesimal calculus had operated on numbers... ? Only because he realized that this was a harmonious way of completing the architecture of the mathematical building' (Hadamard *1945*: 129).

But the topological foundations of the associated function spaces were not sufficiently clear, and thus Volterra's 'generalizing for its own sake' did not affect the development of functional analysis in a very specific way. The large-scale work undertaken by Volterra's student Paul Lévy on generalized differential equations in functional derivatives remained largely unfruitful. In this work, Lévy also spoke of a 'passage from the finite to the infinite'. Here each theorem of the *analyse fonctionnelle* (the term was introduced by Lévy in a book of 1922) is conceived of as the limiting case of a theorem about functions of *n* variables. These variables were construed strictly as values of traditional functions over increasingly finer subintervals. What remained of Volterra's *funzione di linee* was solely the general concept and the recognition of the necessity to analyse the domains of definition of these generalized functions.

4 THE SET-THEORETIC IMPULSE AND FRÉCHET'S *ANALYSE GÉNÉRALE*

This necessity had already been recognized by Volterra's student Cesare Arzelà, who in 1889 attempted to carry over the foundational theorems of Karl Weierstrass on continuous functions to line functions. Arzelà built upon a work by Giulio Ascoli from 1884 in which the Bolzano–Weierstrass accumulation-point theorem (§3.3) – the property of compactness – was carried over to sets of functions that are continuous, of the same order and uniformly bounded. Ascoli's theorem is perhaps the first substantial result on infinite-dimensional function spaces.

With the working out of the difference between the usual (pointwise) convergence and the uniform convergence of sequences of functions, such spaces gradually did emerge, particularly in Fourier analysis. Bernhard Riemann, in his well-known 1854 lecture on geometry (§7.4), gave in an aside the first clear indication (Birkhoff and Kreyszig *1984*: 263): 'But there also exist manifolds in which the determination of location requires not a finite number but either an infinite sequence or a continuum of determinations of quantities. Such a manifold, for instance, is formed by the possible determinations of a *function* for a given domain.'

Weierstrass's approximation theorem of 1885, according to which any continuous function can be represented by a uniformly convergent series over polynomials (Siegmund-Schultze *1988*), was, after Ascoli's theorem, a further important result. It became increasingly clear that Cantor's set theory, developed between 1874 and 1884 (§3.6), was necessary as a foundation for the study of infinite-dimensional sets of functions. This came from a more general methodological view as well as from the generalization of the concept of space-building functions (e.g. Lebesgue integrals: §3.7). The key method in the theorem by Ascoli is, for example, closely related to the so-called second Cantor diagonalization.

Another stimulus for work on infinite sets of functions was concern for the rigorization of Dirichlet's principle in the calculus of variations (§3.17). Arzelà made a particular point of this in 1889. The best-known work in this direction was by Hilbert, around 1900, in which – perfectly analogously to Ascoli's key method – were founded the 'direct methods' of the calculus of variations (where the extreme values of a function of n variables are obtained by taking n to infinity).

Arzelà's work provided an immediate starting-point for the *analyse générale* of the Frenchman Maurice Fréchet. He introduced the concept of the abstract metric space (the term comes from Felix Hausdorff) in his epoch-making dissertation of 1906, investigated abstract (semi-continuous) functionals on it, and formulated the concept of abstract 'compactness'. In the background stands Fréchet's teacher Hadamard, who at the 1897 International Congress of Mathematicians in Zurich had already proposed the making of 'a new chapter in set theory', known today as set-theoretic topology (§3.6). Hausdorff's book of 1914, *Grundzüge der Mengenlehre*, was to be specially influential in this area. In addition, a similar theory was being developed in the USA, under E. H. Moore's name of 'general analysis'.

5 PIONEERING FORAYS WITHOUT EFFECT: AXIOM SYSTEMS OF PEANO AND PINCHERLE FOR INFINITE-DIMENSIONAL VECTOR SPACES

A portrayal of the works of the Italians in early functional analysis would be incomplete if no mention were made of Salvatore Pincherle, who introduced the term *calcolo funzionale* and, in a book of 1901, spoke of a *geometria dello spazio funzionale*. Of all the Italian contributions, Pincherle's works are the closest, with respect to form, to axiomatic functional analysis. In the 1890s he worked out a formal theory of 'distributive operations', that is, linear operators in infinite-dimensional vector spaces. These spaces had already been introduced in a rigorously axiomatic fashion in 1888 by his fellow countryman Giuseppe Peano, who was following Hermann Grassmann's *Ausdehnungslehre* of 1844 (Monna *1973*: 120; see also §6.2). Pincherle, however, worked more in the tradition of the Leibnizian and d'Alembertian abstract operator theory (§4.7). His work had only limited influence on the development of functional analysis. Indeed, the concept of the axiomatic infinite-dimensional vector space was not rediscovered until the 1920s, in the Banach school, when linear functional analysis attained a corresponding level in its development and actually required such general concepts.

6 THE HILBERT THEORY OF INTEGRAL EQUATIONS, AND RIESZ'S SYNTHESIS

With Fréchet's dissertation of 1906, modern functional analysis began to make its appearance. What followed was a filling-out and broadening of its content; above all, it was the Hungarian Frédéric Riesz who, building upon Fréchet's concept of distance, established the connections between the French school in real functions (especially the work of Henri Lebesgue) and the Göttingen school in the theory of linear integral equations (§3.10).

In Göttingen, Hilbert provided a foundation for the theory of integral equations in six famous *Mitteilungen*, or 'communications', between 1904 and 1910. In 1904 he derived for the first time in a rigorous manner the Fredholm resolvents. For symmetric kernels $K(s, t) = K(t, s)$, Hilbert used the analogy with the eigenvalue theory of linear algebra, introducing a parameter in front of the summation in equation (5), and thus generalizing the principal-axes theorem to integral operators with continuous functions.

This method of transition corresponded somewhat to the Volterra–Lévy method. Hilbert's student Erhard Schmidt chose a quite different method in his dissertation of 1905. Using general integral inequalities very similar to those of Friedrich Wilhelm Bessel and K. H. A. Schwarz, he proved the existence of eigenvalues and eigenfunctions, and in particular also provided a proof of the principal-axes transformation. In Schmidt, the orthogonal system of functions (see equation (7) below) that Hilbert had introduced in 1904 in connection with Fourier analysis came clearly to the fore.

In the process the solution of equation (4) led back to the solution of the infinite system of linear equations

$$x_p + \sum_{q=1}^{\infty} K_{pq} x_q = f_p, \quad p = 1, 2, \dots. \tag{6}$$

Here x_p and f_p are no longer viewed, as they are in (5), as function values (note the factor $1/n$ in (5)), but rather as 'Fourier coefficients' defined by an 'orthogonal, normalized and complete system of functions' $\{\phi_p\}$ in the space of continuous functions $C[a, b]$:

$$x_p = \int_a^b \phi_p(t) x(t) \, dt, \qquad \int_a^b \phi_i \phi_j \, dt = \delta_{ij} \tag{7}$$

(using the Kronecker delta). This view of f_p thus underlay the convergence conditions $\Sigma^{\infty} x_p^2 < \infty$ and $\Sigma^{\infty} f_p^2 < \infty$, and with this the space of 'quadratic summable' number series, denoted by ℓ^2, gradually became the object of investigation. This theory was exhaustively developed in Hilbert's fourth *Mitteilung* of 1906, probably the richest of the early work in functional analysis, whose significance was not evident until the end of the 1920s (see Section 9 below). This method of 'transition from the finite to the infinite' broke with the Volterra–Lévy tradition. Lévy criticized bluntly the new level of abstractness that was attained through the interposition of the infinite parameter series, the elements of ℓ^2.

Still, from the standpoint of *analyse générale*, Hilbert theory was not sufficiently abstract, as Fréchet had noted. It was a matter of carrying over the infinite dimensionality into general, axiomatic properties of abstract point sets. Schmidt took a step in this direction with his work of 1908 that introduced geometrical language (projection, decomposition, orthogonality, scalar product) into the 'Hilbert space' ℓ^2, and clearly distinguished two different convergence concepts, the usual convergence and 'strong' convergence.

An important result in this direction was the theorem by Riesz and Ernst Fischer of 1907. One consequence is that, with appropriate generalization of the concepts of convergence and integral, for each element of ℓ^2 there is a function corresponding to it whose Fourier coefficients are the components of the element in ℓ^2. This proof of the isomorphism of ℓ^2 and L^2, the space of quadratic Lebesgue-integrable functions, led to the concept of the Hilbert space, with its scalar-product norm. Up to this time only *functionals* (mostly continuous, frequently linear) on ℓ^2 or C were considered, that is, numerical-valued operators. Even the Hilbert bilinear forms ('functions of infinitely many variables') in ℓ^2 that replaced the integral transformations in (4) could be interpreted as functionals (even continuous). For the moment problems that grew out of probability theory it became highly desirable to address the question of the *general* form of linear and continuous (with respect to the norm) functionals in those spaces.

For given functions $f_n(x)$ (e.g. the power function x^n) and numbers a_n, functions ϕ and α were sought that satisfied

$$\int_a^b f_n(x)\phi(x)\,\mathrm{d}x = a_n \quad \text{or} \quad \int_a^b f_n(x)\,\mathrm{d}\alpha(x) = a_n, \quad n = 1, 2, \ldots . \quad (8)$$

In 1907 Fréchet and Riesz, independently of each other, showed that for the Hilbert space $L^2[a, b]$ each linear and continuous functional may be represented by an integral (8)$_1$, where $f_n(x)$ stands for the argument of the functional. In 1909 Riesz showed in another 'representation theorem' that such functions in $C[a, b]$ may be represented by a 'Stieltjes integral' (8)$_2$ (from the work of the Dutch mathematician T. J. Stieltjes in 1894; §3.7). The moment problems could thus be understood as a search for linear and continuous functionals that are defined on the entire underlying function space and that assume the above values on the finite or infinite subset $\{f_1, f_2, \ldots\}$.

Riesz further generalized the problem in 1910 when he admitted $(p/p - 1)$-fold Lebesgue-integrable functions for the $f_n(x)$ in $(8)_1$, and p-fold Lebesgue-integrable functions for the $\phi(x)$ ($p > 1$). He thereby introduced the new, linear, normed function space L^p that for $p \neq 2$ is not a Hilbert space.

Riesz gave general conditions for the solution of the moment problems, and thus pointed the way to the fundamental Hahn–Banach theorem that was proved by Hans Hahn in 1927 and by Stefan Banach in 1929 for abstract, axiomatic, linear and normed spaces. This theorem, which in general can be established only with the help of the axiom of choice from set theory, ensures the existence of a sufficiently rich theory of the 'dual', of the set of linear and continuous functionals of the relevant space.

Riesz's representation theorem of 1909 and his theory of L^p brought the theory of linear and continuous functionals to a fairly complete state, and at the same time established the functional-analytic approach to integration theory. Riesz's work of 1910, however, signified simultaneously the starting-point of modern operator theory.

7 RIESZ AND THE BEGINNING OF OPERATOR THEORY

Riesz's work of 1910 rests on the fact that L^q can be identified with the dual of L^p if $1/q + 1/p = 1$ ($p, q > 1$). Consequently, for L^2 there does not essentially exist a theory of duality since the space is dual to itself. The generalized representation theory for functionals in L^p enabled Riesz to adjoin transformations (operators) T in L^p, with 'transposed' (now called 'adjoint') transformations T^* in L^q by means of the equation

$$\int_a^b T(f(x))g(x)\,\mathrm{d}x = \int_a^b f(x)T^*(g(x))\,\mathrm{d}x \qquad (9)$$

for f in L^p and g in L^q. In this way the 'function equations' in L^p,

$$T(f(x)) = \phi(x), \quad f(x) \text{ unknown}, \qquad (10)$$

can be converted back to moment problems in L^q. A special role is played by those operators that Riesz, in connection with Hilbert's fourth *Mitteilung* of 1906, called 'completely continuous'. Hilbert had termed 'completely continuous' those functionals $F(x_1, x_2, \ldots)$ in ℓ^2 that were not only continuous with respect to the ℓ^2 norm, but also coordinate-wise continuous. He thus made the property of 'compactness', by means of interposition of the functional concept, fruitful for ℓ^2, since otherwise it does not hold. In 1910 Riesz carried this property over to operators, thereby rendering superfluous the Hilbert method of reducing integral-equation theory to the theory of systems of equations. In L^p, however, the powerful techniques that came out of orthogonality relations were no longer applicable (Birkhoff and Kreyszig *1984*: 293).

In 1918 Riesz came close to an axiomatic theory of complete linear normed spaces and their operators that were later to be named after Banach. In this work the theorem that each locally compact normed space is of finite dimension played a decisive role. It was now possible to carry back the theory of completely continuous operators (in particular the Fredholm alternative) to the theory of finite linear systems of equations ('degenerate operators') and to do it in a 'qualitative' fashion – with regard to existence rather than representation.

8 BANACH AND THE POLISH SCHOOL

An abstract axiomatic presentation was made by Banach in his 1920 Lwów dissertation introducing the 'Banach space' (so named by Fréchet). The principal accomplishment of the Polish school led by Banach in the 1920s and 1930s lay in the production of fundamental, sound principles and theorems of functional analysis under a careful application of set-theoretic methods: the Banach fixed-point theorem, the uniform boundedness theorem (Banach–Steinhaus) and the Hahn–Banach theorem, to name a few. The new quality was especially evident in the open-mapping theorem of 1929 that used the set-theoretic concept of 'category' from René Baire's dissertation of 1899. Here also the 'completeness' of Banach spaces, and thus the convergence of Cauchy series in them, was important.

The fixed-point theorem of Banach's student J. P. Schauder (1930) and work by Schauder and the Frenchman J. Léray of 1934 carried over into infinite-dimensional spaces the topological concepts that came from L. E. J. Brouwer (the fixed-point theorem of 1910, and topological degree; see §7.10). This laid substantial foundations for non-linear functional analysis and its application to non-linear differential and integral equations from the 1950s. The Hahn–Banach theorem has proved fundamental in linear functional analysis, for example in the theory of locally convex spaces since 1935, and in distribution theory.

9 CONCLUSION

At the International Congress of Mathematicians at Bologna in 1928 there were three principal speakers on generalized analysis, Fréchet, Volterra and Hadamard, each of whom took a distinct view of the development of the subject. Banach's works, which had been written without reference to possible applications, were hardly mentioned. The breakthrough to axiomatic functional analysis was made by John von Neumann in work beginning in 1928 that showed the applicability of Hilbert spectral theory to quantum mechanics (§9.15). Von Neumann abstracted from the results of Hilbert's fourth *Mitteilung* (1906) on the theory of integral equations. In typical mathematical generalization for its own sake, Hilbert had considered bounded, not completely continuous 'functions of infinitely many variables' in ℓ^2, even though only the completely continuous ones appeared in applications. Von Neumann extended the results to unbounded operators in Hilbert space which he had defined axiomatically in 1928. With the appearance in 1932 of his *Mathematische Grundlagen der Quantenmechanik* and Banach's *Théorie des opérations linéaires*, functional analysis was established as one of the most important fields of modern analysis, as an independent mathematical discipline.

BIBLIOGRAPHY

Bernkopf, M. *1968*, 'A history of infinite matrices [. . .]', *Archive for History of Exact Sciences*, 4, 308–58.

Birkhoff, G. and Kreyszig, E. *1984*, 'The establishment of functional analysis', *Historia mathematica*, 11, 258–321.

Dieudonné, J. *1981*, *History of Functional Analysis*, Amsterdam: North-Holland.

Hadamard, J. *1945*, *The Psychology of Invention in the Mathematical Field*, Princeton, NJ: Princeton University Press. [Repr., (no date), New York: Dover.]

Heuser, H. *1986*, *Funktionalanalysis. Theorie und Anwendung*, 2nd edn, Stuttgart: Teubner.

Monna, A. F. *1973*, *Functional Analysis in Historical Perspective*, New York and Toronto: Wiley.

Siegmund-Schultze, R. *1982*, 'Die Anfänge der Funktionalanalysis und ihr Platz im Umwälzungsprozess der Mathematik um 1900', *Archive for History of Exact Sciences*, **26**, 13–71.

—— *1988*, 'Der Beweis des Weierstrass'schen Approximationssatzes 1885 vor dem Hintergrund der Entwicklung der Fourieranalysis', *Historia mathematica*, **15**, 299–310.

3.10

Integral equations

J. LÜTZEN

1 ORIGINS

In a paper on potential theory published in 1888, Paul Du Bois-Reymond pointed out that boundary-value problems of partial differential equations often lead to functional equations in which the unknown function appears under a definite integral sign. He named such equations 'integral equations', and proposed them as a worthy area of research since 'in general they pose insurmountable difficulties for present-day analysis'. This programmatic declaration can be considered to mark the birth of integral equations as a separate research area. The development of the theory of linear integral equations was a central part of mathematics from 1890 to 1920 and beyond. In particular, the first 20 years was a period of great progress during which Vito Volterra, Erik Ivar Fredholm and David Hilbert carried the theory to such a degree of perfection that it was considered better understood than the theory of differential equations. This development was also instrumental in the emergence of functional analysis (§3.9).

The integral equations that have played the greatest role are those that Hilbert named as being of the first kind,

$$f(x) = \int_a^b K(x,y)\phi(y)\,\mathrm{d}y, \tag{1}$$

and of the second kind,

$$f(x) = \phi(x) + \int_a^b K(x,y)\phi(y)\,\mathrm{d}y. \tag{2}$$

Here f and the so-called kernel K are given functions, and ϕ is the unknown function.

2 SCATTERED RESULTS BEFORE 1850

During the century preceding the conscious development of the theory of integral equations, special integral equations turned up and were solved in connection with various mathematical and physical problems. However, the results obtained in this period were usually of limited scope and scattered, in the sense that the mathematicians who found them did not see their common features.

The Laplace and Fourier transforms can be considered as furnishing the earliest examples of integral equations, and the inversion formulas for these transforms, found around 1820, provide their solution. For example, let f be the cosine-Fourier transform of ϕ:

$$f(x) = \int_0^\infty \cos(xy)\phi(y)\,dy. \tag{3}$$

This is an integral equation for ϕ, the solution of which is given by the inversion formula

$$\phi(y) = \frac{2}{\pi} \int_0^\infty \cos(xy)f(x)\,dx. \tag{4}$$

Later in the century several other such integral transformations and inversion formulas were studied (§4.8).

A more explicit formulation and solution of a family of integral equations can be found in two publications of 1823 and 1826 by the young Niels Abel. He encountered these equations in his study of the generalized tautochrone problem: to determine the curve AMB so that the time a heavy body takes to slide along the curve from M to A is a given function $f(x)$ of the height $x = \text{MP}$ (Figure 1). By describing the curve by a function $s = \phi(x)$, where s is the length of the curve measured from A, Abel found the time of descent, $f(x)$, to be expressed by the integral

$$f(x) = \frac{1}{(2g)^{1/2}} \int_0^x \frac{\phi'(\xi)\,d\xi}{(x-\xi)^\lambda} \tag{5}$$

for $\lambda = \frac{1}{2}$, and then solved this integral equation for ϕ, for all $\lambda \in (0, 1)$.

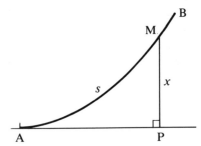

Figure 1 The generalized tautochrone problem

3 LIOUVILLE

A much wider variety of integral equations was encountered and studied by Joseph Liouville in connection with his research on differentiation of arbitrary order, Sturm–Liouville theory, and potential theory. As a student he was influenced by the Laplacian ideas of physics, according to which it should be possible to explain all macroscopic phenomena in terms of infinitesimal elementary interactions. Convinced that these infinitesimal actions could be determined only from their integrated macroscopic consequences, he was faced with integral equations. In his first paper from 1832 on differentiation of arbitrary order (i.e. operators $(d/dx)^\kappa$, where μ is any complex number), he even declared that 'the solution of most physico-mathematical problems basically depend on ... the determination of an arbitrary function placed under an integral sign'. For example, Liouville showed that an experiment performed by J. B. Biot and F. Savart led to an integral equation of the form

$$f(y) = 2 \int_0^\infty \frac{y\phi((s^2 + y^2)^{1/2})}{(s^2 + y^2)^{1/2}} \, ds, \tag{6}$$

the solution of which yields the Biot–Savart law for the interaction between a magnetic pole and an infinitely small conducting element (§9.10). Liouville developed his theory of differentiation of arbitrary order as the first rather general method for solving a variety of integral equations of the first kind.

In connection with his study in 1836–7 of the so-called 'Sturm–Liouville problem' Liouville was also led to consider integral equations of the second kind. In fact he showed that the Sturm–Liouville problem is equivalent to finding a solution $U(z)$ to the equation

$$U(z) = \cos(\rho z) + \frac{h \sin(\rho z)}{\rho} + \frac{1}{\rho} \int_0^z \lambda(z') U(z') \sin(z - z') \, dz'. \tag{7}$$

This was an important step in his proof of the convergence of the 'Fourier' series connected with Sturm–Liouville theory. In this connection, Liouville is often cited as the inventor of the solution by successive approximations of integral equations of the second kind. This is only partially justified. In fact his aim was to solve the differential equation of Sturm–Liouville theory by successive approximations. In so doing he derived an equivalent integral equation with a multiple integral (which Dieudonné *1981*, but not Liouville, pointed out to be equivalent to an equation of the second kind) and solved it by successive approximations.

Finally, in 1845 Liouville published a short paper containing the first general theorems on integral equations. He considered the eigenvalue equation

$$\int_D l(y) T(x, y) \zeta(y) \, dy = m \zeta(x), \tag{8}$$

where D is some vaguely defined n-dimensional manifold, l is a positive function and T is a symmetric kernel. He proved that eigenfunctions ζ_1 and ζ_2 corresponding to two different eigenvalues are 'orthogonal' in the sense that

$$\int_D l(x) \zeta_1(x) \zeta_2(x) \, dx = 0,$$

and he easily deduced that the eigenvalues m are necessarily real. He even stated that the left-hand side of (8), the integral operator, could be replaced by other types of operation as long as they possess a certain symmetry.

These general theorems were only a by-product of an extensive investigation Liouville undertook around 1846 of the spectral theory of an integral operator he encountered in connection with potential theory (Lützen *1990*: 601–35). He was interested in solving the Dirichlet problem (§3.17): given a closed surface S in space and a function $\overline{V}(x)$ on this surface, to find a harmonic function V inside S having $\overline{V}$ as its boundary value. Following Gauss, he sought V in the form of a potential of a single-layer distribution λ on S, and so was led to the integral equation

$$\overline{V}(x) = V(x) = \int_S \frac{\lambda(x')}{|x - x'|} \, dx', \quad x \in S. \tag{9}$$

He saw that he could solve this problem generally if he could determine the 'eigenfunctions' ζ_i on S satisfying

$$\int_S \frac{\zeta_i(x')}{|x - x'|} l(x') \, dx' = m_i \zeta_i(x), \tag{10}$$

where l is the equilibrium density on S. Moreover, he wanted to show that any function on S can be written as a linear combination of the ζ_i. He came a good part of the way to realizing this programme; in particular, he discovered that the ζ_i can be determined by a variational technique which is now often called the Rayleigh–Ritz method. Yet he did not publish these ideas.

4 INTEGRAL EQUATIONS FROM DIFFERENTIAL EQUATIONS

As pointed out by Du Bois-Reymond in the paper quoted at the beginning of this article, a number of integral equations turned up in connection with boundary-value problems of differential equations during the nineteenth century. Liouville's equations (7), (9) and (10) are early examples, and after 1850 many more appeared, particularly in potential theory (§3.17). Many interesting results and techniques were developed, but since they were aimed at a study of the differential equation they can at most be considered implicit results about integral equations. Dieudonné has very aptly called them 'crypto-integral equations'. The most interesting crypto-integral equations arose from the Dirichlet problem. In 1865 A. Beer proposed to solve (9) by successive approximations, but he did not prove the convergence of the process; this was done by Carl Neumann in 1877. Neumann sought to represent the harmonic function V as the potential of a double layer on S, the density of which must satisfy the integral equation of the second kind

$$2\pi\rho(x) + \int_S \rho(x') \frac{\partial}{\partial n} \left(\frac{1}{|x-x'|} \right) dx' = \bar{V}(x). \tag{11}$$

His method of the arithmetic mean can be considered as a solution by successive approximations to this equation and its plane logarithmical analogue. He showed that this process converges when S bounds a convex domain.

The problem of vibrating membranes further led K. H. A. Schwarz (1885) and Henri Poincaré (1895–6) to study eigenvalue problems of integral equations. They both used the idea of a Green's function, a function $G(x, x')$ (x, x' in $\mathbb{R}^3$ or $\mathbb{R}^2$) with given boundary values on a given surface (curve) S, and with the property that a solution V of $\Delta V + f = 0$ having the same boundary values as G, is given by

$$V(x) = \int_S G(x, x') f(x') dx.$$

This may be considered an integral equation for the determination of f if V is known. More to the point, the eigenvalue equations $\Delta V + \lambda V = 0$ (Schwarz) and $\Delta V + \lambda V = f$ (Poincaré) can be translated into the following eigenvalue problems for integral operators:

$$V(x) = \lambda \int_S G(x, x') V(x') dx'$$

and $$\tag{12}$$

$$V(x) = \lambda \int_S G(x, x') V(x') dx' + F(x).$$

By this method Schwarz and Poincaré proved several important theorems about the spectrum, and the eigenfunctions of the differential equations and thus of the integral equations above.

Introduced by George Green in 1828 for the Laplacian in $\mathbb{R}^3$, the idea of a Green's function was generalized around 1880 by Neumann to the Laplacian in $\mathbb{R}^2$, and to ordinary differential operators by Heinrich Burkhardt (1894) and David Hilbert. Hilbert used it to reduce the Sturm–Liouville problem to a problem in spectral theory of integral operators. After Green's functions had been generalized to arbitrary linear differential operators at the beginning of the twentieth century, they provided a general method of translating problems concerning differential operators into problems concerning integral operators.

5 THE FIRST GENERAL THEORY

The study of integral equations with arbitrary kernels was undertaken independently by J. Le Roux and Vito Volterra in 1895 and 1896–7 respectively. They both showed how the method of successive approximations, popularized for differential equations by E. Picard in 1890, could be used to solve integral equations with variable upper limit (i.e. b in equations (1) and (2) replaced by x). This type of integral equation is often named after Volterra, whose research was the most influential and far-reaching of the two.

Motivated by Abel's work, Volterra first discussed equations of the first kind and then turned to equations of the second kind:

$$f(y) = \phi(y) + \int_a^x K(x, y)\phi(x)\,\mathrm{d}x, \tag{13}$$

where f and K are continuous for $x, y \in [a, b]$. He defined the iterated kernels by $K = K_0$ and

$$K_i(x, y) = \int_y^x K_{i-j}(x, \xi)K_{j-1}(\xi, y)\,\mathrm{d}\xi \tag{14}$$

(which is independent of j), and showed that the series $\Sigma_{i=0}^{\infty} K_i(x, y)$ converges uniformly in $[a, b] \times [a, b]$ to a continuous kernel, which he denoted by $\Phi[K(x, y)]$. The solution of (13) can then be found by the inversion formula

$$\phi(y) = f(y) + \int_a^y \Phi[K(x, y)]f(x)\,\mathrm{d}x. \tag{15}$$

(For this reason Hilbert later called $\Phi[K(x, y)]$ the 'resolving kernel'.) As an aside, Volterra remarked that integral equations could be considered as limiting cases of systems of linear equations of the form $b_i = \Sigma_{j=1}^{i} a_{ij}x_i$.

6 THE BEGINNING OF SPECTRAL THEORY: FREDHOLM

Inspired by this analogy, Fredholm undertook in 1900 a study of the general equation

$$f(x) = \phi(x) + \lambda \int_a^b K(x,y)\phi(y)\,\mathrm{d}y \tag{16}$$

in order to simplify the solution of the Dirichlet problem. He imitated an idea of Helge von Koch's on the development of an infinite matrix in a series of 'minors', in order to define a 'determinant' $D(\lambda)$ of the equation (16). According to a theorem by Jacques Hadamard, $D(\lambda)$ represents an entire function in $\mathbb{C}$, and Fredholm showed that it has the same properties as the determinant of a system of linear equations: namely, if $D(\lambda) \neq 0$, then there is one and only one solution of (16), which he expressed in terms of D; if $D(\lambda) = 0$, then there exists a non-zero solution of the homogeneous equation

$$0 = \phi(x) + \lambda \int_a^b K(x,y)\phi(y)\,\mathrm{d}y, \tag{17}$$

that is, λ (or rather $1/\lambda$) is an eigenvalue.

In his second and last paper on integral equations (1903), Fredholm studied the case $D(\lambda) = 0$ in more detail. By considering more general determinants, he could show that if $\psi_1, \psi_2, \ldots, \psi_n$ are the n linearly independent solutions of the transpose of (17) (i.e. with $K(x,y)$ replaced by $K(y,x)$), then (16) has a solution if and only if f is 'orthogonal' to all the ψ_i:

$$\int_a^b f(x)\psi_i(x)\,\mathrm{d}x = 0, \quad i = 1, 2, \ldots, n. \tag{18}$$

This is the so-called Fredholm alternative.

7 HILBERT'S MASTERPIECE

Hilbert became interested in integral equations in 1901, when Erik Holmgren came to Göttingen and gave a talk on Fredholm's ideas. During the period 1904–10 Hilbert published a series of papers, subsequently collected in the classic *Grundzüge einer allgemeinen Theorie der linearen Integralgleichungen* (1912). In the first of these papers, he showed how Fredholm's results could be obtained by a rigorous limit procedure from the system of linear equations

$$f(x_i) = \phi(x_i) + \frac{\lambda(b-a)}{n} \sum_{j=1}^n K(x_i, x_j)\phi(x_j), \quad i = 1, 2, \ldots, n \tag{19}$$

obtained from equation (16) by replacing the integral by a Riemann sum corresponding to a division of the interval $[a, b]$ into n equal parts at the points $x_0 = a, x_1, \ldots, x_n$. This proof was much more clumsy than Fredholm's, but in the case considered by Hilbert of a continuous and symmetric kernel $(K(x, y) = K(y, x))$ it yielded much more precise information on the eigenfunctions. Indeed, Hilbert proved that there are at most countably many eigenvalues $\lambda_1, \lambda_2, \ldots$, counted with multiplicity, and if $\phi_1, \phi_2, \ldots$ are the corresponding normalized eigenfunctions, one has the development:

$$\int_a^b \int_a^b K(x, y) f(x) g(y) \, dx \, dy = \sum_n \frac{1}{\lambda_n} \int_a^b \phi_n(x) f(x) \, dx \int_a^b \phi_n(x) g(x) \, dx, \tag{20}$$

where the series converges uniformly for continuous functions f and g for which $\int_a^b (f(x))^2 \, dx$ and $\int_a^b (g(x))^2 \, dx$ remain below a given limit (i.e. in a ball in $L^2[a, b]$). He considered this to be the equivalent of the principal-axes theorem of quadric forms. Moreover, he showed that the eigenfunctions may be determined by the variational Rayleigh–Ritz method. Finally, he proved that a function of the form

$$f(x) = \int_a^b K(x, y) g(y) \, dy, \quad g \in C[a, b]$$

may be expanded in a Fourier series

$$f(x) = \sum_n a_n \phi_n(x), \quad a_n = \int_a^b f(x) \phi_n(x) \, dx, \tag{21}$$

and this series converges absolutely and uniformly. Hilbert stressed that such developments of arbitrary functions on a set of 'orthogonal' functions was one of the prime reasons for studying integral equations. In fact, Hilbert proved this theorem only under the assumption that any continuous function may be approximated in the sense of mean-square value (in L^2) by functions of the form $\int_a^b K(x, y) g(y) \, dy$. However, in 1905 his student Erhard Schmidt showed that this assumption was superfluous. He was able to prove Hilbert's results without using the cumbersome limit procedure or Fredholm's determinants. Thereby the theory became much more elegant.

After applying his results to differential equations (see Section 4 above) and to function theory, Hilbert gave in his fourth and fifth papers a completely new and very important approach to integral equations. Instead of approximating the integral equation by a finite system of linear equations, he now showed that it is equivalent to the infinite system of equations

$$f_i = x_i + \sum_{j=1}^{\infty} k_{ij} x_j, \quad i = 1, 2, \ldots . \tag{22}$$

To this end, he considered a complete orthonormal system $\Phi_i(x)$ of continuous functions on $[a, b]$, and defined

$$\left.\begin{aligned}
f_i &= \int_a^b f(x)\Phi_i(x)\,\mathrm{d}x, \\[2mm]
x_i &= \int_a^b \phi(x)\Phi_i(x)\,\mathrm{d}x, \\[2mm]
k_{ij} &= \int_a^b K(x, y)\Phi_i(x)\Phi_j(y)\,\mathrm{d}x\,\mathrm{d}y.
\end{aligned}\right\} \tag{23}$$

Then, according to Bessel's identity, $\Sigma_i f_i^2$, $\Sigma_i x_i^2$ and $\Sigma_i \Sigma_j k_{ij}^2$ are all finite, and any solution ϕ of (16) will give a solution x_i of (22); conversely, if these sums are finite, a solution x_i of (22) will give a solution $\phi = \Sigma_i x_i \Phi_i$ of (16).

This made Hilbert study the spectral theory of symmetric infinite quadratic forms, $\Sigma_{i,j} k_{ij} x_i y_j$. Since he assumed only that the form be bounded (in the sense that there is an $M \in \mathbb{R} : |\Sigma_{i,j} k_{ij} x_i y_j| \leqslant M$ for all x_i, y_j with $\Sigma_{i=1}^{\infty} x_i < \infty$ and $\Sigma_{j=1}^{\infty} y_j < \infty$), he was led to the problem of a continuous spectrum, which he ingeniously overcame by using the Stieltjes integral (§3.7). However, for the forms he called *vollstetige* ('compact', in modern terminology), which include those for which $\Sigma_{ij} k_{ij}^2 < \infty$, he showed that the spectrum (Hilbert's term) consists entirely of eigenvalues.

8 A GLANCE AHEAD

Hilbert's work on integral equations was instrumental in the rise of functional analysis. It implicitly contained many general concepts, such as the Hilbert space (both ℓ^2 and L^2), the compact operator and orthogonality. Hilbert himself did very little to conceptualize these ideas, but in the work of his successors, such as Schmidt, Fischer and Riesz, the 'geometric' function-space approach gradually developed. In this process, much of the classical theory of integral equations was generalized to, and at the same time swallowed by, a conceptually simpler theory of bounded operators on Hilbert or Banach spaces (Bernkopf *1966*, Dieudonné *1981*). Singular and non-linear integral equations and integro-differential equations, however, could not be directly subsumed under the general picture, and therefore, after 1920, continued as a separate and active field of research.

BIBLIOGRAPHY

Bateman, H. *1911*, 'Report on the history and present state of the theory of integral equations', *Report of the British Association for the Advancement of Science*, (1910), 345–424.

Bernkopf, M. *1966*, 'The development of function spaces with particular reference to their origins in integral equation theory', *Archive for History of Exact Sciences*, **3**, 1–96.

Bôcher, M. *1909*, *An Introduction to the Study of Integral Equations*, Cambridge: Cambridge University Press.

Dieudonné, J. *1981*, *History of Functional Analysis*, Amsterdam: North-Holland.

Hahn, H. *1911*, 'Bericht über die Theorie der linearen Integralgleichungen', *Jahresbericht der Deutschen Mathematiker-Vereinigung*, **20**, 69–117.

Hellinger, E. and Toeplitz, O. *1927*, 'Integralgleichungen und Gleichungen mit unendlichvielen Unbekannten', in *Encyclopädie der mathematischen Wissenschaften*, Vol. 2, Part C, 1335–597 (article II C 13).

Kline, M. *1972*, *Mathematical Thought from Ancient to Modern Times*, New York: Oxford University Press. [In particular, Chap. 45.]

Lützen, J. *1990*, *Joseph Liouville 1809–1822: Master of Pure and Applied Mathematics*, New York: Springer.

Pincherle, S. *1905*, 'Funktionaloperationen und -Gleichungen', in *Encyclopädie der mathematischen Wissenschaften*, Vol. 2, Part A, 761–817 (article II A 11). [In particular, pp. 803–17.]

3.11

Harmonic analysis

KENNETH I. GROSS

As eloquently expressed by Jean Dieudonné at the outset of his beautiful lecture 'Histoire de l'analyse harmonique' (*1971*), harmonic analysis has a vast and rich history which has inspired a great deal of modern mathematics:

> From the very beginning, the theory that we today call *harmonic analysis* has been a most active catalyst in the development of mathematics, and has constituted an extraordinary crossroads at which almost all the branches of our subject have come together to their mutual enrichment. In the eighteenth century, it was the controversies surrounding trigonometric series that led finally to our modern concept of *function*. In the nineteenth century, the theory of Fourier series was the point of contact at which the theory of functions of a complex variable and the theory of functions of a real variable were cross-fertilized. The problems of convergence that it raised greatly contributed to the birth of topology of the line with Cantor, and to the progressive refinement of the concepts of integral and of measure, which were to culminate in the constructs of Stieltjes, Lebesgue, Denjoy and Schwartz. Fourier theory should also be considered as a source of modern functional analysis, in that it suggested the study of numerous non-trivial *operators*, and introduced (for example through the idea of *orthogonality*) the *geometric point of view* in analysis. Finally, from the beginning of the twentieth century, harmonic analysis has become the location of choice at which the fusion of analysis and algebra, so characteristic of our era, is accomplished.

The aim in this article is to convey the historical flow of these ideas and to display the way in which the modern point of view has evolved. In conformity with this approach, some technical terms have been rendered in the modern way; for example, 'noncommutative' and 'abelian'.

1 FOURIER SERIES

The fundamental ideas of harmonic analysis originate in differential equations, in particular with solutions of the classical partial differential equations of mathematical physics. Illustrative is the one-dimensional wave equation

$$\frac{\partial^2 u}{\partial x^2} = \frac{1}{a^2} \frac{\partial^2 u}{\partial t^2} \tag{1}$$

derived by Jean d'Alembert in 1747 as the law governing the motion of a vibrating string. In this equation, $u = u(x, t)$ represents the displacement from equilibrium at time t of the point x on the string, and the constant a^2 is determined from the physical characteristics of tension T and mass density ρ, according to the formula $a^2 = T/\rho$. The general solution of (1), again due to d'Alembert, is given by the expression

$$u(x, t) = f_1(x + at) + f_2(x - at), \tag{2}$$

where f_1 and f_2 are arbitrary, twice-differentiable functions of a real variable. This solution can be viewed as the superposition of two 'waves', f_1 and f_2, propagating with speed a in opposite directions along the string (§9.8).

If one fixes a positive constant L and imposes upon a solution to (1) the boundary conditions $u(0, t) = u(L, t) = 0$ for all t, as well as the initial conditions $u(x, 0) = \phi(x)$ and $\partial u(x, 0)/\partial t = \psi(x)$ for all x (with ϕ and ψ fixed), the solution (2) then takes the form

$$u(x, t) = f(at + x) - f(at - x), \tag{3}$$

where the function f is periodic with period $2L$ and is uniquely determined, up to an arbitrary constant, as

$$f(x) = \frac{1}{2}\left(\phi(x) + \frac{1}{a} \int_0^x \psi(s)\, ds\right). \tag{4}$$

Of course, the initial data ϕ and ψ should be periodic, with period $2L$. Physically, (3) and (4) describe the motion of a string of length L that is clamped at the end-points, and which at time $t = 0$ is given initial displacement $\phi(x)$ and initial velocity $\psi(x)$.

The notion of harmonic analysis originates with the possibility of superimposing certain of the solutions (3); namely, the nth fundamental modes, or nth harmonics,

$$\sin(n\pi x/L)\cos(n\pi at/L) \quad \text{and} \quad \sin(n\pi x/L)\sin(n\pi at/L),$$

which correspond in (4) to $f(x)$ of the forms $\sin(n\pi x/L)$ and $\cos(n\pi x/L)$, respectively. As expressed most colourfully by Daniel Bernoulli *1753*, 'all sonorous bodies contain potentially an infinity of corresponding ways of making their regular vibrations'. If we rephrase this more mathematically, and for convenience take $L = \pi$, then $f(x)$ should be of the form

$$f(x) = \tfrac{1}{2}a_0 + \sum_{n=1}^{\infty} [a_n \cos(nx) + b_n \sin(nx)]. \tag{5}$$

The series on the right-hand side of (5) is called the Fourier series of f, and the study of such series representations and related generalizations is referred to as Fourier analysis. Although trigonometric series had been considered earlier, by Leonhard Euler and Daniel Bernoulli for example, the attribution to Joseph Fourier is well deserved. Fourier's 1807 investigation of heat flow (summarized in Fourier *1808*) was the first broad examination of the series (5), including the determination of the coefficients in terms of f in the form

$$a_n = \frac{1}{\pi} \int_{-\pi}^{\pi} f(x) \cos(nx) \, \mathrm{d}x \quad \text{and} \quad b_n = \frac{1}{\pi} \int_{-\pi}^{\pi} f(x) \sin(nx) \, \mathrm{d}x \tag{6}$$

(Grattan-Guinness and Ravetz *1972*; see also §9.4). In the words of Dieudonné, Fourier's work 'put the final period to the polemics of the preceding century' and ushered in the beginning of modern real variables.

2 CONVERGENCE OF FOURIER SERIES

From a purely mathematical point of view, as opposed to that of physics, Fourier's work goes far beyond that of Euler and d'Alembert in that, by determining the coefficients in terms of integrals of f, he singled out a particular trigonometric series, the Fourier series, to represent a given function. In retrospect, however, in terms of modern mathematical rigour we can say that Fourier simultaneously assumed both too little and too much. On the one hand, he assumed that the series could be integrated term by term, for which we now know that ordinary pointwise convergence is not sufficient. On the other hand, as we also now know, it is not necessary for the series to converge at every point in order for it to represent the function, since altering a function arbitrarily at a finite number of points will leave the integral and, *a fortiori*, the Fourier series, unchanged. In general, the problem of pointwise convergence of Fourier series is profoundly difficult, and the quest for sufficient conditions for convergence of, as well as necessary conditions for representability by, Fourier series has been a touchstone for two centuries of mathematical innovation.

Of great importance is the seminal paper by J. P. G. Lejeune Dirichlet *1829*, which Dieudonné considers to mark the birth of modern harmonic analysis. There, by methods that satisfy even modern standards of rigour and elegance, Dirichlet settled the problem of pointwise convergence of Fourier series for a large class of functions. Through what we now call the method of Dirichlet summability, he proved that the Fourier series of a function having only a finite number of maxima and minima (in modern terminology, a piecewise monotonic function) converges to the average of its right-hand and left-hand values at each point. It turned out to be extraordinarily difficult to improve on this result, and better results than Dirichlet's were not found for several decades. In fact, it is remarkable that a proof of the almost-everywhere convergence of the Fourier series of a continuous function had to wait until L. Carleson (*1966*) introduced fundamental new and subtle estimates that yielded the convergence.

When the importance of uniform convergence for term-by-term integration of series came to be appreciated in the 1870s and 1880s (§3.3), the impact on the study of pointwise convergence of Fourier series was indeed profound. Because the limiting functions were not necessarily continuous, it became clear that Fourier series could fail to converge uniformly. In turn, since at that time uniform convergence was the only known sufficient condition for term-by-term integration, this fact cast doubt on the validity of the formula for the Fourier coefficients, which seemed to depend on term-by-term integration.

In response to these problems, A. Harnack *1880* introduced a new perspective on convergence. He adopted the point of view that a function $f(x)$ on an interval $[a, b]$ is represented by a series whose sum is $\phi(x)$ if

$$\int_a^b (f(x) - \phi(x))^2 \, \mathrm{d}x = 0. \tag{7}$$

At the same time, Harnack gave the following definition: 'A set of points is *of first kind* if its points can be enclosed in intervals of finite length whose total length can be made arbitrarily small.' Thus, in pointing the way towards the concept of mean-square convergence, Harnack formulated for the first time the definition of a set of measure zero. It should be remarked, however, that, without making a formal definition, Bernhard Riemann *1854* had already used this idea implicitly in his investigation of the uniqueness of Fourier-series expansions.

In particular, Harnack stated the following beautiful theorem, which is entirely correct but which he could not possibly have proved with the tools available at that time:

If $f(x)$ denotes an arbitrary function whose square is integrable [in the sense of Riemann, of course] on the interval from $-\pi$ to $-\pi$, its Fourier series [given by (5) and (6)] represents the value of the function in such a way that the value $\phi(x)$ of the sum of this Fourier series differs from *the value of f(x) by a given amount δ only at the points of a set of first kind. Moreover,*

$$\frac{1}{\pi} \int_{-\pi}^{\pi} [f(x)]^2 \, dx = \frac{1}{2} a_0^2 + \sum_{k=1}^{\infty} (a_k^2 + b_k^2). \tag{8}$$

Since at this early date the set-theoretic and metric structure of the real line was not well understood, Harnack's proof contained, as we can now see, misconceptions and mistakes. Yet the insights provided by this work moved harmonic analysis in directions that remain productive to this day. What was needed was the concept of mean-square convergence, the full power of which became available two decades later with Henri Lebesgue's thesis of 1902. Moreover, it was necessary to decouple the notion of representation by a series from that of pointwise convergence of the series. These notions were clarified a quarter-century later when David Hilbert *1912*, Frédéric Riesz *1907*, Ernst Fischer *1907* and others studied ortho-normal sequences of functions and gave the necessary and sufficient conditions for a sequence of constants to be the Fourier coefficients of a square-integrable function (see §3.9 on functional analysis).

In ways that could not even be imagined at the time, Harnack's original idea embodied in the relation (8) between what we would today call the L^2-norm of a function and the ℓ^2-norm of its Fourier coefficients has evolved in the latter half of the present century into a profound area of research – which lies at the interface of algebra and number theory, modern analysis, geometry and theoretical aspects of modern physics – called infinite-dimensional group representations. At the heart of this subject is the concept of a Plancherel theorem, of which (8) is a motivating example.

3 THE PLANCHEREL THEOREM FOR THE CIRCLE

A modern treatment of Fourier series would recast (5) in complex notation as

$$f(\theta) = \sum_{n=-\infty}^{\infty} c_n e^{in\theta}, \tag{9}$$

where

$$c_n = \hat{f}(n) = \frac{1}{2\pi} \int_{-\pi}^{\pi} f(\theta) e^{-in\theta} \, d\theta. \tag{10}$$

Here, $-\pi \leqslant \theta < \pi$ parametrizes the unit circle T of complex numbers $e^{in\theta}$

of absolute value 1. The right side of (9) is called the Fourier series of f, and $\{\hat{f}(n)\}$ is the sequence of Fourier coefficients of f.

There are many techniques (or, in current terminology, 'methods of summability') for obtaining convergence in (9). The most common and elementary procedure involves the introduction of the inner product

$$\langle f_1 \mid f_2 \rangle = \frac{1}{2\pi} \int_{-\pi}^{\pi} f_1(\theta) \overline{f_2(\theta)} \, d\theta, \tag{11}$$

and applies to the Hilbert space $L^2(T)$ of functions f on T which are square-integrable relative to Lebesgue measure; that is, functions f for which the norm

$$\| f \| = \langle f \mid f \rangle^{1/2} = \left\{ \frac{1}{2\pi} \int_{-\pi}^{\pi} | f(\theta) |^2 \, d\theta \right\}^{1/2} \tag{12}$$

is finite. Thus, the functions e_n on T given by $e_n(\theta) = e^{in\theta}$ form a complete orthonormal system; (9) is just the expansion of f in terms of this orthonormal basis (i.e. $\hat{f}(n) = \langle f \mid e_n \rangle$ for all n); and equality in (9) is interpreted in the mean-square sense as

$$\lim_{N \to \infty} \left\| f - \sum_{n=-N}^{N} c_n e_n \right\| = 0. \tag{13}$$

In other words, the Hilbert space $L^2(T)$ is the orthogonal direct sum

$$L^2(T) = \sum_{n=0}^{\infty} \oplus C e_n \tag{14}$$

of the one-dimensional subspaces spanned by the functions $e_n(\theta)$. Equivalently,

$$\| f \|^2 = \sum_{n=-\infty}^{\infty} | \hat{f}(n) |^2. \tag{15}$$

Equation (15) is precisely Harnack's identity (8), rewritten in the spirit of (9), but it is usually referred to as Parseval's identity or, more commonly, as the Plancherel formula for the circle. Historically, Marc Antoine Parseval *1799* wrote down a primitive version of this formula for the circle, and M. Plancherel *1910* formulated the analogous result, (18) below, for the Fourier transform on the real line.

The previous paragraph describes what is known as the L^2-theory of Fourier series, in which (13) expresses the convergence of the Fourier series of f in L^2-norm. If we let ℓ^2 denote the Hilbert space of all square-integrable sequences $\{\alpha_n : -\infty < n < \infty\}$ of complex numbers, then formulas (9) and (10) can be formalized as follows:

PLANCHEREL THEOREM FOR THE CIRCLE. *The mapping $\mathscr{F}: f \mapsto \hat{f}$ from a function f to its sequence $\hat{f}(n)$ of Fourier coefficients is a unitary operator, called the Plancherel transform on the circle, from the Hilbert space $L^2(T)$ to the Hilbert space ℓ^2.*

As noted above, the extension of the concept of a Plancherel transform to contexts more general than the circle (or real line) is a central and dominant theme of modern harmonic analysis.

4 HARMONIC ANALYSIS ON THE REAL LINE: THE FOURIER INTEGRAL

Historically, harmonic analysis on the real line originates, as was the case for the circle, in the partial differential equations of mathematical physics; in particular, with the heat equation for an object, such as a metal rod, of infinite extent. Once more the crucial idea is due to Fourier *1822*, who replaced the series representation of a solution by an integral representation, and thereby initiated the study of Fourier integrals.

Formally, the Fourier integral, or Fourier transform, $\hat{f}$ of a function f on the real line is defined for all real numbers λ by

$$\hat{f}(\lambda) = \int_{-\infty}^{\infty} f(x) e^{-i\lambda x} \, dx, \tag{16}$$

and f is obtained from $\hat{f}$ by the Fourier inversion formula

$$f(x) = (2\pi)^{-1} \int_{-\infty}^{\infty} \hat{f}(\lambda) e^{i\lambda x} \, d\lambda. \tag{17}$$

Clearly, these formulas bear a strong similarity to formulas (10) and (9), respectively, for Fourier coefficients and Fourier series of a function on the circle.

The L^1-theory of the Fourier integral treats the situation in which f is integrable over the real line, in which case $\hat{f}$ is well defined by the absolutely convergent integral (16). None the less, the integrability of f does not imply the integrability of $\hat{f}$ (even in the Lebesgue sense: see §3.7), and generalized 'methods of summability' (§3.3) are required to give rigorous meaning to the integral in (17).

More to our purposes is the L^2-theory. For if f is both integrable and square-integrable (i.e. if $f \in L^1(\mathbb{R}) \cap L^2(\mathbb{R})$), then $\hat{f}$ is square-integrable. Moreover, in complete analogy to the case of Fourier series, equality in

(17) holds in the mean-square sense, and the Plancherel formula

$$\int_{-\infty}^{\infty} |f(x)|^2 \, dx = (2\pi)^{-1} \int_{-\infty}^{\infty} |\hat{f}(\lambda)|^2 \, d\lambda \tag{18}$$

is valid. The L^2-theory is summarized as follows:

PLANCHEREL THEOREM FOR THE REAL LINE. *The mapping $f \mapsto \hat{f}$, which is originally defined on $L^1(\mathbb{R}) \cap L^2(\mathbb{R})$, extends uniquely to a unitary operator $\mathcal{F}$, called the Plancherel transform, from the Hilbert space $L^2(\mathbb{R})$ to itself.*

As one might expect from the analogy with Fourier series, the Fourier integral is closely connected with the study of the operator d^2/dx^2, which we denote by Δ since it is the Laplacian on $\mathbb{R}$. Indeed, the operator $-\Delta = -d^2/dx^2$ is non-negative, and for each real number λ the function $e_\lambda(x) = e^{i\lambda x}$ is an eigenfunction with eigenvalue λ^2, (i.e. $\Delta e_\lambda = -\lambda^2 e_\lambda$), but these eigenfunctions are not square-integrable. That is, none of the eigenfunctions e_λ lie in $L^2(\mathbb{R})$. In short, the non-compactness of $\mathbb{R}$ is manifested in Δ having a continuous spectrum (the entire real line), and − in contrast to the discrete splitting (14) that occurs for the circle − $L^2(\mathbb{R})$ decomposes as a kind of continuous smearing

$$L^2(\mathbb{R}) = \int_{\mathbb{R}}^{\oplus} \mathbb{C} e_\lambda \, d\lambda \tag{19}$$

of these 'infinitesimal' eigenspaces, called a direct integral. In general, such continuous decompositions, which do not occur in the compact setting, add a challenging level of technical difficulty to harmonic analysis on non-compact domains.

5 FOURIER SERIES REVISITED

As motivation for the higher-dimensional considerations in the next section, we reinterpret the Fourier series (9) in relation to the Laplacian $\nabla^2 = \partial^2/\partial x_1^2 + \partial^2/\partial x_2^2$ on $\mathbb{R}^2$. To remind ourselves that the circle plays the role of the one-dimensional 'sphere' in two-dimensional Euclidean space $\mathbb{R}^2$, we adopt the notation $S^1 = \{x = (x_1, x_2) \in \mathbb{R}^2 : x_1^2 + x_2^2 = 1\}$ for the circle. We then write the Laplacian in polar coordinates as

$$\nabla^2 = \frac{\partial^2}{\partial r^2} + \frac{1}{r}\frac{\partial}{\partial r} + \frac{1}{r^2}\frac{\partial^2}{\partial \theta^2}. \tag{20}$$

If we set $\Delta = d^2/d\theta^2$, which is the Laplacian on the circle, then (20) can be interpreted as the statement that the Laplacian Δ on the circle is the circular part of the Laplacian ∇^2 on $\mathbb{R}^2$. From elementary calculus, the null space of Δ consists of the constant functions, and

$$\Delta(e^{in\theta}) = -n^2 e^{in\theta}, \tag{21}$$

with $n = \pm 1, \pm 2, \pm 3, \ldots$. Thus $-\Delta$ is a non-negative differential operator with eigenvalues n^2, and the Fourier series (9) can be viewed as the expansion of a function f in eigenfunctions of the Laplacian on the circle. From this perspective, the method of Abel summability of Fourier series makes the connection with the Laplacian ∇^2 on the ambient Euclidean plane, and with harmonic functions in two variables.

A function F of two real variables x_1 and x_2 is said to be 'harmonic' if $\nabla^2 F = 0$. For integers $n > 0$, the polynomials

$$(x_1 + ix_2)^n = r^n e^{in\theta} \quad \text{and} \quad (x_1 - ix_2)^n = r^n e^{-in\theta} \tag{22}$$

are harmonic, and they span the two-dimensional space of all harmonic polynomials on $\mathbb{R}^2$ homogeneous of degree n. Fourier series considerations imply that a harmonic function F on $\mathbb{R}^2$ can be expanded in an infinite series

$$F(x_1, x_2) = c_0 + \sum_{n=1}^{\infty} c_n(x_1 + ix_2)^n + c_{-n}(x_1 - ix_2)^n \tag{23}$$

of homogeneous harmonic polynomials. In fact, if we restrict the function F to the unit circle (which corresponds to $x_1^2 + x_2^2 = 1$, or more simply to $r = 1$), then the series (23) reduces to the Fourier series (9). Specifically, we substitute (22) in (23) to obtain

$$F(r, \theta) = \sum_{n=-\infty}^{\infty} c_n r^{|n|} e^{in\theta}, \tag{24}$$

which is precisely Niels Abel's method for summing the Fourier series (9). That is,

$$f(\theta) = \lim_{r \to 1} F(r, \theta). \tag{25}$$

To capture the flavour of the above theory, one might think of the functions $e^{in\theta}$ and $e^{-in\theta}$ on the circle as 'circular harmonics' of degree n, and their extensions defined by (22) as 'solid' circular harmonics on two-dimensional space. These are the special functions that arise from harmonic analysis on $\mathbb{R}^2$ in polar coordinates. We now move on to three-dimensional space.

6 SPHERICAL HARMONICS: AN INTRODUCTION TO THE HARMONIC ANALYSIS OF SPECIAL FUNCTIONS

Harmonic analysis provides an orderly overview of many aspects of the theory of special functions. Illustrative is the study of the harmonics on the

2-sphere in 3-space or, in a phrase apparently first introduced by James Clerk Maxwell in his work on electromagnetism, spherical harmonics.

Spherical harmonics arise historically in the problem – first discussed by Isaac Newton in 1687, and thereafter a dominant theme in astronomy, geometry and analysis during the eighteenth century – of determining the equilibrium shapes of a rotating fluid (e.g. a celestial body). In particular, Adrien Marie Legendre, in a paper 'Recherches sur la figure des planètes' on this subject (*1782*) read to the French Academy of Sciences, expanded in polar coordinates the Newtonian gravitational potential for a point mass (§3.17), thereby introducing into mathematics what are now called the Legendre polynomials (§4.4). Legendre's ideas were immediately picked up by Pierre Simon Laplace. In a seminal paper entitled 'Théorie des attractions de sphéroïdes et de la figures des planètes', Laplace *1782* represented the Newtonian potential in spherical coordinates as a solution of the partial differential equation that now bears his name, solved the equation by the method of separation of variables, and obtained the solution as an infinite series of spherical harmonics that, in his honour, we now call a Laplace expansion.

We outline Laplace's contributions in modern terms. The three-dimensional Laplacian $\nabla^2 = \partial^2/\partial x_1^2 + \partial^2/\partial x_2^2 + \partial^2/\partial x_3^2$ in spherical coordinates is

$$\nabla^2 = \frac{\partial^2}{\partial r^2} + \frac{2}{r}\frac{\partial}{\partial r} + \frac{1}{r^2}\left[\frac{1}{\sin\phi}\frac{\partial}{\partial\phi}\left(\sin\phi\frac{\partial}{\partial\phi}\right) + \frac{1}{\sin^2\phi}\frac{\partial^2}{\partial\theta^2}\right], \qquad (26)$$

where r is the radius and ϕ and θ are latitudinal and longitudinal coordinates, respectively, on the 2-sphere S^2. In particular, the spherical part of the Laplacian – that is, the Laplacian on the sphere – is

$$\Delta = \frac{1}{\sin\phi}\frac{\partial}{\partial\phi}\left(\sin\phi\frac{\partial}{\partial\phi}\right) + \frac{1}{\sin^2\phi}\frac{\partial^2}{\partial\theta^2}. \qquad (27)$$

In analogy to the case of the circle, $-\Delta$ is a non-negative differential operator on the sphere. Its eigenvalues are the numbers $\lambda_n = -n(n+1)$; the eigenspace H_n corresponding to λ_n has dimension $2n+1$; and the functions f_m^n defined for $-n \leqslant m \leqslant n$ by

$$f_m^n(\theta, \phi) = e^{im\theta}(\sin\phi)^m \frac{d^{|m|}P_n}{d\phi^{|m|}}\cos\phi \qquad (28)$$

are an orthogonal basis for H_n (relative to the Lebesgue integral on the sphere). Here, P_n is the nth Legendre polynomial. Thus, in analogy to the Fourier series (9), each square-integrable function f on the sphere has a series expansion of the form

$$f(\theta, \phi) = \sum_{n=0}^{\infty} \sum_{m=-n}^{n} c_{n,m} f_{n,m}(\theta, \phi), \tag{29}$$

called the Laplace series of f. The mean-square convergence of (29) may be summarized by the statement that the Hilbert space $L^2(S^2)$ of square-integrable functions on the sphere decomposes as the orthogonal direct sum

$$L^2(S^2) = \sum_{n=0}^{\infty} \oplus H_n \tag{30}$$

of the eigenspaces H_n of Δ.

The functions in H_n are called the (surface) spherical harmonics of degree n. They are, of course, functions on the sphere. However, each function $f \in H_n$ extends uniquely to a polynomial F on $\mathbb{R}^3$, called a solid spherical harmonic, which is harmonic (i.e. it satisfies Laplace's equation $\nabla^2 F = 0$ on $\mathbb{R}^3$) and homogeneous of degree n (i.e. $F(tx) = t^n F(x)$ for all x in $\mathbb{R}^3$ and positive real numbers t). In particular, in analogy to Abel summability (24) in $\mathbb{R}^2$, any continuous solution F to Laplace's equation in three variables may be represented by a Laplace series

$$F(r, \theta, \phi) = \sum_{n=0}^{\infty} r^n \left\{ \sum_{m=-n}^{n} c_{n,m} f_{n,m}(\theta, \phi) \right\} \tag{31}$$

in solid spherical harmonics. In short, spherical harmonics are the special functions associated with harmonic analysis on $\mathbb{R}^3$ in spherical coordinates.

The remainder of this article is devoted to twentieth-century developments in harmonic analysis. Were we to treat the preceding mathematics from a contemporary point of view, we would represent the sphere S^2 group-theoretically as the homogeneous space $SO(3)/SO(2)$ of the three-dimensional rotation group $SO(3)$ modulo the subgroup $SO(2)$ of rotations in the equatorial plane. In this philosophical context, spherical harmonics may be characterized group-theoretically as the special functions associated with the action of the (noncommutative) group $SO(3)$ on S^2, or alternatively on the ambient Euclidean space $\mathbb{R}^3$. In this way, for example, Legendre polynomials have a group-theoretic definition as the spherical harmonics that remain invariant under rotations in the sub-group $SO(2)$. (Equivalently, as we see from (28), Legendre polynomials are the spherical harmonics which are constant on degrees of latitude.)

Other groups and group actions give rise to most of the familiar classes of special functions (§4.4). For example, Gegenbauer polynomials correspond to the action of the n-dimensional rotation group $SO(n)$ on the

$(n-1)$-dimensional sphere in $\mathbb{R}^n$; Bessel functions are the special functions associated with the action of the Euclidean motion group on the plane; Hermite polynomials are related to the group that arises from the Heisenberg uncertainty principle in physics; and similar characterizations exist for Jacobi polynomials, Legendre functions, the confluent and Gaussian hypergeometric functions, and so on. That all these classes of special functions come under the common rubric of noncommutative harmonic analysis is a modern interpretation of ingenious classical research originally motivated by applications in differential equations and mathematical physics. Classically, as illustrated in the example of spherical harmonics above, group theory is implicit in the symmetries of the underlying geometry.

7 GROUP THEORY AND HARMONIC ANALYSIS

Modern harmonic analysis begins in the 1920s with the confluence of two great streams of mathematical thought which had been developing concurrently but independently.

First there is the role of group theory, rooted in large part in geometry through the philosophy (Felix Klein's *Erlanger Programm* (§6.4), announced in 1872) of studying a space through its group of motions. In particular, the theory of geometric transformation groups, known today as Lie groups in honour of the Norwegian mathematician Sophus Lie (§6.5), who created the subject in the latter decades of the nineteenth century, arose from certain kinds of partial differential equation, in rough analogy to the origin of Galois groups from algebraic equations.

On the other hand, a variety of purely analytic theories were developing in the late nineteenth century from very different aspects, such as boundary-value problems, of differential equations. For example, if in the vibrating-string problem described in Section 1 it is not assumed that the tension and density are uniform across the string, then the wave equation (1) is replaced by the equation

$$\frac{\partial}{\partial x}\left(T(x)\frac{\partial u}{\partial x}\right) = \rho(x)\frac{\partial^2 u}{\partial t^2}. \tag{32}$$

The principle of superposition of solutions applied to more general equations such as (32) leads to eigenfunction expansions more general than Fourier series (of which (29) is an example), to Sturm–Liouville theory, to spectral theory on Hilbert space, and ultimately to modern abstract functional analysis (§3.9).

The genius for bringing together these two seemingly unrelated themes belongs to Hermann Weyl, who should be regarded as the father of modern harmonic analysis. The date of birth is 1927, and the official birth certificate is the remarkable paper *1927* by Weyl and his student F. Peter, in which the structure theory for the representations of a finite group is carried over completely to the context of compact Lie groups (Peter and Weyl *1927*). The decisive analytic idea, originating with Peter and Weyl and fundamental to essentially all the succeeding developments in noncommutative harmonic analysis, is the use of infinite-dimensional representations and their decomposition by means of spectral theory for bounded operators on Hilbert space.

The ancestry of modern harmonic analysis can be traced back to 1896 and Georg Frobenius, who introduced the theory of representations into the study of finite groups (§6.4). Frobenius's student I. Schur refined and simplified the constructs of representation theory, and in 1924 carried over many of the concepts of representation theory to the continuous group $SO(n)$ of rotations on n-space, including the celebrated orthogonality relations that today bear his name. Finally, Schur's student Weyl merged representation theory and operator theory on Hilbert space to found modern harmonic analysis.

In introducing the contemporary perspective in harmonic analysis, we shall be brief and paint the group-theoretic themes in broad strokes. For a more detailed presentation in a historical perspective, see the lovely scholarly essays by G. Mackey (*1963, 1978*).

8 COMMUTATIVE HARMONIC ANALYSIS

The circle T and the real line ℝ are both examples of groups, the former being compact and the latter non-compact, but both being abelian. The constructs of Fourier analysis (Fourier series in the case of the circle and the Fourier transform in the case of the real line) extend to all locally compact abelian topological groups, in which context we speak of (abstract) commutative harmonic analysis.

Historically, the group-theoretic underpinnings of Fourier analysis on the circle and real line were slowly distilled from the classical theory, and commutative harmonic analysis was developed well after the non-commutative theory was under way. In fact, the discovery by A. Haar *1933* of invariant measure — which makes available the analogue of the Lebesgue integral for locally compact topological groups — paved the way for L. Pontrjagin *1934* and André Weil *1940*, among others, to establish commutative harmonic analysis from the abstract point of view as the meeting-ground of classical Fourier analysis and the duality theory borrowed from the subject of finite abelian groups. The idea of fundamental importance, that of a group character, was first defined for finite abelian groups in general by Heinrich

Weber *1882*. (However, in number-theoretic investigations that took place much earlier, Carl Friedrich Gauss used the term 'character' in contexts that we would recognize today as having the nature of harmonic analysis.) The notion of character carries over, *mutatis mutandis*, to the continuous setting, and the structure theory for the so-called group algebra of a finite abelian group becomes a special case of the Plancherel theorem for an abelian, locally compact topological group.

A topological group is a group G which is given a topology such that the group multiplication $(x, y) \mapsto xy$ and inversion $x \mapsto x^{-1}$ are continuous mappings. Clearly, each translation (e.g. right translation, $x \mapsto xa$) is a homomorphism of G, so the topology of G is completely determined by the local behaviour at the identity e. Therefore, we call a group locally compact if there exists some compact neighbourhood of e. The importance of local compactness is that it implies — in fact is essentially necessary and sufficient for — the existence of a Haar integral on G; that is, an integral such that

$$\int_G f(xa)\,dx = \int_G f(x)\,dx, \quad \text{for all } a \in G. \tag{33}$$

(Equation (33) expresses invariance under right translation, but if the group G is abelian there is no distinction between left and right translation.)

Let G be an abelian locally compact topological group. A character λ of G is a continuous homomorphism of G into the circle group T; that is, a complex-valued function λ on G such that $|\lambda(x)| = 1$ for all $x \in G$, and

$$\lambda(xy) = \lambda(x)\lambda(y), \quad \text{for all } x, y \in G. \tag{34}$$

We define the dual group $\hat{G}$ as the collection of all characters of G. $\hat{G}$ itself is a locally compact abelian group relative to pointwise multiplication as the law of composition and the compact–open topology. The classical theory of Fourier analysis, exclusive of the connection with differentiation, has the following extension to this general context. For a function f on G, integrable with respect to Haar measure, we define its Fourier transform

$$\hat{f}(\lambda) = \int_G f(x)\overline{\lambda(x)}\,dx, \quad \text{for all } \lambda \in \hat{G}. \tag{35}$$

Whenever $\hat{f}$ is also integrable, the inversion formula

$$f(x) = \int_{\hat{G}} \hat{f}(\lambda)\lambda(x)\,d\lambda \tag{36}$$

recaptures f from $\hat{f}$, where dx and $d\lambda$ are suitably normalized Haar integrals on G and $\hat{G}$, respectively. The resulting Plancherel theorem takes the following form:

PLANCHEREL THEOREM FOR AN ABELIAN LOCALLY COMPACT GROUP. *The mapping $f \mapsto \hat{f}$, originally defined on $L^1(G) \cap L^2(G)$, extends uniquely to a unitary operator $\mathscr{F}$, called the Plancherel transform, from the Hilbert space $L^2(G)$ to the Hilbert space $L^2(\hat{G})$.*

Of course, when $G = T$ the mapping $e_n \mapsto n$ identifies $\hat{G}$ with the group $\mathbb{Z}$ of integers, and the theory described here coincides with that for the Fourier series on the circle. When $G = \mathbb{R}$, the mapping $e_\lambda \mapsto \lambda$ identifies $\hat{G}$ with $\mathbb{R}$ (so $\mathbb{R}$ is self-dual); and the above theory coincides with that for the Fourier integral on $\mathbb{R}$.

As another example, let G be the multiplicative group $\mathbb{R}^+$ of positive real numbers u. Then the characters of G are of the form $\lambda_t(u) = u^{it}$ for $t \in \mathbb{R}$, and the mapping $\lambda_t \mapsto t$ identifies the dual group $\hat{G}$ with the real line. In this example, the abstract Fourier transform (35) is the integral transform

$$\hat{f}(t) = \int_0^\infty f(u) u^{-it} \, du \tag{37}$$

that arises in number theory and classical analysis, and is known in the classical literature as the Mellin transform. Roughly speaking, the Fourier transform on the real line is a fundamental tool in analysing operations that commute with addition, and the Mellin transform plays the same role for multiplication.

9 HARMONIC ANALYSIS ON A FINITE GROUP

To undertake harmonic analysis on nonabelian groups, one needs to generalize the concept of a character. That characters are not sufficient is seen from the defining relation (34), for $\lambda(xy) = \lambda(yx)$ as long as $\lambda(x)$ and $\lambda(y)$ commute. Hence, if one is limited to 'characters' which are complex-valued functions, then one can never distinguish the harmonic analysis at the element xy from the harmonic analysis at yx. To remedy this situation for finite nonabelian groups, in 1896 Frobenius introduced his concept of 'higher-dimensional character', and shortly thereafter the equivalent notion of 'group representation'. Within a decade, principally through the work of Frobenius, Schur and William Burnside, the major structure theory was in place and representation theory had already established itself as the most powerful technique for penetrating the structure of a finite group.

Briefly, a representation τ of a finite group G is a mapping which associates with each element x of G a linear transformation $\tau(x)$ on a non-zero, finite-dimensional complex vector space V_τ, having the following two properties:

$$\tau(xy) = \tau(x)\tau(y), \quad \text{for all } x, y \in G, \quad \text{and} \quad \tau(e) = I, \tag{38}$$

where e is the identity in G, and I the identity transformation on V_τ. Since $\tau(xx^{-1}) = \tau(x)\tau(x^{-1}) = \tau(e) = I$, it follows that $\tau(x^{-1}) = \tau(x)^{-1}$. Thus, a representation of G is just a homomorphism of G into the general linear group $GL(V_\tau)$ of invertible linear transformations on V_τ. The dimension d_τ of V_τ is called the degree of τ.

It is often useful to realize a representation τ concretely as matrix-valued relative to a choice of a basis for the representation space $V = V_\tau$. Thus, if we choose a basis $e_1, \ldots, e_d$ of V, where $d = d_\tau$, then the equations $\tau(x)e_j = \Sigma_{i=1}^d t_{ij}(x)e_i$ define the matrix $t(x) = (t_{ij}(x))$ of $\tau(x)$, and the mapping $x \mapsto t(x)$ is the matrix-valued realization of τ relative to the chosen basis. The d^2 functions t_{ij} on G are called the matrix entries of τ in the given basis.

The structure theory for representations of a finite group is based upon the concept of 'irreducibility' of a representation, and the property, called 'complete reducibility', that any representation of G may be decomposed into its irreducible constituents. The required technical definitions are as follows. If τ is a representation of a finite group G on a space $V = V_\tau$, we say that a subspace W of V is invariant under τ if $\tau(x)W \subset W$ for all $x \in G$. The restriction of the linear transformations $\tau(x)$ to an invariant subspace W defines a representation τ_W of G on the space W called the subrepresentation of τ on W. A representation τ is called irreducible if it has no proper invariant subspaces. Clearly, either a representation is irreducible or it has a subrepresentation that is irreducible.

For a finite group G, not necessarily abelian, the irreducible representations – that is, Frobenius's higher-order characters – are the appropriate generalization of the concept of group character introduced by Weber for finite abelian groups. Indeed, a character on G in the sense of (34) is nothing more than a representation of degree 1, or in other words a one-dimensional representation. In short, for a nonabelian group there are not enough group characters to do harmonic analysis, but there are sufficiently many irreducible representations. That is, without attempting to give a precise description, the essence of harmonic analysis on a finite group is the statement that any representation of G is the direct sum of its irreducible subrepresentations.

10 THE PETER–WEYL THEOREM

A topological group is called 'locally Euclidean' if there exists a neighbourhood of the identity homeomorphic to an open subset of a Euclidean space $\mathbb{R}^n$. Clearly, a locally Euclidean group is, *a fortiori*, locally compact. In terminology more common earlier this century, a locally Euclidean group in which the group operations are infinitely differentiable functions is called a continuous group. Today, these groups, as noted earlier, are called Lie groups (§6.5).

As the terminology suggests, a topological group G, whether or not it is a Lie group, is called compact if G itself is a compact set. The Peter–Weyl theorem provides a complete structure theory for harmonic analysis on any compact Lie group (Peter and Weyl *1927*). This theory subsumes as special cases not only the representation theory of a finite group, but also Fourier series on the circle and the theory of spherical harmonics. With the discovery of Haar measure a few years later, the results of Peter and Weyl became immediately valid for all compact topological groups.

Before stating the Peter–Weyl theorem, we introduce some notation. Let G be a compact topological group.

1 From the compactness, it follows that Haar measure on G is finite. Let the Haar measure be normalized so that the total volume of G is 1.

2 Next, we augment the definition of a representation τ of G by the condition that τ be continuous. That simply means that the matrix entries t_{ij} of τ, relative to any basis for the representation space, are continuous functions on G.

3 Also from the compactness of G, we assume that a representation of G is unitary. That is, given a representation τ of G, we can introduce an inner product on the space V_τ of τ relative to which $\tau(x)$ is a unitary transformation of V_τ for all $x \in G$. Equivalently, τ being unitary means that, in any matrix-valued realization of τ, the matrix entries satisfy the identity $t_{ij}(x^{-1}) = \overline{t_{ji}(x)}$ for all i, j.

4 Finally, we say that two representations τ_1 and τ_2 are equivalent if there exists an invertible linear transformation $A : V_{\tau_1} \mapsto V_{\tau_2}$ such that $A\tau_1(x) = \tau_2(x)A$ for all $x \in G$. Denote by $[\tau]$ the equivalence class of τ. If λ_1 and λ_2 are two irreducible representations of G, then (by a result known as Schur's lemma) they are equivalent if and only if the vector space spanned by the matrix entries of λ_1 coincides with the space spanned by the matrix entries of λ_2.

We can now define the dual object $\hat{G}$ of G as the collection of all equivalence classes of irreducible representations of G. For each $[\lambda] \in \hat{G}$, we fix once and for all a unitary matrix-valued representation belonging to the class. Thus, for each $x \in G$, $\lambda(x)$ is a $d_\lambda \times d_\lambda$ matrix, where d_λ is the degree of λ.

PETER–WEYL THEOREM. *The normalized matrix entries* $e_{ij}^{(\lambda)}(x) = \sqrt{(d_\lambda)}\lambda_{ij}(x)$ *for* $[\lambda] \in \hat{G}$ *and* $1 \leqslant i,j \leqslant d_\lambda$, *form a complete orthonormal system relative to the inner product*

$$\langle f_1 | f_2 \rangle = \int_G f_1(x)\overline{f_2(x)}\,dx.$$

11 FOURIER SERIES OF A COMPACT GROUP

The Fourier analysis implied by the Peter–Weyl theorem can be outlined as follows. For a function f on G we set

$$f(x) = \sum_{[\lambda] \in \hat{G}} d_\lambda \sum_{i=1}^{d_\lambda} \sum_{j=1}^{d_\lambda} \hat{f}_{ij}(\lambda)\lambda_{ij}(x), \tag{39}$$

where

$$\hat{f}_{ij}(\lambda) = \langle f | \lambda_{ij} \rangle = \int_G f(x)\overline{\lambda_{ij}(x)}\,dx. \tag{40}$$

The right-hand side of (39) is called the Fourier series of f, and (40) defines the Fourier coefficients of f. In analogy to formula (9) for Fourier series on the circle, for a function on G that is square-integrable relative to Haar measure the Fourier series (39) converges to f in the mean-square sense, and the Plancherel formula

$$\|f\|^2 = \sum_{[\lambda] \in \hat{G}} d_\lambda \sum_{i=1}^{d_\lambda} \sum_{j=1}^{d_\lambda} |\hat{f}_{ij}(\lambda)|^2 \tag{41}$$

holds. Equivalently, by analogy with the Fourier series statement (14), the Hilbert space $L^2(G)$ decomposes as the orthogonal direct sum

$$L^2(G) = \sum_{[\lambda] \in \hat{G}} \oplus\, C_\lambda(G) \tag{42}$$

of the subspaces $C_\lambda(G)$ spanned by the matrix entries of λ.

The Peter–Weyl theorem has the following representation-theoretic implication: any representation of a compact group, whether finite-dimensional or not, is completely reducible. More specifically, to within equivalence, the irreducible representations that appear and the multiplicities with which they appear are unique. In fact, if τ is any representation of G, for each $[\lambda] \in \hat{G}$ that appears in τ there exists a unique invariant subspace of V_τ, called the '$[\lambda]$-isotypic' or '$[\lambda]$-primary' component, on which τ is equivalent to a multiple of λ. The resulting decomposition of V_τ is called its 'isotypic' or 'primary' decomposition.

As an example, (42) is the isotypic decomposition for the right regular representation R of G on the Hilbert space $L^2(G)$, defined by

$$(R(a)f)(x) = f(xa), \quad \text{for } a \in G \text{ and } f \in L^2(G). \tag{43}$$

The invariant subspace $C_\lambda(G)$ is the $[\lambda]$-isotypic component. Thus, we have the representation-theoretic formulation of the Peter–Weyl theorem: up to equivalence, each irreducible representation of a compact group appears in the right (or left) regular representation with multiplicity equal to its degree.

As another example, the decomposition (30) is the isotypic decomposition of the natural action by rotation of the compact (nonabelian) Lie group $SO(3)$ on $L^2(S^2)$. In fact, each of the subspaces H_n is invariant under rotation and realizes an irreducible representation of $SO(3)$. Thus, the decomposition of $L^2(S^2)$ into spherical harmonics is multiplicity-free. This multiplicity-free phenomenon is characteristic of what are called compact Riemannian symmetric spaces, the spheres S^n being the commonest examples. By means of the Peter–Weyl theorem, Elie Cartan *1929* described the harmonic analysis of all such spaces.

Note that when G is an abelian compact group, by Schur's lemma an irreducible representation is necessarily one-dimensional, and the dual object $\hat{G}$ coincides with the dual group of all characters. In particular, if $G = T$ then formulas (39)–(42) reduce to formulas (9), (10), (15) and (14), respectively, for Fourier series.

12 THE NONCOMMUTATIVE PLANCHEREL THEOREM

Finally, we introduce the constructs necessary to recast the Peter–Weyl theorem as a noncommutative version of the Plancherel theorem valid for any compact group. Denote by $C^{d_\lambda \times d_\lambda}$ the space of all $d_\lambda \times d_\lambda$ matrices, endowed with the inner product

$$\langle A \mid B \rangle_{d_\lambda} = d_\lambda \operatorname{tr}(AB^*) = d_\lambda \sum_{i=1}^{d_\lambda} \sum_{j=1}^{d_\lambda} a_{ij}\overline{b_{ij}}, \tag{44}$$

which is the standard inner product on $C^{d_\lambda \times d_\lambda}$ but normalized by the factor d_λ. Let $\|A\|_{d_\lambda}$ be the corresponding norm. Next, form the Hilbert space direct sum

$$\ell^2(\hat{G}) = \sum_{[\lambda] \in \hat{G}} \oplus\, C^{d_\lambda \times d_\lambda}. \tag{45}$$

Since the elements of $\ell^2(\hat{G})$ are 'sequences' $\{A_\lambda : [\lambda] \in \hat{G}\}$ such that

$$\sum_{[\lambda] \in \hat{G}} \|A_\lambda)\|_{d_\lambda}^2 < \infty, \tag{46}$$

it is natural that $\ell^2(\hat{G})$ should play the same role for Fourier series on the compact group G as does the space ℓ^2 for Fourier series on the circle T. In fact, the space $L^2(\hat{G})$ reduces to ℓ^2 when $G = T$.

PLANCHEREL THEOREM FOR THE COMPACT GROUP G. *For $f \in L^2(G)$ and* $[\lambda] \in \hat{G}$, *define the $d_\lambda \times d_\lambda$ matrix*

$$\hat{f}(\lambda) = \int_G f(x)\overline{\lambda(x)}\,dx, \qquad (47)$$

the matrix entries of which are the Fourier coefficients of f. Then

$$\|f\|^2 = \sum_{[\lambda] \in \hat{G}} \|\hat{f}(\lambda)\|^2_{d_\lambda} \qquad (48)$$

and the mapping $\mathscr{F}: f \mapsto \hat{f}$ is a unitary operator, called the Plancherel transform for G, from the Hilbert space $L^2(G)$ to the Hilbert space $\ell^2(\hat{G})$.

The Plancherel theorem is simply a restatement of the Peter–Weyl theorem, for (47) and (48) are just matricial reformulations of (40) and (41), respectively. Group-theoretically, the significance of the Plancherel theorem is that it provides an explicit decomposition of the regular representations, both right and left, of G into their isotypic components.

13 NONCOMPACT, NONCOMMUTATIVE HARMONIC ANALYSIS

Here we mention only a few basic phenomena. First, as we have already seen for the real line, for a non-compact group harmonic analysis can involve continuous direct integral decompositions, rather than direct sums. A second major complication that does not arise in the commutative theory is much more serious.

As we have indicated, the irreducible representations of a locally compact abelian group are all one-dimensional. For compact groups we have seen that the irreducible representations are finite-dimensional, and if the group is nonabelian they need not be of dimension 1. *However, for groups which are locally compact, but neither compact nor abelian, one is forced to deal with irreducible representations which are infinite-dimensional.*

Prior to the Second World War, the principal contributions to representation theory and the harmonic analysis of noncompact, nonabelian groups came from sources in modern physics, in particular from quantum mechanics and relativity theory. Among the principal contributors were Weyl, John von Neumann, M. Stone and E. Wigner. During the latter half of this century, infinite-dimensional representation theory and noncompact, noncommutative harmonic analysis have been major research areas in mathematics. Although the structure theories that have evolved are often extremely complicated and major open problems still exist, a solid foundation is in place for both the intrinsic theory and applications. No summary of these developments, no matter how brief, would be complete without mentioning the pioneering work by I. M. Gelfand, and the school that formed around him in the Soviet Union, on many facets of infinite-dimensional representation theory, topological algebras and the structures now known as Gelfand pairs; by George Mackey, who developed the notion of induced representation and put into place the algebraic and measure-theoretic foundations for the infinite-dimensional theory; by Harish-Chandra, whose profound investigations of the representation theory of semi-simple Lie groups, culminating in 1968 with his proof of the Plancherel theorem, still dominate that subject; and by Robert Langlands's efforts to understand p-adic representation theory and its implications for number theory. For more information the reader is referred to Gross *1978*, Helgason *1977* and Mackey *1963*, *1978*, and references therein.

14 CONTEMPORARY CLASSICAL HARMONIC ANALYSIS

Another major theme in harmonic analysis this century, and still pursued actively today, often under the name of 'hard analysis', continues the investigation of the deeper convergence questions in the classical settings of the circle and real line, or more generally the product of circles (i.e. multiple Fourier series) and n-dimensional Euclidean space, respectively. Indeed, the historical interplay between Fourier analysis and differential equations exists in current applications for which Fourier transformation is a fundamental tool in analysing translation-invariant operators (or, more generally, operators that are invariant under certain nonabelian groups) on function spaces more subtle and complicated than L^2 spaces. In some directions, for example the study of singular integral operators, there are strong connections with the group-theoretic perspective, even with infinite-dimensional irreducible representations. On the other hand, a broad range of research focuses on the harmonic analysis of more general domains which exhibit too few symmetries to permit group-theoretic analysis. The reader interested in more detail should consult Ash *1976*, Christ *1990*, Stein and Weiss *1971*, Titchmarsh *1937*, Wiess *1965* and Zygmund *1959*.

ACKNOWLEDGEMENTS

It gives me great pleasure to express my sincere appreciation to Professor Jean Dieudonné, and to my colleagues Professors Roger Cooke and

J. Michael Wilson. To Professor Dieudonné I am especially grateful for sharing his unpublished manuscript *1971* on the history of harmonic analysis with me. Professor Cooke generously and patiently lectured to me on the history of convergence and uniqueness of Fourier series, and Professor Wilson offered his expertise on current research directions in classical harmonic analysis. Their contributions helped to focus the exposition in important ways.

BIBLIOGRAPHY

Ash, J. M. *1976*, *Studies in Harmonic Analysis*, Providence, RI: Mathematical Association of America (Studies in Harmonic Analysis, Vol. 13).

Bernoulli, D. *1753*, 'Réflexions et éclaircissements sur les nouvelles vibrations des cordes exposés dans les mémoires de 1747 et 1748', *Mémoires de l'Académie des Sciences de Berlin*, 147–72.

Carleson, L. *1966*, 'On the convergence and growth of partial sums of Fourier series', *Acta mathematica*, **166**, 135–57.

Cartan, E. *1929*, 'Sur la détermination d'un système orthogonale complète dans un espace de Riemann symmétrique clos', *Rendiconti del Circolo Matematico di Palermo*, **53**, 217–52.

Christ, M. *1990*, *Lectures on Singular Integral Operators*, Providence, RI: American Mathematical Society (Regional Conference Series, Vol. 77, Conference Board of Mathematical Sciences).

Cooke, R. *1993*, 'Uniqueness of Fourier series and descriptive set theory: 1870–1985', *Archive for History of Exact Sciences*, **51**, 251–334.

d'Alembert, J. *1747*, 'Recherches sur la courbe que forme une corde tendue mise en vibration', *Mémoires de l'Académie des Sciences de Berlin*, 214–21.

Dieudonné, J. *1971*, 'Histoire de l'analyse harmonique', unpublished manuscript of a lecture delivered at the 13th International Congress on the History of the Sciences, Moscow.

Dirichlet, P. G. *1829*, 'Sur la convergence des séries trigonométriques qui servent à représenter une fonction arbitraire entre deux limites données', *Journal für die reine und angewandte Mathematik*, **4**, 157–69. [Also in *Werke*, Vol. 1, 117–32.]

Fischer, E. *1907*, 'Sur la convergence en moyenne', *Comptes Rendus de l'Académie des Sciences*, **144**, 1022–4.

Fourier, J. *1808*, 'Mémoire sur la propagation de la chaleur dans les corps solides', *Nouveau Bulletin des Sciences par la Société Philomatique de Paris*, **1**, 112–16. [Also in *Oeuvres*, Vol. 2, 215–21.]

—— *1822*, *Théorie analytique de la chaleur*, Paris: Firmin Didot. [Also *Oeuvres*, Vol. 1.]

Frobenius, G. *1896–1906*, *Gesammelte Abhandlungen*, Vol. 3, 1968, Heidelberg: Springer. [Originally, Frobenius's papers were published in the *Sitzungsberichte der Preussischen Akademie der Wissenschaften*.]

Grattan-Guinness, I. *1970*, *The Development of the Foundations of Mathematical Analysis from Euler to Riemann*, Cambridge, MA: MIT Press.

Grattan-Guinness, I. and Ravetz, J. R. *1972*, *Joseph Fourier, 1768–1830*, Cambridge, MA: MIT Press.

Gross, K. I. *1978*, 'On the evolution of noncommutative harmonic analysis', *American Mathematical Monthly*, **85**, 525–48.

Haar, A. *1933*, 'Der Massbegriff in der Theorie der kontinuierlichen Gruppen', *Annals of Mathematics*, **34**, 147–69.

Harnack, A. *1880*, 'Über die trigonometrische Reihe und die Darstellung willkürlicher Functionen', *Mathematische Annalen*, **17**, 123–32.

Helgason, S. *1977*, 'Invariant differential equations on homogeneous spaces', *Bulletin of the American Mathematical Society*, **83**, 751–74.

Hilbert, D. *1912*, *Grundzüge einer allgemeinen Theorie der linearen Integralgleichungen*, Leipzig and Berlin: Teubner.

Laplace, P. S. *1782*, 'Théorie des attractions des sphéroïdes et de la figure des planètes', *Mémoires de l'Académie Royale des Sciences*, 113–96. [Also in *Oeuvres*, Vol. 10, 339–419.]

Legendre, M. *1782*, 'Recherches sur la figure des planètes', *Mémoires de l'Académie Royale des Sciences*, 170–89.

Lie, S. *1922–60*, *Gesammelte Abhandlungen*, 7 vols, Leipzig: Teubner.

Mackey, G. *1963*, 'Infinite-dimensional group representations', *Bulletin of the American Mathematical Society*, **69**, 628–86.

—— *1978*, 'Harmonic analysis as the exploitation of symmetry – A historical survey', *Rice University Studies*, **64**, 73–228.

Parseval, M. A. *1799*, 'Mémoire sur les séries et sur l'intégration complète d'une équation aux différences partielles linéaires du second ordre à coefficients constans', *Mémoires présentés par Divers Savans*, Series 1, **1**, 638–48.

Peter, F. and Weyl, H. *1927*, 'Die Vollständigkeit der primitiven Darstellungen einer geschlossenen kontinuierlichen Gruppe', *Mathematische Annalen*, **97**, 737–55.

Plancherel, M. *1910*, 'Contribution a l'étude de la représentation d'une fonction arbitraire par des intégrales définies', *Rendiconti del Circolo Matematico di Palermo*, **30**, 289–335.

Pontrjagin, L. *1934*, 'The theory of topological commutative groups', *Annals of Mathematics*, **35**, 361–88.

Riemann, B. *1854*, 'Über die Darstellbarkeit einer Function durch eine trigonometrische Reihe', in *Werke*, 1892, 2nd edn, 227–65.

Riesz, F. *1907*, 'Sur les systèmes orthogonaux de fonctions et l'équation de Fredholm', *Comptes Rendus de l'Académie des Sciences*, **144**, 734–6. [Also in *Oeuvres complètes*, 378–81.]

Rudin, W. *1962*, *Fourier Analysis on Groups*, New York: Wiley.

Schur, I. *1924*, 'Neue Anwendungen der Integralrechnung auf Probleme der Invariantentheorie', *Sitzungsberichte der Akademie der Wissenschaften zu Berlin*, **1**, 189–208.

Schwartz, L. *1945*, 'Généralisation de la notion de fonction, de dérivation de transformation de Fourier et applications mathématiques et physiques', *Annales de l'Université de Grenoble*, **21**, 57–74.

—— *1966*, *Théorie des distributions*, rev. edn, Paris: Hermann.

Stein, E. and Weiss, G. *1971*, *Introduction to Fourier Analysis on Euclidean Space*, Princeton, NJ: Princeton University Press.

Titchmarsh, E. *1937, Introduction to the Theory of Fourier Integrals*, Oxford: Clarendon Press.

Vilenkin, N. J. *1968, Special Functions and the Theory of Group Representations*, Providence, RI: American Mathematical Society.

von Neumann, J. *1931*, 'Die Eindeutigkeiten der Schrödingerschen Operatoren', *Mathematische Annalen*, **104**, 570–78. [Also in *Collected Works*, Vol. 2, 221–9.]

Weber, H. *1882*, 'Beweis des Satzes, dass jede eigentlich primitive quadratische Form unendlich viele Primzahlen darzustellen fähig ist', *Mathematische Annalen*, **20**, 301–29.

Weil, A. *1940, L'Intégration dans les groupes topologiques et ses applications*, Paris: Hermann.

Weiss, G. *1965*, 'Harmonic analysis', in *Mathematical Association of America Studies in Mathematics*, Vol. 3 (ed. I. I. Hirschman, Jr), Englewood Cliffs, NJ: Prentice-Hall, 124–78.

Weyl, H. *1925, 1926*, 'Theorie der Darstellung kontinuierlicher halbeinfacher Gruppen durch linearen Transformationen I, II, und III', *Mathematische Zeitschrift*, **23**, 271–309; **24**, 328–76, 377–95. [Also in *Gesammelte Abhandlungen*, vol. 2, 543–647.]

—— *1950, The Theory of Groups and Quantum Mechanics*, London: Methuen. [German original 1931.]

Zygmund, A. *1959, Trigonometric Series*, New York: Cambridge University Press.

3.12

Three traditions in complex analysis: Cauchy, Riemann and Weierstrass

U. BOTTAZZINI

1 INTRODUCTION

Complex analysis as a proper domain of modern analysis developed during the nineteenth century, especially as a result of the work of Augustin Louis Cauchy, Bernhard Riemann and Karl Weierstrass. 'Imaginary' numbers, however, had been introduced in mathematics since the Renaissance times, when the Italian algebraists found the solution of algebraic equations of the third degree (§6.1). Without asking the question of 'what kind' of numbers he faced, Rafael Bombelli showed in his *Algebra* (1572) how to manipulate these quantities, which later René Descartes (1637) called 'imaginary' as opposed to 'real' quantities (§6.2).

The question of the 'ontological' nature of imaginary quantities was left open until the beginning of the nineteenth century (§6.2). Nevertheless, during the previous century imaginary numbers and imaginary variables were used extensively by Jean d'Alembert and Leonhard Euler, among others. They also discovered important properties of complex functions, like the monogenity equations (as Cauchy was to call them)

$$\frac{\partial u}{\partial x} = \frac{\partial v}{\partial y} \quad \text{and} \quad \frac{\partial u}{\partial y} = -\frac{\partial v}{\partial x} \tag{1}$$

for a function $w = u + v\sqrt{-1}$ of an imaginary variable $z = x + y\sqrt{-1}$. Today we call them the Cauchy–Riemann equations, but they first appeared in a paper by d'Alembert (1752) on hydrodynamics. They were also used by Euler in a series of papers (from 1776 on, published posthumously from 1788 on) in which he showed how one can use complex functions to evaluate real integrals.

The work by d'Alembert and Euler on complex variables led to a considerable extension of the techniques of analysis and important developments in the concept of a function (§3.2). Nevertheless, neither man carried out further investigations of the properties of complex functions: instead, they limited themselves to the study of the real and the complex parts of such functions. This attitude was dominant among mathematicians until the early decades of the nineteenth century.

2 CAUCHY'S WORK

In Cauchy's hands the calculus with complex quantities became an indispensable tool of analysis, losing the aura of mystery that had surrounded complex numbers since their first appearance. This uncertainty can still be found at the end of the eighteenth century, as is witnessed by Carl Friedrich Gauss, who refused to take a public position on the question in his 1799 paper on the fundamental theorem of algebra, preferring to 'complete [his] demonstration without any help of imaginary quantities' (§6.1).

The first paper Cauchy devoted to complex analysis was his 1814 memoir on the calculus of 'definite' integrals, or more precisely, on the evaluation of improper (real) integrals. This had been a subject of research since the time of Euler. According to Cauchy, however, Euler himself, like Pierre Simon Laplace, Siméon-Denis Poisson and Adrien Marie Legendre after him, had all been using 'a type of induction based on the passage from real to imaginary' which needed to be established 'by a direct and rigorous analysis'. This was the aim of the first part of the paper, while in the second part Cauchy discussed the possibility of inverting the order of integration of a double integral. In doing so he obtained his most original results. The paper was actually published only in 1827, enriched with a quantity of added notes in which Cauchy explained his 1814 results in terms of the theory of complex integration he had been elaborating since 1825 (Ettlinger *1922*).

Shortly before, the geometrical interpretation of complex numbers had been discussed among Parisian mathematicians and a series of papers on the topic (including some by Jean Argand) were published in Joseph Gergonne's *Annales* (§6.2). However, in his 1814 paper Cauchy said nothing about the geometrical interpretation of complex numbers (and variables) or the possibility of interpreting his calculus with double integrals as the evaluation of line integrals along the boundaries of a rectangle in the complex plane.

Cauchy set out to explain the foundation of the theory of complex numbers (and complex variables) in his influential textbook *Cours d'analyse* (*1821*). Here imaginary quantities are defined as 'symbolic expressions', such an expression being 'any combination of algebraic signs that do not signify anything in themselves or to which one attributes a value different from that which it naturally has' (p. 153). One can consider, for example, two 'symbolic' expressions like $\cos \alpha + \sin \alpha \sqrt{-1}$ and $\cos \beta + \sin \beta \sqrt{-1}$, and define their product by operating with the usual rule 'as if $\sqrt{-1}$ were a real quantity whose square were equal to -1' (p. 154). After defining the 'conventions' according to which it was possible to give meaning to 'imaginary expressions' and to the operations with them, Cauchy introduced the concept of an 'imaginary' function of a real variable.

In order to define imaginary functions of an imaginary variable like $\sin z$, $\cos z$, A^z and their inverses, Cauchy followed a rigorous though cumbersome procedure, using functional equations (§4.9) and power series $\Sigma_n a_n x^n$ (proving among other things the Cauchy–Hadamard theorem, that such a series converges in a disc $|z| < 1/A$, where $A = \limsup |a_n|^{1/n}$).

This way of regarding complex functions was adopted by Cauchy until the late 1840s. In 1847 he proposed the theory of 'algebraic equivalences' in order to gain clarity by banishing imaginaries and by 'reducing the letter i to no more than a real quantity'. This could be done by reducing calculations with complex numbers to a calculation $(\bmod\, i^2 + 1)$ in the ring $R[i]$, where i is 'a real but indeterminate quantity'. But some years later, 'after new and mature reflections', he decided to adhere to the theory of 'geometric quantities', in other words the usual interpretation of complex numbers (and variables) in the complex plane. In this geometrical setting, Cauchy re-established the results he had obtained since 1825.

In 1825 Cauchy had published a pamphlet on integration which can be considered as a milestone in the history of complex analysis. In it he established his integral theorem that, for a function $f(z)$ which is regular in a (simply connected) domain C,

$$\int_{\partial C} f(z)\, dz = 0. \tag{2}$$

Even though he stated his theorem under the (insufficient) hypothesis that $f(z)$ is 'finite and continuous', in the proof he tacitly assumed that $f'(z)$ did exist and was continuous. (This theorem, however, had been known to Gauss, who announced it in a letter to his friend Friedrich Wilhelm Bessel in 1811.) In 1900, Edouard Goursat was able to demonstrate Cauchy's theorem by assuming merely the existence of $f'(z)$. Some years before, in 1886, G. Morera had proved the converse of Cauchy's integral theorem: that if $f(z)$ is continuous in the domain C, and the integral of $f(z)$ taken along any simply closed path lying entirely in C is equal to zero, then $f(z)$ is holomorphic in C.

If the condition of regularity of $f(z)$ is no longer satisfied, and

$z = a + b\sqrt{-1}$ is (what we today call) a simple pole in C, then Cauchy proved that

$$\int_{\partial C} f(z)\,dz = 2\pi\sqrt{-1}\mathscr{F}, \tag{3}$$

where 'without noticeable error', $\mathscr{F}$ was given by

$$\mathscr{F} = \varepsilon f(a + b\sqrt{-1} + \varepsilon), \tag{4}$$

ε being infinitely small. In this way Cauchy introduced the concept of the residue $\mathscr{F}$ of a function in a simple pole (and in a similar way the concept of residue for multiple poles of order m).

According to Cauchy, the calculus of residues was 'a new kind of calculation similar to the infinitesimal calculus', and he was pleased to apply it to the evaluation of definite and improper integrals such as

$$\int_0^{2\pi} R(\cos q, \sin q)\,dq, \quad \int_{-\infty}^{+\infty} R(x)\,dx \quad \text{and} \quad \int_{-\infty}^{+\infty} R(x)e^{ix}\,dx,$$

R being a rational function.

Starting in the 1820s, Cauchy published a series of papers on the calculus of residues, including refinements of equations (2)–(4), as well as applications to interpolation formulas, summation of series, integration of linear differential equations, resolution of algebraic or transcendental equations, and so on. The wealth of results he was able to obtain using the calculus of residues seems to have led him to overestimate it in comparison with his integral theorem (2), which he did not mention until the early 1850s. Apparently it was only after the publication of Victor Puiseux's paper on the integration of algebraic functions (§4.5) that Cauchy fully realized the importance of his theorem as a fundamental result in the theory of complex functions.

In the early 1830s, during his stay in Italy, Cauchy had presented to the Academy of Turin a memoir inspired by the study of Lagrange series and other series of use in celestial mechanics. The memoir was first published in 1832 in Italian translation, and a summary was published by Cauchy in 1841, but the whole memoir had to wait until 1974 before being printed in the last volume of Cauchy's *Oeuvres*.

In that work Cauchy obtained his celebrated integral formula: in his notation,

$$f(x) = \frac{1}{2\pi}\int_{-\pi}^{\pi} \frac{\bar{x}f(\bar{x})}{\bar{x} - x}\,dp, \tag{5}$$

where $\bar{x} = X\exp(\sqrt{-1}p)$, $-\pi \leqslant p \leqslant \pi$ and $f(x)$ is 'continuous and finite' together with its derivative for $|x| < X$. (From (4) one immediately obtains the usual form

$$f(z) = \frac{1}{2\pi i} \int_c \frac{f(\zeta)}{\zeta - z} \, d\zeta \tag{6}$$

by describing the circumference c of the circle of convergence with $\zeta = re^{i\phi}$.)

Therefore, concluded Cauchy, if $|x|$ is not greater than a value for which $f(x)$ and its derivative ceases to be finite and continuous, the function $f(x)$ is expandable in a power series whose generic term is

$$\frac{1}{2\pi} \int_{-\pi}^{\pi} \frac{x^n}{\bar{x}^n} f(\bar{x}) \, dp = \frac{x^n}{n!} f^{(n)}(0), \tag{7}$$

which by means of the previous substitution $\zeta = re^{i\phi}$ may easily be reduced to the usual form

$$f^{(n)}(z) = \frac{n!}{2\pi i} \int_c \frac{f(\zeta)}{(\zeta - z)^{n+1}} \, d\zeta. \tag{8}$$

Now, if $|x| = \xi$, and $\Lambda f(x)$ denotes the largest value of $|f(x)|$ as p varies, then (7) may be majorized by

$$\left| \frac{1}{2\pi} \int_{-\pi}^{\pi} \frac{x^n}{\bar{x}^n} f(\bar{x}) \, dp \right| \leq \left(\frac{\xi}{X} \right)^n \Lambda f(x). \tag{9}$$

In a similar way Cauchy was able to majorize the remainder of the Maclaurin expansion of the function, and evaluate the 'limits' of the errors introduced by disregarding the terms of the series after the first n. For this reason he called this method *calcul des limites*. This *calcul* (nowadays known as the 'method of majorants') could be applied to functions defined by differential equations, both ordinary and partial, as he showed in a paper published in Prague in 1835, as well as in various notes presented to the Paris Academy of Sciences in 1842 (§3.14).

3 THE EMERGENCE OF THE FRENCH SCHOOL

In the 1840s Cauchy's studies joined with those of Joseph Liouville, Pierre Laurent, Puiseux and Charles Hermite to create a 'French school' in complex analysis. Inspired by Cauchy's work on the *calcul des limites*, in 1843 Laurent presented the Academy with a paper on the representability of a holomorphic function within an annulus centred at point c by the series

$$\sum_{k=-\infty}^{+\infty} a_k (z - c)^k, \tag{10}$$

where the coefficients are given by

$$a_k = \frac{1}{2\pi i} \int_\gamma (z - c)^{-k-1} f(z) \, dz, \tag{11}$$

γ being a closed loop around c within the annulus. While stating in his report to the Academy that this result was 'worthy of notice', Cauchy also added that it was included 'as a special case' in a formula he had published in 1826.

Together with the *calcul des limites*, the calculus of residues seemed to promise the most fruitful applications, as Cauchy showed in 1843 in a series of papers on elliptic functions which inspired Hermite's early work on the same subject (§4.5). At that time elliptic functions were also being studied by Liouville. In a short communication to the Academy in 1844 he set out the following 'general principle' that seemed 'to impress on the study of elliptic functions an uncommon character of unity and simplicity': let $U(z)$ be a single-valued, doubly periodic monodromic function for every value of z. If it is doubly periodic and it never becomes infinite, 'one can, from this alone, affirm that it reduces to a constant'. Cauchy promptly recognized the analogy between this 'principle' and his own previous results, and in less than a year he was able to publish five different proofs of Liouville's theorem.

According to Cauchy, however, his integral theorem and the theory of residues constituted the foundations of the theory of the function of 'geometric quantities' which he developed in the last decade of his life in no less than 15 papers published in his own journal, *Exercices d'analyse et de physique mathématique*. Here he refined his previous results, by reformulating his theorems and by introducing a number of special terms for functions like 'monogenic' (i.e. functions satisfying equations (1)), 'monodromic' and others nowadays completely forgotten. Instead of analyticity, however, in all the papers devoted to the subject he always stressed the importance of continuity, the 'great law' for the functions he had discovered and formulated in his *Cours d'analyse*.

In 1855 Cauchy introduced the concept of 'logarithmic indicator' (*compteur logarithmique*) by considering the integral

$$\frac{1}{2\pi i} \int_{\partial S} \frac{Z'(z)}{Z(z)} \, dz, \tag{12}$$

where $Z(z)$ is a monodromic and monogenic function in S, except at poles, and he proved that (12) equals $N - P$, N being the number of zeros and P the number of poles of Z within S. Thus he was able to obtain from (12) a new proof of the fundamental theorem of algebra, as well as a new proof of the theorem, stated by Liouville and already proved by Hermite, that the number of zeros of a doubly periodic function within the parallelogram of periods equals the number of poles.

Perhaps because he had not been required to write a textbook while teaching at the Ecole Polytechnique, Cauchy never wrote a systematic exposition of the deep results he had obtained in the theory of complex functions throughout his life. After Cauchy's death this was done jointly by C. A. Briot and J. C. Bouquet. The first part of their 1859 treatise *Théorie des fonctions doublement périodiques* [...] gave a complete account of Cauchy's theory and remained until the last decades of the nineteenth century the standard text of the French school in complex analysis.

4 RIEMANN'S GEOMETRIC POINT OF VIEW

In 1851, Riemann graduated from the University of Göttingen with a dissertation on the foundations of complex analysis. This paper was to have a tremendous and lasting influence on the later development of mathematics (§3.3). As L. Ahlfors remarked:

> ... it contains the germ to a major part of the modern theory of analytic functions, it initiated the systematic study of topology, it revolutionized algebraic geometry, and it paved the way for Riemann's own approach to differential geometry. (Ahlfors *1953*: 3)

Riemann's dissertation opens with the following definition: $w = u + iv$ is a function of a complex variable z when w changes with z'in such a way that the value of the derivative dw/dz is independent of the value of the differential dz' (Riemann *1851*: 5). It follows at once that equations (1) are satisfied, as well as the equations $\Delta u = 0$ and $\Delta v = 0$; that is, both u and v are harmonic functions. Moreover, w can be geometrically interpreted as a conformal mapping.

Thus equations (1), which for Cauchy isolated only the particular class of monogenic functions, become in Riemann's eyes the basis of the concept of a function of a complex variable. According to E. Prym, Riemann once said that he had been led to attribute decisive importance to equations (1) since his student days, after he had asked under what conditions it was possible to 'continue' the expansion in series of a function from one domain to another.

After establishing these basic properties of complex functions, Riemann introduced in his dissertation an entirely new geometrical concept, the Riemann surface:

> We choose this covering [*Einkleidung*] so that it will be easy to speak of

superimposed surfaces, thus leaving open the possibility that the place of point O extends many times over the same part of the plane. However, in such a case we assume that the superimposed parts of the surface are not connected along a line, so that a folding of the surface or a division into superimposed parts cannot occur. (Riemann *1851*: 7)

This idea of multiply covered surfaces, which turned out to be one of the deepest achievements in his dissertation, was far from geometrically intuitive, and met with some opposition among Riemann's contemporaries. According to Weierstrass, for example, it was nothing more than a 'geometric fantasy'.

One of the intuitive reasons for introducing such surfaces lies in the study of multi-valued functions. If w is multi-valued, every point of the surface represents a single value of w corresponding to a single value of z. This concept of the space of the variable provides a tool for the treatment of branch-points, the points where the m sheets of the surface are connected in such a way as to return to the point of departure when the variable has completed m loops, passing continuously from one sheet to the next in the neighbour-hood of the point in question. For his surfaces, Riemann introduced topological concepts like cross-cuts, which are systems of curves that go from one boundary point to another, and a topological invariant which he called 'the order of connectivity' of a given surface.

Into this geometrical setting Riemann placed the study of the properties of harmonic functions, including the 'maximum principle' named after him: a function u, harmonic within a given surface T, cannot have either a maximum or a minimum at a point within T or be constant only in part of the surface. The fact that u and v are harmonic functions plays a fundamental role in Riemann's theory and allows him to use the results provided by the researches of Gauss and J. P. G. Lejeune Dirichlet in potential theory (§3.17). As Ahlfors (*1953*: 4) has said, Riemann 'virtually puts equality signs between two-dimensional potential theory and complex function theory'.

In contrast to the traditional concept of a function as an analytic expression, this approach led Riemann to define a complex function in a given domain from its singularities, poles and branch points by means of existence and uniqueness theorems. To this purpose, the essential tool was given to him by a variational 'principle' he named after Dirichlet. According to Dirichlet's principle, there exist, for a given bounded domain T of the plane, one and only one function $u(x, y)$ which is harmonic within the domain and assumes given values on the boundary (§3.17). By means of it, Riemann was able to demonstrate a general existence theorem which opened 'the way to an investigation of determinate functions of complex variables (independent of an expression for them)'. As an example, he treated the following: if T is a simply connected surface, the function $w = u + iv$ of z can be determined according to the following conditions: on the boundary, u coincides with a given continuous function (or one which is at most discontinuous at isolated points); the value of v is arbitrarily given in a point; and the function w must be finite and continuous in every point.

At this point, Riemann remarked:

> Our research has shown that, as a consequence of the general character of a function of a complex variable, in a definition of this kind a part of the determining elements is a result of the rest, namely, the range of the determining elements is traced back to those necessary for determining [the function].
>
> (Riemann *1851*: 37)

Thus, for example, he characterized algebraic functions of a variable z (simply or multiply extended over the entire infinite plane) as the functions with only isolated infinities of finite order (i.e. poles).

In the concluding paragraphs of his dissertation, Riemann tackled the problem of the conformal mapping between two given Riemann surfaces, proving his mapping theorem in a particular case by means of Dirichlet's principle. In modern terms, he stated that every simply connected, bounded domain of the complex plane is biholomorphically equivalent to the unit disc.

According to Riemann, the new method he set out in his dissertation was basically applicable 'to every function which satisfies a linear differential equation with algebraic coefficients'. He gave a proof of this claim first in a paper on hypergeometric functions (§4.4), and then in an important paper on Abelian integrals (§4.6), both published in 1857. Here the Riemann surface was thought to cover the Riemann sphere obtained by adding the point ∞ to the complex plane.

The methods of complex analysis also provided Riemann with powerful tools for fundamental work in number theory (§6.10, Section 4). In a very influential paper on the distribution of primes presented in 1859 on the occasion of his election to the Berlin Academy of Sciences, Riemann started by considering Euler's function

$$\zeta(s) = \sum_{n=1}^{\infty} n^{-s} = \prod \frac{1}{1 - p^{-s}}, \tag{13}$$

where s is supposed to be a complex variable and p runs over the primes. In order to study the behaviour of $\zeta(s)$, Riemann applied Cauchy's residue theorem to $\log \zeta(s)$. This required the preliminary study of the zeros of $\zeta(s)$. In this regard he wrote that 'it is very probable that all the zeros are real. A rigorous proof of this is to be desired' (Riemann *1859*: 148). Thus he conjectured that all the zeros of $\zeta(s)$ in the 'critical strip' $0 \leqslant \mathrm{Re}(s) \leqslant 1$ lie on the 'critical line' $\mathrm{Re}(s) = \frac{1}{2}$. This is the celebrated Riemann's hypothesis, still unanswered today.

Together with the dissertation, which was translated into Italian by Enrico Betti in 1859, the publication of these papers spread Riemann's ideas outside Göttingen. He was, however, well aware of the resistance his methods were finding, and in his lectures on complex function theory he eventually preferred to follow a somewhat eclectic way. Thus, in the last course Riemann gave at Göttingen in summer 1861, he combined the principles set out in his dissertation with the method of analytic continuation as well as with Cauchy's and Laurent's results on residues and series expansions. Following such an approach, he had no longer any need of either the theorem of existence or of Dirichlet's principle, which in fact was not mentioned at all in those lectures.

5 WEIERSTRASS'S RIGOUR

At the very same time that Riemann published his paper on Abelian integrals, Weierstrass, then 40 years old, began his teaching career at the University of Berlin, where he lectured until the late 1880s. His work in analysis (§3.3) as well as his lectures on analytic function theory are characterized by a constant effort towards an absolute rigour, which according to Weierstrass could be achieved only by abandoning the intuitive realm of geometric evidence and taking the arithmetic of natural numbers as the foundation for the whole of analysis. 'The more I think about the principles of function theory – and I do so incessantly', he confessed to his former student K. H. A. Schwarz in October 1875, 'the more I am convinced that *this must be built on the foundation of algebraic truths*'.

Consistent with this point of view, he refused to use the results Cauchy had obtained through complex integration (including the theory of residues and Cauchy's integral theorem). Nor could he accept the geometrical and 'transcendent' methods Riemann had proposed – indeed, it was the criticism of Riemann's ideas and methods that marks the starting-point of Weierstrass's systematic work in complex analysis. He made a decisive move in 1870, when he communicated to the Berlin Academy of Science a counter-example of Dirichlet's principle. Weierstrass considered the integral

$$J = \int_{-1}^{1} x^2 y'^2 \, dx, \qquad (14)$$

where the functions $y(x)$ were supposed to be continuous (together with $y'(x)$) in the interval $(-1, 1)$ such that $y(-1) = a \neq b = y(1)$. It is easily shown that the lower bound of J (as $y(x)$ varies) is zero, and yet there is no minimum since that would imply that $y = $ constant, contrary to the hypothesis. 'With this a large part of Riemann's developments come to naught', Felix Klein once remarked (*1894*: 492). However, he added that, according to Weierstrass, 'Riemann had never laid any particular value on finding his existence proofs with Dirichlet's principle'. In any case, Weierstrass's counter-example gave rise to the problem of establishing Riemann's theorems in a rigorous way. This task was to be taken up by Schwarz, Carl Neumann and others before David Hilbert 'resurrected' Dirichlet's principle in 1900, and Hermann Weyl rigorously re-established the idea of the Riemann surface in 1913 using the concepts provided by modern algebraic topology (§7.10).

Weierstrass presented his own approach to the theory of analytic functions in his lectures. This theory was intended to provide the foundation of the theory of elliptic and Abelian functions, set out in the programme of lectures: introduction to the theory of analytic functions, elliptic functions, Abelian functions and the calculus of variations (or applications of elliptic functions). Searching for an increasing rigour, he was apparently never satisfied with the level he had attained and consequently never published a version of his lectures on analytic functions, nor authorized his students to publish their lecture notes. Nevertheless, we can use works of this type (like Pincherle *1880* or Weierstrass *1988*) to get an idea of his approach to the theory of analytic functions.

In his lectures, Weierstrass first established the fundamental principles of arithmetic (including the theory of real numbers) as well as some theorems on 'magnitudes in general' (like the Bolzano–Weierstrass theorem). In order to introduce the concept of analytic function, he then developed the theory of power series by proving the 'classic' theorems on convergence and uniform convergence now named after him.

A basic role in Weierstrass's theory was played by the principle of analytic continuation. Given a power series $P(x, a)$ converging in a circle C of centre a and radius r, for any point b within the circle one can deduce from the primitive series a series $P(x, a, b)$ with respect to the point b, converging in a circle C' of centre b and radius R'. If C' extended outside C, Weierstrass said that $P(x, a, b)$ was the analytic continuation of the element $P(x, a)$. In this way, given an element of an analytic function, by analytic continuation one can obtain the analytic function in its totality. Weierstrass distinguished between monodromic and polydromic functions according to whether or not, starting from a first element $P(x, a)$, one always arrives at a unique expansion $P(x, a, b, \dots, y)$ with respect to every point y of the domain of validity of the function. A singular point a for a function $f(z)$ was called by Weierstrass either an 'essential singularity' or an 'inessential' one (a pole), according to whether or not there existed an integer n such that $(z - a)^n f(z)$ was regular in a.

In a very influential paper published in 1876, Weierstrass proved that in the neighbourhood of an essential singularity c a function $f(z)$ 'can come arbitrarily close to any given value; consequently it has no determinate value for $x = c$'. This is now known as the Casorati–Weierstrass theorem, because the Italian mathematician F. Casorati had included it in his 1868 treatise on complex analysis (Neuenschwander *1978*). In the same paper Weierstrass proved his factorization theorem: that, for any sequence of complex numbers a_n ($a_n \to \infty$ for $n \to \infty$), there is an entire function $f(z)$ which has its zeros at the points a_n and which can be represented as a (possibly infinite) product of linear factors $(1 + z/a_n)$ and exponentials. He also proved representation theorems for functions having n prescribed poles or essential singularities. One year later, Weierstrass's results were extended to meromorphic functions by G. Mittag-Leffler. Weierstrass's paper was translated into French by Emile Picard, who extended the Casorati–Weierstrass theorem with the additional result that in the neighbourhood of an essential singularity a function takes on every value except at most one.

At the end of the nineteenth century, Weierstrass's approach to complex analysis became dominant, and the German word *Funktionenlehre* became synonymous with the theory of functions of a complex variable according to his principles. However, this exclusive attitude has now been abandoned and most modern treatises take the Cauchy–Riemann approach as a starting point. The modern theory of complex integration provides the most elegant methods for developing the theory, and also for showing the equivalence of the Cauchy–Riemann and Weierstrass approaches.

BIBLIOGRAPHY

Ahlfors, L. V. *1953*, 'Development of the theory of conformal mapping and Riemann surfaces through a century', *Annals of Mathematical Studies*, **30**, 3–13.

Bottazzini, U. *1986*, *The Higher Calculus. A History of Real and Complex Analysis from Euler to Weierstrass*, Heidelberg: Springer.

Cauchy, A.-L. *1821*, *Cours d'analyse de l'Ecole Polytechnique. 1re partie* [and only]: *Analyse algébrique*, Paris: De Bure. [Repr. with introduction by U. Bottazzini, 1992, Bologna: CLUEB. Also in *Oeuvres complètes*, Series 2, Vol. 3. German transl.: 1828, Königsberg: Borntrager.]

Dugac, P. *1973*, 'Elements d'analyse de Karl Weierstrass', *Archive for History of Exact Sciences*, **10**, 41–176.

Ettlinger, H. J. *1922*, 'Cauchy's paper of 1814 on definite integrals', *Annals of Mathematics*, Series 2, **23**, 255–70.

Klein, F. *1894*, 'Riemann und seine Bedeutung für die Entwicklung der modernen Mathematik', *Jahresbericht der Deutschen Mathematiker-Vereinigung*, **4**, 71–87. [Also in *Gesammelte mathematische Abhandlungen*, Vol. 3, 482–97.]

Neuenschwander, E. *1978*, 'The Casorati–Weierstrass theorem. (Studies in the history of complex function theory I)', *Historia mathematica*, **5**, 139–66.

Osgood, W. F. *1901*, 'Allgemeine Theorie der analytischen Funktionen a) einer und b) mehrerer complexen Grössen', in *Encyklopädie der mathematischen Wissenschaften*, Vol. 2, Part 2, 1–114 (article II B 1).

Pincherle, S. *1880*, 'Saggio di una introduzione alla teoria delle funzioni analitiche secondo i principi del Prof. C. Weierstrass', *Giornale di matematiche*, **18**, 178–254, 313–57.

Remmert, R. *1983*, *Funktionentheorie*, Vol. 1, Heidelberg: Springer. [English transl.: 1991, Heidelberg: Springer.]

Riemann, G. F. B. *1851*, *Grundlagen für eine Theorie der Funktionen einer veränderlichen complexen Grösse*, Inauguraldissertation, Göttingen. [Also in *Werke*, 2nd edn, 1892, 3–48.]

—— *1859*, 'Über die Anzahl der Primzahlen unter einer gegebenen Grösse', *Monatsberichte der Berliner Akademie der Wissenschaften*, 671–80. [Also in *Werke*, 145–55.]

Weierstrass, K. *1988*, *Einleitung in die Theorie der analytischen Funktionen* (ed. P. Ullrich), Braunschweig: Vieweg.

3.13

Geometry in complex function theory

J. J. GRAY

1 INTRODUCTION: THE WORK OF GAUSS

Geometrical complex function theory deals with those parts of the subject where the domain of a function carries a natural geometrical structure and where the many-valued nature of complex expressions might otherwise cause difficulties. The first to adopt this approach was Carl Friedrich Gauss, who used it in the first decade of the nineteenth century to obtain elliptic functions with arbitrary periods. In modern terms, an elliptic function (§4.5) is a function f of a complex variable z which has two periods which, without loss of generality, may be taken to be σ and τ:

$$f(z) = f(z + \sigma) = f(z + \tau). \tag{1}$$

Here σ/τ is a complex number with positive imaginary part, called the modulus. The problem is to show that there is an elliptic function with precisely these periods.

If the periods are σ and τ, then any number of the form $m\sigma + n\tau$ is also a period. These numbers fill out the points of a lattice in the plane of complex numbers. The same lattice of periods, and so the same function, is obtained on replacing σ and τ by any expression of the form

$$\frac{\alpha\tau + \beta\sigma}{\gamma\tau + \delta\sigma},$$

where

$$\begin{pmatrix} \alpha & \beta \\ \gamma & \delta \end{pmatrix}$$

is a matrix with integer entries and determinant 1 (call this Condition 1). So, the problem reduces to finding an elliptic function for every different lattice of periods. The range of essentially distinct values of the modulus σ/τ is

432

therefore given by any region of the plane consisting of complex numbers such that every set

$$\left\{ \frac{\alpha\tau + \beta\sigma}{\gamma\tau + \delta\sigma}, \text{ all matrices } \begin{pmatrix} \alpha & \beta \\ \gamma & \delta \end{pmatrix} \text{ satisfying Condition 1} \right\}$$

has exactly one representative in the region. Such a set is conveniently given by the semi-infinite region ABCD depicted in Figure 1,

$$R_1 := \{z : |z| > 1, \ -\tfrac{1}{2} < \text{Im}(z) < \tfrac{1}{2}\}, \tag{2}$$

together with the boundary points for which $\text{Re}(z) \geqslant 0$. The basic approach of geometric function theory consists of drawing such a region and considering it rather than the whole complex plane.

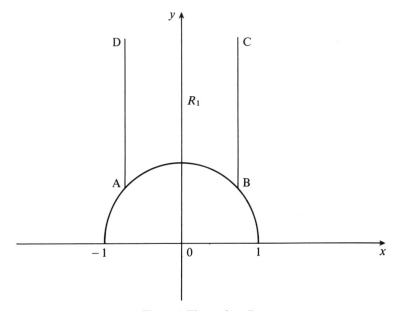

Figure 1 The region R_1

Since Gauss wrote an elliptic function in terms of three simpler functions (p, q and r), each of which had no poles, and each of which depended in a complicated way on the modulus τ, he had to investigate how these functions behaved as functions of τ. Their behaviour was a little more complicated, but he was able to show that what mattered was the behaviour of p^2/q^2. If this quotient could take any prescribed value, then the problem

433

was solved. This quotient was invariant under all the transformations of τ of this form:

$$\tau \to \left\{ \frac{\alpha\tau + \beta\sigma}{\gamma\tau + \delta\sigma}, \text{ where } \begin{pmatrix} \alpha & \beta \\ \gamma & \delta \end{pmatrix} \text{ is a matrix with integer entries,} \right.$$

of determinant 1, and such that α and δ are odd

and β and γ are even$\Big\}$.

Call this Condition 2.

Accordingly, Gauss sought a region of the complex plane consisting of complex numbers such that every set

$$\left\{ \frac{\alpha\tau + \beta\sigma}{\gamma\tau + \delta\sigma}, \begin{pmatrix} \alpha & \beta \\ \gamma & \delta \end{pmatrix} \text{ runs through the matrices satisfying Condition 2} \right\}$$

has exactly one representative in the region. Such a set is conveniently given by the region R_2 or ABCDE, lying between the two verticals and the two semicircular arcs in Figure 2. Having isolated the crucial region, Gauss was then able to show that there are indeed functions p and q such that the square of their quotient takes any prescribed value.

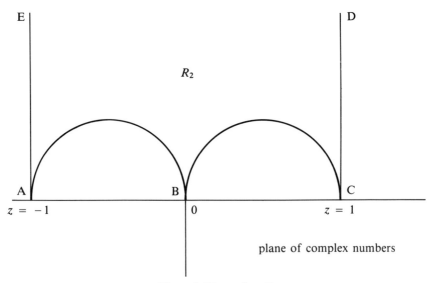

Figure 2 The region R_2

This was a dramatic achievement, for Gauss had now created a family of new functions which not only displayed a new phenomenon, double periodicity, but for which the ratio of the periods could take any non-real value. Remarkably, Gauss did not publish his findings, leaving it to Niels Abel and Carl Jacobi a generation later to discover elliptic functions, and then only when one of the periods was real (Gauss *1876*). One way in which Gauss's approach surpassed those of his successors was that he thought of complex numbers geometrically, whereas in the late 1820s they took a purely formal view.

2 THE LATER NINETEENTH CENTURY

By the 1850s, when Gauss's work was published, geometric-function theory had received a great stimulus from the work of Bernhard Riemann. In his lectures, Riemann advocated studying elliptic functions by thinking of them as defined on a torus and then looking at a corresponding region of the complex plane. In his celebrated and difficult memoir on Abelian functions, he proposed that any algebraic curve of higher genus (later called a Riemann surface in his honour) be cut by curves into a polygon which could be thought of as being in the complex plane. Paths on the surface were then to be thought of as paths in the polygon whose sides were glued together along the cuts. However, Riemann's ideas were.difficult to follow in any given case, and for a time they were rejected in favour of more rigorous algebraic approaches, until Felix Klein and Richard Dedekind independently sought to revive them.

Dedekind's motivation was to elucidate the theory of elliptic modular functions, which concerns such questions as how the periods of an elliptic function depends on its modulus. He wished to free this theory from the theory of elliptic functions themselves, which was up to this time (1878) the only way in which it had been approached. He succeeded, and showed how the theory depended only on a function j which had simple invariance properties:

$$j\left(\frac{\alpha\tau + \beta}{\gamma\tau + \delta}\right) = j(\tau),$$

where

$$\begin{pmatrix} \alpha & \beta \\ \gamma & \delta \end{pmatrix}$$

is a matrix with integer entries and determinant 1.

It follows that the function j takes every value once and only once on the region R_1, called its fundamental domain. On the basis of this discovery,

Dedekind gave an elegant account of how two elliptic functions whose moduli are simply related are themselves related, explaining, for example, what happened if the ratio of their periods is doubled or multiplied by any prime number. These were traditional questions in the subject, but Dedekind's approach was the first direct one and rapidly became standard.

Klein's contribution was to stress the group-theoretic aspect of all of this material. Matrices of the form

$$\begin{pmatrix} \alpha & \beta \\ \gamma & \delta \end{pmatrix}$$

with integer entries and determinant 1 (those satisfying Condition 1) form a group (called the modular group and denoted by Γ), and there are many sub-groups to be obtained by imposing simple congruence conditions on the integer entries. For example, Gauss's matrices satisfying Condition 2 form the sub-group of matrices

$$\begin{pmatrix} \alpha & \beta \\ \gamma & \delta \end{pmatrix} \equiv \begin{pmatrix} 1 & 0 \\ 0 & 1 \end{pmatrix} (\mathrm{mod}\, 2).$$

He also discussed how the matrices act geometrically on the complex plane. The effect of the modular group is to move points around as described by the formula above. It therefore moves its fundamental domain around *en bloc*, and tessellates the upper half-plane with it. Klein described in some detail how sub-groups have fundamental domains which are made up of several copies of the fundamental domain for the function j, forming a new tile which is moved *en bloc* by the sub-group.

Klein's best discovery relates to the subgroup of matrices

$$\Gamma_7 := \text{all } \begin{pmatrix} \alpha & \beta \\ \gamma & \delta \end{pmatrix} \equiv \begin{pmatrix} 1 & 0 \\ 0 & 1 \end{pmatrix} (\mathrm{mod}\, 7), \tag{3}$$

an important case in the theory of elliptic modular functions. It led him in 1879 to a figure consisting of 168 copies of the fundamental domain for the function j (Figure 3). This 14-sided polygon is moved in such a way that one can infer a pairing on the sides and so think of it as defining a Riemann surface (as it happens, of genus three). This was the first time that such an explicit description of a Riemann surface had been obtained, and Klein was able to find an equation for it as a curve in projective space ($x^3 y + y^3 z + z^3 x = 0$). It is unexpectedly symmetric, and the group of its symmetries is the quotient group Γ/Γ_7, which has 168 elements. It consists of all the matrices

$$\begin{pmatrix} \alpha & \beta \\ \gamma & \delta \end{pmatrix}$$

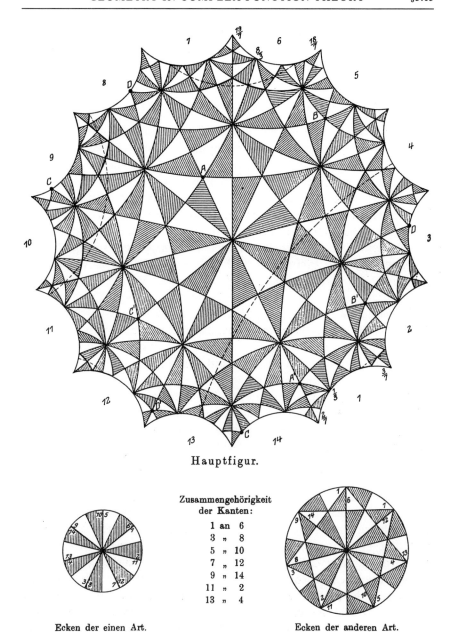

Hauptfigur.

Zusammengehörigkeit
der Kanten:

1 an 6
3 „ 8
5 „ 10
7 „ 12
9 „ 14
11 „ 2
13 „ 4

Ecken der einen Art.

Ecken der anderen Art.

Figure 3 The first explicit tessellation of a Riemann surface
(from Klein *1923*: 126)

437

whose entries lie in the finite field of integers mod 7, and where two matrices are also considered equivalent if their corresponding entries are equal and opposite (Beardon *1983*).

3 ESSENTIAL SINGULARITIES

Geometric function theory speedily proved its worth in other aspects of complex function theory. Its most celebrated success was in the elucidation of an essential singularity (Caratheodory *1960*). A singularity of a function was originally loosely regarded as a point at which it took an infinite value; for example, $f(z) = 1/z$ has a singularity at the point $z = 0$. The singularity is essential if the reciprocal function $1/f$ is not zero there. The singularity of $1/z$ is not essential, but the singularity at the point $z = 0$ of the function $e^{1/z}$ is essential. This distinction between types of singularity was introduced by Karl Weierstrass in the 1860s (§3.12) and shown to be of fundamental importance. In particular he showed, as did F. Casorati, that in any neighbourhood of any of its essential singularities a function comes arbitrarily close to any pre-assigned value (the Casorati–Weierstrass theorem). This means that no value can be attached by a limiting argument to the reciprocal of the function at the essential singular point. Then, in a very short paper of 1879, Emile Picard showed, using the techniques of geometric function theory, that in any neighbourhood of an essential singularity a function took every possible finite value except possibly one. For example, the function $e^{1/z}$ takes every finite value except 0.

Another indication of the power of geometric function theory was Henri Poincaré's increasingly profound elucidation of the theory of algebraic curves, culminating in his conjecture that all but the simplest of them arise, like Klein's curve mentioned above, from decompositions of the upper half-plane into a tessellation by polygons. By simultaneously introducing new classes of complex functions defined on the upper half-plane and taking the same value at all corresponding points of the tessellation, Poincaré showed how the corresponding algebraic curve could be parametrized. This extended the theory of elliptic functions to cover every curve. This conjecture, the so-called uniformization theorem, was eventually proved by Poincaré and P. Koebe in 1907.

BIBLIOGRAPHY

Beardon, A. F. *1983*, *The Geometry of Discrete Groups*, New York: Springer.

Carathéodory, C. *1960*, *Theory of Functions* (transl. F. Steinhart), 2 vols, New York: Chelsea.

Gauss, C. F. *1876*, *Werke*, 2nd edn, Vol. 3, Göttingen: Academy.

Gray, J. J. *1986*, *Linear Differential Equations and Group Theory from Riemann to Poincaré*, Boston and Basel: Birkhäuser.

Klein, F. *1923*, *Gesammelte mathematische Abhandlungen*, Vol. 3, Leipzig: Teubner.

3.14

Ordinary differential equations

CHRISTIAN GILAIN

1 THE INVERSE PROBLEM OF TANGENTS AND FIRST-ORDER EQUATIONS

In the strict sense, ordinary differential equations began with the birth of the calculus of Isaac Newton and Gottfried Wilhelm Leibniz, which led to the study of the inverse problem of differentiation: given a relation between two quantities and their differentials (or fluxions), how to find a relation between the quantities (or fluents) themselves (§3.2). However, this analytic problem of the integration of first-order differential equations corresponds to a geometric problem formulated previously, that of the inverse method of tangents; that is, how to find a curve characterized by a given property of its tangent (Florimond Debeaune and René Descartes: §3.1).

Beginning with the correspondence between Newton and Leibniz in 1676–7, a debate arose in connection with the question of methods of integration, about what it meant to solve a differential equation (Scriba *1963*). By utilizing expansions of expressions in power series, Newton showed that the inverse problem of tangents was fully solvable. Leibniz, however, expressing his desire to achieve solutions giving the nature of curves, was not satisfied with the systematic use of series and thought that, generally speaking, not enough was known yet about the inverse method of tangents. His procedure was essentially to change variables in order to attempt to transform the given differential equation into an equation with variables separated,

$$f(x)\,dx = g(y)\,dy, \qquad (1)$$

whose solution is immediately obtained by quadratures. Even before the end of the seventeenth century, the works of, especially, Leibniz, and John I and James Bernoulli, led towards the integration (i.e. the reduction to quadratures) of the homogeneous differential equation and of the first-order linear equation.

However, even having achieved such a separation of variables, which is not always the case, there remained the problem of reducing the quadratures thus obtained into other, simpler ones, particularly into those of the circle or the hyperbola (a question on which Newton also did much work). Also, as John Bernoulli remarked in his *Lectiones mathematicae* in 1691, the separation of the variables can also mask the nature of the sought-for curve. Thus, for example, solving the equation

$$y \, dx = 2x \, dy, \tag{2}$$

written with variables separated as

$$dx/x = 2 \, dy/y, \tag{3}$$

would involve logarithmic curves when in fact the solution of equation (2) is algebraic. By multiplying equation (2) written in the form $2x \, dy - y \, dx = 0$ by the factor y/x^2, the left-hand side of the equation so obtained is an exact differential, namely the differential of y^2/x. Hence the solution of equation (2) is $y^2 = bx$, the equation of a parabola. Here one recognizes the use of the so-called method of integrating factors, which however did not come into widespread use until 1740 (Alexis Clairaut, Leonhard Euler; see §4.3 on methods in the calculus).

2 SECOND-ORDER DIFFERENTIAL EQUATIONS AND RICCATI EQUATIONS

Various geometric or mechanical problems very soon caused mathematicians to start thinking about differential equations of order greater than one. In particular, the search for a plane curve characterized by a property of its curvature depends on a second-order equation. An important case was presented by Jacopo Riccati in 1723 to solve the second-order differential equation

$$x^m \, d^2x = d^2y + dy^2. \tag{4}$$

He reduced it, by means of a change of variables, to the first-order equation

$$x^m \, dq = du + (u^2/q) \, dx. \tag{5}$$

Then, by letting $q = x^n$, he raised the problem of finding the values of the exponent n that would allow the equation to be integrated by separation of the variables. Daniel Bernoulli was the first to publish a solution, in 1725.

Considering the differential equation in the simplified form

$$ax^n \, dx + u^2 \, dx = b \, du \tag{6}$$

(in which a and b are constants), he showed that separation of the variables is possible if $n = -4c/(2c \pm 1)$ (where c is a whole number) or $n = -2$.

The differential equation

$$\frac{dy}{dx} = Ay^2 + Bx^n \tag{7}$$

came to be called the Riccati equation, and stimulated much work. Euler's efforts yielded the main results. Considering the generalized Riccati equation

$$\frac{dy}{dx} = Py^2 + Qy + R \tag{8}$$

(where P, Q and R are functions of x), he showed in 1763 that the knowledge of a particular integral v reduced (8) by the change of function $y = v + (1/z)$ to a first-order linear differential equation, and therefore to two quadratures. He also established that the knowledge of two particular solutions M and N led, in stipulating that $(y - M)/(y - N) = z$, to an equation with separated variables z and x whose integral is obtained by a single quadrature. However, Euler does not appear to have realized that three particular solutions furnished the general integral of (8) directly, without quadrature, which is true because the anharmonic ratio of four solutions is constant. This last result was brought to light by Emile Picard in his thesis (1877) in connection with the determination of the asymptotic lines on a ruled surface by the Riccati equation.

By a change of function, Euler also showed the equivalence of the generalized Riccati equation and the second-order homogeneous linear differential equation. In particular, the equivalence of the equation

$$\frac{du}{dx} + u^2 + ax^n = 0 \tag{9}$$

and the linear equation

$$\frac{d^2 y}{dx^2} + ax^n y = 0 \tag{10}$$

enabled him, by expanding the integral of equation (10) in power series, to obtain the expression in finite terms of the solutions of the Riccati equation in the cases given by Daniel Bernoulli in which the variables are separable.

442

3 nTH-ORDER LINEAR DIFFERENTIAL EQUATIONS

The researches of Daniel Bernoulli and Euler on small oscillations of a horizontal elastic bar, one of whose ends is fixed, led to a linear differential equation

$$k^4 \frac{d^4 y}{dx^4} = y. \tag{11}$$

Euler first obtained a solution of this equation in the form of an infinite series (1735). Four years later he gave the solution in finite terms, as an application of the general method he had just created for the integration of nth-order homogeneous linear differential equations with constant coefficients. At the start of the important paper in which he presented this method in 1743, Euler laid the general framework of the theory of nth-order differential equations. He introduced the concepts of the complete integral (later referred to as the general integral), in which there are n arbitrary constants, and of the particular integral. For an nth-order homogeneous linear equation, he stated that knowledge of n particular integrals implied that the complete integral could be found as a linear combination of them. Although Euler sought no general criterion, he saw that these particular integrals had to be linearly independent, and he approached this in a practical manner.

Thus, for the equation with constant coefficients

$$0 = Ay + B \frac{dy}{dx} + C \frac{d^2 y}{dx^2} + \cdots + N \frac{d^n y}{dx^n}, \tag{12}$$

he sought particular integrals having the form $y = e^{px}$, where the quantity p satisfies the nth-order algebraic equation

$$0 = A + Bp + Cp^2 + \cdots + Np^n. \tag{13}$$

In the various cases of equation (13) for which the roots are simple or multiple, real or imaginary, he gave n linearly independent particular integrals of (12) and from them deduced the complete integral. Euler thus directly obtained integrations in finite terms of the members of an extended class of nth-order differential equations. A little later, in a paper published in 1753, he dealt (not without a few errors) with the case of the inhomogeneous linear equation.

The bases of the general theory of nth-order linear differential equations with variable coefficients were developed in 1765 by Joseph Louis Lagrange and Jean d'Alembert. Using two different methods, they showed that n particular integrals of the homogeneous equation determine the complete integral of the inhomogeneous equation with the help of n quadratures

(Demidov *1983*). In 1776, Lagrange noticed that this result could also be demonstrated using the so-called 'method of the variation of constants', which he gave the status of a general method.

4 DIFFERENTIAL EQUATIONS OF HIGHER DEGREE AND SINGULAR SOLUTIONS

In geometric problems in particular, there appeared first-order differential equations unsolved with respect to y'. These included equations which may be written in the form $F(x, y, y') = 0$, in which F is a polynomial of degree $p > 1$ in y'. These so-called algebraic differential equations of degree p highlighted a new phenomenon: the possible existence of a solution which is not included in the complete integral.

In 1715, Brook Taylor had already come across such a solution in the case of a second-degree equation, and noticed its singular character. In 1734, Clairaut was confronted with the same situation, arising out of the problem of how to find a curve enveloped by a right angle which slides in the plane and whose vertex describes a given curve. The equations indicated by Clairaut, which the solution of the preceding problem must satisfy, can be written in classical form as

$$y = xp + f(p), \quad p = \frac{dy}{dx}. \tag{14}$$

Proceeding by differentiation, we obtain on the one hand $dp = 0$ which, by integrating, gives the complete integral $y = ax + f(a)$ (which depends on the arbitrary constant a); and on the other hand, a relation between x and p which, combined with the initial equation, furnishes another solution. This latter solution, Clairaut emphasized, is actually the solution of the geometric problem posed.

In 1758, Euler stressed the dual paradox of such singular solutions in the integral calculus. They are obtained not by integration, but by differentiation of the differential equation, and they are not included in the so-called complete integral. All the great mathematicians of the time attempted to study this phenomenon. Significant was Lagrange's contribution, the results of which, published from 1776 onwards, were developed at length in his textbook *Leçons sur le calcul des fonctions*, first published in 1801. In particular, he demonstrated that if $F(x, y, a) = 0$ is the complete integral of the differential equation, a singular integral must satisfy the equation $D = 0$ obtained by eliminating a between the equations $F = 0$ and $\partial F/\partial a = 0$. From this he deduced that, geometrically, a singular integral corresponds to the envelope of curves defined by the complete integral.

However, many gaps remained in his study. The theory was to evolve in the direction of the development of a local viewpoint, in which particular attention was paid to the study of the singular points of solutions, and reached maturity towards the end of the nineteenth century. In particular, it appeared that the existence of a singular integral is the exception rather than the rule, and that the equation $D = 0$ often leads to a set of singular points of the integral curves which does not constitute a solution of the differential equation (Painlevé *1910*, Ince *1927*).

5 THE GENERAL PROBLEM OF INTEGRATION IN FINITE TERMS

Towards the middle of the eighteenth century, many differential equations had still not been reduced to quadratures by separation of the variables, and quadratures had not been reduced to those of the circle or the hyperbola. It was widely accepted at the time, on an empirical basis, that the class of elementary functions (algebraic, logarithmic, trigonometric) was not sufficient to express all solutions of the problems of integral calculus.

Following the line of thought in Fontaine's works, the Marquis de Condorcet, after a period marked by erroneous results, managed in the 1770s to derive a general formulation of the dual problem of recognizing in advance whether or not a differential equation is integrable in finite terms (i.e. with the help of elementary functions alone) and, if so, how to calculate the integral. In addition, he stated some important results. For example, if the integral of the first-order equation $P\,dx + Q\,dy = 0$, where P and Q are algebraic functions, is elementary, then it necessarily has the general form $G + L + C = 0$, in which C is a constant, G is an algebraic function of x and y, and L is a linear combination of logarithms of algebraic functions. This corresponds to the theorem demonstrated by D. Mordukhay-Boltovskoy at the beginning of the twentieth century. Statements of this kind, the demonstrations of which were certainly incomplete, also enabled Condorcet, in the case of quadratures, to obtain results of non-integrability in finite terms, in particular for a class of integrals including the classic integral $\int e^x/x\,dx$ (Gilain *1988*).

Starting in the 1830s, Joseph Liouville was to provide solid foundations for the theory. In particular, he showed that the Riccati equation (except for the cases pointed out by Daniel Bernoulli) is integrable neither in finite terms nor even with the help of quadratures (Lützen *1990*).

6 THE SITUATION APPROACHING 1800

In addition to works on exact integration, the theory of differential equations included at the time a set of results for the analytic expression of the solutions with the aid of infinite processes (power series, continued fractions, etc.) which formed the domain then called integration by approximation. Use of the method of series, present from the start, was developed particularly in the second half of the eighteenth century. This was done not only to express explicitly the solutions of certain types of differential equations not integrable in finite terms, but to demonstrate the general results of the theory as well. Thus, thanks to expansions in Taylor series, it was thought at the time to have been demonstrated that the complete integral, whose existence was not doubted, of an nth-order differential equation depends on n arbitrary constants.

7 EXISTENCE THEOREMS

The critical attitude of Augustin Louis Cauchy with respect to the high status then granted to the expansion of functions in Taylor series (§3.2) led him to formulate the problem of the existence of solutions of a general differential equation. Thus, in his analysis course of 1824 at the Ecole Polytechnique (Cauchy *1981*), he demonstrated the first existence theorem of a function $y = F(x)$ satisfying the differential equation $dy = f(x, y)\, dx$ such that $y_0 = F(x_0)$ (the solution of the so-called Cauchy problem) under the assumption that $f(x, y)$ and $\partial f(x, y)/\partial y$ are continuous in a neighbourhood of the initial values x_0 and y_0. Cauchy's procedure is based on the transformation of the algorithm of approximate calculation, utilizing the equation of finite differences, $\Delta y = f(x, y)\Delta x$ (already known to Euler in 1768), into a method of demonstrating existence by studying the convergence of the process in the limit. In addition to this theorem of local existence, Cauchy also took up, in his course, the question of global existence with the study of the extension of the solution over a maximal interval. Using the same hypotheses, Cauchy established the local uniqueness of the solution, then extended the existence theorem to a system of first-order differential equations. With the weaker hypotheses given by Rudolf Lipschitz in 1868 ($f(x, y)$ being continuous and satisfying a condition now named after him), this result of existence and uniqueness was to become the so-called Cauchy–Lipschitz theorem.

Later, in 1835, Cauchy presented a second demonstration of existence. This utilized the expression of the solutions in the form of power series, generally not entire, whose convergence he demonstrated using majorant series obtained via the so-called Cauchy inequalities relative to functions of a complex variable. This is not, in fact, a second demonstration of the same result, but rather a second local existence theorem, less general than the first since it involved the hypothesis that $f(x, y)$ be analytic. But this was not the opinion of Cauchy, who did not make this distinction and preferred the second method, in the context of the increasing role played by the theory of functions of complex variables. In 1856, C. A. Briot and J. C. Bouquet demonstrated, for the equation $du/dz = f(z, u)$ in the complex domain, with precise hypotheses concerning the function f, the existence and uniqueness of a holomorphic solution $u(z)$ in a disc whose centre is z_0. This demonstration, utilizing expansions in power series, has become a classic.

Towards the end of the nineteenth century, Picard transformed the so-called method of successive approximations then known into a general method of demonstrating existence (1890). In his *Traité d'analyse* (1893) one finds the first consistent exposition of the results of existence in which a clear distinction is made between the different theorems and fields of application of the methods.

The hypotheses of the theorems indicated thus far assured not only the existence but also the local uniqueness of the solution of the Cauchy problem. In 1886, Giuseppe Peano established an existence theorem for a first-order differential equation in the real domain with the sole hypothesis that $f(x, y)$ be continuous. In 1890, he emphasized the absence of a unique solution in this case, giving the famous counter-example of the solutions of the equation $dx/dt = 3x^{2/3}$ with the initial condition $x(0) = 0$.

8 SINGULAR POINTS AND THE COMPLEX DOMAIN

In the second half of the nineteenth century, we find a considerable number of works on differential equations in which the theory of functions of a complex variable, then in full sway, is used. In 1856, Briot and Bouquet undertook the study of solutions of the first-order equation $du/dz = f(z, u)$ in the neighbourhood of the singular points (z_0, u_0) – points where f no longer satisfies the conditions of regularity which assure the existence and local uniqueness of a holomorphic solution.

In 1866, L. I. Fuchs presented the foundations of the theory of linear differential equations in the complex domain (Gray *1986*). For an nth-order homogeneous linear equation

$$\frac{d^n y}{dx^n} + p_1 \frac{d^{n-1} y}{dx^{n-1}} + \cdots + p_n y = 0, \tag{15}$$

he showed that the solutions are holomorphic as long as the p_i are holomorphic. Then, in the neighbourhood of a singular point x_0 of the coefficients, he looked for a system of n fairly 'regular', linearly independent solutions of (15), where the condition of independence is expressed by the non-vanishing of the determinant that Thomas Muir in 1882 was to call the 'Wronskian' (§6.6). Fuchs highlighted the important case in which the singular point x_0 is a pole whose order is at most i for each coefficient

p_i (which was later to be called a 'regular singular' point), and he defined the class of linear differential equations all of whose singular points (assumed to be of finite number) and the point at infinity are of this type. This class of equations of so-called Fuchsian type in particular includes the hypergeometric equation. In the early 1880s Henri Poincaré, systematically taking up the global study of the solutions, managed by constructing new transcendental functions to express analytically the integrals of some fairly large classes of linear differential equations (§3.16).

For non-linear equations the situation is even more complicated, since the singular points of the solutions are no longer fixed and determinable in advance from the differential equation, as they are for linear equations, but generally depend on the constant of integration. In 1884, works by Fuchs and Poincaré on algebraic differential equations whose solutions have no other mobile singular points than poles showed that first-order equations of this kind reduce to quadratures or to the Ricatti equation. Going on to second-order equations, P. Painlevé obtained equations whose solutions are new transcendental functions (1900). Thus, for example, the differential equation $y'' = 6y^2 + x$ has meromorphic solutions in the complex plane, even though the equation cannot be reduced either to quadratures or to linear equations (Ince *1927*).

9 QUALITATIVE STUDIES

In 1836, J. C. F. Sturm published a paper in which he looked at second-order linear differential equations from a new, qualitative viewpoint. Having accepted that one cannot integrate most of these equations exactly, and having noticed that, even though one might know the analytic expression of the solution, this does not mean that one can determine its essential properties, he proposed finding these properties directly from the differential equation. Thus, for the equation in the self-adjoint form

$$\frac{d}{dx}\left(K\frac{dV}{dx}\right) + GV = 0 \qquad (16)$$

(with $K(x) > 0$ on an interval $[a, b]$), he studied how the zeros of a solution vary when one varies the initial conditions or the coefficients K and G of the equation. In particular, Sturm showed that if V_1 and V_2 are two independent solutions of the same equation (16), then one zero of V_2 lies exactly between two zeros of V_1 (the so-called Sturm separation theorem: §4.10). Then, by varying the coefficients of equation (16), he obtained 'comparison theorems', including this fundamental one: if K decreases and G increases, then the number of zeros of the differential equation increases.

This theory of Sturm had many applications, particularly in the study of so-called Sturm–Liouville systems (Lützen *1990*).

On the basis of such questions as the stability of n bodies in celestial mechanics, Poincaré, beginning in 1880, stressed the importance of the global qualitative properties of the real solutions of general differential equations and the lack of pertinence of the analytic methods then preferred for solving them. He developed the direct study of these properties in a long paper entitled 'Sur les courbes définies par une équation différentielle', thus creating the geometric or qualitative theory of differential equations.

For a first-order, first-degree equation $dx/X = dy/Y$ (where X and Y are real polynomials), Poincaré proposed to determine the form of the integral curves over the whole extent of a sphere onto which he projected them. He made no assumptions about the analytic expression of the solutions. The principle of his methods was to consider the family of all integral curves and to study the geometric relations between these, the singular points and certain particular algebraic curves. In particular, he arrived at the following classification theorem: any integral curve which does not end at a singular point is a cycle or a spiral winding round a cycle (called a limit cycle). This theorem, fundamental for the general theory of differential equations on the sphere or the plane, came to be called the Poincaré–Bendixson theorem, in recognition of the Swedish mathematician Ivar Bendixson, who in 1901 published a more rigorous demonstration with weakened hypotheses.

Poincaré then went on to consider first-order algebraic differential equations of higher degree, $F(x, y, dy/dx) = 0$. His essential idea here was to associate this equation with the surface whose equation is $F(x, y, z) = 0$. An integral curve then appears as the projection onto the plane of a curve traced on this surface which locally satisfies a differential equation of the type $du/U = dv/V$ (where U and V are sufficiently regular functions). For the curves situated on the surface (in modern language, the integral curves of a field of tangent vectors), one finds the same local properties on which the results obtained previously were based. But the same is not true of the global behaviour, which seemed to depend on the genus of the surface $F(x, y, z) = 0$. Thus in the case of the torus (a surface of genus one) dealt with by Poincaré, there may be integral curves everywhere dense in the surface and no singular point. The role of the topology thus appeared to be essential in changing from the local study to the global study of the integral curves of non-linear differential equations. The wish to develop the geometric theory of the differential equations of order higher than two, particularly with applications to celestial mechanics in mind, was indeed one of the main elements which led Poincaré to found algebraic topology shortly afterwards (§7.10).

10 THE SITUATION APPROACHING 1900

Towards the end of the nineteenth century, the theory of ordinary differential equations can be considered to have attained its classic maturity. The various existence theorems for first-order equations or systems of such equations formed by then a consistent whole, providing a solid foundation for general theory both in the real domain and in the complex domain. On this basis, research developed principally in three directions corresponding to the algebraic, analytic and qualitative viewpoints (although with some overlap).

The approach of algebraic integration (formal integration, in the language of the time) was to study the reducibility of the differential equations to simpler types. In particular, Picard and E. Vessiot, drawing on the work by Sophus Lie on continuous groups of transformations (§6.5), developed for homogeneous linear differential equations a theory analogous to Evariste Galois's theory for algebraic equations, including solutions in finite terms (Vessiot *1910*). For an irreducible differential equation, analytic integration seeks to represent the integral − locally or globally − by an infinite process, in particular by series. This approach involves essential relations with the theory of functions of a complex variable. Lastly, the qualitative study of certain properties then gave rise to a few works; one of the most influential studies was by Aleksandr Lyapunov, on the general problem of the stability of solutions (1892) (see §8.4 on feedback control).

11 FUNCTIONAL ANALYSIS AND MODERN THEORY

In the twentieth century, the study of differential equations has developed largely along the three main lines just indicated (Pommaret *1983*, Hille *1976*, Hartman *1982*). One essential characteristic is the importance of the role that functional analysis (§3.9) and topology (§7.10) have come to play in the general theory. Thus, in 1927, J. Schauder gave a new demonstration of the existence of solution to the Cauchy problem for an equation $dy/dx = f(x, y)$, showing that the operator which associates the function

$$y_0 + \int_{x_0}^{x} f(s, y(s)) \, ds \tag{17}$$

with the function y has a fixed point. For this, he used a generalization of Brouwer's fixed-point theorem in infinite-dimensional spaces, obtained by G. D. Birkhoff and O. D. Kellogg in 1922.

Complete normed vector spaces, the axiomatization of which was being pursued by Stefan Banach in the 1920s (§3.9), were indeed to become the general framework of the modern theory of differential equations. Shortly after the Second World War, there appeared Bourbaki's *Eléments de mathématique*, a classic presentation of the general theory of differential equations in Banach spaces. Although many methods and theorems valid in finite-dimensional spaces can be generalized to infinite-dimensional spaces, some cannot. The former is the case for Peano's theorem, for example: the counter-example given by J. Dieudonné in 1950 of a differential equation $y' = f(x, y)$ in a space of sequences showed that the continuity of f is no longer sufficient to ensure the existence of a solution in infinite-dimensional spaces.

Today, after more than three centuries of work, the theory of ordinary differential equations still appears to be one of the most active branches of mathematics.

BIBLIOGRAPHY

Cauchy, A. L. *1981*, *Equations différentielles ordinaires*, Paris: Etudes Vivantes, and New York: Johnson. [Photoreprint of 1824 printed lecture notes, with editorial introduction by C. Gilain.]

Demidov, S. S. *1983*, 'On the history of the theory of linear differential equations', *Archive for History of Exact Sciences*, **28**, 369–87.

Gilain, C. *1988*, 'Condorcet et le calcul intégral', in *Sciences à l'époque de la Révolution Française*, Paris: Blanchard, 85–147.

Gray, J. *1986*, *Linear Differential Equations and Group Theory from Riemann to Poincaré*, Boston: Birkhäuser.

Hartman, P. *1982*, *Ordinary Differential Equations*, 2nd edn, Boston: Birkhäuser. [Particularly for qualitative theory.]

Hille, E. *1976*, *Ordinary Differential Equations in the Complex Domain*, New York: Wiley.

Ince, E. L. *1927*, *Ordinary Differential Equations*, London: Longmans, Green. [Repr. 1944, New York: Dover. Many historical notes.]

Kline, M. *1972*, *Mathematical Thought from Ancient to Modern Times*, New York: Oxford University Press. [Particularly Chaps 21 and 29.]

Lützen, J. *1990*, *Joseph Liouville 1809–1882: Master of Pure and Applied Mathematics*, New York: Springer.

Mawhin, J. *1988*, 'Problème de Cauchy pour les équations différentielles et théories de l'intégration: Influences mutuelles', *Cahiers du Séminaire d'Histoire des Mathématiques*, **9**, 231–46.

Painlevé, P. *1910*, 'Existence de l'intégrale générale. Détermination d'une intégrale particulière par ses valeurs initiales', in *Encyclopédie des sciences mathématiques*, Tome 2, Vol. 3, 1–57 (article II 15).

Pommaret, J. F. *1983*, *Differential Galois Theory*, New York: Gordon & Breach.

Scriba, C. J. *1963*, 'The inverse method of tangents: A dialogue between Leibniz and Newton (1675–1677)', *Archive for History of Exact Sciences*, **2**, 113–37.

Vessiot, E. *1910*, 'Méthodes d'intégration élémentaires. Etude des équations différentielles ordinaires au point de vue formel', in *Encyclopédie des sciences mathématiques*, Tome 2, Vol. 3, 58–170 (article II 16).

3.15

Partial differential equations

J. LÜTZEN

1 INTRODUCTION

Originating in the middle of the eighteenth century in the theory of the mechanics of continuous bodies, partial differential equations (PDEs) have continued to draw much of their lifeblood from this area as well as from heat conduction, fluid mechanics, electromagnetism, quantum mechanics and other parts of physics. There exists no general method of solving PDEs, but from the end of the eighteenth century various classes of equations have found their own general solution. During the nineteenth century the search for general solutions was overshadowed by initial- and boundary-value problems. The idea emerged that it was important to show the existence and uniqueness of solutions of such problems, even if no explicit method of solution was within reach. Existence theorems have dominated much of pure mathematical research on PDEs during the twentieth century.

Although general theorems, concepts and methods have emerged since the eighteenth century, the field of PDEs is still not united under one all-embracing, unified point of view. It has always been characterized by a great variety of ideas, methods and tricks drawn from many diverse areas of mathematics such as ordinary differential equations, real and complex function theory, series expansions, the calculus of variations, difference equations, group theory, differential geometry, integral equations and functional analysis. Also, PDEs have been a prime motivation for developments in these fields. In Morris Kline's words, in the nineteenth century 'partial differential equations became and remain the heart of mathematics' (Kline *1972*: Chap. 28).

2 PRELUDE: FAMILIES OF CURVES

When Jean d'Alembert solved the first PDE in the 1740s, he drew heavily on earlier concepts and techniques on partial differentiation. Some historians of mathematics have seen the origin of partial derivatives in Isaac

Newton's procedure for finding the radius of curvature. Indeed, Newton considered an algebraic curve, whose equation we can write as $f(x, y) = 0$. He then introduced the symbols $\mathcal{X}$ and $\mathcal{X}$ to denote expressions which in special cases can be identified with our $x \, \partial f / \partial x$ and $y \, \partial f / \partial y$, respectively. However, as pointed out by Engelsman *1984*, the conceptual novelty of partial differentiation did not become apparent until problems with two independent variables were encountered. Interestingly enough, this first occurred not in the study of surfaces, but in problems concerning families of curves.

After an isolated appearance in Gottfried Wilhelm Leibniz's papers on envelopes from 1692 and 1694, partial differentiation came to the fore in the new differential calculus in work produced from 1697 by John I and James I Bernoulli, Leibniz, Nicholas I and II Bernoulli, and Leonhard Euler on orthogonal trajectories and other trajectories to families of curves. While Leibniz's original calculus was limited to two variables x and y related to one curve, study of the new problems yielded the insight that a similar formalism would allow differentiation with respect to the parameter a which determines the curve of the family. The families in question were either defined by an algebraic equation $V(x, y, a) = 0$ (or $y = f(x, a)$) or given transcendentally by $y = \int_{x_0}^{x} p(x, a) \, dx$. Many different techniques were devised to study these questions. Here are mentioned a few key results and ideas that eventually led Euler to a sort of partial differential equation (for more details see Engelsman *1984*).

In 1697 John I Bernoulli discovered that the a and the x of an orthogonal trajectory are related by the differential equation

$$(1 + p^2) \, dx + pq \, da = 0, \tag{1}$$

where $p = \partial y / \partial x$ and $q = \partial y / \partial a$. (The notation $\partial f / \partial x$ used here was suggested by Adrien Marie Legendre, although it was not generally accepted until the second half of the nineteenth century.) The relation between x and y can then be found by eliminating a from a solution of this equation and the equation of the family of curves. However, if the family of curves was given in the form $y = \int_{x_0}^{x} p(x, a) \, dx$, this was no simple matter. The first problem, and the only one considered here, was to determine q. Leibniz discovered that this could be done by interchanging integration and differentiation with respect to a parameter

$$q = \frac{\partial}{\partial a} \int_{x_0}^{x} p(x, a) \, dx = \int_{x_0}^{x} p_a(x, a) \, dx. \tag{2}$$

As reformulated by Nicholas I Bernoulli around 1719, the problem is this: given a partial differential $d_x y = p(x, a) \, dx$, in which a must be considered constant, find the corresponding total differential

$$dy = p(x, a)\, dx + q(x, a)\, da, \tag{3}$$

where both y and a can vary freely. Around 1730, the young Euler pointed out that the interchangeability result (2) does not allow one to consider a as a free variable unless the integral $\int_{x_0}^{x} p_a(x, a)\, dx$ can be evaluated explicitly. The reason for this seems to be the confusion of the dummy variable of integration and the upper limit. Indeed, in Euler's notation, it is of course impossible to let a be a function, $V(x)$, say, because $\int_{x_0}^{x} p_a(x, V(x))\, dx$ is different from the result one obtains by making the substitution $a = V(x)$ after the integration has been performed. Yet Euler discovered that, if $p(x, a)$ is a homogeneous function of degree $n - 1$ (and if $x_0 = 0$), then the total differential corresponding to $d_x = p(x, a)\, dx$ may be written as

$$dy = p(x, a)\, dx + (ny - p(x, a)x)\, da/a. \tag{4}$$

This follows from Euler's theorem stating that $ny = px + qa$, where y is a homogeneous function of x and a. Euler called a total differential of the first or higher order with algebraic coefficients a 'modular equation'.

In an 'Additamentum' to his paper 'De infinitis curvis eiusdem generis' (1734–5, published in 1740) Euler then set out to find other classes of functions for which $q(x, a)$ has a simple relation to $p(x, a)$ of the form

$$F(x, y, a, p, q) = 0. \tag{5}$$

This is a partial differential equation. Euler's analyses depend crucially on the theorem that a differential form $M(x, y)\, dx + N(x, y)\, dy$ is exact (i.e. the total differential of some function of x and y) if and only if $\partial M/\partial y = \partial N/\partial x$. This theorem was discovered by Euler (1734–5), and independently by Alexis Clairaut (1739–40), also for differentials of more than two variables.

Euler's treatment of the equation

$$q = - px/a \tag{6}$$

is typical. By way of (6) he eliminated q from the total differential

$$dy = p\, dx + q\, da \tag{7}$$

and found that

$$dy = p\, dx - (px/a)\, da = p(dx - x/a\, da). \tag{8}$$

The differential in the last parenthesis is not an exact differential, but Euler noticed that it can be made exact by multiplication by a so-called integrating factor $1/a$ so that

$$\mathrm{d}y = pa(\mathrm{d}x/a - x/a^2\,\mathrm{d}a) = pa\,\mathrm{d}(x/a).\tag{9}$$

From this equation he concluded that pa must be an arbitrary function of x/a: $p = 1/a\phi(x/a)$ (i.e. p is a homogeneous function of degree -1). This result is in accordance with Euler's theorem mentioned above. Euler was able to treat rather complicated equations such as $q = R(x, a)p + F(a)y$. In all cases he followed the same method: using the given equation and some ingenious multipliers and substitutions of variables, he transformed the total differential (7) into the form $\mathrm{d}y = t\,\mathrm{d}u$, and concluded that t must be some arbitrary function of u.

Did Euler solve partial differential equations here? Certainly the equations he analysed were PDEs, and he had a uniform method of treating them. However the problem he set himself was not the solution of the equation – that is, the determination of y – but the determination of p. Moreover, Euler's programme of modular equations led nowhere, and so the breakthrough to PDEs had to wait another 10 years, until d'Alembert introduced them in his work on mechanics.

3 THE FIRST SOLUTIONS OF PDEs: THE VIBRATING STRING

A continuing tradition in partial differential equations began with d'Alembert's work from the 1740s on continuum mechanics (Demidov *1982a*). He formulated the first PDE in his *Traité de dynamique* (1743), but was unable to solve it. Three years later d'Alembert won a prize at the Berlin Academy of Sciences for his 'Réflexions sur la cause générale des vents' in which he solved a number of PDEs. His formulation of the problems was different from the modern one. For example, he posed the problem of determining $\alpha(u, s)$ and $\beta(u, s)$ so that the differentials

$$\alpha(u, s)\,\mathrm{d}s + \beta(u, s)\,\mathrm{d}u\tag{10}$$

and

$$\rho\alpha(u, s)\,\mathrm{d}u + \nu\beta(u, s)\,\mathrm{d}s + A(u, s)\,\mathrm{d}u + \Gamma(u, s)\,\mathrm{d}s\tag{11}$$

(where ρ and ν are constants) are both exact differentials of some functions z and v, respectively. Using the Clairaut–Euler condition for the exactness of a differential form, d'Alembert wrote the problem in the form

$$\frac{\partial\alpha}{\partial u} = \frac{\partial\beta}{\partial s}, \qquad \nu\frac{\partial\beta}{\partial u} = \rho\frac{\partial\alpha}{\partial s} + \phi(u, s),\tag{12}$$

where $\phi(u, s) := \partial A/\partial s - \partial\Gamma/\partial u$. From these equations it is easy to deduce that z must satisfy the PDE

$$\nu\frac{\partial^2 z}{\partial u^2} - \rho\frac{\partial^2 z}{\partial s^2} = \phi(u, s) \tag{13}$$

and the corresponding PDE for v. But d'Alembert did not write these equations explicitly; his method of solution followed Euler's method in the 'Additamentum'. He introduced the new variables $m = \alpha + \beta(\nu/\rho)^{1/2}$ and $t = s + u/(\nu/\rho)^{1/2}$ in a suitable combination of (10) and (11), and thus arrived at a new differential:

$$m\,dt + \Psi(s, t)\,dt + \Pi(s, t)\,ds. \tag{14}$$

This must also be exact, and so

$$\frac{\partial m}{\partial s} + \frac{\partial\Psi(s, t)}{\partial s} = \frac{\partial\Pi(s, t)}{\partial t}, \tag{15}$$

from which d'Alembert concluded that

$$m = -\Psi(s, t) + \int\frac{\partial\Pi(s, t)}{\partial t}\,ds - \phi(t), \tag{16}$$

where ϕ is an arbitrary function. Similarly, he introduced the variables

$$\mu = \beta(\nu/\rho)^{1/2} - \alpha \quad\text{and}\quad y = u/(\nu/\rho)^{1/2},$$

and determined μ. From the expressions for m and μ, he determined α and β, and thus z and v. The essential feature of the new variables t and y is that $t = c_1$ and $y = c_2$ represent what Gaspard Monge later called 'characteristic curves'.

D'Alembert's most influential work on PDEs was his solution in 1747 (published in 1749) of the problem of the vibrating string. Brook Taylor had argued in 1713 that the accelerating force of a point on a homogeneous string was proportional to the radius of curvature of the string. We would now write this as the wave equation

$$\frac{\partial^2 y(x, t)}{\partial t^2} = \frac{\partial^2 y(x, t)}{\partial x^2}, \tag{17}$$

where $y(x, t)$ is the displacement of the string at the point x at time t, and the speed of propagation is unity.

However, Taylor immediately made some restrictive assumptions on the behaviour of the string and thus missed this equation. In fact, he only determined the fundamental sine-function solution of the problem. Unlike Taylor, d'Alembert immediately sought the solution in the form of a function y of both x and t, and arrived at the equation

$$dp = \alpha\,dt + \nu\,dx \quad \text{and} \quad dq = \nu\,dt + \alpha\,dx \tag{18}$$

for the partial derivatives $p = \partial y/\partial t$ and $q = \partial y/\partial x$. With the methods used above (i.e. introducing $t + x$ and $t - x$ as new variables), he easily obtained

$$y(x, t) = \Psi(t + x) + \Gamma(t - x), \tag{19}$$

where Ψ and Γ are arbitrary functions. Moreover, he showed how to determine Ψ and Γ so as to satisfy the initial-value conditions

$$y(x, 0) = f(x) \quad \text{and} \quad \partial y/\partial t(x, 0) = g(x) \tag{20}$$

and the boundary conditions

$$y(0, t) = y(\pi, t) = 0. \tag{21}$$

D'Alembert's paper triggered a lengthy discussion between d'Alembert himself, Euler and Daniel Bernoulli, and later Joseph Louis Lagrange. There were two points at issue. First, how general can the arbitrary functions Ψ and Γ be? D'Alembert initially held that they must be analytical expressions, but later relaxed the condition almost to a modern-looking concept of a solution: they must be twice differentiable. Euler, on the other hand, argued that physical necessity forced mathematicians to consider Ψ and Γ corresponding to any hand-drawn curve. Some of his arguments can be read as attempts to generalize the concept of a solution to a PDE. This part of the discussion marks an important stage in the development of the concept of function. Second, Daniel Bernoulli solved the problem by a trigonometric series, representing a superposition of the fundamental and higher modes. This raised the question of whether an arbitrary function can be developed in a trigonometric series. Euler, d'Alembert and, later, Lagrange agreed that a series of the form

$$y = \sum A_n \sin(nx)\cos(nt) \tag{22}$$

would solve the wave equation with the boundary conditions (21), but they also agreed that this is only a special solution because the initial-value equation

$$f(x) = \sum A_n \sin(nx) \tag{23}$$

would be satisfied only for very special functions f. This remained the general belief until Joseph Fourier (1807, 1822) (see Section 5).

4 MODELS OF PHYSICAL PHENOMENA

After the initial success with the vibrating string, many new second-order linear PDEs were suggested as mathematical descriptions of various physical systems. First, Euler and d'Alembert, and later Lagrange, studied generalizations of the wave equation in an attempt to describe other vibratory phenomena such as vibrations of inhomogeneous strings, hanging cords, drums, and so on. Laplace's equation,

$$\Delta V := \frac{\partial^2 V}{\partial x^2} + \frac{\partial^2 V}{\partial y^2} + \frac{\partial^2 V}{\partial z^2} = 0, \tag{24}$$

was studied first by d'Alembert (1752) (in two variables), then by Euler (1754) in connection with non-viscous fluid dynamics, and from the 1780s by Lagrange, Legendre, Pierre Simon Laplace and others in work on potentials due to Newtonian forces. Indeed, this equation has such a rich history that it is considered separately in §3.17.

During the first quarter of the nineteenth century a new area of physics, heat conduction, was subjected to a mathematical treatment (§9.4). Fourier gave the fundamental PDE, the heat equation

$$k^2 \frac{\partial u}{\partial t} = \Delta u, \tag{25}$$

first in a prize-winning essay (1807) and later in his *Théorie analytique de la chaleur* (1822). His initial work was soon taken up by Siméon-Denis Poisson and by the younger generation of French mathematicians. Only slightly later the system of PDEs describing a viscous fluid was obtained by Claude Navier (1821), Poisson (1829) and George Stokes (1845). Other systems of PDEs were derived by Navier (1821) to describe the vibration of isotropic elastic media, and these were generalized the following year by Cauchy to anisotropic solids. In the 1860s, James Clerk Maxwell derived the eponymous Maxwell equations governing electric and magnetic fields, and in 1895 D. J. Korteweg and J. de Vries set up the equation named after them as a means of describing the propagation of long water waves.

Even the mechanics of point masses and rigid bodies, once the prime domain of ordinary differential equations, came under the reign of PDEs. This happened in two steps. First, William Rowan Hamilton (1834–1835) and Carl Jacobi (1837) showed that the solution of the equations of motion, the Hamilton equations, was equivalent to the solution of one (Hamilton had two) first-order PDE, the so-called Hamilton–Jacobi equation (§8.1). Second, the quantum-mechanical revolution in principle transformed all mechanics into the study of a PDE, namely the Schrödinger equation, derived in 1926 by the physicist after whom it is named (§9.14).

5 SEPARATION OF VARIABLES; HEAT CONDUCTION; SPECTRAL THEORY, SERIES AND SPECIAL FUNCTIONS

Unable to find the general solutions to the PDEs of the more complicated vibratory phenomena, Euler and d'Alembert returned to the method of separation of variables. This method was used with great success in the nineteenth century and gave rise to important development in several other areas of analysis.

The classical form of the method of separation of variables was formulated by Fourier in his work on heat conduction. For example, stationary heat conduction in a semi-infinite bar, held at zero degrees along its horizontal edges and at a fixed temperature $u_0(y)$ at the left boundary, led him to the boundary-value problem shown in Figure 1:

$$\frac{\partial^2 u}{\partial x^2} + \frac{\partial^2 u}{\partial y^2} = 0 \tag{26}$$

with

$$u(x, 0) = u(x, \pi) = 0, \tag{27}$$

$$u(x, y) \to 0 \quad \text{as } x \to \infty \tag{28}$$

and

$$u(0, y) = u_0(y), \quad 0 < y < \pi. \tag{29}$$

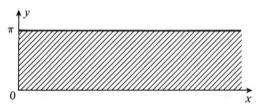

Figure 1 Boundary-value problem: heat conduction in a semi-infinite bar

Following an idea of d'Alembert's, Fourier separated the variables, looking for solutions that can be written as a product of a function of x and a function of y:

$$u(x, y) = F(x)f(y). \tag{30}$$

In this case equation (26) reduces to

$$\frac{F''(x)}{F(x)} = -\frac{f''(y)}{f(y)}. \tag{31}$$

Fourier argued that since the left-hand side of (31) is only a function of x and the right-hand side is only a function of y, then they can be equal only if they are equal to a constant, say m^2, so that

$$F''(x) = m^2 F(x), \tag{32}$$

$$f''(y) = -m^2 f(y). \tag{33}$$

According to (27), f must satisfy the boundary condition

$$f(0) = f(\pi) = 0, \tag{34}$$

which implies that

$$f(y) = A \sin(my). \tag{35}$$

In modern terms we would say that the eigenvalue problem (33) and (34) has eigenvalues $m = 1, 2, 3, \ldots$, and the corresponding eigenfunctions are given by (35). Moreover, taking (28) into account, Fourier concluded from (32) that $F(x) = e^{-mx}$, so that

$$u(x, y) = A e^{-mx} \sin(my). \tag{36}$$

Because of the linearity of the problem (26)–(28), the infinite linear combination

$$u(x, y) = \sum_{m=1}^{\infty} A_m e^{-mx} \sin(my) \tag{37}$$

is also a solution. Fourier was then faced with the problem of determining the coefficients A_m so that (29) is fulfilled,

$$u_0(y) = \sum_{m=1}^{\infty} A_m \sin(my), \quad 0 < y < \pi. \tag{38}$$

Thus the problem of expanding an arbitrary function in a trigonometric series turned up again. Unlike Euler, d'Alembert and Lagrange, Fourier argued that an arbitrary function can be expanded in a Fourier series (the first rigorous proof was given by J. P. G. Lejeune Dirichlet in 1829; §3.3), and was therefore of the opinion that the method of separation of variables gave the general solution to the problem.

Most boundary-value problems led to eigenvalue problems of ordinary differential equations more complicated than (33). For example, Euler's study of the vibrating circular membrane had led to eigenfunctions that are Bessel functions, and in the study of potentials of spheroids Laplace and Legendre had found spherical functions as eigenfunctions. The reason for this complexity is that in order for the method of separation of variables to work it is necessary for one of the coordinates to be constant on the surface where the boundary condition is given. This forced the mathematicians to use polar, spherical, cylindrical and other curvilinear coordinate systems. Expressed in these coordinates, even the simple PDEs of mathematical physics become complicated with variable coefficients.

Fourier also encountered such complicated eigenvalue problems when he studied heat conduction in a sphere and a cylinder. During the 1830s, in the hands of Gabriel Lamé, the method of separation of variables in curvilinear coordinate systems became a versatile method of solving boundary-value problems for PDEs, in particular the Laplacian. Lamé functions originate in this research. Later in the nineteenth century, this technique gave rise to many other classes of special functions (§4.4).

A conceptually even more significant use of the method of separation of variables appeared in the mid-1830s when J. C. F. Sturm and Joseph Liouville undertook the study of heat conduction in a heterogeneous, unequally polished bar. This phenomenon is described by a linear PDE which has coefficients that are variable, and not even explicitly known. After separation of variables, they arrived at an eigenvalue problem whose eigenvalues and eigenfunctions cannot be found explicitly. Still, by arguing directly from the equation they were able to deduce many interesting qualitative properties of the eigenfunctions. Thus they initiated the general spectral theory of differential operators.

In order to show that separation of variables leads to the general solution of various boundary-value problems, it was important to show that 'arbitrary' functions may be expanded in a series of eigenfunctions. This problem of completeness became a central one in the nineteenth century, both for the many concrete classes of special functions and for the Sturm–Liouville problem and other general eigenvalue problems.

Until the middle of the nineteenth century, separation of variables led only to eigenvalue problems of ordinary differential operators. However, when one variable is separated out in a PDE with three or more independent variables, the result is an eigenvalue problem of a PDE. For example, if the time variable is separated out in the two- or three-dimensional wave equation, the result is $\Delta V + k^2 V = 0$, where k is a parameter which must often be chosen so that there exist solutions V satisfying given boundary conditions. This equation is often called the Helmholtz equation because Herman von Helmholtz initiated the general study of it in 1860 in a work on oscillations of air in an organ pipe (§9.8). The two-dimensional problem was subsequently studied by Gustav Kirchhoff, Heinrich Weber, K. H. A. Schwarz, Henri Poincaré and other mathematicians in connection with vibrating membranes.

6 GENERAL SOLUTIONS: FIRST-ORDER PDEs

Alongside the study of physically interesting PDEs, mathematicians also tried to design general methods to solve various classes of PDEs. First, d'Alembert and Euler tried to systematize and generalize their methods based on exact differentials and multipliers, and produced what may be called the first treatises on the subject: d'Alembert's *Recherche de calcul intégral* (1768) and the third volume of Euler's *Institutiones calculi integralis* (1770). In particular, Euler's work standardized the concept of a general integral of an nth-order PDE: it is a solution which includes n arbitrary functions (Engelsman *1980*). Euler's idea was that any solution of the equation could be found by specifying the arbitrary functions. Therefore, when in 1758 Euler was faced with singular solutions, ones that could not be thus obtained from the general solution, he considered them to be a 'paradox in the integral calculus'.

In his work on first-order PDEs (1776), Lagrange overcame this problem by defining the complete solution of

$$f(x, y, z, p, q) = 0, \quad p = \frac{\partial z}{\partial x}, \quad q = \frac{\partial z}{\partial y} \tag{39}$$

to be a solution containing two arbitrary constants:

$$V(x, y, z, a, b) = 0. \tag{40}$$

Lagrange's complete solution does not yield all solutions when a and b are specified, but elimination of a and b from (40) and its partial derivatives with respect to x and y will lead to the given PDE. Moreover, Lagrange pointed out that Euler's general solution and the singular solution can both be found by varying the arbitrary constants. More precisely, he showed that if ϕ is an arbitrary function, then eliminating a from the equations

$$\left. \begin{array}{c} V(x, y, z, a, \phi(a)) = 0 \\ \frac{\partial}{\partial a} V(x, y, z, a, \phi(a)) + \phi'(a) \frac{\partial}{\partial b} V(x, y, z, a, \phi(a)) = 0 \end{array} \right\} \tag{41}$$

would yield a general solution in Euler's sense, while eliminating a and b from the equations

$$V(x, y, z, a, b) = 0, \quad \frac{\partial}{\partial a} V(x, y, z, a, b) = 0, \quad \frac{\partial}{\partial b} V(x, y, z, a, b) = 0 \tag{42}$$

gives a singular solution of (39). With this new concept of a complete solution at hand, Lagrange was able to solve the general first-order PDE with two independent variables, (39). By 1774 Lagrange had already, in effect, reduced it to a linear first-order PDE. He assumed the equation to be given in the form $q = f(x, y, z, p)$, and sought to determine p as a function P of x, y and z, so that the differential equation

$$dz - p\,dx - q\,dy = 0 \qquad (43)$$

is integrable in the sense that the differential form can be made exact by multiplying by a factor $M(x, y, z)$. According to a theorem of Euler's, this is so if and only if

$$\frac{\partial p}{\partial y} + \left(f - p\frac{\partial f}{\partial y}\right)\frac{\partial p}{\partial z} - \frac{\partial f}{\partial p}\frac{\partial p}{\partial x} - \frac{\partial f}{\partial x} - p\frac{\partial f}{\partial p} = 0. \qquad (44)$$

This is a first-order linear PDE in p, but Lagrange did not write it down explicitly. If p is a solution of (44) containing one arbitrary constant a, then the integration of (43) will give a solution of (39) which will contain, in addition to a, the arbitrary constant of integration; it will therefore be a complete solution of (39).

In 1779 and 1785 Lagrange turned to linear first-order PDEs:

$$\sum_{i=1}^{n} P_i \frac{\partial z}{\partial x_i} + R = 0, \qquad (45)$$

where P_i and R are functions of x, y and z. He showed that it may be reduced to the integration of the system of ordinary differential equations

$$\frac{dx_1}{P_1} = \frac{dx_2}{P_2} = \cdots = \frac{dx_n}{P_n} = \frac{dz}{R}. \qquad (46)$$

Strangely, Lagrange never combined his two methods, and it was P. Charpit who in 1784 pointed out that together they provide a complete integration of the general first-order PDE of two variables. Here it should be remarked that since Euler's time a PDE has been considered solved if it has been reduced to the integration of ordinary differential equations.

Lagrange's theory was mostly algebraic and formal. He did remark that the singular integral could be interpreted as the envelope of the solutions in the complete integral, but otherwise the geometrical theory was left for Gaspard Monge. Having begun his research in 1770, Monge published it in a number of papers after 1780, and in his *Feuilles d'analyse appliquée à la géométrie* (1795) and its later editions entitled *Application de l'analyse à la géométrie* (1807). In these works Monge introduced the concepts of a characteristic curve and a characteristic cone. The characteristic curve is analytically determined by a solution of equation (46) corresponding to the PDE (45), and is defined geometrically as the curve in the x, y, z-space along which two consecutive solutions cut each other tangentially. Monge showed how solutions can be built up from characteristic curves in the sense that a surface is a solution surface if the characteristic curve through any point of the surface is contained in the surface. Monge also applied his geometric approach to second-order PDEs, but his ideas were not appreciated by his contemporaries. They were later taken up by Augustin Louis Cauchy, and became central to the development of the subject in the last half of the nineteenth century in the works of Gaston Darboux, Paul Du Bois-Reymond, and others.

Charpit is reported to have tried to extend his and Lagrange's method to first-order PDEs of n variables ($n > 2$), but J. F. Pfaff was the first to succeed. In 1815 he transcribed this problem to a special case of a so-called Pfaffian equation in total differentials,

$$A_1(y_1, y_2, \ldots, y_k) \, \mathrm{d}y_1 + \cdots + A_k(y_1, y_2, \ldots, y_k) \, \mathrm{d}y_k = 0, \quad k = 2n. \quad (47)$$

He then showed that, by solving a system of ordinary differential equations, he could determine a substitution of variables which would transform the equation (47) into one of the same kind but having one variable fewer. By successive applications of this idea, he reduced the solution of the general first-order PDE with n independent variables to the solution of n systems of ordinary differential equations of order $2n - 1, 2n - 3, \ldots, 3, 1$, respectively. Subsequent research lowered the number and order of these equations (Demidov *1982b*). Thus, Cauchy (1819), independently of Pfaff, reduced the solution of (47) to one system of ordinary differential equations.

Jacobi discovered two different methods of integrating first-order PDEs. The first, which he published in 1827, was an improvement on Pfaff's method. He discovered the second method when, inspired by Hamilton's work on mechanics, he returned to the subject in the 1830s. It rests on the important theorem stating that if

$$f_i(x_1, x_2, \ldots, x_n, p_1, p_2, \ldots, p_n) = k_i, \quad i = 1, 2, \ldots, n - 1 \quad (48)$$

are $n - 1$ independent integrals of the first-order PDE

$$f_0(x_1, x_2, \ldots, x_n, p_1, p_2, \ldots, p_n) = 0, \quad (49)$$

and one uses (48) and (49) to determine the p_i as functions of the x_i, then the differential form $\Sigma_i p_i \, \mathrm{d}x_i$ is exact if and only if the Poisson brackets

$$[f_k, f_l] := \sum_{i=1}^{n} \frac{\partial f_k}{\partial p_i} \frac{\partial f_l}{\partial x_i} - \frac{\partial f_k}{\partial x_i} \frac{\partial f_l}{\partial p_i} \quad (50)$$

are zero for $k \neq l$, $k, l = 0, 1, 2, \ldots, n - 1$. Since every first-order PDE can be reduced to the form (49), which does not explicitly involve z, and since the general solution can easily be deduced from the exact differential $\Sigma_i p_i \, dx_i$, Jacobi had thus reduced the problem to determining the f_k successively so that

$$[f_k, f_l] = 0, \quad l = 0, 1, 2, \ldots, k - 1. \tag{51}$$

This is in fact a system of linear homogeneous PDEs for f_k, and as a result of Jacobi's identity for Poisson brackets he succeeded in reducing it to a system of ordinary differential equations.

In the theories formulated by Euler, Legendre, Pfaff and Jacobi, there was a growing tendency to treat the derivatives p_i as independent variables. Combining this idea with his geometric studies of contact transformations, Sophus Lie began to investigate, in the period 1871–3, what can be said about the solution of a given PDE if it is known that a certain group of contact transformations leaves the equation invariant in a certain way. This 'Galois' theory of PDEs was the origin of the theory of Lie groups (see Hawkins *1991*, and §3.16 and §6.5). By taking this approach, Lie could successfully treat first-order PDEs. His attempts to grapple with second-order equations were followed up by E. Cartan, H. Goldschmidt and S. Sternberg, whose work gave rise to results on smooth manifolds.

7 THE WAVE EQUATION: GREEN'S FUNCTIONS

In connection with optics, the three-dimensional wave equation

$$\Delta V = \frac{1}{a^2} \frac{\partial^2 V}{\partial t^2} \tag{52}$$

attracted much interest in the nineteenth century. Poisson studied it in a number of papers from 1808 to 1819. In 1818 he solved the initial-value problem by showing that the value of the solution u at a point p and time t can be found by integrating the initial values of u and $\partial u / \partial t$ over a sphere of radius equal to ta, where a is the wave velocity.

A new approach was opened up by Bernhard Riemann in his 1858–9 paper on the propagation of sound waves of finite amplitude. He considered an equation of the form

$$\frac{\partial^2 u}{\partial x \partial y} + A(x, y) \frac{\partial u}{\partial x} + E(x, y) \frac{\partial u}{\partial y} + F(x, y)u = 0, \tag{53}$$

and showed how to find its solution at a particular point, given u and $\partial u / \partial n$ along a given curve. Using a generalized version of Green's theorem (§3.17), he reduced this to the problem of finding a solution (often called the Riemann function) to a simpler initial-value problem for the so-called adjoint equation. A different use of Green's theorem led Helmholtz (1860) and Kirchhoff (1882) to an important solution of the Helmholtz equation for simple harmonic waves. It expresses the solution in terms of its value and the value of its normal derivative on a given surface surrounding the point, but taken at an earlier time. This generalization of Poisson's result is sometimes called 'Huygens' principle'.

These solutions prompted Du Bois-Reymond (1889) and Darboux to seek a generalization of Green's theorem to general linear second-order PDEs. Moreover, it became clear that it was desirable to find a generalization of the Green's function and the Riemann function which, once found, would solve the boundary-value problem for arbitrary boundary values. Several methods of defining such functions were suggested at the end of the nineteenth and the beginning of the twentieth century; the definitive one, given by Jacques Hadamard in 1932, also applied to higher-order equations. The problem was that the type of singularity one has to require of the solution was different for each equation. This was overcome in Laurent Schwartz's theory of distributions (1950), according to which a fundamental solution with respect to a point a and a differential operator L is a distribution E_a satisfying $L(E_a) = \delta_a$, where δ_a is the δ-distribution concentrated at the point a. In this context, a Green's function can be defined simply as a fundamental solution satisfying certain boundary conditions (Lützen *1982*).

8 EXISTENCE THEOREMS; BOUNDARY VALUE PROBLEMS; AND CLASSIFICATION

During the eighteenth century the method of undetermined coefficients was often used to express solutions of ordinary and partial differential equations in terms of power series. However, the question of convergence was generally overlooked until 1842, when Cauchy, using the method of majorants, proved that for quasi-linear PDEs the series will converge within specified limits. He explicitly stressed that he had thereby given the first general existence proof for solutions of PDEs. Cauchy's results were generalized in 1875 by Sonya Kovalevskaya to PDEs on the so-called 'normal form'. She emphasized that if an nth-order equation is on normal form with respect to one of its variables, say x, one can prescribe the value of the solution and its $n - 1$ first derivatives with respect to x at $x = 0$ (this is the so-called Cauchy problem). On the other hand, she showed by an example that the Cauchy problem need not have an analytic solution if the equation is not on normal form (Cooke *1984*).

Around the same time, Carl Friedrich Gauss, George Green, William Thomson, Dirichlet and others began to study the existence of solutions of Laplace's equation in a given domain with given values or given normal derivatives on the boundary (§3.17). Some of the methods worked out in these studies were also used to show the existence of a countable infinity of eigenvalues of Helmholtz's equation $\Delta u + \lambda u = 0$, whose eigenfunctions have required boundary values such as $u = 0$ (Poincaré *1894*). It gradually became clear that the type of initial or boundary values one could impose depended on the type of equation under investigation.

This feeling was made precise by Du Bois-Reymond. In a book of 1864, he clarified Monge's theory of characteristics and this led him (in a post-humous paper of 1889) to classify second-order PDEs into hyperbolic equations (e.g. the wave equation), parabolic equations (e.g. the heat-conduction equation) and elliptic equations (e.g. the Laplace equation). Hyperbolic equations have real and distinct characteristics, parabolic equations have real and coinciding characteristics and elliptic equations have imaginary characteristics (this distinction had earlier been blurred by the liberal use of complex variable transformations). This classification allowed a generalization to general classes of PDEs of the results obtained previously on the boundary values of the physically important PDEs. In particular, Du Bois-Reymond showed that there exists a solution of a hyperbolic PDE with prescribed values and normal derivative on a curve that is cut no more than once by any characteristic. Moreover, in 1890 Emile Picard began a general study of boundary-value problems of elliptic PDEs, showing that many of the results concerning Laplace's equation could be carried over to the general setting.

These ideas were generalized at the beginning of the twentieth century, in particular by Hadamard, who introduced the concept of a well-posed initial-value or boundary-value problem. This is a problem which has a unique solution that depends continuously on the initial values. During the 1930s, I. G. Petrovsky extended Hadamard's research to a systematic study of the well-posedness of the Cauchy problem for rather general systems of PDEs, and he generalized Du Bois-Reymond's classification to such systems.

The regularity of the solutions turned out to depend on the type of equation. For elliptic equations the solutions are typically smooth, even though the boundary conditions are not smooth, whereas hyperbolic equations typically possess singularities that 'spread' along the characteristics. The study of the regularity of solutions and the propagation of singularities was taken up during the second half of the nineteenth century, and has remained an active area of research during the twentieth century (Bermelmans *et al.* *1990*).

9 CONCLUSION

At the beginning of the twentieth century, research on PDEs was dominated by boundary-value problems for hyperbolic, parabolic and elliptic equations. As late as 1962, Richard Courant declared that a general theory of PDEs, like the one for first-order equations summarized in Section 6, would be impossible for higher-order equations. However, since then many deep theorems have been established for general differential operators, irrespective of their type, and – as mentioned in Section 6 – Lie's and Cartan's geometric theory was carried further (Pommaret *1978*). More recently, there has been increasing interest in non-linear PDEs.

During the nineteenth century the problem of finding solutions to PDEs was gradually supplemented by an abstract study of the structure of the equations and of the possible types of solution. This tendency continued into the twentieth century, when the field was strongly influenced by the introduction of mathematical structures, in particular the development of functional analysis. However, since partial differential operators are unbounded in the usual topologies, many special tricks and methods were still called for. Thus, from the 1930s the consideration of PDEs in connection with various function spaces, like Sobolev spaces, became a major topic. A strong influence from functional analysis came indirectly through the theory of distributions (Lützen *1982*), created by Laurent Schwartz in the period 1945–50 (and anticipated by L. S. Sobolev in 1936).

In the hands of Lars Hörmander and others, this new framework completely revolutionized the subject after 1960. By generalizing the concept of a solution, many problems can be overcome. The uniform definition of the Green's function has been mentioned above. Another example is the study of the propagation of singularities, which was earlier hindered by the fact that the 'solution' did not satisfy the PDE in the singular points. In the theory of distributions this problem disappears. For example, any distribution T, and in particular any integrable function, will give solutions $T(x + t)$ and $T(x - t)$ to the two-dimensional wave equation. This, in a way, provides a rigorous underpinning of Euler's idea of a solution. Even classical existence theorems have been simplified and strengthened by first showing the existence of a distribution solution, and then using some regularity theorems to establish that they are in fact classical solutions.

BIBLIOGRAPHY

Bermelmans, J., Hildebrandt, S. and von. Wahl, W. *1990*, 'Partielle Differentialgleichungen und Variationsrechnung', in *Ein Jahrhundert Mathematik 1890–1990. Festschrift zum Jubiläum der DMV*, Braunschweig: Vieweg, 149–230.

Cooke, R. *1984*, *The Mathematics of Sonya Kovalevskaya*, New York: Springer.

Demidov, S. S. *1982a*, 'Création et développement de la théorie des équations différentielles aux dérivées partielles dans les travaux de J. d'Alembert', *Revue d'Historie des Sciences*, **35**, 3–42.

—— *1982b*, 'The study of partial differential equations of the first order in the 18th and 19th century', *Archive for History of Exact Sciences*, **28**, 325–50.

Engelsman, S. B. *1980*, 'Lagrange's early contributions to the theory of first order partial differential equations', *Historia mathematica*, **7**, 7–23.

—— *1984*, *Families of Curves and the Origins of Partial Differentiation*, Amsterdam: North-Holland.

Hawkins, T. *1991*, 'Jacobi and the birth of Lie's theory of groups', *Archive for History of Exact Sciences*, **42**, 187–278.

Kline, M. *1972*, *Mathematical Thought From Ancient to Modern Times*, New York: Oxford University Press.

Lützen, J. *1982*, *The Prehistory of the Theory of Distributions*, New York: Springer.

Pommaret, J. F. *1978*, *Systems of Partial Differential Equations and Lie Pseudogroups*, New York: Gordon & Breach.

Sommerfeld, A. *1900*, 'Randwertaufgaben in der Theorie der partiellen Differentialgleichungen', in *Encyclopädie der mathematischen Wissenschaften*, Vol. 2, Part A, 504–70 (article II A 7c).

Weber, E. *1900*, 'Partielle Differentialgleichungen', in *Encyclopädie der mathematischen Wissenschaften*, Vol. 2, Part A, 294–399 (article II A 5).

3.16

Differential equations and groups

J. J. GRAY

1 THE PRINCIPAL EXAMPLE: THE HYPERGEOMETRIC EQUATION

The study of differential equations goes back to Isaac Newton and Gottfried Wilhelm Leibniz (§3.2), while the study of groups is essentially nineteenth century in origin (§6.4); and it is not obvious that they are related. But a fertile area for the growth of group-theoretic ideas was the study of transformations, and the interrelation of the solutions to a differential equation came to be studied as the way in which one solution could be transformed into another. The simplest case was that of linear differential equations, where every solution is expressible as a sum of certain arbitrarily chosen basic ones. The first intimate and detailed connection between the theories of linear differential equations and groups to emerge concerned the hypergeometric differential equation:

$$z(1 - z)\frac{d^2 w}{dx^2} + [c - (a + b + 1)z]\frac{dw}{dz} - abw = 0. \tag{1}$$

This equation had been studied by many authors (§4.4) when Bernhard Riemann investigated it in 1857. He showed that a (many-valued) complex function can be defined so as to have a prescribed behaviour under analytic continuation around the points $z = 0$, $z = 1$, and at infinity, suitable linear relations holding between any three of its branches; moreover, such a function necessarily satisfies a suitably chosen hypergeometric equation. This was the first time the equation had been studied from the complex point of view, apart from an unpublished study by Carl Friedrich Gauss. By introducing the idea of analytic continuation, Riemann shed light on the earlier studies by Gauss and Ernst Kummer.

Group theory remained implicit in Riemann's presentation, but was soon to be made explicit. Since the hypergeometric equation is of second order, its solutions are expressible as a sum of two basic solutions. The analytic continuation of these two around each singular point yields more branches of the solution that depend linearly on the original choice. This dependence is captured by a matrix of constants (called a 'monodromy matrix' by Charles Hermite). The group generated by these matrices is called the monodromy group of the equation; it is a quotient group of the fundamental group of the Riemann sphere punctured at the three singular points. Riemann's work therefore established that the monodromy group determines the differential equation and its solutions.

What was true of the hypergeometric equation could be asked more generally. In his address to the International Congress of Mathematicians in 1900, David Hilbert raised a number of problems which he felt should guide the future direction of mathematical research. To a remarkable degree he was successful, and Hilbert's problems continue to generate real interest to this day. The Riemann–Hilbert problem, posed by Hilbert as the 21st of his problems, asks, given a quotient group of a fundamental group of a punctured Riemann surface, whether there is a differential equation having its singular points only at the punctures and the given group as its monodromy group. It was answered affirmatively by G. D. Birkhoff in 1913.

A further twist to Riemann's ideas came when they were taken up by Henri Poincaré, to whom explicit mention of group theory is due (Gray *1986*). Poincaré was responding to papers by L. I. Fuchs, who had generalized Riemann's theory to a class of differential equations with solutions that have no essential singularities. He became interested in the behaviour of the quotient of two solutions to a second-order linear differential equation, in particular in finding when it has a single-valued inverse on some domain. For the hypergeometric equation, Poincaré could show that this happened whenever the quantities a, b and c determined monodromy matrices of suitable finite orders. The inverse of the quotient function then defined maps of a triangular region onto a half-plane, and analytic continuation showed that, depending on their angles, the triangles tessellated either the sphere, the plane or the non-Euclidean disc. Poincaré showed that more general polygons arise in the same way from differential equations with more singular points. Poincaré's introduction of the tools of non-Euclidean geometry led him to recognize that every Riemann surface should arise from the action of a suitable discrete group on one of those three surfaces. (Discrete means that the polygon is moved around *en bloc* in such a way as never to overlap itself, except neatly along its edges.) By identifying the edges that overlap in this way, the polygon is made into a Riemann surface. This result, conferring an intrinsic geometry on every Riemann surface, was first rigorously proved by Paul Koebe and Poincaré independently in 1907.

Another connection between group theory and differential equations forms the subject of Picard–Vessiot theory, named after two pioneers who attempted to use Evariste Galois's ideas about polynomial equations to this new setting. Galois theory can be thought of as describing how solutions to polynomial equations (whose coefficients have already been constructed) can be successively adjoined to the set of rationals. Emile Picard and E. Vessiot associated with a differential equation the group of all linear transformations of its solutions (in general, this group is larger than the monodromy group). They then investigated how this group changes as solutions to simpler types of differential equation are assumed known. In thus establishing a hierarchy of functions defined by differential equations, they were following the earlier example of Joseph Liouville, who in 1834 had shown in the same way that elliptic functions were not expressible in terms of the simpler logarithmic and trigonometric functions. Liouville's ideas were later taken up by J. F. Ritt and E. R. Kolchin, and more recently by M. Singer.

2 THE WORK OF SOPHUS LIE

The most fruitful connection to have been discovered between the theories of differential equations and groups has also the most obscure of origins, recently elucidated in Hawkins *1991*. It arose in the work of Sophus Lie, who had formed the hope that problems in partial differential equations could likewise yield to a theory of Galois type. Partial differential equations (PDEs) involve several independent variables, and are much harder to understand than are ordinary ones (§3.15). In the 1860s a number of mathematicians caught up with earlier but unpublished ideas of Carl Jacobi about how to solve them, and difficulties in understanding their work motivated Lie to attempt a geometrical theory (Hawkins *1991*). Lie concentrated on the way in which certain transformations of the variables can make a PDE easier to solve. This led him to ask when two PDEs are related by a change of variables: are there any invariant features by which PDEs can be distinguished? He also looked at Jacobi's account of how, given some solutions to a (certain kind of) PDE, other solutions can be found by a process of combination. Lie then investigated how the existence of transformations of a PDE affects the solution procedure discovered by Jacobi. He found that instead of starting, as others had done, with a PDE, he could work backwards. He showed that a particular type of transformation, called a contact transformation and known to him from earlier work on a purely geometric problem, led to a PDE. In this way, the process of combining solutions transferred to a process of combining these transformations. So he began to speak of the family of all transformations appropriate to a given PDE, and since this family was closed under composition, he called it a group.

Lie had come to these realizations by the end of 1873. At that time the notion of a group was still imprecise; later developments crystallized the concept in such a way that Lie's families were not groups, but rather what we might now best call Lie algebras. The idea of a Lie group (§6.5) was to come much later. He succeeded in showing that there are distinct systems of such 'groups' that could not be transformed into one another by changes of variable. Indeed, if there was only one variable, there were just three distinct types of such 'group'. If the number of variables was fixed, there would correspond only a finite number of distinct transformation 'groups'. Elucidating this realization that such 'groups' formed a practicable object of study, albeit requiring a lifetime's work, was to take Lie away from differential equations altogether.

Lie himself sought to reach an audience committed to Jacobi's analytic style, and so he increasingly suppressed his geometrical approach without, unhappily, achieving adequate lucidity. His later, and rich, theory of 'groups' of transformations was therefore not widely understood. His old friend Felix Klein encouraged Lie to work in Germany for a while and to receive bright students capable of presenting his ideas; several books resulted from these collaborations. Independently, and for other reasons, Wilhelm Killing came to analyse the groups of transformations that Lie had found. Finally, the French mathematician Elie Cartan took up their work. By 1900, the culmination of all this endeavour was a profound classification theorem for the 'groups', but the original connection with the theory of partial differential equations was almost completely obliterated.

Lie algebras and Lie groups arise naturally where there is symmetry, and for this reason they are useful even when no differential equation can be written to describe the situation. But they also have a significant role to play in the study of differential equations with invariants (Olver *1986*). For example, in problems involving bodies in motion, the total energy of the system is often a constant, and other quantities, such as total angular momentum, may also be preserved. Knowledge of these invariants can be used to simplify the differential equations describing the motion. The central theorem here is due to Emmy Noether, and it asserts that every conservation law associated with a system arising from a variational principle (which is the usual case in physics) comes from a symmetry property. In dynamics, for example, if the variational principle leading to the so-called Euler–Lagrange equations (§8.1) is invariant under translations of the time variable, there is conservation of energy; invariance under a group of spatial transformations yields conservation of momentum. Noether's theorem is not merely a deep theoretical connection between conservation laws and symmetry properties; it is also very useful since new symmetries are often to be found by combining known ones, while conservation laws are useful but often hard to spot.

BIBLIOGRAPHY

Gray, J. J. *1986, Linear Differential Equations and Group Theory from Riemann to Poincaré*, Boston and Basel: Birkhäuser.

Hawkins, T. *1991*, 'Jacobi and the birth of Lie's theory of groups', *Archive for History of Exact Sciences*, **42**, 187–278.

Ince, E. L. *1927, Ordinary Differential Equations*, London: Longmans, Green. [Repr. 1944, New York: Dover.]

Lie, S. with Engel, F. *1888, 1890, 1893, Theorie der Transformationsgruppen*, 3 vols, Leipzig: Teubner.

Olver, P. J. *1986, Applications of Lie Groups to Differential Equations*, New York: Springer.

Poincaré, J. H. *1916, Oeuvres*, Vol. 2, Paris: Gauthier-Villars. [Repr. 1952.]

Riemann, B. *1990, Gesammelte mathematische Werke*, 3rd edn (ed. R. Narasimhan), New York: Springer.

Schlesinger, L. *1895, 1897, Handbuch der Theorie der linearen Differentialgleichungen*, 2 vols, Leipzig: Teubner.

3.17

Potential theory

J. J. CROSS

1 ELEMENTARY POTENTIAL THEORY

The roots of potential theory lie in the work of Gottfried Wilhelm Leibniz, Christiaan Huygens and the Bernoullis on mechanics and extremization in the late seventeenth century (Cross *1983*). The basic Galilean idea that velocities are proportional to the square root of the distance fallen, and its Leibnizian form in the work of Huygens, the Bernoullis and Jakob Hermann on mechanics, came to Leonhard Euler and Daniel Bernoulli in a developed form. From the theory of maxima and minima as practised by the Bernoullis and by Leibniz, the idea of comparison with neighbouring curves was seized on by Alexis de Fontaine and Euler (among others), and developed by them into the basis of a calculus of several variables; the idea of fields of tangents on the plane arose in the study of curves and of fluid mechanics with this new generation.

Nicholas I and Daniel Bernoulli, Euler, Alexis Clairaut and Fontaine began working in the period 1720–30 (Engelsman *1984*). They shared work on the geometry of curves, the calculus of variations and the study of mechanics. Clairaut learnt geometry from his father and mechanics with Pierre de Maupertuis; Euler learnt from both the elder Bernoullis (James I and John I) and the younger ones. Fontaine studied the work of the Bernoullis on various motions on curves, and extended it in his own work on the foundations of the calculus of several variables (Cross *1985a*). The younger Bernoullis worked on fluid mechanics and functions of several variables. Fontaine and Clairaut in Paris, Nicolas I Bernoulli in Basel and Euler in Saint Petersburg developed the calculus of functions of several variables in various forms (Greenberg *1984*; compare §3.2); Daniel Bernoulli first in 1721, Euler in 1734, Fontaine in 1740, when Clairaut published the conditions for the integration of 1-forms in two and three variables.

These are the key ideas. A 1-form, or differential, is an expression

$$dS = P\,dx + Q\,dy + R\,dz, \tag{1}$$

where P, Q and R are functions of x, y and z. Such a form is complete or integrable if

$$\frac{\partial P}{\partial y} = \frac{\partial Q}{\partial x}, \quad \frac{\partial Q}{\partial z} = \frac{\partial R}{\partial y}, \quad \frac{\partial R}{\partial x} = \frac{\partial P}{\partial z}, \tag{2}$$

and not integrable otherwise. If the variable z is absent, then setting $R \equiv 0$ yields the corresponding condition of integrability for functions of the two variables x and y. Clairaut used these in 1743 to find the shape of the rotating, gravitating earth (§8.14). But Jean d'Alembert raised an objection: for functions in the punctured plane which satisfy the commutativity condition, there need be no function from which they are derived. He cited the arctangent (inverse tangent) function for the first time: $y = \tan^{-1}(x/a)$ is a function defined on the plane slit from the origin out to infinity. It would reappear to haunt and destroy all the general assertions about potentials and principles.

Fontaine had written and published his work on the calculus of several variables which included the commutation of second-order mixed partial derivatives. Once the result for the mixed derivative was known, it was quite easy for Clairaut to give a simple proof based on the (infinite) series in two independent variables directly or with an integrating factor, and presenting an algorithm to do the integration. Clairaut used these ideas in his *Théorie de la figure de la terre* published in 1743; he used a field of force and the idea of a complete differential to integrate the force equations, where the conservative nature of the forces acting on the fluid was derived from his model of the Earth and of its equilibrium as a mass of rotating fluid. D'Alembert acted as devil's advocate, saying that there were differentials which were complete in Clairaut's sense in that they satisfied the mixed derivative test (as did M and N above), but in fact they were not integrable in that the 'function' which the integration gave was not really a function, but had a discontinuity: his ϕ, given by

$$M = \frac{\partial \phi}{\partial x} = -\frac{y}{x^2 + y^2} \quad \text{and} \quad N = \frac{\partial \phi}{\partial y} = \frac{x}{x^2 + y^2}, \quad \phi := \tan^{-1}\left(\frac{y}{x}\right) \tag{3}$$

is the first use of the inverse tangent as a counter-example. So the distinction between the two types of 1-form was born. D'Alembert was to use the idea of the complete differential to get a potential for the fluid velocity at the end of the decade in his solution of the partial differential equations of vibrating cords, and in his fluid mechanics. In the mid-1740s he developed the two-dimensional velocity field, with the Cauchy–Riemann equations as its integrability conditions and the complex velocity potential as its integral (in a fashion similar to his solution of the wave equation) (see §3.13 on complex-variable analysis).

476

D'Alembert showed that the velocity (p, q) for fluid flow in two dimensions can be integrated under complicated assumptions that he simplified to

$$dp = M\,dx + N\,dz \quad \text{and} \quad dq = N\,dx - M\,dz, \tag{4}$$

and these give a complex potential for the velocity:

$$\phi = \phi(x + iz), \qquad p = \frac{\partial \phi}{\partial x}, \quad q = \frac{\partial \phi}{\partial z}, \quad p + iq = \frac{d\phi}{dw}, \qquad w = x + iz. \tag{5}$$

Equations (5) are the so-called Cauchy–Riemann equations (§3.12) with

$$\frac{\partial p}{\partial x} = M = -\frac{\partial q}{\partial z} \quad \text{and} \quad \frac{\partial p}{\partial z} = N = \frac{\partial q}{\partial x}. \tag{6}$$

This is matched by Euler's potential for irrotational fluid flow, using integration criteria similar to Clairaut's: $\text{curl}(u, v, w) = 0$, or

$$\frac{\partial w}{\partial y} - \frac{\partial v}{\partial z} = 0, \qquad \frac{\partial u}{\partial z} - \frac{\partial w}{\partial x} = 0, \qquad \frac{\partial v}{\partial x} - \frac{\partial u}{\partial y} = 0. \tag{7}$$

When Euler provided the basic equations of motion for fluid mechanics (the continuity equation for conservation of mass, and the momentum equation linking the motion to the forces; see §8.1), he noted the usefulness of potentials, both for the motion (1755) and for the forces (1769). He also showed that the velocity potential satisfied Laplace's equation for a liquid where the continuity equation reduced to

$$\frac{\partial u}{\partial x} + \frac{\partial v}{\partial y} + \frac{\partial w}{\partial z} = 0, \tag{8}$$

with the velocity vector (u, v, w) given by the potential ϕ:

$$u = \frac{\partial \phi}{\partial x}, \qquad v = \frac{\partial \phi}{\partial y}, \qquad w = \frac{\partial \phi}{\partial z} \tag{9}$$

and the density of the fluid constant (§8.5). But he also demonstrated that there were perfectly good fluid motions for which no potential existed; for example, when the fluid rotates as a rigid body about an axis, or more generally when

$$u = -f((x^2 + y^2)^{1/2})y, \qquad v = f((x^2 + y^2)^{1/2})x, \qquad w = 0, \tag{10}$$

where the differential for the velocity becomes

$$u\,dx + v\,dy = -f((x^2 + y^2)^{1/2})y\,dx + f((x^2 + y^2)^{1/2})x\,dy. \tag{11}$$

This is integrable in a certain sense when

$$f((x^2 + y^2)^{1/2}) = 1/(x^2 + y^2); \tag{12}$$

and the potential is $\phi = \tan^{-1}(y/x)$, which is not a function in the plane or the punctured plane, as d'Alembert had stated. Joseph Louis Lagrange was to assume the same potentials for the velocity and the force in 1782.

The work of Clairaut, d'Alembert and Euler may be characterized as 'elementary potential theory'. It draws upon the postulation of a potential, but is based on many doubtful arguments or on outright assumption. The main contributor to its propagation was Euler. With Fontaine and Clairaut, he helped to develop a logical, well-founded calculus of several variables in a clear notation; he transformed, with Daniel Bernoulli and Clairaut, the energy equation for a particle falling under gravity into a general principle applicable to continuous bodies and general forces; and he founded – after the attempts of the Bernoullis, d'Alembert and, especially, Clairaut – the modern theory of fluid mechanics, using complete differentials for forces and velocities. His work was fruitful: the theories of Lagrange grew from his writings on extremization, fluids and sound, and mechanics; the work of Pierre Simon Laplace followed.

2 TRANSITION: FUNCTIONS AND INTEGRALS

In correspondence with Euler in 1759–60, Lagrange developed the theory of sound extensively (§9.8) and published an article in 1761 in which he integrated the equations of motion via an elementary Fourier transform and a volume-to-surface integral transformation. The year 1770 sees a change in potential theory. In 1771, Lagrange began to apply the Bernoulli–Euler small-element method to the attraction of spheroids. The initial results of the integration, in Cartesian and spherical polar coordinates, of the expressions for the attractions under the law $R = 1/r^2$, involved terms

$$\frac{R(x-a)}{r}\,dx\,dy\,dz, \qquad \frac{R(y-b)}{r}\,dx\,dy\,dz, \qquad \frac{R(z-c)}{r}\,dx\,dy\,dz, \quad (13)$$

useful but not spectacular. But in these formulas the direction cosines $(x-a)/r$, $(y-b)/r$ and $(z-c)/r$, the small element $dx\,dy\,dz$ and the Newtonian force proportional to $1/r^2$ have been brought together; r is the distance of the attracting particle from the attracted point.

The next step appears in Lagrange's prize essay on the secular equation of the Moon in 1773. He had to take into account the non-spherical shapes of the Moon and the Earth. When he came to resolve the forces of the Earth on the Moon, he used not the coordinates $(x-a, y-b, z-c)$ of the attracting point A, but the coordinates (α, β, γ) of the attracted point B centred on the attracting point A. So the direction cosines of the ray between the points became $dr/d\alpha$, $dr/d\beta$, $dr/d\gamma$. With the force F proportional to $1/r^2$, the axial components X, Y, Z became the coefficients of the differential d(M/r) with respect to α, β, γ:

$$d\left(\frac{M}{r}\right) = -\frac{M}{r^2}\frac{\partial r}{\partial \alpha}d\alpha - \frac{M}{r^2}\frac{\partial r}{\partial \beta}d\beta - \frac{M}{r^2}\frac{\partial r}{\partial \gamma}d\gamma. \tag{14}$$

So the sum of quantities $\Sigma\, M/r$ was seen to be important, and the force was obtained by differentiating such a sum. It was a small step for Lagrange from his previous paper, in taking the whole mass as the limit of a set of point masses, to make M a differential dm for a continuous body of arbitrary shape, and hence to write the sum Σ as an integral, with the force as its derivative:

$$\frac{M}{r} + \frac{M'}{r'} + \frac{M''}{r''} + \cdots = \Sigma = \int\frac{dm}{r}, \quad \text{with forces} \quad \frac{\partial \Sigma}{\partial \alpha}, \frac{\partial \Sigma}{\partial \beta}, \frac{\partial \Sigma}{\partial \gamma}. \tag{15}$$

Differentiation under the integral sign does not appear because series expansions sufficed for his purpose.

At this time (1772–6) Laplace began a series of researches which impinged on potential theory: on the shape of the Earth as a rotating mass of fluid, on gravity and on waves. In general he used series expansions, with systems of equations for each level of approximation, but eventually double and triple integrals for the force components appear. These are by no means recognizable yet as Lagrange's integrals; they are crude first steps, for spheroids which are almost true spheres. But his discussion on waves used the small parallelepiped idea, mentioned only Newton, and obtained the Laplace equation

$$\frac{\partial^2 z}{\partial x^2} + \frac{\partial^2 z}{\partial y^2} = 0 \tag{16}$$

in two independent variables, x and y, which he solved, following d'Alembert, as

$$z = \phi(x + y\sqrt{-1}) + \psi(x - y\sqrt{-1}), \tag{17}$$

where $\phi(t)$ and $\psi(t)$ are arbitrary functions of t. The Laplace equation for ϕ also appeared.

Laplace's important paper on the attraction of spheroids in 1782 begins by expressing the attraction of a spheroid via triple integrals in spherical polar coordinates, and then notes that these attractions (A, B, C) are the derivatives of

$$V = \int\frac{\partial M}{r}. \tag{18}$$

The Laplace equation for V appears in polar coordinates, but the name 'Lagrange' does not. By 1787, when Laplace wrote about the rings of

Saturn and the observed orbital inequalities of Jupiter and Saturn, the integral for V in Cartesian and polar coordinates was well established, as was the equation it satisfied: Laplace's equation (16) for three independent variables. The treatment is now at the second stage: potential theory with integrals and not just functions.

3 INTEGRAL TRANSFORMS

Integral transforms begin with Lagrange (1759–60), with the change of a triple integral over the volume of a rectangular box to double integrals over pairs of faces, by integration by parts with respect to the variable perpendicular to the pair of faces. This method is used repeatedly and, coupled with the change of area from plane to curved surface, it is now standard. Integrals over curved surfaces occur in Euler and Lagrange involving a cosine ($\cos \gamma$) to relate elements dS of area of the curved surface to the element $dx\,dy$ in the plane. Carl Friedrich Gauss (1813) used this to show that closed surfaces projected zero area onto planes perpendicular to any axis:

$$\int \cos \alpha \, dS = 0, \qquad \int \cos \beta \, dS = 0, \qquad \int \cos \gamma \, dS = 0, \qquad (19)$$

and hence his 'divergence theorem':

$$\int (T(y, z) \cos \alpha + U(z, x) \cos \beta + V(x, y) \cos \gamma) \, dS = 0. \qquad (20)$$

Note the absence of derivatives.

In connection with his work on electrodynamics (§9.10), A. M. Ampère (1826) was the first to transform a line integral (an integral along a curve in space) into an integral over a curved surface (a surface integral) whose boundary is the curve, and he also did the reverse; Augustin Louis Cauchy (1825) and Mikhail Ostrogradsky (1826) had formulated such transformations for curves and areas in the plane. Ampère's expressions do not reflect a 'natural' transformation: there is no pattern relating derivative, tangent, and normal to curve and surface. The first to do this was Siméon-Denis Poisson (1826), who linked a volume integral for the potential of a body to a surface integral over its boundary, using integration by parts to obtain the integral over the curved surface (Cross *1985b*). It must be noted that Poisson was one of the referees of one of Ostrogradsky's papers for the Académie des Sciences in 1826, and so had the chance to read Ostrogradsky's ideas (Grattan-Guinness *1990*: 1171–3). Ostrogradsky, in a paper written in 1827 and published in 1829–31, uses a fully developed surface-to-volume transform for partial differential operators which had the divergence theorem as a special case:

$$\iint (P(x,y,z)\cos\alpha + Q(x,y,z)\cos\beta + R(x,y,z)\cos\gamma)\, dS$$

$$= \iiint \left(\frac{\partial P}{\partial x} + \frac{\partial Q}{\partial y} + \frac{\partial R}{\partial z}\right) dV. \quad (21)$$

This included Gauss's versions as the simplest cases, where the derivatives on the right are zero. In modern terms, we have

$$\int_S \mathbf{v}\cdot\mathbf{n}\, dS = \int_V \operatorname{div}\mathbf{v}\, dV. \quad (22)$$

George Green did neither of these things in his *Essay* (1828). He introduced volume-to-surface integrals to transform the integral

$$\int dx\, dy\, dz\left(\frac{\partial U}{\partial x}\frac{\partial V}{\partial x} + \frac{\partial U}{\partial y}\frac{\partial V}{\partial y} + \frac{\partial U}{\partial z}\frac{\partial V}{\partial z}\right) \quad (23)$$

in two ways, again by integration by parts and the cosine projection as well as the directional derivative along the normal to the surface to make it equal to the two sides of

$$\int dx\, dy\, dz\, U\delta V + \int d\sigma\, U\left(\frac{dV}{dw}\right) = \int dx\, dy\, dz\, V\delta U + \int d\sigma\, V\left(\frac{dU}{dw}\right), \quad (24)$$

where δ is the Laplace operator (from (16)), $d\sigma$ represents the element of surface area, and dV/dw is the derivative along the interior normal to this element. He used this in applications to gravity, electrostatics and magnetism. Gauss (1840) re-derived a special case of this equation where $U = V$; in this case Green's integral is called the 'energy integral'. Green, Gauss, Rudolph Clausius, Arthur Cayley and P. Humbert all talk about potentials, potential functions and prepotentials. These are all names for the same thing; some people regard the potential as the integral and the potential function as the integrand, and a prepotential as the same integral but with the integrand having a power other than that determined by the dimension of the space (Humbert *1936*).

4 GREEN'S FUNCTION AND THE DIRICHLET PRINCIPLE

Green facilitated the calculations of his *Essay* (Green *1871*) by introducing a function U which had the following properties: (1) it is harmonic; (2) at one point p' it is equal to the simple singularity $1/r$; and (3) on the given closed surface surrounding the point p' the function is zero:

$$\frac{\partial^2 U}{\partial x^2} + \frac{\partial^2 U}{\partial y^2} + \frac{\partial^2 U}{\partial z^2} = 0, \quad U(p') = \frac{1}{r}, \quad U(p) = 0, \quad \text{for all } p \in S. \quad (25)$$

Green gave a physical argument for the existence of his function, just as Henri Poincaré was to do sixty years later for his similar fundamental solutions. The Green's function was picked up and popularized by Carl Neumann, and now forms a central pillar in the theory of differential equations, in potential theory and, indeed, in most of physics (Sologub *1975*).

The next key year is 1839. J. P. G. Lejeune Dirichlet had begun his journey through mathematical physics several years earlier when he started lecturing; now he discovered potential theory and began to deal with it from a global viewpoint, with series expansions and their convergence an outcome of his deliberations rather than the central method, as they were for Laplace and Lagrange. In his treatment the Dirichlet problem and its concomitant Dirichlet principle received wide publicity – he lectured on it in Berlin in 1842, 1846, 1848 (to Bernhard Riemann), 1852 and 1855, and later in Göttingen (to Richard Dedekind), and his listeners spread the message. It was known well before Riemann's death that the principle had a 'hole' in it, and Karl Weierstrass used the inverse tangent function to highlight this gap.

In modern terms, Dirichlet's *problem* is as follows: we look at a domain D in Euclidean space $\mathbb{R}^n$, $n \geqslant 2$; a domain is an open connected set. We assume that its boundary S is compact (closed and bounded). We let ϕ be some continuous function on this boundary S. The problem is to find the potential f which is twice differentiable on D, satisfies Laplace's equation $\nabla^2 f = 0$, and coincides with the given function ϕ on S:

$$\text{given } \phi \in C^1(S), \text{ find } f \in C^2(D), \; \nabla^2 f = 0, \; f|_{\partial D} = \phi \in C^0(S),$$
$$S = \partial D, \; D \subseteq \mathbb{R}^n, \; n \geqslant 2. \quad (26)$$

The Dirichlet *principle* (first with William Thomson, 1847) uses the Green–Gauss energy integral

$$\int dx\,dy\,dz\left[\left(\frac{\partial U}{\partial x}\right)^2 + \left(\frac{\partial U}{\partial y}\right)^2 + \left(\frac{\partial U}{\partial z}\right)^2\right] \equiv \iiint |\,\text{grad } U\,|^2\,dV. \quad (27)$$

This time D is a bounded domain in $\mathbb{R}^n$, $n \geqslant 2$. We look at all functions g which are piecewise continuously differentiable in D, with finite energy integral

$$\iiint |\,\text{grad } g\,|^2\,dV < \infty. \quad (28)$$

For a given function ϕ which is continuous on the boundary S of the domain D, we assume that g coincides with ϕ on $S = \partial D$:

$$g\,|_{\partial D} = \phi \quad \text{in } C^0(\partial D). \tag{29}$$

Then Dirichlet (and William Thomson) asserted that the solution f of the Dirichlet problem has minimum energy integral:

$$\iiint |\operatorname{grad} f|^2 \, dV \leqslant \iiint |\operatorname{grad} g|^2 \, dV \tag{30}$$

for all g.

Many attempts have been made to prove the Dirichlet principle (Malossini 1990, Monna 1975). David Hilbert had an idea for a proof in 1899–1900, but it fails under conditions on the space: in 1909 S. Zaremba showed that it was not solvable for the punctured ball, and Henri Lebesgue in 1913 showed that smoothness of the boundary was essential by constructing an irregular point. Later attempts were to succeed by changing the function framework and admitting the so-called 'weak' solutions (Monna 1975).

The story of integral transforms continues with the Stokes–Thomson theorem, the informative version of Ampère's result. Again, we are in three dimensions, and we transform the integral along a curve in space to an integral over a surface; if C is a curve which is the boundary of a surface S so that $C = \partial S$, then we have the relation

$$\int (\alpha \, dx + \beta \, dy + \gamma \, dz)$$

$$= \iint \left[l\left(\frac{\partial \beta}{\partial z} - \frac{\partial \gamma}{\partial y}\right) + m\left(\frac{\partial \beta}{\partial z} - \frac{\partial \gamma}{\partial y}\right) + n\left(\frac{\partial \beta}{\partial z} - \frac{\partial \gamma}{\partial y}\right) \right] dS, \tag{31}$$

where l, m and n denote the direction cosines of a normal to the element dS of the surface. In modern terms,

$$\int_C \mathbf{v} \cdot d\mathbf{r} = \int_S (\operatorname{curl} \mathbf{v}) \cdot (\mathbf{n} \, dS). \tag{32}$$

Even more informative is to write the boundary operation taking the surface S to its boundary C as a derivative, and to represent the derivatives grad, div and curl as a derivative d whose definition in coordinate terms varies with the objects on which it works, after the ideas of Josiah Willard Gibbs, Gregorio Ricci-Curbastro, Tullio Levi-Città and their followers

(§3.4). We then have the symmetric relation for all the integral formulas we have seen so far:

$$\int_{\partial R} \omega = \int_R d\omega, \qquad (33)$$

where R is the region of integration (volume or surface), ∂R is its boundary of one dimension lower (surface or line), ω is a 'form' and $d\omega$ is its derivative (which also is a form, but of higher degree). The forms include the element of volume (or area, or length) of the region of integration. The symmetry shows that the 'derivative' moves from the geometric object to the form as we go from left to right, and the 'dimension' increases from left to right. In fact, with the correct definitions of boundary and derivative for regions and forms, this is the theorem of Poincaré, G. de Rham and W. Hodge.

5 BERNHARD RIEMANN

Riemann introduced Dirichlet's principle to a wider audience in the 1850s and 1860s through his articles and lectures. By 1861 it was clear to him, to Dedekind and to others attending his lectures that there was indeed a 'hole' in the principle, a hole they hoped Gauss had closed (he had not), or they themselves could close (they could not). The method of proof used by Dirichlet was destroyed by Weierstrass's counter-example – based once again on the inverse tangent function.

But Riemann's work took potential theory away from the applications of Euler, Laplace, Green, Thomson and Stokes into function theory, the (complex) harmonic functions satisfying Laplace's equation in two real variables and the surfaces or manifolds on which they are defined (§3.11). Riemann's ideas on surfaces led directly to the theories of Felix Klein, Emile Picard, Poincaré, Hermann Weyl, Paul Koebe, Richard Courant and Elie Cartan: automorphic functions, homology and homotopy, conditions on integrability for complex functions of several variables, manifolds, the classification of manifolds, conformal mappings, flows on manifolds (be they of fluid or electromagnetism) and the singularities of such flows, the corresponding integral transforms and derivatives in spaces of dimension greater than three. The interplay of function and geometry culminated in Poincaré's duality theorem of 1895 and the Hodge–de Rham theory of 1930–40, linking homology (geometry) and cohomology (differentiable functions) (see §7.10 on topology).

These theories generalized the earlier transforms of Ampère, Ostrogradsky and Thomson to higher dimensions and more general spaces. There are three vector operators in three dimensions (§6.2), which are summarized in terms of Cartesian coordinates. First, the gradient operator takes scalar functions $f(x, y, z)$ to a vector or its corresponding 1-form:

$$\text{grad } f = v \frac{\partial f}{\partial x^1} \mathbf{e}^1 + \frac{\partial f}{\partial x^2} \mathbf{e}^2 + \frac{\partial f}{\partial x^3} \mathbf{e}^3, \qquad df = \frac{\partial f}{\partial x^1} dx^1 + \frac{\partial f}{\partial x^2} dx^2 + \frac{\partial f}{\partial x^3} dx^3. \tag{34}$$

In the 1-form in equation $(34)_2$, the vectors dx^i are regarded as the tangent vectors to the coordinate lines.

Second, the operator curl, the rotation operator, takes vectors $\mathbf{u}$ to vectors or their corresponding 2-forms:

$$\text{curl } \mathbf{u} = \left(\frac{\partial u_2}{\partial x^3} - \frac{\partial u_3}{\partial x^2} \right) \mathbf{e}^1 - \left(\frac{\partial u_1}{\partial x^3} - \frac{\partial u_3}{\partial x^1} \right) \mathbf{e}^2 + \left(\frac{\partial u_1}{\partial x^2} - \frac{\partial u_2}{\partial x^1} \right) \mathbf{e}^3, \tag{35}$$

$$d\mathbf{u} = \left(\frac{\partial u_2}{\partial x^3} - \frac{\partial u_3}{\partial x^2} \right) dx^2 \wedge dx^3 - \left(\frac{\partial u_1}{\partial x^3} - \frac{\partial u_3}{\partial x^1} \right) dx^1 \wedge dx^3$$

$$+ \left(\frac{\partial u_1}{\partial x^2} - \frac{\partial u_2}{\partial x^1} \right) dx^1 \wedge dx^2. \tag{36}$$

The vector products $dx^i \wedge dx^j$ are skew-symmetric so that

$$dx^i \wedge dx^j = -dx^j \wedge dx^i. \tag{37}$$

Thus there are only three such basis vectors when the dimension of the space is three, and $\frac{1}{2} n(n - 1)$ in general.

The third operator in three dimensions takes a vector to a scalar function, or the corresponding 3-form:

$$\text{div } \mathbf{v} = \frac{\partial v^1}{\partial x^1} + \frac{\partial v^2}{\partial x^2} + \frac{\partial v^3}{\partial x^3}, \tag{38}$$

$$d\mathbf{v} = (\text{div } \mathbf{v}) \, dx^1 \otimes dx^2 \otimes dx^3. \tag{39}$$

It is easy to show that, for all functions f and vectors $\mathbf{u}$,

$$\text{curl grad } f \equiv \mathbf{0} \quad \text{or} \quad ddf \equiv d^2 f = 0, \quad \text{curl } \mathbf{u} = 0 \quad \text{or} \quad dd\mathbf{u} \equiv d^2\mathbf{u} = 0. \tag{40}$$

So, we have four vector spaces of functions, scalar or vector, linked by a derivative (or a set of derivatives) with the property that $d^2 = 0$. Similarly, we have already seen that we have a set of geometric objects – bodies, surfaces, curves and points – and a boundary operator denoted by ∂, linked as follows: for a body V we assume that it is bounded by a closed surface

S, where this surface has no 'holes' in it, no boundary, so that the boundary operator gives

$$\partial V = S, \qquad \partial S = 0, \qquad \partial\partial = \partial^2 = 0. \tag{41}$$

Similarly, for surfaces S bounded by curves C,

$$\partial S = C, \qquad \partial C = 0, \qquad \partial\partial = \partial^2 = 0. \tag{42}$$

6 CULMINATION

After Riemann, potential theory and the Green's function were developed by Hermann Schwarz and by Neumann.

Schwarz, in 1870, proved the existence of a solution to the Laplace equation, established by solving the boundary-value problem for a domain of the plane where it can be solved easily, say a circle, and hence completing the boundary values for another domain, say a square; the new boundary-value problem for the second domain is solved, which gives a new set of boundary values for the first domain. Thus one forms sequences of solutions of each domain which converge to a single solution on the union of the two domains, where both the sequences agree on the intersections in the limit, as, for example, on the non-empty intersection of the circle and the square.

In 1877, Neumann codified nearly twenty years of his work: his book covered elementary potential theory, distributions of mass and charge, double distributions or the double layer of magnetism, and the use of the method of arithmetic means to establish the existence of a solution to the first Dirichlet boundary-value problem for convex surfaces (avoiding the error in the Dirichlet principle). His work extended that of his father, Franz, and was carried on by his nephew Ernst.

There were many other developers of potential theory to particular cases: E. Betti, V. J. Boussinesq, E. Mathieu, E. Beltrami, V. Bäcklund, V. Bjerknes, V. G. Robin, Hilbert and Poincaré. In 1890, Poincaré expounded the *méthode de balayage*, the method of sweeping out mass from space onto a given surface. The method dates back to Poisson and Michel Chasles; it gives a unique potential, but not a unique mass distribution because of the selection process involved. Poincaré proved by physical arguments (which he expected could be made rigorous, and suggested how) that there is a sequence of fundamental functions, mutually orthogonal, which will solve a sequence of boundary-value problems and so provide a series for the potential – the Green's functions in full flower.

With the new century came Weyl, Charles de la Vallée Poussin, O. Frostman and many others. A change came over potential theory, a change not fully reflected in the work of Oliver Kellogg's widely read book *Foundations of Potential Theory* (1929). The theory became more precise, with the arrival of the exact conditions for the existence of solutions to the problems of potential theory and the associated theory of real and complex functions. But the theory became remote from its applications, more abstract and somehow less real. The introduction of measure theory (§3.7) and function spaces (§3.9) removed the immediacy of the elementary and classical theory. This was regrettable, no matter how necessary it might have been. But potential theory spawned many important other branches such as differential geometry and topology, homology theory, manifold theory, measure theory with capacitance, regular and irregular points and boundaries. One may see it in full flower in the Hodge–de Rham theorem in homology, finally explaining the details of d'Alembert's original objections to Clairaut, and in the proof of the Dirichlet principle (M. Brelot, J. Deny, S. Zaremba and O. Nikodym), both of these coming in the 1930s and 1940s (Malossini *1990*, Monna *1975*).

BIBLIOGRAPHY

Burkhardt, H. and Meyer, W. F. *1900*, 'Potentialtheorie', in *Encyclopädie der mathematischen Wissenschaften*, Vol. 2, Part A, 464–503 (article II A 7b).

Cross, J. J. *1983*, 'Euler's contributions to potential theory, 1730–1755', in J. J. Burckhardt, E. A. Fellmann and W. Habicht (eds), *Leonhard Euler. Beiträge zu Leben und Werk. Gedenkband des Kantons Basel-Stadt*, Basel: Birkhäuser, 331–43.

—— *1985a*, 'Potential theory', in J. W. Dauben (ed.), *The History of Mathematics from Antiquity to the Present: A Selective Bibliography*, New York and London: Garland, 353–63.

—— *1985b*, 'Integral theorems in Cambridge mathematical physics, 1830–1855', in P. Harman (ed.), *Wranglers and Physicists*, Manchester: Manchester University Press, 113–48.

Engelsman, S. *1984*, *Families of Curves and the Origins of Partial Differentiation*, Amsterdam: North-Holland.

Grattan-Guinness, I. *1990*, *Convolutions in French Mathematics, 1800–1840*, 3 vols, Basel: Birkhäuser.

Green, G. *1871*, *Mathematical Papers* (ed. N. M. Ferrers), London: Macmillan. [Repr. 1970, New York: Chelsea.]

Greenberg, J. L. *1984*, 'Alexis Fontaine's route to the calculus of several variables', *Historia mathematica*, 11, 22–38.

Humbert, P. *1936*, *Potentiels et prépotentiels*, Paris: Gauthier-Villars.

Malossini, M. *1990*, 'Il principio di Dirichlet: Storia e sviluppi nella seconda metà dell'ottocento', Laureate thesis, Università degli Studi di Trento.

Monna, A. F. *1975*, *Dirichlet's Principle: A Mathematical Comedy of Errors and its Influence on the Development of Analysis*, Utrecht: Oosthoek, Scheltema & Holkema.

Sologub, V. S. *1975, Razvitie teorii elliptickeskikh uravneniy v xviii i xix stoletiyakh*, Kiev: Naukovo dumka. [Differential equations and potential theory, 1740–1900.]

Part 4
Functions, Series and Methods in Analysis

4.0

Introduction

As was mentioned in §3.0, this Part complements its predecessor by dealing largely with methods and techniques. §4.2 performs the special role (promised in §0) of emphasizing and binding together the basic notions of trigonometry and trigonometric functions as they arise in this and several other Parts.

The next four articles treat either special kinds of function or important methods. Then in §4.9 come functional equations, which have turned up in a wide variety of contexts in mathematics; in the period covered by this book they were valued mostly for the solutions as such.

The remaining topics of this Part are oriented around the theory of equations and numerical methods, with §4.12 providing a concluding general survey. Of articles elsewhere, §5.11 on calculating machines is relevant, as are a variety of articles in Parts 8 and 9 on mechanics and mathematical physics, in which fields the methods described in this Part were deployed.

4.1

The binomial theorem

M. PENSIVY

---

1 THE CASE OF INTEGRAL EXPONENTS

Two results of a different nature go under the name of the 'binomial theorem'. The first, which is purely algebraic, gives the expansion of $(a + b)^n$ when the exponent is an integer:

$$(a + b)^n = a^n + na^{n-1}b + \cdots + nab^{n-1} + b^n. \tag{1}$$

Since the time of ancient Greece, and in all regions of the world where traces of mathematical activity have been found, we find texts containing this expression for the first values of the integer n.

From China, around AD 1100, there is evidence of a triangular layout permitting the closer and closer calculation of binomial coefficients using the additive formula we write today as

$$nC_r = (n - 1)C_{r-1} + (n - 1)C_r. \tag{2}$$

The most famous is Zhu Shijie's triangle (1303), which gives results for up to $n = 8$ (§1.9). In this and other Chinese texts, the binomial expansion appears in the context of the search for solutions of polynomial equations, using a method close to Horner's method.

Whereas in India in the sixth century AD (§1.12) Āryabhaṭa stopped at the exponent $n = 3$ (as did Brahmagupta in 628), 'Umar al-Khayyāmī (*circa* 1100) wrote the formula for $n = 4$, 5 and 6, and stated that this formula could be generalized for all integers. Al-Kāshī (1427), and then Girolamo Cardano (1570), stated the general formula, and Blaise Pascal made the final contribution: in his 'Treatise on triangular arithmetic' (1654), he demonstrated the theorem after having proved by recurrence the multiplicative relation

$$nC_r = \frac{n - r + 1}{r} nC_{r-1} \tag{3}$$

492

and given two different expressions of the binomial coefficients (see §10.1, on probability).

In the eighteenth century, mathematicians rarely mentioned Pascal and often attributed the formula to James Bernoulli, who effectively demonstrated it by enumeration in the *Ars conjectandi* (1713). Thus in no case where the exponent is an integer can the formula be called 'Newton's binomial theorem'. More detailed information can be found in Edwards *1987*: Chap. 5, and in other chapters of this book, a history of the combinatorial aspects of the coefficients of Pascal's triangle; see also §7.13.

2 THE GENERAL BINOMIAL THEOREM

In the remainder of this article, by the 'binomial theorem' we mean the expansion in infinite series

$$(1 + x)^\alpha = 1 + \alpha x + \alpha C_2 x^2 + \cdots, \tag{4}$$

where the exponent α is not an integer. In modern beginners' textbooks, this is a banal statement, a mere consequence of Brook Taylor's formula (§4.3). However, this theorem played a central role in constructing analysis from 1665 (when, as a young man, Isaac Newton wrote a first version of it in the margin of his copy of John Wallis's *Arithmetica infinitorum*) until 1826, when Niels Abel gave a near-perfect demonstration of it (Section 4).

A few facts suffice to highlight the importance of the formula at this time. First, Augustin Louis Cauchy's *Cours d'analyse* (1821) seems organized so as to converge towards the demonstration of an important result: the binomial theorem. This is therefore considered a key result in the reformulation of analysis undertaken by Cauchy in this book. Second, throughout the entire eighteenth century, the binomial theorem was not regarded as a mere expansion in series like others: it was, in effect, virtually the first series in all treatises on analysis, from which all the other necessary expansions in series were obtained. Third, during this period of approximately one hundred and fifty years, the theorem resisted all attempts at demonstration. One can point to more than fifty different attempts in all the European scientific reviews and reports of the various academies (Pensivy *1986*). Every mathematician of some repute had at least one try!

It can be said that the hesitations and errors committed in trying to demonstrate the binomial theorem increasingly emphasized the need to make a clearer distinction between algebra and differential calculus, and led to a preferred structure for treatises. After circulation of Leonhard Euler's *Introductio in analysin infinitorum* (1748), almost all textbooks adopted the following order: arithmetic, algebra, binomial theorem, expansions in series, differential calculus, Taylor's formula.

Examination of all these demonstrations modifies accepted ideas concerning the attitude of eighteenth-century mathematicians to mathematical rigour. If Euler returned to the subject on five occasions, it was in an attempt to provide a solid basis for his conception of the organization of analysis.

3 NEWTON, EULER, CAUCHY AND THE BINOMIAL THEOREM

3.1 Newton's discovery

The first statement of the formula with a rational exponent is found in a letter which Isaac Newton sent to Henry Oldenburg in 1676, replying to a question asked by Gottfried Wilhelm Leibniz about infinite series. In the absence of all notations using indices, the coefficients are deduced one from the other by a sort of algorithmic cascade:

$$(P+PQ)^{m/n} = P^{m/n} + (m/n)AQ + \frac{m-n}{2n}BQ + \frac{m-2n}{3n}CQ + \cdots, \quad (5)$$

with

$$A = P^{m/n}, \qquad B = mAQ/n, \qquad C = (m-n)BQ/2n, \ldots . \quad (6)$$

The formula was stated without demonstration, and was followed by many examples of applications. In another letter in 1676, Newton located the origin of his ideas on the question near the time when in 1665, aged around 22 years, he was studying Wallis's works on the quadratures of curves of equations $y = (1-x^2)^0$, $y = (1-x^2)^1$, $y = (1-x^2)^2, \ldots$. Newton explained how by observing the coefficients he intuitively divined the binomial formula for the following expressions: $(1-x^2)^{1/2}$, $(1-x^2)^{3/2}$, $(1-x^2)^{1/3}$, ..., then the general result. In fact, we know that Newton arrived at the idea of the coefficients with the help of a complicated generalization of Pascal's triangle written in 1665 in the margin of one of his books (Whiteside *1961*).

Independently of Newton, James Gregory used the binomial theorem in 1670, in camouflaged form, to interpolate logarithms. Before then, around 1620, Henry Briggs, who calculated his logarithms using the square roots of numbers close to 1 (§2.5), was already using the first four coefficients of the binomial expansion of $(1 + \alpha)^{1/2}$.

From the period of Newton's discovery onwards, the binomial formula appeared to be an essential result. It is through its intermediary that interpolation techniques led to the production of a very great number of expansions in series (§4.3). After Newton's manuscript *De analysi* (1669) was

494

communicated to English mathematicians, the movement gathered momentum: many series were developed for the lengths of arcs of ellipses, the areas of the zones of circles, the trigonometric functions and their reciprocals, and so on. The binomial theorem also enabled Newton to establish the first formulas of his 'method of fluxions': indeed, it made possible the calculation of the 'fluxion' (in other words, the derivative: see §3.2) of x^n when n is a fraction.

In the quarrel that ensued, it is symbolic that Newton rejected Leibniz's dissociation of the method of series from the method of fluxions. They form the basis of his analysis of the infinite, and the the fact that both methods rest on the binomial theorem shows the importance of the method of fluxions in the eyes of Newton's contemporaries.

3.2 Euler's treatment

Euler's *Introductio* (1748), founded on the notion of a function (§3.2), is the first treatise in which the exponential is dealt with before logarithms. His contribution was to bring new order to the presentation of analysis. He did not demonstrate the binomial theorem in this book, but made the general statement of it; and he deduced the expansions in series of all the usual functions, including, for the first time, those of trigonometric functions. The book had a great influence on the mathematicians of the time and, after its circulation, when other authors attempted to demonstrate the binomial theorem, they no longer limited themselves to the case of fractional exponents.

Eighteenth-century attempts to demonstrate the binomial theorem contributed to the development of rigour in analysis. We note Euler's logical error in the *Institutiones calculi differentialis* (1755), in seeking to derive the binomial theorem from Taylor's formula, when that theorem had already been used. This attempt was not renewed until Cauchy, to whom we turn.

3.3 Cauchy's demonstration

The first six chapters of his *Cours d'analyse* (1821) seem to be aimed at rigorously establishing the expansion of $(1 + x)^\mu$, where x and μ are any real numbers. Indeed, at the end of Chapter 6 he defined

$$g(\mu) = 1 + \mu x + \mu C_2 x^2 + \cdots. \qquad (7)$$

He went on to show, mainly using results established by him previously, that $g(\mu)$ is defined for $-1 < x < 1$, that g is continuous (his definition of continuity enabled him to 'demonstrate' that the sum of a series of continuous functions is continuous; Abel pointed out the error in 1826), and that g satisfies the functional equation (§4.9)

$$g(\mu)g(\nu) = g(\mu + \nu) \tag{8}$$

(thanks to his results on the product of two infinite power series). Since he had previously resolved this functional equation, it followed that

$$g(\mu) = A^\mu = [g(1)]^\mu = (1 + x)^\mu. \tag{9}$$

Cauchy obtained from this the expansions of the exponential and of the logarithm, using the formulas

$$e^x = \lim_{\alpha \to 0}(1 + \alpha x)^{1/\alpha}, \ \ln(1 + x) = \lim_{\mu \to 0}\{[(1 + x)^\mu - 1]/\mu\}. \tag{10}$$

(In Halley *1695*, a calculation of logarithms using the binomial theorem based on the same principle can already be found.) In the last three chapters of his book, Cauchy again took up calculations for complex values of x, the exponent μ remaining real.

The *Cours d'analyse* indubitably drew attention to the binomial theorem: most of the above results were used only to demonstrate this theorem. This is particularly clear for the two long chapters dealing with the interpolation of functions, which serve solely to prove that g satisfies the functional equation. Five years later, Abel's demonstration *1826* filled in the gaps in Cauchy's demonstration and generalized it to the case of a complex exponent.

4 METHODS OF DEMONSTRATION, FROM NEWTON TO ABEL

Reasoning by interpolation is usually found in the first statements of the formula (by Newton and Gregory). It is more a method of 'discovery'; the 'real' demonstrations can be divided into four categories.

4.1 Integral powers of a multinomial

In an article which received quick acclaim, Abraham De Moivre *1697* indicated how to raise a 'multinomial': that is, an infinite series $az + bz^2 + cz^3 + dz^4 + \cdots$ to a power of an integer.

J. Castillon was the first to use this result to obtain the expansion of a binomial with a rational exponent (Castillon *1744*: Preface). He defined

$$(p + q)^{r/n} = Ap^{r/n} + Bp^{(r/n)-1}q + Cp^{(r/n)-2}q^2 + \cdots, \tag{11}$$

then raised both sides of the equation to the power n. The known expansion of $(p+q)^n$ then permitted, by means of a term-by-term identification (writing relations between A, B, C, and so on) these coefficients to be calculated. (Castillon gives only the first four, for the relations found do not yield a simple recurrence.) This result only touched upon the case of fractional exponents. It can still be found in textbooks from the end of the eighteenth century.

De Moivre's method (improved by Leibniz) gave rise to the German combinatorial school grouped around K. F. Hindenburg towards 1800 (§11.2). In Hindenburg *et al. 1796* the multinomial theorem is presented as 'the most important theorem of all analysis'.

4.2 Differential methods

The first example is found in Colson *1736*. The principle (simplified) consisted first in defining

$$(a + x)^m = A + Bx + Cx^2 + \cdots \tag{12}$$

with m a fraction; hence $A = a^m$ for $x = 0$. By equating the 'fluxions' of the two sides,

$$m(a + x)^{m-1}\dot{x} = B\dot{x} + 2Cx\dot{x} + \cdots, \tag{13}$$

one obtains $B = ma^{m-1}$ by dividing by x and taking $x = 0$; and so on. Colson immediately realized the vicious circle and wrote that the derivative of x^m when m is a fraction was itself calculated using the binomial theorem!

Many authors proceeded likewise, and although some were able to skirt round the difficulty (for instance Colin Maclaurin in 1742), it is clear that they did not wish to do so: in the eighteenth century, expanding in series was a purely algebraic operation and the mathematicians wished to establish all the expansions before taking up differential calculus (§4.3).

4.3 Algebraic methods

J. Landen *1758* attempted a purely algebraic demonstration based on an original procedure; later, Euler, Joseph Louis Lagrange and others did the same. None of their calculations used multiplication of series, and they all involved a rational exponent. However, they furnished a recurrence relation linking two consecutive coefficients, which enabled them all to be calculated.

4.4 Functional equations

Some demonstrations led to a functional equation satisfied by the second coefficient of the binomial expansion (for example, n in equation (1)), the other coefficients being derived from this one. (Several original ideas are found in Aepinus *1757.*) Most demonstrations led to a functional equation satisfied by the sum of the series. The original idea came from Euler *1775*, who overturned the accepted way of thinking by reasoning about the second term of the series. In Euler's view the functional equation was poorly justified; it could be resolved only for the rationals, and its domain of validity was not specified. But his idea led to the first satisfactory demonstration when first Cauchy and then Abel elucidated all the steps. After that the binomial theorem lost its position as the cornerstone of analysis, and took a more modest place. Functional equations are dealt with in more detail in §4.9.

BIBLIOGRAPHY

Abel, N. H. *1826*, 'Untersuchungen über die Reihe $1 + (m/1)x + [m(m-1)/2]x^2 + \dots$', *Journal für die reine und angewandte Mathematik*, **1**, 311–39. [French transl. in *Oeuvres complètes*, 2nd edn, 1881, Vol. 1, 219–50.]

Aepinus, F. U. T. *1757*, 'Demonstratio generalis theorematis Newtoniani [...]', *Novi commentarii Academiae Scientiarum Petropolitanae*, **8**, 169–80.

Castillon, J. *1744*, [...] *Newton*, [...] *opuscula mathematica* [...], Lausanne and Geneva: Bousquet.

Colson, J. *1736*, *The Method of Fluxions* [...], London: Nourse.

De Moivre, A. *1697*, 'A method of raising an infinite multinomial to any given power [...]', *Philosophical Transactions of the Royal Society of London*, (230), 619–25.

Edwards, A. W. F. *1987*, *Pascal's Arithmetical Triangle*, London: Griffin.

Euler, L. *1775*, 'Demonstatio theorematis newtoniani [...]', *Novi commentarii Academiae Scientiarum Petropolitanae*, **19**, 103–11. [Also in *Opera omnia*, Series 1, Vol. 15, 202–16.]

Halley, E. *1695*, 'A most compendious and facile method for constructing the logarithms [...]', *Philosophical Transactions of the Royal Society of London*, (215), 58–67.

Hindenburg, K. F., Klügel G. S., Kramp C. and Tetens, J. N. *1796*, *Der polynomische Lehrsatz* [...], Leipzig: Fleischer.

Landen, J. *1758*, *A Discourse Concerning the Residual Analysis*[...], London: Nourse.

Pensivy, M. *1986*, 'Jalons historiques pour une épistémologie de la série infinie du binôme', *Sciences et Techniques en Perspective* (Nantes), **14**.

Whiteside, D. T. *1961*, 'Newton's discovery of the general binomial theorem', *The Mathematical Gazette*, **45**, 175–80.

4.2

An overview of trigonometry and its functions

I. GRATTAN-GUINNESS

1 EXPLANATION

The purpose of this article is rather special in the encyclopedia, due to the peculiar position that trigonometry occupies in the history of mathematics. Today planar trigonometry is (or should be) a staple part of school curricula; maybe in the final school years or at university level spherical and even spheroidal trigonometry will also be studied. While in fact quite a respectable branch of mathematics, trigonometry is too often seen merely as a bank of results and techniques. In line with this reputation, the history of trigonometry is not well represented in the literature: there has been no general study since von Braunmühl's excellent *1900–03*, although Tropfke *1923* profited from it greatly in his general history of elementary mathematics. Later Zeller *1946* took the story forward nicely to around 1700.

By contrast, the place of trigonometry in the past is far more substantial, and indeed too great to be treated in a single article; instead, it appears here in several articles in a number of Parts. Further, the trigonometric functions are prominent in algebra and in calculus and its applications. Here some general features of the history of trigonometry and its functions are 'pulled together', with cross-references to other articles.

2 THE TRIGONOMETRY OF GEOMETRY

This article is placed here because in the encyclopedia trigonometry and its functions arise mostly in connection with the calculus and its uses. However, this situation is itself the result of historical processes. During the Renaissance, trigonometry developed between algebra and geometry, being concerned with the properties of angles and sides of the (planar or spatial) triangle, and with the analysis of shadows (when cast, for example, by a sundial).

The principal motivation was astronomy; thus planar trigonometry was developed as a part of its more important spherical counterpart. In Greek mathematics, for example, Ptolemy devoted much of the *Almagest* (second century AD) to a summary of trigonometric results, including a table of chords, which correspond to sine tables (see §1.3–1.4; and Berggren *1987*). Islamic mathematicians continued this interest (see §1.6–1.7; and King *1988*); so did the Indians (§1.12), who also studied mathematical properties of the pertaining 'functions' (see e.g. Gupta *1967* on approximating to the sine). However, almost none of these achievements were known in the medieval West; indeed, they were not rediscovered until the twentieth century.

The word 'sine' made it debut in the West in rather curious circumstances. Around 1140 Robert of Chester prepared his Latin translation of a treatise by Abū Ja'far Muḥammad ibn Mūsā al-Khwārizmī, thereby bringing in the word 'algebra' as a rendering of *al-jabr* (§1.6); he used the Latin word 'sinus' as his translation of the Arabic word *jayb*, which means 'bay' or 'inlet' as well as serving as the rendition of the Sanskrit word *jiva* for 'sine'.

By the Middle Ages not all the interest lay in astronomy; planar trigonometry came sometimes to be handled separately, for purposes such as calculating the heights of buildings and mountains. It played an important part in the flowering of Western mathematics; conversely, trigonometry was becoming sufficiently significant for it to be viewed as a 'subject' and therefore treated in books (Wolfenbüttel *1989*: Chap. 9). For example, the chapters of Nicolaus Copernicus's *De revolutionibus orbium coelestium* (1543) dealing with planar and spherical trigonometry were first published the year before by his editor, Georg Joachim Rheticus; it served as a compendium of the trigonometry pertinent to the astronomy of the day (§2.7). Another important book was *De triangulis omnimodis* ('On Triangles of All Kinds', 1543) by Johann Müller (Regiomontanus); he gave a good coverage of planar and spherical trigonometry, with some emphasis on the sine and its inverse (Hughes *1967*).

The word 'trigonometry' was introduced in another text, the *Trigonometria* (1595) by B. Pitiscus, which was significant also for introducing the formulas for $\frac{\sin}{\cos}(A \pm B)$. (At that time substantial tables produced by Rheticus were posthumously published (compare §2.5 on logarithms); he had also found the formula for $\frac{\sin}{\cos} 2A$ and $\frac{\sin}{\cos} 3A$.) The word 'goniometry' was introduced to refer to that part of the subject dealing solely with angles.

3 ALGEBRIZATION AND THE TRIGONOMETRIC FUNCTIONS

The formulas are stated above anachronistically in that, while algebra was now 'around', a full-scale algebraic treatment was not yet available. A pioneer here was François Viète, and it is noteworthy that as part of his advocacy of the 'new art' (§2.3) he wrote several tracts on trigonometry in the late sixteenth century, including (versions of) formulas for $\sin nA$ as functions of $\sin A$ and $\cos A$. He also extended the tables of Rheticus. Now 'trigonometry was treated by most of the mathematicians of the Renaissance' (Rouse Ball *1908*: 520), and a range of notations was gradually introduced (Cajori *1929*: 142–79). In addition, the words 'cosine', 'tangent', and so on were introduced over the course of the seventeenth century (see the table at the end of Zeller *1946*); 'co' came in to abbreviate terms such as 'sinus complementari'.

As algebra developed into function theory in the seventeenth and eighteenth centuries, the range of formulas for both planar and spherical trigonometry increased greatly (von Braunmühl *1903*). Important topics included the relationship between the main functions and between powers and multiples of the angles, series (with Isaac Newton, infinite ones) and continued-fraction expansions, and related questions such as the binomial series (§4.1) and the evaluation of π (and see §4.3 on series). James and John Bernoulli found several formulas in the early eighteenth century, and Leonhard Euler presented a large selection (including several new results) in his textbook *Introductio ad analysin infinitorum* (1748); he followed a strictly algebraic style, with no reference to the geometrical dimensions of the variables.

At that time complex variables began to come into mathematics in a general way (§6.2), and formulas such as

$$(\cos A + i \sin A)^n = \cos nA + i \sin nA \quad \text{and} \quad \cos A + i \sin A = e^{iA}, \quad (1)$$

found by Abraham De Moivre and Euler, led to further properties. It also helped to stimulate interest in the hyperbolic functions: in the 1760s Johann Heinrich Lambert gave a very nice presentation in terms of a parametrization of the hyperbola, by analogy with such a treatment of the sine and cosine on the circle.

With the functions well established, series of the form $\Sigma_r \, a_r {\sin \atop \cos} rx$ gained attention, in two main contexts. One concerned Euler's expansion (1747) of the distance function in mathematical astronomy as such a series; this became a major method (§8.8). The second was the Fourier-series expansion, which was also aired in the 1750s; but Euler was opposed to it, and it gained acceptance only after Joseph Fourier's advocacy from the early

501

1800s (§3.11). Then they were recognized as a staple technique for solving partial differential equations (§3.15), and found applications in a wide range of problem areas in mechanics and mathematical physics; they were also used in trigonometric interpolation (Burkhardt *1904*) and in probability theory as characteristic functions (§10.2).

4 MORE MAP- AND TABLE-MAKING

From the mid-eighteenth century onwards the geometrical side received fresh impetus from geodesy, as the spheroidicity of the Earth came to be better understood (§8.14). In connection with the measurement of the meridian during the 1790s, Adrien Marie Legendre proved in 1799 this elegant result: to a small triangle PQR on the sphere there corresponds a planar triangle XYZ with sides of the same length and angles given by

$$X := P - w/3, \qquad Y := Q - w/3, \qquad Z := R - w/3, \qquad (2)$$

where w is the excess of the sum of its angles over $180°$. Geodesy and cartography (§8.15) benefited from equation (2) and related results for spheroids; Puissant *1805* is an influential treatise on geodesy in this tradition.

In addition, more elaborate tables of values of the functions were produced, culminating in a monstrous project of the 1790s run by Gaspard de Prony, where values to many places were calculated but the final product was never published (Grattan-Guinness *1990*). However, more modest books of tables were produced until the pocket calculators of our times rendered them obsolete.

BIBLIOGRAPHY

Berggren, J. L. *1987*, 'Mathematical methods in ancient science: Spherics' and '[. . .]: Astronomy', in I. Grattan-Guinness (ed.), *History in Mathematics Education* [. . .], Paris: Belin, 14–32, 33–49.

Burkhardt, H. K. F. L. *1904*, 'Trigonometrische Interpolation', in *Encyklopädie der mathematischen Wissenschaften*, Vol. 2, Part 1, 642–94 (article II A 9a).

Cajori, F. *1929*, *A History of Mathematical Notations*, Vol. 2, Chicago, IL: Open Court.

Grattan-Guinness, I. *1990*, 'Work for the hairdressers: The production of de Prony's logarithmic and trigonometric tables', *Annals of the History of Computing*, 12, 177–85.

Gupta, R. C. *1967*, 'Bhāskara's approximation to the sine', *Indian Journal for the History of Science*, 2, 121–136.

Hughes, B. (ed.) *1967*, *Regiomontanus on Triangles* [. . .], Madison, WI: University of Wisconsin Press.

King, D. *1988*, 'Universal solutions to problems of spherical trigonometry from Mamluk Egypt and Syria', in F. Kazemi and R. D. McChesney (eds), *A Way Prepared* [...], New York: New York University Press, 153–84.

Puissant, L. *1805*, *Traité de géodésie* [...], 1st edn, Paris: Courcier. [2nd edn, 2 vols, 1819–27.]

Richard of Wallingford *1976*, *An Edition of his Writings*, 3 vols (ed. J. D. North), Oxford: Clarendon Press. [Pertinent texts with commentary.]

Rouse Ball, W. W. *1908*, *A Short Account of the History of Mathematics*, 4th edn, Cambridge: Cambridge University Press. [Repr. 1960, New York: Dover.]

Tropfke, J. *1923*, *Geschichte der Elementar-Mathematik*, 2nd edn, Vol. 5, Berlin: de Gruyter.

von Braunmühl, A. *1900–03*, *Vorlesungen über Geschichte der Trigonometrie*, 2 vols, Leipzig: Teubner. [Repr. 1971, Wiesbaden: Steiner.]

—— *1908*, 'Trigonometrie. Polygonometrie. Tafeln', in M. B. Cantor (ed.), *Vorlesungen über Geschichte der Mathematik*, Vol. 4, Leipzig: Teubner, 403–50. [See also Cantor in his Vol. 2 (1892), especially Chaps 68 and 73.]

[Wolfenbüttel] *1989*, *Mass, Zahl and Gewicht* [...], Weinheim: VCH, *Acta humaniora*. [Catalogue for an exhibition held at the Herzog August Bibliothek, Wolfenbüttel, by M. Folkerts, K. Reich and E. Knobloch, with substantial commentaries.]

Zeller, M. C. *1946*, *The Development of Trigonometry from Regiomontanus to Pitiscus*, Ann Arbor, MI: Edwards.

4.3

Infinite series and solutions of ordinary differential equations, 1670–1770

L. FEIGENBAUM

From the infinitesimal calculus of Newton and Leibniz emerged several new branches of mathematics in the late seventeenth and eighteenth centuries, among them infinite series, differential equations, the calculus of variations and complex variables. Two of these are surveyed here: infinite series, used as an analytical tool from Newton to Lagrange; and the early solutions to ordinary differential equations, first developed by Leibniz and Newton in the 1670s and further explored by the Bernoullis and Taylor in the early eighteenth century.

1 INFINITE SERIES

1.1 Introduction

Although infinite series are taught today as a crucial part of the calculus, they are not considered to be as fundamental to the subject as they were in the late seventeenth and eighteenth centuries. Isaac Newton paved the way by making infinite series the primary tool in his fluxional calculus for handling both the transcendental functions and more difficult algebraic functions (§3.2). He always underlined their importance, as in the following letter to Henry Oldenburg in 1676:

> From all this it is to be seen how much the limits of analysis are enlarged by such infinite equations: in fact by their help analysis reaches, I might almost say, to all problems, the numerical problems of Diophantus and the like excepted. (Turnbull *1960*: 39)

Gottfried Wilhelm Leibniz, the Bernoullis, Brook Taylor and their contemporaries all employed infinite series to deal with differential equations and to represent most of the elementary functions. Only later, largely through the work of Leonhard Euler, did mathematicians begin to handle closed-form expressions for the elementary functions and to begin to address some of the theoretical issues that had begun to be troubling: how to define the notion of convergence rigorously; whether series were no more than infinite polynomials to be treated by the normal algebraic rules; and whether every function could be expressed, as both Euler and Joseph Louis Lagrange believed, as an infinite series.

1.2 Newton

In two letters of 1676, the so-called 'Epistola prior' and the 'Epistola posterior' written to Henry Oldenburg, the Secretary of the Royal Society, to be transmitted to Leibniz (Turnbull *1960*: 20–47, 110–61), Newton revealed not only his discovery of the binomial series, which he had found by 1665 (§4.1), but also some of his results on infinite series in general.

In the summer of 1669 Newton had written up his investigations on series in the treatise *De analysi per aequationes infinitas*, which was circulated privately but not published until 1711. He was probably motivated to set down his ideas after the appearance of Nicolaus Mercator's *Logarithmotechnia* (1668), in which Mercator had derived his famous result

$$\log(1 + x) = x - \frac{x^2}{2} + \frac{x^3}{3} - \cdots. \tag{1}$$

In *De analysi* Newton gave several examples of his own method. To find the area under the hyperbola $y = a/(b + x)$, he first used long division to obtain the expansion

$$y = \frac{a^2}{b} - \frac{a^2 x}{b^2} + \frac{a^2 x^2}{b^3} - \cdots, \tag{2}$$

and then integrated term by term to find the area (Mercator's result), adding that the 'first few terms will be of some use and sufficiently exact provided x be considerably less than b'. For $y = (a^2 - x^2)^{1/2}$ he used the technique of root extraction 'in the same way as arithmeticians in decimal numbers ... extract roots' (Whiteside *1968*: 213) to obtain an infinite series which he then integrated to find the circle's area.

Newton then showed how to invert a series by using successive approximations, a method called the 'reversion of series' and discussed more fully by Abraham De Moivre in the *Philosophical Transactions* in 1698. Applying it to the series for the area under the hyperbola,

$$y = x - \frac{x^2}{2} + \frac{x^3}{3} - \cdots, \tag{3}$$

505

he solved for x in terms of y to obtain the exponential series

$$x = 1 + y + \frac{y^2}{2} + \frac{y^3}{6} + \cdots. \tag{4}$$

By further use of these techniques he obtained the series for $\sin^{-1} x$, $\tan^{-1} x$, $\sin x$ and $\cos x$, and then used them to calculate the areas under the cycloid and the quadratrix.

For Newton the normal algebraic rules also extended to infinite series. As he commented in *De analysi*,

> And whatever common analysis performs by equations made up of a finite number of terms ... this method may always perform by infinite equations: in consequence, I have never hesitated to bestow on it also the name of analysis. To be sure, deductions in the latter are no less certain than in the other, nor its equations less exact .... (Whiteside *1968*: 241)

In the 'Epistola posterior', Newton described a rule for finding a variable y, expressed implicitly as a function of x, in terms of a series of powers of x. This method, now called 'Newton's parallelogram method', allows one to determine in advance the form of the series and then to compute it using the method of undetermined coefficients. The proof, which Newton did not provide, was finally given several decades later by A. G. Kästner and Gabriel Cramer.

In Section 2.2 we consider Newton's use of infinite series to solve differential equations.

1.3 Leibniz and the method of undetermined coefficients

Leibniz also found series expansions for $\sin x$, $\cos x$ and $\tan^{-1} x$, in 1673. In 1674 he obtained his famous arithmetical quadrature of the circle,

$$\frac{\pi}{4} = 1 - \frac{1}{3} + \frac{1}{5} - \frac{1}{7} + \cdots, \tag{5}$$

using a general 'transmutation theorem' for dividing an area into infinitely small parts which are then reassembled to form equivalent areas. In response to hearing about Newton's work on the same subject, he communicated these results and other series to Oldenburg in August 1676 for transmission to Newton (Turnbull *1960*: 65–8).

Although the method of undetermined coefficients for finding series solutions to differential equations was already known privately (see Section 2.2), Leibniz was the first to publish it, in an article in the *Acta eruditorum* of April 1693. Remarking that Mercator had found series by division and Newton by extraction, Leibniz felt his own method would provide results 'more easily and universally'. Thus he showed how to find either the logarithm y or the number x, given

$$y = \int \frac{a\,dx}{a+x} \tag{6}$$

or the corresponding differential equation

$$a\,dy + x\,dy - a\,dx = 0. \tag{7}$$

By substituting either

$$y = bx + cx^2 + ex^3 + \cdots \quad \text{or} \quad x = ly + my^2 + ny^3 + \cdots, \tag{8}$$

he obtained

$$y = x - \frac{x^2}{2a} + \frac{x^3}{3a^2} - \cdots \tag{9}$$

or

$$x = y + \frac{y^2}{2a} + \frac{y^3}{6a^2} + \cdots, \tag{10}$$

thereby obviating the need for the technique of reversion of series to obtain the logarithmic from the exponential expansion, and vice versa.

To find the series for $\sin y$ from its differential equation

$$a^2(dx)^2 = a^2(dy)^2 + y^2(dx)^2, \tag{11}$$

he substituted

$$y = bx + cx^3 + ex^5 + \cdots. \tag{12}$$

Assuming dx to be constant, he differentiated the original equation to obtain

$$a^2\,d^2y + y(dx)^2 = 0, \tag{13}$$

and then substituted the series expressions for y, dy/dx and d^2y/dx^2. Equating the coefficients to zero under the assumption that $b = 1$, he obtained

$$y = x - \frac{x^3}{2\cdot 3a^2} + \frac{x^5}{2\cdot 3\cdot 4\cdot 5a^4} - \cdots. \tag{14}$$

1.4 The Bernoulli series

In the *Acta eruditorum* of November 1694, John Bernoulli announced a new theorem for quadratures that he considered to be an improvement over Leibniz's method of undetermined coefficients. He then introduced his own 'series universalissima' which 'expresses generally all quadratures, rectifications, and integrals of other differentials':

$$\int n\,dz = nz - \frac{z^2}{2}\frac{dn}{dz} + \frac{z^3}{2\cdot 3}\frac{d^2n}{dz^2} - \frac{z^4}{2\cdot 3\cdot 4}\frac{d^3n}{dz^3} + \cdots. \tag{15}$$

Applying this to the differential equation for the sine, $dy = (a^2 - y^2)^{1/2}\,dx/a$, he set

$$n = \frac{(a^2 - y^2)^{1/2}}{a} \quad \text{and} \quad dz = dx. \tag{16}$$

Then

$$y = \int dy = \frac{(a^2 - y^2)^{1/2}}{a}\,x + \frac{yx^2}{2a^2} - \frac{(a^2 - y^2)^{1/2}}{2\cdot 3a^3}\,x^3 - \frac{yx^4}{2\cdot 3\cdot 4a^4} + \cdots, \tag{17}$$

or

$$\frac{y}{(a^2 - y^2)^{1/2}} = \frac{x - \dfrac{x^3}{2\cdot 3a^2} + \dfrac{x^5}{2\cdot 3\cdot 4\cdot 5a^5} - \cdots}{a - \dfrac{x^2}{2a} + \dfrac{x^4}{2\cdot 3\cdot 4a^3} - \cdots}. \tag{18}$$

Calculating the logarithm as Leibniz did, John Bernoulli found that

$$y = \frac{ax}{a+x} + \frac{ax^2}{2(a+x)^2} + \frac{ax^3}{3(a+x)^3} + \cdots, \tag{19}$$

and noted that it was a series which, although 'different from that of Mr Leibniz, nevertheless has the same value'.

It is easily shown that the Bernoulli series is equivalent to, but not the same as, the Taylor series, by generating both from the same integral using repeated integration by parts:

$$\int_0^x f'(t)\,dt = \begin{cases} f'(t)t\,\big|_0^x - \displaystyle\int_0^x f''(t)t\,dt \; \ldots & \text{Bernoulli series with} \\ & n = f' \text{ in (15)} \\[2mm] f'(t)(t-x)\,\big|_0^x - \displaystyle\int_0^x f''(t)(t-x)\,dt \; \ldots & \text{Taylor series.} \end{cases} \tag{20}$$

1.5 The Taylor series

Taylor was the first to publish the theorem that now bears his name, but he was not the first to discover it. At least five mathematicians anticipated him: James Gregory (1671), Newton (1691), Leibniz (1670s), John Bernoulli (1694) and De Moivre (1708) (Feigenbaum *1985*: 72–96). Of the five, only

508

Bernoulli published his result, (20)$_1$; however, he does not seem to have been aware that his series was equivalent to Taylor's. In any case, it is clear that Taylor was the first to have appreciated the fundamental significance and the applicability of the result (Feigenbaum *1985*: 40–47, 96–129).

The Taylor series was published as Proposition 7, Corollary 2 of Taylor's *Methodus incrementorum* (1715), the first text to discuss the theory of finite differences and their applications to difference equations, to differential equations and to problems in mechanics. To prove the theorem, he used the elementary properties of finite increments to develop the formula, which in modern notation amounts to:

$$x(z + n\,\Delta z) = x(z) + \frac{n}{1}\Delta x(z) + \frac{n}{1}\frac{n-1}{2}\Delta^2 x(z) + \frac{n}{1}\frac{n-1}{2}\frac{n-2}{3}\Delta^3 x(z)$$

$$+ \cdots. \tag{21}$$

Then, setting

$$v = n\,\Delta z, \qquad \frac{v - \Delta z}{2\Delta z} = \frac{n-1}{2}, \tag{22}$$

and so on, he obtained

$$x(z + v) = x(z) + \frac{\Delta x(z)v}{\Delta z} + \frac{\Delta^2 x(z)}{(\Delta z)^2}\frac{v(v - \Delta z)}{2!} + \cdots$$

$$+ \frac{\Delta^n x(z)}{(\Delta z)^n}\frac{v(v - \Delta z)\ldots[v - (n-1)\Delta z]}{n!} + \cdots, \tag{23}$$

which in essence is the Gregory–Newton interpolation formula, first enunciated by Newton in Book III of the *Principia* (Lemma 1, Case 5). Both Gregory and Newton used the formula for interpolating the values of functions expressed as infinite series (Goldstine *1977*: 68–84; Edwards *1979*: 281–7), while Gregory also used it to prove the binomial theorem.

Taylor then substituted for the finite increments the fluxions proportional to them, and made v, $v - \Delta z$, $v - 2\Delta z$, and so on, all equal to v, obtaining the famous series

$$x(z + v) = x(z) + x'(z)v + \frac{x''(z)}{2!}v^2 + \frac{x'''(z)}{3!}v^3 + \cdots. \tag{24}$$

Taylor did not address the questions of convergence, of a remainder term, or of the validity of expressing a function in such a series, and he implicitly assumed that

$$\lim_{\substack{n \to \infty \\ \Delta z \to 0}} (v - n\,\Delta z) = v,$$

prompting Felix Klein (*1932*: 233) to characterize his argument as 'a passage to the limit of unexampled audacity'. In fact, Taylor was no more audacious than either his contemporaries or his immediate successors: Euler proved the Taylor series in virtually the same way in 1736 (Feigenbaum *1985*: 44).

Euler also praised Taylor for being the first to make the world aware of the series and for demonstrating its usefulness. In Section 2.6 we examine how Taylor employed the theorem to find series solutions to differential equations. He also used it in approximating the roots of equations, including those with transcendental expressions; however, he did not explicitly emphasize employing it as we do today to expand well-known functions in power series, although he certainly knew that this was possible.

1.6 Stirling and Maclaurin

It was James Stirling, not Taylor, who published the first explicit use of the Taylor theorem to generate a power-series expansion for a well-known function, namely $\cos x$. In his *Lineae tertii ordinis Newtonianae* (1717), Stirling derived the Taylor series about zero by using the method of undetermined coefficients, as most modern texts do. He then applied it to express y as a power series in x, where x and y were related either by a fluxional or by an ordinary algebraic equation.

Colin Maclaurin also developed the Taylor series about zero, using the method of undetermined coefficients (*Treatise of Fluxions*, 1742). Although he fully attributed the theorem to Taylor, ironically this is the chief result to which his name is universally attached. Carrying Taylor's work further. Maclaurin applied the Taylor series to determine the conditions for which local maxima and minima exist (Struik *1969*: 340–41), and also devised the integral test for the convergence of an infinite series.

1.7 Euler

In his three great textbooks, the *Introductio in analysin infinitorum* (1748), *Institutiones calculi differentialis* (1755) and *Institutiones calculi integralis* (1768–70), Euler shaped the modern subject of analysis. In the *Introductio* we find the logarithmic and exponential functions both defined by power series and treated as inverses of each other. Each is derived using the binomial theorem and a limiting argument involving both infinitely small and infinitely large quantities. For example, to derive the series for a^x he observed that, for ε infinitely small (and therefore $a^0 = 1$),

$$a^\varepsilon = 1 + k\varepsilon. \tag{25}$$

Then, for $x/\varepsilon = N$, an infinitely large number,

$$a^x = (a^\varepsilon)^N \tag{26}$$

$$= [1 + (kx/N)]^N \tag{27}$$

$$= 1 + kx + \frac{N(N-1)}{2!\,N^2} k^2 x^2 + \frac{N(N-1)(N-2)}{3!\,N^3} k^3 x^3 + \cdots \tag{28}$$

$$= 1 + kx + \frac{k^2 x^2}{2!} + \frac{k^3 x^3}{3!} + \cdots. \tag{29}$$

The number e is defined as the value of a for which $k = 1$:

$$e = 1 + 1 + \frac{1}{2!} + \frac{1}{3!} + \cdots \quad \text{and} \quad e^x = (1 + (x/N))^N. \tag{30}$$

The trigonometric functions were introduced as ratios, and their series expansions derived also using the binomial series and a limiting argument (Edwards *1979*: 272–7). Moreover, Euler unified the theory involving all these concepts by means of the idea of a function, the algebraic treatment of infinite series and special formulas which brought several notions together, such as

$$e^{iv} = \cos v + i \sin v. \tag{31}$$

When Euler derived the series expansions for the elementary transcendental functions, he did so without relying on the calculus. Later, in the *Institutiones calculi differentialis*, he used these expansions to find the differential of the elementary functions. For example, to find $d(\log x)$ he wrote it as

$$\log(x + dx) - \log x = \log\left(1 + \frac{dx}{x}\right) = \frac{dx}{x} - \frac{(dx)^2}{2x^2} + \frac{(dx)^3}{3x^3} - \cdots, \tag{32}$$

and then rejected all differentials of second order and higher to obtain $d(\log x) = dx/x$. Similarly,

$$d(\sin x) = \sin(x + dx) - \sin x \tag{33}$$

$$= \sin x \cos dx + \cos x \sin dx - \sin x \tag{34}$$

$$= \sin x\left(-\frac{(dx)^2}{2!} + \frac{(dx)^4}{4!} - \cdots\right) + \cos x\left(dx - \frac{(dx)^3}{3!} + \cdots\right), \tag{35}$$

so that

$$d(\sin x) = \cos x \, dx. \tag{36}$$

Euler devoted much effort to the summation of infinite series (Hofmann

1959). From a modern viewpoint some of these results are not legitimate, since the series are divergent (§3.3); but he probably saw himself as examining relationships between series and functions. One lasting result was the 'Euler constant', C, which arose in a study of 1734 of the harmonic series $\Sigma_r\ a/(b + rc)$, where a, b and c are constants:

$$C := \lim_{n \to \infty} \left(\sum_{r=1}^{n} \frac{1}{r} - \ln n \right) = 0.577\,21 \ldots. \tag{37}$$

It is still not known whether C is rational or irrational.

1.8 Lagrange and the calculus as algebra

In his *Théorie des fonctions analytiques* (1797), Lagrange attempted to provide a sound foundation for the calculus by reducing it to algebra (§3.2). Rejecting the Newtonian notion of limit, he tried to prove Taylor's theorem algebraically and then develop the calculus on the basis of the theorem. Assuming that any function could be expanded in a power series,

$$f(x + i) = f(x) + pi + qi^2 + ri^3 + \cdots, \tag{38}$$

he called the coefficients p, q, r, ... the 'derived functions', f', f'', f''', ... (the source of our modern notation and term for derivative), since they arose from the original function $f(x)$. He then observed that, with a little understanding of the differential calculus, one can see that the derived functions coincide with

$$\frac{dy}{dx}, \quad \frac{d^2y}{dx^2}, \quad \frac{d^3y}{dx^3}, \quad \ldots.$$

For the first time we also find an expression for the remainder term, the well-known Lagrange form of the remainder for the Taylor series (Edwards *1979*: 297–9).

2 EARLY SOLUTIONS TO ORDINARY DIFFERENTIAL EQUATIONS

2.1 Introduction

The subject of ordinary differential equations arose in the attempts by the early practitioners of the calculus to describe and solve physical problems, especially those originating in mechanics and astronomy: gravitational attraction, pendulum motion, elasticity, projectile motion in a resisting medium, orthogonal trajectories, and curves such as the isochrone, the brachistochrone and the catenary. Leibniz named them 'differential equations' because he thought of them as functions involving parts of the characteristic triangle (dx, dy, ds), while Newton called them 'fluxional equations', expressing a relationship between fluxions, which he defined as the velocities with which flowing quantities (fluents) increase. Newton stated the fundamental problem of the subject in the 'Epistola posterior' of 1676 (see Section 1.2), which was to play a role several decades later in the calculus priority controversy. It was concealed in the anagram

$$6accdae13eff7i3l9n4o4qrr4s8t12vx,$$

whose meaning he revealed in 1693: 'given an equation involving any number of fluent quantities to find the fluxions, and conversely' (Turnbull *1960*: 153).

Most of the elementary methods for solving first-order equations and many of the *ad hoc* techniques for solving particular equations of higher order were devised during this period. However, the systematization of the subject did not begin until the work of Euler (§3.14).

2.2 Newton's fluxional equations

In Newton's treatise 'De methodis serierium et fluxionum' of 1671 we find the first substantial discussion of fluxional equations. However, his results were known only to those few who saw or learned about the manuscript privately, for it was not published until 1736 (Whiteside *1969*: 32–353). In it Newton classified first-order equations into three types: those in which two fluxions and one of their fluents are involved; those in which two fluxions and their fluents are involved; and those in which more than two fluxions are involved (i.e. partial differential equations). While suggesting particular solutions in finite form for special examples, he proposed solving them generally through infinite series. For example, in the equation

$$\dot{y}^2 = \dot{x}\dot{y} + \dot{x}^2 x^2, \tag{39}$$

he solved algebraically for $\dot{y}/\dot{x}$ and then extracted the root to obtain

$$\dot{y}/\dot{x} = \tfrac{1}{2}[1 + (1 + 4x^2)^{1/2}] = 1 + x^2 - x^4 + 2x^6 + \cdots. \tag{40}$$

After integrating term by term, he arrived at

$$y = x + \tfrac{1}{3}x^3 - \tfrac{1}{5}x^5 + \tfrac{2}{7}x^7 + \cdots. \tag{41}$$

Newton also used the method of undetermined coefficients to express y as an infinite series in x: for

$$\dot{y}/\dot{x} = 1 - 3x + y + x^2 + xy, \tag{42}$$

he found that

$$y = a + (1 + a)x + (-1 + a)x^2 + (\tfrac{1}{3} + \tfrac{2}{3}a)x^3 + \cdots, \qquad (43)$$

observing that, since a is arbitrary, one obtains an infinity of solutions.

2.3 Leibnizian differential equations

Although Newton and Leibniz developed the calculus at about the same time, Leibniz was the first to publish his ideas in two now celebrated papers in the *Acta eruditorum* of 1684 and 1686. In the latter we find the first published use of the integral sign, while in the former the first published differential equation, namely the problem that Florimond Debeaune had proposed years before to René Descartes: to find the ordinate w of a curve whose subtangent is a constant a. Leibniz set down and solved the resulting equation

$$\mathrm{d}w/\mathrm{d}x = w/a \qquad (44)$$

by noting that if the xs form an arithmetic proportion, then the ws form a geometric proportion and the curve will be logarithmic.

In 1691 Leibniz implicitly discovered the technique of separating the variables in a differential equation; however, the general method was not formulated until 1694, by John Bernoulli. Leibniz also showed how to integrate a homogeneous first-order equation, and reduced the first-order linear equation to quadratures. As was mentioned in Section 1.2, he also introduced the method of undetermined coefficients to obtain series solutions, although he failed to include the arbitrary constant which provides the general solution.

2.4 Leibniz and the Bernoullis

During the 1690s Leibniz and his two chief disciples, the brothers James and John Bernoulli, applied the calculus to a host of problems in mathematical physics that involved differential equations. For example, Leibniz proposed the isochronous curve in 1687, the curve along which a body falls with uniform velocity. In James Bernoulli's solution (*Acta eruditorum*, May 1690), which leads to the differential equation

$$\mathrm{d}y/(b^2y - a^3)^{1/2} = \mathrm{d}x(a^3)^{1/2}, \qquad (45)$$

we find the first use of the word 'integral' in equating the antiderivatives of both sides. In 1690 James proposed finding the catenary, the curve formed by a chain of uniform weight hung freely from its ends. Galileo had incorrectly assumed the curve to be a parabola; however, John Bernoulli showed that it satisfied the equation

$$\frac{\mathrm{d}y}{\mathrm{d}x} = \frac{s}{a}, \tag{46}$$

where s is the arc length and a is constant. Leibniz and Christiaan Huygens also published correct solutions.

In 1695 James Bernoulli proposed the problem of separating the indeterminates x and y in the equation now known as the 'Bernoulli equation':

$$a\,\mathrm{d}y = yp\,\mathrm{d}x + by^n q\,\mathrm{d}x, \tag{47}$$

where a and b are constants, and p and q are both functions of x. Leibniz showed that equation (47) can be reduced to a linear equation by substituting $z = y^{1-n}$.

John Bernoulli offered a different substitution that was the basis of a new method: letting $y = mz$ and $\mathrm{d}y = m\,\mathrm{d}z + z\,\mathrm{d}m$, he obtained

$$am\,\mathrm{d}z + az\,\mathrm{d}m = mzp\,\mathrm{d}x + bm^n z^n q\,\mathrm{d}x, \tag{48}$$

with m, z, p, q and x all variable. Then, exercising the freedom of choice this allowed him, he set

$$am\,\mathrm{d}z = mzp\,\mathrm{d}x \quad \text{and} \quad az\,\mathrm{d}m = bm^n z^n q\,\mathrm{d}x. \tag{49}$$

From equation $(49)_1$ he obtained z in terms of x; separating the variables in $(49)_2$, he found m and then y in terms of x.

2.5 John Bernoulli's integral calculus

In addition to his ingenious solutions to special differential equations like that of the brachistochrone (the curve of swiftest descent, which he determined to be a cycloid), John Bernoulli made several contributions to the general theory of differential equations. As mentioned above, he was the first to formulate the method of separating the variables, and the first to apply a 'universal series', named after him (see Section 1.3), to obtain series solutions to differential equations.

When, in 1691, the Marquis de l'Hôpital asked the 24-year-old John Bernoulli to tutor him in the differential and integral calculus, Bernoulli wrote down his lessons, but not for publication. In a financial arrangement unprecedented in the history of mathematics (Truesdell 1958: 61), John agreed to surrender his discoveries to the Marquis, who published those on the differential calculus in his *Analyse des infiniment petits* (1696). The lessons on the integral calculus were finally published by John in 1742, long after l'Hôpital's death. In them we find, for example, the first use of integrating factors. To solve

$$ax\,dy - y\,dx = 0, \tag{50}$$

Bernoulli multiplied by y^{a-1}/x^2, so that $d(y^a/x) = 0$; therefore the integral, y^a/x, will be equal to a constant. Note that although this equation is easily separable, John did not yet know how to handle the quantity $\int dx/x$ (he learned this soon enough); thus the integrating factor nicely allowed him to circumvent the difficulty.

2.6 Taylor and the method of increments

As we saw in Section 1.5, Taylor wrote about the theory and solutions of differential equations in his treatise *Methodus incrementorum* (1715). Four of his contributions deserve mention here (for further details see Feigenbaum *1985*). Using his now-famous series, he showed how to find a solution to the fluxional equation

$$\ddot{x}z + n\ddot{x}x - \dot{x} - \dot{x}^2 = 0, \quad \text{where } \dot{z} = 1. \tag{51}$$

Computing the first few fluxions of x, namely

$$\ddot{x} = \frac{\dot{x} + \dot{x}^2}{z + nx}, \qquad \dddot{x} = (2 - n)\frac{\ddot{x}\dot{x}}{z + nx}, \qquad \ddddot{x} = (3 - 2n)\frac{\dddot{x}\dot{x}}{z + nx}, \tag{52}$$

he observed the pattern for calculating the rest. Then, setting

$$x(a) = c, \qquad \dot{x}(a) = \dot{c}, \tag{53}$$

and so on, the solution will be expressed by means of his series

$$x(a + v) = c + \dot{c}v + \frac{\ddot{c}v^2}{2} + \frac{\dddot{c}v^3}{2 \cdot 3} + \cdots, \tag{54}$$

where $z = a + v$ and $\ddot{c}$, $\dddot{c}$, and so on will be given in terms of c and $\dot{c}$, the two arbitrary constants. He then showed that this solution could be expressed in closed form.

Taylor also discovered the first singular solution to a differential equation when he considered

$$4x^3\dot{z}^2 - 4x^2\dot{z}^2 = (1 + z^2)^2\dot{x}^2. \tag{55}$$

Substituting $x = v^\theta y^\lambda$, he transformed the equation into two equations, one of which had the solution

$$x = \frac{1 + z^2}{[a + (1 - a^2)^{1/2}z]^2} \tag{56}$$

and the other reduced to $x = 1$, which he noted is 'a certain singular solution of the problem', there not being any choice of a in equation (56) that would give the solution $x = 1$.

Although Newton had noted earlier in passing that the general solution to an nth-order differential equation would contain n arbitrary constants, only Taylor devoted considerable attention to the initial and boundary conditions that can accompany a differential equation. Thus he described in great detail not only the number and nature of the conditions necessary for any equation, but also the manner in which to apply them, namely the various restrictions on the distribution of conditions.

Taylor's fourth contribution concerns the formulas for the derivatives of the inverse function, which in modern notation amount to expressing

$\mathrm{d}x/\mathrm{d}z$, $\mathrm{d}^2x/\mathrm{d}z^2$, $\mathrm{d}^3x/\mathrm{d}z^3, \ldots$ in terms of $\mathrm{d}z/\mathrm{d}x$, $\mathrm{d}^2z/\mathrm{d}x^2$, $\mathrm{d}^3z/\mathrm{d}x^3, \ldots$.

He used these in particular to transform certain differential equations that might be too difficult or impossible to solve in closed form into ones that might be more easily treated. Thus equation (51) is the transformed version of

$$\ddot{z}z + \dot{z}^2 + n\ddot{z}x + \dot{z}\dot{x} = 0, \qquad (57)$$

which according to Taylor did not lend itself to a closed-form solution, as (51) did.

3 CONCLUDING REMARKS

The formulas for the derivatives of the inverse function are directly related to the problem of specifying the independent variable in a differential equation. In his differential equations Leibniz specified the 'progression of the variable' by indicating which of the first-order differentials was to be assumed constant (as in equation (11)). The Newtonians referred to the same problem as choosing the variable which would increase uniformly (for example, in equation (51)). Both practices correspond to the modern concept of denoting the independent variable in a differential equation. According to Bos (*1974*: 43–7), by the middle of the eighteenth century mathematicians became less interested in this special issue, looking instead at the more general problem of transforming differential equations in which one differential is assumed constant into those in which no differential is assumed constant. First introduced by John Bernoulli, this problem influenced the work of Euler and Jean d'Alembert.

Concerning solutions themselves, although more general ones were to be found later in the eighteenth century, several notable problems were treated in the early eighteenth century which led to second- and third-order differential equations, and more often than not also led to disputes between some of the solvers. Three deserve note here: the isoperimetric problem proposed by James Bernoulli in 1696, which led to a bitter feud between the Bernoulli brothers; the orthogonal-trajectory problem proposed by Leibniz in 1716 to 'test the pulse of the English'; and finally the path of a projectile in a medium resisting as the square of the velocity, proposed by John Keill in 1717 and which erupted into a battle in which almost all the famous mathematicians of the time became involved (Whiteside *1981*: Part 2).

Infinite series and elementary methods of solving ordinary differential equations were only two of the new branches of analysis which grew out of the calculus. The eighteenth century also saw the development of several others, which are treated in Part 3 of the encyclopedia: the general theory of these equations (§3.14), partial differential equations (§3.15), the calculus of variations (§3.5) and differential geometry (§3.4). In addition, they feature in various aspects of mechanics (Part 8).

BIBLIOGRAPHY

Bos, H. J. M. *1974*, 'Differentials, higher-order differentials and the derivative in the Leibnizian calculus', *Archive for History of Exact Sciences*, **14**, 1–90.

Burkhardt, J. J., Fellmann, E. A. and Habicht, W. (eds) *1983*, *Leonhard Euler 1707–1783: Beiträge zu Leben und Werk. Gedenkband des Kantons Basel-Stadt*, Basel: Birkhäuser.

Cantor, M. *1898*, *Vorlesungen über Geschichte der Mathematik*, Vol. 3, Leipzig: Teubner. [Repr. 1965.]

Edwards, C. H. Jr *1979*, *The Historical Development of the Calculus*, New York: Springer.

Feigenbaum, L. *1985*, 'Brook Taylor and the method of increments', *Archive for History of Exact Sciences*, **34**, 1–140.

Goldstine, H. H. *1977*, *A History of Numerical Analysis from the 16th through the 19th Century*, New York: Springer.

Hofmann, J. E. *1956*, *Über Jakob Bernoullis Beiträge zur Infinitesimalmathematik*, Geneva (*L'Enseignement Mathematique*: pamphlet No. 3).

—— *1959*, 'Um Eulers erste Reihenstudien', in K. Schröder (ed.), *Sammelband der zu Ehren des 250. Geburtstages Leonhard Eulers*, Berlin: Akademie Verlag, 139–208.

Ince, E. L. *1927*, 'Historical note on formal methods of integration', in his *Ordinary Differential Equations*, London: Longmans, Green, 529–39. [Repr. 1944, New York: Dover.]

Klein, F. *1932*, *Elementary Mathematics from an Advanced Standpoint*, Vol. 1, New York: Macmillan. [Transl. from the German edition of 1924 by E. R. Hedrick and C. A. Noble. Repr. 1945, New York: Dover.]

Kline, M. *1972*, *Mathematical Thought from Ancient to Modern Times*, New York: Oxford University Press. [Repr. in 3 vols, 1990.]

Maseres, F. (ed.) *1791–1807, Scriptores logarithmici*, 6 vols, London: White. [Includes reprints of early papers on series.]

Struik, D. J. (ed.) *1969, A Source Book in Mathematics, 1200–1800*, Cambridge, MA: Harvard University Press.

Truesdell, C. *1958*, 'The new Bernoulli edition', *Isis*, **49**, 54–62.

Turnbull, H. W. (ed.) *1960, The Correspondence of Isaac Newton*, Vol. 2, Cambridge: Cambridge University Press.

Walker, H. E. *1983*, 'Taylor's theorem with remainder: The legacy of Lagrange, Ampère, and Cauchy', Doctoral Dissertation, New York University.

Whiteside, D. T. *1961*, 'Patterns of mathematical thought in the later 17th century', *Archive for History of Exact Sciences*, **1**, 179–388.

—— (ed.) *1968, 1969, 1976, 1981, The Mathematical Papers of Isaac Newton*, Vols. 2, 3, 7, 8, Cambridge: Cambridge University Press.

4.4

Special functions

I. GRATTAN-GUINNESS

1 JOB DESCRIPTION

The name 'special functions' has become attached to a group of functions which arose out of a variety of questions in pure and applied mathematics, but which often came to be associated with solutions to certain differential equations. As one of their forms is usually an expansion in an infinite power series, they are often also called 'transcendental'. There are mathematical connections between some of them, but these links were not realized at once, and the histories are somewhat separate. This article follows that characteristic by treating, in turn, the beta and gamma functions, Legendre functions and Bessel functions. The order is roughly the chronology of their rises to importance, which occurred up to the mid-nineteenth century. These first examples are functions of real variables; the hypergeometric function (Section 5) runs forward to the late nineteenth century, when complex-variable mathematics (§3.12) was furnishing many new properties for all special functions. Some of these properties for Legendre and Bessel functions are noted in Section 6, which finishes with a brief survey of the era of textbooks and treatises.

Certain other functions often called 'special' are treated in other articles: elliptic and Abelian functions in §4.5–4.6, while Fourier series form part of §3.11 on harmonic analysis. To some extent this article is a continuation of the survey in §4.3 of early techniques in the calculus and differential equations; it also relates to §4.1 on the binomial expansion.

Many of the main publications on these functions were concerned with the applications from which the functions were generated; but only the 'pure' properties are stressed here. These are (where applicable): the differential equation which a function satisfies; its expression as a (power) series; the reality of its zeros; conversion of the series to a finite, or 'closed' form; involvement of parameters which define different 'orders' of the function, and maybe 'recurrence relations' between them; a 'generating function' $f(q)$, in which some or all of the functions of these various orders are defined as coefficients of the appropriate powers of q; and orthogonality of some set(s) of functions over an interval of values, and the possibility of expressing 'any' function as a sum of functions from this set.

The order followed here is not definitive: indeed, to offer one at all is rather anachronistic in that the emergence of the functions *as such* is part of the story. Emphases of this kind will be pointed out on occasion, although space prevents a complete record.

The amount of material involved is quite massive; only a few of the equations and formulas can be quoted here. Grattan-Guinness *1990* provides *passim* much contextual information for the period 1780–1840. Wangerin *1904* gives a valuable overview of knowledge at the time of writing. Whittaker and Watson *1927* is a classic textbook in this area, which not only contains an excellent survey but also furnishes many historical references.

2 BETA AND GAMMA FUNCTIONS

These functions are usually defined respectively as

$$\Gamma(x) = \int_0^\infty e^{-t} t^{x-1} \, dt \quad \text{and} \quad B(x,y) = \int_0^1 p^{x-1} (1-p)^{y-1} \, dp. \quad (1)$$

Leonhard Euler was the first major contributor (Davis *1959*), defining $\Gamma(x)$ in 1731 in order to extend the concept of the factorial

$$n! = n(n-1)(n-2)(n-3)\ldots 3\cdot 2\cdot 1 \quad (2)$$

to cases where n is not an integer, thereby expressing various results in interpolation. Later he also considered $B(x, y)$, and found the fundamental relationship

$$\Gamma(x)\Gamma(y) = \Gamma(x+y)B(x,y) \quad (3)$$

between the two functions.

The next major student was Adrien Marie Legendre. He introduced the 'Γ' notation and gave the definition $(1)_2$ (Euler's was different in form), and the name 'gamma function'; he also found several further properties of both functions. He gave a comprehensive presentation in his treatise *Exercices du calcul intégral* (1811–17), which was mainly concerned with (the related topic of) elliptic integrals (§4.5). By then both functions were well recognized by mathematicians, and used in a variety of contexts in geometry, analysis and mechanics.

Along with the beta function, but slower in its emergence, was the indefinite integral which has become known as the 'incomplete beta function', defined as this function of z:

$$B_z(x, y) = \int_0^z p^{x-1}(1 - p)^{y-1} dp, \quad x, y > 0 \text{ and constant.} \qquad (4)$$

The motivations came almost entirely from probability and statistics. John Wallis, Thomas Bayes and Pierre Simon Laplace experimented at finding approximate values for this function for large x and y (§10.2), and much later the function was studied further when the beta distribution became of interest (Dutka *1981*); but these developments took place largely outside those of the special functions.

3 LEGENDRE FUNCTIONS AND POTENTIAL THEORY

From 1830, following a suggestion by William Whewell, the names 'Laplace's functions' or 'coefficients' became adopted; however, the name 'Legendre functions' became current from the late nineteenth century. This change evinced Legendre's major contributions to the theory; it also reflects an unlovely competition between the two men over the period 1775–1800 (Todhunter *1873*: Vol. 2). The noun 'polynomial' is more common but less accurate, as for some orders these functions have no representation by *finite* power series.

The context for Laplace and Legendre was the attraction to the spheroidal Earth of an exterior mass-point. The potential V satisfied Laplace's equation, which in the spherical polar coordinates (r, ω, θ) appropriate for the problem took the form

$$((1 - \mu^2) V_\mu)_\mu + V_{\omega\omega}/(1 - \mu^2) + r(rV)_{rr} = 0, \quad \mu := \cos\theta. \qquad (5)$$

When the method of separating variables was applied, the function $H(\mu)$ satisfied 'Legendre's equation',

$$(1 - \mu^2)H''(\mu) - 2\mu H'(\mu) + [n(n + 1) - m^2/(1 - \mu^2)]H(\mu) = 0, \qquad (6)$$

where m and n are constants. Its solutions were the Legendre functions, now written as $P_n^m(\mu)$ and usually found in power-series form; the case where $m = 0$ attracted the most interest (for which the notation is $P_n(\mu)$). Another motivation from astronomy was this expansion of the inverse of the 'distance function', which stated the length of the side of a triangle whose other sides, of lengths 1 and q, subtended the angle θ:

$$(1 - 2\mu q + q^2)^{-1/2} = \sum_r P_r(\mu) q^r. \qquad (7)$$

Laplace and Legendre worked around equations (5)–(7), finding many properties especially of $P_r(\mu)$ for r an integer (and when the polynomial form arises); in particular, they found that the functions are orthogonal over $[-1, 1]$ of μ, so that expansions of polynomials as series could be effected. They also examined the forms $\cos m\phi P_n^m(\mu)$ and $r^n \cos m\phi P_n^m(\mu)$ because of their importance in potential theory as solutions of equation (5); the names 'surface' and 'solid harmonics' have become attached to them. Other aspects, such as convergence of the series and the generality of the solutions of equation (6), were less fully handled.

Laplace deployed the theory well in the first two volumes of his *Mécanique céleste* (1799), and from then on they were well-established tools. Various results were added, of which an important one was proved in 1816 in a Paris doctorate of the Université de France, 'On the attraction of spheroids': the formula

$$P_n^m(\mu) = \{(1 - \mu^2)^{m/2} d^{m+n}[(1 - \mu^2)^n]/d\mu^{m+n}\}/2^n n!, \quad n \geqslant m \qquad (8)$$

derived by the young Jewish mathematician Olinde Rodrigues. Among established figures, Siméon-Denis Poisson used the functions in appropriate problems in heat diffusion, and in a paper summarizing current knowledge in 1823.

4 BESSEL FUNCTIONS UP TO BESSEL

The very title of this section shows that a misnomer is at hand, for, while Friedrich Wilhelm Bessel contributed, he was not the first or the main figure. As with the Legendre functions, they have their origins in the eighteenth century, although it is a much more scrappy story. Daniel Bernoulli and Euler created fragments on occasion, in connection with problems in oscillation (Truesdell *1960: passim*), but without establishing a theory. The strongest vision came from Euler in a study (1759) of the vibrating membrane, where he formed the differential equation, found the series expansion and claimed that 'any' function could be expanded in a series of Bessel functions $J_0(x)$ (to use the standard notation) defined from its assumed real zeros $\{a_n\}$:

$$f(x) = \sum_n k_n J_0(a_n x). \qquad (9)$$

His optimism was in contrast to his reserve about such expansion by Fourier series (Bôcher *1893*).

This is a cue to bring in Joseph Fourier. The Bessel functions come naturally out of differential equations like Laplace's cast in cylindrical polar coordinates, for procedures similar to (5) and (6) produce them as the functions of the radial variable r (for this reason they are often given the name 'cylinder functions'); yet we owe this approach only to Fourier's initiation of heat diffusion during the 1800s (§9.4). Having found his series for diffusion in the bar and lamina, he cast his equation in appropriate coordinates for the cylinder and ran into $J_0(r)$. His treatment was brilliant: from the relevant ordinary differential equation

$$rJ_0''(r) + J_0'(r) + rJ_0(r) = 0 \tag{10}$$

he found the power-series expansion and its conversion to an integral, and proved that it has an infinity of real zeros. Then − in imitation of the orthogonality of trigonometric series, but in ignorance of Euler's contribution − he showed that equation (9) obtained, and found the formula for the coefficients k_n (Grattan-Guinness and Ravetz *1972*: Chaps 15 and 16).

This work was restricted to $J_0(x)$. Interest in higher orders (of which fragments were known in the eighteenth century) was stimulated by Bessel's contribution of some recurrence relations (motivated by problems in astronomy) and Poisson's use of them in his work on heat diffusion in the 1820s. Progress was similar to that with the Legendre functions, although the generating function (for functions of integral order) seems to have played a much smaller role.

5 THE HYPERGEOMETRIC FUNCTION

Definitions of varying generality have been given at different times; the most appropriate one here is the series definition

$$F(a, b, c; x) := \sum_n [(a)_n(b)_n/(c)_n n!] x^n, \tag{11}$$

where a, b and c are constants, and the shifted factorial $(a)_n$ is defined by

$$(a)_0 := 1, \quad (a)_n := a(a+1)(a+2)\ldots(a+n-1), \quad n \geqslant 1. \tag{12}$$

The first to study a series of this form was Wallis in 1655; he was followed by Euler and then J. F. Pfaff in 1797 (Dutka *1984*). A significant step forward was taken by Carl Friedrich Gauss in 1813, when he showed that the series converged for complex x satisfying $|x| < 1$. (This was done at a time when convergence was not normally considered (§3.3): his criterion is the origin of 'Gauss's test'.) In unpublished work at the same time Gauss showed that the function satisfied the differential equation

$$x(1-x)F''(x) + [c - (a+b+1)x]F'(x) - abF(x) = 0. \tag{13}$$

His study of the series yielded numerous identities and applications, including several to Euler's theory of gamma functions. The first to publish the connection between the function and the differential equation (although

524

only for real x) was Ernst Kummer, over twenty years later. He also derived a famous list of the 24 solutions of the equation.

International, especially German, interest in this series increased, the main reason being the rise of complex-variable analysis (§3.16). In the late 1850s Bernhard Riemann re-derived the differential equation (13) (and extended Kummer's results) from a consideration of a complex function with specified behaviours at the singular points ($x = 0$, 1 and ∞). Riemann made another contribution when, while working in number theory, he introduced the 'zeta function' $\zeta(s)$ of the complex variable s (§3.12). These extensions and their consequences led to a new range of 'confluent' forms for the function, and at last clarified the point that several of the special functions are particular cases of it. One case, studied by K. H. A. Schwarz, a follower of Karl Weierstrass, indicated how to define new classes of functions, called automorphic functions, later and independently adopted by Henri Poincaré in his study of the differential equation. By 1893 Klein was stressing the great importance of the hypergeometric function for mathematics (Gray *1986*), and soon the Indian mathematician Srinivasa Ramanujan was working intensively on it (Askey *1980*).

6 FURTHER RESULTS FOR SPECIAL FUNCTIONS

Meanwhile, and not entirely independently, much more work was being carried out on the special functions; sometimes the work was inspired 'internally' (including properties of the hypergeometric function), but quite often an application was the source. Again, complex-variable analysis provided a huge stimulus for results, especially expressions (and even new definitions) involving contour integrals; and emphasis on the general solutions of the relevant differential equations highlighted the additional forms of these functions which would arise as second solutions. In addition, greater attention was lavished on definability over values of variables, ranges of convergence of infinite series, uniqueness, the reality (or not) of zeros, and so on.

The gamma function even received a new definition, from Weierstrass in 1856, in the terms favoured by him of an infinite product. Regarding the other two main functions, just a few features involving orthogonality are noted here, as this property stimulated many of the main extensions and applications.

Legendre functions, especially in polynomial form, helped inspire the introduction of other polynomials $\{g_r(x)\}$ (Szegö *1958*). A weighting function $w(x)$ was involved in the orthogonality condition: that is, over some interval of (real) values $[a, b]$,

$$\int_a^b g_r(x)g_s(x)w(x)\,\mathrm{d}x = 0, \quad r \neq s. \tag{14}$$

(For the Legendre functions over $[0, 1]$, $w(x) = 1$.) Names have been given to the originators of these polynomials, although as usual partial predecessors are sometimes evident. Here are some of the main types, with first (significant) dates, one notable predecessor (if any), and specifications of $[a, b]$ and $w(x)$:

Jacobi	1859	$[-1, 1]$	$w(x) = (1 - x)^a(1 + x)^b$, $a, b > -1$,	
				(15)

| Chebyshev | 1859 | $[-1, 1]$ | $w(x) = (1 - x^2)^{\pm 1/2}$, | (16) |

| Hermite (Laplace) | 1864 | $[-\infty, \infty]$ | $w(x) = \exp(-x^2)$, | (17) |

| Laguerre (Euler) | 1879 | $[0, \infty]$ | $w(x) = \mathrm{e}^{-x}x^a$. | (18) |

For Bessel functions, valuable expansion theorems analogous to equation (9) were found; for example:

| Schlömilch | 1857 | $f(x) = \sum_n k_n J_m(a_n x), \quad n > 0$ | (19) |

| C. Neumann | 1867 | $f(x) = \sum_n k_n J_{m+n}(a_n x), \quad n \geqslant 0$, | (20) |

| Lommel | 1868 | $f(x) = \sum_n k_n J_m(a_n x), \quad n > 0$, | (21) |

| Kapteyn | 1893 | $f(x) = \sum_n k_n J_{m+n}[(m + n)x], \quad n \geqslant 0$. | (22) |

In these (and other) cases, complex values of x were allowed.

In addition, two other important functions took inspiration from the Bessel, and also from the use of orthogonal curvilinear coordinates (§3.4). Gabriel Lamé introduced his function in 1857 by transforming Laplace's equation into elliptic coordinates (i.e. coordinates whose constant values define ellipses in the plane), separating these variables and forming the solutions (compare equation (5)). Emile Mathieu formed his function in similar manner in 1868 by taking the differential equation for the vibrating membrane in elliptic–hyperbolic coordinates and separating variables.

From the 1870s, special functions were sufficiently well established, not only to be treated at some length in many works on applied mathematics and/or differential equations, but even to warrant special texts of their own (see the bibliographies in Brunel *1899* and Wangerin *1904*). German mathematicians naturally led the field: for example, Carl Neumann wrote up Bessel functions in 1868, with E. Lommel following a year later; Neumann's father Franz treated 'spherical functions' in 1878, while E. Heine issued the second edition of his handbook in two volumes (1878–81), following a pioneering first edition of 1861. Among other nationalities, English mathematicians were alert: Isaac Todhunter ran through *The Functions of Laplace, Lamé and Bessel* in 1875, and the latter functions received an account twenty years later from A. Gray and G. B. Matthews. Then in 1902, E. T. Whittaker surveyed the field in *Modern Analysis* (1902); Whittaker and Watson *1927*, cited in Section 1, is a successor. In 1922 G. W. Watson himself weighed in with a magnificent treatise on Bessel functions, of which the second edition is cited here as Watson *1944*. Extensive tables of values of some of these functions, and of their zeros, were also compiled.

For the last hundred years the special functions have been a staple diet in mathematics, and new properties and relationships are still being found. Erdelyi *1953–5* reports upon the continuing interest in a vast compendium of results; the references mostly cite recent work. Askey *1988* recounts the history of this remarkable project.

BIBLIOGRAPHY

Askey, R. *1980*, 'Ramanujan's extensions of the gamma and beta functions', *American Mathematical Monthly*, **87**, 346–59.

—— *1988*, 'Handbooks of special functions', in P. Duren (ed.), *A Century of Mathematics in America*, Part 3, Providence, RI: American Mathematical Society, 369–91.

Bôcher, M. *1893*, 'A bit of mathematical history', *Bulletin of the New York Mathematical Society*, **2** (1892–3), 107–9.

Brunel, G. *1899*, 'Bestimmte Integrale', in *Encyklopädie der mathematische Wissenschaften*, Vol. 2, Part A, 135–88 (article II A 3).

Burkhardt, H. F. K. L. *1908*, 'Entwicklungen nach oscillirenden Functionen und Integration der Differentialgleichungen der mathematischen Physik', *Jahresbericht der Deutschen Mathematiker-Vereinigung*, **10**, Part 2, xii + 1804 pp. [Valuable references *passim*.]

Davis, P. J. *1959*, 'Leonhard Euler's integral [...]', *American Mathematical Monthly*, **66**, 849–69.

Dutka, J. *1981*, 'The incomplete beta function – A historical profile', *Archive for History of Exact Sciences*, **24**, 11–29.

—— *1984*, 'The early history of the hypergeometric function', *Archive for History of Exact Sciences*, **31**, 15–34.

Erdelyi, A. (ed.) *1953–5*, *Higher Transcendental Functions*, 3 vols, New York: McGraw-Hill. [Based on manuscript notes by H. Bateman. Includes elliptic functions.]

Grattan-Guinness, I. *1990*, *Convolutions in French Mathematics, 1800–1840* [. . .], 3 vols, Basel: Birkhäuser, and Berlin: Deutscher Verlag der Wissenschaften.

Grattan-Guinness, I. and Ravetz, J. R. *1972*, *Joseph Fourier 1768–1830* [. . .], Cambridge, MA: MIT Press.

Gray, J. J. *1986*, *Linear Differential Equations and Group Theory from Riemann to Poincaré*, Boston, MA: Birkhäuser.

Molina, E. C. D. *1930*, 'The theory of probability: Some comments on Laplace's *Théorie analytique*', *Bulletin of the American Mathematical Society*, **36**, 369–92. [Laplace and the Hermite polynomials.]

Papperitz, E. *1889*, 'Über die historische Entwickelung der Theorie der hypergeometrischen Functionen', *Abhandlungen der naturwissenschaftlichen Gesellschaft Isis in Dresden*, 61–73.

Szegö, G. *1958*, 'An outline of the history of orthogonal polynomials', in his *Collected Papers*, Vol. 3, 1982, Basel: Birkhäuser, 857–69. [Includes historical notes by the editor, R. Askey.]

Todhunter, I. *1873*, *A History of the Mathematical Theories of Attraction and Figure of the Earth* [. . .], 2 vols, London: Macmillan. [Repr. 1962, New York: Dover.]

Truesdell, C. A. III *1960*, *The Rational Mechanics of Flexible or Elastic Bodies 1638–1788*, Zurich: Orell Füssli. [As L. Euler, *Opera omnia*, Series 2, Vol. 11, Part 2.]

Wangerin, A. *1904*, 'Theorie der Kugelfunktionen und der verwandten Funktionen [. . .]', in *Encyklopädie der mathematischen Wissenschaften*, Vol. 2, Part A, 695–795 (article II A 10).

Watson, G. N. *1944*, *The Theory of Bessel Functions*, 2nd edn, Cambridge: Cambridge University Press. [Contains many references to original works.]

Whittaker, E. T. and Watson, G. N. *1927*, *A Course of Modern Analysis*, 4th edn, Cambridge: Cambridge University Press. [Various reprints.]

4.5

Elliptic integrals and functions

ROGER COOKE

1 INTRODUCTION

The introduction of infinitesimal methods and symbolic algebra into geometry during the seventeenth century made it feasible to study new curves such as the cycloid and the lemniscate, and a large number of previously intractible geometric problems such as the quadrature and rectification of the conic sections (§7.2). The result was an increase in the number of geometric curves accessible to study and useful in solving the problems of mechanics. The new mechanics was able to describe accurately the motion of the planets and falling bodies, and by the eighteenth century investigation was begun into more complicated motions, such as those of a vibrating string, a rotating rigid body or a pendulum. By the year 1700 the integral calculus had been developed to the extent that a substantial body of techniques existed for expressing integrals ('fluents', in the notation of the British) in terms of elementary functions where possible, and various non-elementary integrals began to appear repeatedly in different contexts.

The simplest of the latter integrals are those later named elliptic functions by Adrien Marie Legendre. This term was originally applied to an indefinite integral of the form $\int R[x, (p(x))^{1/2}]\,dx$, where $p(x)$ is a polynomial of degree 3 or 4 and $R(x, y)$ is a rational function of two variables, but since the work of Niels Abel and Carl Jacobi in the early nineteenth century it has been used to denote the function inverse to such an indefinite integral. The indefinite integral that originally bore this name is now known as an elliptic integral. The non-elementary character of such integrals was proved rigorously by Joseph Liouville in 1840. The word 'elliptic' here refers to the geometric interpretation of one class of such integrals, namely

$$\int [(a^2 - e^2 x^2)/(a^2 - x^2)]^{1/2}\,dx$$

as the length of an arc of an ellipse. On the history, see Fricke *1913* and Houzel *1978*.

The problems posed by such integrals have been viewed differently at different times, as the prevailing ideas about mathematics have changed. In the seventeenth and eighteenth centuries the subject now known as calculus was more geometrical and less analytic than it is today, in the sense that mathematicians of the earlier era sought to express the solutions of their problems geometrically rather than by means of a formula; the occurrence of an elliptic integral in the solution of both a rectification problem and a mechanical problem meant that the solution of the mechanical problem could be expressed by an arc of a conic section. The period of a pendulum, for example, was expressed in this way (§8.13). It was realized, of course, that more information about the rectification of the conic sections was desirable. In modern mathematics the integral is regarded as the fundamental concept, in terms of which both the conic arc and the period of the pendulum are expressed.

Being the simplest integrals not expressible in terms of elementary functions, elliptic integrals began to be studied in the late seventeenth century. In the nineteenth century elliptic functions were an important application of many of the concepts of complex analysis, such as contour integrals, conformal mapping and the description of analytic functions in terms of their singularities, providing illustrations of the principles of abstract function theory and connections with number theory. In the context of algebraic geometry and algebraic number theory they continue to be an object of study today.

The focus of attention in the study of elliptic functions has naturally shifted over the centuries as old problems have been solved and new ones posed. In the history of the subject one can conveniently distinguish four periods: early encounters with particular elliptic integrals (1650–1750); isolation of elliptic integrals as an object of study, leading to the classification of elliptic integrals and their properties (1750–1825); discovery and exploration of the fundamental properties of elliptic functions (1825–75); and complete integration of elliptic functions into the setting of modern abstract mathematics (1875 to the present). The extraordinarily full development of this subject can be explained by the fact that in all four periods elliptic functions and integrals have attracted the interest of some of the most distinguished mathematicians of the day.

2 EARLY ENCOUNTERS WITH ELLIPTIC INTEGRALS, 1650–1750

Attempts were made to solve the rectification problem for the ellipse, which gave its name to elliptic integrals and functions, as early as 1655. That year John Wallis began to study arc length on various cycloids, and from the form of the expression for these arcs deduce relations between the lengths of arcs on these curves and the lengths of arcs of an ellipse. In 1669 Isaac Newton gave an infinite series for the length of an arc of an ellipse; this result was published by Wallis in 1685 and appeared also in Newton's work. James Bernoulli encountered elliptic integrals in 1679 in connection with the rectification of a spiral; he remarked that the resulting expression implied that two arcs of different shape would have the same length. Later he encountered elliptic integrals in studying the shape of an elastic rod and the length of the lemniscate. In 1694 both James and John Bernoulli encountered the integral $\int x^2(a^2 - x^4)^{-1/2}\,dx$, and John expressed the opinion that this integral could not be written in terms of the arc length of a conic section. In 1698 John Bernoulli studied the transformation

$$(x, y) \mapsto (x\,dy^3 : dx^3, 3x\,dy^2 : 2\,dx^2),\tag{1}$$

under which arc length transforms in a particularly simple way. For the cubic curve $y = 2x^{1/3}$, which is invariant under this transformation, the length of an arc and the length of its image differ by an algebraic expression. A few years earlier John Bernoulli had posed the general problem of studying a pair of arcs (generally on two different curves) whose difference equals an arc of a circle or a line segment. This type of problem was investigated in great detail by G. Fagnano in 1714 in connection with the rectification of the lemniscate. All these investigations, as well as other investigations of the rectification problem by Blaise Pascal, Christopher Wren, Jean d'Alembert, Leonhard Euler and Colin Maclaurin, were concerned with particular geometric and mechanical problems. No unified, systematic investigation of integrals of the particular form now known as elliptic had yet been undertaken.

3 ELLIPTIC INTEGRALS AS AN OBJECT OF STUDY, 1750–1825

In 1750 Fagnano sent his works to the Berlin Academy of Sciences, and Euler was appointed to study them. The relations that Fagnano had established between arcs of the lemniscate and hyperbola attracted Euler's attention, and he sought to clarify the principles underlying such theorems. Interpreting Fagnano's relations as particular integrals of a differential equation, Euler posed the problem of finding the general solution in terms of a parameter in such a way that the particular solutions found by Fagnano would correspond to special values of the parameter. The result was the famous addition theorem for elliptic integrals, published in 1757, which expresses the general integral of the differential equation

$$\frac{dx}{(1-x^4)^{1/2}} = \pm \frac{dy}{(1-y^4)^{1/2}} \tag{2}$$

as the relation

$$x^2 + y^2 = c^2 + 2xy(1-c^4)^{1/2} - c^2x^2y^2, \tag{3}$$

where c is the arbitrary parameter. This paper of Euler's established elliptic integrals as an object of independent interest. It was not, however, as systematic as Euler would have wished. In fact, he said that he had arrived at the result essentially by guessing, and posed the problem of deriving the result in a more natural manner. In 1764 he published a paper unifying the earlier work on rectification of the ellipse and hyperbola by Maclaurin and d'Alembert, and pointing out the analogy between the elliptic integrals and the integrals for the inverse trigonometric functions. This observation was the basis of the systematic investigations undertaken later by Legendre. (Indeed, it seems to have been this paper of Euler's that inspired Legendre's lifelong interest in elliptic integrals, and a quotation from this paper adorns the *avertissement* to Legendre's treatise on elliptic integrals.)

The related differential equation

$$\frac{dx}{(\alpha + 2\beta x + \gamma x^2 + 2\delta x^3 + \varepsilon x^4)^{1/2}} = \frac{dy}{(\alpha + 2\beta y + \gamma y^2 + 2\delta y^3 + \varepsilon y^4)^{1/2}} \tag{4}$$

was studied by Joseph Louis Lagrange in 1766. His idea was to differentiate the equation and combine the resulting equation with the original so as to obtain a simpler equation of the same order. For this particular equation his technique was to denote the common value of the two sides by dt/T. The idea of making the integral itself the independent variable and letting the limit of integration be the dependent variable, which is implied by this technique, was to prove crucial in the development of the theory, though nearly two generations were to pass before this idea was developed.

In his *Théorie des fonctions analytiques* (1797), Lagrange found the solution of the differential equation

$$\frac{dz}{(1-k^2\sin^2 z)^{1/2}} = \frac{du}{(1-k^2\sin^2 u)^{1/2}} \tag{5}$$

in the form

$$\cos z \cos u + \sin z \sin u \cos M = \cos m, \tag{6}$$

where m is a constant of integration and $\sin M = k \sin m$. The first of these equations is of course the law of cosines for a spherical triangle, and leads to a geometric interpretation of u, z and m as the sides of a spherical triangle in which k is the ratio of the sine of each angle to the sine of the opposite side.

In reducing integrals to a standard form, Euler, Maclaurin and d'Alembert had frequently expressed such integrals as the difference between an arc of a hyperbola from the vertex to a given point P and the projection of the line segment from the centre of the hyperbola to P onto the tangent at P. This difference, however, assumes the indeterminate form $\infty - \infty$ when the point P is infinitely distant. To remove this indeterminacy, J. Landen showed in 1771 that the indeterminate quantity could be expressed as the sum of two such finite differences decreased by the projection of a line from the centre of an ellipse onto the tangent to the ellipse. The techniques developed in solving this problem enabled Landen to express an arc of the hyperbola

$$\frac{x^2}{m-n} - \frac{y^2}{2(mn)^{1/2}} = 1 \tag{7}$$

in terms of an algebraic function of position and the arcs of two ellipses, one with semi-axes m and n, the other with semi-axes $\frac{1}{2}(m + n)$ and $(mn)^{1/2}$; that is, the semi-axes of the second ellipse are the arithmetic and geometric means of the axes of the original ellipse. In discovering this relationship, Landen implicitly gave the first example of the technique known as transformation of elliptic integrals, which was to form an essential part of both the theoretical and the computational aspect of elliptic functions.

An elliptic integral obviously depends on the coefficients of the polynomial appearing under the radical sign as well as on the limits of integration, and the nature of this dependence is a matter of considerable interest. In particular, one would like to know which elliptic integrals can be obtained from a given elliptic integral by a change of variable, as Landen had essentially done. He had obtained his result by using a fractional linear change of variable. In 1784 Lagrange showed how to reduce elliptic integrals to the form $\int Z[(1 \pm r^2z^2)(1 \pm s^2z^2)]^{-1/2} dz$, thereby providing a rapidly convergent series for the computation of elliptic integrals and removing the defect inherent in the use of arcs of conic sections for the solution of mechanical problems. The transformation technique was used by Legendre to shorten the labour of compiling tables of elliptic integrals.

The most visible progress on elliptic integrals in the years from 1785 until 1825 was the work by Legendre, who established the fundamental properties of these integrals and classified them into three simple types:

$$
\left.
\begin{aligned}
F(\phi, c) &= \int_0^\phi \frac{d\phi}{(1 - c^2 \sin^2\phi)^{1/2}}, \qquad E(\phi, c) = \int_0^\phi (1 - c^2 \sin^2\phi)^{1/2}\, d\phi \\
\Pi(n, c, \phi) &= \int_0^\phi \frac{d\phi}{(1 + n \sin^2\phi)(1 - c^2 \sin^2\phi)^{1/2}}.
\end{aligned}
\right\} \tag{8}
$$

He showed that every elliptic integral could be reduced to a sum of integrals of the three basic types and deduced some very subtle properties of each type, the most remarkable of which is perhaps the law of interchange of amplitude (ϕ) and modulus (c) in an integral of the third kind.

By the time Legendre published the last volume of his treatise on elliptic integrals (*1825–8*), the topics mapped out by Euler had been thoroughly explored. These topics involved studying the algebraic transformation of one elliptic integral into another, the addition theorems for these integrals, and the differential equations satisfied by the integrals as functions of the upper limit when the coefficients of the polynomial are fixed, and as functions of the coefficients of the polynomial when the upper limit is fixed.

For some thirty years, Carl Friedrich Gauss had been conducting his own investigations of elliptic integrals and making profound discoveries, many of which unfortunately were not published during his lifetime. Allusions to them occur in his *Disquisitiones arithmeticae* (1801), where he mentions that a problem similar to the construction of a regular 17-sided polygon can be solved also for the lemniscate (i.e. the lemniscate can be divided into 17 equal arcs). As this problem was solved by Abel using the double periodicity of elliptic functions, it seems that Gauss must have known a great many of the properties of elliptic functions by the year 1800. According to Schering's notes in Gauss's collected works, Gauss had studied the lemniscatic elliptic integral $\phi = \int (1 - x^4)^{-1/2}\, dx$ by 1797 and had discovered its second period. He later expressed the periods as arithmetic–geometric means of simple functions of the parameter of the integral, and obtained a direct inversion of the elliptic integral. He also studied the function $\sum_{k=-\infty}^\infty \exp[-a(k + \omega)^2]$, now known as a theta function, which was to become a basic tool in the investigation of elliptic functions. Only a little of this work was published, mainly his researches on the relation between the arithmetic–geometric mean and the lemniscate.

Legendre paid special attention to the transformation problem, both as a way of shortening the labour of compiling tables of elliptic integrals and as a way of gaining insight into the relations between integrals of differing parameter. In particular he was interested in the parameters of the integrals that could be obtained from a given elliptic integral by iterating a rational transformation of specified order. He grouped all such parameters into a single 'scale'. In 1824, just as he was about to publish the last edition of his treatise on elliptic integrals, he discovered a new scale based on a transformation of order 3; but by that time Abel and Jacobi had already discovered the new transformation, and Jacobi had boldly announced the existence of such transformations of all orders in the form $y = U(x)/V(x)$, where U and V are relatively prime polynomials of degree $2k + 1$ and $2k$, respectively. Jacobi found eventually that the number of distinct transformations of this type is equal to the sum of the divisors of $2k + 1$.

4 ELLIPTIC FUNCTIONS AS FUNCTIONS OF A COMPLEX VARIABLE, 1825–1875

The researches of Abel and Jacobi represent a continuation of the theory begun by Euler and based on the extensive foundation laid by Legendre (Königsberger *1879*). The idea of inverting the integrals leads to the problem of solving equations of the type

$$t = \int_0^x \frac{du}{[(1 - u^2)(1 - k^2 u^2)]^{1/2}} \qquad (9)$$

for x in terms of t, and this process reveals both the double periodicity of the inverse function and the need for complex values of the argument. On these insights Abel constructed a systematic theory of such functions, including a study of their zeros and discontinuities (poles, in later terminology), an expression for the inverse function at a point mz as a rational function of its value at z and, conversely, an expression for the value of the function at z as an algebraic function of its value at mz. For prime values of m Abel investigated this problem separately, and was led to the 17-fold division of the lemniscate as a result. Abel gave a variety of both single and double series and product representations of his elliptic functions, and found the relation between the parameters of two elliptic integrals transformable into each other by general transformations.

Abel died in 1829, before he could perfect a systematic treatise in which he would have attempted to find all possible cases in which a sum of elliptic integrals whose limits of integration are algebraically related can be expressed by an algebraic function of the limits. Some of his ideas on the transformation problem and on the inversion of elliptic integrals had been independently discovered by Jacobi, however, who continued the work after Abel's death. Jacobi gave explicit names to the inverse functions, which have been preserved to the present day: 'sine amplitude', 'cosine amplitude' and 'Δ amplitude', now more familiar as sn, cn and dn. Jacobi also studied the multiplication problem and the relation between the modulus k of the integral of the first kind and the ratio of the complete elliptic integrals of the first kind corresponding to this value of the modulus and the conjugate modulus $k' = (1 - k^2)^{1/2}$. His greatest achievement, however, was to discover a simple global expression for the elliptic functions not subject to restrictions on the radius of convergence, as a quotient of the functions now known as theta functions. These functions turn out to be the simplest elements from which elliptic functions can be constructed, and their generalization to more than one variable was crucial to the understanding of algebraic integrals more complicated than elliptic integrals.

The use of complex arguments to study elliptic functions in the work of Abel and Jacobi occurred just as the theory of analytic functions of a complex variable was making strong advances in the hands of Gauss, Augustin Louis Cauchy and others. In particular, the phenomenon of double periodicity could be clearly understood in the light of the contour integrals developed by Cauchy. Gauss had earlier remarked on the multi-valuedness of the integral $\int dz/z$, but Cauchy extended this observation to integrals $\int f(z) dz$, where $f(z)$ is itself a multi-valued (algebraic) function. The systematic development of the properties of analytic functions using the Cauchy integral to find power series and infinite products led to the realization that such functions are characterized by their zeros and poles. The result of these new developments was a new way of looking at algebraic functions in general and elliptic functions as a special case. During the 1840s, the new possibilities attracted the attention of many talented mathematicians, who were able to give definitive solutions to many old problems of the theory and find new perspectives from which to look at the theory as a whole, leading to new questions to be answered. Among them were Liouville, who established the fundamental fact that the number of zeros of a doubly periodic, meromorphic function within a period parallelogram equals the number of poles (counting multiplicity); Victor Puiseux, who exhibited the periods of the elliptic functions as integrals around contours enclosing the singularities of the integrand; and Charles Hermite, who made a systematic study of transformations of theta functions. A full treatise on the theory was published by C. A. Briot and J. C. Bouquet in 1859.

The systematic exploitation of the poles to characterize an analytic function was applied to elliptic functions by Ferdinand Gotthold Eisenstein, who studied series of the form

$$\sum_{m,n} \frac{1}{(x + \alpha m + \beta n)^g}, \quad g > 0, \tag{10}$$

which he denoted by (g, x). In this approach he anticipated some later work by Karl Weierstrass and Henri Poincaré. Eisenstein showed that the function $y(x) = (2, x) - (2^*, 0)$ (where the asterisk indicates summation only over pairs (m, n) for which $m^2 + n^2 > 0$) must satisfy the differential equation

$$(\mathrm{d}y/\mathrm{d}x)^2 = p(x), \tag{11}$$

with $p(x)$ a cubic polynomial. Thus he deduced that any function representable by a series of the indicated type must represent an elliptic function, a sort of converse to the fact that elliptic functions are doubly periodic.

As a young man Weierstrass had investigated the second integrals of the squares of four Jacobi elliptic functions and had expressed the Jacobi functions sn u, cn u and dn u as quotients of these primitives. In a later, polished version of the theory (1882), Weierstrass deduced all his results from a single function $\sigma(u)$, the exponential of the second primitive of the famous $\wp$-function. Although Weierstrass's exposition essentially sums up the theory along the lines of the Legendre–Jacobi approach, new ideas from geometry were already penetrating the theory of elliptic functions even as the theory seemed to be reaching perfection along the established lines. Among these ideas was the Riemann surface, applied in the more general setting of Abelian integrals, but contributing essentially to an understanding of the mapping properties of elliptic functions. In the future, however, the theory of elliptic functions would be able to contribute to analysis, geometry and number theory as much as it took from them, and the interaction of the theories would be of mutual benefit.

The earliest example of such new ideas occurs in the problem of complex multiplication – the problem of determining the complex numbers w such that $f(wz)$ is a doubly periodic function having the same periods as a doubly periodic function $f(z)$. Abel had considered the case $w = i\sqrt{n}$, and Jacobi had found that there are transformations of order p corresponding to the two values $w = a \pm ib\sqrt{n}$ if $p = a^2 + nb^2$ with a odd and b even. Eisenstein noted in 1846 that if a transformation leaves the ratio of the periods invariant, it corresponds to a complex multiplication by w, and w is an algebraic integer in a quadratic extension of the rational numbers. Through this connection, Eisenstein found a proof of the law of biquadratic reciprocity. This example is only one of many in which elliptic functions contribute to number theory.

An application to geometry occurs in the theory of elliptic curves, whose history goes back at least to 1676, when Newton showed how to study a general cubic plane curve via a curve $y^2 = ax^3 + bx^2 + cx + d$. In 1861 S. H. Aronhold found a parametric representation of the general plane cubic in which each of the three homogeneous coordinates of the general point is a quadratic polynomial plus the square root of a cubic polynomial in a variable λ, which in turn is a fractional linear function of x and y. This representation was given an interpretation in terms of elliptic functions by

Alfred Clebsch in 1863. The result was a description of the cubic curve as the path followed by the point of intersection of two variable lines. This description provided a great deal of valuable information about the curve, for example the number (nine) and location of the points of inflection, and the existence of four tangents issuing from a given point of the curve and an expression for their points of contact. Clebsch later found other curves representable parametrically by elliptic functions. These curves are nowadays called elliptic curves.

An application in algebra, noted by Leopold Kronecker in 1853, rests on the fact that the equations with coefficients having rational real and imaginary parts and commutative Galois group are the equations for division of the periods of the lemniscatic elliptic functions. A conjecture based on this observation, known as 'Kronecker's *Jugendtraum*', laid the foundation for class field theory (§6.4). It was listed by David Hilbert as his twelfth problem in 1900, and its solution was achieved by the 1930s as a result of the combined efforts of H. Weber, R. Fueter, T. Takagi and H. Hasse.

5 ELLIPTIC FUNCTIONS IN MODERN MATHEMATICS, 1875 TO THE PRESENT

The role played by the theory of elliptic functions in modern mathematics, providing insight into other areas and simultaneously applying theorems from other areas to gain more information about elliptic functions, can be seen from the three examples above. To them one could add many other important interactions. Two outstanding instances of these interactions are the solution of the Dirichlet problem for the lune-shaped region bounded by two circles of different radius, achieved by H. Villat in 1920, and an essential role they played in the proof of the Picard theorem that the image of a non-constant entire function is the entire plane with the exception of at most one point, a splendid example of the application of properties of a particular function to prove theorems about an entire class of functions. It can also be shown that the general theory of automorphic functions arose from the study of elliptic functions.

Elliptic functions are nowadays studied in the context of modern abstract algebraic geometry. In the 1940s F. Conforto, C. L. Siegel and A. Weil found new existence proofs in this context for functions (theta functions) having the invariance properties needed to serve as building blocks for a unified theory of such functions. The natural context for this theory is an Abelian variety, and the theory is built upon a broad foundation of topology, geometry and algebra. The central ideas discovered in the nineteenth century have not been lost, however, only generalized, and Weil has recently reworked the basic ideas of Eisenstein and Kronecker in order to call attention to the fact that their implications have not yet been exhausted.

BIBLIOGRAPHY

Brill, A. and Noether, M. *1893*, 'Die Entwicklung der Theorie der algebraischen Functionen in älterer und neuerer Zeit', *Jahresbericht der Deutschen Mathematiker-Vereinigung*, 3, 109–566.

Dieudonné, J. A. *1985*, *History of Algebraic Geometry* (transl. from the French by J. Sally), Monterey, CA; Wadsworth.

Fricke, R. *1913*, 'Elliptische Funktionen', in *Encyklopädie der mathematischen Wissenschaften*, Vol. 2, Part 3, 177–348 (article II B 3).

Houzel, C. *1978*, 'Fonctions elliptiques et intégrales abéliennes', in J. Dieudonné (ed.), *Abrégé d'histoire des mathématiques*, Vol. 2, Paris: Hermann, 2–113.

Königsberger, L. *1879*, *Zur Geschichte der elliptischen Transcendenten in den Jahren 1826–1829*, Leipzig: Teubner.

Legendre, A.-M. *1825–8*, *Traité des fonctions elliptiques*, 3 vols, Paris: Huzard-Courcier.

Shafarevich, I. R. *1974*, *Basic Algebraic Geometry* (transl. from the Russian by K. A. Hirsch), New York: Springer.

4.6

Abelian integrals

ROGER COOKE

An Abelian integral is a complex contour integral of the form $\int_\gamma R(x,y)\,dx$, where $R(x,y)$ is a rational function of the complex variables x and y and, when the integral is evaluated over a contour γ, the function y is expressed locally as a function of x by means of an underlying assumed algebraic relation $f(x,y) = 0$, where $f(x,y)$ is a polynomial in x and y. The simplest non-elementary Abelian integral is an elliptic integral (§4.5); for example $\int [(1-x^2)(1-k^2x^2)]^{-1/2}\,dx$, which can be thought of as $\int dx/y$, when x and y are related by the equation

$$f(x,y) = y^2 - (1-x^2)(1-k^2x^2) = 0. \tag{1}$$

The history of Abelian integrals has been intensively studied; see, for example, Dieudonné *1985*, Houzel *1978*, Krazer and Wirtinger *1920*, and Brill and Noether *1893*.

1 EARLY STAGES

The study of general algebraic integrals was given a firm foundation by Niels Abel in a paper sent to the Paris Academy of Sciences in October 1826, but unfortunately it was not appreciated at the time and not published until 1841, twelve years after his death. In 1828, despairing of the appearance of his great paper, and terminally ill, he wrote a shorter paper *1828* explaining the essence of his ideas through the example of hyperelliptic integrals.

Abel's great paper *1841* contains many important facts about such integrals. Perhaps the most significant is the existence of an algebraic law of addition: the recognition that the transcendental relation

$$\int_{a_1}^{x_1} R(x,y)\,dx + \cdots + \int_{a_n}^{x_n} R(x,y)\,dx = 0 \tag{2}$$

can be equivalent to an algebraic relation $p(x_1,\ldots,x_n) = 0$ (p a polynomial) when the domains of the variables are suitably restricted. This fact

540

had been appreciated earlier for elementary and elliptic integrals, but in those cases a simpler statement of the property is possible, so that the general property remained obscured. In particular, Abel showed that the underlying algebraic relation $f(x, y) = 0$ determines a number p such that the sum of any number of indefinite integrals can be expressed as the sum of p such integrals whose upper limits are algebraic functions of the upper limits in the original sum. For elliptic integrals $p = 1$, and so the problem is completely determinate; but for more complicated hyperelliptic integrals, in which y is the square root of a polynomial of degree greater than 4, the number p is greater than 1, and the problem thus becomes indeterminate.

The indeterminacy was not explicitly pointed out by Abel *1828*, but it was recognized by Carl Jacobi *1832*. He eliminated the indeterminacy by considering a set of transcendental equations involving enough different integrands to make the problem determinate; the result is known as the Jacobi inversion problem. As an example, consider the algebraic relation

$$f(x, y) = y^2 - (x - r_1)(x - r_2)(x - r_3)(x - r_4)(x - r_5), \tag{3}$$

which leads to hyperelliptic integrals. Let the variables u_1 and u_2 be defined by the equations

$$u_1 = \int_{a_1}^{x_1} \frac{1}{y} dx + \int_{a_2}^{x_2} \frac{1}{y} dx \quad \text{and} \quad u_2 = \int_{a_1}^{x_1} \frac{x}{y} dx + \int_{a_2}^{x_2} \frac{x}{y} dx. \tag{4}$$

The Jacobi inversion problem is to invert these equations and express x_1 and x_2 as functions of u_1 and u_2.

The solution of this problem was gradually achieved over the next quarter-century, and gave rise to a great many parts of complex analysis (§3.12). In particular, the study of algebraic functions provided guidelines for the more general topic of analytic functions of a complex variable. The depth of the subject of Abelian integrals lies in the exceptional values of x near which the equation $f(x, y) = 0$ does not define y uniquely as a function of x. The 'local' point of view is thus inadequate for an understanding of the subject, which thereby becomes an important stimulus and testing laboratory for the global point of view. (The distinction between local and global, of course, would not have been made by Abel, whose point of view was formalistic and concerned with finding exact differentials to express the integrals explicitly.) The technique necessary to solve the Jacobi inversion problem involves theta functions of several variables, in particular the study of the places where such functions vanish. Both Bernhard Riemann and Karl Weierstrass developed their different approaches to analytic function theory in connection with the study of the Jacobi inversion problem. In particular, this problem seems to have been the single most important motivating factor in the invention of Riemann surfaces (Brill and Noether *1893*: 265–6).

The fundamental complication involved in dealing with multi-valued analytic functions arises at the points known as branch points. For example, although the function $y = \sqrt{x}$ can be represented locally by a power series, it is found that when x comes back to its starting-point having traversed a circle enclosing the point $x = 0$, y approaches the negative of its starting-point. Riemann *1857* handled problems of this type by imagining that x ranges over two copies of the complex plane glued together along a ray from the point $x = 0$ in such a way that x passes from one copy of the plane to the other whenever it crosses this ray. On one of the copies $y = \sqrt{x}$ has one of its two possible values, and at the corresponding point of the other copy y takes on the negative of this value. In this way y becomes an analytic function of x on the resulting manifold of two copies of the complex plane. Such manifolds, known as Riemann surfaces, can be constructed for any function x defined implicitly by a polynomial equation $f(x, y) = 0$. Weierstrass, on the other hand, defined analytic functions as those representable locally by power series. At a branch point the function y defined by the equation $f(x, y) = 0$ is not representable by a power series in x, but it has one or more representations as power series in variables t that can be thought of as roots of x. The collection of all such power series was taken by Weierstrass as the definition of the structure (*Gebilde*) of the function defined by $f(x, y) = 0$. Both these approaches exhibit the algebraic function as a unified whole and show that some of its properties can be inferred from its singularities.

2 THE RIEMANN–ROCH THEOREM

The global approach was a striking success in the construction of theta functions of several variables. These functions had been introduced by A. Rosenhain and J. G. Göpel in the 1840s in a partial solution of the Jacobi inversion problem. Riemann's construction of them through their formal properties rather than by direct power-series definition enabled him to give a complete solution of the Jacobi inversion problem. Riemann's work was clarified and perfected by Roch *1864* in what is now known as the 'Riemann–Roch theorem'. To explain this fundamental theorem, as we shall now do, requires a bit of technical apparatus.

The global structure of an algebraic function determines a connection between the underlying polynomial equation $f(x, y) = 0$ and an Abelian integral $\int \eta(x, y) \, dx$ based on this equation. The integral is interpreted as $\int \eta(x(t), y(t)) x'(t) \, dt$, where t is the variable used for power-series representation at a given point x_0 ($t = x - x_0$, except when x_0 is a branch point, in which case t is a root of $x - x_0$). The order of the integral is then the smallest power of t that occurs in the expansion of $\eta(x(t), y(t))x'(t)$. (The order of a rational function $\xi(x, y)$ is defined similarly.) This order is zero except at a finite number of points, say $P_1, \ldots, P_s$, where the orders are, say, $l_1, \ldots, l_s$, respectively. The sum of the orders at these exceptional points, it turns out, is always an even integer $2(p - 1)$, where p, known as the genus of the function defined by $f(x, y) = 0$, is independent of the particular integrand η, being determined entirely by the function $f(x, y)$.

There is always a rational function $\xi(x, y)$ having orders at least as large as a prescribed set of values, say $j_1, \ldots, j_s$, at the points $P_1, \ldots, P_s$, since the identically zero function has order infinity at every point. If the maximal number of linearly independent functions satisfying these conditions is, say, $M(j_1, \ldots, j_s)$, it turns out that

$$M(-l_1 - j_1, \ldots, -l_s - j_s) - M(j_1, \ldots, j_s) = p + j_1 + \cdots + j_s - 1. \quad (5)$$

This equation, which is the Riemann–Roch theorem, limits the possible behaviour of a rational function. For example, it is known that there is no rational function having a zero of order 2 at infinity, and only one simple pole in the finite plane. The Riemann–Roch theorem provides an analogue of this fact for general algebraic functions. By making different choices of the integrand $\eta(x, y)$, one can obtain a variety of results on the number of independent rational functions $\xi(x, y)$ having prescribed orders at prescribed places.

Although Felix Klein, writing early in the twentieth century, lamented the loss of interest in Abelian integrals (the ironic result of the perfection the topic seemed to have achieved), the subject experienced a renewal after the Second World War. In the late 1940s C. L. Siegel, F. Conforto and A. Weil discovered new proofs for the existence of theta functions and placed the entire topic in the context of the study of Abelian varieties, where it continues to be a thriving part of mathematics.

BIBLIOGRAPHY

Abel, N. H. *1841*, 'Mémoire sur une propriété générale d'une class très étendue de fonctions transcendantes', *Mémoires Présentés par Divers Savants*, **7**, 176–264. [Also in *Oeuvres complètes*, 2nd edn, 1881, Vol. 1, 145–211.]

——*1828*, 'Remarques sur quelques propriétés d'une certaine sorte de fonctions transcendantes', *Journal für die reine und angewandte Mathematik*, **3**, 313–23. [Also in *Oeuvres complètes*, Vol. 1, 444–56.]

Brill, A. and Noether, M. *1893*, 'Die Entwicklung der Theorie der algebraischen Functionen in älterer und neuerer Zeit', *Jahresbericht der Deutschen Mathematiker-Vereinigung*, **3**, 109–566.

Dieudonné, J. A. *1985, History of Algebraic Geometry* (transl. from the French by J. Sally), Monterey, CA: Wadsworth.

Houzel, C. *1978*, 'Fonctions elliptiques et intégrales abéliennes', in J. Dieudonné (ed.), *Abrégé d'histoire des mathématiques*, Vol. 2, Paris: Hermann, 2–113.

Jacobi, C. G. J. *1832*, 'Considerationes générales de transcendentibus Abelianis', *Journal für die reine und angewandte Mathematik*, 9, 394–403. [Also in *Gesammelte Werke*, Vol. 2, 5–16.]

Krazer, A. and Wirtinger, W. *1920*, 'Abelsche Funktionen und allgemeine Theta-funktionen', in *Encyklopädie der mathematischen Wissenschaften*, Vol. 2, Part 3, 604–873 (article II B 7).

Riemann, B. *1857*, 'Theorie der Abelschen Functionen', *Journal für die reine und angewandte Mathematik*, 54, 115–55. [Also in *Gesammelte mathematische Werke*, 2nd edn, 1892, 86–144.]

—— *1866*, 'Über das Verschwinden der Thetafunctionen', *Journal für die reine und angewandte Mathematik*, 65, 161–72. [Also in *Werke*, 212–24.]

Roch, G. *1864*, 'Über die Anzahl der willkürlichen Constanten in algebraischen Functionen', *Journal für die reine und angewandte Mathematik*, 64, 372–6.

Shafarevich, I. R. *1974, Basic Algebraic Geometry* (transl. from the Russian by K. A. Hirsch), New York: Springer.

4.7

Operator methods

I. GRATTAN-GUINNESS

1 FRENCH INITIATIVES

In this article the word 'operator' refers to operations of the calculus when they are treated as mathematical objects. The principal operator is differentiation, thought of as d/dx and written 'D'; in an algebra of D's, D^2 is d^2/dx^2, the integral operator $\int$ is D^{-1}, and so on. Now, $D^n y \neq (Dy)^n$ (*n* an integer); however, a powerful (and to critics, regrettable) analogy between powers and orders of differentiation was invoked to allow $D^n = (D)^n$.

The operator idea has pedigree in Gottfried Wilhelm Leibniz's notion of differential d as an operator on x to produce the infinitesimal dx (§3.2). But the algebra of D's has French origins, mainly from Joseph Louis Lagrange's algebraic conception of the calculus (§3.2). He thought of derivation as an operator, indicated by the prime symbol, on the function $f(x)$ to produce $f'(x)$, and used the analogy to cast Taylor's series in the form

$$\Delta f(x) := f(x+h) - f(x) = \exp(hDf(x)) - 1, \tag{1}$$

thereby relating D to the forward difference operator Δ. This led in turn to methods of summing series, since summation was viewed as an operation inverse to differencing, and so the operator $\Sigma = \Delta^{-1}$. Iterated summations were also allowed, based upon $\Sigma^n = \Delta^{-n}$, $n > 0$. He gave other results in an operator style; for example, $d\delta x = \delta dx$ in the calculus of variations (§3.5).

Lagrange's approach led to a range of methods of summing series and solving differential and difference equations. Among his followers, the Alsatian mathematician L. F. A. Arbogast went a step beyond the analogy in the book *Calcul des dérivations* (1800) by *separating* the operator symbol from that of the function; thus, for example, equation (1) became

$$\Delta f(x) := f(x+h) - f(x) = (e^{hD} - 1)f(x), \quad \text{and thus} \quad \Delta = e^{hD} - 1, \tag{2}$$

an operator equation in which '1' was the identity operator. Working with an operator called *dérivation*, which possessed the same basic properties (such as linearity) as D, he obtained many series expansions and extended versions of the binomial theorem (§4.1), and also results useful for combinatorics (§7.13).

In 1807, B. Brisson pioneered the idea of a differential equation as a differential operator on the function, by using the algebra of operators to produce solutions. For example, a linear partial equation was written as follows, with 'L' as the pertaining operator:

$$y + Ly = g(x);$$
$$\therefore y = (1 + L)^{-1}g(x) = (1 - L + L^2 - \cdots)g(x) = \cdots. \tag{3}$$

Lagrange's approach also encouraged the study of functional equations (§4.9). These 'new' algebras are of importance for the history of algebra in general, as they were the first ones in which the objects of study were not numbers or geometrical sizes. Partly for this reason, however, differential operators met with reservations, although most of the results obtained with them could be shown to be correct. For example, in the 1820s Augustin Louis Cauchy founded procedures such as Brisson's in his own form of solution using Fourier integrals with complex-variable integrands (§3.11; see also Section 3 below).

2 ENGLISH AND IRISH ENTHUSIASMS

During the early nineteenth century methods of solution by differential operators attracted considerable interest, especially in England and Ireland, when British mathematics at last began to come out of a long slumber (§11.7). Lagrange's approach was encouraged, and both the new algebras gained popularity in England and Ireland (Koppelman *1971*). In particular, building upon the work in the late 1830s by R. Murphy and D. F. Gregory, George Boole examined in a paper of 1844 the basic principles of the D-operator method. He noted that functions F of D could obey three properties: commutativity and distributivity (words introduced in 1814 by Lagrange's follower F. J. Servois, in connection with functional equations; §4.9), and the 'index law'

$$F^m F^n = F^{m+n}, \quad m \text{ and } n \text{ positive integers.} \tag{4}$$

There was an explicit similarity between these laws and those of his (later) algebra of logic, described in §5.1; he was aware of the limitations imposed by commutativity.

Many mathematicians used D methods to solve differential equations.

Ordinary equations were the principal target, especially one due to Pierre Simon Laplace referring to the shape of the Earth:

$$x^2(D^2 + n^2)y = m(m + 1)y, \quad n \text{ constant, } m \text{ a positive integer.} \quad (5)$$

In the 1840s some partial equations attracted attention, especially (the so-called) Laplace's equation. Much emphasis was laid upon the forms of the inverse operator(s) from which solutions were usually found, especially their existence, uniqueness and generality. A principal hope was to obtain in the end solutions in finite form rather than cumbersome ones in infinite series (Panteki 1992).

In the explication of the methods, techniques from the theory of (orthodox) analysis were deployed, such as power-series and partial-fraction expansions, and factorization; for example, Murphy's result

$$D(D - 1)(D - 2) \ldots (D - n) = x^{n+1}D^{n+1}, \quad n > 1. \quad (6)$$

As an example, here is the start of a treatment of (5) by T. Gaskin, which was published in Hymers 1839. He first used a change of dependent variable

$$u := yx^m \quad \text{to convert (5) to} \quad x(D^2u + n^2u) = 2mDu. \quad (7)$$

Applying $(D^2 + n^2)$ to equation $(7)_2$ yielded, after separating the symbols,

$$x(D^2 + n^2)^2u = 2m(m - 1)D(D^2 + n^2)u; \quad (8)$$

and $m - 1$ further iterations furnished the equation

$$(D^2 + n^2)^{m+1}u = 0, \quad \text{with solution} \quad u = A(x)\cos nx + B(x)\sin nx, \quad (9)$$

where A and B are polynomials of degree m. Equation $(9)_2$ was amenable to further analysis.

However, the 'mystery' of operator methods was apparent: not only the analogy but also the rather liberal rigour (such as concerning the convergence of infinite series). To address these questions, the non-interpretability of formulas was permitted: for example, a differential equation like (5) and its solution could have meaning as (say) potential theory, but the intervening lines of reasoning (such as (7)–(8)) were not required to refer to anything. British concerns of this time with the philosophy of algebras (§6.9) were influenced by such issues.

In some competition with methods based on Boole's laws, non-commutative operators, used already by Murphy in the 1830s and furthered especially by Boole himself, C. Graves and B. Bronwin, gained favour. Applied to ordinary equations were operators L and M which obeyed the law

$$LM - ML = 1, \quad (10)$$

where '1' again denoted the identity operator; partial equations received the attention of

$$xd/dx + yd/dy + zd/dz + \cdots, \tag{11}$$

a form significant for homogeneous functions. These alternative developments had consequences for Boole's hope that his methods would be general, for it became clear that different classes of equations required different methods.

In addition to applications, interactions with other branches of mathematics occurred from the 1850s. Arthur Cayley and J. J. Sylvester made use of operators in the theory of invariants (§6.8), and when determinants became well known (§6.6) they were used to facilitate the technical manipulations. By then enough work had been produced for the results to be included in further textbooks, such as R. D. Carmichael's *A Treatise on the Calculus of Operations* (1855) and Boole's *A Treatise on Differential Equations* (1859) and *A Treatise on the Calculus of Finite Differences* (1860). They also received treatment in encyclopedia articles (Russell *1873*) and survey papers (Glaisher *1881*).

However, overall success was limited; many important differential and difference equations did not respond to any of these methods, or at least other methods were more efficacious. So the movement rather fell away during the 1860s, and the results obtained attracted less attention than they deserved.

3 THE MYSTERIES OF HEAVISIDE

From the 1890s there was a revival in interest in operator methods when attempts were made to bring rigour to Oliver Heaviside's operator methods of solving the telegraph equation and various other equations (Cooper *1952*, Petrova *1987*). He expressed his basic conception in very general terms: given a force $f(t)$ dependent on time t acting on a connected system, to find the effects $F(t)$ of some given kind of its action. For him the answer took the form of an operator equation

$$f(t) = Y(p)F(t), \tag{12}$$

where Y was a function of the operator $p := d/dt$. He applied this approach mainly to electromagnetic theory, based on Maxwell's equations (§9.10); their mathematical form led him to power-series and basic functions in p, which he freely manipulated in order to solve them. In particular, he obtained an important 'expansion theorem' (as he called it) of the partial-fraction type found by some predecessors (especially Cauchy (1825, 1827) on the solutions of ordinary differential equations), of which he was unaware.

Heaviside published his results especially in his three-volume book *Electromagnetic Theory* (1893–1912). He made no attempt to find a basis for them, and some of the results found by him and his successors were incorrect. The growing importance of telecommunications (§9.11) gave urgency to the clarification of his mysterious procedures. Today the most satisfactory version of his theory is the symbolic calculus developed in the late 1950s by the Polish mathematician J. Mikusiński, based upon the convolution of functions (Freudenthal *1969*). But at that time the main consequence for mathematics was the emergence of a rigorous theory of the Laplace transform, to which we now turn.

BIBLIOGRAPHY

Carslaw, H. S. and Jaeger, J. C. *1941*, *Operational Methods in Applied mathematics*, Oxford: Oxford University Press. [Classic survey; 2nd edn 1947.]

Cooper, J. L. B. *1952*, 'Heaviside and the operational calculus', *The Mathematical Gazette*, **36**, 5–19.

Freudenthal, H. *1969*, 'Operatorenrechnung – von Heaviside bis Mikusiński', *Überblicke Mathematik*, **2**, 131–49.

Glaisher, J. W. L. *1881*, 'On Riccati's equation [. . .]', *Philosophical Transactions of the Royal Society of London*, **172**, 759–828.

Hymers, J. *1839*, *Treatise on Differential Equations and on the Calculus of Finite Differences*, 1st edn, Cambridge: Deighton. [2nd edn 1860.]

Koppelman, E. *1971*, 'The calculus of operations and the rise of abstract algebra', *Archive for History of Exact Sciences*, **8**, 155–242.

Panteki, M. *1992*, 'Relationships between algebra, differential equations and logic in England: 1800–1860', Council for National Academic Awards (London), Doctoral Dissertation.

Petrova, S. S. *1987*, 'Heaviside and the development of the symbolic calculus', *Archive for History of Exact Sciences*, **37**, 1–23.

Russell, W. H. L. *1873*, 'Symbols, calculus of', in *The English Cyclopaedia*, Suppl. Vol. 4, London: Knight, cols. 2020–26.

4.8

The Laplace transform

M. A. B. DEAKIN

An integral transform is a map from one function, $F(t)$, to another, $f(s)$, by means of a definite integral:

$$f(s) = \int_a^b K(s, t)F(t)\,dt. \tag{1}$$

The initial interpretation of equation (1) has s and t as real variables, but, more generally, they may be taken as complex, with the integral taken over some specified contour in the t-plane. Indeed, such an extension is needed for a full theory of the Laplace transform.

Nowadays the Laplace transform is defined by

$$f(s) = \int_0^\infty e^{-st}F(t)\,dt, \tag{2}$$

provided the integral converges. Here it is mainly the application of the transform to the solution of differential equations that is considered, although it has other uses, for example as a generalization of the Dirichlet series in analytic number theory.

1 THREE METHODS

Method 1. Consider the differential equation

$$tF''(t) - tF'(t) + F(t) = 0 \tag{3}$$

as an example. To solve this by means of the Laplace transform, multiply throughout by e^{-st} and integrate from $t = 0$ to $t = \infty$. This yields the first-order equation

$$s(s - 1)f'(s) + 2(s - 1)f(s) = F(0), \tag{4}$$

whose solution is

$$f(s) = F(0)\,[s^{-1} + s^{-2}\ln(s - 1)] + As^{-2}. \tag{5}$$

The problem is thus solved provided we can invert the transform (2) – that is, recover $F(t)$ from a knowledge of $f(s)$. In the present case, this yields

$$F(t) = At + B[e^t - t\,\mathrm{Ei}(t)], \tag{6}$$

where $B = F(0)$ and $\mathrm{Ei}(t)$ is the exponential integral function.

Method 2. This technique would not have been used last century. One approach would have been to seek a solution to equation (3) in the form

$$F(t) = \int_a^b e^{st}\Phi(s)\,\mathrm{d}s. \tag{7}$$

Substituting this into equation (3) yields (after an integration by parts)

$$[e^{st}s(s-1)\Phi(s)]_a^b - \int_a^b e^{st}\left\{\frac{\mathrm{d}}{\mathrm{d}s}[s(s-1)\Phi(s)] - \Phi(s)\right\}\mathrm{d}s = 0. \tag{8}$$

If we now choose $a = -\infty$ and $b = 1$, the first, integrated, term will vanish, and we may now set

$$\frac{\mathrm{d}}{\mathrm{d}s}[s(s-1)\Phi(s)] = \Phi(s), \tag{9}$$

which is in fact equation (4) in the special case $F(0) = 0$. We thus have $\Phi(s) = As^{-2}$, and so

$$F(t) = A\int_{-\infty}^1 (e^{st}/s^2)\,\mathrm{d}s. \tag{10}$$

Although this integral is divergent, a formal integration by parts yields

$$F(t) = A[e^t - t\,\mathrm{Ei}(t)]. \tag{11}$$

This is a partial solution. The second solution, Bt, may be made to emerge if the definite integral (7) is replaced by a contour integral, one relevant contour being a circle about the origin. Alternatively, if tediously, our nineteenth-century problem-solver could have – indeed would have – used equation (11) and reduction of order to achieve the full solution.

Method 3. Any such hypothetical solvers would almost certainly have lived in Continental Europe. Had they lived in Britain, they would probably have proceeded differently, writing equation (3) in the form

$$t(\mathrm{D}^2 - \mathrm{D})F(t) + F(t) = 0, \tag{12}$$

where D stands for $\mathrm{d}/\mathrm{d}t$ and 'symbolically' represents an 'operator', an instruction to operate on (in this case, by differentiating) the function $F(t)$ (§4.7).

These operators were treated as if they obeyed the same algebraic rules as numerical quantities, and formal expressions and rules for their interpretation were devised. Such a formal expression, applied to equation (12), would yield

$$F(t) = \frac{1}{D^2}\left[\frac{1}{t}\left(\frac{D}{D-1}\right)0\right]. \tag{13}$$

The expression $D/(D-1)$ would then be expanded in powers of D^{-1}, where D^{-1} was interpreted as an integral. Thus $D^{-1}0 = a$, $D^{-2}0 = at + b$, $D^{-3}0 = \frac{1}{2}at^2 + bt + c$, and so on; so that

$$\left(\frac{D}{D-1}\right)0 = Ae^t, \tag{14}$$

where $A = a + b + c + \cdots$. The calculation would then proceed with

$$F(t) = A \iint \frac{e^t}{t} dt\, dt = A\,[t\,\mathrm{Ei}(t) - e^t] + Bt + C. \tag{15}$$

Substitution back into the original equation would then yield $C = 0$.

2 EARLY WORK, 1737–1837

Method 2 is the one that, for many years, had the most direct bearing on the subject. Some have seen papers by Leonhard Euler, dating from as early as 1737, as being relevant to the development of the Laplace transform. However, the first clear case is the discussion occurring in the second volume of Euler's *Institutiones calculi integralis* (1769), which presents a generalization of Method 2 applicable to some cases where the equation for solution is non-homogeneous. It is also the first account of integral transforms, properly so-called, apart from a brief fragment. For more on this history, see Deakin *1980, 1985.*

Pierre Simon Laplace also considered the matter, notably in a lengthy paper on asymptotic formulas published in the *Mémoires* of the Paris Academy of Sciences (1785), and in his *Théorie analytique des probabilités* (1812, 1820). The form (2) was exhibited and discussed along with other possibilities, and (somewhat vague) reference was made to prior work by Euler. Twice attempts were made to invert the transform – a task that eluded Laplace, as the required formula gives $F(t)$ in the plane of complex s (§3.12). In fact, much more attention is devoted in Laplace's work to a related integral transform (today named the Mellin transform) than to the form (2) which now bears his name. In particular, he used the Mellin transform to provide a derivation of Stirling's formula.

The question of the inversion of an integral transform is raised most dramatically by the Fourier integral theorem. This shows (in one version) that the two integral transforms

$$f(s) = \int_0^\infty F(t) \cos st \, dt \quad \text{and} \quad F(t) = \frac{2}{\pi} \int_0^\infty f(s) \cos st \, ds \qquad (16)$$

are inverse to one another. Fourier gave his result, and a heuristic proof of it, in his *Théorie analytique de la chaleur* (1822) (§9.4); a complete and rigorous demonstration was not available, however, till 1907 (§3.11). This is not to say that the result was not accepted prior to this; rather it became a fruitful source of problems for research, and ultimately contributed to the rise of functional analysis (§3.9).

Central to the proof of the theorem is a passage to the limit, in which a finite conducting bar is extended until it is semi-infinite. This impinges directly on the story of the Laplace transform, for in the course of one discussion of the matter (by Siméon-Denis Poisson) integrals of the form (2) were produced. This circumstance has led to Poisson being wrongly credited with the Laplace transform.

Of the various forms of the Fourier-transform pair (16), a complex version, due to Augustin Louis Cauchy, is perhaps the simplest and most elegant. Certainly it relates most directly to the theory of the Laplace transform, for the factor $\cos st$ in equations (16) is replaced by e^{-ist} in the first and by e^{ist} in the second, so that equation $(16)_1$ becomes, apart from minor notational change, equation (2). This is the key to the inversion problem which had eluded Laplace himself.

Cauchy was quite capable of producing the result, but his work took him in other directions and the opportunity was missed. He is, however, relevant to the story in another way as well. Although symbolic approaches, such as that of Method 3, may be traced back to Gottfried Wilhelm Leibniz (Koppelmann *1971*), Cauchy investigated them more thoroughly (and with more rigour) than any previous author. Most notably, he proved a result that was in fact a generalized version of what became known as the Heaviside expansion theorem.

The remaining major figure from the early period is Joseph Liouville, whose researches into fractional differentiation led him to seek to express functions in forms like that of equation (7). This allows ready extension of the formulas for m-fold differentiation to the case of non-integral m. One study applied these results to the solution of differential equations (Deakin *1981*).

3 THE VIENNESE SCHOOL

It was Liouville whose work had the most direct influence on Josef

Petzval, a Czech mathematician who became Professor of Higher Mathematics at the University of Vienna. Essentially, Petzval rediscovered equation (7) and its use to solve differential equations, omitting the detour Liouville had taken – the reference to fractional derivatives. He was in his mid-twenties when he began this work, which resulted eventually in a long paper and a two-volume treatise. Much of his output was concerned with the properties and applications of equation (7), especially in the context of ordinary linear differential equations whose coefficient functions are linear.

The treatment is thorough, tedious in fact, and turns into an unsatisfactory melange of special cases. The basic reason for this is Petzval's inability to use contour integration (§3.12). Because the inversion of equation (2) requires this, the attempt to solve equations by substituting (in essence) that inverse directly calls for a complex analogue of equation (7). (This is why the solution (11) is incomplete.) The main theoretical advance was his generalization of the relation between equations (3) and (9). They are adjoint, and this property applies universally.

Petzval is remembered today for his work on lens systems rather than his research into the Laplace transform. In the early 1850s he fell out with one of his students, Simon Spitzer, a forgettable figure best known for a book of compound-interest tables. Petzval described Spitzer as not 'a mathematician' but 'a calculator'. This judgement, while certainly correct, came to Spitzer's attention and caused a prolonged and public quarrel. Spitzer's revenge took the form of his clamorously asserting that Petzval had not only been anticipated by Laplace, but had in fact plagiarized him.

The claim was false, but the attribution to Laplace was adopted by George Boole and (subsequently) Henri Poincaré, and has stuck. Spitzer continued to publish trivia, almost till his death. The only real advance made during the period 1860–80 was Bernhard Riemann's inversion of the Mellin transform.

4 POINCARÉ AND COMPLEX ANALYSIS

It was Poincaré who first provided the key to further advance by using a complex form for equation (7) and thus extending Method 2. This work, part of an essay 'Non inultus premor' ('I Will Be Vindicated'), was submitted for the Paris Academy's prize in 1880, when Poincaré was 26 years old; but it failed to win, and remains unpublished even today.

However, the method itself was later published, by Poincaré himself in 1885, with the acknowledgement that the integral form (7) had been discussed before. He attributed it initially to Bessel (did he mean Petzval?), later to Laplace. Camille Jordan's *Cours d'analyse* (1887) gave it wide dissemination.

The most prolific of the researchers who followed was Salvatore Pincherle, a pupil of Karl Weierstrass. It was Pincherle who gave the complex inverse of equation (2) and greatly popularized the method. L. Schlesinger, R. H. Mellin and H. Bateman extended the theory of the adjoint property, Emil Picard applied Method 2 to partial differential equations, and M. Lerch proved the uniqueness of the inverse (Deakin *1982*).

5 OPERATIONAL CALCULUS

Method 3 was widely explored by the English school, notably by Boole, who was the main influence on the self-taught Oliver Heaviside. In an engineering context that included transoceanic telegraphy, Heaviside made great use of operational methods (§4.7). They became very popular and were widely taught in engineering schools; J. R. Carson and B. van der Pol promulgated them. The Heaviside expansion theorem and other such results were widely used. Attempts were made to rigorize the approach, which many mathematicians continued to regard as unsatisfactory, but which led eventually to such modern developments as the Mikusiński's calculus (§4.8). However, it was the modern Laplace transform and not these that displaced the operational techniques from undergraduate syllabuses (Lützen *1979*).

6 THE MODERN THEORY

Method 1 was first used by Bateman in 1910, but was explored more fully by S. N. Bernstein and, especially, G. Doetsch. It will be seen from the example given that this approach avoids the loose logic and *ad hoc* methodology of Methods 2 and 3. Doetsch devoted his life to this cause, his *Theorie und Anwendung der Laplace-Transformation* appearing in 1937. Raised in the rigorous Weierstrassian tradition, Doetsch showed an understandable (if uncharitably expressed) impatience with Heaviside's eccentric amateurism.

When, in the 1950s, *Theorie und Anwendung* was updated, these passages were removed – though they were never retracted. One might wish that (the Jewish) Schlesinger's name might have been restored to the reference list also at that time, but it was not. The reasons for omitting it in 1937 were perhaps compelling, and Doetsch did show some courage in documenting his early collaboration with Bernstein.

Bernstein was part of that scientific diaspora that so benefited the intellectual life of the USA. Dover Publications reprinted Doetsch's *Theorie und Anwendung* under wartime abrogation of copyright agreements, so on both grounds promulgation was swift. Texts by M. F. Gardner and J. L. Barnes

(1942, USA) and by H. C. Carslaw and J. C. Jaeger (1944, UK) influentially promoted the new technique. By the mid-1950s it was standard fare. Rarely has research mathematics so rapidly and thoroughly achieved curricular acceptance (Deakin *1992*).

BIBLIOGRAPHY

Deakin, M. A. B. *1980*, 'Euler's version of the Laplace transform', *American Mathematical Monthly*, **87**, 264–9.

—— *1981*, *1982*, 'The development of the Laplace transform 1737–1937', *Archive for History of Exact Sciences*, **25**, 343–90; **26**, 351–81.

—— *1985*, 'Euler's invention of integral transforms', *Archive for History of Exact Sciences*, **33**, 307–19.

—— *1992*, 'The ascendancy of the Laplace transform [...]', *Archive for History of Exact Sciences*, **44**, 265–86.

Koppelmann, R. *1971*, 'The calculus of operations [...]', *Archive for History of Exact Sciences*, **8**, 155–242.

Lützen, J. *1979*, 'Heaviside's operational calculus [...]', *Archive for History of Exact Sciences*, **21**, 161–200.

4.9

Functional equations

I. GRATTAN-GUINNESS

1 INTRODUCTION

A functional equation is one in which the function itself is the unknown. To take two simple (but important) examples: in

$$f(x+y) = f(x) + f(y) \quad \text{or} \quad f(x+y) = f(x)f(y) \tag{1}$$

we seek the function(s) f which is (are) satisfied for all x and y. In some cases only certain ranges of values of x are specified; special conditions may be imposed upon f. More than one function can occur in a single equation; the task is then to find relationships between them, or to solve for one when particular functions are assigned to the others. Simultaneous systems of these equations may also be considered. Difference equations are a class of special cases.

The methods of solution are different from those for ordinary algebraic equations or differential equations, although similar broad questions arise (general and particular solutions, and conditions for their existence). Single- and multi-valued functions, and inverse(s) of a function raise other important questions. In addition, equations in one variable are distinguished from those in several variables: equations (1) are examples of the latter kind, for $f(x+y)$ is a function of x and y. For substantial bibliographies of original works in each category, see respectively Kuczma *1968* and Aczel and Dhombres *1989*, and Aczel *1966* for both categories.

Some older literature included integral equations and (contemporary versions of) Laplace-transform theory under 'functional equations', but here those topics are treated separately in §3.10 and §4.8. The subject was then often called 'the calculus of functions', referring to the determination of functions where perhaps no equation was involved. The ubiquity of functions in mathematics means that functional equations can arise in a wide variety of contexts; only a small selection of cases can be considered here.

2 EARLY SOURCES: THE CALCULUS, AND THE FOUNDATIONS OF MECHANICS

Some early thinking in the direction of functional equations occurred in the development of logarithms (§2.5); for example, the inverse logarithm satisfies equation $(1)_2$. Similarly implicit thought can be detected in early mechanics: in particular, some medieval concepts of types of motion (§2.6) can be expressed in terms of this equation for the 'velocity' or 'acceleration':

$$\Delta f(x_3)/\Delta f(x_2) = \Delta x_3/\Delta x_2 \tag{2}$$

for any trio of distances x_1, x_2 and x_3. Further implicit procedures of this kind can be found in some later thinkers on mechanics, although formulations such as (2) are grossly anachronistic. Explicit thought of this kind first arose in the eighteenth century with the concern to prove elementary properties in statics. Daniel Bernoulli, Jean d'Alembert, Pierre Simon Laplace and others formed and then solved functional equations to prove certain laws of statics. For example, solutions to

$$2f(x) = f(x - A) + f(x + A), \quad A \text{ constant, and}$$
$$[g(y)]^2 + [g(90° - y)]^2 = 1 \tag{3}$$

produced respectively $f(x) = Kx$ for the law of moments and $g(y) = \cos y$ for the law of angular displacement y (or composition) of forces.

The solution of functional equations within the context of differential equations was pioneered by Gaspard Monge in the 1770s, when in effect he tackled equations such as

$$f[g(x, h(x))] = k(x) \tag{4}$$

for f, with g, h and k known. His methods were soon extended by Laplace; a study of difference equations (1776) related to probability theory involved

$$F[G(x)] = H(x)F[K(x)] + X(x) \tag{5}$$

for F, with G, H, K and X known. Solutions usually involved substitutions which converted the equation to a known one, maybe a difference equation (Panteki *1992*: Chap. 1).

In connection with the foundations of analysis, Leonhard Euler proved the binomial theorem (§4.1) by showing that its series satisfied a version of equation $(1)_2$ and solving the latter (Dhombres *1987*). This approach was extended still further in the programme of Joseph Louis Lagrange to found the calculus only upon algebraic manipulations, working out from the Taylor expansion of 'any' function (§3.2):

$$f(x+z) = f(x) + g(x)z + h(x)z^2/2! + \cdots . \tag{6}$$

One strategy was to use the properties of f to find a functional equation satisfied by g (or h, etc.), and then solve it. Lagrange and his followers were able to find the derivatives of some simple functions by such means. Typically (from Poinsot *1815*),

for $f(x) = x^m$, since $(ax)^m = a^m x^m$, then $g(x, m) = G(x)x^{m-1}$; (7)

since $x^{m+n} = x^m x^n$, then $G(m+n) = G(m) + G(n)$ for some G. (8)

From equation (8)$_2$, for a multiple integral K of m,

$$G(Km) = KG(m) = mG(K), \tag{9}$$

by symmetry. Now, when $K = 1$, $G(m) = mG(1) = 1$. So, in equation (7)$_3$, $g(x, m) = mx^{m-1}$, as required.

Function letters have been used several times here in order to stress the possibility that the function itself is the object under study. Lagrange's followers L. F. A. Arbogast, F. J. and J. F. Français, and F. J. Servois explicitly took this step by 'separating the symbol' from its argument(s). From this move two new algebras were furthered: functional equations, and differential operator methods using the operator D for d/dx (§4.7). An enduring contribution to algebra was made by Servois in 1814; seeking the fundamental properties of both algebras, especially functions, he characterized f as 'distributive', and f and g as 'commutative with each other' if, respectively,

$$f(x+y+\cdots) = f(x) + f(y) + \cdots \quad \text{and} \quad f[g(x)] = g[f(x)]. \tag{10}$$

3 SOME ENGLISH FREEWHEELING, SOME CONTINENTAL RIGOUR

The methods of Monge and Laplace, and also the Lagrangian programme, proved successful exports to England when British mathematics revived in the early nineteenth century (§11.7); the young Charles Babbage and John Herschel published several papers between 1813 and 1827 in this area. Many of their results dealt with functions of one variable (though often involving several functions of it); for example, Babbage studied

$$f(x) = f[g(x)] \quad \text{and also} \quad F^n(x) = G(x), \quad n \text{ an integer} \tag{11}$$

to solve for F in terms of a given G. (Both equations had important special cases: when $g(x) := x + 1$, (11)$_1$ became the difference equation $\Delta f(x) = 0$;

when $G(x) := x$, F was 'periodic' of order n.) The most general equation studied was

$$H[x, f(x), f^2(g_1(x)), f^3(g_2(x)), \ldots, f^n(g_{n-1}(x))] = 0 \qquad (12)$$

to solve for f, given H and the $\{g_r\}$. Babbage also tried a few differential functional equations and integral equations.

The methods of solution were pretty freewheeling manipulations of functions, self-substitutions and elimination of functions in equations, cunning changes of variable, and so on. However, the concerns of the solvers included the place of symmetric functions (where $h(x, y) = h(y, x)$), and especially the inverse function. Babbage also studied the form $g^{-1}fg$ (perhaps the first example of this significant 'conjugate' in mathematics), while John Herschel was an important founder of the type of notation 'f^{-1}' for a general function f and also for the inverse trigonometric functions.

Their work did not make a great impact, unfortunately; only one major follower emerged, namely Augustus De Morgan, who wrote the first systematic survey *1836* of the subject, for both one and several variables, including some methods of his own for building up special symmetric-function solutions. His piece appeared as an encyclopedia article, and was itself largely ignored by mathematicians (and historians also – see Panteki *1992*: Chaps 2 and 3). However, there was an important similarity between this work and his later introduction of the logic of relations (§5.1).

Meanwhile, Augustin Louis Cauchy had taken up functional equations around 1820, as part of his founding of mathematical analysis (§3.3). In particular, he refined the treatment of the binomial series made by his predecessors by obtaining it via equation (1)$_2$ without use of the calculus (§4.1). More significantly, in tune with his desire for rigour, he brought to bear considerations of the continuity of a function and convergence of a series. His follower Niels Abel soon extended the treatment to cover complex values of variables; he also gave a penetrating analysis of

$$f(x) + f(y) = g[xh(y) + yh(x)], \qquad (13)$$

in which he noted the plurality of solutions that could eventuate.

Unfortunately even this study gained little attention; the subject lay rather in the doldrums for most of the nineteenth century, although it continued to arise in various contexts of pure and applied mathematics. For example, in the 1870s Ernst Schröder studied a variant of equation (11)$_1$:

$$Af(x) = f[g(x)], \quad A \neq 0 \text{ or } 1 \qquad (14)$$

and g known, in connection with iterations of the function f; for him x was real, but complex-variable cases also came to be studied (Aczel *1966*:

Chap. 9). In the next decade Gaston Darboux considered afresh equation $(1)_1$, and its continuous solution Ax (A a constant), in connection with the law of decomposition of forces.

Perhaps the major worker of that time was the Italian Salvatore Pincherle; in the 1880s he applied the Laplace transform (§4.8) to solve a complicated linear equation in one variable, and later he wrote the article on the subject for the *Encyklopädie der mathematischen Wissenschaften* (Pincherle *1903*). He laid emphasis upon distributive properties in all current branches of mathematics, and emphasized connections between functional equations and operator methods (and their algebras) in general.

4 TWENTIETH-CENTURY POPULARITY

In his famous lecture of 1900 on 'mathematical problems', David Hilbert *1900* could fairly refer to functional equations as a 'wide and not uninteresting field', while describing his fifth problem (on Lie groups); he also praised the contributions of Abel. Interest increased when in 1905 (and using the newly revealed axioms of choice; see §5.4) G. Hamel found criteria under which the solution of equation $(1)_1$ would be continuous in the vicinity of a value of x. This result led to various constructions of discontinuous solutions, in which tools of set topology and measure theory (§3.6 and §3.7) were deployed (Dhombres *1986*: Section 10).

Only over the past fifty years has activity increased, although the change has been very substantial; leading figures include J. Aczel and M. Kuczma. On the foundational side, the concern is now more with continuous, discontinuous, regular and other functions defined over sets of points or within certain function spaces (see §3.9 on functional analysis), conditions for their (non-)uniqueness, solutions under special conditions such as monotonicity or convexity on the function(s), equations for operators on functionals, the algebraic properties of the solution functions themselves, extensions such as vectorial and matricial functional equations, and the place of functional inequalities (Aczel *1966*).

In addition, the realm of empirical applications has grown, both in traditional branches such as analysis, geometry and probability theory, and in the newer areas. For example, their use in mathematical economics has been fruitful for expressing properties of such functions as production and price indexing (Guerraggio *1985*), while in recent times information theory has been an important new recruit to their ambit (Aczel *1984*). By now the field is very wide, and not at all uninteresting.

BIBLIOGRAPHY

Aczel, J. *1966, Lectures on Functional Equations and their Applications*, New York: Academic Press.

—— (ed.) *1984, Functional Equations: History, Applications and Theory*, Dordrecht: Reidel.

Aczel, J. and Dhombres, J. G. *1989, Functional Equations in Several Variables, with Applications to Mathematics, Information Theory and to the Natural and Social Sciences*, Cambridge: Cambridge University Press.

De Morgan, A. *1836*, 'Calculus of functions', in *Encyclopaedia Metropolitana*, Vol. 2, 305–92. [Date of article publication; volume carries date 1845 of completion.]

Dhombres, J. G. *1986*, 'Quelques aspects de l'histoire des équations fonctionnelles [...]', *Archive for History of Exact Sciences*, **36**, 91–181. [Only tackles equations of several variables.]

—— *1987*, 'Les Présupposés d'Euler dans l'emploi de la méthode fonctionnelle', *Revue d'Histoire des Sciences*, **40**, 179–202.

Guerraggio, A. *1985*, 'Le equazione funzionali nei fondamenti della matematica finanzaria', *Rivista di Matematica per le Scienze Economiche e Sociali*, **9**, 33–52.

Hilbert, D. *1900*, 'Mathematische Probleme', in *Gesammelte Abhandlungen*, Vol. 3, 290–329. [Various reprints and translations.]

Kuczma, M. *1968, Functional Equations in a Single Variable*, Warsaw: Polish Scientific Publishers.

Panteki, M. *1992*, 'Relationships between algebra, differential equations and logic in England: 1800–1860', Council for National Academic Awards (London), Doctoral Dissertation.

Pincherle, S. *1903*, 'Functionaloperationen und -gleichungen', in *Encyklopädie der mathematischen Wissenschaften*, Vol. 2, Part 4, 761–817 (article II A 11). [Extended version in *Encyclopédie des sciences mathématiques*, Tome 2, Vol. 5 (1912), 1–81 (article II 26).]

Poinsot, L. *1815*, 'Des principes fondementaux et des règles générales du calcul différentiel', *Correspondance sur l'Ecole Polytechnique*, **3** (1814–16), 111–23.

4.10

The roots of equations: Detection and approximation

I. GRATTAN-GUINNESS

This article deals with aspects of the theory of equations which relate to the calculus, rather than to resolution and elimination, the fundamental theorem of algebra, and so on (see §6.1). This distinction was not often drawn in major textbooks and treatises, although separate parts were devoted to them. Very many methods have been proposed since the advent of algebra (§2.3–2.4), in a variety of contexts. Not all are efficacious, and some have a narrow compass of use; only certain principal results or interesting cases can be treated here. Particular cases of our topic are treated elsewhere: for example, latent roots of matrices are noted in §6.7, and the zeros of special functions in §4.4. For convenience, no distinction is made between the roots of an equation $f(x) = 0$ and the zeros of the function $f(x)$.

1 DETECTION OF ROOTS: FROM DESCARTES TO STURM

René Descartes's *Géométrie* (1637) has left its mark, among other reasons, for the 'rule of signs' known after his name, which states that the number of positive/negative roots of a polynomial equation is not more than the number of variations/permanences of sign in its sequence of coefficients (that is, respectively the pair $+ -$ or $- +$, or $+ +$ or $- -$). He did not prove the result in a general way; he probably intuited from special cases of the linear and quadratic equations. (So did the Englishman Thomas Harriot, who came to the same result a little earlier.) Proofs were offered by various authors in the eighteenth century: the clearest statement of the proof by induction upon the degree of the polynomial was given by the young Joseph Fourier in the late 1780s. More importantly, he came to a generalization which offered upper bounds for the number of roots within any interval $[a, b]$ of values: take the polynomial $f(x)$ and all its derivatives, and note the signs of the values of each function when $x = a$ and $x = b$; then the number of real roots is not less than that of the difference between the two sequences. A priority dispute erupted with an obscure French mathematician, Ferdinand Budan de Boislaurent, who had developed a very similar method independently (but later); sometimes the result is known after him.

Like Descartes's rule, Fourier's theorem gives only an upper bound because it is insensitive to the presence of complex roots, which occur in conjugate pairs. While pondering upon this defect in 1829, the young J. C. F. Sturm came up with a variant theorem in which, after $f(x)$ and $f'(x)$, the rest of the sequence was defined from the process of Euclidean division. When this sequence of values was formed, the exact number of real roots was found. His result found rapid favour and was soon in textbooks (Sinaceur *1991*: Part 1).

At that time Augustin Louis Cauchy had found a quite different theorem on the number of roots, and indeed other functions of the roots; they were based on his complex-variable residue calculus, then being developed (§3.12). Sturm and his friend Joseph Liouville were among those who popularized these results from the 1830s onwards, and the name 'winding numbers' became attached to them; but complex analysis was progressing slowly, and much time passed before they came into prominence.

A theorem for cubics C, due to Joseph Louis Lagrange, used the equation whose roots were the differences of the squares of those of C. If the discriminant of this equation were negative/zero/positive, then the roots of C were real and unequal/included two equal ones/included a pair of complex conjugates. Lagrange's treatise *1798* is not only a major work on equations in general, but also quite a useful historical source (Hamburg *1976*).

2 APPROXIMATION: FROM NEWTON TO HORNER

Fourier's results appeared principally in a posthumous book *1831*, which became well known. In addition to theorems on detection and other considerations such as linear programming (§6.11), he also reviewed, partly historically, various methods of approximating to real roots. Graphical methods had been used from Descartes's time onwards (Boyer *1945*); Isaac Newton and several of his followers refined them with techniques using the calculus. The best known is that named after him and Joseph Raphson: if the guess c_1 is made for a root, then a closer estimate c_2 is given by

$$c_2 = c_1 - [f(c_1)/f'(c_1)]. \tag{1}$$

(This form, using the calculus, is actually due to Thomas Simpson (1740).)

But the method was known not to be foolproof, and also to be insensitive to closely neighbouring roots. Fourier considered the rates of convergence, and the possibility of approximating from above and below by variants of equation (1).

Another method of approximating is named after the English mathematician William Horner. It was similar to Budan's version of Fourier's theorem in that it involved the evaluation of $f(x - a)$, and could be evaluated in a better way than straight insertion of the value $x - a$. If a, an integer between 0 and 9 or a multiple of 10, is the first guess at a root, and the next decimal place is found to be b, then the method is used again to evaluate $f(x - b)$ (for example, $a = 40$, $b = 43$, $c = 43.6$, and so on). One homes in on the required root by this process (see §1.9 on Chinese anticipations).

The Swiss mathematician C. H. Graeffe introduced another method in the 1830s which received much attention. He noticed that $f(x)f(-x)$ was a function of x^2, with roots the squares of those of $f(x)$ itself; iteration of the process produced a sequence of equations, with roots the 2^r-th powers of the originals. Drawing on an approximation to the largest root in absolute value, proposed a century earlier by Daniel Bernoulli, he generated a powerful means of approximating to all the roots of the original equation (§4.13).

3 LATER WORK

Interest in detection and approximation continued, with determinants (§6.6) being deployed on occasion, and various numerical methods being proposed. Most of the results and techniques taught today, however, were known by Sturm's time. Nevertheless, new areas of application of mathematics sometimes provoked ideas; one of these was aerodynamics (§8.12): in 1914 the Englishman Leonard Bairstow extended Graeffe's method and extensions by Carl Runge and others to study the presence of many complex pairs of roots in polynomials of fairly high degree (octics and above). Although obscurely published (and so cited here as Bairstow *1914*), it became circulated.

BIBLIOGRAPHY

Bairstow, L. *1914*, 'Appendix' to 'Investigations relating to the stability of the aeroplane', *Advisory Committee for Aeronautics. Reports and Memoranda*, No. 154, London, 51–64.

Boyer, C. B. *1945*, 'Early graphical solutions of polynomial equations', *Scripta mathematica*, **11**, 5–19.

Burnside, W. S. and Panton, A. W. *1881, The Theory of Equations*, 1st edn, Dublin: Hodges, Figgis, and London: Longmans, Green. [Also later edns. Influential textbook; contains historical notes.]

Fourier, J. B. J. *1831, Analyse des équations indéterminées* (ed. C. L. M. H. Navier), Paris: Firmin Didot. [German transls: 1846, Braunschweig: Meyer; 1902, Leipzig: Engelsmann (Ostwalds Klassiker, No. 127).]

Hamburg, R. R. *1976*, 'The theory of equations in the 18th century: The work of Joseph Lagrange', *Archive for History of Exact Sciences*, **16**, 17–36.

Lagrange, J. L. *1798, Traité de la résolution des équations numériques* [. . .], Paris: Duprat. [2nd edn 1808; also in his *Oeuvres*, Vol. 8.]

Runge, C. *1898*, 'Separation und Approximation der Wurzeln', in *Encyklopädie der mathematischen Wissenschaften*, Vol. 1, 404–448 (article I B 3a).

Sinaceur, H. *1991, Corps et modèles* [. . .], Paris: Vrin.

4.11

Solving higher-degree equations

U. BOTTAZZINI

1 INTRODUCTION

The proof first given by Paolo Ruffini *1799* and, independently of him, by Niels Abel *1826* of the impossibility of algebraic solution of the general quintic equation settled a question which had been opened by the work of the Italian algebraists of the Renaissance (§6.1). For some three hundred years mathematicians had thought that the solution of the fifth-degree equation could be given 'by radicals', as was the case for the equations of degree 2 to 4. In such cases the solutions were expressed by formulas involving only rational operations on the coefficients and extractions of roots. Their guiding idea was that a similar method must exist (and therefore had to be found) for the general algebraic equation of degree n. In other words, it had to be always possible to reduce an nth-degree equation to the simple form $x^n = A$ by means of auxiliary equations of degree (at the most) $n - 1$ (Pierpont *1895*).

Joseph Louis Lagrange *1770–71* proved that for $n > 4$ the auxiliary equation (the 'resolvent') was of a degree strictly greater than n (§6.1). Using Lagrange's result, Ruffini showed that the quintic could not be reduced to the form $x^5 = A$ by solving algebraic equations of lower degree. It was possible, however, to reduce the equation to the form $f(x, A) = 0$, f being a fifth-degree polynomial with coefficients given by rational functions of a parameter A. Therefore, x could be algebraically expressed as $x = g(A)$, and the general quintic equation could be solved by means of radicals involving the function g.

The reduction of the fifth-degree equation to the form $f(x, A) = 0$ had actually been found independently by the Swedish mathematician E. S. Bring in 1786. Some fifty years later, in 1834, the Englishman G. B. Jerrard showed that it was always possible to put an algebraic equation of degree n into such a form that it did not contain terms of degree $n - 1$, $n - 2$ or

567

$n - 3$. To this purpose he used the following method of transformation, introduced by Ehrenfried Tschirnhaus in a paper of 1683: given the equation

$$x^n + a_1 x^{n-1} + \cdots + a_{n-1} x + a_n = 0, \tag{1}$$

and its auxiliary equation

$$x^{n-1} = b_1 x^{n-2} + \cdots + b_{n-2} x + b_{n-1} + y, \tag{2}$$

eliminate x between equations (1) and (2) to obtain a new equation of degree n,

$$y^n + c_1 y^{n-1} + \cdots + c_{n-1} y + c_n = 0, \tag{3}$$

where the coefficients c_k depend on the b_k. Tschirnhaus had thought (wrongly) it was always possible to determine the b_k in such a way that $c_k = 0$ for $k = 1, 2, \ldots, n - 1$, thus reducing (after the transformation) equation (3) to the form $y^n + c_n = 0$. For $n = 5$ Jerrard showed that, by using Tschirnhaus's method (i.e. solving equations of degree 2 and 3 only), the general equation could be reduced to the form

$$x^5 + x + a = 0. \tag{4}$$

This is the Bring–Jerrard form, which was the starting-point of the researches of Charles Hermite.

2 HERMITE'S SOLUTION OF THE QUINTIC EQUATION

Before Hermite published his result in 1858, an important step had been made by Enrico Betti in 1854 using the following reduction form:

$$x^5 + 5x^3 = y \tag{5}$$

(y being a parameter). By differentiating equation (5) he found that

$$5x^2(x^2 + 3)\,dx = dy. \tag{6}$$

After eliminating x between equation (5) and $x^2 + 3 = 0$ and some more calculations, he obtained the equation

$$\frac{dy}{5(y^2 + 108)^{1/2}} = \frac{x^2\,dx}{(x^6 + 4x^4 - 8x^2 + 12)^{1/2}}. \tag{7}$$

The left-hand side of equation (7) is easily integrable by elementary functions, while the right-hand side can be transformed into an elliptic differential by means of the change of variable $x^2 = z$. Therefore, Betti said, the equation could be solved (at least in principle).

Independently of him, the same idea was pursued by Hermite *1858*, who actually found a way of expressing the roots of the quintic equation by means of elliptic functions. He considered the reduced form

$$x^5 - x - a = 0 \tag{8}$$

and asked whether it was possible to represent each solution of it by means of single-valued functions of new variables. The analogy of the trigonometric solution of the cubic equation suggested to him that he should consider transcendental functions analogous to them (i.e. elliptic functions).

Hermite considered the moduli k and k' of the elliptic integrals (§4.5)

$$K = \int_0^{\pi/2} \frac{d\phi}{(1 - k^2 \sin^2\phi)^{1/2}} \quad \text{and} \quad K' = \int_0^{\pi/2} \frac{d\phi}{(1 - k'^2 \sin^2\phi)^{1/2}}. \tag{9}$$

As Jacobi had shown, $\sqrt[4]{k}$ and $\sqrt[4]{k'}$ can be expressed as quotient of power series of $q = \exp(-\pi K'/K)$. Hermite then considered

$$q = \exp(i\pi\omega), \qquad \varphi(\omega) = \sqrt[4]{k}, \qquad \psi(\omega) = \sqrt[4]{k'}. \tag{10}$$

If n is a prime number, then $v = \varphi(n\omega)$ and $u = \varphi(\omega)$ are related by an equation of degree $n + 1$ (modular equation) which, as Evariste Galois had stated in 1832 in his testamentary letter on group theory (§6.4), reduces to an equation of degree n when $n = 5, 7$ or 11. For $n = 5$ the modular equation is

$$u^6 + v^6 + 5u^2v^2(u^2 - v^2) + 4uv(1 - u^4v^4) = 0. \tag{11}$$

Hermite then considered the expression

$$\phi(\omega) = [\varphi(5\omega) + \varphi(\omega/5)]$$
$$\times \left[\varphi\left(\frac{\omega + 16}{5}\right) - \varphi\left(\frac{\omega + 4 \cdot 16}{5}\right)\right]\left[\varphi\left(\frac{\omega + 2 \cdot 16}{5}\right) - \varphi\left(\frac{\omega + 3 \cdot 16}{5}\right)\right]. \tag{12}$$

Now, $\phi(\omega + m \cdot 16)$, $m = 0, 1, \ldots, 4$, are actually the roots of the following quintic equation whose coefficients can be rationally expressed by means of $\varphi(\omega)$:

$$\phi^5 - 2^4 \cdot 5^3 \phi\varphi^4(\omega)\psi^{16}(\omega) - 2^6\sqrt{(5^5)}\varphi^3(\omega)\psi^{16}(\omega)[1 + \varphi^8(\omega)] = 0. \tag{13}$$

Hermite was able to show that equation (13), under the assumptions he made for $\phi(\omega)$, $\varphi(\omega)$ and $\psi(\omega)$, did coincide with the resolvent of fifth degree of the modular equation (11); and also that it could be reduced to the Jerrard form (8) by setting

$$\phi = \sqrt[4]{(2^4 \cdot 5^3)}\varphi(\omega)\psi^4(\omega)x \tag{14}$$

and therefore

$$a = \frac{2}{\sqrt[4]{(5^5)}} \frac{1 + \varphi^8(\omega)}{\varphi^2(\omega)\psi^4(\omega)}. \tag{15}$$

By assuming $\varphi^4(\omega) = k$ as an unknown, and recalling that $\psi^4(\omega) = k'$ (with $k'^2 = 1 - k^2$), he obtained the quartic in k

$$k^4 + A^2k^3 + 2k^2 - A^2k + 1 = 0, \quad A = \sqrt[4]{(5^5)}a/2, \tag{16}$$

which allowed him to determine k and consequently the roots of the quintic (8).

3 THE WORK OF BRIOSCHI AND KRONECKER

Inspired by Hermite's paper, in the same year of 1858 Francesco Brioschi published a second method for the solution of the quintic equation. Instead of using the modular equation, as Hermite did, he started by considering the 'equation of the multiplicator', which had been introduced by Carl Jacobi in his work on the transformation of elliptic integrals. In 1827–8 Jacobi had solved the problem of determining a rational function y such that

$$\frac{\mathrm{d}y}{(P_4(y))^{1/2}} = \frac{\mathrm{d}x}{(P_4(x))^{1/2}}, \tag{17}$$

where $P_4(x)$ is a fourth-degree polynomial (§4.5). He was able to show that, for every odd prime n and any module k, the problem reduced to the determination of a rational function $y = U(x)/V(x)$ (U being a polynomial of degree n, and V a polynomial of degree $n - 1$) such that

$$\frac{\mathrm{d}x}{[(1 - x^2)(1 - k^2x^2)]^{1/2}} = \frac{M\mathrm{d}y}{[(1 - y^2)(1 - \lambda^2y^2)]^{1/2}}, \tag{18}$$

where the unknowns are the coefficients of U and V, the new module λ, and M (called the 'multiplicator' of the transformation of order n). For $n = 5$, $z = 1/M$ satisfied Jacobi's 'equation of the multiplicator'

$$(z - 1)^6 - 4(z - 1)^5 + 2^8k^2k'^2z = 0. \tag{19}$$

Just as Hermite had done, Brioschi was able to obtain from equation (19) a fifth-degree resolvent which could be reduced to the Jerrard form (8) after some more calculations based on the properties of Jacobi's equation (19). Following a suggestion by Hermite, Brioschi quickly realized that equation (19) has a simpler resolvent whose existence, according to a method elaborated by Leopold Kronecker, was to be ensured *a priori*.

For some years, independently of Hermite and Brioschi, Kronecker had been working on the same problem. Two months after Hermite had published his solution, Kronecker communicated his own method in a letter to him published in the *Comptes Rendus* of the Paris Academy of Sciences. Like Hermite, Kronecker had first followed the idea of using the modular equation and the reduction of the quintic to the form (8). He then decided instead to adopt a more direct procedure. he considered a cyclic (rational) function $f(x_0, \ldots, x_4)$ of the five roots $x_0, \ldots, x_4$ of the quintic as

$$\sum_{m=0}^{4} \sum_{n=0}^{4} \sin\left(\frac{2n\pi}{5}\right) (x_m x_{m+n}^2 x_{m+2n}^2 + \nu x_m^3 x_{m+n} x_{m+2n}), \tag{20}$$

and thought that ν could be determined in such a way that the 12 transformed functions obtained by means of even permutations of the x's could be written as $\pm f$, $\pm f_0$, $\pm f_1$, $\pm f_2$, $\pm f_3$, and $\pm f_4$ in order to satisfy both the relation

$$f^2 + f_0^2 + f_1^2 + f_2^2 + f_3^2 + f_4^2 = 0 \tag{21}$$

and the equation

$$f^{12} - 10\varphi f^6 + 5\varphi^2 = \psi f^2, \tag{22}$$

where φ and ψ are rational functions of both the coefficients and the square root of the discriminant of the quintic. Then he observed that equation (22) could be reduced to Jacobi's 'equation of the multiplicator', and therefore solved by means of elliptic functions.

Kronecker was not entirely satisfied with his result because of the necessity of introducing what he called an 'accessorial irrationality', the extraction of the square root of the discriminant of the quintic in order to ensure the condition (21). As Abel had done for the equations of lower degree (2 to 4), Kronecker wanted to use only 'natural irrationalities', auxiliary quantities which depended on the roots of the quintic in a rational way. In other words, he was looking for a method in which the auxiliary quantities he needed remained in the field of the roots of the equation. But later he discovered that it was impossible to obtain an equation like (22) without introducing 'accessorial' quantities, nor to obtain a resolvent of the equation depending on one parameter only. In other words, it was impossible to reduce the general quintic equation to a form depending on one parameter only without leaving the field of the roots. Kronecker's claim was to be proved by Felix Klein in 1879.

The method, which Kronecker communicated in a cryptic way in a letter to Hermite, was explained in full detail by Brioschi in 1858 and by Kronecker himself in 1861. A complete discussion of the different methods (including all the calculations they entailed) was eventually given by Hermite in 1866.

4 LATER DEVELOPMENTS

Kronecker had elaborated his method of tackling a more general problem concerning seventh-degree equations. This problem was eventually solved in 1878 by Klein, who showed that it was possible to reduce all seventh-degree equations, whose Galois group was a certain simple subgroup of order 168 of the alternate group A_7, to the modular equation of order 7, and then to solve them by means of elliptic functions.

In the late 1870s Klein's researches on higher-degree equations intertwined with Brioschi's before they found their natural setting in the theory of automorphic functions (§3.16). Klein himself focused on the deep connections between the algebraic aspects of the theory of equations and the geometrical ones in his book *1884* on the icosahedron, and gave an historical account of the previous developments and presented in geometrical terms the main concepts of the theory of the quintic equation (Slodowy *1986*).

BIBLIOGRAPHY

Abel, N. H. *1826*, 'Beweis der Unmöglichkeit algebraische Gleichungen von höheren Graden als dem vierten allgemein zu lösen', *Journal für die reine und angewandte Mathematik*, **1**, 65–85. [French transl. in *Oeuvres complètes*, 2nd edn, 1881, Vol. 1, 66–87.]

Hermite, C. *1858*, 'Sur la résolution de l'équation du cinquième degré', *Comptes Rendus de l'Académie des Sciences*, **46**, 508–12. [Also in *Oeuvres*, Vol. 1, 5–12.]

Klein, F. *1884*, *Vorlesungen über das Ikosaeder und der Gleichungen vom fünften Grade*, Leipzig: Teubner. [Repr. 1992, Basel: Birkhäuser.]

Lagrange, J.-L. *1770–71*, 'Réflexions sur la résolution algébrique des équations', *Nouveaux Mémoires de l'Académie Royale des Sciences de Berlin*, (1770), 134–215; (1771), 138–253. [Also in *Oeuvres*, Vol. 3, 203–421.]

Pierpont, J. *1895*, 'Zur Geschichte der Gleichung des V. Grades (bis 1858)', *Monatshefte für Mathematik und Physik*, **6**, 15–68.

Ruffini, P. *1799*, *Teoria generale delle equazioni in cui si dimostra impossibile la soluzione algebraica delle equazioni generali di grado superiore al quarto*, Bologna: S. Tommaso d'Aquino. [Also in *Opere matematiche*, Vol. 1, 1–324.]

Slodowy, P. *1986*, 'Das Ikosaeder und die Gleichungen fünften Grades', in H. Knörrer *et al.*, *Arithmetik und Geometrie*, Basel: Birkhäuser, 71–113.

4.12

Nomography

H. A. EVESHAM

1 THE NATURE OF NOMOGRAPHY

Nomography is the study of those computational techniques which use results from geometry. To illustrate the method, consider two very simple cases. Figure 1 shows a set of rectangular hyperbolas $xy = z$ plotted for some values of x and y between 1 and 10; the value of the parameter z for each curve is shown on the curve. Normally we are interested in such curves for their geometric properties, and take their equation $xy = z$ as an algebraic description of their shape. However, if our interest is computational then we observe that the equation expresses the fact that the product of x and y is z. In this context Figure 1 becomes a computational device:

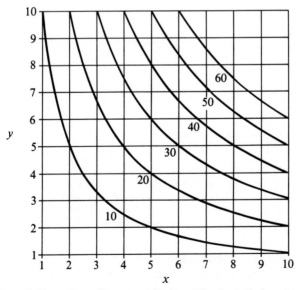

Figure 1 Nomogram for a set of rectangular hyperbolas $xy = z$

the product of x and y is seen to be the value of the parameter of the curve at which the appropriate vertical and horizontal lines intersect.

This particular case is too restricted to be of any real value, but it is an illustration of the type of thinking on which nomography is based. It is in fact an elementary example of an intersection nomogram, so called because the result of the computation is given by the intersection of curves.

An example of a second type of nomogram in which points are aligned is shown in Figure 2. This example is based on the geometric fact that a line joining the points having ordinates y_1 and y_2 of the parabola $y^2 = x$ cuts the axis of the parabola at the point $y_1 y_2$. Thus the line ABC demonstrates the fact that $2 \times 3 = 6$. Note that values of y below zero have had their negative signs removed in order to give results for positive multiplication. This idea was first expressed by August Ferdinand Möbius in 1841.

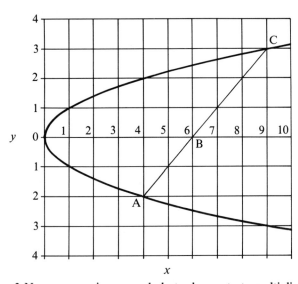

Figure 2 Nomogram using a parabola to demonstrate multiplication

2 EARLY HISTORY

No published history of the subject exists, although there is at least one unpublished Ph.D. thesis (Evesham *1982*); d'Ocagne *1928* contains an informative section.

It is possible to cite certain developments which were prerequisites to the development of nomography. There was the development of coordinate geometry by Pierre de Fermat and René Descartes (§7.1), without which nomography would have been impossible. The representation of three variables on a two-dimensional plane is a simple enough idea, but fundamental. In 1752 Philippe Buache described how submarine channels and the slope of the sea bed in coastal waters could be represented in one plane; he demonstrated it for the English Channel, recording depths from 10 fathoms increasing by units of 10 fathoms. The development of statistics in the early years of the nineteenth century also led to graphical representation of data (Royston *1956*).

However, the impetus necessary to initiate the serious development of geometric computation seems to have been the introduction of a new system of weights and measures in France at the end of the eighteenth century. Initially the conversion from one system to another had been facilitated by the use of tables, but in the spring of 1795 the authorities published a law which stated that 'in place of tables of relationship between old and new measures, which had been provided by the order of 8 May 1790, will be graphical scales to estimate these relationships without having need to any calculation'. A publication giving these graphical scales appeared in 1795, prepared by Louis Pouchet. This work had an appendix called 'Arithmétique linéaire' which gave graphical methods for addition, subtraction, multiplication, division, squaring and the extraction of roots. The method for multiplication was similar to that represented in Figure 1, but with values of z ranging from 2 to 100 in steps of 2.

Other examples of graphical computation appeared later: by I. Didion in 1839 in ballistics, in which the ability to perform speedy calculations is of great value (§8.11); and by L. Lalanne in meteorology (§9.7). The French and English editions of *A Complete Course in Meteorology* by L. F. Kaemtz appeared in 1843 and 1845, respectively. These contained an appendix by Lalanne in which graphical representations of some of the numerical tables were given. It was suggested that graphical methods were preferable to tabular ones on the grounds that interpolation was easier to carry out. The preface to this work suggests that Lalanne was the first to generalize the representation of three coordinates in one plane.

The problem of constructing diagrams which contained curves must have been a tedious task in an age when there were few mechanical aids (§5.11). It therefore comes as no surprise that one of the early advances in the emerging discipline was to simplify the constructions required. In 1846 Lalanne made the observation that there was no reason why the frame of a computational diagram should be graduated in a regular manner, and went on to propose that an irregular scale which replaced curves by straight lines would have considerable advantages. To this new approach he gave the name 'anamorphosis'. A simple illustration of the technique is shown in Figure 3, which is an anamorphic version of Figure 1. The anamorphosis

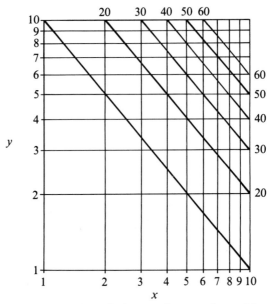

Figure 3 Anamorphic version of Figure 1: linear scales and hyperbolas are now logarithmic scales and straight lines

has been achieved by the use of logarithmic scales. The equation $xy = z$ has been written as $\log x + \log y = \log z$ to give the linear form. The values of z appear at either end of the anamorphic lines.

Lalanne was a civil engineer who in 1846 worked on the construction of railways from Paris. It was the construction in France and Belgium of railways and their associated structures that provided the second thrust in the development of nomography. There is no evidence that similar undertakings in Britain led to similar developments. Lalanne produced what he called a 'universal calculator', a nomogram which performed a wide range of calculations, including ratios for chemical equivalents. He advocated that it should replace the slide rule, 'the use of which is so common in England'.

3 THE EMERGENCE OF NOMOGRAPHY AS A DISTINCT BRANCH OF MATHEMATICS

An indication of the insights to be gained from a nomogram is well illustrated by Lalanne's nomogram for the cubic equation, which can always be reduced to the form

$$x^3 + px + q = 0. \tag{1}$$

From the nomographic standpoint this equation has two variables, p and q, and therefore represents a straight line for a given value of x. Figure 4 shows straight lines for values of p and q from -1 to $+1$. The corresponding values of x are marked on the lines. The diagram shows immediately that some equations will have only one real root, while others will have three. The curve which separates the two regions has the form

$$4p^3 + 27q^2 = 0. \tag{2}$$

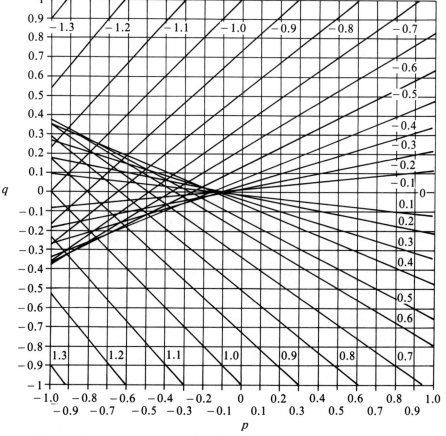

Figure 4 Nomogram for the cubic $x^3 + px + q = 0$, with straight lines for values of p and q between -1 and 1

An early indication that there were interesting mathematical undertones to the subject came in 1867, when there arose the question of under what circumstances an equation in three variables $F(x, y, z) = 0$ can be transformed into an equation of the form

$$f_1(z) = f_2(x) + f_3(y). \tag{3}$$

The point of such a transformation is that, for any fixed value of z, a suitable change of variable for x and y will produce the linear form. This transformation was achieved in the case of Figure 3 by inspection to obtain $\log x + \log y = \log z$ from $xy = z$. Most cases, however, will not permit of such simple treatment. The problem was posed and solved by Paul de Saint-Robert in 1871. It is of interest to note that, although he produced a result of importance to nomography, he was actually investigating the construction of a slide rule. The result of his investigations is known as Saint-Robert's criterion, and may be expressed thus: an equation $F(x, y, z) = 0$ can be reduced to the form

$$f_1(z) = f_2(x) + f_3(y) \tag{4}$$

if

$$[\ln (F_x/F_y)]_{xy} = 0, \tag{5}$$

where subscripts denote partial differentiation. An extension to this result explains how, if the criterion is satisfied, the functions f_1, f_2 and f_3 can be found.

Two important contributors emerged during the 1880s. One was J. Massau, a Belgian engineer also working on the construction of railways. In general, his work builds on that of Lalanne and is concerned with making practical improvements to intersection nomograms. The purpose of these improvements was to reduce the time required for construction, and to do this the concept of anamorphosis was of great value. He introduced determinants (§6.6) into the subject. Suppose that a straight-line intersection nomogram in the variables x, y and z is constructed of three sets of lines:

$$\left.\begin{array}{ll} f_1 p + g_1 q + h_1 = 0, & \text{where } f_1, \ g_1, \ h_1 \text{ are functions of } x, \\ f_2 p + g_2 q + h_2 = 0, & \text{where } f_2, \ g_2, \ h_2 \text{ are functions of } y, \\ f_3 p + g_3 q + h_3 = 0, & \text{where } f_3, \ g_3, \ h_3 \text{ are functions of } z. \end{array}\right\} \tag{6}$$

Then it is necessary that the following condition should hold if lines from each set are to intersect:

$$
\begin{vmatrix}
\dfrac{f_1(x)}{h_1(x)} & \dfrac{g_1(x)}{h_1(x)} & 1 \\[2ex]
\dfrac{f_2(y)}{h_2(y)} & \dfrac{g_2(y)}{h_2(y)} & 1 \\[2ex]
\dfrac{f_3(z)}{h_3(z)} & \dfrac{g_3(z)}{h_3(z)} & 1
\end{vmatrix} = 0
\quad \text{or equivalently} \quad
\begin{vmatrix}
f_1(x) & g_1(x) & h_1(x) \\
f_2(y) & g_2(y) & h_2(y) \\
f_3(z) & g_3(z) & h_3(z)
\end{vmatrix} = 0.
\tag{7}
$$

Such determinants are occasionally referred to as 'Massau determinants'.

In itself this is not a great step forward since it is the condition for non-trivial solutions to the equations, but Massau recognized that there is associated with it a problem which is certainly not simple. He supposed that, for the equations given above, $f_1, f_2, f_3, \ldots, h_3$ represent functions of the second degree in x, y and z. Then the determinant will yield an equation of the sixth degree in x, y, z, x^2, y^2 and z^2. Now, the starting-point for the construction of any nomogram will be some algebraic equation arising from a practical situation. The nomogram can be constructed if the Massau determinant holds. Therefore the problem that Massau faced was this: given an equation of the sixth degree in x, y, z, x^2, y^2 and z^2, how can it be identified with the Massau determinant? Massau did not solve this problem, although he did arrive at results for certain cases.

4 FURTHER DEVELOPMENTS

The other important figure to emerge during the 1880s was Maurice d'Ocagne. In 1884, at the age of 22 years, he published a paper which described 'a new method of graphical calculation'. The new method was alignment. Although Möbius had already noted that the alignment of certain points on a parabola and its axis could be interpreted as multiplication (as illustrated in Figure 2), this was little more than an interesting observation. d'Ocagne turned it into a sophisticated tool. He did this by converting lines to points and points to lines. This was achieved through the use of a parallel coordinate system. The basis of such a system is a pair of parallel lines AU and BV, and a transversal AB (Figure 5). It is convenient to make AB perpendicular to AU and BV. The coordinates u and v are measured from, respectively, A along AU and B along BV, so the pair (u, v) represents the straight line MN. The equation $v + au + b = 0$ represents a point.

This coordinate system is well suited for the construction of an alignment nomogram for the cubic equation $x^3 + px + q = 0$. If p is measured along

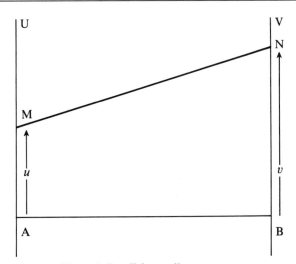

Figure 5 Parallel coordinate system

the u-axis and q along the v-axis, then for every line in Lalanne's solution (Figure 4) we have a point in d'Ocagne's solution (Figure 6). The points are particularly easy to plot, and lie on the curve referred to by d'Ocagne as C_3. Figure 6 is based on d'Ocagne's original nomogram for the equation $x^3 - 7x + 6 = 0$. Taking $p = -7$ and $q = 6$, the line $(-7, 6)$ intersects C_3 at 1 and 2, these being two of the roots. To get the third root, observe that, since the sign of 6 is positive and the two roots found are both positive, the third root must be negative. Replacing x by $-x$ in the equation gives $x^3 - 7x - 6 = 0$. The line $(-7, -6)$ gives the third root, -3.

At about the same time that d'Ocagne was developing his alignment nomogram, another Frenchman, Charles Lallemand, was also occupied with the simplification of calculation. He was employed by the Nivellement Général de la France, a topographical organization of which he later became head. Once again, it is clear that developments in engineering and applied science spurred the development of nomograms. Lallemand is particularly remembered for his hexagonal nomogram, which is based on the following geometric result: the algebraic sum of the projection of a segment of a line on two axes having an angle of 120° between them is equal to the projection of the same segment onto the internal bisector of the angle between these axes. The hexagonal nomogram appears to be independent of any earlier work, and is the first nomogram to use an oriented transparency as an essential part. An oriented transparency has the effect of extending a nomogram by placing part of it on a transparent sheet which can be positioned in a variety of ways.

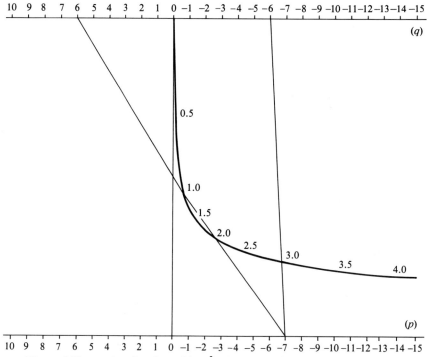

Figure 6 Nomogram for the cubic $x^3 - 7x + 6 = 0$ in parallel coordinates, based on d'Ocagne's original diagram

The period 1840–90 saw the development of techniques in geometric computation. Eventually, these techniques became public knowledge, usually as the result of papers published by learned societies. In the case of Lallemend's hexagonal nomogram a brochure was published for the use of the Nivellement Général, but was not made available to the public. It was d'Ocagne who collected these techniques together and published them in 1891 in his *Nomographie – Les calculs usuels effectués au moyen des abaques*. This was the first text on nomography, and also the first work to describe nomography by that name. It not only describes most of the methods known at that time, but also gives their underlying mathematical principles.

The publication of d'Ocagne's book marks the point at which nomography ceased to be a collection of interesting techniques and became a proper branch of mathematics. At the 1893 International Mathematical Congress at Chicago, d'Ocagne read a paper on nomography.

5 MAJOR PUBLICATIONS

The most important work on nomography is d'Ocagne's *Traité de nomographie* (1899), with a second edition in 1921. This work is in French, as is his *1928* and also other major works such as R. Soreau's *Nomographie, ou traité des abaques* (1921). Perhaps the first published account in English of nomographic methods was a series of articles by John B. Peddle which appeared in *The American Machinist* during 1908. In the following year the *Journal of the Royal Artillery* published the first of many items by R. K. Hezlet illustrating the application of nomography to ballistics. An early book to appear in English was *A First Course in Nomography* (1920) by Selig Brodetsky, Reader in Mathematics at the University of Leeds. It was not until 1932 that anything approaching d'Ocagne's *Traité* was published in Britain. This was *The Nomogram* by M. J. Allcock and J. M. Jones, which through its various editions became the standard British text (Allcock *et al. 1963*).

In general those countries which had some special link with France, either through language or history, tended to make greater use of nomography. Thus in Egypt in the early 1900s nomograms were used by the Irrigation Service and by Egyptian Railways. Some important theoretical advances which resulted in beautiful nomograms were made by J. Clark, Professor of Mathematics at the Ecole Polytechnique at Cairo, and published in 1907. Clark's work is interesting because he reversed the tendency to have as many straight lines as possible in a nomogram and produced some having conics or cubics as their main features. His work is not as well known as it deserves to be. In Canada in 1906, E. Deville, a land surveyor, proposed the use of nomography to find the celestial coordinates of the pole star. Nomograms were in use in Italy before the end of the nineteenth century, and in Spain, Holland, the Ukraine and Russia in the early twentieth century.

Nomograms have now been largely superseded by the computer and the pocket calculator. Indeed, nomograms can in many cases be constructed without too much effort by using graphics software and a personal computer. They continued to be used in some disciplines after computers had become popular.

In Russia, interest in all aspects of nomography was such that in 1959 an entire issue of the journal *Vychislitel'naya matematika* ('Computational Mathematics') was devoted to the subject. One of the contributors, I. N. Denisyak, presented a nomogram for the calculation of generalized Laguerre polynomials. At about the time that Denisyak was producing his nomogram, the Manchester University Mark 1 computer was also computing Laguerre polynomials (Lavington *1980*) – a coincidence which is both a comment on the relative technological attainments of the two countries at that time, and an indication of when nomograms began to give way to electronic methods.

6 THEORETICAL PROBLEMS

Of enduring interest have been the theoretical problems which nomography has presented. At the Paris International Congress of 1900, David Hilbert presented his 23 outstanding mathematical problems. Problem 13 was the 'impossibility of the solution of the general equation of the seventh degree by means of functions of only two arguments'. This is not at first sight a nomographic problem, but Hilbert elaborated it in terms of nomography. The problem is really about the representation of a function as a superposition of functions of fewer variables. It has been investigated by the Russians V. I. Arnol'd and A. N. Kolmogorov among others, most recently by the American R. C. Buck. Their contributions lead into other branches of mathematics.

The theoretical problem of more direct interest to nomography is the one first expressed by Massau. In its most general form, it applies to both intersection nomograms with anamorphosis and to alignment nomograms and can be expressed as follows; to find under what conditions $F(x, y, z) = 0$ can be expressed as

$$\begin{vmatrix} f_1(x) & g_1(x) & h_1(x) \\ f_2(y) & g_2(y) & h_2(y) \\ f_3(z) & g_3(z) & h_3(z) \end{vmatrix} = 0. \tag{8}$$

There are four different approaches to the problem. First there is the line, suggested by d'Ocagne, through partial differential equations. Then there is an algebraic approach which starts from the recognition that the determinant must have the form

$$f_1(x)R_1(y, z) + g_1(x)R_2(y, z) + h_1(x)R_3(y, z) = 0 \tag{9}$$

if expanded along the x row, with parallel expressions if expanded along the y or z row. A third approach is to set up a classification system which will, for any given function, lead to the appropriate determinant, where possible, or to a statement of the impossibility of a determinant form. The fourth approach is based on an approximation method for constructing nomograms. If the function is nomographic, then the method leads to an exact nomogram; if it is non-nomographic, the method will produce an approximate nomogram.

A considerable time gap exists between the first two approaches and the second two. T. H. Gronwall published his work on the partial-differential approach in 1912, followed three years later by O. D. Kellogg's paper on the form of the expanded determinant. However, it was not until 1959 that the other two appeared: from Poland came the classification scheme of M. Warmus, and from Russia came the approximation method of G. E. Dzhems-Levi.

The four papers reveal the different philosophical positions of the authors, ranging from the classical mathematical tradition to a constructionist approach, the attitudes in part reflecting the times when the work was carried out. Perhaps the most satisfactory for the would-be nomographer is the work of Warmus, but there is no easy way to produce a nomogram for a very complicated function. It may be this as much as the electronic revolution which has brought about the demise of nomography.

BIBLIOGRAPHY

Allcock, M. J., Jones, J. R. and Michel, J. G. L. *1963, The Nomogram*, 4th edn, London: Pitman.

Evesham, H. A. *1982*, 'The history and development of nomography', Ph.D. Thesis, University of London.

—— *1986*, 'Origins and development of nomography', *Annals of the History of Computing*, **8**, 324–33.

Lavington, S. *1980, Early British Computers*, Manchester: Manchester University Press.

d'Ocagne, M. *1911*, 'Calculs numériques', in *Encylopédie des sciences mathématiques*, Tome 1, Vol. 4, 196–452 (article I 23). [See especially arts. 42–55.]

—— *1928, Le Calcul simplifiée*, 3rd edn, Paris: Gauthiers-Villars. [English transl.: *Le Calcul simplifiée: Graphical and Mechanical Methods for Simplifying Le Calculation* (transl. J. Howlett and M. R. Williams), 1986, Cambridge, MA: MIT Press.]

Otto, E. *1963, Nomography*, Oxford: Pergamon Press.

Royston, E. *1956*, 'A note on the history of graphical presentation of data', *Biometrika*, **43**, 241–6. [Also in E. S. Pearson and M. G. Kendall (eds), 1970, *Studies in the History of Statistics and Probability*, Vol. 1, London: Griffin, 173–82.]

4.13

General numerical mathematics

PETER SCHREIBER

1 INTRODUCTION AND SURVEY

Up to the end of the eighteenth century, mathematics existed largely only in its union with astronomy, geodesy, mechanics, architecture and other disciplines. Distinction between 'pure' and 'applied' mathematics at this time is possible for individual results, but never for the whole of a mathematician's work. The general shift in the position of mathematics and mathematicians in society since the early nineteenth century produced what we may call the first pure mathematicians, such as Niels Abel, Evariste Galois, Bernard Bolzano and János Bolyai, as well as the first applied mathematicians, such as Gaspard Monge, Gaspard de Prony, Charles Babbage and Adolphe Quetelet. We find fewer mathematicians who were strong in both directions; those who were include Augustin Louis Cauchy, Joseph Fourier, Carl Friedrich Gauss and Pafnuty Chebyshev. Some further preliminary remarks are necessary.

It was not the case, as one might assume, that progress in applied mathematics has been concentrated at engineering schools and that of pure mathematics at universities; in many cases it was the converse. Second, at the end of the nineteenth century applied mathematics was still considered to include parts of astronomy, geodesy, mechanics, demography and social statistics, and to the 1920s it included all graphical methods of solving numerical problems. This state of affairs remained after the founding of the first special institutes and professorships for applied mathematics – for example, at Göttingen in 1904 (Carl Runge), at Munich in 1910 (H. Liebmann) and at Berlin in 1920 (Richard von Mises) – and this is reflected in the contents of the first textbooks on numerical mathematics and in the *Zeitschrift für angewandte Mathematik und Mechanik*, founded in 1921 by von Mises. So it is true that as a special topic numerical mathematics became established in the nineteenth century, although this may not have been so evident at the time.

Third, even if we (in an ahistorical manner) limit numerical mathematics in a narrower sense to its applications in astronomy, geodesy, and so on, in contrast to the body of graphical methods, the contents of numerical mathematics within the period treated here is still essentially different from what it is today. Algorithmic descriptions of procedures, explicit studies of the propagation and estimation of errors and of the numerical stability of procedures, and all considerations of the efficiency of algorithms can be found as implicit hints at most. Perhaps only de Prony, Gauss and Babbage were true pioneers of modern numerical mathematics.

Fourth, the constitution of numerical mathematics and its progress in the nineteenth century is not adequately reflected in most textbooks on the history of mathematics, and hitherto insufficiently studied in historical research. The only fairly recent monographs are Goldstine *1977* and Brezinski *1990*, Goldstine ends at 1900 and concentrates on a few central persons; the book is recommended, although in some aspects it reflects the special interests of its author more than the 'spirit of an age'.

2 NUMERICAL SOLUTIONS OF EQUATIONS AND SYSTEMS OF EQUATIONS

In his Latin treatise of 1810 on the orbit of the asteroid Pallas, Gauss published his famous elimination procedure for systems of linear equations, now familiar as the 'Gaussian algorithm'. His motive was the need to solve the normal equations which arise when the least-squares method is used. It is interesting how slowly it took a general and suitable presentation of this algorithm to appear in the textbooks. Even in Runge and König *1924* we find a very cautious treatment of small-scale examples, albeit enriched with valuable remarks about accuracy and other practical aspects of calculating 'by hand'.

Gauss himself remarked upon the insufficiency of his elimination procedure from the point of view of practical computing with many variables and rounded coefficients. In a letter to his friend C. L. Gerling in 1823, he described for the first time the iterative, self-correcting solution procedure which was independently rediscovered in 1879 by P. L. von Seidel and is now known as the Gauss–Seidel–Southwell algorithm. In 1884 and 1892 P. A. Nekrasov gave the first exact estimations of the error of this method. Gauss's *Theoria motus corporum* (1809) implicitly contains a generalized *regula falsi* rule for systems of non-linear equations: if the function f of n variables in $n+1$ places, $x_1^i, \ldots, x_n^i$ with $i = 1, \ldots, n+1$, has the values $f(x_1^i, \ldots, x_n^i)$ which are not all of the same sign, then these values together determine a linear function of n variables which has the value zero along a linear manifold. By cutting such manifolds for several functions, one finds approximately the roots of the system of equations $f_j(x_1, \ldots, x_n) = 0$.

The main problem area in this topic to be addressed in the nineteenth century was that of solving non-linear (especially polynomial) equations in one variable, which corresponds to finding the roots of a polynomial or other type of function. Many mathematicians were concerned with the further study of Isaac Newton's iteration method (1669), among them Fourier who, in a posthumous book of 1831, defined for the first time the exact conditions under which this method will work (§4.10). Fourier's work also contains the germ of J. C. F. Sturm's method of separation of the zeros of a polynomial; Sturm himself had published his method (without proof) in 1829. In connection with the numerical computation of values of polynomials necessary for finding the roots, Paolo Ruffini in 1813 and, independently, William Horner in 1819 published the so-called 'Horner scheme', reducing the number of steps of the computation by using the term

$$(\ldots(a_n x + a_{n-1})x + a_{n-2})x + \cdots + a_1)x + a_0 \tag{1}$$

instead of

$$a_n x^n + a_{n-1} x^{n-1} + \cdots + a_1 x + a_0. \tag{2}$$

With the coefficients $a_n, a_{n-1}, \ldots, a_1, a_0$ laid out in a row, one completes the scheme as illustrated in Figure 1.

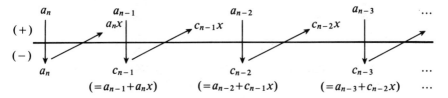

Figure 1 The Horner scheme

This is one of the earliest examples of thinking in terms of the complexity and efficiency of algorithms. As far as I know, it is the first time in modern mathematics that a suitable arrangement of data in the plane was explicitly used for economical computing (Cajori *1911*).

Another way to compute the zeros of a polynomial is now known as

C. H. Graeffe's root-squaring method (Graeffe *1837*). From the given polynomial

$$f(x) := \sum_{i=1}^{n} a_i x^i = a_n(x - x_1)(x - x_2)\ldots(x - x_n), \tag{3}$$

we go to its 'Graeffe square'

$$f^g(x) = f(x) \cdot a_n(x + x_1)(x + x_2)\ldots(x + x_n) \tag{4}$$

$$= a_n^2(x - x_1^2)(x - x_2^2)\ldots(x - x_n^2) = \sum_{i=1}^{n} b_i x^{2i}, \tag{5}$$

where the computation of the new coefficients b_j from the a_i is simple because of

$$(x + x_1)(x + x_2)\ldots(x + x_n) = \sum_{i=1}^{n} (-1)^{n-1} a_i x^i. \tag{6}$$

By iteration of Graeffe squaring, the differences between the roots become so large that, after m steps, according to Viète,

$$b_1^{(m)} = x_1^{2^m} + x_2^{2^m} + \cdots + x_n^{2^m} \tag{7}$$

will be approximately the largest of these summands. Complications arise if f has double roots or some pairs of conjugate complex roots. Many authors (e.g. E. Routh in 1877, A. Hurwitz in 1895, and S. Brodetsky and G. Smeal in 1924) tried to extend Graeffe's idea to such cases. Today we know that Graeffe had several forerunners in other countries (Edward Waring in 1770, Pierre Dandelin in 1826 and Nikolai Lobachevsky in 1834). Such multiplicity of progress in science is a phenomenon typical of the nineteenth century, when the transition to publishing in vernacular languages was not matched by a rapid spread of learning vernacular languages, and the growth of the scientific community was not followed by the prompt founding of review journals.

3 INTERPOLATION AND APPROXIMATION OF FUNCTIONS; NUMERICAL QUADRATURE

As is well known, in the *Principia* (1687) Newton set out his method for finding a polynomial of degree $\leq n$ so that it has given values at $n + 1$ interpolation nodes. This method was further developed by Joseph Raphson in 1690, and by many others (§4.10). Although Newton's method has many advantages (e.g. the number of interpolation nodes may be increased within the procedure) many people did further work in this field. Perhaps Joseph Louis Lagrange's method, published in 1798, now has more popularity because it may be described as more elegant or concise, and in line with modern thinking on linearity. The function

$$f(x) = \frac{(x - x_1)(x - x_2) \ldots (x - x_{n-1})}{(x_n - x_1)(x_n - x_2) \ldots (x_n - x_{n-1})} \tag{8}$$

has the value 0 for $x = x_1, x_2, \ldots, x_{n-1}$, and the value 1 for $x = x_n$. By linear combination of such functions one obtains a polynomial of degree $\leqslant n - 1$ which has given values at the n places $x_1, \ldots, x_n$. Most textbooks today mention the priority of Waring (1779) to his interpolation formula, but in fact Lagrange communicated it in 1778 at a session of the Berlin Academy of Sciences (Kowalewski *1917*: 90).

Exactly 100 years after Lagrange, in 1878, Charles Hermite generalized Lagrange's method so that derivatives of arbitrary order may be taken at the interpolation nodes. Interpolation by functions other than polynomials was studied first by Cauchy (in his *Cours d'analyse* (1821), Note V) with fractional rational functions, by de Prony with exponential functions, by Lagrange and Gauss with trigonometrical series and, later on, by Adrien Marie Legendre and Franz Ernst Neumann with spherical functions.

A different direction, approximation instead of interpolation, goes back to Jean Victor Poncelet and Chebyshev and their study of mechanisms. The idea of minimizing

$$\max |f(x) - g_n(x)|, \quad a \leqslant x \leqslant b \tag{9}$$

by suitable approximation functions g_n (especially polynomials of given degree) had a rapid and deep influence, especially in France (e.g. Edmond Nicolas Laguerre and Hermite) and in the Saint Petersburg mathematical school founded by Chebyshev (§11.6). To approximate by minimization of

$$\int_a^b |f(x) - g_n(x)|^p \, dx \quad \text{for some } p \tag{10}$$

instead of (9) comes in principle from Gauss and Fourier, but it was only in 1904 that David Hilbert began to clarify the background of such an approach, proceeding from the point of view of what was to become functional analysis (§3.9). The approximation by 'area' is strongly connected with Fourier analysis of functions (§3.11). Recently it has been found that as early as 1805 Gauss treated the orbits of celestial bodies by Fourier approximation and invented the so-called Cooley–Tukey algorithm for the fast Fourier transform (Heideman *et al. 1985*).

A special aim for interpolation and approximation of functions has been to realize the numerical quadrature of a complicated function by integration of a simple approximate function. The roots of such procedures are the rules of Johannes Kepler, Thomas Simpson and Newton. In 1814 Gauss generalized these methods by suitable choice of different distances of the nodes, and in 1826 Carl Jacobi devoted a deep study to the superiority of this Gaussian method over increase in the number of equidistant nodes (Kowalewski *1917*).

4 CALCULUS OF OBSERVATIONS: CALCULUS OF ERRORS

Today we know that Gauss possessed his method of least squares from about 1795 (§10.5), although he published it for the first time in 1809 in *Theoria motus corporum*. Legendre had the priority with his 'méthode des moindres carrés' in *Nouvelles méthodes pour la détermination des orbites des comètes* (1806), a priority which Gauss explicitly acknowledged. It is remarkable (and hitherto not sufficiently clarified in its relation to the publications of Legendre and Gauss) that, as early as 1803 in the English journal *Leybourn's Mathematical Repository*, there appeared the geodetical problem of finding that point for which the sum of the squares of its distances from three given straight lines is a minimum. (Later this minimum point became well known as 'Lemoine's point'; see Mackay *1892*.) This problem amounts to finding the destination of the locus (e.g. of a ship) by compensation of the errors of three astronomical observations, each giving an (approximately straight) line for the desired locus.

The titles of these works by Gauss and Legendre clearly tell us the motive of compensating observational errors. In 1821 and 1823 Gauss again treated his method, this time in connection with his geodetical work. In 1806 Legendre justified his method only by the remark that also the arithmetic mean m of numbers $a_1, \ldots, a_n$ minimizes $\Sigma_i (m - a_i)^2$. In 1809 Gauss gave a derivation from the hypothesis of normal distributed errors. Laplace then justified this hypothesis by the central limit theorem. A new probabilistic justification (the one usual today) was given by Andrei Markov in 1898 (§10.6).

The theory of probable errors from these roots was developed by Gauss, Pierre Simon Laplace, Chebyshev and their students in connection with general probability theory; however, we have insufficient historical insight into the development of the theory of maximal errors, which in principle lies at the root of modern interval mathematics. From the idea that

$$f(M_1, \ldots, M_n) = \{f(x_1, \ldots, x_n) \,|\, x_1 \in M_1, \ldots, x_n \in M_n\} = M_0, \quad (11)$$

this subject takes the sets $M_0, M_1, \ldots, M_n$ of real numbers or other mathematical objects which are described by a finite number of parameters, and investigates how the parameters of M_0 depend on the parameters of $M_1, \ldots, M_n$. In the simplest case, which gives its name to the topic, the sets

M_i are intervals of real numbers, their parameters are the lower and upper bounds, and f is a simple arithmetic function (e.g. addition or multiplication). If M_1 and M_2 are given by mean values x and y, and error bounds Δx and Δy as $M_1 = [x - \Delta x, x + \Delta x]$ and $M_2 = [y - \Delta y, y + \Delta y]$, then

$\Delta(x + y) = \Delta x + \Delta y$, i.e.

$$M_1 + M_2 = [x + y - (\Delta x + \Delta y), x + y + (\Delta x + \Delta y)]. \quad (12)$$

Up to the time of Lagrange, there seems to have been little interest in exact boundaries for the difference between exact and approximate values. Instead of the notions of 'δ' and 'ε', there was mostly only a good feeling for the rapidity of approximations and the goodness of values. First investigations in the direction of a general error propagation law

$$\Delta f(x_1, \ldots, x_n) = \sum_{i=1}^{n} \left| \frac{\partial f}{\partial x_i} \right| \Delta x_i \quad (13)$$

came from Roger Cotes (in a posthumous work, 1722), Nicolas Louis de Lacaille (1741), Pierre Bouguer (1749), J. J. Marioni (1751) and others, and were confined to the special case of the dependence of a side or an angle of a triangle on the other sides and angles (Kaunzner *1987*).

Explicit rules for rounding and the application of a linear Taylor approximation of functions of several variables to the law (13) were given by Gauss in his *Theoria motus corporum* (1809) (Goldstine *1980*: 258–60). F. Wolff in *Theoretisch-practische Zahlenlehre* (1832–9, 2 vols) and J. M. L. Vieille in *Théorie générale des approximations numériques* (1852) seem to have been the first to discuss the inverse problem: how to designate Δx_i such that $\Delta f < \varepsilon$ for a given ε. Lagrange's estimation of the remainder in the Taylor series is in his *Théorie des fonctions analytiques* (1797), but not very clearly (§3.2). It was followed by the more useful remainder formulas of Cauchy (1823) and the more general ones of E. A. Roche and O. Schlömilch (1847). Chebyshev then founded in Saint Petersburg a remarkable school of 'approximation mathematics' with many prominent followers, for example S. N. Bernstein and A. N. Krylov (§11.6). Krylov developed the applications of the error propagation theory in physics and technology, and showed the role of the physical units of measures in appropriate rounding and estimating errors.

Seemingly without direct forerunners, Felix Klein in his *Anwendung der Differential- und Integralrechnung auf Geometrie* (*1902*) developed a remarkable point of view: namely, that there is an exact (pure) and an approximative mathematics, and only the latter is directly useful in practice. The propositions of exact mathematics have an indirect usefulness, being 'abbreviations' of propositions of approximative mathematics. For example, the proposition 'sin x is a continuous function' leads to the more useful proposition: that 'for given x_0 and ε, there is a δ such that

$$|\sin x - \sin x_0| < \varepsilon \quad \text{if} \quad |x - x_0| < \delta. \tag{14}$$

5 NUMERICAL SOLUTION OF DIFFERENTIAL EQUATIONS

The nineteenth century saw significant advances in the theory of differential equations (especially the formulation and proof of existence theorems) as well as in the mathematical modelling of physical laws and processes, but only a small number of numerical methods. For initial-value problems the dominant technique was the method of broken lines, which in principle dates back to Leonhard Euler: from the given equation $y' = f(x, y)$ and the given initial values x_0 and y_0, one goes forward a small step to the 'new initial values' $x_1 = x_0 + h$ and $y_1 = y_0 + hf(x_0, y_0)$, and so on. This method was improved by Cauchy, who introduced around 1824 (§3.14) the iterative correction of solutions, so that the whole problem now is often named 'Cauchy's problem': if $y = g_n(x)$ is an approximate solution of $y' = f(x, y)$ with $g_n(x_0) = y_0$, the next approximation is

$$g_{n+1}(x) = y_0 + \int_{x_0}^{x} f(x, g_n(x)) \, dx. \tag{15}$$

Independently of Cauchy, Joseph Liouville in 1837 and Rudolf Lipschitz in 1868 rediscovered Cauchy's methods. New ideas in this field came later from F. Bashforth and J. C. Adams (1883) and then from Runge (1895), Wilhelm Kutta (1901) and E. J. Nyström (1925). The well-known Runge–Kutta procedure is based on the idea of Simpson's quadrature rule: to calculate the increment $k(h)$ of y resulting from the increment h of x, the first coefficients of the approximate solution are compared with the coefficients of the Taylor series for $y(x_0 + h)$.

In the domain of boundary-value problems, the substitution of the given differential equation by a corresponding difference equation ruled until 1908, when W. Ritz invented the method of setting up the solution as a linear combination of suitable fundamental solutions.

6 DEVELOPMENT OF REMEDIES

Calculating machines (§5.11), nomography (§4.12) and other graphical methods, tables of functions and also of simple operations like multiplication, of decomposition of numbers in primes, and so on, attained a rapidly growing importance for the practice (and also for parts of the theory) of numerical mathematics within the period treated here. Those concerned with computing by machines saw these useful methods as not primarily in the enormous raising of productivity and speed (as we do today), but in the safeguards they afforded against human error. Babbage was stimulated to the first thoughts on programmable computers by the many errors in mathematical and nautical tables of his time. From as recently as 1924, we may read: 'However, the big advantage in using calculating machines is not in the accuracy but in the exclusion of errors, assuming a good construction' (Runge and König *1924*: 22).

The introduction of a decimal measurement of angles after the French Revolution initiated the great project of the fresh computation of all logarithmic–trigonometric tables in France, directed by de Prony. In its organization, he was influenced by the British economist Adam Smith in dividing the enormous amount of work among many people, and in turn he influenced Babbage to the last step, the automation of the government process (Grattan-Guinness *1990*). However, the whole work of 17 volumes was never published completely. By contrast, the logarithmic–trigonometric manual of the Austrian officer G. von Vega, first issued in 1793, ran through some ninety editions to the end of the nineteenth century; and the computing tables (only for multiplication and division) edited by the German engineer August Leopold Crelle came out in eight editions. In most textbooks on numerical mathematics from the beginning of the twentieth century we find chapters on the appropriate design and the correct use of tables (see e.g. Bruns *1903*). Prominent mathematicians were very interested in this 'profane' side of mathematics; Gauss carefully reviewed the logarithmic tables edited by F. Callet (1802). M. von Prasse (1881), J. Pasquich (1817), Babbage (1828), F. K. Hassler (1831) and von Vega (1840, 1851) (Gauss, *Werke*, Vol. 3).

BIBLIOGRAPHY

Note the special entry under '*Encyklopädie*'.

Biermann, O. *1905*, *Vorlesungen über mathematische Näherungsmethoden*, Braunschweig: Vieweg.
Braunmühl, A. von *1908*, 'Trigonometrie, Polygonometrie und Tafeln', in M. Cantor (ed.), *Vorlesungen über Geschichte der Mathematik*, Vol. 4, Leipzig: Teubner, 403–50.
Brezinski, C. *1990*, *History of Continued Fractions and Padé Approximants*, New York: Springer.
Bruns, H. *1903*, *Grundlinien des wissenschaftlichen Rechnens*, Leipzig: Teubner.
Cajori, F. *1911*, 'Horner's method of approximation anticipated by Ruffini', *Bulletin of the American Mathematical Society*, **17**, 409–14.

Encyklopädie der mathematischen Wissenschaften articles:

Bauschinger, J. *1901a*, 'Interpolation', Vol. 1, Part 2, 799–820 (article I D 3).

—— *1901b*, 'Ausgleichsrechnung', Vol. 1, Part 2, 769–98 (article I D 2).

Burkhardt, H. *1904*, 'Trigonometrische Interpolation', Vol. 2, Part A, 643–94 (article II A 9a).

Mehmke, R. *1902*, 'Numerisches Rechnen', Vol. 1, Part 2, 938–1079 (article I F 1). [Extensively revised by M. d'Ocagne (1911) in *Encyclopédie des sciences mathématiques*, Tome 1, Vol. 4, 196–452 (article I 23).]

Runge, C. *1898*, 'Separation und Approximation der Wurzeln', Vol. 1, Part 1, 405–48 (article I B 3a).

Goldstine, H. H. *1977, A History of Numerical Analysis from the 16th through the 19th Century*, New York: Springer.

Graeffe, C. H. *1837, Die Auflösung der höheren numerischen Gleichungen*, Zurich.

Grattan-Guinness, I. *1990*, 'Work for the hairdressers: The production of de Prony's logarithmic and trigonometric tables', *Annals of the History of Computing*, **12**, 177–85.

Heideman, M. T., Johnson, D. H. and Burrus, C. S. *1985*, 'Gauss and the history of the fast Fourier transform', *Archive for History of Exact Sciences*, **34**, 265–77.

Heinrich, H. *1978*, 'Über Gauss' Beiträge zur numerischen Mathematik', *Abhandlungen der Akademie der DDR (Berlin), Mathematik-Naturwissenschaft-Technik*, **3**, 109–22.

Kaunzner, W. *1987*, 'Über eine Entwicklung in der Fehlerrechnung', *Österreichische Akademie der Wissenschaften, Mathematisch-naturwissenschaftliche Klasse*, **196**, 435–61.

Klein, F. *1902, Anwendung der Differential- und Integralrechnung auf Geometrie*, lithograph. [2nd edn, 1907, Leipzig: Teubner.]

Kowalewski, A. *1917, Newton, Cotes, Gauss, Jacobi. Vier grundlegende Abhandlungen über Interpolation und genäherte Quadratur*, Leipzig; Veit. [There is another book with a very similar title by G. Kowalewski, brother of A. Kowalewski.]

Mackay, J. S. *1892*, 'Early history of the symmedian point', *Proceedings of the Edinburgh Mathematical Society*, **9**, 92–103.

Richenhagen, G. *1985, Carl Runge (1856–1927): von der reinen Mathematik zur Numerik*, Göttingen: Vandenhoeck und Ruprecht.

Runge, C. and König, H. *1924, Vorlesungen über numerisches Rechnen*, Berlin: Springer.

Part 5
Logics, Set Theories and the Foundations of Mathematics

5.0

Introduction

The Part on foundational aspects of mathematics is placed here because, as emerges in §5.2–5.6, the dominating questions arose largely from problems in mathematical analysis, especially of the kind discussed in §3.3 and §3.6. (The important topics of (in)completeness are handled mainly in §5.5 and §5.8.) However, Boolean algebra and its consequences, which took their main inspiration from algebra, have priority, as is acknowledged by their location at §5.1.

All these articles are oriented towards the history of modern mathematicized logics and of foundational questions in mathematics, as they started to be studied in the mid-nineteenth century. The controversies of logicism versus formalism versus intuitionism of the early twentieth century, and the important Polish contributions, arise in §5.2–5.8; the current situation in the philosophy of mathematics is reviewed in §5.9. Next come rather broader philosophical questions, with discussion of algorithmic thought. The articles on calculating machines and (the early types of) computer

(§5.11–5.12) are placed here and in this Part, because they should go together and the links between computing and metamathematics are rather close.

Logic is of course a very ancient subject, although before the development covered in §5.1 its connections with mathematics were too slender for their (fascinating) logical and philosophical contexts to be explored here. Similarly, a variety of notations were brought into logical theories and schemes of various kinds (Cajori *1929*: 280–314), but they are not discussed here.

The articles in this Part are more heterogeneous than is usual in this encyclopedia. One reason is that the philosophical component is larger than for any other Part (although most other Parts have articles of philosophical import for their own contexts), and a wide range of issues is thereby engaged. In addition, the presence of philosophy means that a greater proportion of modern considerations is described here than in other Parts, which also adds to the variety.

This point has more consequences than may be immediately apparent. One might suppose that mathematics and logic have always been closely linked, but in fact the connections (of various kinds) noted here are exceptions rather than the rule in the practice of mathematics and of logic (for examples, see Grattan-Guinness *1988*). Thus, even the most 'rigorous' of modern mathematics is often pretty sloppy to a modern logician (Corcoran *1973*). This Part of the encyclopedia could have didactic as well as nostalgic consequences for the mathematician-reader.

Some histories of logic have been written, usually not restricted to its mathematicized kinds; Kneale and Kneale *1964* is a well-known example, while van Heijenoort *1967* is a valuable source-book for mathematical logic. Fraenkel *1953* is graced not only with fine discussions of set theory, but also with an outstanding bibliography.

Rand *1905* is a sadly neglected bibliography of philosophy and related subjects, including a good section on logic. Risse *1965–79* is an eclectic but more general collection of bibliographies on logics (of all brands). Among journals, it is worth mentioning the review section of the *Journal of Symbolic Logic*, *History and Philosophy of Logic* and *Noûs* for the philosophy of mathematics (usually in the modern sense, reviewed in §5.9). In addition, the *Annals for the History of Computing* regularly contains articles relating to logic (and also to numerical methods related to Part 4).

BIBLIOGRAPHY

Burkhardt, H. and Smith, B. (eds) *1991, Handbook of Metaphysics and Ontology*, 2 vols, Munich: Philosophia. [Many pertinent bibliographical or topical entries.]

Cajori, F. *1929, A History of Mathematical Notations*, Vol. 2, Chicago, IL: Open Court.

Corcoran, J. *1973*, 'Gaps between logical theory and mathematical practice', in M. Bunge (ed.), *The Methodological Unity of Science*, Dordrecht: Reidel, 23–50.

Fraenkel, A. A. *1953, Abstract Set Theory*, Amsterdam: North-Holland.

Grattan-Guinness, I. *1988*, 'Living together and living apart: On the interactions between mathematics and logics from the French Revolution to the First World War', *South African Journal of Philosophy*, 7, (2), 73–82.

Kneale, W. and Kneale, M. *1964, The Development of Logic*, Oxford: Clarendon Press.

Rand, B. (ed.) *1905, Bibliography of Philosophy, Psychology and Cognate Subjects*, 2 parts, New York: Macmillan. [Constitutes Vol. 3 of J. M. Baldwin, *Dictionary of Philosophy and Psychology*. Some reprints.]

Risse, G. *1965–79, Bibliographica logica*, 3 vols, Hildesheim: Olms.

van Heijenoort, J. (ed.) *1967, From Frege to Gödel. A Source Book in Mathematical Logic*, Cambridge, MA: Harvard University Press.

5.1

Algebraic logic from Boole to Schröder, 1840–1900

NATHAN HOUSER

1 INTRODUCTION

The original algebraic logic, usually said to be the first modern logic, was George Boole's algebra of logic, which evolved into Boolean algebra as we know it today. Boole held that traditional logic, developed from Aristotle's theory of syllogism and assumed by such an eminent thinker as Immanuel Kant to be final, was inadequate as a general theory of inference: it was too incomplete and not sufficiently fundamental. Influenced by developments towards greater abstraction in algebra and geometry, Boole supposed that all deductive processes must instantiate a common formal structure. A few others before Boole, most notably Gottfried Wilhelm Leibniz, had entertained a similar thought, but it was left to Boole to be the first to have the simple yet profound insight that, by limiting ordinary quantitative algebra to two values, 1 and 0, he could come very close to obtaining a fully fledged algebra of logic.

Modifications of Boolean algebra, primarily by William Stanley Jevons, Charles Sanders Peirce and Ernst Schröder, transformed it first into a more felicitous model for the logic of classes and also of terms, then into a fully working predicate/function logic with quantifiers, and finally into (or at least well along the way towards becoming) a logic of mathematics. The influence of algebraic logic reached its zenith with Schröder's monumental three-volume *Vorlesungen über die Algebra der Logik* (*1890–1905*). At the end of the nineteenth century, Alfred North Whitehead developed his universal algebra, with which he hoped to establish algebraic logic as a branch of mathematics by identifying its abstract algebraic structure with that of other special algebras, including William Rowan Hamilton's quaternions and Hermann Grassmann's calculus of extension (§6.2). In effect, this was a return to Boole's own idea of developing his algebra as an uninterpreted system. In this century there have been a number of contributors to the development of algebraic logic, most notably Alfred Tarski.

2 PRECURSORS

Although Boole should be given full credit for the creative insight that set logic on its modern course, he was acquainted to a degree with the work of others who sought to extend the range of syllogistic logic and to put it on a more mathematical foundation. This is borne out by the many references in Boole's writings to the work of other innovators, including even Chrysippus, the third-century BC Stoic logician whose work represents the culmination of the Megarian–Stoic school(s) of logic. Bochenski *1951* credits that ancient school, the rival of Aristotle's Academy, with several marked achievements, including the development of a complete and consistent axiomatic system of propositional logic, the invention of truth tables and the investigation of truth-functional connectives (including a thorough discussion of the meaning of implication), the formulation of a sharp distinction between logical laws and metalogical rules of inference, and the recognition of the difference between intension and extension.

The rise of experimental science in the sixteenth and seventeenth centuries signalled the decline of the almost two-thousand-year reign of Aristotelian (syllogistic) logic. The astonishing growth of knowledge that resulted from the application of experimental method called for a radical expansion of logic to encompass inductive and extended deductive reasoning. To some extent the key to this transformation of logic was foreshadowed by the emphasis of Johannes Kepler and Galileo Galilei on the importance of mathematics and mathematical models. But it was perhaps due more to the urging of Francis Bacon and the successes of René Descartes and Isaac Newton that traditional logic came to be regarded as decidedly inadequate. Modern logic was born of the marriage of logic and mathematics.

The first 'modern' logician to contribute to the extension of logic towards its eventual algebraization was Leibniz. He forged an intimate linkage between logic and mathematics and noticed, probably for the first time, the benefits of representing logical expressions algebraically. He was aware of an important peculiarity about his algebraically expressed logic: 'we ... take no account here of repetition; AA is the same as A for us' (Leibniz *1976*). This principle would later be put forward by Boole as the distinguishing difference between the algebra of logic and ordinary numerical algebra. Leibniz envisaged a universal language based on a common alphabet of thought (the *characteristica universalis*) and a general calculus of reasoning (the *calculus ratiocinator*). A key idea was that logical symbols can, and should, be treated formally, without recourse to special interpretations.

Between Leibniz and Boole there were a number of mathematicians and logicians who made important contributions to the groundwork, if not the theory, of algebraic logic, although frequently in indirect ways. In the early nineteenth century, algebraists such as George Peacock, François-Joseph Servois and William Rowan Hamilton among others, clearly set the stage for Boole (§4.7, §6.2), but there is no indication that part of their purpose was to provide a model for deductive logic. At the same time, logicians such as Richard Whately and Sir William Hamilton launched powerful programmes to expand syllogistic logic (Van Evra *1977*), and even though they did not seek to establish logic on a strictly mathematical basis, they built bridges between logic and mathematics. The latter was influential in the continuing formalization of logic and promoted the conception of logic as a theory of laws of thought.

Another major figure in the modernization of logic was Augustus De Morgan. Some of his results, in his *Formal Logic* (1847) and a sequence of papers, merely prefigured more effective treatment at the hands of Boole, but others dealt with matters that would have to wait for Peirce and Schröder for fuller treatment. Such, in particular, was De Morgan's remarkable creation in 1860 of the logic of relations and relative terms, which helped pave the way for quantification theory and important advances in the foundations of arithmetic, but which only through Boole's followers became part of algebraic logic (Merrill *1990*).

Another important contribution of De Morgan was his 'generalized copula', which he represented as transitive, commutative and reflexive (in effect defining an equivalence relation). Such refinements helped prepare the way for the mathematicization of logic.

Some of De Morgan's other logical extensions and innovations include his discovery and discussion of various non-syllogistic forms of inference, the development of the theory of the quantified predicate (introduced first by George Bentham in 1827, but also discussed by Sir William Hamilton), the isolation of the logical relations between compounds and aggregates now known as 'De Morgan's laws', the isolation of the extended set of compound propositions known as 'De Morgan's eight', and the introduction of the concept of a universe of discourse.

Among the non-syllogistic forms of reasoning discussed by De Morgan, we find his famous example:

Man is an animal,
therefore, the head of a man is the head of an animal;

his numerically definite syllogism:

At least three-quarters of a company wear hats,
and at least three-quarters wear low shoes,
therefore, at least two-thirds of those of the company who wear low shoes
wear hats;

and his important syllogism of transposed quantity:

Some X is Y,
for every X there is something neither Y nor Z,
therefore, something is neither X nor Z.

The principal significance of De Morgan's syllogism of transposed quantity
is that it is valid only when the referent universe is finite, something Peirce
would emphasize some years later.

De Morgan's eight compound propositions can be represented in several ways, but in one of his later notations they appear as follows (with equivalent expressions in words):

A. X))Y   Every X is Y.
a. X((Y   Every Y is X.
E. X)·(Y   No X is Y.
e. X(·)Y   Everything is X or Y or both.
I. X()Y   Some Xs are Ys.
i. X)(Y   Some things are neither Xs nor Ys.
O. X(·(Y   Some Xs are not Ys.
o. X)·)Y   Some Ys are not Xs.

In all of these ways, he extended and improved the logic of his predecessors. Most of his extensions had a generalizing effect and helped prepare logicians to conceive of logic as essentially mathematical.

3 BOOLE'S ALGEBRA OF LOGIC

Boole, a man well known for his independence of mind, saw in mathematics the possibility for far more than the mere enhancement of traditional logic. He envisaged a new comprehensive logic, modelled on algebra, which would subsume the logic of syllogism as a minor part. Boole's many contributions, together with those of De Morgan, but especially with the expansions and developments of his followers, have revealed the correctness of his view.

Boole's results were principally set out in two books: *The Mathematical Analysis of Logic* (1847) and *An Investigation of the Laws of Thought* (1854). He regarded logic as a normative science of reasoning, and was much influenced by the operator methods of solving differential equations (§4.7). He sought to isolate algebraic laws that would model the mental operations involved in sound reasoning. Although his approach was non-axiomatic, it was essentially a development from a few basic principles.

The basis of Boole's algebra of logic consists of lower-case letters (e.g. x, y, z), which he called 'literal symbols'; three binary signs of operation, $+$, $\times$ and $-$; the sign of identity, $=$; and two constants, 0 and 1. In practice, contiguity is taken as a sign synonomous with $\times$. In Boole's basic (uninterpreted) algebra of logic, which is merely a restricted part of ordinary quantitative algebra, all these signs take their standard meanings, but within a context of only two values. As a binary operation, $-$ represents what is usually called the operation of exception, but sometimes Boole seemed to use it as a unary operator to represent complementation or negation, and exception can be regarded as the compound operation 'and not', where the unary $-$ is assumed to be preceded by $\times$. He also introduced a fourth operation into his calculus, division, but he declined to use $\div$ as a sign in his basic alphabet. This was because formulas involving division were not interpretable and were only used in the process of working towards expressions restricted to the basic alphabet.

Boole isolated the following equivalences as having special importance for logic:

1 $xy = yx$.
2 $x^2 = x$.
3 $x + y = y + x$.
4 $z(x + y) = zx + zy$.
5 $x - y = -y + x$.
6 $z(x - y) = zx - zy$.
7 If $x = y + z$, then $x - z = y$.
8 If $x = y$, then $zx = zy$; $z + x = z + y$; $x - z = y - z$.

1, 2, and 4 are the principal laws. All these equivalences — except for the second, $x^2 = x$, which Boole called the fundamental law of duality or the index law — apply to ordinary numerical algebra. But $x^2 = x$ is special, as Leibniz had noticed, and applies to algebra only when it is limited to two values, 1 and 0 (or 1 and ∞, as Peirce would later point out). Another way to represent Boole's laws is to use subtraction to move all the letters to one side of the equals sign with the resulting expressions set equal to 0. The index law, for example, can be expressed as $x - x^2 = 0$ or as $x(1 - x) = 0$.

Interpreted as a logic of classes (what Boole called the logic of primary propositions, i.e. propositions relating to things), the literal symbols stand for simple names of qualities, the constants 0 and 1 represent the empty class and the universe (the two limits of class extension), and the operators represent class union, class complementation and exclusive class intersection. The equals sign stands for identity. In addition, in his logic of primary propositions, he used a symbol, v, in conjunction with a literal symbol to represent an indefinite class about which we can say only that it contains *some* of its members; it therefore serves somewhat as an existential quantifier. In his early work Boole called his literal symbols 'elective symbols' and conceived of them (when not prefixed by v) as selecting (or denoting) all the members of a class.

Interpreted as a logic of propositions (what Boole calls the 'logic of secondary propositions', i.e. propositions relating to propositions), the literal symbols represent propositional variables (more precisely, the times when a designated proposition is true), the constants 0 and 1 represent 'never true' and 'always true', and the operators represent negation, conjunction and exclusive disjunction; the equals sign stands for equivalence. Boole introduced capital letters (e.g. X, Y, Z) to represent actual propositions. He employed his indefinite symbol, v, in representing what he called 'the eight fundamental types of proposition':

1 All Y's are X's, $y = vx.$
2 No Y's are X's, $y = v(1 - x).$
3 Some Y's are X's, $vy = vx.$
4 Some Y's are not-X's, $vy = v(1 - x).$
5 All not-Y's are X's, $1 - y = vx.$
6 No not-Y's are X's, $1 - y = v(1 - x).$
7 Some not-Y's are X's $v(1 - y) = vx.$
8 Some not-Y's are not-X's $v(1 - y) = v(1 - x).$

These can be seen to correspond (in a different order from that given above) to the De Morgan eight.

Boole also developed an interpretation of his algebra as a calculus of probability, where the literal symbols represent probabilities of events, and the constants 1 and 0 represent certainty and impossibility (Hailperin *1986*). In his day, this was regarded by some as his most innovative interpretation.

In working with his logical calculus in *The Laws of Thought* (1854), Boole set aside whatever interpretation was operative and applied ordinary algebraic procedures. 'We may in fact lay aside the logical interpretation of the symbols in the general equation; convert them into quantitative symbols, susceptible only of the values 0 and 1; perform upon them as such all the requisite processes of solution; and finally restore to them their logical interpretation.' Boole's general method consisted of three algebraic procedures: development, elimination and reduction. By development, expressions could be transformed into equivalent expressions but with the elements rearranged; by elimination, the number of terms in an expression could be reduced, thus yielding a more specific result; and by reduction, two or more expressions could be combined into a single compound expression which could then be analysed by development and elimination. It is easy to see that the method for drawing simple syllogistic conclusions was to reduce the premises to a compound expression and then use development and elimination until a desired conclusion was obtained.

4 JEVONS AND ALGEBRAIC LOGIC AFTER BOOLE

Algebraic logic evolved rapidly in the final third of the nineteenth century, especially with the work of Jevons, Peirce and Schröder. The general outline of this development was, first, the streamlining of algebraic logic to make it more fully interpretable as (isomorphic with) a system of logic structures, operations and procedures; second, the extension of algebraic logic to include relations; third, the focusing on mathematical reasoning both for analysis and for discovery. Another contributor was John Venn, whose diagrams extended those of Leonhard Euler and are now standard devices for modelling Boolean expressions. Venn was a close follower of Boole, insisting even on keeping Boole's exclusive addition; but while he helped to propagate algebraic logic, he contributed little to its development.

Jevons was the first to raise well-considered objections to Boole's system, although quite clearly with the intention of improving what he took to be the logic of the future (Grattan–Guinness *1991*). Jevons's most strenuous objection (*Pure Logic*, 1864) was to Boole's addition, which *excluded* whatever elements the addends had in common. Jevons was convinced that logic required that addition (disjunction/union) be *inclusive*, so that $a + a = a$ would count as a legitimate law of logic. This was an important modification to Boole's system for, even though Jevons may not have realized it (as Peirce and Schröder would), it opened the way to a thoroughgoing duality for the laws of logic (i.e. when for every law involving logical addition there is a corresponding law where every instance of addition is replaced by multiplication, and vice versa).

Jevons's more general objection was that Boole's logic simply was not the logic of common thought, and this led him not only to modify Boole's addition but also to reject Boole's subtraction and division, for which he found no analogue in logic. The principal thrust of Jevons's reaction to Boole was to weaken the analogy between logic and algebra. Jevons did not, however, believe that mathematics and logic were independent, rather that logic is the foundation for mathematics, a view soon to be given much weight by the logicists (§5.2).

Of the three major contributors to nineteenth-century post-Boolean algebraic logic, only Jevons opted primarily for an intensional interpretation of his logical terms. The expressions in his system were equations, and the main rule of inference was what he called 'the substitution of similars'. Jevons was the first to realize that a systematic and rule-governed Boolean approach to logic was consistent with its mechanization, and in 1869 he constructed a logic machine to perform simple Boolean operations. The idea of building logic machines seems to have especially interested the algebraic logicians. One of Peirce's students, Allan Marquand, soon constructed a slightly more advanced logic machine and, in 1886, Peirce outlined for the first time how electricity could be used for logical addition and multiplication.

5 PEIRCE

Charles Sanders Peirce, who was much influenced by his father, the American algebraist Benjamin Peirce, was the second logician to make significant modifications to Boole's algebra of logic. In 1868, in his first published writings on algebraic logic, Peirce recognized that logical operations should be regarded as distinct from their mathematical counterparts. At first he followed Boole in holding that it was legitimate to employ strictly mathematical operations in logic, so for a time his logic employed both mathematical and logical operations and procedures. But by 1870 Peirce had substituted inclusion (or predication) for equivalence (or identity) as the primary copula (primitive) of logical statements (he used $\prec$, a modified $\leq$, as his sign of inclusion), and by 1879 he agreed with Jevons that it had been a mistake to push the analogy with mathematics so far. Peirce believed that algebraic logic, for the logician, should be for the purpose of 'analyzing inferences and showing precisely upon what their validity depends'. This purpose was best served by weakening the logic–mathematics analogy. Peirce acknowledged that it might be different for the mathematician, for whom algebraic logic was mainly of value for solving problems and for finding the conclusions to be drawn from given premises.

Peirce's algebra of logic is in some respects closer to modern mathematical logic than to Boole's algebra of logic. In addition to his rejection of Boole's exclusive addition and of logically uninterpretable operations and procedures, his contributions include the discovery of full duality for algebraic logic and the first assertion of the full principle of distribution, an algebraic treatment of relations, the development out of algebraic logic of a systematic propositional logic with implication as the principal connective, an effective axiomatization of propositional logic, the demonstration that 'neither–nor' is a sufficient single connective for the classical propositional calculus, the introduction of truth values and truth-function analysis into algebraic logic, the discovery of the logical law known since Jan Łukasiewicz as Peirce's law $(((p \prec q) \prec p) \prec p)$, the development of quantification theory for algebraic logic, and his rather neglected treatment of the theory of individuals and of sets. Among Peirce's more mathematically motivated contributions are his perhaps inadvertent isolation, in 1880, of the basis for lattice theory and, in 1881, his successful axiomatization of the arithmetic of natural numbers.

In 1880 Peirce published a somewhat non-rigorous axiomatization of algebraic logic in which he gave the first definitions of logical addition and multiplication suitable for modern Boolean algebra. He defined these operations as follows:

If $a \prec x$ and $b \prec x$,	If $x \prec a$ and $x \prec b$,
then $a + b \prec x$;	then $x \prec a \times b$;
and conversely,	and conversely,
if $a + b \prec x$,	if $x \prec a \times b$,
then $a \prec x$ and $b \prec x$	then $x \prec a$ and $x \prec b$.

These definitions were included in Peirce's 1880 paper in a section on the logic of non-relative terms, which is often regarded as the first presentation of Boolean algebra in its modern form. In his classic monograph on lattice theory, Garrett Birkhoff *1948* gives the basic principles of lattice theory as deriving from this section of Peirce's paper.

The axioms of Peirce's full 1880 system, as isolated by Prior *1964*, are:

1 $(b \prec c) \prec ((a \prec b) \prec (a \prec c))$.
2 $(a \prec (b \prec c)) \prec (b \prec (a \prec c))$.
3 $(a \prec a)$.
4 $(\bar{\bar{a}} \prec a)$.
5 $(0 \prec a)$.
6 $(a \prec c) \prec ((b \prec c) \prec ((a + b) \prec c))$.
7 $((a + b) \prec c) \prec (a \prec c)$.
8 $((a + b) \prec c) \prec (b \prec c)$.
9 $(a \prec b) \prec ((a \prec c) \prec (a \prec (b \times c)))$.
10 $(a \prec (b \prec c)) \prec ((a \times b) \prec c)$.

Prior has shown that with only slight enhancements this system gives a complete basis for the classical propositional calculus. It is somewhat uncertain, however, to what extent Peirce's 1880 logic should be regarded as a propositional calculus, for in Peirce's opinion his algebraic logic was highly abstract and could be interpreted as a class calculus, predicate calculus or propositional calculus. His primary connective $\prec$ was defined simply as a transitive and reflexive binary relation. It is noteworthy that Peirce was familiar with the work of Hugh MacColl, who had already (*1877–8*) begun to express the propositional calculus as a Boolean algebra.

Between 1880 and 1885 Peirce developed a conception of truth values (a sentence has the value **v** if it is true or **f** if it is false) and created a semantics for his algebraic logic. In 1885 he used his truth values in giving the first decision procedure for theoremhood. The first theorem proved by this method was $(((p \prec q) \prec p) \prec p)$, which (we now know) is needed as an axiom to make complete the propositional calculus without negation.

The axioms (sometimes excluding 4) for Peirce's 1885 first-order logic are:

1 $(p \prec p)$.
2 $(p \prec (q \prec r)) \prec (q \prec (p \prec r))$.
3 $(p \prec q) \prec ((q \prec r) \prec (p \prec r))$.
4 $(0 \prec p)$.
5 $(((p \prec q) \prec p) \prec p)$.

To this basis he added his quantifiers: '$\Sigma_i x_i$ means that x is true of some one of the individuals denoted by i or $\Sigma_i x_i = x_i + x_j + x_k +$ etc.'. In the same way, '$\Pi_i x_i$ means that x is true of all these individuals, or $\Pi_i x_i = x_i x_j x_k$, etc.'. Peirce also applied his quantifiers to relations (greater than monadic), and cautioned that quantified formulas are not sums and products, strictly speaking, 'because the individuals of the universe may be innumerable'. His 1885 system provides a complete quantification theory with identity (Zeman *1986*).

Peirce's theory of quantification was an important enhancement of Boole's algebra, transforming it into an algebraic logic far more suitable for a general theory of inference, but also more useful for mathematics. Another major expansion of Boole's algebra of logic was Peirce's, and then Schröder's, development of the algebraic logic of relations. The classic calculus of relations, according to Tarski *1941*, consists of the Boolean calculus of classes with the addition of the relative (or Peircean) constants: 1' (or I for the identity relation), 0' (or **T** for the diversity relation), ˘ (for the converse), † (or † for relative addition) and ; (or, simply, contiguity for relative multiplication). Peirce developed the logic of relations both as a basic calculus of relations and as a quantified theory of relations.

Peirce made a careful study of algebraic notations for logic, and concluded that a fully adequate notation must use three different kinds of sign: iconic signs, which represent relational forms; indexical signs, which designate individuals in a selected universe; and symbolic signs, which are general terms with agreed (or conventional) meanings. He studied the benefits of specially designed notations that instantiate the relational structures they represent, and found that where this is done it is possible to minimize the semantic content of transformation rules by developing purely mechanical rules to physically manipulate iconic signs. For example, he designed a notation for the 16 binary connectives that accommodates a purely mechanical procedure for identifying tautologies from the substitution sets of expressions with up to three term (proposition) variables and five connective variables (see G. Clark and S. Zellweger in Houser *et al.* *1994*). It is regrettable that Peirce left so much interesting work unpublished; his logical programme was carried on for a time, and in some areas extended, by a few students, most notably C. Ladd-Franklin and O. H. Mitchell.

6 SCHRÖDER AND THE CONTINUATION OF ALGEBRAIC LOGIC

Although Peirce advanced the theory of algebraic logic further than any other of its founders, Schröder is the one who brought algebraic logic to its historical zenith. First in his *Operationskreis des Logikkalkuls* (1877), but more emphatically in his *Vorlesungen über die Algebra der Logik* (*1890–1905*), he brought a systematic thoroughness and rigour to algebraic logic – which, unlike Boole and Peirce, he never extended beyond the sphere of deductive reasoning. In his *Operationskreis*, Schröder emphasized the duality of the properties of conjuction (intersection) and disjunction (union) by printing dual laws in parallel columns. He asserted the 'law of duality', which states that for any theorem, if 0's and 1's are interchanged, complemented and non-complemented terms are interchanged, and if signs of disjunction and conjunction are interchanged, then the resulting dual expression is also a theorem. Peirce had already noticed this but did not systematically employ his duality principle – extended to include inclusions – until nearly two years later, after having read Schröder's *Operationskreis*. In his *Vorlesungen*, Schröder gave a comprehensive and detailed account of algebraic logic which organized much of the best work up to that time and was expected to inaugurate a research programme perhaps comparable in scope to that launched in 1910 with Russell and Whitehead's *Principia mathematica* (§5.2).

Schröder's mature system (based on inclusion/subsumption, denoted by $\nleqslant$, and where addition is non-exclusive) may be characterized by the following set of axioms (as isolated by Dipert *1978*):

1 $A \nLeftarrow A$	Identity or tautology axiom.
2 $(A \nLeftarrow B)(B \nLeftarrow C) \nLeftarrow (A \nLeftarrow C)$	Subsumption rule.
3 $(A \nLeftarrow B)(B \nLeftarrow A) \nLeftarrow (A \nLeftarrow B)$	Equality definition.
4 $0 \nLeftarrow A$	Zero postulate.
5 $A \nLeftarrow 1$	One postulate.
6 $1 \nLeftarrow 0$	Existence postulate.
7 $(X \nLeftarrow A)(X \nLeftarrow B) = (X \nLeftarrow AB)$	Product definition.
8 $(A \nLeftarrow X)(B \nLeftarrow X) = (A + B \nLeftarrow X)$	Sum definition.
9 $(A + X)(\bar{A} + X) = X = AX + \bar{A}X$	Negation or distribution principle.

It should be noted that Schröder gave only one of these formulas explicitly as an axiom. Of the others, there are two sets of duals, 4 and 5, and 7 and 8, and duals are combined in 9.

The *Vorlesungen* is a massive work, completed after Schröder's death by a follower, E. Müller. It contained many techniques and proofs, and posed many questions that had a continuing influence, especially on logic in Germany, well into the twentieth century. As with Peirce, Schröder's contributions extend far beyond the realm of logic proper. For mathematicians, it is probably Schröder's development of his conception of *Dualgruppe* that stands out as having lasting importance, for it was with that work that he, rather than Peirce, initiated the movement that resulted in modern lattice theory (§6.4, Section 3). Schröder also made significant contributions to model theory, and some to set theory and the theory of individuals. He systematized and extended Peirce's theory of relations and, with his presentation in *Vorlesungen*, provided a basis similar to Russell's mathematical theory of relations, and also the source from which Tarski developed the modern algebraic theory (see Grattan-Guinness *1975* on Norbert Wiener's comparison of the two logics of relations).

Even though Schröder, far more than Peirce, kept in mind what contributions his algebraic logic might make to mathematics, he continued to regard logic as an objective theory of deductive inference. He may thus be said to have remained, along with Peirce, philosophical in his approach to algebraic logic. Those who followed Schröder, however, tended to focus on the mathematical character of algebraic logic almost exclusively. Whitehead, in *A Treatise on Universal Algebra* published in 1898 – before he gave up the algebraic tradition to join forces with Russell – explicitly sought to integrate algebraic logic, as an abstract algebra, with the rest of mathematics. Others who contributed to the reconsecration of algebraic logic as a branch of mathematics were E. V. Huntington, G. D. Birkhoff, H. M. Sheffer, R. A. Bernstein and C. I. Lewis.

It was mainly in Germany and Poland, but also in Russia, that the achievements of the algebraic logicians were studied and developed and, in some cases, integrated into modern mathematical logic. Leopold Löwenheim and Thoralf Skolem based their research on Schröder's *Vorlesungen*, and in Poland, algebraic logic, often directly from Peirce, was influential in the work of Łukasiewicz, Stanisław Leśniewski, Tarski and M. Wajsberg (§5.7). According to Hiż *1994*, Peirce's influence on logic in Poland was fairly wide-ranging and was felt even in such areas as the study of many-valued logics. In Russia, the chief representative of this line of development was P. Poretsky (Styazhkin *1969*). He is probably best known for the law named for him: $A = 0$ is equivalent to $T = A - T + (-AT)$. This allows for the introduction of an arbitrary term T into any expression (Lewis *1932*). But of all who followed in the algebraic tradition, it was Tarski more than anyone else who carried forward the legacy from Schröder to the present day, and who has been most instrumental in the resurgence of interest in this approach to logic.

In the twentieth century, Boolean algebra has been regarded principally as a logic of classes. As refined by Huntington, Birkhoff and others, Boolean algebra is usually characterized as a closed algebraic system of elements $A, B, C, \ldots$, and their complements, $\bar{A}, \bar{B}, \bar{C}, \ldots$, all of which are subject to the operations of union, symbolized by $\cup$ (corresponding to Boole's $+$) and intersection, symbolized by $\cap$ (corresponding to Boole's $\times$). Four dual postulates (or their equivalents) are commonly asserted:

1 $A \cup B = B \cup A$ and $A \cap B = B \cap A$.
2 $A \cup (B \cap C) = (A \cup B) \cap (A \cup C)$ and
  $A \cap (B \cup C) = (A \cap B) \cup (A \cap C)$.
3 $A \cup 0 = 0 \cup A = A$ and $A \cap 1 = 1 \cap A = A$.
4 $A \cup \bar{A} = 1$ and $A \cap \bar{A} = 0$.

In mathematics this system is called a Boolean lattice, and the resulting theory is called Boolean algebra.

7 ALGEBRAIC LOGIC AND MATHEMATICAL LOGIC

Quite obviously, algebraic logic is mathematical logic in one sense of that expression: it is logic treated by the mathematical method (Church *1956*). However, its precise relationship to mathematics is difficult to determine. Some of the difficulty stems from the origins of algebraic logic. It was Boole's brilliant idea to isolate a part of ordinary algebra that could serve as an algebra of logic, but from the beginning he understood that the resulting algebra could be studied either in the abstract as a special branch of mathematics, or under various interpretations as different kinds of logic.

Boole's main interest was the application of his algebra to logic but, unlike most of his followers (Venn excepted), he was reluctant to let the requirements of logic completely shape his algebra. He believed that universal interpretability was too great a restriction on his calculus. This liberal marriage of algebra and logic did not last long (although Peirce's 1870 paper on 'The logic of relations' is a monument to the free application of mathematics to logic); the process of refining Boole's algebra to fit logic more closely accelerated to such a degree with Boole's successors, especially Jevons, Peirce and Schröder, that Boolean algebra may almost be regarded as a branch of mathematics that evolved in the service of logic. But it can also be regarded as a foundational branch of mathematics that was discovered by logicians in the search for an abstract structure that could support and extend (and perhaps even reveal) their science. In this sense we might call algebraic logic the mathematics of logic. After Schröder, it was mainly as a branch of mathematics that algebraic logic survived.

It is sometimes supposed that algebraic logic may be distinguished from mathematical logic proper by its limitations, but when the full development of Peirce and Schröder and later algebraic logicians is taken into account that no longer seems feasible. Of course, there are differences which, from one side or the other, may be viewed as limitations or deficiencies. Algebraic logic tends to emphasize logical laws (especially the laws of commutativity, association and distributivity) and multiple interpretability, while mathematical logic emphasizes axioms and formal properties, such as minimality and completeness (§5.5). Duality plays an important role in algebraic logic, but not in mathematical logic. Also, the theory of collections that supports algebraic logic is generally a 'part–whole theory' as opposed to the Cantorean set theory of mathematical logic (Grattan-Guinness *1994*). But these comparisons are not decisive, for there are significant exceptions. For example, Peirce's set theory, though not Cantorean, was mathematically sophisticated and was developed with attention to Cantor's work, and both Peirce and Schröder advanced theories of individuals that in some respects were more advanced than competing theories in mathematical logic.

Probably a better approach to the distinction between these two traditions is historical. The mathematical roots of algebraic logic can be traced to Joseph Louis Lagrange (§3.2) and the Abbé de Condillac, and this tradition is marked by its emphasis on the theories of functions and differential operators. On the other hand, the mathematical roots of mathematical logic can be traced to Augustin Louis Cauchy and Karl Weierstrass, and this tradition is marked by its emphasis on mathematical analysis (Grattan-Guinness *1988*; see also §3.3).

Another way to distinguish these two logical traditions is by their different programmatic characters. Algebraic logic was never conceived of as a logic research programme primarily in the service of mathematics. It is true that both Peirce and Schröder were mindful of the contributions their work could make to mathematics, especially their logic of relations, and they made some outstanding contributions to the foundations of mathematics. But the benefits for mathematics were largely incidental to their main purpose, which continued to be the further development of the general theory of reasoning. Of course, one branch of the theory of reasoning is the theory of mathematical reasoning (reasoning *in* mathematics, or mathematical inference), which was of interest to most of the algebraic logicians. This part of logic may be regarded as the logic of mathematics. But the study of reasoning, whether in mathematics or in any other discipline, is the traditional mission of logic, and the purpose is to improve and extend the theory of inference, not to ground mathematics. Quite simply, algebraic logic is mathematics in the service of logic. Mathematical logic, on the other hand, the logic deriving from Gottlob Frege, Giuseppe Peano, Whitehead and Russell, was conceived of as a logic research programme primarily intended to establish and work out the foundations of mathematics. Mathematical logic, then, is logic in the service of mathematics, as we shall now see.

BIBLIOGRAPHY

Birkhoff, G. *1948*, *Lattice Theory*, Providence, RI: American Mathematical Society.

Bochenski, I. M. *1951*, *Ancient Formal Logic*, Amsterdam: North-Holland.

Couturat, L. *1905*, *L'Algèbre de la logique*, Paris: Gauthier-Villars. [2nd edn 1914. English transl. by L. G. Robinson, with preface by P. E. Jourdain, 1914, Chicago, IL: Open Court.]

Church, A. *1956*, *Introduction to Mathematical Logic*, Princeton, NJ: Princeton University Press.

Dipert, R. *1978*, 'Development and crisis in late Boolean logic: The deductive logics of Peirce, Jevons, and Schröder', Doctoral Dissertation, Indiana University.

Grassmann, R. *1872*, *Die Begriffslehre oder Logik*, Vol. 2 of *Formenlehre oder Mathematik*, Stettin: R. Grassmann.

Grattan-Guinness, I. *1975*, 'Wiener on the logics of Russell and Schröder [. . .]', *Annals of Science*, **32**, 103–32.

—— *1982*, 'Psychology in the foundations of logic and mathematics: The cases of Boole, Cantor and Brouwer', *History and Philosophy of Logic*, **3**, 33–53.

—— *1988*, 'Living together and living apart. On the interactions between mathematics and logics [. . .]', *South African Journal of Philosophy*, **7**, (2), 73–82.

—— *1991*, 'The correspondence between George Boole and Stanley Jevons, 1863–1864', *History and Philosophy of Logic*, **12**, 15–35.

—— *1994*, 'Peirce between logic and mathematics', in Houser *et al. 1994* (to appear).

Hailperin, T. *1986*, *Boole's Logic and Probability*, 2nd edn, Amsterdam: North-Holland.

Hamilton, Sir William *1859–60*, *Lectures on Metaphysics and Logic* (ed. H. L. Mansel and J. Veitch), 4 vols, Edinburgh and Boston: Gould & Lincoln.

Hiż, H. *1994*, 'Peirce's influence on logic in Poland', in Houser *et al*. *1994*.

Houser, N. with Roberts, D. D. and Van Evra, J. (eds) *1994*, *Studies in the Logic of Charles S. Peirce*, Bloomington, IN: Indiana University Press.

Huntington, E. V. *1904*, 'Sets of independent postulates for the algebra of logic', *Transactions of the American Mathematical Society*, 5, 109–115.

Johnson, W. E. *1892*, 'The logical calculus', *Mind*, n.s., 1, 3–30, 235–50, 340–57.

Kneale, W. and Kneale, M. *1962*, *The Development of Logic*, Oxford: Oxford University Press.

Leibniz, G. *1976*, 'A study in the logical calculus', in L. E. Loemker (ed.), *Philosophical Papers and Letters*, 2nd edn, Dordrecht: Reidel.

Lewis, C. I. *1918*, *A Survey of Symbolic Logic*, Berkeley, CA: University of California Press. [Revised edn 1960, New York: Dover.]

Lewis, C. I. and Langford, C. H. *1932*, *Symbolic Logic*, New York: The Century Co. [Revised edn 1959, New York: Dover.]

MacColl, H. *1877–8*, 'The calculus of equivalent statements', *Proceedings of the London Mathematical Society*, 9, 9–20, 177–86.

—— *1880*, 'Symbolic reasoning', *Mind*, n.s., 5, 45–60.

Merrill, D. D. *1990*, *Augustus De Morgan and the Logic of Relations*, Dordrecht: Kluwer.

Peckhaus, V. (ed.) *1990–91*, Schröder issue, *Modern Logic*, 1, 113–247. [Papers by I. Anellis, R. Dipert, N. Houser, Peckhaus and C. Thiel.]

Peirce, C. S. *1883*, *Studies in Logic, by Members of The Johns Hopkins University*, Boston, MA: Little, Brown. [Edited by Peirce with papers by him and his students: A. Marquand, C. Ladd(-Franklin), O. H. Mitchell and B. I. Gilman. Repr. 1983, Amsterdam and Philadelphia, PA: John Benjamins.]

—— *1932–3*, *Collected papers of Charles Sanders Peirce*, Vols 2–4 (ed. C. Hartshorne and P. Weiss), Cambridge, MA: Harvard University Press. [These three volumes contain the best known of Peirce's logic papers, but see also *The New Elements of Mathematics*, 4 vols (ed. C. Eisele), 1976, The Hague: Mouton.]

Prior, A. N. *1964*, 'The algebra of the copula', in E. Moore and R. Robin (eds), *Studies in the Philosophy of Charles Sanders Peirce*, Amherst, MA: The University of Massachusetts Press, 79–84. [See also A. Turquette in this volume.]

Schröder, E. *1890–1905*, *Vorlesungen über die Algebra der Logik*, 3 vols, Leipzig: Teubner. [Repr. 1966, New York: Chelsea.]

Styazhkin, N. I. *1969*, *From Leibniz to Peano: A Concise History of Mathematical Logic*, Cambridge, MA: MIT Press.

Tarski, A. *1941*, 'On the calculus of relations', *The Journal of Symbolic Logic*, 6, 73–89.

Van Evra, J. *1977*, 'A reassessment of George Boole's theory of logic', *Notre Dame Journal of Formal Logic*, **18**, 363–77.

—— *1984*, 'Richard Whately and the rise of modern logic', *History and Philosophy of Logic*, **5**, 1–18.

Venn, J. *1881*, *Symbolic Logic*, 1st edn, London: Macmillan. [2nd edn 1894. Repr. 1971, New York: Burt Franklin.]

Zeman, J. J. *1986*, 'Peirce's philosophy of logic', *Transactions of the Charles S. Peirce Society*, **22**, 1–22.

5.2

Mathematical logic and logicism from Peano to Quine, 1890–1940

F. A. RODRÍGUEZ-CONSUEGRA

1 INTRODUCTION

What is known today as mathematical logic is a very recent creation, which can be attributed to Giuseppe Peano, Gottlob Frege and Bertrand Russell. However, some other names stand in the background. First there is George Boole, who in 1847 introduced the first mathematical treatment of logic, by stating a symbolic calculus able to be interpreted in many ways, for instance in terms of propositions and classes. William Stanley Jevons simplified the calculus, and Hugh MacColl proposed the clear interpretation as a calculus of propositions in a series of writings from 1877 onwards. In America, Charles Sanders Peirce extended the resulting logic to relations (following Augustus De Morgan), and added other technical improvements. In Germany, Ernst Schröder began a great synthesis from 1890 (§5.1).

All of them regarded logic as standing in a certain relationship to mathematics, which was progressively clarified. For Boole, MacColl and Peirce we have to regard logic and mathematics as complementary, but for Schröder pure mathematics was a branch of logic in the sense that it can be developed without introducing particular categories or primitive notions (e.g. that of number). This position – that mathematics can be reduced to logic – is usually referred to as 'logicism', and can be traced to some extent in Peano and his school, and more deeply in Frege and Russell.

2 PEANO: THE CREATOR OF CONTEMPORARY MATHEMATICAL LOGIC

The title of this section is no exaggeration. Although Frege (discussed in the next section) is today admitted as having the priority for some principal concepts in contemporary mathematical logic, he exerted little influence on the actual development of the field. On the other hand, Peano's writings from 1889 to 1901 (appearing in the journal *Revue de Mathématiques* and the several collective volumes of *Formulaire de mathématiques*, both of which he edited), which made Russell's logic possible, are now very little known; thus the rather recent discovery of Frege's basic works, together with the traditional lack of interest of most scholars in reading Peano's works (written in Latin, French, Italian and 'Latino sine flexione'), constitutes a real historical paradox (Nidditch *1962*).

Peano's main goal was to achieve a system for rigorously expressing all propositions of mathematics in such a way as to divide them into two classes: primitive propositions (Pp's), which are to be accepted as indemonstrable axioms; and derivative propositions, which have to be shown to be logically deduced from the primitive ones. At the same time, the concepts or ideas involved in these propositions were to be divided into two classes as well: primitive ideas (Pi's), which are to be presented as the result of an analysis of the relevant ideas and are not to be defined; and derivative ideas, which have to be defined in terms of the primitive ones. This goal led him to the need for developing the instrument of mathematical logic, including a suitable and flexible notation, a logic of propositions and/or classes based upon a set of connectives and rules of inference, and a device for suitably representing mathematical existence and generality (the two quantifiers), together with other technical means later employed in axiomatizing arithmetic, analysis and geometry (which then constituted 'pure' mathematics).

The main devices to make possible the mutual conversion of the logics of propositions and classes were, first, a clear distinction between membership (ϵ) and inclusion ($\supset$; originally a reversed 'c') – the former has the transitive property, the second does not; and, second, the idea of 'such that' ($\ni$). Thus, for example, an operation between classes can be expressed in terms of an operation between propositions:

$$a \cap b = x \ni (x \in a \cdot x \in b), \tag{1}$$

which can be read as 'the intersection of classes a and b is equal to those x such that they belong to a and to b'. As for the duality class-proposition itself, derived from the double interpretation of the symbol $\supset$ as implication and inclusion, Peano never completely decided for any of them, although the propositional interpretation had more logicist possibilities, apart from other advantages (i.e. implication between propositions, $p \supset q$, is a proposition, while inclusion between classes, $a \supset b$, is not a class, but another proposition). Besides, the distinction between membership and inclusion can make the subsequent distinction between a class and the set of its members easier, making it possible to distinguish between an individual and the class of which it is the only member (the unit class), which is extremely important in mathematics (see Rodríguez-Consuegra *1991* for dates and details of these achievements and the following ones).

But there are two kinds of special proposition in mathematics, those expressing existence and generality, which require special treatment. The first one is easier: we need only a symbol to 'translate' the ordinary word. Thus, Peano expressed the existence of a class a as '$\exists a$', which is still the standard way to express existential quantification, and was interpreted as technically meaning that the class a has at least one member (i.e. it is not equal to the null class). The second is a little more difficult, but already had an important precedent: the conversion of general expressions, 'every a is b', into hypothetical ones, 'if x is an a, then x is a b', which was maintained by MacColl and Peirce as well.

Peano introduced universal quantification by combining membership, which was needed for the 'hypothesis' fixing the meaning of the variables, with subscripts in the signs for implication and identity (and dots instead of brackets):

$$p, q \in P. \supset : .p =_{x,y} q. = :p \supset_{x,y} q.q \supset_{x,y} p, \tag{2}$$

which can be read as 'p and q being propositions, then to say that p is equal to q, for all undetermined constituents x and y of them, is to say that p implies q and q implies p, for the same constituents'. Through these quantificational devices, Peano was also able to differentiate between bound and free variables ('apparent' and 'real' in his terminology) which, respectively, do not fall or do fall under a quantifier.

The above-mentioned distinction between the unit class and its only member gave rise to another important distinction, also in combination with membership, which was reached through an analysis of the symbol for identity:

$$(a = b) = (a \text{ is equal to } b) = (a \in \iota b). \tag{3}$$

Thus, the symbol ι immediately made it possible to distinguish between a class in itself (a) and the same class regarded as an individual (ιa), which one needs to do, for instance, in geometry: if a and b are straight lines, $a \cup b$ designates the whole set of points, whereas $\iota a \cup \iota b$ designates the pair of straight lines as individuals. In this way a new symbol of a function was obtained which, written before any individual x, gives rise to a class (ιx), the class of the individuals which are equal to x. Thus, the distinction between the 'is' of membership and the 'is' of identity becomes clear, and the definitions of the notions of individual and class become unnecessary: they will be, respectively, all that which appears to the left or to the right of ϵ. Finally, ιx was defined as the class formed by any *one* object $(\iota x = y \ni (y = x))$, and, through the idea of its converse, ιa was regarded as the only individual of the class $(x = \iota a \cdot = \cdot a = \iota x)$. The new symbol ι('the') made the definition of many others possible, especially through its capacity to isolate an entity.

One of the most famous applications of Peano's techniques was devoted to what is now known as 'Peano's postulates', which from 1889 onwards provided an axiomatization for arithmetic based upon three Pi's: zero (0), number (N) and successor (+), the notion of class, denoted by Cls, and five Pp's:

1 $0 \in N$.
2 $a \in N \cdot \supset \cdot a^+ \in N$.
3 $a, b \in N \cdot a^+ = b^+ \cdot \supset \cdot a = b$.
4 $a \in N \cdot \supset \cdot a^+ - = 0$.
5 $s \in Cls \cdot 0 \in s : x \in s \cdot \supset_x \cdot x^+ \in s : \supset \cdot N \supset s$.

These postulates, which had been introduced the year before by Richard Dedekind in a somewhat different way, are usually read as follows: (1) '0 is a number'; (2) 'the successor of any number is a number'; (3) 'no two numbers have the same successor'; (4) '0 is not the successor of any number'; and (5) 'any property that belongs to 0, and also to the successor of every number having the property, belongs to all numbers'. Also, Peano provided symbolic axiomatizations of several kinds of Euclidean geometry based on only a few Pi's and Pp's, which were improved by M. Pieri in 1898 (and later), starting before David Hilbert presented his famous work of 1899 in this field (which, by the way, was written without resorting to symbolic logic: §7.4).

As for logicism, it is true that in the above postulates Peano introduced three primitive (and then undefined) arithmetic ideas, and that he regarded number (N) as that which can be abstracted from his axioms (unlike Dedekind, who said that arithmetic can be reduced to logic). Thus, N could be defined only 'by abstraction' (i.e. by defining equality between numbers), and the same can be said for his geometric axiomatizations (e.g. by introducing vectors). However, Pieri had provided in 1898–9 a method for transforming definitions by abstraction into nominal definitions (which can explicitly define concepts); Cesare Burali-Forti had succeeded in nominally defining the arithmetic indefinables of Peano's five postulates in logical terms; and in 1901 Peano published a logical definition of the number of a class as the class of classes similar to that class. In addition, Peano had shown the possibilities of Moritz Pasch's methods of nominally constructing real numbers in terms of classes of rationals; and Pieri had already constructed several geometries combining only a few geometrical indefinables with logical ideas, by insisting that these indefinables had no special meaning, but were characterized only by the logical relations represented by the postulates. Thus, we can say that Peano and his school constituted the main source of Russell's logicism, for Russell came to Frege's works only after having developed his own logicist philosophy (see Section 4).

3 FREGE: THE GREAT UNKNOWN

Many of the ideas presented in the last section had been anticipated by Frege. His *Conceptual Notation* (1879) had provided the first systematization of logic (i.e. an axiomatic presentation), with the aim of preparing a later foundation for arithmetic, in the logicist sense. Likewise, his *Foundations of Arithmetic* (1884) provided an impressive definition of number in logical terms, after having criticized several empiricist, formalist and psychologistic approaches to mathematics. The definition was constructed in terms of properties of concepts rather than through classes. Thus, the number of a class was introduced as the number which applies to a given concept, and this last as the extension of the concept 'equinumerous with the given concept', which can be defined in terms of bijective correspondence between sets. Then the class of such numbers is introduced by means of a definition 'in use': to say 'n is a number' is to say 'there is a concept to which the number n can be applied'. This is more or less equivalent to Peano's definition of the number of a class as the class of all classes which can be put into bijective correspondence with a given class, but Russell did not know Frege when he formulated his famous definition following Peano.

The Fundamental Laws of Arithmetic (1893–1903) constituted the final stage of Frege's efforts to provide arithmetic with an axiomatic foundation from a logicist point of view – that is, by presenting a few logical ideas and logical axioms through which arithmetical ideas can be defined and from which arithmetical truths can be deduced by explicit rules of inference. This was supposed: (a) to improve the axiom system of 1879; (b) to clarify the quantificational devices; (c) to add rules to ensure that definitions are well formed; and (d) to provide a series of rigorous (and difficult) conceptual distinctions in an attempt to disentangle the philosophical foundations of logic. Among these distinctions we can mention the following pairs: object–function, aimed at providing an ontological division of all entities into two exclusive classes (things and concepts/relations); function–argument, which was destined to replace the old distinction between predicate and subject; and sense–reference, aimed at solving certain problems in the philosophy of logic, for instance the fact that 'the morning star' and 'the evening star' – two senses – both designate Venus – only one reference, which gave rise to many difficulties in (opaque) contexts involving belief and other propositional attitudes (Klemke *1979*).

In addition, Frege introduced a new symbol for a function to represent existence and uniqueness (\), similar to Peano's ι (see above), as well as the concept of 'course of values' of a function, which corresponded to classes and were intended to represent them through the notion of 'extension of a concept' (i.e. the set of objects falling under a given concept). The difficult notion of course of values was presented by means of an axiom stating that two courses of values are equal if and only if the corresponding functions coincide in their values for all their possible arguments. But, as Russell discovered in 1901, this axiom was vulnerable to his famous paradox of the class of all classes which are not members of themselves (see Section 4; and §5.3), which finally caused Frege to abandon mathematical logic as a suitable foundation for mathematics, after some attempts to save his axiom system. Besides, Frege's notation, in spite of its high precision, was constructed in two dimensions, which constituted a great disadvantage as compared with the linear one, employed by Peano, following the Boolean tradition. That, together with a difficult, extremely abstract and very specialized language made Frege's influence very limited in this crucial foundational period.

4 RUSSELL: THE CLIMAX OF LOGICISM

4.1 Russell's initial forays

Russell first attempted to found mathematics philosophically before he had any knowledge of Frege or even of Peano. His main goal at this period (1898–1900) was to analyse actual mathematics in search of the absolute and indefinable concepts which compose its main ideas, as well as of the indemonstrable propositions which would serve as a ground for the assertions of accepted mathematics. This seems to be close to the axiomatic approach, but Russell, following G. E. Moore, believed that the analysis of ordinary language, as employed to express mathematical propositions, had to provide the required material, which led him into endless complications. It was only at the Congress of Paris in the summer of 1900 that Russell met Peano and his school and became acquainted with his writings. Then he adopted their notation, their methods and their underlying logicist philosophy of mathematics, which he himself was destined to develop to its ultimate consequences.

The first step was constituted by *The Principles of Mathematics* (1903) where Russell expounded a logic based upon a series of Pi's and Pp's, which are extended until they cover the three fields of his logic: propositions, classes and relations. The main novelties can be found in the logic of relations, originally introduced in former papers by him (particularly in one of them published in 1901 in Peano's *Revue*). The list of Pi's is rather obscure, both because of the difficulty of fixing a clear set of components and because of the problems in understanding their meanings. One reason was that for Russell the Pi's included not only the standard logical constants, but also a series of ontological or linguistic notions. Thus, the official list includes implication (material – between propositions – and formal – between propositional functions), membership, 'such that', propositional function (a function whose argument is a subject and whose value is a proposition) and relation, which together were used to define all derivative ideas. But there were also Pi's such as truth, constancy of form, denotation, class, assertion and variable, which gave rise to a lot of unsolved problems. The list of Pp's is rather long, and was expressed without using symbolic logic: it contained ten for propositions (including the rules of inference, in the Peanian way), plus two for classes and six for relations.

The main problems that arose concerned denotation and class. The main difficulty with denotation was that the concept of variable was hardly distinguishable from the many ways in which a concept can ambiguously denote other concepts. Thus Peano's quantifiers were located within a list of six 'denoting concepts' (including the descriptor): *the, a, some, all, every* and *any*, whose respective status was hardly clarified (even taking into consideration the appendix devoted to Frege's doctrines). As for the problem with class, the reason was of course the paradox of the class of all classes not members of themselves. When we ask whether or not this class is a member of itself, each answer led to the contrary one (§5.3). This led Russell to collect other similar paradoxes but, although a second appendix was devoted to a first theory of types, which introduced a hierarchy of propositional functions according to the types of the 'range of significance' of the objects which can be taken as their arguments, no satisfactory solution was found to the numerous problems involved.

However, despite these obvious difficulties Russell managed to achieve his main objective: the construction of a logicist philosophy of mathematics through the reduction of mathematics to logic by means of nominal definitions of mathematical entities in logical terms (Rodríguez-Consuegra *1990b*). Thus, he first offered a natural transition from logic to arithmetic through the famous definition of number in logical terms (following Peano): the number of a class is the class of all classes similar to it; that is, those that can be put into bijective correspondence with the given class. Then he explained the transition to mathematical analysis through the reduction of reals to rationals (following Pasch and Peano): a real number is a class of rational numbers under certain conditions. Finally, he avoided the introduction of new primitive ideas in geometry (i.e. point) through the construction of geometrical spaces as general concepts based on certain relations and a set of axioms, which were regarded as properties of those concepts (following Pieri), in spite of the fact that these kinds of definition can hardly be regarded as nominal.

4.2 Russell's definitive position

After two preliminary contributions to these problems in 1905 and 1908, Russell wrote the three great volumes of *Principia mathematica* (1910–13) in collaboration with Alfred North Whitehead, his old master, who had joined the project in 1903. The first of these contributions helped to resolve the problem of denoting concepts. The new theory of descriptions permitted the elimination of the definite article, needed to individuate important mathematical entities, which can give rise to problems of 'apparent' existence. Thus, the celebrated sentence 'the present King of France is bald' can be reduced, by means of propositional functions, to 'there is an entity which is now King of France and is bald', which is false and no longer presents the descriptor − the symbol introduced by Peano − as a constituent. In symbols:

$$\psi(\imath x)(\phi x)\cdot = \cdot (\exists b):\phi x\cdot \equiv_x\cdot x = b:\psi b. \tag{4}$$

The other denoting concepts were then reduced to the two standard Peanian quantifiers, which were introduced through the corresponding Pi's. Also, the Fregean distinction between meaning and denotation (sense–reference) was rejected, under the argument that both notions are unable to be clearly separated. Ultimately, the theory also provided a method which was later applied to classes (and relations) as entities, which were finally eliminated through a similar procedure and declared to be 'incomplete symbols'. However, these eliminations were not enough to solve all paradoxes (Rodríguez-Consuegra *1990a*).

The second contribution provided a much more sophisticated theory of types, after several alternative theories were discarded, in order to solve the 'semantic' paradoxes as well, such as the paradox of the Liar who says 'I am lying' (§5.3). The main idea was the so-called vicious-circle principle (from Henri Poincaré), according to which we have to avoid the assumption of illegitimate totalities − those involving themselves as other members of a totality they try to define. The theory of logical types avoids such possibilities by requiring that a propositional function be regarded as belonging to a higher logical type than its arguments, a requirement which has to be applied to classes, predicates, relations and propositions as well. Thus the paradox of the class of all classes not members of themselves is removed because the proposition that a class is a member of itself is meaningless, and the paradox of the Liar disappears because the truth of the original sentence of the Liar has to be located in a higher level of language, from where it cannot be applied to itself (Cocchiarella *1980*, Rodríguez-Consuegra *1989*).

With these resources, the new edifice could already be built by following the axiomatic architecture. The new list of Pi's is now somewhat simplified: elementary proposition (p) and propositional function (ϕx), assertion ($\vdash$); negation ($\sim$); disjunction ($\vee$) and implication ($\supset$). As for the list of Pp's, after two of them stating that anything implied by a true proposition is true, and that, for real variables, the assertion of an implication between propositional functions and of its antecedent lead us to the assertion of the consequent, we find the following five (all them 'asserted'):

1 $p \vee p \cdot \supset \cdot p$.
2 $q \cdot \supset \cdot p \vee q$.
3 $p \vee q \cdot \supset \cdot q \vee p$.
4 $p \vee (q \vee r) \cdot \supset \cdot q \vee (p \vee r)$.
5 $q \supset r \cdot \supset : p \vee q \cdot \supset \cdot p \vee r$.

However, in later parts of the work a few additional primitive ideas and propositions appear: the two quantifiers, together with their corresponding propositions, the axiom of reducibility (see Section 5) and others.

Concerning logicism, the philosophical position is the same, and the level of sophistication in the constructions is now considerably higher, although great parts of the original construction of 1903 vanished, mainly most of the mathematical analysis and geometry (the latter intended to be developed in a fourth volume, which never appeared). However, philosophical problems were avoided, and the old Platonism that forced numbers and other mathematical objects to be regarded as genuine entities was also discarded in favour of the theory of descriptions. But the move towards a position where propositional functions were the main basis for the reduction of classes, relations, propositions and mathematical entities, all of which were declared to be 'incomplete symbols', was never clarified.

5 LIMITATIONS AND CRITICISMS

The logic of *Principia mathematica* was soon shown to be faulty: one of the five main Pp's was redundant; the theory of types was complicated and obscure; the status of classes and relations, together with the strange role of propositions, led to a rather peculiar view of the Pi's of logic. In addition, there was no clear distinction between Pp's and rules of inference, which went back to Peano and was caused by a lack of metalogical concepts (i.e. the possibility of taking into consideration the whole logic as an object of study). Regarded as a logicist work, *Principia mathematica* also had some flaws, particularly the admission of certain axioms which can hardly be regarded as logical, so that the reduction of mathematics to logic was rather doubtful. The most conspicuous of them was the axiom of reducibility, designed to reduce propositional functions of higher orders to 'predicative functions' (i.e. to functions whose arguments can be regarded as individuals), which was needed to ensure the general properties of numbers, such as mathematical induction.

Unfortunately the second edition of *Principia mathematica* (1925–7) did not solve these problems, despite Russell's attempt to take into consideration some of Ludwig Wittgenstein's criticisms in his *Tractatus* (1922), mainly in the sense of introducing a more extensional system which avoided the admission of entities not reducible to their constituents. Also, as Russell's student Frank Ramsey pointed out in 1926, there is a need to simplify Russell's theory of types by regarding semantic paradoxes (e.g. the Liar) rather as linguistic, for which we need only admit the 'simple' types – that is, those considering mathematical (set-theoretical) entities (Grattan-Guinness *1981*, *1984*). However, the possibility of completely abandoning logical types was later demonstrated by Willard Quine, the most distinguished contemporary follower of the spirit of Russell's logicism, who in 1937 invented a technique, stratification, with which paradoxes could be avoided (the main goal of the theory of types) by suitably indexing the various elements of a symbolic expression. In addition, Quine, among others, showed that the 'logic' of *Principia mathematica* included set theory as well, but that set theory has to be regarded as belonging to mathematics (§3.4), so that Russell's logicism was no longer defensible in the original sense.

However, the definitive proof of the untenability of logicism came from Kurt Gödel's incompleteness theorems (§5.5). One of the assumptions of the original logicist programme was to show that *all* mathematical truths could be deduced from the axioms of logic. But Gödel proved in 1931 that any complete description of mathematical truths is impossible: it is always possible to find at least one proposition which cannot be demonstrated despite being true in the system. (However, the possibility of maintaining a certain kind of logicism can still be discussed; see Sternfeld *1976*, Musgrave 1977, and Hellman *1981*.) Of course, Gödel's results affect any other foundation for mathematics which presupposes completeness. Strangely enough, this seems not to have been noted by most mathematicians, who have shown no interest in these limitations on their mathematical practice.

BIBLIOGRAPHY

Cocchiarella, N. *1980*, 'The development of the theory of logical types and the notion of a logical subject in Russell's early philosophy', *Synthèse*, **45**, 71–115.

—— *1989*, 'Russell's theory of logical types and the atomistic hierarchy of sentences', in C. Wade Savage and C. Anthony Anderson (eds), *Rereading Russell*, Minneapolis, MN: University of Minnesota Press, 41–62.

Gödel, K. *1944*, 'Russell's mathematical logic', in Schilpp *1944*, 123–54.

Grattan-Guinness, I. *1981*, 'On the development of logics between the two world wars', *American Mathematical Monthly*, **88**, 495–509.

—— *1984*, 'Notes on the fate of logicism from *Principia mathematica* to Gödel's incompletability theorem', *History and Philosophy of Logic*, **5**, 67–78.

Hellman, G. *1981*, 'How to Gödel a Frege–Russell: Gödel's incompleteness theorems and logicism', *Noûs*, **15**, 451–68.

Jourdain, P. E. B. *1910*, 'The development of theories of mathematical logic and the principles of mathematics', *Quarterly Journal of Pure and Applied Mathematics*, **41**, 324–52; **43** (1912), 219–314; **44** (1913), 113–28. [Also in *Selected essays* [...] (ed. I. Grattan-Guinness), 1991, Bologna: CLUEB, 101–244.]

Kennedy, H. *1980*, Peano. *Life and Works of Giuseppe Peano*, Dordrecht: Reidel.

Klemke, E. D. *1979*, 'Frege's philosophy of logic', *Revue Internationale de Philosophie*, **33**, 666–93.

Moore, G. H. *1988*, 'The emergence of first-order logic', in W. Aspray and P. Kitcher (eds), *History and Philosophy of Modern Mathematics* (Vol. XI of Minnesota Studies in the Philosophy of Science), Minneapolis, MN: University of Minnesota Press, 95–135.

Musgrave, A. *1977*, 'Logicism revisited', *British Journal for the Philosophy of Science*, **28**, 99–127.

Nidditch, P. H. *1962*, *The Development of Mathematical Logic*, London: Routledge & Kegan Paul.

Quine, W. V. *1963*, Set Theory and its Logic, Cambridge, MA: Harvard University Press. [2nd edn 1969.]

Reichenbach, H. *1944*, 'Bertrand Russell's logic', in Schilpp *1944*, 21–54.

Rodríguez-Consuegra, F. A. *1988*, 'Elementos logicistas en la obra de Peano y su escuela', *Mathesis*, **4**, 221–99.

—— *1989*, 'Russell's theory of types, 1901–1910: Its complex origins in the unpublished manuscripts', *History and Philosophy of Logic*, **10**, 131–64.

—— *1990a*, 'The origins of Russell's theory of descriptions according to the unpublished manuscripts', *Russell: The Journal of the Bertrand Russell Archives*, **9**, (2), 98–132.

—— *1990b*, 'El logicismo Russelliano: su significado filosófico', *Crítica*, **67**, 15–39.

—— *1991*, *The Mathematical Philosophy of Bertrand Russell: Origins and Development*, Basel: Birkhäuser.

Schilpp, P. A. (ed.) *1944*, *The Philosophy of Bertrand Russell*, La Salle, IL: Open Court.

Sternfeld, R. *1976*, 'The logicist thesis', in M. Schirn (ed.), *Studien zu Frege/Studies on Frege*, 3 vols, Stuttgart and Bad Canstatt: Fromman, Vol. 1, 139–60.

Tichý, P. *1988*, *The Foundations of Frege's Logic*, Berlin: de Gruyter.

Ullian, J. S. *1986*, 'Quine and the field of mathematical logic', in L. E. Hahn and P. A. Schilpp (eds), *The Philosophy of W. V. Quine*, La Salle, IL: Open Court, 569–89.

5.3

The set-theoretic paradoxes

ALEJANDRO R. GARCIADIEGO

1 CANTOR'S PARADOXES

Sometime between 1883 and 1899, Georg Cantor came across what we now call the paradoxes of the great cardinal and ordinal numbers. How, when and why did he discover them? Cantor himself claims that he had *intuitively* understood them since 1883. In spite of his own assertions, it is difficult to establish conclusively from the surviving sources the events that caused him to discover the paradoxes, his reasoning, or even the date of the discovery. First, based on cogent grounds, some historians have speculated that Cantor *formally* encountered the paradoxes while working on one of the following questions: (a) an attempt to prove that every cardinal number was an aleph; (b) an investigation of the comparability of the powers of all transfinite sets; or, perhaps, (c) an enquiry into the more general question of the well-ordering of all sets. Second, others have thought that the discovery of the paradoxes arose from Cantor's 1891 theorem asserting that the cardinality of a power set is always greater than the cardinality of the set itself. A third possible explanation rests on his composition of a clear and final presentation of his arithmetic of transfinite cardinal and ordinal numbers.

Historians have suggested that Cantor discovered the paradoxes in one of the following years: 1883, 1890, 1892, 1895, 1896 and 1899. There is evidence both for and against all of these dates. The clues contained in Cantor's public works and private correspondence are brief and scarce. From his correspondence with Richard Dedekind (Grattan-Guinness *1974*: 126–8; van Heijenoort *1967*: 113–17), we learn that Cantor concluded that it was necessary to distinguish between two kinds of multiplicity: 'absolutely infinite' (or 'inconsistent') multiplicities and 'consistent' multiplicities (or 'sets'). He realized this limitation of size was mandatory when he became aware that, for some multiplicities, 'the assumption that *all* of its elements "are together" leads to a contradiction' (van Heijenoort *1967*: 114). Examples of inconsistent multiplicities were the totality of everything thinkable and the system of all cardinalities. Cantor used an argument similar to the paradox now associated with the concept of the greatest of all ordinal numbers in his attempts to show that Ω (the system of all ordinal numbers) was an inconsistent multiplicity. Cantor's argument ran as follows:

> If Ω' [Ω' results from adding 0 to Ω] were consistent, then, since it is a well-ordered set, there would correspond to it a number δ greater than all numbers of the system Ω; but the number δ also occurs in the system Ω, because this system contains *all* numbers; δ would be greater than δ, which is a contradiction. (van Heijenoort *1967*: 115)

Contrary to popular belief, the discovery of the paradoxes did not result in the logical refutation of Cantorian set theory, nor was it imperative to re-examine the theory's foundations. Cantor, for example, regarded the paradoxes as a positive contribution to his studies, from the mathematical, philosophical and theological points of view. In fact, his theological convictions allowed him to metamorphose the (theological) Absolute into his mathematical absolute of collections too large to be sets (Dauben *1979*: 294–7).

Almost immediately afterwards, perhaps coincidentally, other philosophers and mathematicians came across similar paradoxes – that is, arguments that derive self-contradictory conclusions by valid deductions from apparently acceptable premises. In particular, in an unpublished letter of 1898 to Cantor, E. H. Moore found the paradox expressible in terms of the greatest of all ordinal numbers. Edmund Husserl discovered the paradox of the greatest cardinal number, and Ernst Zermelo came across the paradox known today as Russell's contradiction (see Section 2). The investigations leading to these discoveries were not undertaken to criticize or discredit Cantor's thought. Prior to 1903, David Hilbert, Dedekind, Zermelo, E. H. Moore, Husserl and Giuseppe Peano, among others, were acquainted with these paradoxes or others like them. Nevertheless, none of these philosophers and mathematicians published any comment on the paradoxes, and the mathematical community at large ignored them.

2 RUSSELL'S CONTRADICTIONS

When Bertrand Russell arrived at Cambridge University in 1890, at the age of 18, he was already interested in studying the foundations of mathematics. There he studied mathematics (the first three years) and philosophy (a fourth year). Still under the influence of some of his philosophy professors, Russell made his first unfulfilled attempt at writing a book on the principles of arithmetic in a manuscript of 1898. There were two more unsuccessful endeavours before his meeting with Peano at the First International Congress of Philosophy at Paris in July 1900.

In October 1900, Russell sat down to begin writing what would eventually turn out to be the book *The Principles of Mathematics* (*1903*). But even before he had finished the manuscript, Russell thought that he had found a 'fallacy' in Cantor's theory of transfinite numbers. In Russell's words:

> There is a greatest of all infinite numbers, which is the number of things altogether, for every sort and kind. It is obvious that there cannot be a greater number than this, because, if everything has been taken, there is nothing left to add. Cantor has a proof that there is no greatest number, ... But in this one point, the master has been guilty of a very subtle fallacy .... (Russell *1901*: 95)

Over time, slowly and unconsciously, Russell metamorphosed Cantor's apparent mistake into inescapable difficulties (Coffa *1979*, Grattan-Guinness *1978*).

There are sources, both unpublished and published, showing that around May and June 1901 Russell was involved intellectually with not one but three potential 'contradictions' (as he often called them). Between May 1901 and May 1903, while writing the final version of *The Principles of Mathematics*, he would come to think of two of these three arguments as irresoluble *contradictions* (i.e. statements containing propositions one of which denies or is logically at variance with the other).

More precisely, between January and June 1901 Russell had shifted his attention from his original difficulty involving Cantor's supposed mistake to an argument concerning the class of all classes which are not members of themselves. For Russell, perhaps because of its technical simplicity and also because he was unaware that he was discovering the other contradiction, the most important of the first two contradictions was the one described in terms of the class of all classes which are not members of themselves. The contradiction arose when he asked whether this class was a member of itself, or not. Each alternative led to its opposite.

The second contradiction also arose from Russell's original concern. Russell's final version reads:

> Briefly, the difficulty may be stated as follows. Cantor has given a proof that there can be no greatest cardinal number, and when this proof is examined, it is found to state that, if u be a class, the number of classes contained in u is greater than the number of terms of u, or (what is equivalent), if α be any number, 2^{α} is greater than α .... The difficulty arises whenever we try to deal with the class of all entities absolutely or with any equally numerous class .... (Russell *1903*: 362)

As mentioned above, Russell did not realize that he was discovering this second and independent contradiction. The process of revelation was mostly unconscious. Perhaps he thought this second contradiction was too obvious, and Cantor could not have missed it.

A third contradiction eventually emerged from *The Principles of Mathematics*, which has come to be known as the Burali-Forti paradox. However, Russell originally considered it as two different contradictions. According to him, on the one hand, a form of the Burali-Forti paradox involved contradictory theorems by Cantor and Cesare Burali-Forti on the comparability of transfinite ordinal numbers. On the other hand, in the same paragraph (Russell *1903*: 323), he also indicated the existence of yet another contradictory result discovered by Burali-Forti. Nevertheless, he also initially thought that there was a way to deny a premiss of Burali-Forti's original argument. In time, partly due to the works of Henri Poincaré and Philip E. B. Jourdain, among others, Russell's second form of the contradiction materialized as the paradox of the greatest ordinal number, and it is known today essentially in the same form as used by Burali-Forti in his *reductio ad absurdum* proof of the incomparability of transfinite ordinal numbers published in 1897 (Copi *1958*: 281, 284).

3 OTHER PARADOXES

In the mathematical literature it has been suggested that the three paradoxes described above led directly to many others. This claim, however, has recently been disputed; there may not have originally been such a direct and strong connection.

In December 1904, Russell received a letter from G. G. Berry, then a part-time librarian at the Bodleian Library at Oxford, communicating his dissatisfaction with *The Principles of Mathematics*. In the first place, Berry rejected Russell's solution to the Burali-Forti paradox, claiming that it was easy to prove that the series of all ordinal numbers was a well-ordered set (and that Cantor had actually done it). Second, he described a contradiction (which he thought was too obvious to have passed unnoticed) in terms of 'the least ordinal number not definable in a finite number of words'. But this expression defined that specific number in and of itself.

Soon after, in a letter of June 1905 addressed to the editor of the *Revue Générale des Sciences Pures et Appliquées*, the French mathematician Jules Richard described what he labelled a 'contradiction' in the roots of set theory. Richard had read some notes published by Jacques Hadamard. These notes mentioned the possible existence of contradictions between, on the one hand, results obtained by Julius König (presented at the Third International Congress of Mathematicians, 1904) and by Zermelo (published in *Mathematische Annalen* in 1904) and, on the other hand, theorems on ordinal numbers by Cantor and Burali-Forti. Richard's contradiction was also formulated in terms of a number not definable in a finite number of words. Richard's finding has some remarkable features. He explicitly claimed to have discovered a contradiction – not a paradox, nor an antinomy. He claimed to have found the contradiction in the mere roots of the concept of set. Later, he suggested that the source of the contradiction was the notion of infinite set. Finally, and most importantly, Richard provided a construction (using Cantor's own diagonal procedure) of such a number.

At about the same time, other mathematicians discovered additional findings involving numbers not definable in a finite number of words. König attempted to prove that the continuum was not a well-ordered set. In the process of his proof, by *reductio ad absurdum*, he supposed that the continuum was a well-ordered set and arrived at a contradiction. Then, he rejected his original hypothesis and concluded the opposite. Simultaneously, Alfred C. Dixon attempted to prove that it was impossible to well-order a transfinite set by a finite set of rules. He too attempted to carry out his demonstration by a *reductio ad absurdum* argument, and his proof also involved a contradiction concerning a number not definable in a finite number of words. But he too rejected his original hypothesis, and concluded the opposite.

Nevertheless, in 1908 it was Russell, once again, who took König's and Dixon's proofs, detached the submerged contradictions in the arguments by *reductio ad absurdum* and recast the contradictions in their own terms. Russell's formulation runs as follows:

> Among transfinite ordinals some can be defined, while others can not; for the total number of possible definitions is $\aleph_0$, while the number of transfinite ordinals exceeds $\aleph_0$. Hence there must be indefinable ordinals, and among these there must be a least. But this is defined as 'the least indefinable ordinal', which is a contradiction. (Russell *1908*: 223)

Finally, it has recently been disputed whether these paradoxes did lead Zermelo to his axiomatization of set theory (Moore *1978*: 307). On the contrary, unquestionably the paradoxes had a great effect upon the final presentation of Russell and Whitehead's *Principia mathematica* (1910–13), and upon various later researches.

BIBLIOGRAPHY

Coffa, A. *1979*, 'The humble origins of Russell's paradox', *Russell: The Journal of the Bertrand Russell Archives*, (33/34), 31–8.
Copi, I. *1958*, 'The Burali-Forti paradox', *Philosophy of Science*, **25**, 281–6.

Dauben, J. W. *1979, Georg Cantor: His Mathematics and Philosophy of the Infinite*, Cambridge, MA: Harvard University Press.

Garciadiego, A. *1985*, 'The emergence of the non-logical paradoxes of the theory of sets, 1903–1908', *Historia mathematica*, **12**, 337–51.

—— *1992a, Bertrand Russell and the Origins of the Set Theoretic Paradoxes*, Basel: Birkhäuser.

Grattan-Guinness, I. *1972*, 'Bertrand Russell on his paradox and the multiplicative axiom. An unpublished letter to Philip Jourdain', *Journal of Philosophical Logic*, **1**, 103–10.

—— *1974*, 'The rediscovery of the Cantor–Dedekind correspondence', *Jahresbericht der Deutschen Mathematiker-Vereinigung*, **76**, 104–39.

—— *1978*, 'How Bertrand Russell discovered his paradox', *Historia mathematica*, **5**, 127–37.

Menzel, C. *1984*, 'Cantor and the Burali-Forti paradox', *The Monist*, **67**, 92–106.

Moore, G. H. *1978*, 'The origins of Zermelo's axiomatization of set theory', *Journal of Philosophical Logic*, **7**, 307–29.

Moore, G. H. and Garciadiego, A. *1981*, 'Burali-Forti's paradox: A reappraisal of its origins', *Historia mathematica*, **8**, 319–50.

Purkert, W. *1986*, 'Georg Cantor und die Antinomien der Mengenlehre', *Bulletin de la Société Mathématique de Belgique*, **38**, 313–27.

Russell, B. *1901*, 'Recent work on the principles of mathematics', *International Monthly*, **4**, 83–101.

—— *1903, The Principles of Mathematics*, Cambridge: Cambridge University Press. [2nd edn 1937, London: Allen & Unwin, with new introduction.]

—— *1908*, 'Mathematical logic as based on the theory of types', *American Journal of Mathematics*, **30**, 222–62.

Thomas, W. and Rang, B. *1981*, 'Zermelo's discovery of the Russell paradox', *Historia mathematica*, **8**, 15–22.

Van Heijenoort, J. (ed.) *1967, From Frege to Gödel: A Source Book on Mathematical Logic, 1879–1931*, Cambridge MA: Harvard University Press.

5.4

Logic and set theory

GREGORY H. MOORE

1 THE BOUNDARY BETWEEN LOGIC AND SET THEORY

The boundary between logic and set theory is a porous one, encouraging exchanges in both directions, since the notion of class (or set) falls naturally on both sides of the boundary. These exchanges were increased by the fact that mathematical logic and set theory originated during the same period, the late nineteenth century.

At first it was not clear that the main ingredients of mathematical logic are a separate syntax and semantics. It was surprisingly difficult for logicians to distinguish the words used to define a mathematical object (linguistic entities within the syntax) from the mathematical object itself (part of the semantics). Nor was it clear, until 1930, that there was any point in the distinction, apart from the purely grammatical one. It was a common assumption of the late nineteenth and the early twentieth century that mathematical syntax and semantics were completely interchangeable. David Hilbert, for example, held the view that the existence of a mathematical object satisfying an axiom was equivalent to the consistency of the axiom. Only with the work of Kurt Gödel in 1930 did it become clear that Hilbert's view, which he held dogmatically, was true for some kinds of logic but not for logic in general. More specifically, Gödel proved the two most important theorems in mathematical logic: the completeness theorem for first-order logic and the incompleteness theorems for any axiom system containing arithmetic (Grattan-Guinness *1979*; §5.5, §5.8).

Set theory influenced logic, both through its semantics, by expanding the possible models of various theories and by the formal definition of a model; and through its syntax, by allowing for logical languages in which formulas can be infinite in length or in which the number of symbols is uncountable (Moore *1980*). Logic influenced set theory by treating it axiomatically, giving it an underlying logic, and studying its models.

2 WHAT IS MATHEMATICAL LOGIC?

Mathematical logic crystallized gradually over the period 1850–1950. Although there are many kinds of mathematical logic, they all share certain features. First, each kind is an artificial language, somewhat like a computer programming language, with a syntax and a semantics. The syntax fixes the 'alphabet' of signs of the logic, as well as any additional non-logical signs taken as primitive. Expressions, called formulas, are built up from these logical and non-logical signs by means of recursive rules. Certain formulas are taken as axioms of the logic. A proof is a finite list such that each item in the list is an axiom, or is obtained from previous items by a rule of inference. A theorem is the last item on such a list.

About 1880, when Gottlob Frege and Charles Sanders Peirce independently separated quantifiers ('for all x', 'there exists x') from the Boolean operations such as 'and', 'or', 'if...then', the road was opened for a logic whose principal goal was to be adequate for mathematics. This fundamental shift, carried out by Peirce within the algebraic Boolean tradition (§5.1) and by Frege outside it (§5.2), began to create a truly mathematical logic from the philosophical logic bequeathed by Aristotle, the Stoics and the medieval scholastics.

From today's standpoint, Frege and Peirce should have invented first-order logic; but they did not. First-order logic extends propositional logic (that is, Boolean logic) by the quantifiers 'there exists an x' and 'for every x', by variables for individuals, and by symbols for relations and functions. The semantics of first-order logic considers structures $\langle A, F, R \rangle$, where F contains a function for each function symbol (and R contains a relation for each relation symbol) in the first-order syntax.

During the 1920s, before first-order logic became central, Bertrand Russell's theory of types dominated mathematical logic (§5.2). But during that period neither the theory of types nor any other system was universally accepted by logicians, and philosophical controversies affected the system of logic used.

First-order logic began to emerge only around 1920. What Frege and Peirce had invented earlier was second-order logic, which includes first-order logic as a subsystem. The syntax of second-order logic has quantifiers that range over relations (or functions). Its semantics extends that of first-order logic in the most direct way. In second-order logic, the usual proofs for the categoricity of the natural numbers ℕ and of the real numbers ℝ can be expressed (§5.8). But in first-order logic, neither ℕ nor ℝ nor any other infinite set has a categorical axiomatization.

After 1930, first-order logic gradually became the dominant logic. Its proponents, such as Thoralf Skolem in the 1920s and, later, Willard Quine, saw it as adequate to express all mathematical ideas. Its opponents, such as Ernst Zermelo, saw it as inadequate even to capture the real numbers, much less set theory. But by 1950, first-order logic was well established as the fundamental logic system.

The dominance of first-order logic continues today; every student of mathematical logic now learns the fundamental facts about first-order logic, but, in general, very little about other logics. Yet in the mid-1950s there emerged officially a kind of logic that had existed in embryonic form for decades: infinitary logic. This is a whole family of logics, generally including first-order logic as a subsystem, and each of them has sentences, or proofs, that are infinitely long. They can express concepts that first-order logic cannot, such as the notion of finite set.

3 THE ORIGINS OF SET THEORY

In essence, set theory was created by a single man, Georg Cantor, who formulated the fundamental concepts and proposed the major problems (§3.6). He did the first part of this work while establishing that a function is represented by a unique trigonometric series. Having proved this for when the series converged everywhere, he extended it to the case when convergence failed for infinite sets having a special form. If P is a point set, then the derived set of P, called $P^{(1)}$, was the set of all limit points of P. Iterating this operation gave $P^{(2)}$, $P^{(3)}$, and so on for every finite index. By 1872 he had considered infinite indices $P^{(\infty)}$, $P^{(\infty+1)}$, $P^{(\infty+2)}$, and so on. These infinite indices, which he refrained from publishing at the time, were the origin of his infinite ordinal numbers, which he developed a decade later. At this early stage, his set-theoretic ideas were closely tied to point set topology (Cavaillès *1962*, Dauben *1979*, Purkert and Ilgauds *1986*).

In 1872 Cantor published a rigorous foundation for the real numbers, and Richard Dedekind did likewise. Cantor constructed the real numbers as equivalence classes of Cauchy sequences of rational numbers, while Dedekind used 'cuts' in the rationals. Cantor's approach was later used to complete any metric space, while Dedekind's was employed to complete any partial order.

Cantor next initiated the theory of infinite cardinal numbers. He defined two sets to have the same cardinal number, or power, if they can be put in 1–1 correspondence. A set is 'denumerable' if it has the same cardinal number as the positive integers. His first major result was that the set of real numbers is not denumerable, thus establishing that there are at least two infinite cardinal numbers. In 1878 he showed that a line and a plane have the same cardinal number, called the 'power of the continuum'. He then proposed that there is no cardinal number between the denumerable and the power of the continuum. This was his famous continuum hypothesis (CH), which has stimulated the development of set theory up to the present (Moore *1989*).

In 1883 Cantor published his theory of infinite ordinal numbers, and modelled its development on that of the positive integers. He arranged his ordinals in number classes, thereby connecting them to his infinite cardinal numbers. The first number class consisted of the positive integers, and its cardinal was denumerable. The second number class consisted of the denumerable ordinals, and its cardinal number (later called $\aleph_1$) was the first one larger than the denumerable. He defined the nth number class for each finite n. But he clearly had in mind that there are many larger cardinals; for he took it as true that there are as many cardinals as there are ordinals, and this gives many number classes with infinite index. Indeed, his statement was the first of what was later called a 'large cardinal' axiom (see Section 6).

Also in 1883, Cantor introduced the notion of well-ordered set. A set is 'well-ordered' if it is linearly ordered and if every non-empty subset has a least element. He argued for the truth of the well-ordering principle, which states that every set can be well-ordered. Thus every infinite set has the cardinal number of some number class, and the 'trichotomy of cardinals' holds: that is, for any cardinals A and B, $A < B$ or $A = B$ or $B < A$. (Here, $A < B$ means that A has the cardinal of some subset of B, but no subset of B has the cardinal of A.) Finally, in 1891 he showed that, for any set A, the set of all subsets of A has a cardinal larger than that of A; hence there is no largest infinite cardinal.

By 1900, despite opposition from constructivist mathematicians such as Leopold Kronecker (§5.6), set theory had proved its value in other branches of mathematics, especially analysis. Hilbert treated a proof of CH, and of the well-ordering principle applied to the real numbers, as the first of his famous 23 problems.

In August 1904 Julius König claimed to refute both the well-ordering principle and CH. But a gap was found in König's argument, and Zermelo soon offered a proof for the well-ordering principle, known henceforth as the well-ordering theorem.

This theorem provoked intense opposition from mathematicians in England, France and Germany. Much of that opposition was directed against the axiom of choice (AC), which Zermelo introduced as the basis for his proof. This axiom states that, given a set A, there is some function f such that, for each non-empty subset B of A, $f(B)$ belongs to B. Thus the function 'chooses' an element from each non-empty subset. It turned out that many propositions, including the trichotomy of cardinals, are equivalent to AC (Moore *1982*).

In 1908, wishing to refute the opponents of AC and realizing that he must find a way round the paradoxes, Zermelo gave a Hilbert-style axiomatization for set theory and included AC. The axiomatization began with the empty set, and possibly certain non-sets or individuals, and generated further sets by the operations of pairing, union and power set. It also assumed the existence of an infinite set. Especially important was the axiom of separation, which stated that, given a set A, all its elements satisfying a 'definite' property also form a set.

4 LOGIC AND THE AXIOMATIZATION OF SET THEORY

In 1922 Abraham Fraenkel in Germany and Skolem in Norway independently proposed to modify Zermelo's system. First, they introduced the axiom of replacement, which had also been proposed by Dmitri Mirimanov in Switzerland in 1917: if A is a set, then any collection in 1–1 correspondence with A is also a set. Fraenkel and Skolem realized that this axiom was necessary to get cardinals as large as $\aleph_\omega$. It was also necessary, as John von Neumann recognized in 1923, to show that every well-ordered set is isomorphic to some ordinal. Second, Fraenkel and Skolem modified, in quite different ways, the axiom of separation, which Zermelo had based on the vague notion of 'definite' property. Skolem's way was the most influential in the long run, for he insisted that a 'definite' property be one expressible in first-order logic by means of the primitive symbols ϵ and $=$.

5 THE INFLUENCE OF SET THEORY ON LOGIC

Within the Boolean tradition, the border between set theory and logic was hazy (§5.1). Indeed, one of George Boole's interpretations for his calculus was in terms of classes or sets. The set-theoretic operations of union, intersection and complementation were part of the semantics of Boole's logic. And Peirce developed the existential quantifier in terms of the union of a class of sets. When Ernst Schröder continued Peirce's work, the ambiguity between the logical and the set-theoretical viewpoints became quite fruitful, for it opened the door to an infinitary logic – that is, a logic in which some sentences were infinitely long. Whether Schröder himself intended a quantifier to be construed as an infinite sentence (e.g. 'for every whole number, $P(x)$' to be construed as '$P(1)$ and $P(2)$ and $P(3)$ and . . .'), his successors such as Leopold Löwenheim certainly understood him to do so.

Nor was it an accident that the first important result in model theory was due to Löwenheim and arose within the Peirce–Schröder logical tradition. That tradition blended set theory, algebra and logic. Löwenheim asked when his linguistic expressions could have a model that is countable (i.e. finite or denumerable). Certainly not always, since Cantor had shown that the set of real numbers is uncountable. But Löwenheim distinguished clearly, as his predecessors had not, between a first-order sentence, where the quantifiers range only over individuals, and a second-order sentence, where the quantifiers range over sets of individuals. In 1915 he answered that any first-order sentence has a countable model, if it has a model at all.

The Peirce–Schröder tradition was continued by Skolem, who first argued that first-order logic is all of logic, and not merely a subsystem of a stronger logic (Moore *1980*). Whereas Löwenheim had used infinitely long sentences in his version of first-order logic, Skolem soon dropped any use of infinitely long sentences. In 1920 Skolem expressed his central result, now known as the Löwenheim–Skolem theorem, as follows: in first-order logic, every set of sentences with an uncountable model has a countable model.

Skolem had a philosophical aim: to diminish the importance of set theory as a foundation for mathematics. Thus he formulated the Skolem paradox: set theory has a countable model, even though set theory asserts that there are uncountable sets. This paradox was troubling at the time because mathematicians were not used to thinking that a mathematical system, such as set theory, should be considered only within some previously given logic; in first-order logic, a model of set theory can be countable, but inside the model some sets are seen as uncountable.

6 LARGE CARDINAL AXIOMS

Felix Hausdorff hoped to solve Cantor's continuum problem by studying order types. In 1907 he introduced the fundamental notion of cofinality: an ordinal β is 'cofinal' with an ordered set A if there is a sequence of length β of members of A such that each a in A is less than some member of that sequence. The cofinality of A is the least ordinal cofinal with A. He used cofinality to introduce the notions of regular and singular cardinal, on which much further progress depended. An infinite cardinal is 'singular' if it is the union of a smaller set of smaller cardinals; otherwise, it is 'regular'. Each $\aleph_{\alpha+1}$ is regular, and if β is a limit ordinal, then $\aleph_\beta$ is singular for all known cardinals. If such an $\aleph_\beta$ is regular, it is said to be a 'weakly inaccessible' cardinal. Such 'exorbitantly' large cardinals, he insisted, could have nothing to do with set theory as it had been developed until then. But he could not have been more mistaken. The existence of certain kinds of inaccessible cardinal affects the projective sets (rather simple sets of real numbers), and whether CH holds for them. At the same time, Hausdorff extended Cantor's perspective by formulating the generalized continuum hypothesis (GCH): that for every ordinal α, $2^{\aleph_\alpha} = \aleph_{\alpha+1}$.

Meanwhile, a problem in analysis was influencing set theory. In 1902

RHEADHenri Lebesgue had put forward the measure problem: to find a non-negative function m on all the subsets of the interval $E = [0, 1]$ such that (1) $m(E) = 1$; (2) if A and B are congruent subsets of E, then $m(A) = m(B)$; and (3) the measure m of a finite or denumerable union of disjoint sets is the sum of the measures of those sets (§3.7). Lebesgue hoped that his Lebesgue-measurable sets would solve the measure problem, but in 1905 Giuseppe Vitali showed that they did not.

In 1929 Stefan Banach dropped the geometric condition (2) and replaced it with a weaker set-theoretic condition (2'): that $m(A) = 0$ if A contains one element. He recognized that, with (2'), the problem really concerned the cardinality of the set E, and so he considered an arbitrary set E rather than the set of real numbers. The cardinal of such an E is called a 'real-valued measurable cardinal'. By using GCH, he showed that any real-valued measurable cardinal is weakly inaccessible. In 1930 Stanisław Ulam strengthened this result by proving it without recourse to GCH. He also introduced the important notion of a measurable cardinal – that is, a real-valued measurable cardinal whose measure m takes only the values 0 and 1. By using AC, he established that every measurable cardinal is strongly inaccessible.

For the next thirty years, one of the major unsolved problems of set theory was whether the first strongly inaccessible cardinal is measurable. Only in 1960, with the help of William Hanf's results on infinitary logic, did Alfred Tarski prove that the first measurable cardinal $\varkappa$ is so large that there are $\varkappa$ strongly inaccessible cardinals below it.

7 MODELS OF SET THEORY

The first work on models of set theory was by Fraenkel, who in 1922 gave a model of Zermelo's axioms and the negation of AC. The model was constructed by beginning with a denumerable set of pairs of individuals and closing under the operations of union, pairing and power set. Unfortunately, this model said nothing about whether AC, when restricted to sets of real numbers, followed from the other axioms, nor did such models say anything about CH. Such models could be used in second-order as well as first-order logic.

In 1930 Zermelo gave the first understanding of what the models of set theory are in second-order logic. To do so, he introduced the cumulative-type hierarchy of sets (Zermelo–Fraenkel set theory), which has been the basis of set theory ever since, via his axiom of foundation. This axiom states that there is no infinite descending sequence of sets $\cdots \in A_2 \in A_1 \in A_0$, and it splits the universe V of sets into levels. The lowest level is the set of all individuals. Each higher level is the set of all subsets of the previous level.

If a level has no immediate predecessor, it is the union of all previous levels. Zermelo established that a model is determined by its 'width' (the cardinal of its set of individuals) and its 'length' (the least ordinal not in the model). He found that the only possible lengths are the strongly inaccessible cardinals, whereas any cardinal can be a width.

Gödel's incompleteness theorems showed that such inaccessible cardinals cannot be proved consistent, except from some stronger assumption (§5.5). He suggested an idea, which proved very important after 1960, that the existence of large cardinals could decide questions about the natural numbers or the real numbers (Fraenkel *et al. 1973*: 110–13).

After 1931, Gödel's work was concerned, like Fraenkel's, with AC, but with its relative consistency rather than its independence. Gödel followed Paul Bernays in using first-order logic as the basis for set theory, but used Russell's ramified theory of types (extended to transfinite orders) as a model of set theory. In other words, he invented what he called the 'constructible sets': those sets definable in first-order logic from those already defined at a lower level. In effect, he took Zermelo's cumulative-type hierarchy of sets and restricted it.

In 1938 Gödel used the class L of all constructible sets to establish that if Zermelo–Fraenkel set theory (ZF) is consistent, then so is ZF together with AC and GCH. To do so, he introduced the axiom of constructibility stating that $V = L$: that is, every set is constructible.

Gödel's work on set theory stood as an isolated landmark until the 1950s. In Poland, Andrzej Mostowski had worked on L during the Second World War, but his work was lost. In the 1950s, models of set theory were investigated in England, Hungary, Israel and, above all, in the USA. These logicians extended both Fraenkel's and Mostowski's independence results and Gödel's relative consistency results.

Only in 1961 was a connection made between large cardinals and constructible sets. Dana Scott showed that if there is a measurable cardinal, then Gödel's axiom of constructibility is false. Soon this axiom came to be viewed as dubious.

In 1963 Paul Cohen made what Gödel later called 'the most important progress in set theory since its axiomatization'. Cohen discovered the method of 'forcing', which, given the consistency of ZF, showed the independence of AC from ZF as well as the independence of CH from ZF including AC (Moore *1988*, Mostowski *1966*). In particular, Cohen was able to show that very weak forms of AC, restricted to any denumerable family of pairs of real numbers, are not provable in ZF.

Cohen, who received the Fields Medal for his work, intended it to be an isolated landmark, but it was applied before it saw print. At Berkeley, even graduate students soon obtained new independence results by forcing, and models of set theory remained throughout the 1960s the most exciting part of mathematical logic.

Forcing was soon combined with large cardinals. The most important problem studied was this: what are the possible ways in which GCH can be violated? It was shown in 1964 that GCH can be violated easily at all regular cardinals. Singular cardinals are another matter. Robert M. Solovay soon established that if there is a supercompact cardinal (one much larger than a measurable cardinal), then GCH holds at many singular cardinals. A recent landmark result, due to Donald A. Martin and John R. Steel, is the following: if there is a supercompact cardinal, then CH holds for all the projective sets (i.e. any projective set is countable or has the power of the continuum).

Today the most important results being obtained in set theory depend on the interplay of forcing, large cardinals and first-order logic.

BIBLIOGRAPHY

Cavaillès, J. *1962, Philosophie mathématique*, Paris: Hermann.

Dauben, J. W. *1979, Georg Cantor: His Mathematics and Philosophy of the Infinite*, Cambridge, MA: Harvard University Press.

Fraenkel, A. A. *1954, Abstract Set Theory*, Amsterdam: North-Holland. [Excellent bibliography.]

Fraenkel, A. A., Bar-Hillel, Y. and Levy, A. *1973, Foundations of Set Theory*, Amsterdam: North-Holland.

Grattan-Guinness, I. *1979, 'In memoriam* Kurt Gödel: His 1931 correspondence with Zermelo on his incompletability theorem', *Historia mathematica*, **6**, 294–304.

Hallett, M. *1984, Cantorian Set Theory and Limitation of Size*, Oxford: Clarendon Press.

Moore, G. H. *1980*, 'Beyond first-order logic: The historical interplay between mathematical logic and axiomatic set theory', *History and Philosophy of Logic*, **1**, 95–137.

—— *1982, Zermelo's Axiom of Choice: Its Origins, Development, and Influence*, New York: Springer.

—— *1988*, 'The origins of forcing', in F. R. Drake and J. K. Truss (eds), *Logic Colloquium '86*, Amsterdam: North-Holland, 143–73.

—— *1989*, 'Toward a history of Cantor's continuum problem', in D. E. Rowe and J. McCleary (eds), *The History of Modern Mathematics*, Vol. 1, Boston, MA: Academic Press, 78–121.

Mostowski, A. *1966, Thirty Years of Foundational Studies*, New York: Barnes & Noble.

Purkert, W. and Ilgauds, H.-J. *1986, Georg Cantor*, Basel: Birkhäuser.

5.5

Metamathematics and computability

STEWART SHAPIRO

1 COMPUTABILITY

A function f is said to be computable if there is an algorithm (or mechanical procedure) A that computes f. For every m in the domain of f, given m as input, A would produce $f(m)$ as output. Put this way, computability applies to functions on things that can be processed by algorithms or machines, such as marks on paper or bits of magnetic media. Idealizing a little, the domain of computability consists of functions on strings – finite sequences of characters on a finite alphabet. Since strings are structurally similar to natural numbers, it is common to think of computability as applying to number-theoretic functions. A set S of strings or numbers is 'effectively decidable', or simply 'decidable', if there is an algorithm B for deciding membership in S. Given s as input, B would produce 'yes' if s is in S and B would produce 'no' if s is not in S.

In a paper published in 1936, Alonzo Church proposed that computability be 'defined' as recursiveness. This became known as 'Church's thesis'. In the same paper, it is noted that recursiveness is coextensive with λ-definability. The equivalence was established by Church's student, Stephen Kleene, in the same year. (Definitions of recursiveness and λ-definability are provided in Section 7.) Also in 1936, Alan Turing published his own characterization of computability, introducing the celebrated notion of the Turing machine. Turing's work was independent of Church's and Kleene's, but on learning of the latter, Turing showed that a function is recursive (or λ-definable) if and only if there is a Turing machine that computes it. In 1936, Emil Post also published a characterization of computability remarkably similar to Turing's. Post's work was independent of Turing's, but not of the activity of Church and his students at Princeton.

Two themes are raised here. First, what led to the development of computability in the 1930s? Algorithms have been known since Antiquity, and computation was occasionally discussed in the history of mathematics and philosophy, but the period in question here produced the first attempts at a characterization of computability *per se*. Second, what convinced the major contributors – Turing, Kurt Gödel, Post, Church and Kleene – of Church's thesis (or its equivalent)? What led each of them to believe that the characterizations at hand are successful?

2 PRECURSORS

Gandy *1988* contains a description of the algorithms behind Charles Babbage's proposed computing device (1837). Gandy shows that, suitably idealized and suitably programmed, a Babbage machine can compute any recursive function. Thus, given Church's thesis, a function is computable if and only if it can be computed on a Babbage machine. But it does not appear that Babbage was interested in the limits of computability *per se*, nor did he make a claim like Church's thesis. He did assert that his work shows that 'the whole of the development and operations of analysis are now capable of being executed by machinery', but this is not the issue here (§5.11). A statement that a certain area is capable of mechanization is not a statement on the overall limits of computation.

Another possible precursor is Gottfried Wilhelm Leibniz's aborted *characteristica universalis*, an attempt to reduce all of mathematics, science and philosophy to algorithms. Again, the concern is with the extent of these fields, not with the extent of computation. In Leibniz's programme the expressions of a language are regarded as uninterpreted inputs to algorithms, and the language itself is regarded as a formal object, capable of mathematical treatment. Thus the programme is a forerunner of modern mathematical logic, proof theory in particular, and it is this field that gave birth to computability.

Today, there is a consensus that a function is an arbitrary correspondence between collections of mathematical objects. For a given function f, there need be no rule, or even an independent description, that specifies, for each x in the domain of f, its value $f(x)$. In the set-theoretic foundation, for example, a function is defined to be a set of ordered pairs with a certain property and, again, a set is an arbitrary collection. One reason for the relatively late development of computability may be that this notion of arbitrary correspondence, and thus the very possibility of a non-computable function, has emerged only fairly recently. Stein *1988* shows that what is today called 'classical mathematics' is scarcely a century old. Before that, there was no widespread understanding of such notions as function and set. This point is well illustrated in the early debates over the axiom of choice (§5.4). Many opponents of the axiom, notably René Baire, Emil Borel and

Henri Lebesgue, raised doubts about the intelligibility of functions and sets that are not uniquely specifiable. It is not a big step from this to question the intelligibility of non-computable functions of natural numbers. To this day, there are intuitionists and constructivists who make this step (§5.6).

Once the notion of an arbitrary function or set is available, then perhaps it becomes natural to ask for an algorithm to compute a particular function or to decide membership in a particular set. And such problems were formulated and pursued early this century. For example, some of David Hilbert's 23 problems, posed in 1900, are what may be called 'decision-by-algorithm' problems. Consider the once-elusive tenth, in number theory (§6.10):

> Determination of the Solvability of Diophantine Equations. Given any Diophantine equation with any number of unknown quantities and with rational integer coefficients: to devise a process according to which it can be determined by a finite number of operations whether the equation is solvable in rational integers.

Clearly, the best way to provide a positive solution to a decision problem is to give an algorithm and show that it does the required task. There is rarely a question over whether a purported algorithm really is an algorithm (although the subsequent debate over Church's thesis produced a number of exceptions; see §4.10 on algorithms in general). However, a negative solution to a decision problem amounts to a theorem that *no* algorithm accomplishes the task at hand. To establish this, one must first identify a property shared by *all* algorithms, or all computable functions. One straightforward way to do this would involve a precise characterization of computability. Thus, unsolved algorithm problems may have been a motivation behind the development of computability. In short, it became important to make coherent and useful statements about all algorithms, or all computable functions, in order to show that a certain function is not computable.

If this is correct, then in a sense, Hilbert himself foreshadowed the development of computability. In the aforementioned 'mathematical problems' lecture, he recognized the possibility of negative solutions to some of his problems: 'Occasionally it happens that we seek the solution ... and do not succeed. The problem then arises: to show the impossibility of the solution ...' (Hilbert *1900*: 444). This is illustrated with historical instances of problems which eventually found 'fully satisfactory and rigorous solutions, although in another sense than originally intended'. He mentions the problems of squaring the circle and proving the axiom of parallels. The solutions of these problems involved the development of new theory to provide mathematical characterizations of pre-formal notions. It is curious that, although many unsolved algorithm problems originated from Hilbert's thinking, the development of computability did not come from his school.

3 THE HILBERT PROGRAMME

At the beginning of Hilbert *1925*, Karl Weierstrass is praised for eliminating the 'infinitely large' and the 'infinitely small' from mathematical analysis, but Hilbert notes that the infinite remains in analysis in the 'infinite number sequences that define the real numbers' (e.g. Cauchy sequences of rationals or, equivalently, Dedekind cuts; see §3.3). Moreover, the real number system is itself infinite and it is conceived 'to be an actually given totality, complete and closed'. This is manifest in the use of unrestricted universal and existential quantifiers. Hilbert suggests that intellectual history provides good reason to regard any such use of infinite collections as problematic. Thus, the aim of the programme is 'to endow mathematical method with the definitive reliability that the critical era of the infinitesimal calculus did not achieve; thus it shall bring to completion what Weierstrass . . . endeavored to do and toward which he took the first necessary and essential step' (van Heijenoort *1967*: 370).

According to Hilbert, there is a central 'finitary' core of mathematics that is unquestionably reliable. Its subject matter is strings of characters on a finite alphabet or, equivalently, natural numbers. There are, of course, infinitely many strings and natural numbers, but Hilbert did not regard them as a 'complete and closed' totality. The domains are merely 'potentially infinite', in the sense that there is no upper bound on the size of strings that can be considered – given any string, one can always produce a larger one. As indicated, unrestricted quantifiers are banned from finitary mathematics; every quantifier must be restricted to a finite domain.

To be sure, mathematics goes well beyond the finitary and, unlike the intuitionists and constructivists, Hilbert is not out to restrict available methodology. The idea is that the non-finitary parts of mathematics be regarded as meaningless, akin to the ideal 'points at infinity' sometimes introduced into geometry. The purpose of non-finitary systems is to streamline inferences leading to finitary conclusions. With a view like this, we need some assurance that employing the non-finitary methods will not lead to results that are refuted on finitary grounds; that is, we need a guarantee that the non-finitary system is consistent with finitary mathematics. To achieve this, the Hilbert programme called for the discourse of each branch of mathematics to be cast in a rigorously specified deductive system. These deductive systems are to be studied syntactically, with the aim of establishing their consistency. For this metamathematics, only finitary methods are to be employed. Thus, if the programme were successful, finitary mathematics would establish that deductive systems are consistent, and can be used to derive finitary results with full assurance that the latter are correct: 'We shall carefully investigate those ways of forming notions and those modes of inference that are fruitful; we shall nurse them, support them, and make them usable... No one shall be able to drive us from the paradise that Cantor created for us' (Hilbert *1925*: 375–6). Even though we are still enjoying Georg Cantor's paradise, Hilbert's programme was not to achieve its goal. But it did bear fruit of its own.

In the early 1920s, Thoralf Skolem published a paper developing 'the recursive mode of thought' in order to formulate a portion of elementary arithmetic' without the use of apparent variables ranging over infinite domains'. This and other work established the usefulness of recursively defined functions for foundational studies. The technique of definition by recursion is also natural in formulating deductive systems. The subsequent research on recursion and proof theory culminated in Gödel's *1931* paper on the incompleteness of arithmetic, to which we now turn.

4 GÖDEL'S INCOMPLETENESS THEOREM

The main result of Gödel *1931* is that the axiomatization of Russell and Whitehead's *Principia mathematica* does not meet Hilbert's criterion of completeness: there is a formula in the relevant language which is neither derivable nor refutable in the given deductive system. To establish this, Gödel provided a rigorous account of the aforementioned connection between the structure of the natural numbers and the structure of syntactic items like formulas and formal derivations. Each character of the formal language in question is assigned a number, later called its Gödel number or Gödel code. Then one gives a uniform procedure for determining a unique Gödel number for each formula and each sequence of formulas from the codes of its constituent elements. In this way, relations on the items in a language and deductive system correspond to relations on natural numbers. Gödel established that many common syntactic properties and relations correspond to primitive recursive relations on natural numbers (see Section 7 for definitions). Examples include 'x is the Gödel number of a well-formed formula', 'x is the Gödel number of an axiom' and, ultimately, 'x is the Gödel number of a derivation of the formula whose Gödel number is y'. He also showed that every primitive recursive function is definable in the language of *Principia mathematica*. Together, these results indicate that there is a formula Prf(x, y) such that, for every pair of numerals m and n, Prf(m, n) is derivable in the deductive system in question if m denotes the Gödel number of a derivation of the formula whose Gödel number is denoted by n; and the negation of Prf(m, n) is derivable otherwise.

In a sense, then, Gödel showed how a formal system of arithmetic can 'talk about' its own language and deductive system. Through an ingenious technique of self-reference, Gödel then showed that there is a sentence G of the formal language that 'says' that there is no derivation of G. The sentence G is an analogue, of sorts, of the infamous Liar: 'This sentence is false.' In place of the predicate 'false', G contains the arithmetic analogue of the predicate 'non-derivability in the deductive system of *Principia mathematica*'. At this point, one can establish that if the deductive system is, in fact, consistent, then G is not derivable in it. Moreover, if the deductive system enjoys a certain stronger consistency property, called 'ω-inconsistency', then the negation of G is not derivable either. Thus, the deductive system is not complete.

There is a surprising corollary of this result, which has profound consequences for the Hilbert programme. Consider the statement just above: 'If the deductive system is consistent, then G is not derivable in it.' This has an analogue *in* the formal language of arithmetic, via the Gödel coding. Let 'Con' be a formula asserting that 'for every x, x is not the Gödel number of a derivation of a contradiction'. That is, Con is an analogue of the statement that the deductive system is consistent. Recall that the phrase 'G is not derivable' is equivalent to G. So the above assertion that 'if the deductive system is consistent, then G is not derivable' corresponds to 'if Con then G'. The latter *is* derivable in the deductive system in question. Assuming consistency, it follows that Con is *not* derivable in the deductive system. If it were, then G would be derivable as well. In other words, Gödel showed that the consistency of this deductive system cannot be established *in that system*, much less in its finitary core. This is usually regarded as showing that the Hilbert programme cannot be carried out (but see Detlefsen *1986*).

Gödel's theorem is reviewed again in §5.8, in the context of model theory.

5 TOWARDS A THEORY OF COMPUTATION

The methods of Gödel *1931* appear to be general. Any of the straightforward extensions of the deductive system in question will also be incomplete (if consistent). At least with hindsight, the obvious generalization of Gödel *1931* is that *no* suitable axiomatization of arithmetic is complete. In order to clarify and, perhaps, prove this generalization, the concept of 'suitable axiomatization' *vis-à-vis* the Hilbert programme had to be developed, explicated and studied.

Clearly, not every collection of sentences of a given language is a suitable axiomatization. For example, one cannot simply declare that a sentence A is an 'axiom' if and only if A is true (in some intended interpretation of the language). Or at least one cannot do this in the context of the Hilbert programme.

In the clarification of 'suitable axiomatization', considerations of computability arise. Axiomatic deductive systems are to represent or codify actual mathematical discourse, one purpose of which is to communicate proofs. This suggests that in a suitable axiomatization one should be able to determine, 'using only finite means', whether an arbitrary string of characters is a formula and whether an arbitrary sequence of formulas is a derivation. In short, the syntax of a suitable axiomatization should be effectively decidable. So the proposed generalization of Gödel's 1931 result is this: that there is no consistent, *decidable* axiomatization of a theory that includes a certain amount of arithmetic in which every sentence is either derivable or refutable. Here we have a conjecture of a negative theorem about algorithms: *no* algorithm decides the syntax of a consistent, complete axiomatization of (a sufficiently rich) arithmetic.

Gödel's 1934 lecture notes (published in Davis *1965*) contain an elaboration and extension of the incompleteness result. They open with an explicit statement that the syntax of an axiomatic deductive system must be decidable. He states that the condition of primitive recursiveness 'in practice suffices as a substitute for the imprecise requirement' of effective decidability. It does not follow, however, that no suitable axiomatization is complete because, as was known at the time, there are computable functions that are not primitive recursive. After mentioning that the primitive recursive functions are all computable, Gödel adds a footnote that 'the converse seems to be true, if ... recursions of other forms ... are admitted'. The lecture notes contain an extension of primitive recursiveness, equivalent to recursiveness, and it is shown that every recursive axiomatization of arithmetic is either incomplete or is not ω-consistent. This work motivated Kleene's 1935 detailed study of recursiveness, which was soon followed by Church's and Post's characterizations of computability. As noted, Turing's work also appeared almost simultaneously (Shapiro *1983*).

This line of thought is not the whole story behind the development of computability, for Turing and Church seem to have been interested in another aspect of the Hilbert programme. One of the desiderata was that, for each axiomatization, there should be an algorithm that enables one to decide, for any formula, whether it is derivable in the deductive system. This is a decision problem, similar to Hilbert's tenth problem. Another important decision problem is that of finding an algorithm to decide whether a given sentence is a logical truth. If attention is restricted to first-order languages, the Gödel completeness theorem shows that the latter is, in fact, the decision problem for the first-order predicate calculus, a well-known deductive system. Hilbert characterized it as *the* fundamental problem of mathematical logic. This decision problem remained open, despite the best efforts of many great mathematical minds. Some may have conjectured that it does not have a positive solution; again, such a conjecture could lead to a characterization of computability. Turing and Church both show how their characterizations result in a negative solution to the decision problem for first-order languages. The set of first-order logical truths is not recursive – this is Church's theorem, called an 'application' of the studies of computability.

6 THE ACCEPTANCE OF CHURCH'S THESIS

It is straightforward that every λ-definable function is computable. The same goes for recursive functions and, especially, Turing computable functions (see Section 7). The converses of these statements are another story. In a number of places, Kleene has provided first-hand recollections of the events that led him and Church to hold that every computable function is λ-definable (see e.g. Kleene *1979*). After the notation was first formulated, Church and his students began to investigate individual functions to determine whether they are λ-definable. The goal of this activity was to explore the extent of the newly defined property of functions. Every computable function that was 'tested' in this way was shown to be λ-definable, some more easily than others. A major breakthrough was Kleene's proof that the predecessor function is λ-definable. This result surprised Church, who had come to speculate that the predecessor function might not be λ-definable. When Church finally proposed the connection between computability and λ-definability, Kleene 'sat down to disprove it by diagonalizing out of the class of λ-definable functions. But quickly realizing that the diagonalization cannot be done effectively, I [Kleene] became overnight a support of the thesis.'

In Church's 1936 paper, the identification of computability with recursiveness is proposed as a 'definition'. Of course, he did not intend to introduce a new word with the equation. Computability is an imprecise concept from ordinary language, and the proposal is that recursiveness (or λ-definability) be substituted for it. If Church's thesis is construed in this way, there is no question of an attempted 'proof' of it. How does one establish that a vague notion is coextensive with a precise mathematical one? Indeed, from this perspective, Church's thesis does not have a truth value. It is a pragmatic matter, depending on how useful the identification is, for whatever purpose is at hand. One can, of course, demand that the precise notion correspond more or less with the vague, pre-formal counterpart, but

this was amply confirmed by the extensive study of examples. The usefulness of the definition was also bolstered with the discovery that different
formulations of computability, different 'definitions', come to the same
thing. Such results show that independently motivated attempts to characterize the same notion converge on a single class – a good indication that
all of them are on target.

It seems that the other main players did not share this attitude to
Church's thesis. As we have seen, Gödel suggested that there may be a
connection between recursiveness and computability. In fact, he proposed
the identification of the two as a 'heuristic'. But at the time he remained
sceptical, to say the least. Kleene *1979* reports that in a letter of November
1935, Gödel wrote of how he regarded Church's proposed 'definition' as
'thoroughly unsatisfactory'. Church 'replied that if [Gödel] would propose
any definition of effective calculability, [Church] would undertake to prove
that it was included in lambda-definability'. This, of course, would be more
of the same kind of evidence that Church relied on in proposing the 'definition'. But there was already plenty of that, and Gödel was unconvinced by
it. Gödel was not simply being more stubborn than Church here: rather, he
saw the thesis differently. He preferred the rigour of conceptual analysis
to the wealth of examples and the impressive convergence of various
efforts. Presumably, one would attempt to formulate basic premises, or
axioms, about algorithms or computation, and *derive* Church's thesis from
those. Perhaps, for Church, this 'rigour' is not needed. One cannot precisely analyse a vague concept from ordinary language, only propose that
a precise one be substituted for it (Davis *1982*).

It was not long before Gödel was convinced of the truth of Church's
thesis, and it was Turing's work that did it. Turing carefully considered the
possible actions of a human who was following a previously specified
algorithm, and he showed that every such action can be carried out by a
Turing machine.

Post also rejected the idea that Church's thesis be accepted as a 'definition', calling it a 'working hypothesis'. Moreover, in an early article (not
published until 1965), he wrote that 'Establishing [the thesis] is not a
matter of mathematical proof but of psychological analysis of the mental
processes involved in combinatory mathematical processes' (Davis *1965*:
418). I presume that Turing's work would satisfy this demand as well, or
at least contribute to it. Later writings make it clear that Post did accept
Church's thesis. In fact, he proposed that the identification of computability with recursiveness go beyond heuristic, and be used to develop a fully
fledged theory of computability, perhaps along the lines of Rogers *1967*. In
such a framework, one does not constantly think in terms of an individual
formulation, such as λ-definability or Turing computability; rather, the

existence of an algorithm for computing a function is both necessary and sufficient to conclude that it is recursive. Such inferences, sometimes called 'argument by Church's thesis', dominate the field today.

7 ADDENDUM – DEFINITIONS

In this section, characterizations of recursiveness and λ-definability are provided. For more detail, see Rogers *1967* and the papers in Davis *1965*. Let f be an n-place function, and g an $(n+2)$-place function of natural numbers. Then the *primitive recursion of f and g* is the $(n+1)$-place function h such that, for all $x_1, \ldots, x_n$,

$$h(0, x_1, \ldots, x_n) = f(x_1, \ldots, x_n) \tag{1}$$

and

$$h(x+1, x_1, \ldots, x_n) = g(x, h(x, x_1, \ldots, x_n), x_1, \ldots, x_n). \tag{2}$$

Notice that h is computable if f and g are. Indeed, given $m_1, \ldots, m_n$, one can successively compute $h(0, m_1, \ldots, m_n)$, $h(1, m_1, \ldots, m_n)$, $\ldots$, using the algorithms for f and g. Let f be an n-place function and $g_1, \ldots, g_n$ be m-place functions of natural numbers. Then the *composition of f* with $g_1, \ldots, g_n$ is the m-place function k such that, for all $x_1, \ldots, x_m$,

$$k(x_1, \ldots, x_m) = f(g_1(x_1, \ldots, x_m), g_2(x_1, \ldots, x_m), \ldots, g_n(x_1, \ldots, x_m)). \tag{3}$$

Again, k is computable if $g_1, \ldots, g_n$ and f are all computable. Define an $(n+1)$-place function f to be *regular* if, for every $x_1, \ldots, x_n$, there is at least one y such that $f(y, x_1, \ldots, x_n) = 0$. If f is regular, then the *minimalization of f* is the n-place function whose value at any $x_1, \ldots, x_n$ is the smallest y such that $f(y, x_1, \ldots, x_n) = 0$. If f is regular, then its minimalization is computable.

Define a number-theoretic function f to be *primitive recursive* if there is a sequence $f_1, \ldots, f_n$ such that f_n is f, and for each $i \leqslant n$, either f_i is the successor function, f_i is the identically 0 function (i.e. for all x, $f_i(x) = 0$), f_i is a projection function (i.e. for all $x_1, \ldots, x_m$, $f_i(x_1, \ldots, x_m) = x_j$) or f_i is obtained from previous functions in the sequence by primitive recursion or composition. Define a number-theoretic function f to be *recursive* if there is a sequence $f_1, \ldots, f_n$ such that f_n is f, and for each $i \leqslant n$, either f_i is primitive recursive, f_i is obtained from previous functions in the sequence by composition or f_i is the minimalization of a previous regular function in the list. This is standard terminology today, but in some of the early work on computability the term 'recursive' is used for what is today called 'primitive recursive', and 'general recursive' is used for 'recursive'.

Let $a, b, \ldots$ be a collection of variables. Define a string to be a λ-*term*, and a variable that occurs in a λ-term to be *free* or *bound*, as follows: a variable x by itself is a λ-term, and the occurrence of x in x is free. If F and X are λ-terms, then $\{F\}(X)$ is a λ-term, and every occurrence of a variable x as free (respectively bound) in F and X is an occurrence of x as a free (respectively bound) variable of $\{F\}(X)$. If F is a single symbol, then $\{F\}(X)$ can be abbreviated as $F(X)$. Finally, if M is a λ-term that contains a variable x as free, then $\lambda x[M]$ is a λ-term. Every occurrence of x in $\lambda x[M]$ is bound, and if y is a variable other than x, then every occurrence of y as free (respectively bound) in M is free (respectively bound) in $\lambda x[M]$. Each natural number is associated with a λ-term as follows:

$$0: \quad \lambda a[\lambda b[a(b)]],$$
$$1: \quad \lambda a[\lambda b[a(a(b))]], \tag{4}$$
$$2: \quad \lambda a[\lambda b[a(a(a(b)))]],$$

and so on. If n is a natural number, then let $l(n)$ be the corresponding λ-term.

A *renaming* of a λ-term in the form $\lambda x[M]$ is a λ-term $\lambda y[M']$ such that y does not occur in M and M' is the result of replacing each free occurrence of x in M with y. An *evaluation* of a λ-term in the form $\{\lambda x[M]\}(N)$ is the λ-term obtained from M by replacing each occurrence of x with N, provided there is no bound occurrence of x in M and no variable has both a free occurrence in N and a bound occurrence in M. A λ-term M is a *reduction* of a λ-term N if there is a sequence $M_1, \ldots, M_n$ of λ-terms such that M_1 is N, M_n is M, and, for each $i < n$, M_{i+1} is obtained by replacing a part of M_i that is itself a λ-term with a renaming or an evaluation of that part.

Finally, a unary function f of natural numbers is *λ-definable* if there is a λ-term F such that, for each pair m and n of natural numbers, if $f(m) = n$ then $l(n)$ is a reduction of $\{F\}(l(m))$. A binary function g of natural numbers is *λ-definable* if there is a λ-term G such that, for all natural numbers m, n and p, if $g(m, n) = p$ then $l(p)$ is a reduction of $\{\{G\}(l(m))\}(l(n))$. The λ-definability of n-place functions is defined similarly. Again, if a function is λ-definable, then it is computable. The process of reduction provides an algorithm for computing the function. It is interesting to note that λ-terms are used to denote both numbers and functions. The system is sometimes called the *type-free λ-calculus*.

BIBLIOGRAPHY

Two major source-books of the original texts are available. Davis *1965* contains papers by Gödel, Turing, Church, Kleene and Post; van Heijenoort *1967* includes papers by Gödel, Hilbert and Skolem in English translation. Gödel's collected works *1986* contain the original papers, English translations and many insightful introductory notes by eminent scholars.

Davis, M. *1958, Computability and Unsolvability*, New York: McGraw-Hill.
—— *1965* (ed.), *The Undecidable*, Hewlett, NY: The Raven Press.
—— *1982*, 'Why Gödel didn't have Church's thesis', *Information and Control*, **54**, 3–24.
Detlefsen, M. *1986, Hilbert's Program*, Dordrecht: Reidel.
Gandy, R. *1988*, 'The confluence of ideas in 1936', in R. Herken (ed.), *The Universal Turing Machine*, New York: Oxford University Press, 55–111.
Gödel, K. *1931*, 'Über formal unentscheidbare Sätze der *Principia mathematica* und verwandter Systeme I', *Monatshefte für Mathematik und Physik*, **38**, 173–98. [Translated as 'On formally undecidable propositions of the *Principia mathematica*', in Davis *1965*: 4–35; also in van Heijenoort *1967*: 596–616.]
—— *1986, Collected Works*, Vol. 1 (ed. S. Fefferman *et al.*), New York: Oxford University Press.
Hilbert, D. *1900*, 'Mathematische Probleme'. [Various publications; English transl. as 'Mathematical problems', *Bulletin of the American Mathematical Society*, 1902, **8**, 437–79.]
—— *1925*, 'Über das Unendliche', *Mathematische Annalen*, **95**, 161–90. [Transl. as 'On the infinite', in van Heijenoort *1967*: 369–92 (cited here).]
Kleene, S. *1979*, 'Origins of recursive function theory', in *Twentieth Annual Symposium on Foundations of Computer Science*, New York, Institute of Electrical and Electronics Engineers, 371–82.
Rogers, H. *1967, Theory of Recursive Functions and Effective Computability*, New York: McGraw-Hill.
Shapiro, S. *1981*, 'Understanding Church's thesis', *Journal of Philosophical Logic*, **10**, 353–65.
—— *1983*, 'Remarks on the development of computability', *History and Philosophy of Logic*, **4**, 203–20.
Stein, H. *1988*, '*Logos*, logic, and *Logistiké*: Some philosophical remarks on the nineteenth century transformation of mathematics', in W. Aspray and P. Kitcher (eds), *History and Philosophy of Modern Mathematics*, Minneapolis, MN: University of Minnesota Press, 238–59.
Tait, W. *1981*, 'Finitism', *The Journal of Philosophy*, **78**, 524–46.
van Heijenoort, J. (ed.) *1967, From Frege to Gödel*, Cambridge, MA: Harvard University Press.

5.6

Constructivism

MICHAEL DETLEFSEN

1 INTRODUCTION

In the philosophy of mathematics, the term 'constructivism' is generally associated with a class of views placing certain characteristic restrictions on the kinds of reasoning taken to be admissible in mathematics, and a class of figures ranging from solid and celebrated citizens of the mathematical establishment (e.g. Leopold Kronecker, Henri Poincaré, Emil Borel and David Hilbert) to talented eccentrics (chiefly L. E. J. Brouwer, though, to a lesser extent, also Hermann Weyl and Brouwer's student A. Heyting) who not only held what many have regarded as curious philosophical views, but also were quite serious in their efforts to make mathematics conform to them.

Among the restrictions mentioned, perhaps the best known are the refusals to make use of (if not, in some cases, to reject in principle) any knowledge- or evidence-transcendent notion of truth, and of any mind- or intellect-transcendent notion of existence (though, as we shall see, both Kronecker and Hilbert are to some extent exceptions to this). By a knowledge- or evidence-transcendent notion of truth, we mean a notion of truth according to which a proposition can be true (false) even if we do not, and perhaps cannot, have any knowledge or evidence for (or against) it. Similarly, by a knowledge- or evidence-transcendent notion of existence, we mean a notion of existence according to which a thing can exist even though we do not, and perhaps cannot, have any knowledge or evidence that acquaints us with it.

From the refusal to make use of any evidence-transcendent notion of truth derive such familiar constructivist tenets as the disjunction principle, which requires that a proof of a disjunction prove one of its disjuncts, and the even more familiar ban on the unrestricted use of the principle of excluded middle. Likewise, from the refusal to allow any evidence-transcendent notion of existence arise such characteristic standards as the existence principle (which requires that an existence claim be accompanied by a description of how mentally to construct an exemplar, and thus prohibits the use of reasoning *reductio ad absurdum* as a means of establishing existence), and the general binding of sets to rules of generation (with the associated proscription of reasoning concerning actual infinities – that is, reasoning which results from treating infinite collections as cases of Being rather than cases of Becoming, or, in other words, as if they have an existence transcending that of the rules of generation by which they are given to thought).

2 EVIDENTIAL CONSTRUCTIVISM

Philosophically, it is possible to distinguish three broad types of constructivism, which we shall term, respectively, evidential, meaning-theoretic and idealistic. The first of these is based on a sorting of proofs into evidensory kinds (i.e. according to the kind and quality of evidence that they provide for their conclusions). Among these evidensory kinds, one is identified as superior, and there then arises a programme aimed at placing all of mathematics on this evidentially favoured basis – either by retaining its theorems (albeit perhaps in a 'translated' form) and finding proofs of the superior kind to replace proofs of inferior kind, or by cutting back mathematics to the point where something is counted as a theorem only if it is proved by the superior means.

2.1 Kronecker

At least in his stereotypical guise (on which more in a moment), Kronecker is commonly taken to have favoured such a position. In his view, elementary arithmetic reasoning sets the evidensory standard for the rest of mathematics. There, objects are defined and proved to exist by providing a means of constructing them in a finite number of steps. In the stereotypical Kronecker's view, proof of this sort represents the ultimate in objective justification, a view which he is said to have stated in a lecture of 1886 in the immortal words, 'God made the whole numbers, all the rest is human creation.'

In his only philosophical essay, 'Über den Zahlbegriff', Kronecker *1887* claimed to have found a similar view expressed by Carl Jacobi in a letter of 1846 to Alexander von Humboldt. Included in this letter was a parody of a Schiller poem, the closing lines of which are:

Was Du im Kosmos erblickst, ist nur der Göttlichen Abglanz,
In der Olympier Schaar thronet die ewige Zahl.

(What you glimpse in the cosmos is only the divine reflection of the immortal Number, enthroned amidst the Olympian throng.)

Kronecker then attributed this view to Carl Friedrich Gauss, citing as evidence Gauss's remark that 'Mathematics is the queen of the sciences and arithmetic the queen of mathematics.' He concurred, and further expressed the hope that in the not-too-distant future the 'arithmetization' of mathematics (i.e. the translation of all concepts into arithmetical concepts, together with the arithmetic proof of the resulting translations) would actually be carried out.

What comes next, however, is surprising. For, paraphrasing another remark by Gauss (given below), Kronecker characterized the principal difference between the arithmetic and non-arithmetic parts of mathematics as consisting in the fact that the numbers are purely a product of the human intellect, while space and time have an 'extra-mental' reality which prohibits a complete *a priori* determination of their laws. This seems to reverse the position of the 1886 remark, for instead of a realistic view of the arithmetic of the whole numbers we now find what would appear to be either an idealistic view like that of Brouwer and Weyl, or an instrumentalist view like that of Hilbert. Indeed, many years later, we find Weyl (*1949*: 22) stating that 'the numbers are to a far greater measure than the objects and relations of space a free product of the mind and therefore transparent to the mind'. This remark is strikingly similar to Gauss's original: 'wir mussen in Demuth zugeben, dass, wenn die Zahl *bloss* unsers Geistes Product ist, der Raum auch *ausser* unserm Geiste eine *Realität* hat, der wir *a priori* ihre Gesetze nicht vollständig vorschreiben können' (Kronecker *1887*: 265). In the light of these remarks, in which the numbers are presented more as 'human creation' and space and time as the work of God, it is difficult to embrace whole-heartedly the famous 1886 dictum as representing Kronecker's settled philosophical views.

2.2 Hilbert

Hilbert represents a different variety of evidential constructivism. For though Kronecker, at least in his stereotypical guise, may have regarded elementary arithmetic as evidentially privileged *vis-à-vis* the rest of mathematics, he gave little indication that he regarded the non-arithmetic portion as literally meaningless. This, however, is exactly what Hilbert did in his division of mathematics into a real and an ideal part (Hilbert *1922*, *1926*, *1927*; Detlefsen *1986*).

In this scheme, ideal mathematics, which forms the bulk of classical mathematics (in particular, analysis and set theory), is literally meaningless and has merely instrumental value; a value which, in this case, consists in its superior efficiency as a means of identifying truths of real mathematics (which efficiency Hilbert, in turn, attributed to the fact that classical logic is the psychologically natural logic of human reasoning). Real mathematics, on the other hand, is literally meaningful and evidentially genuine. It is comprised of the most elementary forms of arithmetic reasoning (termed 'finitary' by Hilbert), the epistemologically salient features of which are that it consists in the syntactical appearances of finite configurations of concrete symbols and thus, owing to the supposedly elemental (and elementary) character of judgments concerning such appearances, that it justifies the maximum attainable in rational confidence.

Since, in Hilbert's view, finitary evidence is the only genuine evidence in mathematics, it follows that the existence of mathematical objects (which he took to be symbolic expressions) can be justifiedly asserted only when syntactical appearances for them are made manifest (either by actually producing them, or by saying enough about some process for producing them as to have clarified what their appearances would be were they actually to be produced by this process). Hence one aspect of Hilbert's finitism is a constructivistic conception of existence claims. Its other aspect (as noted in Gödel *1958*) is that the objects with which one deals be 'concrete', or spatio-temporally 'inspectable'.

One also finds reflected in Hilbert's view of induction a commitment to the typical constructivist conception of the infinite as a form of potentiality. In his view, a universal sentence is not really a proposition at all (specifically, not the 'logical product' of its instances), but rather a proposition-schema (i.e. an expression which comes to express a proposition only by being instantiated). Similarly, an inductive proof is really not a proof, but rather a proof-schema, indicating how to prove any of the specific propositions $F(n)$ which form instances of the proposition-schema $(x)Fx$ for which it is a proof-schema. Specifically, it directs one to prove $F(n)$ by beginning with a (sub)proof of $F(1)$, adding to it a (sub)proof of $F(1) \rightarrow F(2)$, inferring $F(2)$, adding a (sub)proof of $F(2) \rightarrow F(3)$, inferring $F(3)$, and so on, to $F(n)$.

3 MEANING-THEORETIC CONSTRUCTIVISM

Unlike evidential constructivism, meaning-theoretic constructivism, which is our term for the so-called 'anti-realist' position of Michael Dummett (*1975*, *1977*), does not consist essentially in preference for a particular kind of evidence. Rather, it presents constructivism as the consequence of a proper account of meaning for mathematical language. The concept of meaning adopted is the well-known meaning-as-use doctrine associated with

Ludwig Wittgenstein's *Philosophical Investigations*, according to which to know the meaning of an expression is to know the conditions of its proper usage. Meaning, in this view, is something 'public' – that is, something shared by the members of a linguistic community – and not something 'private' within the mind of the user. It cannot transcend the publicly perceivable evidence for correct usage. Thus, since a proposition is 'used' by being either asserted or denied, knowledge of its meaning consists in a knowledge of the conditions under which these acts are proper; and since, in mathematics, an assertion of a proposition is proper only when it is accompanied by a proof, it follows that a sentence is assertable only if it is provable.

Clearly, such a position reproduces the traditional constructivist repudiation of the law of excluded middle. It is, however, less clear how or whether the remaining traditional constructivist principles (e.g. the full disjunction and existence principles) can be recovered. Dummett (*1975*: 20–21) says that their recovery is assured by the fact that the meaning-as-use conception makes inevitable the adoption of a general 'picture' of mathematical reality as being the product of our own thought, and that the adoption of this picture brings with it the full stable of constructivist principles. This, however, fails to make clear whether the inevitability in question reflects some kind of rational connection between the meaning-as-use doctrine and an idealist metaphysics, or is only a psychological compulsion to join what are rationally unrelated matters.

4 IDEALIST CONSTRUCTIVISM

4.1 Brouwer

We come now to idealist constructivism, which is perhaps the main variety and that represented by the intuitionism of Brouwer (*1907, 1933, 1948, 1955*). Unlike meaning-theoretic constructivism, it is based on more traditional epistemological and metaphysical considerations. And, unlike Kronecker's and Hilbert's evidential constructivism, it does not attempt to provide a constructivist basis for non-constructivist thought, but rather recommends the removal of such thought from mathematics altogether. It begins with the 'observation' that, for mathematicians, all that matters is how the objects of enquiry are given, any reality going beyond this being of no concern to them. This metaphysically indecisive epistemological point then acquires metaphysical momentum by being paired with what we shall call the transparency principle: to wit, that what the mind itself creates is especially accessible to it. (A version of this put forward by Weyl *1949*: 22, and closely similar to attitudes expressed by Gauss and Kronecker, was noted in Section 2 above.) With the transparency principle (germs of which can be found in Plotinus) supplying the epistemological motive, the move to an idealist metaphysics (according to which mathematical entities simply *are* the products of our thought, so that there *is* nothing to know about them aside from that which is 'given' in thought) becomes quite natural.

As remarked above, it is probably Brouwer's name that is most closely associated with this view. He described mathematics as 'an autonomous interior constructional mental activity which . . . neither in its origin nor in the essence of its method has anything to do with language or an exterior world' (*1955*: 551–2). He thus repudiated any truth or existence in mathematics 'existing independently of human thought', and from this follow such typical constructivist principles as the rejection of excluded middle and the demand that objects claimed to exist be 'exhibited'.

Brouwer distinguished himself from other constructivists (and even from other intuitionists such as Heyting and Weyl), however, by going well beyond this in his opposition to the use of logic in classical mathematics. His criticism of the classical mathematician's use of logic, unlike that of the constructivist generally, was not chiefly a local criticism of particular principles (e.g. excluded middle), but rather a criticism of the very role given to logical reasoning in classical mathematics (see Detlefsen *1990* for a sustained argument).

In Brouwer's view, logical reasoning follows regularities that are induced by the linguistic representation of our mathematical thought, and not regularities of that thought itself (which he described as 'an essentially languageless activity of the mind'; *1951*: 4). Because of this, we should not expect the development of mathematical thought to proceed according to the principles of any logic (i.e. the principles of any scheme of regularities induced by the linguistic representation of our thought). Consequently, it is a fundamental mistake to believe 'in the possibility of extending one's knowledge of truth by the mental process of thinking, in particular thinking accompanied by linguistic operations independent of experience called "logical reasoning"' (*1955*: 551).

4.2 Weyl

Like Brouwer, Weyl too was an idealistic constructivist. He regarded the numbers as 'a free product of the mind', and was explicit in his advocacy of the transparency principle. He also offered a description of that in virtue of which a mathematical item is 'given' or takes on mental existence: namely, by its being grafted into a scheme of manipulations (partially) determinative of that mental activity which constitutes (its part of) mathematics. (Broadly and roughly speaking, a scheme of mental manipulations involves the relation of presentations of the phenomena treated as 'canonical' or 'preferred', with surrounding families of presentations viewed as non-canonical. The canonical presentations are those that are manifest in the phenomenon itself, while the non-canonical presentations represent characteristics that are not manifest, but which are none the less determined by the canonical manifestation through the possible actions of the mind upon it. The 'given' thus consists of both the manifest and the non-manifest presentations of the phenomenon.) Thus the emphasis on computational control, which is a typifying characteristic of constructivism, is presented as a consequence of the fact that such control is that by virtue of which an entity comes to be 'given'.

Thus, too, the rejection of the actual infinite and reasoning which passes from ideal to some transcendent form of existence for sets. Generally speaking, a set is 'given' by a property that is characteristic of its elements, and not by the possession of individual descriptions of its various elements. Indeed, only in the case of finite sets is the latter a possibility. Hence, even if the elements of a set were to exist in a transcendent sense, an infinite collection of such elements would not; and this should make it clear that it is not, in general, permissible to proceed from the ideal existence of a set to a cognition-transcendent mode of existence (Weyl *1949*: 37–8 describes this as the chief mistake made by Gottlob Frege, Bertrand Russell and Richard Dedekind).

5 THE MATHEMATICAL DEVELOPMENT OF CONSTRUCTIVISM

Among mathematicians, the chief concern about constructivism has always been its strength – that is, whether it permits a body of mathematical thought as rich and interesting as classical mathematics. Hilbert (*1927*: 475–6) argued for the superiority of his form of constructivism over Brouwer's, precisely on the grounds that it would allow one to retain the rich 'paradise' of classical mathematics, while Brouwer's would require all but a fragment of it to be scrapped. Brouwer countered by arguing that much of what the classical mathematician counts as riches really is not, and that, just as the classicist can prove things that the intuitionist cannot, so too can the intuitionist prove things that the classicist cannot (e.g. the so-called continuity principle, which says that every total function defined on a unit continuum is uniformly continuous).

No final resolution of this issue has yet been reached. However, in the early 1930s Brouwer's student Heyting (*1930a, 1930b, 1934, 1971*) embarked upon a programme aimed at resolving it. His project called for the development of formal axiomatizations of various areas of intuitionistic mathematics so that their strength could be compared with that of their classical counterparts. Since then, several of Heyting's students have vigorously advanced this work; indeed, it is fair to say that this has become the major focus of contemporary work on intuitionism and constructivism (see the excellent surveys by Beeson *1985* and by Troelstra and van Dalen *1988* for an up-to-date account of this, along with extensive bibliographical references). Along similar lines, and having similar motives, is the project initiated by Bishop *1967* and continued by his students and followers (Bishop and Bridges *1985*, Richman *1981*).

BIBLIOGRAPHY

Beeson, M. *1985*, *Foundations of Constructive Mathematics*, Berlin: Springer.

Bishop, E. *1967*, *Foundations of Constructive Analysis*, New York: McGraw-Hill.

Bishop, E. and Bridges, D. *1985*, *Constructive Analysis*, Berlin: Springer.

Bridges, D. *1987*, *Varieties of Constructive Mathematics*, Cambridge: Cambridge University Press.

Brouwer, L. E. J. *1907*, 'Over de grondslagen der Wiskunde', Doctoral Thesis, University of Amsterdam. [Portions translated in Brouwer *1975*: 11–101.]

—— *1933*, 'Volition, knowledge and language', in Brouwer *1975*: 443–6.

—— *1948*, 'Consciousness, philosophy and mathematics', in Brouwer *1975*: 480–94.

—— *1951*, *Brouwer's Cambridge Lectures* (ed. D. van Dalen), Cambridge: Cambridge University Press.

—— *1955*, 'The effect of intuitionism on classical algebra of logic', in Brouwer *1975*: 551–4.

—— *1975*, *Collected Works*, Vol. 1 (ed. A. Heyting), Amsterdam: North Holland.

Detlefsen, M. *1986*, *Hilbert's Program*, Dordrecht: Reidel.

—— *1990*, 'Brouwerian intuitionism', *Mind*, **99**, 501–34.

Dummett, M. A. E. 1975, 'The philosophical basis of intuitionistic logic', in H. E. Rose and J. C. Shepherdson (eds), *Logic Colloquium '73*, Amsterdam: North-Holland, 5–40.

—— *1977*, *Elements of Intuitionism*, Oxford: Oxford University Press.

Gödel, K. *1958*: 'Über eine bisher noch nicht benützte Erweiterung des finiten Standpunktes', *Dialectica*, **12**, 280–87. [English transl. by W. Hodges and B. Watson in *Journal of Philosophical Logic*, 1980, **9**, 133–42.]

Heyting, A. *1930a*, 'Die formalen Regeln der intuitionistischen Logik', *Sitzungsberichte der Preussischen Akademie der Wissenschaften, Physikalisch–Mathematische Klasse*, 42–56.

—— *1930b*, 'Die formalen Regeln der intuitionistischen Mathematik', *Sitzungsberichte der Preussischen Akademie der Wissenschaften, Physikalisch–Mathematische Klasse*, 57–71, 158–69.

—— *1934*, *Mathematische Grundlagenforschung. Intuitionismus. Beweistheorie*, Berlin: Springer. [Repr. 1974.]

—— *1971*, *Intuitionism*, 3rd edn, Amsterdam: North-Holland.

Hilbert, D. *1922*, 'Neubegründung der Mathematik', *Abhandlungen aus dem mathematischen Seminar der Hamburgschen Universität*, 1, 157–77. [Also in *Gesammelte Abhandlungen*, Vol. 3, 157–77.]

—— *1926*, 'Über das Unendliche', *Mathematischen Annalen*, **95**, 161–90. [English transl.: van Heijenoort *1967*: 367–92.]

——*1927*, 'Die Grundlagen der Mathematik', *Abhandlungen aus dem mathematischen Seminar der Hamburgschen Universität*, **6**, 65–85. [English transl.: van Heijenoort *1967*: 464–79.]

Kronecker, L. *1887*, 'Über den Zahlbegriff', *Journal für die reine und angewandte Mathematik*, **101**, 337–55. [Also in *Philosophische Aufsätze Eduard Zeller* [. . .], Leipzig: Teubner, 261–74; and *Werke*, Vol. 3, Part 1, 249–74.]

Richman, F. (ed.) *1981*, *Constructive Mathematics*, Berlin: Springer.

Troelstra, A. and van Dalen, D. *1988*, *Constructivism in Mathematics*, Vols. 1 and 2, Amsterdam: North-Holland.

van Heijenoort, J. (ed.) *1967*, *From Frege to Gödel*, Cambridge, MA: Harvard University Press.

Weyl, H. *1949*, *Philosophy of Mathematics and Natural Science*, Princeton, NJ: Princeton University Press. [Repr. 1963, New York: Atheneum Press.]

5.7

Polish logics

PETER SIMONS

1 PLACES, PERSONS AND TIMES

Logicians from Poland have made a disproportionately significant contribution to modern logic, especially between the wars. This article outlines the historical and institutional background, and surveys the main figures and achievements of Polish logicians up to 1939. A comprehensive account of the Lwów–Warsaw school can be found in Woleński *1989*; a collection of outstanding papers in English translation is McCall *1967*.

1.1 Twardowski and Lwów

In 1895, Kazimierz Twardowski (1866–1938), who had studied in Vienna under Franz Brentano, became Professor of Philosophy at Lwów in Galicia, and proceeded to catalyse Polish philosophy through his organizational and didactic powers, and his example of clear analysis backed by sound historical knowledge. His students before 1914 included the first generation of those philosophers and logicians who were to dominate Polish thought. While not an original logician, Twardowski taught from 1898 a course surveying modern attempts to reform logic. His first major pupil, Jan Łukasiewicz, began to teach mathematical logic in Lwów from about 1905. His criticism of psychologism in logic influenced all subsequent Polish logicians. In 1910 Łukasiewicz published the monograph *O zasadzie sprzeczności u Arystotelesa* ('On the Principle of Contradiction in Aristotle', not yet translated), which questioned the principle's self-evidence and suggested it was a theorem rather than an axiom of logic. In an appendix he discussed the algebra of logic, and Russell's and other antinomies in the foundations of mathematics. This work made the basic ideas of mathematical logic easily available in Polish, and prompted another Twardowski student, Stanisław Leśniewski, to embark upon a lifelong mission to provide a logical foundation for mathematics.

1.2 The Warsaw centre

Despite the disruption of the First World War, Warsaw University was reconstituted in 1915, and Łukasiewicz was appointed Professor of Philosophy. He was followed in 1919 by Leśniewski, who had continued working on foundations in Russia. They were the principal teachers of the following generations of logicians. Though trained as philosophers, they were both (Łukasiewicz from 1920) professors in the Faculty of Mathematical and Natural Sciences. Warsaw mathematicians such as Z. Janiszewski, W. Sierpiński and S. Mazurkiewicz were sympathetic to logic and to sct theory (§3.6), and the opportunistic specialization of Polish mathematicians in set theory and topology was also fortunate, as these turned out to have close ties with mathematical logic. The cooperation between philosophers and mathematicians in teaching and in running journals was excellent, so it is not surprising that many of the most talented mathematics students specialized in logic. Foremost among these was Alfred Tarski, who quickly became the third principal of the Warsaw group.

1.3 Elsewhere

Although Warsaw was the unrivalled centre, logic also flourished elsewhere. At Kraków, then Lwów, the mathematician-painter Leon Chwistek developed a system of the foundations of mathematics combining extreme nominalism with criticism of *Principia mathematica*, becoming the first to advocate dispensing with the ramified in favour of the simple type hierarchy. In the 1930s, a Kraków group around Father Jan Salamucha (1903–44) attempted to modernize Catholic dogma using mathematical logic.

Though many Polish logicians survived the Second World War, 1939 marked the end of Polish logic's golden age. Leśniewski died in May; Tarski was fortunately in America when war broke out; Salamucha and the Jewish logicians Adolf Lindenbaum, Mordechaj Wajsberg and Mojżesz Presburger were murdered; and survivors such as Łukasiewicz, Bolesław Sobociński, Józef Bocheński and Czesław Lejewski did not return after the war.

2 THEMES

2.1 Classical sentential calculi

The intensive study of sentential calculi was initiated by Łukasiewicz. He invented a bracket-free notation, in which, for example

$$((p \wedge \sim q) \supset (r \vee \sim \sim p)) \quad \text{becomes} \quad CKpNqArNNp. \qquad (1)$$

In 1929 he produced a calculus based on implication and negation, with the axioms $CCpqCCqrCpr$, $CCNppp$ and $CpCNpq$. Łukasiewicz also invented an extremely compact yet exact way of presenting proofs. In 1936 he discovered the single axiom

$$CCCpqCCCNrNstrCuCCrpCsp. \qquad (2)$$

Polish logicians, with a highly developed sense of logical aesthetics, were constantly seeking shortest axioms and optimizing other virtues. Another of Łukasiewicz's innovations, later also used in characterizing Aristotelian syllogistic, was the introduction of rules of rejection in addition to rules of assertion: whereas assertion, $\vdash A$, means that all substitution instances of A are theorems, rejection, $\dashv A$, means that not all substitution instances of A are theorems.

Another Warsaw speciality was partial calculi, based on one or more primitives, such as pure implicational and pure equivalential calculi. Further innovations in sentential logic were Leśniewski's addition of quantifiers and variable functors of higher order, a system he called 'protothetic', Łukasiewicz's use of variable functors, and Stanisław Jaśkowski's system of natural deduction. Metalogical properties such as completeness and independence of axioms were also studied.

2.2 Many-valued logics

The ground for Łukasiewicz's invention of three-valued logic was broken between 1910 and 1913, in his Aristotle studies, and when Tadeusz Kotarbiński and Leśniewski debated the truth of propositions about the future. Łukasiewicz created his first three-valued system in 1917. The third truth value – he called it 'possibility' – characterized propositions about indeterminate future contingents. In the 1920s, n-valued and infinite-valued systems were introduced. Initially, many-valued logic was investigated using truth-value matrices. Later, Łukasiewicz's three-valued system was axiomatized by Wajsberg, and the first functionally complete system was given by Jerzy Słupecki. Attitudes to many-valued logic differed widely: for Łukasiewicz they were characteristic of creative freedom; for Leśniewski they were false and fictitious; for others they were simply interesting objects of investigation.

2.3 Other non-classical logics

Łukasiewicz linked modality to many-valuedness, but after seeing that three values would not capture all the intuitively correct laws of modal inference, he eventually (1953) opted for a four-valued system. This view remained idiosyncratic. Several logicians studied the metalogic of intuitionistic sentential calculi: in 1936 Jaśkowski constructed infinite truth matrices adequate for intuitionistic logic. Normally Polish logicians were extensionalists, so they had little sympathy for the philosophy behind intensional logic.

2.4 Leśniewski's logic

Repelled by the inexactness of Russell and Whitehead's *Principia mathematica* (1910–13), Leśniewski initially presented his work on the foundations of mathematics in a regimented Polish; only after 1920 did he use symbols. He rejected formalism and insisted that logical systems should consist of meaningful, true sentences about the real world. His systems employ languages constructed according to a rigorous scheme of semantic categories, later (1935) formalized by Kazimierz Ajdukiewicz. The most fundamental system, though the last to be axiomatized, was protothetic. Next came ontology, a calculus of terms combining aspects of the Boole–Schröder algebra with Fregean quantifiers and higher-order categories. The basic form of atomic sentence, '$a \in b$', meaning roughly '(the) a is a b', employs a singular copula between two names, which may, unlike the singular terms of predicate logic, designate not just one object, but also several, or none. Mereology, chronologically the first system (1916), was a calculus of part, whole and aggregates, intended as a substitute for standard set theory, which Leśniewski found incomprehensible.

The nominalist Leśniewski regarded his logical systems as expandable collections of concrete inscriptions, formulating his rules of expansion, called 'directives', with unparalleled exactness. He gave precise requirements for definitions and showed the need for scrupulous observance of the use–mention distinction. His publications lagged behind his research, and after his death in 1939 Sobociński was to publish his *Nachlass*, but it was destroyed in 1944 and the results had to be reconstructed by Leśniewski's students. Though the most forceful personality among Warsaw logicians, Leśniewski did not establish his systems as the norm.

2.5 Semantics and metamathematics

Truth was a constant issue for Twardowski and his pupils from before 1914.

Leśniewski's view that logical theses are truths was threatened by the mathematical and semantic paradoxes, and Tarski's semantic theory of truth (formulated around 1929, published 1933), the most famous and influential product of the Warsaw school, required the object language to be expressively weaker than its meta-language to allow a formally consistent and materially adequate bivalent truth theory. Prompted by Kurt Gödel's work, Tarski showed that the class of true sentences of formalized arithmetic is not definable in arithmetic. In the 1930s Tarski broadened his concept from truth theory to semantics. Gödel's and Tarski's limitation results showed that a syntactic concept of logical consequence was too weak, and Tarski replaced it by a semantic concept, laying a basis for later work in model theory. Metamathematics in Poland was not tied to a formalist or finitist programme like David Hilbert's. Tarski axiomatized logical consequence, investigated deductive systems, the algebra of logic and (with Lindenbaum) the theory of definability. Together with Andrzej Mostowski, they co-operated with mathematicians like Stefan Banach, K. Kuratowski and Sierpiński on the foundations of mathematics, especially in set theory and decidability.

2.6 History of logic

Using formal methods, Łukasiewicz revolutionized research into the history of logic. His 1910 monograph on Aristotle was just the beginning. He showed the importance of sentential logic in the Stoics and medievals. He and Słupecki formally investigated and revitalized syllogistic, Salamucha analysed Thomas Aquinas's proof *ex motu* (1934) and demonstrated William of Ockham's pre-eminence in medieval sentential logic, while both Tadeusz Czeżowski (1936) and Bocheński (1938) wrote on the history of modal logic.

BIBLIOGRAPHY

Jordan, Z. *1945, The Development of Mathematical Logic and of Logical Positivism in Poland Between the Two Wars*, Oxford: Oxford University Press. [Partly repr. in McCall *1967*: 346–406.]
McCall, S. (ed.) *1967, Polish Logic 1920–1939*, Oxford: Clarendon Press.
Tarski, A. *1983, Logic, Semantics, Metamathematics. Papers from 1923 to 1938*, 2nd edn (ed. J. Corcoran), Indianapolis, IN: Hackett. [1st edn 1956, Oxford: Clarendon Press.]
Woleński, J. *1989, Logic and Philosophy in the Lvov–Warsaw School*, Dordrecht: Kluwer.

5.8

Model theory

GEORGE WEAVER

1 INTRODUCTION

Model theory is that branch of mathematical logic which studies relationships between mathematical systems and sets of propositions. Literally, the term 'model theory' applies to any discussion of the relation 'is a model of'. This relation holds between a system and a set of propositions provided every member of the set is true on the system.

The term 'theory of models' was introduced by Tarski *1954*. Model theory emerged as a separate branch of logic in the late 1940s or early 1950s. Readers interested in the major ideas and influences in research should consult Chang *1974*; Vaught *1973* and Morley *1973* present representative methods and results; Chang and Keisler *1990* provide an up-to-date introduction to the field and give an extensive bibliography and historical notes.

Model theory originated in the study of metalogic. Roughly, the study of metalogic began with the publication of Löwenheim *1915* and reached maturity in the mid-1930s. The reader interested in this period in the development of model theory should consult Vaught *1974* and Corcoran *1983*; many of the important papers from this period can be found in van Heijenoort *1967*.

Prior to 1915, model-theoretic concepts played an important role in the foundations of mathematics. These concepts were applied to various mathematical theories, and general principles involving them were articulated in the late nineteenth and the early twentieth century. Discussions of the model theory of this period can be found in Corcoran *1980*, *1981* and Scanlan *1988*, *1991*. Papers from this period can be found in van Heijenoort *1967*.

The mathematics of the eighteenth century was naturally seen as the abstract study of nature, but in the nineteenth century concepts were introduced which had no clear connection with physical reality (e.g. quaternions and non-Euclidean geometry). These developments led to attempts in the late nineteenth century to rethink the purpose(s) of mathematics.

2 PROPOSITIONS AND SYSTEMS

By the turn of the century, workers in the foundations of mathematics had recognized a distinction between mathematical systems and their theories. A system was thought of as a collection of objects (the domain of the system) together with a finite sequence of distinguished properties, relations and functions on that domain and elements of that domain. The theory of a system was thought of as a collection of propositions about that system.

There were several views of the nature of propositions and their relations to systems. According to one, the non-syntactic propositions are the meanings of sentences. In this view, propositions are tied to particular systems and cannot be interpreted on others; Gottlob Frege appears to have arrived at something like this view by 1906. According to another view, the syntactic propositions are sentences. In this view, in contrast to the non-syntactic view, propositions are not themselves true or false, but true on a system or false on a system. Both David Hilbert *1899* and Oswald Veblen *1904* maintained something like this second view.

The notion of interpreting a proposition on a system is fundamental to the syntactic view of propositions. To get a flavour of what is involved here, consider an example. Let (N, 0, s) be that mathematical system whose domain (N) is the set of non-negative integers, together with distinguished element zero (0) and successor function (s). Let (E, 2, s') be that system whose domain (E) is the set of positive even integers together with distinguished element (2) and distinguished function (s'), where s' (2n) = 2n + 2. Finally, let (N, 0, I) be that system where N and 0 are as above, and the distinguished function (I) is the identity function on N (i.e. I(n) = n). Now, consider the principle of mathematical induction, formulated as follows: 'for all sets S, if S(*0*), and for all x, if S(x), then S(*s*(x)), then for all y, S(y)'. This proposition, which can be interpreted on all three of the above systems, presupposes a grammatical analysis of the proposition. On this analysis, '*0*' is an individual constant (or numeral), '*s*' is a functional (or operational) constant, 'S' and 'x' are variables, 'for all' is a universal quantifier, and 'if...then' and 'and' are sentential connectives. Terms like '*0*' and '*s*' are called non-logical terms.

The principle of induction is interpreted on a system as follows: the individual constant denotes the distinguished object, the functional constant denotes the distinguished function, the variables 'x' and 'y' range over members of the domain, the variable 'S' ranges over subsets of the domain, and the quantifier and sentential connectives have their usual meanings. Given a value for 'x', '*s*(x)' denotes that member of the domain which is the image of the value of 'x' under the distinguished function. A given value for 'S' satisfies 'S(*0*)' only when the distinguished element is a member of the value for 'S'. In the same way, given values for 'S' and 'x', 'S(x)' is satisfied by these values only when the value of 'x' is a member of the value of 'S', and 'S(s(x))' is satisfied by these values only when the denotation of 's(x)' is a member of the value for 'S'.

Given this understanding of how the principle of induction is interpreted on a system, it is easily seen that it is true on (N, 0, s) and (E, 2, s'), but false on (N, 0, I). Obviously, the principle of induction can be interpreted on any system of the form (A, a, f) where A is a set, a is an element in A and f is a function from A to A. Note that in interpreting the principle of induction on a system, the meanings of '0' and 's' are ignored.

This understanding of interpreting a proposition in a system was accepted without explicit formulation in the late nineteenth and the early twentieth century. Similar remarks apply to the concepts of being true on a system and being false on a system. Precise formulations of all these notions were presented in Alfred Tarski *1935*. While some writers were explicit about the underlying grammatical analysis (e.g. Peano *1889*), others were not (e.g. Hilbert *1899*).

It appears that a third view of propositions can be found in the work of Eugenio Beltrami and J. Hoüel from the late 1860s. In this view, propositions, while tied to a given system, could be reinterpreted on at least some others provided (in contrast to the above) the meaning of the non-logical terms remained fixed (Scanlan *1988*).

3 CATEGORICITY

From a model-theoretic point of view, one of the most interesting features of foundational studies was the project of characterizing mathematical systems uniquely up to isomorphism. Richard Dedekind *1887* appears to have been the first to suggest that the axioms of the theory of a system should determine that system uniquely up to isomorphism. He seems to have believed that mathematics should study the structure of system (i.e. what is grasped when we ignore the identity of the objects in the domain, and focus only on how many of them there are and how they are related to one another by the distinguished properties, functions and relations of the systems (art. 73)); and that when a system is determined uniquely up to isomorphism we have succeeded in capturing its structure (i.e. that isomorphic systems have the same structure).

A set of axioms is called 'categorical' if all models of the axioms are isomorphic. For example, the set consisting of the principle of mathematical induction and propositions to the effect that the successor function is 1–1 and that zero is not the successor of any member of the domain is categorical. Systems (A, a, f) and (B, b, g) are isomorphic provided there is a function, ψ, from A to B which is 1–1 and onto B, sends a to b and preserves the distinguished functions (i.e. for all a' in A, $\psi(f(a')) = g(\psi(a'))$). (N, 0, s) and (E, 2, s') are isomorphic, but neither is isomorphic to (N, 0, I).

The term 'categorical' was introduced by Veblen in *1904*. By the end of that decade, several important systems had been given categorical axiomatizations; many of these results are reported in Huntington *1905–06*. It was common practice to refer to these systems by referring to their domains: thus, the system (N, 0, s) is the system of the non-negative integers. Following this practice, each of the following can be said to have been given categorical axiomatizations by 1905: the non-negative integers, the integers, the rationals, the reals, the complex numbers and Euclidean space.

4 CATEGORICITY AND COMPLETENESS

In the course of these investigations it was realized that categorical axiomatizations of a system provide a complete codification of the propositions true on that system in the sense that every proposition true on the system is a logical consequence of the axioms. Dedekind (*1887*: art. 134) observed that isomorphic systems make the same propositions true; Veblen *1904* concluded from this observation that every proposition true on a system was a logical consequence of a categorical set of axioms for that system.

The thesis that categoricity implies completeness was merely stated; no proof of it was offered. The notion of logical consequence presupposed by Veblen is precisely that formulated in Tarski *1936*: if c is a proposition and P a set of propositions, then a system is called a counter-interpretation for (P, c) provided it is a model of P but not a model of c. Veblen presupposed that having no counter-interpretation was a necessary and sufficient condition for c being a logical consequence of P. He explicitly distinguished between logical consequence and provability, and raised the completeness question for the underlying logic: whether or not every logical consequence of P could be proved from P.

Counter-interpretations were used at the turn of the century to show that sets of axioms were independent in the sense that no one of them was provable from the others. Let c be an axiom for the theory of a system, and let P be the axioms of that system other than c. Padoa *1900* observed that if c is provable from P, then (P, c) has no counter-interpretation. Hence, if there is a model of P which is not a model of c, then c is not provable from P. This principle, which Padoa states for all theories, had been used by Hilbert *1899* in geometry and Giuseppe Peano *1889* in arithmetic.

5 DEFINITION

Model-theoretic ideas can also be found in the nineteenth-century literature on the role of definition in the axiomatization of the theory of a system. Padoa *1900* articulated a method for showing that a non-logical term was not definable on the basis of others. No attempt was made to prove the correctness of the method, but Tarski *1934* did announce a theorem which justifies the method. In the axiomatization of $(N, 0, s)$ discussed in Section 3, '*0*' can be defined in terms of s. To see this, note that the following is a consequence of the axioms: 'for all x, $x = 0$ iff for all y, $x \neq s(y)$'. Now, consider whether or not 's' is definable in terms of '*0*'. According to Padoa, such a definition expresses a relation between '*0*' and 's' which is sufficient to determine uniquely the denotation of 's' in any model of the axioms. Thus, if (A, a, f) and (A, a, f') are models of the above axioms, and $f \neq f'$, then 's' is not definable in terms of '*0*'. Note that the system $(N, 0, f)$ is a model of these axioms, where $f(0) = 2$, $f(2) = 1$, $f(1) = 3$ and $f(n) = n + 1$ for $n \geqslant 3$. Thus, 's' is not definable in terms of '*0*'.

6 FROM 1915 TO 1930

While investigators in the foundations of mathematics articulated several model-theoretic principles, they did not attempt to prove them, perhaps because they saw themselves as merely articulating principles which they understood to be implicit in practice. The work of Tarski and others in the 1920s and 1930s can be seen as giving these principles mathematically precise formulations and justifications.

By the mid-1930s, the grammars, semantics (model theory) and deductive systems for a variety of logics had been given mathematically precise formulations; the model-theoretic notions of satisfaction, truth and logical consequence had been given precise formulations; various mathematical theories had been formulated in these logics; and several model-theoretic results had been established (e.g. that isomorphic systems make the same propositions true; and that categoricity implies completeness).

Between 1915 and 1930 a distinction emerged between first-order logics and logics of second and higher order, based on syntactic differences between the languages of these logics. The languages of first-order logics (first-order languages) contained but one kind of variable, called individual variables, which, when interpreted in a system, range over members of the domain of that system. In contrast, the languages of second-order logics (second-order languages) contain, in addition to individual variables, set variables ranging over subsets of the domain, relational variables ranging over relations on the domain, and functional variables ranging over functions on the domain. Thus, second-order languages contain first-order sublanguages. The reader interested in differences between first- and second-order languages should consult Boolos *1975* and Shapiro *1985*.

Second-order logics were common in the late nineteenth and the early twentieth century. For example, the principle of induction as described in Section 2 is a second-order proposition and all the categorical axiomatizations mentioned at the end of Section 3 contain second-order propositions. Model-theoretic results about first-order languages appeared as early as 1915. For example, Leopold Löwenheim *1915* gave a proof that any set of propositions in a first-order language with a model has a model whose domain is either finite or countably infinite. This result, generalized further by Thoralf Skolem *1920* and now known as the Löwenheim–Skolem theorem, implies that no set of propositions in a first-order language can provide a categorical axiomatization of the system of the real numbers.

First-order languages were initially seen as interesting sublanguages of second-order languages. Gradually, however, research in metamathematics and model theory came to focus on first-order logics. The reader interested in these developments should consult Moore *1987*, and §5.4.

Kurt Gödel *1930* showed that first-order logics were complete in the sense that, if P is any set of first-order propositions, then any first-order proposition which is a consequence of P is provable from P. Since this result involves both the semantics and the deductive system of first-order logic, it is not, strictly speaking, model-theoretic. However, it does have model-theoretic consequences: in particular, it implies that, if every finite subset of a set of first-order propositions has a model, then the whole set has a model. This result, which has come to be called the compactness theorem, is widely regarded as one of the most fundamental results in the model theory of first-order languages.

Another fundamental result in this area was discovered before 1930, although it was not published until later. From 1926 to 1928, Tarski held a seminar in logic at the University of Warsaw. In 1928 he presented in that seminar the theorem that, if a set of first-order propositions has a model whose domain is infinite, then it has a model in each infinite power (i.e. for β, any infinite cardinal, the set has a model whose domain is of cardinality β). This result, a version of what is called the upward Löwenheim–Skolem theorem, implies that no set of propositions in a first-order language can provide a categorical axiomatization of any system with an infinite domain. Hence it is impossible to capture the structure of any system which has an infinite domain with a set of first-order propositions.

Interest in obtaining complete axiomatizations continued after 1915. However, the focus of investigation shifted to first-order logic. Here, the first-order theory of a system is that set of first-order propositions true on the system. Axioms are chosen from the first-order theory; a set of such axioms is said to be complete provided every first-order proposition true on the system is a consequence of the axioms.

By the late 1920s, several first-order theories were shown to have complete axiomatizations. For example, the first-order theory of the rational numbers, together with the 'less than' relation, was shown to be complete in 1928. Other results were presented at Tarski's seminar in Warsaw. Since no set of first-order propositions can provide a categorical axiomatization for a system with an infinite domain, the methods used to obtain these results were fundamentally different from those used at the turn of the century. The development of techniques for establishing the completeness of first-order theories remained an important area of research in model theory well into the 1950s.

7 GÖDEL'S INCOMPLETENESS THEOREM

The first incompleteness result was published by Gödel in *1931*. This result, called Gödel's incompleteness theorem, is widely considered to be one of the most profound results in logic. Consider the system N whose domain is the set of non-negative integers, together with a single distinguished object (zero), the successor function, and the operations of addition, multiplication and exponentiation. A categorical axiomatization for N is provided by the principle of induction, propositions to the effect that zero is not the successor function of any number and that the successor function is 1–1, and the recursive definitions of addition, multiplication and exponentiation. Of these axioms, all but the proposition of induction are first order.

An axiomatization for the first-order theory of this system can be obtained from the above by replacing the principle of induction by the collection of its first-order 'instances'. Consider a first-order language whose non-logical vocabulary consists of the numeral zero ('*0*') and operation symbols for the successor function, addition, multiplication and exponentiation. The numerals of this language are '*0*', '*s(0)*', '*s(s(0))*', and so on. For each non-negative integer n, let n be the numeral which denotes n in N. Let $\phi(n)$ be a first-order principle in which the numeral n occurs. Let $\phi(x)$ denote the result of replacing every occurrence of n in $\phi(n)$ by an occurrence of the individual variable 'x', where 'x' does not occur in $\phi(n)$; $\phi(x)$ is called a first-order condition. For $\phi(x)$, a first-order condition 'In($\phi(x)$)' denotes the following first-order proposition: 'if $\phi(0)$ and for all x, if $\phi(x)$ then $\phi(s(x))$, then for all x, $\phi(x)$'.

In($\phi(x)$) is a consequence of the principle of induction, and hence is true on N. Consider the set of propositions obtained by adding In($\phi(x)$) to the first-order axioms above, for all first-order conditions $\phi(x)$. This set is infinite.

Gödel showed that the elements of the vocabulary of the language could be associated with non-negative integers in such a way that the collection of finite sequences of members of the vocabulary are associated in a 1–1 way with the non-negative integers. The number associated with a sequence is called its Gödel number. Thus, corresponding to each set of finite sequences there is a set of non-negative integers: the set of Gödel numbers of the sequences in the set. A set of finite sequences is said to be represented by the first-order condition $\psi(x)$ provided, for all n, $\psi(n)$ is true on **N** iff n is the Gödel number of a member of the set.

The Gödel numbering can be done in such a way that those sets of finite sequences which are characterized by their outward appearances or form are represented by a first-order condition. A set of sequences has a formal characterization provided there is a formal procedure (i.e. one that utilizes only the form of the sequences) for determining, for any sequence and in a finite amount of time, whether or not it is in the set. Sets of sequences which have a formal characterization in this sense are said to be recursive. For example, the set of first-order propositions and the set of axioms are both recursive; furthermore, any finite set is recursive.

Since the set of axioms is recursive, the set of proofs from these axioms is also recursive. It follows that the set of first-order propositions provable from the axioms is representable by a first-order condition. It is possible to enumerate the first-order conditions in such a way that there is a first-order condition $\psi(x)$ such that, for all n, $\psi(n)$ is true on **N** provided $\phi_n(n)$ is not provable from the axioms, where $\phi_n(x)$ is the n-th member of the enumeration. Since $\psi(x)$ is a first-order condition, there exists m such that $\psi(x)$ is the m-th member of the enumeration. Thus, $\psi(m)$ is true on **N** provided $\psi(m)$ is not provable from the axioms. Since **N** is a model of the axioms, any proposition provable from the axioms is true on **N**; $\psi(m)$ is either provable from the axioms or not provable from the axioms, but in either case $\psi(m)$ is true on **N**. Thus, $\psi(m)$ is true on **N**, but not provable from the axioms. Since first-order logic is complete, $\psi(m)$ does not follow from the axioms, and the axioms are incomplete.

The reasoning outlined above can be extended to any set of axioms for the first-order theory of **N**, if that set is recursive. Thus, since every first-order proposition true on **N** is in the first-order theory of **N** and hence provable from the theory, the first-order theory of **N** is not recursive.

Infinite sets of axioms first appeared in the early 1920s in connection with the attempt to axiomatize particular first-order theories. There appears to have been no recognition prior to the early 1920s that sets of axioms might be infinite – in fact, as late as 1930 Tarski appears to have required all sets of axioms to be finite. This requirement seems to have been removed in Tarski *1935*, to be replaced by the requirement that they be recursive, which appears to have gained general acceptance by the early 1950s. Thus, accepting that all sets of axioms are recursive, it follows from the above that there is no complete set of axioms for the first-order theory of N.

BIBLIOGRAPHY

Several works are cited from the collections Tarski *1983* and van Heijenoort *1967*.

Boolos, G. *1975*, 'On second order logic', *The Journal of Philosophy*, 72, 509–27.
Chang, C. C. *1974*, 'Model theory 1945–1971', in Henkin *et al.* *1974*: 173–86.
Chang, C. C. and Keisler, H. J. *1990*, *Model Theory*, 3rd edn, Amsterdam: North-Holland.
Corcoran, J. *1980*, 'Categoricity', *History and Philosophy of Logic*, 1, 187–207.
—— *1981*, 'From categoricity to completeness', *History and Philosophy of Logic*, 2, 113–19.
—— *1983*, 'Editor's introduction to the revised edition', in Tarski *1983*: xv–xxvii.
Dedekind, R. *1887*, *Was sind und was sollen die Zahlen?*, 1st edn, Braunschweig: Vieweg. [English transl. by W. W. Berman, 1901, *Essays on the Theory of Numbers*, Chicago, IL: Open Court; repr. 1963, New York: Dover.]
Gödel, K. *1930*, 'The completeness of the axioms of the functional calculus of logic', in van Heijenoort *1967*: 582–91.
—— *1931*, 'On formally undecidable propositions of *Principia mathematica* and related systems', in van Heijenoort *1967*: 592–617.
Henkin, L. *et al.* (eds) *1974*, *Proceedings of the Tarski Symposium*, Providence, RI: American Mathematical Society.
Hilbert, D. *1899*, *Die Grundlagen der Geometrie*, 1st edn, Leipzig: Teubner. [English transl. by E. J. Townsend as *The Foundations of Arithmetic*, 1902, La Salle, IL: Open Court.]
Huntington, E. V. *1905–06*, 'The continuum as a type of order', *Annals of Mathematics*, Series 2, 6, 151–84; 7, 15–43.
Löwenheim, L. *1915*, 'On possibilities in the calculus of relatives', in van Heijenoort *1967*: 228–51.
Moore, G. *1987*, 'The emergence of first-order logic', in W. Aspray and P. Kitcher (eds), *History and Philosophy of Modern Mathematics*, Minneapolis, MN: University of Minnesota Press, 95–135.
Morley, M. D. (ed.) *1973*, *Studies in Model Theory*, Providence, RI: American Mathematical Society.
Padoa, A. *1900*, 'Logical introduction to any deductive theory', in van Heijenoort *1967*: 118–23.
Peano, G. *1889*, 'The principles of arithmetic presented by a new method', in van Heijenoort *1967*: 83–97.
Scanlan, M. *1988*, 'Beltrami's model and the independence of the parallel postulate', *History and Philosophy of Logic*, 5, 13–34.

—— *1991*, 'Who were the American postulate theorists?', *Journal of Symbolic Logic*, **56**, 981–1002.

Shapiro, S. *1985*, 'Second order languages and mathematical practice', *Journal of Symbolic Logic*, **50**, 714–42.

Skolem, T. *1920*, 'Logico-combinatorial investigations in the satisfiability or provability of mathematical propositions: A simplified proof of a theorem by L. Löwenheim and generalizations of the theorem', in van Heijenoort *1967*: 252–63.

Tarski, A. *1934*, 'Some methodological investigations on the definability of concepts', in Tarski *1983*: 296–319.

—— *1935*, 'The concept of truth in formalized languages', in Tarski *1983*: 152–278.

—— *1936*, 'The concept of logical consequence', in Tarski *1983*: 409–20.

—— *1954–5*, 'Contributions to the theory of models I, II, III', *Indagationes Mathematicae*, **16**, 572–88; **17**, 56–64.

—— *1983*, *Logic, Semantics, Metamathematics*, 2nd edn (ed. J. Corcoran), Indianapolis, IN: Hackett.

van Heijenoort, J. (ed.) *1967*, *From Frege to Gödel*, Cambridge, MA: Harvard University Press.

Vaught, R. L. *1973*, 'Some aspects of the theory of models', *American Mathematical Monthly*, **80**, 3–37.

—— *1974*, 'Model theory before 1945', in Henkin *et al. 1974*: 152–72.

Veblen, O. *1904*, 'A system of axioms for geometry', *Transactions of the American Mathematical Society*, **5**, 343–84.

5.9

Some current positions in the philosophy of mathematics

NICOLAS D. GOODMAN

1 SET-THEORETIC PLATONISM

Since the Second World War, the dominant position on the foundations of mathematics has been the view that mathematics should be thought of as formalized in some extension of Zermelo–Fraenkel axiomatic set theory ZFC (including the axiom of choice). Sometimes this position is held as an empirical thesis: that is, it is maintained that as a matter of fact all known mathematics can be formalized in ZFC or in some suitable extension of ZFC. More frequently, the position is asserted in a normative sense: that is, it is maintained that formalizability in ZFC or in some suitable extension of ZFC is a criterion of correctness for mathematical arguments and results.

In either case, the set-theoretic interpretation of mathematics is taken as canonical. Ordered pairs really are unordered pairs of a unit set and an unordered pair: (a, b) is $\{\{a\}, \{a, b\}\}$. Functions really are sets of ordered pairs. Natural numbers really are sets of preceding natural numbers. Real numbers really are sets of rational numbers, which are equivalence classes of pairs of integers, which are equivalence classes of pairs of natural numbers. Points in Euclidean three-dimensional space really are ordered triples of real numbers. So pervasive has this orthodoxy become that there have even been widespread attempts to teach set theory to schoolchildren on the grounds that it is more fundamental than the traditional elementary mathematical curriculum.

A number of authors have attacked this set-theoretic reductionism (see e.g. Paul Benacerraf on 'What numbers could not be', reprinted in Benacerraf and Putnam *1983*). It continues to be popular, nevertheless, because it appears to give such a smooth and uniform account of mathematical ontology. Mathematicians, moreover, still feel a strong allegiance to the Euclidean ideal of deducing everything from a few axioms stated at the beginning. The axioms of ZFC admirably fill that role without intruding very much on ordinary mathematical practice.

If mathematics is set theory and all mathematical objects are sets, then the main metaphysical issue raised by mathematics is the question of the nature and status of sets. The most popular answer to the question was articulated forcefully and clearly by Kurt Gödel (see his essays reprinted in Banacerraf and Putnam *1983*). According to him, sets are eternal, non-material objects whose reality is no more problematic than is the reality of physical objects. Just as we must assume the existence of physical objects to understand and systematize our sense experience, so we must assume the existence of sets in order to understand and systematize our experience with numbers and computation.

The most important difference, for Gödel, between physical objects and sets is epistemological. In his view, we know the correctness of the axioms of ZFC primarily because their correctness is evident in a proper under-standing of the meaning of set theory. The standard interpretation of set theory, often called the 'iterative conception of sets', is expounded by, among many others, George Boolos (in his paper reprinted in Benacerraf and Putnam *1983*). The view that mathematical axioms become self-evident once properly understood was defended in a number of influential works by Georg Kreisel, who acknowledged his indebtedness to Gödel. In this view, then, mathematics is an *a priori* science, founded on self-evident truths, and concerned with non-material objects existing independently of human thought (a view referred to as Platonism).

There has been extensive philosophical criticism of set-theoretic Platonism (see Section 2). The main problem that it has faced, however, has been internal. Specifically, it follows from well-known results obtained by Gödel, Paul Cohen, and a multitude of other authors that the axioms of ZFC do not suffice to settle even very fundamental questions about the nature of the set-theoretic universe. In particular, they leave open the cardinality of the set of real numbers. This is the famous 'continuum problem' (§5.4). The search for new axioms that will resolve this and other questions had led to extensive discussions of the nature of sets and of the sources of our knowledge of set theory. (For a survey of this work and some of its philosophical implications, see Maddy *1988*.) Many mathematicians feel that the failure of this research to lead to conclusive results undermines their confidence in a Platonistic interpretation of set theory.

2 TWO VERSIONS OF CONSTRUCTIVISM

Set-theoretic Platonism may perhaps be seen as the contemporary heir to the logicism of Gottlob Frege and Bertrand Russell. The intuitionism of

L. E. J. Brouwer has led to two main successor views that have been advocated recently. Among mathematicians the more influential of these has been the constructivism of Errett Bishop (Bishop and Bridges *1985*). He and his followers accept the traditional intuitionistic criticism of classical mathematics (§5.6), but without accepting Brouwer's subjectivistic views. Like Brouwer, they hold that a correct proof of the existence of a number should, at least in principle, show us how to construct that number. A non-constructive existence proof which claims to establish the existence of a number without giving us any idea of how to produce the number is just fraudulent. It is like a conjuring trick which seems to produce a rabbit when there is actually no rabbit there. From Bishop's point of view, then, classical mathematics is not so much metaphysically wrong-headed as insufficiently rigorous. Mathematicians have been tricked by logicians into accepting arguments that are, as a matter of fact, not correct.

Instead of trying to articulate a sophisticated philosophical position, Bishop and his followers have attempted to redo as much of classical mathematics as possible in a way that will be constructively acceptable. All their concepts and arguments are classically correct, but they avoid the non-constructive use of the law of the excluded middle. They also eschew set-theoretic reductionism; instead, they view mathematics as the study of effective procedures operating, directly or indirectly, on the natural numbers.

Among philosophers the most influential contemporary version of constructivism is the intuitionism of Michael Dummett *1977*. His view is founded on an essentially Wittgensteinian philosophy of language: to understand mathematics is primarily to understand mathematical speech, the meaning of which must be constituted by its use. It follows for Dummett that an account of the semantics of mathematical language cannot rely on a transcendent notion of mathematical truth since there is no way that such a notion could be taught. What we teach in the classroom is not what it means for a mathematical assertion to be true, but rather what constitutes a proof of a mathematical assertion. Therefore the meaning of a mathematical assertion can only consist in the rules for determining what is to count as a proof of that assertion.

Following Frege, Dummett insists that the meaning of a complex assertion must be composed of the meanings of its parts. Thus for Dummett the meaning of a disjunction, for example, is to be specified by requiring that a proof of the disjunction must consist of a proof of one of its disjuncts. Since not every statement can be either proved or refuted, Dummett follows Brouwer in rejecting the universal applicability of the law of the excluded middle.

For Dummett, the contrast between Platonism and intuitionism is a special case of the more general contrast between realistic and anti-realistic views about almost any subject. That contrast will always consist ultimately in a disagreement about how the language used to talk about the subject is to be interpreted. The realist will interpret it using a Tarskian truth-conditional semantics; the anti-realist, some sort of justification or operational semantics. Thus, where physical objects are concerned, a realist should accept a straightforward Tarskian account of the semantics of sentences like 'the book is on the table'; it suffices to explain the sentence by saying that it is true simply if the book is on the table. One common form of anti-realism about physical objects is phenomenalism. The phenomenalist, according to Dummett, must interpret the sentence 'the book is on the table' by explaining what sense experiences would justify the assertion of the sentence. Of course, it is perfectly possible that I have neither experiences to justify the assertion that the book is on the table, nor experiences to justify the assertion that it is not. On Dummett's version of phenomenalism, then, I would not be justified in making either assertion.

For Dummett the problem of justifying Platonism comes down to the problem of founding a realistic semantics for mathematical language in our actual use of that language. A realistic interpretation of talk about physical objects, for example, can be justified by pointing out that we deal with physical objects constantly, and that our dealing with those objects forms part of the context of our use of these linguistic expressions. Words denoting physical objects can be taught by ostension. Since, in Dummett's view, we have no direct contact with infinite sets, there is no way for our talk about such objects to be justified in any analogous fashion. A theory like ZFC lacks any comprehensible interpretation.

3 EMPIRICISM

Recent years have seen at least one major development not arising directly from the great schools that flourished between the wars. The instigation for much of this work was the very original dissertation of Imre Lakatos, republished as Lakatos *1976*. Lakatos tried to show that the general framework of Karl Popper's scientific methodology can be applied to mathematics. In doing this he was influenced by the beautiful work of George Polya on mathematical heuristics (see especially Polya *1954*). Lakatos's essay consists primarily in a detailed account of the historical development of Augustin Louis Cauchy's proof of Leonhard Euler's formula that, for all polyhedra,

$$V - E + F = 2, \tag{1}$$

683

where V is the number of vertices, E the number of edges and F the number of faces. He emphasizes the way in which the concept of polyhedron evolved so as to preserve the correctness of the theorem and its proof, despite the appearance of a series of counter-examples.

One reason for the popularity of Lakatos's ideas has been their congruence with the anti-foundational spirit of so much recent Anglo-American philosophy. Perhaps the most important effect of his influence has been to lead philosophers of mathematics away from their fascination with the increasingly esoteric details of logical investigations, and back to a concern with studying what mathematicians actually do. As the study of 'the foundations of mathematics' has declined, we have seen a flourishing of philosophical investigations of mathematical history and methodology.

Lakatos has provided the stimulus for considerable work on concept formation in mathematics and on non-deductive elements in mathematical epistemology (see the representative essays in Tymoczko *1985* and in Aspray and Kitcher *1988*). This work had led to a revival of empiricism in the philosophy of mathematics (for a general exposition of this view, see Kitcher *1984*). One of the most striking conclusions from it is that older mathematical theories are constantly being reinterpreted so as to make them fit contemporary standards of rigour and contemporary patterns of conceptualization. The illusion that mathematical theorems, once established, will be accepted for ever is fostered by such reinterpretations. For example, when it turned out, towards the beginning of the twentieth century, that Euclidean geometry applied to physical space only locally, mathematicians reinterpreted that geometry as being about 'Euclidean space', rather than as being about physical space, as had always been thought before. Nevertheless, they often think that the theorems proved by the ancient Greek geometers are still accepted today.

The fundamental position underlying the new mathematical empiricism is that the deductive model of mathematical methodology is an oversimplification. Mathematicians characteristically develop their theories by emphasizing the properties of striking examples, by exploiting the power of methods whose justification may not yet be clear, by adapting their definitions to the results they find intuitively or aesthetically satisfying. They employ a complex mixture of heuristics which cannot be described as deductive. It is only after the work has been done and the theory has arrived at a more-or-less final form that mathematicians seek to give it a deductive structure by deducing it from axioms which arise as the culminating generalizations of the subject. Thus mathematics is expounded in an order almost the reverse of that in which it was discovered.

A striking recent example of the non-deductive character of mathematical research is provided by the increasing use of computers as an aid in constructing proofs too complex to be assembled without their aid (compare §5.10 on algorithms). The resulting unsurveyable deductive structures are very different from traditional arguments intended to give insight into the reasons for the truth of a theorem. (For a discussion see Tymoczko's essay, 'The four-color problem and its philosophical significance', reprinted in Tymoczko *1985*.)

From the present point of view, set-theoretic Platonism gives a distorted account of the nature of mathematical knowledge. Since mathematics is not derived from the abstract contemplation of the properties of concepts held antecedently, but rather arises from computational and conceptual experimentation, it is not natural to describe mathematics as *a priori*. The axioms of ZFC are neither evident nor permanent: they represent a temporary codification of a particular stage in the development of mathematics. Set-theoretic reductionism gives a misleading account of the nature of mathematical concepts, since how a particular concept is to be reduced to set-theoretic language will change as the concept evolves in response to our experience. From this point of view, indeed, there is no reason to believe that all mathematical concepts will always be reducible to set theory or that all the arguments mathematicians will rationally accept will always be formalizable in set theory.

The essence of the new empiricism in the philosophy of mathematics is to emphasize the extent to which mathematics is a historical product, and the extent to which it is similar to the other sciences. The traditional dichotomy between the natural sciences as consisting of changing empirical laws, on the one hand, and mathematics as consisting of unchanging *a priori* principles, on the other, is being undermined by closer attention to the actual historical practice of mathematicians.

BIBLIOGRAPHY

Aspray, W. and Kitcher, P. (eds) *1988*, *History and Philosophy of Modern Mathematics*, Minneapolis, MN: University of Minnesota Press.

Benacerraf, P. and Putnam, H. (eds) *1983*, *Philosophy of Mathematics: Selected Readings*, 2nd edn, Englewood Cliffs, NJ: Prentice-Hall.

Bishop, E. and Bridges, D. *1985*, *Constructive Analysis*, Berlin: Springer.

Dummett, M. E. *1977*, *Elements of Intuitionism*, Oxford: Clarendon Press.

Kitcher, P. *1984*, *The Nature of Mathematical Knowledge*, Oxford: Clarendon Press.

Lakatos, I. *1976*, *Proofs and Refutations* (ed. J. Worrall and E. Zahar), Cambridge: Cambridge University Press.

Maddy, P. *1988*, 'Believing the axioms', *Journal of Symbolic Logic*, **53**, 481–511, 736–64.

Polya, G. *1954*, *Mathematics and Plausible Reasoning*, 2 vols, Oxford: Clarendon Press.

Tymoczko, T. (ed.) *1985*, *New Directions in the Philosophy of Mathematics*, Boston, MA: Birkhäuser.

5.10

Algorithms and algorithmic thinking through the ages

PETER SCHREIBER

1 PROBLEMS AND SOLUTIONS

From the beginnings of human activity, mathematics has developed out of the necessity to solve problems, especially those which recur frequently with different 'input objects' (sets, numbers, areas, lines, points, functions, etc.). Both generally and from a modern point of view, a problem is given by a list $x_1, \ldots, x_n$ of (names of) given objects, together with some assumptions about them, a list $y_1, \ldots, y_m$ of (names of) required objects, and some conditions or restrictions pertaining to $x_1, \ldots, x_n$ and $y_1, \ldots, y_m$. Such a problem is to be solved by an algorithm (procedure, program) which constructs such $y_1, \ldots, y_m$ from the given $x_1, \ldots, x_n$ in a uniform manner, not depending on the actual values of the variables $x_1, \ldots, x_n$. In ancient Greece, from as early as about 400 BC, the solution of a problem was completed by a proof of the correctness of the algorithm: a proof that it is applicable to all values of $x_1, \ldots, x_n$ which satisfy the stated assumptions, and that the values $y_1, \ldots, y_m$ produced, together with the given values of $x_1, \ldots, x_n$, satisfy the stipulated conditions (Schreiber *1984*). Problems of geometric construction are important examples of such a treatment of problems, while Euclid (*circa* 300 BC) also applies such a treatment to number theory (*Elements*: Books VII–XI) (Schreiber *1987a*).

Designing algorithms, proving their correctness, searching for algorithms which solve problems in a small number of steps and with high accuracy, searching for instruments and machines to execute the individual steps of algorithms, transforming objects into other sorts of objects more suitable for a particular algorithm (coding) and, last but not least, searching for proofs of the impossibility of solving problems by special sorts of algorithms – most of these procedures have assumed a central place within mathematics, but sometimes they were displaced by an excessively deductive view of mathematics ('the only aim of mathematics is to prove difficult and

687

beautiful theorems'). However, in the twentieth century a general definition of the notion of an algorithm has been given, and general explicit studies of this notion have been made. The computer is a relatively new but superb instrument for executing algorithms, so it is quite natural that the questions about problems and algorithms listed above should once again return to the centre stage of mathematics.

2 DEVELOPMENT OF COMMUNICATION TECHNIQUES FOR ALGORITHMS

An important problem, which nevertheless has been explicitly studied only in the twentieth century, is the development of linguistic means for the formulation and communication of algorithms. The first to use letters as variables (better, 'addresses') for given and required quantities was Jordanus de Nemore in his treatise *De numeris datis* (*circa* 1200), but this had little influence (§2.3); the next was François Viète (second half of the sixteenth century), whose work was of lasting influence (§6.9). Before then, algorithms could in general be communicated only by instructive numerical examples. (So it was also with the Egyptians and Babylonians, as well as medieval mathematicians.)

Occasionally there have been attempts to formulate general rules, using several techniques which we now may characterize as 'hidden addresses'. Those quantities about to be joined by an operation were named according to their specific mathematical context (e.g. produce the new 'denominator' by multiplying the 'denominator of the first and the second fraction'), or the stage of the calculation at which they appear (e.g. add the 'number given in the beginning' to the 'last computed number'), or how they were being computed (e.g. add the two 'square roots'). Especially remarkable is the ancient Chinese 'Computational Prescriptions in Nine Chapters (*Jiuzhang suanshu*)', in which addresses for the purpose of formulating general rules are taken from an example treated before (§1.9). For instance, the 'Gaussian algorithm' for solving three linear equations with three unknowns is explained in terms of an example by denoting the unknowns as 'sheaf of a good harvest', 'sheaf of an average harvest' and 'sheaf of a poor harvest'. In a similar manner, Leonardo of Pisa in his famous *Liber abbaci* (1202) denoted the unknowns in linear equations generally as 'money of the first buyer', 'money of the second buyer', and so on, again with respect to a specific example (§2.4).

A remarkable exception from the usages outlined above is the description of geometrical constructions since the time of the ancient Greeks. We find single letters being used as addresses of variable points, and combinations of letters as addresses of straight lines or curves. A virtually modern 'problem-oriented programming language' is used to describe construction algorithms in Euclid's *Elements*. (In addition, Euclid proves to be an early and surprising master in describing branched or cyclic procedures.) The reason for this exception is clearly that geometric objects, in contrast to numbers and other mathematical objects, are bounded at a fixed place, and therefore it seems to be impossible to communicate geometrical construction algorithms by substituting special data instead of variables in a linguistic description of algorithms.

3 ORIGIN AND EARLY HISTORY OF THE WORD 'ALGORITHM'

Abū Ja'far Muḥammad ibn Mūsā al-Khwārizmī (i.e. a man from Khwarizm, a middle-eastern empire) was born about 783 in Khiva (in present-day Uzbekistan). He worked under the rule of 'Abbāsid caliphs in Baghdād, and died there about 850 (§1.6). His 'arithmetic' treated (probably for the first time) the handling of the elementary arithmetic operations with respect to the Hindu–Arabic decimal numerals. The Arabic original of this work, of fundamental importance for the history of mathematics, seems to be lost. Preserved are only a Latin translation entitled *Algorithmi de numero indorum* (perhaps by Gerhard of Cremona or Magister A – i.e. Adelard of Bath, both twelfth century) and some early Latin compilations of this text. They show the rapid dissemination of al-Khwārizmī's in Western Europe.

The word 'algorithm' (originally 'algorism') is a corruption of al-Khwārizmī's name, and the word was originally used in Western Europe to mean the body of arithmetical methods he advanced. The adherents of these new methods of calculating 'by pencil and paper' were known as 'algorithmicians', as opposed to the 'abacists', the adherents of traditional calculating on the abacus, a heritage from ancient Rome. The word 'algorist' is now often used.

From this time until the end of the sixteenth century, a rich algorithm-oriented computational tradition developed, exemplified by the 'reckoning masters' like Robert Recorde in England, Adam Ries and Ulrich Wagner in Germany, and Niccolò Tartaglia in Italy (§2.3). Since about 1600 the notion of algorithm, and also the algorithmic description of calculating procedures, have been gradually displaced by new mathematical ideas and influences. One of these new ideas may be traced back to the *Ars magna* of the thirteenth-century Spanish theologian and logician Ramón Llull (Raimundus Lullus). His great concept of artificial languages of universal expressive power, and of machines or systems of rules applicable to expressions of such languages and capable of deciding the truth of propositions or solving classes of problems, influenced many thinkers, especially René

Descartes and Gottfried Wilhelm Leibniz. Descartes's 'rational method of reasoning' (1637) and Leibniz's *Calcul* (1684) are some of the 'children' of this concept. A very significant factor in the decline of the algorithmic tradition is that, after the end of the Renaissance, mathematical works, books and papers were addressed, not as before to the lay-user of mathematical knowledge, but to the learned colleague. Authors began to write with the expectation that their readers would be able to make the correct inferences from ideas presented only in their brief essentials – that they would 'read between the lines' – and that they could go on to apply the knowledge thus learnt. Further, there grew among mathematicians a competitiveness that for some amounted to a burning ambition to achieve fame through mathematical discovery; in consequence, this led some mathematicians to keep secret their procedures for problem-solving.

A further factor in the decline of algorithmic thinking was, following the invention of the differential and integral calculus, the growing attachment to 'closed analytical expressions' which, to the mathematicians of the eighteenth century, seemed to offer the best means of solving problems (§3.14, §4.3). For example, even Leonhard Euler had no intuitively correct understanding of algorithmic procedures, and failed in describing simple algorithms like that for solving systems of linear equations (see his treatment of Pell's equation in Euler *1770*: 232, 380–87).

4 SOME CAUSES FOR THE INSUFFICIENT UNDERSTANDING OF ALGORITHMS

A linear algorithm (i.e. a simple sequence of assignments) may often be communicated more briefly and elegantly as a sequence of terms (functional expressions), provided it is not intended to be read and executed by a computer. For instance, the well-known formula

$$x_1, x_2 = -\tfrac{1}{2}p \pm [(\tfrac{1}{2}p)^2 - q]^{1/2} \tag{1}$$

for solving the equation $x^2 + px + q = 0$ conveys to a human reader in essence the same information as the program

$$p, q \ \rightarrow \ z_1 := \tfrac{1}{2}p \ \rightarrow \ z_2 := z_1^2 \ \rightarrow \ z_3 := z_2 - q \ \rightarrow$$
$$z_4 := \sqrt{z_3} \ \rightarrow \ x_1 := -z_1 + z_4 \ \rightarrow \ x_2 := -z_1 - z_4 \ \rightarrow \ x_1, x_2. \tag{2}$$

The presentation (1), however, says nothing about the succession of steps or the optimal design of the calculation.

An essential step in the correct understanding of the algorithms is to be able to tell easily whether solutions of problems can (or cannot) be cast in the form of a linear program; that is, whether each correct solution of a certain problem must be branched by some tests or even be cyclic (with loops back to previous steps). Such situations indeed occur in very simple problems: for example, in branching in the solution of two linear equations, if one makes suitable assumptions about the given coefficients which ensure a uniquely determined solution, or cycles for computing the greatest common divisor of two numbers. But inspection of the mathematical literature of more than 2000 years shows that (besides Euclid and a few other forerunners) branching and cyclic behaviour have never been recognized as natural and often necessary properties of algorithms.

The lack of attention paid to cyclic behaviour may be because typical cyclic algorithms often approximate to a result; so it seems to be sufficient to describe one step of approximation. It is 'clear', then (but only to a human), that one has to repeat that step until the required accuracy is achieved. With respect to branching, before the rise of first programming languages (and, in some classical fields of mathematics, even after it!), mathematicians often worked out different linear algorithmic solutions for all possible cases of a problem, then decided (often unconsciously) which case was currently requiring execution. Such decisions, however, were not accepted as a part of the procedure, and consequently their effect was not properly considered. Also, in branched algorithms there is often a 'probable' regular case, and one or more singular or exceptional cases.

From the seventeenth century to the mid-nineteenth, most mathematicians were primarily scientists, and from the natural sciences they knew that most natural laws have a bounded field of validity, but may permit exceptional cases. Thus they found it quite natural in mathematics to concentrate on the regular cases and to neglect the exceptional ones; this is confirmed by the fact that of the few mathematicians of this epoch who occasionally discussed different cases correctly (e.g. Johann Heinrich Lambert and Bernard Bolzano), most were philosophers or logicians as much as scientists.

Since Antiquity, the treatment of geometric constructions has been a special domain − sometimes a refuge − of algorithmic thinking. Only such thinking could motivate the search for further solutions of solved problems, especially new solutions achieved with other instruments, with a better accuracy, by more practicable means in a bounded part of the plane. In this field only were made the first studies on the unsolvability of problems by special instruments (i.e. by special classes of algorithms) and the first attempts to measure, compare and optimize the complexity of different algorithms for solving the same problem. These attempts came from E. Lemoine and his adherents, and his 'geometrography' (*1902*).

However, this new discipline had three remarkable defects which caused its premature end around 1920. First, the adherents of the 'geometrography' never could agree on a generally accepted measure of complexity (Coolidge *1916*: 166). Second, they held, incorrectly, that solutions must always be linear algorithms, and therefore that solutions would certainly be found after the required steps were taken. And third, they had no method of *proving* the optimality of a solution. Thus the title 'geometrographic solution' of a problem was rather like a prize awarded for the best solution achieved.

5 MODERN HISTORY OF THE WORD AND NOTION OF 'ALGORITHM'

Until about 1800, almost nobody used the word 'algorithm'; knowledge of the heritage and meaning of this word had been lost. In the *Mathematisches Wörterbuch* by G. S. Klügel we read:

> Algorithmus, also Algorismus, in the Middle Ages when calculating by the decadic ciphers was introduced into Europe, meant this new mode of computing. The word is composed from the Arabic '*Al*' and the Greek '*Arithmos*', like Almagestum... (Klügel 1803: 68–9)

The French orientalist J. T. Reinaud proposed in *1849* that 'algorithm' may come from the name of a learned Islamic man. In 1857 the Italian historian of mathematics Baldassarre Boncompagni rediscovered and published two of the Latin versions of the 'arithmetic' mentioned in Section 3. Since then, the life and works of al-Khwārizmī and the spread and influence of his methods in Western mathematics have attracted the interest of mathematicians and historians of mathematics (Zemanek *1981*). However, the use of the word 'algorithm' has been diffused in the international mathematical literature only since about 1950. Without doubt this happened because of the influence of Russian mathematics, which have been translated and extensively studied in Western countries.

As has been mentioned, both in classical mathematics and, especially, in areas of its application, there have been down the ages many occasions for contemplating the notion of algorithm, and the qualities and properties of algorithms. But, because of the failing of classical mathematics on this matter, the first attempts to define and to investigate notions like (effective) computability, decidability and enumerability (which are strongly connected with algorithms, although the word 'algorithm' was not used) came from the metamathematicians and the logicians (David Hilbert 1900, H. Behmann 1922, Paul Finsler 1925, and Alonzo Church, Emil Post and Alan Turing in the 1930s) in the context of proving the undecidability of the truth of propositions from the axioms and rules of inference of formal languages

(§5.5). For some time these investigations remained wholly separate from the first attempts to design computers and programming languages. Today the precise notion of algorithmic (or recursive) computability in discrete mathematics seems in principle to have made it possible to answer all questions concerning algorithms; but it cannot be used to address the algorithmic problems of classical mathematics because it is not applicable to non-finite or non-discretizable objects. A more general and generally accepted definition of the notion of algorithm does not yet exist.

BIBLIOGRAPHY

Coolidge, J. L. *1916*, *A Treatise on the Circle and the Sphere*, Oxford: Clarendon Press.

Euler, L. *1770*, *Vollständige Anleitung zur Algebra*, Saint Petersburg: Academy. [Also in *Opera omnia*, Series 1, Vol. 1 (cited here).]

Klügel, G. S. *1803*, *Mathematisches Wörterbuch*, Part 1, Section 1, Leipzig: Schwickert.

Lemoine, E. *1902*, *La Géométrographie, ou art des constructions géométriques*, Paris: Naud.

Reinaud, J. T. *1849*, *Mémoir sur l'Inde*, Paris: Imprimerie Nationale.

Schreiber, P. *1984*, 'Constructivity in arbitrary mathematical theories', in G. Wechsung (ed.), *Frege Conference 1984*, Berlin: Akademie, 96–100.

—— *1987a*, *Euklid*, Leipzig: Teubner.

—— *1987b*, 'Zur Geschichte des Verfahrensaspektes der Mathematik', *Mitteilungen der Mathematischen Gesellschaft der DDR*, Heft 3.

—— *1991*, 'Generalized construction problems', *Zeitschrift für mathematische Logik und Grundlagen der Mathematik*, **37**, 57–62.

Zemanek, H. *1981*, *Al-Khorezmi: His Background, His Personality, His Work and His Influence* (Lecture Notes in Computer Sciences, No. 122), Heidelberg: Springer.

5.11

Calculating machines

D. D. SWADE

This article provides a brief survey of calculating machines under the following general headings: early physical aids, instruments, mechanical calculators, automatic calculating machines and pre-electronic twentieth-century developments. The concern is to identify significant or interesting developments, rather than to exhaust the rich variety of devices and designs in the prehistory of computing. Punched-card machines are not covered: these are regarded as part of the data-processing movement in which the preoccupation is less with numerical computation than with information management (Aspray *1990*). Automata and logic machines are also outside the brief of the survey.

1 EARLY PHYSICAL AIDS

The prehistory of mechanical calculation features a number of elementary physical aids to counting and simple calculation – tokens, counters, pebbles, knotted cords and tally sticks. A development in the use of tokens or counters is found in the reckoning board or table abacus; markings on the flat surface assigned numerical value to the position of loose tokens placed by the user.

An elementary form of mechanization can be found in the wire-and-bead abacus. Unlike tokens on the reckoning board, the beads are physically constrained by the wires or dowels held in a frame. In the hands of a skilled operator, the abacus is a rapid and accurate aid for the four basic operations of arithmetic as well as for more advanced use such as the extraction of roots. The most familiar forms of the wire-and-bead abacus are the Chinese *suan-pan*, the Japanese *soroban* and the Russian *stchoty*.

An aid to calculation that attracted widespread interest was 'Napier's rods' otherwise known as 'Napier's bones', supposedly so-called because the more expensive sets were made from bone or ivory. The rods, devised to aid multiplication, are numbered sticks or oblong blocks, with the face of each marked with multiples of one of the counting digits. Multi-digit multiplication is carried out by laying rods alongside each other to make up the digits of the multiplicand, reading off and recording the partial products, and summing them. The layout of the numbers allowed the carries to be performed mentally as the partial products are read off, and the system reduced multi-digit multiplication to a series of look-up operations and simple additions. The use of the rods was described by John Napier in *Rabdologia* (1617) (Horsburgh *1914*). Division was assisted by using the bones for trial multiplications; there were separate bones for calculating square and cube roots (see §2.5 on logarithms).

Extensions of Napier's rods took various forms. Gaspard Schott described sets of bones devised for special purposes including geometry, and calculating planetary movements in astronomy. Various physical forms were devised: versions with rotatable cylinders by Wilhelm Schickard in 1623 (see Section 3), and probably independently by Schott (1668); a version using circular discs by Samuel Morland (described in 1673); and one by Charles Cotterell (1667) which included a wire-and-bead abacus for adding the partial products. A final development took the form of a series of rulers devised by Henri Genaille, demonstrated in 1891. The Genaille rulers eliminated the need to carry digits when looking up partial products, and the rulers were extended to aid division.

2 INSTRUMENTS

The devices described so far are essentially digital in that the representation of value is discrete. Analogue devices, in which value is rerepresented continuously, form an important separate class of aids. The proportional compass, sector, quadrant and slide rule are examples of devices using graduated scales on which intermediate values between number-markings are valid. The proportional compass, dating from at least as early as Roman times, consists of two scales in scissor-like arrangement with a sliding pivot. It was used for a variety of calculations, including simple multiplication, dividing the circumference of a circle into equal parts, and finding square and cube roots. A related device is the sector, developed in the late 1500s and widely used until the middle of the nineteenth century. It consists of two arms with scale markings, hinged at one end, and was usually used in combination with a compass. Sectors were produced with scales for basic multiplication and division as well as for a range of special applications including gunnery, surveying and navigation.

Edmund Gunter plotted logarithms on a straight scale (*circa* 1620) to allow multiplication and division to be performed by the addition or subtraction of lengths using a divider. The crucial step from the Gunter's 'line of numbers' to the slide rule is attributed to William Oughtred who claims to have used, as early as 1621, two Gunter scales sliding against each other to perform the addition and subtraction of logarithms without the need for dividers. The first slide rule in which the slide moved between fixed blocks is attributed to Robert Bissaker (1654).

The slide rule was developed in a variety of physical forms primarily to extend the effective length of the scales to improve range and accuracy: circular and cylindrical versions were first described in print by Richard Delamain (1630); a circular version with spiral scales was made by Thomas Brown (1633). A cylindrical version with helical scales designed by George Fuller (1878) achieved a working length of nearly 13 metres and four-figure accuracy in a device about 30 centimetres in length. The use of slide rules became increasingly widespread, particularly among engineers with their need for a versatile and convenient aid for fast calculations requiring no more than a few figures of accuracy. Slide rules rapidly disappeared with the availability of electronic pocket calculators in the 1970s.

3 MECHANICAL CALCULATORS

There were several attempts in the seventeenth century to mechanize calculation, notably by Schickard, Blaise Pascal and Gottfried Wilhelm Leibniz. Though their calculators represent significant landmarks they were, by and large, serious-minded curiosities of questionable reliability, and unsuited to routine use.

Schickard's calculator (1623) consisted of a cylindrical version of Napier's rods and a mechanical adder which accumulated, taking account of any carry, the partial products manually entered by turning dials. Until the discovery of Schickard's work was publicly revealed in 1957, the credit for the first calculator went to Pascal, who conceived his first machine in 1642. Pascal's carry mechanism was ingenious but its reliability doubtful. Addition was performed by dialling numbers using a stylus; since dials could not be turned backwards, subtraction was accomplished using complements (e.g. for $7 - 3$, take $7 + 6 = 13$, add 1, and discard the tens column to obtain 4). Multiplication and division were by repeated addition and repeated subtraction, respectively.

Leibniz's machine, completed in 1674, incorporated a new device, the stepped drum, which dominated calculator design for the next two hundred years. His calculator performed multiplication by generating partial products by repeated addition and accumulated the final result by the mechanical addition of products. Propagating a carry across several digits was not automatic and required intervention by the operator. Several calculators based on the Leibniz stepped drum were devised in the eighteenth century. The better known of these include devices by Matthaus Hahn (*circa*

1774), Johann Müller (1784) and Viscount Mahon (later Earl Stanhope, 1775). None of these calculators was manufactured in commercial quantities.

The first commercially successful four-function calculator was introduced by Thomas de Colmar in the early 1820s. The 'arithmometer', also based on the Leibniz drum, was widely used towards the end of the nineteenth century. A major breakthrough in the physical construction of calculators followed from the invention of the variable-toothed gear, patented in 1875 by Frank Baldwin in the United States, and later, with improvements, by Willgodt Odhner. The Baldwin–Odhner designs led to a range of desktop calculators under various names, among which Brunsviga and Facit are perhaps the best known (Eames and Eames *1973*).

Key-set and key-driven machines form a separate class of devices. The first key-driven machine was devised by Dorr Felt in 1885; the Felt 'comptometer' enabled addition and subtraction to be performed by pressing typewriter-like keys. A key-set adder–subtractor was patented by William Burroughs in 1892. The number to be added was entered on numbered keys and a lever pulled to perform the calculation. Baldwin–Odhner-type machines were widely used for scientific calculations, while the Felt and Burroughs key machines were favoured for commercial and financial applications. Desktop mechanical calculators were superseded by the electronic desktop or pocket calculators in the 1970s.

4 AUTOMATIC CALCULATING MACHINES

None of the calculators so far described was automatic: their operation required the continuous, informed intervention of the operator. The design of the first automatic calculator was conceived in 1821 by Charles Babbage, with the aim of producing error-free tables. Babbage's difference engines were designed to tabulate polynomial functions using the method of finite differences. The advantage of the method is that it allows successive tabular values of the function to be generated by repeated addition, which is comparatively easy to mechanize, and eliminates the need for the multiplications and divisions ordinarily required to evaluate each term in the polynomial expression. Except for a few partial assemblies, Babbage failed to complete any of his engines in physical form (Bromley *1987*).

About one-seventh of his Difference Engine No. 1 was assembled in 1832. This is capable of three orders of difference, and represents the first successful attempt to incorporate a mathematical rule in a mechanism. Difference Engine No. 2, designed between 1847 and 1849, has seven orders of difference and 31-figure accuracy (see Figure 1, which shows the construction of this machine, completed in 1991). Though Babbage was the first to

Figure 1 Charles Babbage's Difference Engine No. 2, built by the Science Museum to original designs for the bicentenary, in 1991, of Babbage's birth. Pictured with the engine is the author who, as Curator of Computing, led the project. This is the first complete Babbage engine to be realized in physical form. It was designed between 1847 and 1849, measures over 3 metres long, over 2 metres high and half a metre deep, and weighs over $2\frac{1}{2}$ tonnes. It consists of 4000 parts made from bronze (gunmetal), cast iron and steel. The printing unit, which would have stood on the left of the machine, was not built

produce a complete design of an automatic machine, he was not the first to propose the use of finite differences as the basis for such a device. This distinction goes to Johann Müller, who recorded the suggestion in 1784.

Difference engines were designed and built by Georg and Edvard Scheutz, a Swedish father-and-son team, stimulated by a description of Babbage's work (Lindgren *1987*). The Scheutz engines, like Babbage's, were decimal digital machines. The Scheutzes built three difference engines, a prototype (1843) and two 'production models' (1853, 1859). Other difference engines were built by Martin Wiberg in Sweden (1860), by Barnard Grant in the United States (1876) and (a model) by Alfred Deacon in London. The Scheutz and Wiberg engines were used to produce published tables. However, the nineteenth-century movement to mechanize table-making using difference engines largely failed.

These difference engines were dedicated devices: their use was confined to a fixed set of operations determined by their wheelwork. The first design for a general-purpose calculating machine was Babbage's analytical engine, conceived of by 1834. The engine had as its repertoire the four basic arithmetic functions, and could be programmed, using punched cards, to perform these in any sequence. A sequence of operations could be repeated a given number of times (i.e. the machine was capable of iteration). Babbage saw the engine as a universal calculating machine capable of evaluating almost any algebraic expression. The engine designs were developed to an advanced stage, though only a small experimental model was built during Babbage's lifetime. For all its pioneering ingenuity, the analytical engine represents an isolated episode with apparently little direct influence on subsequent design (Babbage *1989*).

5 PRE-ELECTRONIC TWENTIETH-CENTURY DEVELOPMENTS

Percy Ludgate described in 1909 an original design for a mechanical, digital, general-purpose program-controlled calculator. Ludgate's design for his analytical engine represents another isolated episode which had no discernible influence on subsequent developments. During the 1930s, Vannevar Bush developed large mechanical analogue machines at the Massachusetts Institute of Technology for the solution of differential equations. His 'differential analyser' used gears for multiplication, differential gears for addition and subtraction and wheel-and-disc integrators. Bush-type differential analysers saw useful service, particularly in the calculation of ballistics tables.

In Germany, Konrad Zuse built a series of general-purpose automatic digital machines using mechanical and electromechanical relay components as switching elements. His Z1 (1938), prototype, was entirely mechanical and used, as did his later machines, the binary rather than the decimal number system. His Z2 was a hybrid machine with a mechanical memory and electromechanical arithmetic unit. His Z3 (1941), entirely electromechanical, was the first successful general-purpose program-controlled calculator.

The wartime conditions of the 1940s stimulated the development of purely electronic calculators and computers based on vacuum tubes. These eventually superseded electrically driven mechanical devices and electromechanical relay-based systems. Notable landmark electronic projects include the ENIAC, developed in the United States for ballistics calculations, and Colossus, built in England for cryptanalysis. To such topics we now turn.

BIBLIOGRAPHY

Aspray, W. (ed.) *1990, Computing before Computers*, Ames, IA: Iowa University Press. [Includes chapters on logic machines, analogue computing machines and punched-card machinery.]

Babbage, C. *1989, Works* (ed. M. Campbell-Kelly), Vol. 2, *The Difference Engine and Table Making*, and Vol. 3, *The Analytical Engine and Mechanical Notation*, London: Pickering.

Baxandall, D. *1926, Catalogue of the Collections in the Science Museum: Mathematics – Calculating Machines and Instruments*, London: HMSO.

Bromley, A. G. *1987*, 'The evolution of Babbage's calculating engines', *Annals of the History of Computing*, **9**, 113–36.

Eames, C. and Eames, R. *1973, A Computer Perspective*, Cambridge, MA: Harvard University Press. [Excellent illustrations.]

Horsburgh, E. M. (ed.) *1914, Handbook of the Exhibition of Napier Relics and of Books, Instruments and Devices for Facilitating Calculation*, Edinburgh: Royal Society of Edinburgh. [Repr. 1982, Los Angeles and San Francisco: Tomash.]

Lindgren, M. *1987, Glory and Failure: The Difference Engines of Johann Müller, Charles Babbage and Georg and Edvard Scheutz*, Stockholm: Royal Institute of Technology Library. [2nd edn 1990, Cambridge, MA, and London: MIT Press.]

d'Ocagne, M. *1911*, 'Calculs numériques', in *Encyclopédie des sciences mathématiques*, Tome 1, Vol. 4, 196–452 (article I 23). [See especially arts. 56–61.]

Williams, M. R. *1985, A History of Computing Technology*, Englewood Cliffs, NJ: Prentice-Hall.

5.12

Computing and computers

M. CAMPBELL-KELLY AND S. B. RUSS

Computers can be classified as special-purpose or general-purpose, and digital or analogue. A special-purpose computer is one which can perform only one task (such as integrating differential equations), whereas a general-purpose machine can be programmed to perform a wide variety of mathematical and other tasks. An analogue computer is one which uses some physical process to emulate a mathematical one; for example, an analogue electronic computer might emulate the process of integration by the charging of a capacitor. In contrast, a digital computer handles numbers and nothing more: a computation proceeds by operating directly on those numbers, without regard to any physical meaning the numbers might have. The difference between analogue and digital computation is well conveyed by the distinction between a slide rule and a personal electronic calculator. It is only since the Second World War that computation has come to be dominated by general-purpose digital machines (Aspray *1990*).

1 THE THEORETICAL BACKGROUND

The modern computer was *not* the inevitable outcome of technological advance. The crucial prerequisite for the useful application of technology to computing was the development of notation, or language systems, sufficiently comprehensive to satisfy both the need for representation, and the need to express and implement mechanisms for the transformation of expressions in the language. The natural context today for discussing such features is that of a formal system (see §5.5 on metamathematics). Ordinary arithmetic is a special, but familiar example of a formal system in which we can apply the rules, or procedures, for transforming strings of symbols without regard to their interpretation. The expressive power, or generality of a formal system is determined by our ability to represent a problem and its solution within the system. An important purpose of such a representation is to allow for an effective, general procedure, or algorithm, to transform an instance of a problem into a solution. These themes appear in a fully mathematical setting for the first time in what is probably the most important single paper in the theory of computation, namely Turing *1936*. We now sketch the background leading up to that publication.

It is natural to trace the theoretical origins of computing to the work of Gottfried Wilhelm Leibniz because the themes of generality and the transformation of expressions are united in him in a powerful motivating vision (§5.1). In *De arte combinatoria* (published in 1666, when he was 19) he outlined his ideas for a *characteristica universalis*, a notation in which the structure of compound symbols should reflect the structure of the objects symbolized. Associated with this universal language he imagined a means of reasoning which would be as effective as calculation in arithmetic. At this stage, at least, he expected questions of physics, metaphysics, ethics, law and medicine to be capable of resolution in his system. Given a 'calculus', or procedure for solving problems, Leibniz was well aware of the possibility of mechanical implementation. In the early 1670s, he designed and built a working mechanical calculator to perform multiplications directly. Leibniz derived from his study of Aristotelian logic the notion of formal proof, and he was the first to explore and exploit the formal affinity between logic and mathematics. He suggested that logic could be developed into 'a sort of universal mathematics', and he applied mathematical methods to derive some syllogistic forms using the algebraic methods of symbolism, forming equations and making substitutions.

It was not until well over a century after Leibniz's death that George Boole completed this application of mathematics to logic in such a way that all the valid syllogisms could be produced by means of the appropriate interpretation, in terms of classes, of an algebraic system (§5.1). In the opening paragraph of *The Mathematical Analysis of Logic* (1847), he wrote that, provided it is consistent, 'every system of interpretation is equally admissible'. Boole's work represents an end of one era, that of classical syllogistic logic, which thus became a branch of mathematics. At the same time it is an example of the beginning of a new era in a more formal, abstract approach to mathematics. Ironically, as this movement towards greater abstraction led away from the intuitive roots of mathematics, so the new trend to formalization and axiomatics pioneered later in the century by, among others, Richard Dedekind, Giuseppe Peano and David Hilbert, led some (such as Gottlob Frege and Bertrand Russell) to turn back to logic to seek satisfactory foundations (§5.2). Without pursuing the many foundational issues here, we note that Hilbert himself sought to base mathematics, not on logic, but on the idea of a formal system which had certain desirable properties such as consistency and completeness.

After Kurt Gödel's famous results of 1930–31, there still remained a further outstanding question, to which Hilbert had sought a positive answer for over thirty years: does predicate logic admit a 'decision procedure' – that is, is there an algorithm to determine whether a given formula is valid or not? This question was the direct motivation for the remarkable paper by Alan Turing *1936*. The answer, because it turned out to be negative and therefore required a proof that *no* algorithm could do the job, required a formal definition of the notion of algorithm. Surprisingly, this was provided independently in four different ways, in papers by Turing (using a machine model), by Alonzo Church (using λ-calculus), by Stephen Kleene (using recursive functions) and by Emil Post (using production systems) which all appeared during 1936 (Gandy *1988*). Turing gave a definition of an algorithm, or of what was 'effectively computable', in terms of a machine model of computation that was promptly christened by Church a Turing machine; a function, or number, is effectively computable precisely if there is a Turing machine which computes it (§5.5).

Turing *1936* laid the foundation for the theory of computation, and while this paper undoubtedly influenced the design and realization of high-speed digital computers, its influence is hard to evaluate precisely. In any case, the real intellectual origin of the modern computer has much deeper roots in the themes of representation and of automatic methods of symbolic transformation.

2 REALIZATION

The first major step in the practical realization of an automatic computer was taken by Charles Babbage (§5.11). In 1822 he produced the design for a machine he called the difference engine, which was intended to calculate and set the type for mathematical tables using the method of differences. The difference engine was a special-purpose digital machine which was limited solely to the production of mathematical tables (which is not to understate the importance of this task). However, in about 1834 Babbage abandoned the still-unbuilt difference engine for his analytical engine. The analytical engine was a general-purpose digital computer which was to be programmed in much the same manner as a modern computer; a key feature of the design was the separation of the store, which held the numbers during a computation, from the mill, which performed arithmetic operations on them. In arriving at this design and terminology, Babbage was drawing a parallel between the process of computation and the organization of a factory.

For several reasons – including funding and personal temperament – Babbage never built his analytical engine; nor did anyone succeed in building a computer until over a century after Babbage's original conception. During this period such large-scale computation as took place was often performed by teams of human computers working with desktop calculating machines. In the late 1920s, the New Zealand mathematician L. J. Comrie, working at the Nautical Almanac Office in England, mechanized the method of differences for table-making by linking together a number of commercial accounting machines. Other computing laboratories adapted punched-card machines for mathematical computation. Various analogue instruments, ranging from Lord Kelvin's tide predictor of 1876 to Vannevar Bush's differential analyser of the 1930s, were also developed.

The first fully automatic calculating machine, the IBM Automatic Sequence Controlled Calculator, was constructed at Harvard University between 1937 and 1943 with funding provided by the business-machine company IBM. The calculator was based on electromechanical-relay technology, and was capable of around two hundred arithmetic operations per minute. This machine, generally known as the Harvard Mark I, encouraged a number of similar electromechanical computer developments in the years around the Second World War.

What could claim to be the first programmable electronic computer was the Colossus, built in 1943 under the direction of M. H. A. Newman, but with the close cooperation of Turing, as part of the English code-breaking work carried out at Bletchley Park (Hodges *1983*). This entailed using symbolic combinatorial computation of the kind that would have general processing applications and not be restricted to numerical calculations. In Turing's *ACE Report* (Turing *1947*) he gives his own account of the connections between his ideas of 1936 and the design of the ACE machine developed while he was working at the UK's National Physical Laboratory.

A parallel and much more important electronic computer development, the ENIAC, took place at the Moore School of Electronics of the University of Pennsylvania. The ENIAC (Electronic Numerical Integrator and Calculator) was developed for ballistics calculations between 1943 and 1945. Although of rather limited flexibility, the machine was capable of a startling 5000 arithmetic operations per second, making it more than 1000 times faster than any previous non-electronic machine.

In 1945, the distinguished mathematician John von Neumann associated himself with the Moore School group. At that time he was seeking a computational solution to partial differential equations in connection with the Manhattan Project to develop the atomic bomb. The ENIAC turned out to be unsuitable for this type of work, but the realization of its shortcomings led von Neumann and other members of the Moore School team to evolve the design of the modern stored-program computer, called the EDVAC (Electronic Discrete Variable Automatic Computer). This design is now frequently referred to as the 'von Neumann architecture', although this

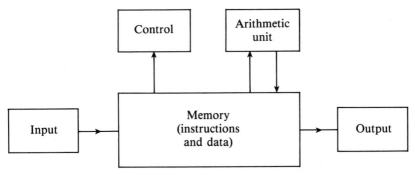

Figure 1 Stored-program computer

does considerable injustice to von Neumann's collaborators, notably J. Presper Eckert and John W. Mauchly.

The stored-program computer, as described in von Neumann's classic 'First draft of a report on the EDVAC' (*1945*) has five functional parts: a control unit, an arithmetic unit, a memory, and input and output 'organs' (Figure 1). In using this structure and terminology, von Neumann was drawing a metaphor between the computer and the brain (much as Babbage, a century earlier, had made an analogy between his analytical engine and factory organization). The modern computer is termed a stored-program machine because the program is held in the memory along with the numerical data. Instructions are fed in rapid succession to the control unit, which causes them to be executed and the appropriate arithmetic,control and input–output operations to be obeyed. In spite of the marked similarity between Babbage's analytical engine and the stored-program computer, the latter was developed in complete ignorance of Babbage's designs.

3 IMPACTS

The first practical computer of the new type, the EDSAC (Electronic Delay Store Automatic Calculator) began operation at Cambridge University in May 1949, and during the next two years around a dozen further machines came into operation in the United States and Britain. The stored-program computer has been the archetype of virtually all modern computers in the post-war era, and by the 1960s special-purpose and analogue computers had largely been superseded.

The stored-program computer has had a profound impact on numerical mathematics. One immediate impact was to raise the status of numerical analysis from something of an unfashionable backwater to a major and well-funded research area. On the operational level, the economics of

705

computation was entirely transformed by the electronic computer. The superstructure of practical computation had been developed on the basis of cheap storage (typically by writing intermediate results on paper, or using punched cards) and the avoidance of multiplication (which was expensive in time and machinery). But with the new digital computer, storage was expensive (a few thousand numbers at most) whereas multiplication was cheap, and could be performed at the rate of several hundred operations per second.

To take advantage of the new possibilities for automatic computation, numerical analysis underwent a renaissance. One major research area arose, for example, because existing methods of matrix computation had never been tested with really large matrices. Similarly, little work had been done on error analysis or the stability of iterative processes. As new numerical techniques were developed they had a major impact on engineering design and simulation – especially in the aircraft industries and weather forecasting – which provided an important customer-base for the high-speed and giant computers that emerged in the 1960s.

The digital computer, besides being a superb instrument for numerical mathematics, also proved adaptable to business accounting and automatic control, and rapidly swept away all the existing mechanical and analogue technologies. Indeed, by about 1970 the use of computers for numerical purposes had become a somewhat specialized activity within computer science. Today, the computer is seen as an information-processing machine of much wider scope than mathematical computation. However, the persistence of the name 'computer' serves as a vestigial reminder of the origins of this most important development of the twentieth century.

BIBLIOGRAPHY

The *Annals of the History of Computing* (1980–) is the principal source for this topic.

Aspray, W. (ed.) *1990, Computing before Computers*, Ames, IA: Iowa University Press.
Bowden, B. V. (ed.) *1953, Faster than Thought. A Symposium on Digital Computing Machines*, London: Pitman. [Contemporary and also historical appraisals.]
Gandy, R. *1988*, 'The confluence of ideas in 1936', in R. Herken (ed.), *The Universal Turing Machine*, Oxford: Oxford University Press, 55–111. [A fine survey of the major approaches to the notion of effective calculability, with full references. See also other papers in this very useful compendium.]
Goldstine, H. H. *1972, The Computer from Pascal to von Neumann*, Princeton, NJ: Princeton University Press.

Hodges, A. *1983*, *Alan Turing: The Enigma*, London: Burnett Books.

Randell, B. *1982*, *The Origins of Digital Computers: Selected Papers*, Berlin: Springer.

Turing, A. M. *1936*, 'On computable numbers, with an application to the *Entscheidungsproblem*', *Proceedings of the London Mathematical Society*, Series 2, **42**, 230–65.

—— *1947*, 'The automatic computing engine', in *A. M. Turing's ACE Report of 1946 and Other Papers* (ed. R. E. Carpenter and R. W. Doran), 1986, Cambridge, MA: MIT Press, 106–24. [Lecture given to the London Mathematical Society in February 1947.]

von Neumann, J. *1945*, 'First draft of a report on the EDVAC', Moore School of Electrical Engineering, University of Pennsylvania. [Repr. in *Papers of John von Neumann on Computing and Computer Theory*, Vol. 12 of the Charles Babbage Institute Reprint Series for the History of Computing (ed. W. F. Aspray and A. Burks), Cambridge, MA: MIT Press, and Los Angeles: Tomash.]

Part 6
Algebras and Number Theory

6.0

Introduction

The plural 'algebras' in the title of this Part is deliberate: over the centuries more and more of them have sprouted, with an ever wider range of laws, structures, interpretations and ontologies. Up to the early nineteenth century they usually handled numbers or geometrical sizes, but then operator methods (§4.7) and functional equations (§4.9) began to broaden the scope, which widened further to the conception of abstract algebras, discussed without reference to interpretations.

This last stage is rehearsed in §6.4; prior to that various specific algebras dating back to the time of Girolamo Cardano (in a pick-up from §2.2–2.3) are considered. Among other algebras, linear algebra and its relatives are treated in §6.6–6.8, and the general philosophical situation is apraised in §6.9. The final three topics relate to algebras without properly belonging to them: §6.10, on number theory, treats its 'analytical' branches (in which the calculus is central) as well as the algebraic ones, while §6.11–6.12 take two connected and relatively modern topics (although linear programming has a lengthy prehistory), interdisciplinary in character, in which algebras are significant.

In addition to §4.7 and §4.9, other articles in which algebras and other things work together include §5.1 on algebras in logic; §7.9 on early modern algebraic geometry; §7.12–7.13 on finite vector spaces and graph theory, respectively; §8.1, in which algebras are prominent in mechanics; §9.17 on crystallography, where group theory lurks; and §10.1–10.3, where algebraic aspects of probability theory are manifest.

BIBLIOGRAPHY

Scholz, E. (ed.) *1990*, *Geschichte der Algebra. Eine Einführung*, Mannheim: Wissenschaftsverlag.
van der Waerden, B. L. *1985*, *A History of Algebra* [...], Berlin: Springer.

6.1

The theory of equations from Cardano to Galois, 1540–1830

L. TOTI RIGATELLI

It is well known that the method for solving the general second-degree algebraic equation was known in Antiquity; there is certainly evidence of it in Babylonian mathematics (§1.1). In contrast, the methods for solving the general equations of the third and the fourth degree (cubics and quartics) are much more recent.

1 THE SOLUTION OF CUBIC AND QUARTIC EQUATIONS

The work of the Tuscan abbacists of the fourteenth century gave rise to some rather interesting partial results, but the actual solutions of cubic and quartic equations were found only in the sixteenth century (§2.4). They were first published in the treatise *Ars magna* (1545) by Girolamo Cardano. It is for this reason that the solution to cubic equations is known as 'Cardano's formula'. Take the equation

$$ax^3 + bx^2 + cx + d = 0, \tag{1}$$

and suppose that $x = y - b/3a$. Then (1) takes the form

$$y^3 + py + q = 0, \tag{2}$$

of which one root is

$$y = \left[-\frac{q}{2} + \left(\frac{q^2}{4} + \frac{p^3}{27} \right)^{1/2} \right]^{1/3} + \left[-\frac{q}{2} - \left(\frac{q^2}{4} + \frac{p^3}{27} \right)^{1/2} \right]^{1/3}. \tag{3}$$

Actually, Cardano (to whom we owe the method which allows us to eliminate the second-degree term from every complete cubic equation) was not alone in working out the formula: Scipione Del Ferro and Niccolò Tartaglia shared in its discovery.

In a manuscript recording lessons given by P. Bolognetti at the University of Bologna between 1554 and 1558, reference is made to a method for solving the equation $ax^3 + bx = c$ that attributed it to Del Ferro, who is said to have discovered it earlier in the sixteenth century. The fact that negative numbers were not used, and that zero was not considered to be a number, led to the idea of distinct cases for the cubic equations. In fact, among those free of the second-degree term the following were also examined:

$$ax^3 + c = bx \quad \text{and} \quad ax^3 = bx + c. \tag{4}$$

Tartaglia found the solutions for these last two equations in 1535, along with the one that Del Ferro had already identified, but which Tartaglia claimed to have found independently. Tartaglia did not publish his findings right away but tried to keep them secret, as was normal at that time. Finally, giving in to Cardano's insistent demands, he showed them to the latter, who subsequently published them in his *Ars magna*. In this work Cardano analysed every distinct case of the complete cubic equation, transforming each equation into one free of the second-degree term, giving a numerical example for each case, and proving geometrically the validity of the single solutions.

Lodovico Ferrari discovered the method for solving the general quartic equation. It is also included in the *Ars magna*, where it was illustrated by the equation $x^4 + 6x^2 + 36 = 60x$.

A problem related to the method of solving cubic equations led to the invention of the so-called imaginary numbers. Cardano had already realized the problem in (2) when $q^2/4 + p^3/27 < 0$, although he was not able to apply the solution since he would have had to calculate the root of a negative number. None the less, by resorting to strategems, it is possible to find the three roots of the equation, which are all real numbers.

This particular type of cubic equation, later to be known as the 'irreducible case', was fully analysed, along with the others, by Rafael Bombelli in his work *Algebra* (1572). In this book Bombelli made a detailed examination of what we now call the arithmetic field containing quadratic and cubic irrationals, and observed that the addition of quadratic irrationals is enough to solve quadratic equations, while the addition of cubic irrationals serves to solve cubic equations. For cubics, Bombelli added that, because there are no real numbers able to represent the square root of negative numbers, it is necessary to add some other 'irrational particulars' to represent those numbers. In Bombelli's text the locution *più di meno* stands for the expression $+\sqrt{-1}$, while *men di meno* indicates $-\sqrt{-1}$. For example, the formula

$$\text{R.c.L } 4 \text{ p. di m. R.q.11l} + \text{R.c.L } 4 \text{ m. di m. R.q.11l} \tag{5}$$

represents the modern expression

$$(4 + i\sqrt{11})^{1/3} + (4 - i\sqrt{11})^{1/3}. \tag{6}$$

Having established the formal rules of calculation for imaginary numbers, Bombelli applied them to numerous examples. It is in his work that we find the first complete and comprehensive treatment of the quartic equation. Interestingly, he claimed that Diophantus had taught the method of solving those equations in the lost books of his *Arithmetic* (§6.10).

The history of the cubic and quartic equations had been reconstructed by P. Cossali in 1799 and by E. Bortolotti in the 1920s. Nevertheless, lack of knowledge of the sources led these scholars to neglect the contributions of the numerous Italian abbacists of the fourteenth and fifteenth centuries (§2.4). At the Centro Studi della Mathematica Medioevale of the University of Siena, the algebraic treatises of this period are now being edited with the aim of making possible the compilation of a more complete history of algebra. Meanwhile, see Franci and Toti Rigatelli *1979, 1985*.

2 THE QUINTIC EQUATION

From the end of the sixteenth century, once the method of solving the general cubic and quartic equations had been found, it was only natural that mathematicians should turn to the problem of finding a method for solving the general quintic equation. This method, following the example of those already known, would presumably make use of a finite number of rational operations and extractions of roots on the coefficients of the equation. Thus they set to work to find a solution 'by radicals' or, as it was later called, an 'algebraic' solution. This frequently led algebraists to re-examine the methods already known and to seek solutions different from those of Cardano and Ferrari for solving cubic and quartic equations. Among them was the French algebraist François Viète. In his *In artem analyticem isagoge* (1591), he introduced the use of a vowel to indicate that quantity in an equation which is presumed unknown, and a consonant to represent a quantity or number which is presumed to have been given (Klein *1968*; see also §6.9). Furthermore, he discovered a new method for solving cubic equations as well as a particularly elegant method for solving quartic equations of the type $x^4 + 2ax^2 = c - bx$.

The search for a method for solving the general quintic equation using radicals turned out to be fruitless, although it often led to some interesting results. Among these was the discovery by Ehrenfried von Tschirnhaus (1683) of a transformation by which it is possible to eliminate from a given algebraic equation some of the intermediate terms between that of the highest degree and the known term. Tschirnhaus obviously hoped to use his transformations to develop an algorithm capable of reducing the general equation of degree n to a binomial equation also of degree n: that is, an equation which contains only terms of degree n and degree zero and is therefore, as is shown below, solvable by radicals. Actually Tschirnhaus was able to reduce only the cubic equation to binomial form, while he reduced the quartic to one of the type $ax^4 + bx^2 + c = 0$. Using his transformations he also demonstrated that an equation of degree $n > 2$ can be reduced to a form in which the coefficients of the terms of degree $n - 1$ and $n - 2$ are both zero.

In 1786 the Swedish mathematician E. S. Bring demonstrated that it was possible to find a Tschirnhaus transformation which reduces the quintic equation to the form $y^5 + py + q = 0$, although this still does not lead to its resolution. A single method for solving the equations of the first four degrees was proposed by Leonhard Euler in 1732. He even hypothesized that any equation of degree n would admit a resolvent of degree $n - 1$, as in fact do equations of second, third and fourth degree. Therefore he proposed, for the roots of an equation of degree n, the form

$$x = \sqrt[n]{A_1} + \sqrt[n]{A_2} + \cdots + \sqrt[n]{A_{n-1}}, \tag{7}$$

where the A_i are the roots of the resolvent, although he never did the calculations for $n = 5$. For further stages in this development, see §4.11.

Joseph Louis Lagrange undertook a more careful examination of the whole question, not just the search for the solution, in his long memoir 'Réflexions sur la résolution algébrique des équations', published in 1770 (Rider Hamburg *1976*). He concentrated on creating auxiliary equations (now known as 'resolvents', but which he called 'reduced equations') whose roots would be rational expressions of the roots of the given equation and of the roots of unity. He demonstrated that in general the resolvent of an equation of degree n is $n!$, and that for $n = 3$ or 4 the rational expression of the roots can be chosen so that the respective resolvents, of degree 6 and 24, can be reduced to an equation of degree less than n. For $n = 5$, the degree of the resolvent, which is 120, can be reduced to 6; he was unsure about the possibility of any further reduction. His reflections on resolvents led him to work out a unitary method for solving equations of the first four degrees, a method which for a quartic equation consists in looking for a rational function of the four roots x_1, x_2, x_3 and x_4 which, however they permute among themselves, assumes only three values. The function considered is $\phi = x_1 x_2 + x_3 x_4$.

The effective calculation of a resolvent of sixth degree for the general quintic equation was carried out by G. Malfatti in his memoir *1771*. This calculation was important because the resolvent of sixth degree allowed Malfatti to solve some quintic equations using radicals, including

$$x^5 - 5x^3 + 10x^2 - (35/4)x + 3 = 0 \quad \text{and} \quad x^5 + 20x^2 - 48 = 0. \quad (8)$$

3 THE THEOREMS OF RUFFINI AND ABEL ON THE IRRESOLVABILITY OF THE QUINTIC

The resolution of the general quintic equation was proving to be much harder. Paolo Ruffini overturned the belief of scholars of algebra when he demonstrated in 1799 that equations of the fifth degree and higher were not solvable by radicals. His results are contained in a very long memoir of 516 pages entitled *Teoria generale delle equazioni, in cui si dimostra impossibile la soluzione algebraica delle equazioni generali di grado superiore al quarto*. The proof, which carried to its extreme consequences the path traced out by Lagrange, really proved only the impossibility of solving the quintic equation by the transformation–reduction method (Cassinet *1988*).

It is worth pointing out that Ruffini's work contains, in an implicit form, some of the concepts that Evariste Galois would later use. What would later become the concept of the group of substitutions, Ruffini called simply 'permutation', and the word, used in the singular, meant the group of all the substitutions for which a function remained unchanged.

In his memoir *Riflessione intorno alla soluzione delle equazioni algebraiche generali* of 1813, Ruffini gave a different demonstration of his theorem. It largely coincides with what, in later treatises of classical algebra, is called 'Wantzel's modification' of the theorem of Abel; however, this modification is already in Ruffini. It was in fact the Norwegian mathematician Niels Abel who in 1824 demonstrated Ruffini's theorem on the impossibility of solving the general equation of degree $n \geqslant 5$ by means of radicals. Of particular importance was Abel's paper 'Mémoire sur une classe particulière d'équations résolubles algébriquement' (1829), in which he showed that, in those equations which were solvable by radicals, all roots could be expressed as rational functions of any other root, and that these functions were permutable with respect to the four arithmetical operations. That is, if F_1 and F_2 are any two corresponding functional operations, then $F_1F_2x = F_2F_1x$. In his *Disquisitiones arithmeticae* of 1801, Carl Friedrich Gauss had thoroughly analysed the equation $x^n = 1$, giving the method of solving by radicals for every natural number n, and carrying out the calculations for $n = 17$ and 19. The case for $n = 11$ had already been examined by Alexandre Vandermonde in 1774. On the next algebraic stages, see §6.4, Section 1.

4 THE WORK OF GALOIS

After the demonstration of the theorem of Ruffini–Abel and the claim that there exist an infinite number of particular equations of any degree n which are solvable by means of radicals, it still remained to find the necessary and sufficient conditions under which an algebraic equation of any degree n could be solved by means of radicals. These conditions were given in 1831 by Galois (who died in a duel a year later at only 20 years of age) in the memoir 'Sur les conditions de résolubilité des équations par radicaux'. Galois may be considered the founder of modern algebra since he was the first to introduce the concept of a group of substitutions, the first example in history of an algebraic structure (§6.4, Section 1). He was the first to express clearly the idea of a group of substitutions which is closed with respect to the product. In his memoir he affirmed that:

> When we want to group some substitutions together we will take them all from a single permutation. ... if in such a group there are the substitutions S and T, it is sure that there will also be the substitution ST'.

For every algebraic equation of degree n, Galois took an appropriate group made up of a certain number, divisor of $n!$, of substitutions of the n roots of the equation. These are exactly those substitutions that leave numerically unchanged every rational expression of the roots which belong to what is now known as the field generated by the coefficients of the equation. (The group was later called by O. Hölder the 'Galois group' of the equation.) It is important to note that Galois expressed the condition of solvability by radicals, not in terms of the property of the equation, but rather the property of the group:

> The group G whose equation is solvable by radicals must be divided into a prime number of similar and identical groups H; this group H into a prime number of similar and identical groups K, and so on, until a certain group M does not contain other than a prime number of substitutions. Reciprocally, if group G satisfies the previous conditions the equation will be solvable by radicals.

Nowadays we express this idea by saying that an equation is solvable by radicals if and only if its Galois group is a 'solvable group'. This method of investigation − to associate with objects of various sorts others quite different which translate, wholly or in part, the property of those which one wants to study – has become typical of modern algebra (Edwards *1984*, Toti Rigatelli *1989*).

In his memoir Galois examined in particular the case of an irreducible equation whose degree is a prime number, in this way obtaining the following theorem:

> *Because an irreducible equation of first degree can be solved by radicals, it is necessary and sufficient to note any two of the roots; the others can be rationally deduced from them.*

Galois used not only the concept of group, but also those of coset, subgroup, normal subgroup and factor group. However, he did not define these concepts, indicating each of them only by the word 'group'. The theory of fields also has its origins in the Galois theory, although the concept of field would be rigorously defined only fifty years later, by Richard Dedekind (§6.4).

5 TRIGONOMETRIC SOLUTIONS

There are equations of the second, third and fourth degree which can be solved by trigonometric functions as well as by radicals. The first attempt at a trigonometric solution for equations of second degree appears in Bonaventura Cavalieri's *Compendio di regole trigonometriche* (1638), and the subject was well developed in the treatise *Trigonometry plane and spherical* (1748) by the English mathematician Thomas Simpson.

At first mathematicians limited themselves to finding trigonometric solutions for equations with real solutions. Later, from the first half of the eighteenth century, after the introduction of imaginary numbers in trigonometry (largely as a result of the work of James Bernoulli and Abraham De Moivre), they also found trigonometric solutions for equations with real coefficients but with complex roots.

It is worth remembering that in 1593 Viète used the formulas for multiple angles, equivalent to what we would nowadays write as

$$\cos nx = \cos^n x - \frac{n(n-1)}{1\cdot 2}\cos^{n-2}x\sin^2 x$$

$$+ \frac{n(n-1)(n-2)(n-3)}{1\cdot 2\cdot 3\cdot 4}\cos^{n-4}x\sin^4 x + \cdots \qquad (9)$$

and

$$\sin nx = n\cos^{n-1}x\sin x - \frac{n(n-1)(n-2)}{1\cdot 2\cdot 3}\cos^{n-3}x\sin^3 x + \cdots, \qquad (10)$$

to solve the following equation of the 45th degree:

$$x^{45} - 45x^{43} + 945x^{41} - \cdots - 3795x^3 + 45x = k. \tag{11}$$

To Viète we also owe the trigonometric solution for the irreducible case of the cubic equation.

6 THE FUNDAMENTAL THEOREM OF ALGEBRA

Various studies described above related to this important existence theorem, which states that every polynomial equation $f(x) = 0$ of degree n with real coefficients has n real or complex roots or, equivalently, that it can be fully factorized into real-valued linear and quadratic terms in x.

The importance of the result was soon recognized after a discussion given by Jean d'Alembert in 1746; but the proof turned out to be formidably difficult. One problem was the logical danger of using a technique which assumed the theorem; for example, Lagrange's analysis of equations described in Section 4 above handled the roots as if they were real or complex numbers. In addition, various conceptual matters impeded progress (Loria *1891*, Gilain *1991*). Some attempts, such as that by Etienne Bézout in 1779, required more knowledge of determinants than was then available (§6.6), although he found valuable properties of elimination between equations; while others, such as Euler's, drew on complex variables and integrals, which were then little developed (§3.12) or made assumptions about topology, which had hardly begun (§7.10).

The first major contributor was Gauss, who produced four proofs between 1799 and 1850 (van der Waerden *1985*: Chap. 5). The first two were defective, but their successors, which used complex-variable integrals, were more rigorous and set the standard for later proofs. In this context, algebra, geometry, analysis and topology were to interact, with mutually beneficial consequences.

BIBLIOGRAPHY

Cassinet, J. *1988*, 'Paolo Ruffini (1765–1822): La résolution algébrique des équations et les groupes de permutations', *Bollettino di storia delle scienze matematiche*, **8**, 21–69.

Edwards, H. M. *1984*, *Galois Theory*, New York: Springer.

Franci, R. and Toti Rigatelli, L. *1979*, *Storia della teoria delle equazioni algebriche*, Milan: Mursia.

—— *1985*, 'Towards a history of algebra from Leonardo of Pisa to Luca Pacioli', *Janus*, **72**, 17–82.

Gilain, C. *1991*, 'Sur l'histoire du théorème fondemental de l'algèbre [...]', *Archive for History of Exact Sciences*, **42**, 91–136.

Klein, J. *1968*, *Greek Mathematical Thought and the Origins of Algebra*, Cambridge, MA: MIT Press.

Loria, G. *1891*, 'Il teorema fondamentale della teoria delle equazione algebriche', *Rivista di Matematica*, **1**, 185–248.

Malfatti, G. *1771*, 'De aequationibus quadrato–cubicis dissertatio analytica', *Atti della Accademia delle Scienze di Siena*, **4**, 129–85.

Maracchia, S. *1979*, *Da Cardano a Galois*, Milan: Feltrinelli.

Matthiessen, L. *1878*, *Grundzüge der antiken und modernen Algebra*, Leizpig: Teubner.

Novy, L. *1973*, *Origins of Modern Algebra*, Leyden: Noordhoff.

Rider Hamburg, R. *1976*, 'The theory of equations in the 18th century: The work of Joseph Lagrange', *Archive for History of Exact Sciences*, **16**, 17–36.

Todhunter, I. *1867*, *An Elementary Treatise on Theory of Equations*, 2nd edn, Cambridge: Cambridge University Press.

Toti Rigatelli, L. *1989*, *La mente algebrica. Storia dello sviluppo della teoria di Galois nel XIX secolo*, Busto Arsizio: Bramante.

van der Waerden, B. L. *1985*, *A History of Algebra*, Berlin: Springer.

Wussing, H. *1969*, *Die Genesis des abstrakten Gruppenbegriffes*, Berlin: VEB Deutscher Verlag der Wissenschaften. [English transl.: *The Origins of Group Theory*, 1984, Cambridge, MA: MIT Press.]

6.2

Complex numbers and vector algebra

ALBERT C. LEWIS

1 INTRODUCTION

Historical accounts have generally taken the extension of the number concept as a key indicator of the growth of the whole of mathematics itself. Thus the recognition of irrational numbers by the Classical Greeks, and the incorporation of complex numbers by European mathematicians by the middle of the nineteenth century, and of the number systems which came out of Georg Cantor's transfinite numbers of the late nineteenth century (§3.6), are viewed as major signs of progress. The detailed history of each of these episodes also shows, however, that what are today regarded as 'natural' extensions of the number concept were far from readily accepted as valid mathematical entities when first introduced.

With imaginary and complex numbers, records of the world's major cultures in the first millennium AD show that, while general classes of concrete problems were recognized, there was no recognition of correspondingly general solutions of these problems. For example, there was a class of geometric problems of the type 'Find the dimensions of the rectangle enclosing an area of x square units with a perimeter of length y units.' Solutions for given x and y were found but, along with these, cases were noticed which were regarded as having no solution, such as $x = 40$ and $y = 20$. We might regard this case as a formally valid problem, in that it follows the same pattern as the others, but one which has no real or intuitional content in that no such area of land is possible.

The story of the acceptance of these new types of number is much more difficult to extract than the stories of the discoveries themselves by individual mathematicians. Undoubtedly, the mutual influences of various developments in algebra, geometry, analysis and applied mathematics eventually put complex numbers on the same footing as real numbers.

2 IMAGINARY NUMBERS IN EARLY ALGEBRA

As European algebra came into existence in the sixteenth century, Rafael Bombelli and Girolamo Cardano recognized 'impossible' solutions (§6.1). Bombelli's solution of a cubic used properties of conjugate imaginaries. Glushkov *1977* has shown that François Viète constructed operations on right-angled triangles which anticipated the direct and inverse operations in the multiplicative group of complex numbers, but without explicitly recognizing such numbers. In the seventeenth century René Descartes (who used the term *imaginaire*) and others gave formal recognition to complex numbers, at least to the extent of writing square roots of negative numbers as outcomes of algebraic manipulations. It was recognized from at least the second half of the seventeenth century that each real number has n nth roots, of which two at most are real. Negative numbers themselves – let alone complex numbers – were not, however, regarded as having the same status as the positive real numbers, but were generally taken as symbolizing the absence of a solution. In 1629 Albert Girard posited, in effect, the possibility that the number of roots of an algebraic equation with integral coefficients is equal to its degree (a result first rigorously proved by Carl Friedrich Gauss in the early nineteenth century). The desire to satisfy such a generalization helped to put negative and imaginary numbers on the same footing as the real numbers.

3 COMPLEX NUMBERS IN THE CALCULUS

Complex analysis was one of the fields which made formal use of complex numbers and thus further helped to extend the notion of number to include them (§3.12). The use of complex numbers is essentially algebraic, but the motivating contexts are problems in analysis.

The logarithms of imaginary numbers were discussed in correspondence between Gottfried Wilhelm Leibniz and John Bernoulli in 1702 in connection with the integration of rational functions. The relation between such logarithms and the arcs of circles was also the subject of correspondence in the same year, between John Bernoulli and Pierre Varignon.

In his *Miscellanea analytica* (1730), Abraham De Moivre presented further analytical trigonometric results (some formulated as early as 1707), making use of complex numbers. Although he did not state what is now known as De Moivre's theorem, it is clear that he was making use of it. He gave

$$\cos B = \tfrac{1}{2} \sqrt[n]{(\cos nB + \sqrt{-1} \sin nB)} + \tfrac{1}{2} \sqrt[n]{(\cos nB - \sqrt{-1} \sin nB)}, \qquad (1)$$

which relates to expressions of the type

$$\sqrt[n]{(a + \sqrt{-b})} + \sqrt[n]{(a - \sqrt{-b})}, \tag{2}$$

which in turn came from early attempts at solving cubic equations. This last expression had been something of a paradox for some time since it appears to be complex but in fact is real.

Extensive use was made of complex numbers, especially in the form of logarithms of negative numbers, in the works of Leonhard Euler. He used the symbol 'i' for $\sqrt{-1}$ and was the first to prove, for all numbers n, the formula named after him:

$$(\cos \phi + i \sin \phi)^n = \cos n\phi + i \sin n\phi. \tag{3}$$

He also proved that

$$\sin \phi = (e^{\phi i} - e^{-\phi i})/2i, \qquad \cos \phi = (e^{\phi i} + e^{-\phi i})/2. \tag{4}$$

4 GEOMETRIC REPRESENTATION

In 1797, at the age of 52, the Norwegian surveyor Caspar Wessel presented a paper to the Royal Academy of Denmark in which he demonstrated essentially the same geometric representation of complex numbers used today. Then, in 1806, the 38-year-old Jean Argand, Swiss and apparently self-educated, while working in Paris as a bookseller published a graphical representation of complex numbers as *lignes en direction*, the first such to be printed. In the period 1813–15 he, as well as J. F. Français and F. J. Servois, attempted an extension of complex numbers that would allow a three-dimensional representation. As with earlier mathematicians, Argand regarded the truth or falsity of imaginaries as irrelevant: he was, he claimed, simply proposing that his notation could be of use in demonstrating certain truths.

In his study of these efforts, Cartan *1908* makes a fine but rather crucial historical distinction between mere geometrical representation and the actual definition of complex numbers as directed lines. In his view the latter can occur only if all the laws of the calculus are explicit. It is not sufficient simply to illustrate addition and multiplication, say, using the geometrical representation. As the first work to meet this rigorous standard, he credits C. V. Mourey's *La Vrai Théorie des quantités négatives et des quantités prétendues imaginaires* (1828).

Gauss mentioned the subject of the graphical representation of complex numbers in 1798, but did nothing further with it. Apparently independently of Wessel, he presented the equivalent form in 1831. Gauss used complex numbers, in particular Gaussian integers ($a + ib$, where a and b are integers), in this 1831 work on biquadratic residues (§6.10). In order to distinguish between $a\sqrt{-1}$ and $a + b\sqrt{-1}$ he gave the name 'complex number' to the latter. He also coined the term 'norm' of a complex number $a + ib$ to describe $a^2 + b^2$.

Current historical opinion seems to be that complex numbers reached their watershed of acceptance as a result of this work by Gauss: the imprimatur of the greatest mathematician of the time served as a mark of official mathematical acceptance. But it should be noted that the two most important discoveries next made in the field, in Germany by Hermann Grassmann and in Ireland by William Rowan Hamilton, were made without knowledge of Gauss's work on complex numbers.

5 GENERALIZED COMPLEX NUMBERS

The next step beyond complex numbers, already anticipated by Wessel and Argand, was to numbers that would extend to three dimensions the geometrical properties of complex numbers in the plane (i.e. a representation as rotations about an axis). The discoveries of such entities, principally by Grassmann and Hamilton, led to questions of how they interacted algebraically with the system of complex numbers.

Grassmann's *Ausdehnungslehre* (1844) purported to lay the foundations of a new mathematical science, but it was not until the end of his life in 1877 that mathematicians began to recognize that he had in fact not just founded a new branch (multilinear algebra) but created a new way of doing mathematics (Lewis *1977*). Part of his method was to posit quite general properties of operations which would apply not only to ordinary arithmetic and algebra, but also to combinatorics (compare §7.13).

The *Ausdehnungslehre* was to be that branch of mathematics which applied such operations to extensive (i.e. geometrical) quantities. Although originally motivated by work on the theory of tides, Grassmann's presentation of his theory in 1844, when he was a 35-year-old Prussian schoolteacher, proceeded along rather abstract lines. (Detailed descriptions of this algebra are given in Study *1898* and Crowe *1967*.) In his version *1862* of the *Ausdehnungslehre*, Grassmann justified calling the abstract objects of his algebra 'quantities' by pointing out that mathematicians were now used to referring to elements generated from a basis of 1 and $\sqrt{-1}$ as 'quantities', albeit imaginary or complex quantities. Grassmann's extensive quantities were to be similarly generated from two or more units:

> But ... I go further, since I call these not just quantities but *simple* quantities. There are other quantities which are themselves compounded quantities and whose characteristics are as distinctive relative to each other as

the characteristics of the different simple quantities are to each other. These quantities come about through addition of higher forms and especially through the consideration of quotients and functions.

One result is what is today termed the 'Grassmann algebra'.

The 38-year-old Hamilton discovered quaternions in 1843; he presented part of his theory in 1844 and the first complete treatment in 1853. Hamilton, when he first read Grassmann's work around the beginning of 1853, recognized how close Grassmann had come to quaternions. The crucial difference is that Grassmann's exterior product of two base elements results in a member of a new order or step, whereas the product of two base elements for Hamilton can be expressed as a linear combination of the base elements. This is accomplished by having the base elements, 1, i, j and k, obey the following multiplication laws:

$$ijk = i^2 = j^2 = k^2 = -1. \tag{5}$$

Grassmann interested himself in the general question of what sorts of hypercomplex number were possible. Although his methods could be regarded as precursors of group-theory methods, his efforts were not immediately appreciated. Quaternions, on the other hand, were a major influence on the work of the Americans Benjamin Peirce and his son Charles Sanders Peirce in their generalizations, linear associative algebras, in the 1870s (Pycior *1979*). Both quaternions and Grassmannian algebra were an influence on the English mathematician William Clifford's work in 1873 on biquaternions. Grassmann eventually came to have his major influence on modern mathematics mainly through Hermann Hankel *1867*, Giuseppe Peano, Alfred North Whitehead (in his 'universal algebra' of 1898) and Josiah Willard Gibbs (in the 1880s and 1890s in his spreading of the importance of 'multiple algebras').

In the 1880s the Englishman J. J. Sylvester published a number of papers which explored the correspondence between quaternions and their generalizations on the one hand and matrices on the other. His countryman Arthur Cayley had in 1858 noted the similarity in behaviour between quaternions and 2×2 matrices in his own writings on hypercomplex numbers. The general connection between hypercomplex numbers and matrices is evident when the multiplication table of base elements is written as

$$e_i e_j = \sum_{k=1}^{n} a_{ij}^k e_k, \quad i, j = 1, \ldots, n. \tag{6}$$

For each hypercomplex number $\boldsymbol{\alpha} = \Sigma_{i=1}^{n} \alpha_i e_i$ there is an associated matrix,

726

namely the matrix of coefficients in the system of equations

$$\boldsymbol{\alpha} e_j = \sum_{i=1}^{n} \sum_{k=1}^{n} \alpha_i a_{ij}^k e_k. \tag{7}$$

This discovery connected hypercomplex numbers to the nascent theory of transformation groups, and thus formed one of the strands in the history of representation theory in the early decades of the twentieth century (Hawkins *1972*).

6 VECTOR ALGEBRA

In much the same way that complex numbers came from solving algebraic equations, and in turn broadened the scope of algebra, so hypercomplex numbers were generated algebraically and at the same time broadened the notion of algebraic operation itself. In the earliest stages of development, for example, properties of multiplication became an object of change because the new 'numbers' did not follow the usual multiplication rules of real numbers.

Addition and subtraction of the new entities presented less of a problem, mainly, perhaps, because of the long history of use of the parallelogram 'law' determining the combination of velocities and of forces. Algebraically this form, in modern terms, corresponds to the components of the sum of two vectors as the sum of the components of the addends:

$$\boldsymbol{\alpha} + \boldsymbol{\beta} = \sum_{i=1}^{n} (\alpha_i + \beta_i) e_i. \tag{8}$$

However straightforward and empirically satisfactory it seems to think of the sum as one of the diagonals of a parallelogram, it has been a stumbling block for some who found it difficult to accept a sum which did not 'consist' of its addends except in the special case of forces or velocities acting along the same straight line. The mathematical justification for this definition of addition lies, as Grassmann expressed it, in the definition of multiplication as an operation which is distributive over addition. Though defined by Grassmann for any finite dimension, in three-dimensional geometric terms such multiplication of two directed line segments is the signed area of the parallelogram formed by them. This corresponds to the cross or vector product,

$$\boldsymbol{\alpha} \times \boldsymbol{\beta} = \begin{vmatrix} e_1 & e_2 & e_3 \\ \alpha_1 & \alpha_2 & \alpha_3 \\ \beta_1 & \beta_2 & \beta_3 \end{vmatrix} = \sum_{i=1}^{3} \sum_{j=1}^{3} \sum_{k=1}^{3} \alpha_j \beta_k \varepsilon_{ijk} e_i \tag{9}$$

in modern notation (where $\varepsilon_{ijk} = 1$, 0 or -1 according as the indices are cyclic, anticyclic or repetitive). The fact that these properties imply a non-commutative and non-associative multiplication was not a problem for Grassmann. In his 1844 work he also described a commutative product, the modern dot or scalar product:

$$\alpha \cdot \beta = \sum_{i=1}^{3} \alpha_i \beta_i. \tag{10}$$

Both the cross and dot products appeared as parts of Hamilton's definition of quaternion multiplication, in fact as the vector and scalar components respectively of the quaternion product. Both products were also given by Clifford in *Elements of Dynamics* (1878), a textbook which he was not able to complete before his death in 1879 at the age of 34.

The first reasonably fully fledged vector algebras incorporating these new multiplications were developed by the American physicist Gibbs and the self-taught Englishman Oliver Heaviside. In much the same way that Grassmann had been motivated by the theory of tides and a desire to improve on the tools provided in Joseph Louis Lagrange's *Mécanique analytique* (1811–15), so Gibbs and Heaviside were influenced by a desire to provide a mathematical treatment of James Clerk Maxwell's field theory (§9.10). Just as Grassmann and Hamilton were initially unacquainted with Gauss's work on complex numbers and with each other's work, so Gibbs and Heaviside initially developed their ideas independently of Grassmann and of each other (though not independently of Hamilton, whose ideas came to them via Maxwell and through P. G. Tait's *Elementary Treatise on Quaternions* (1873)). All also had difficulties in getting their ideas accepted, though each for different reasons.

The greatest usefulness of these algebraic products lay in the new analytic operators which resulted when they were combined with analytic notions. Hamilton in 1846 noted the possible usefulness of a partial differential operator based on quaternions which he wrote as

$$\triangleleft = \frac{i\mathrm{d}}{\mathrm{d}x} + \frac{j\mathrm{d}}{\mathrm{d}y} + \frac{k\mathrm{d}}{\mathrm{d}z}. \tag{11}$$

In *Elements of Vector Analysis* (1881, second part 1884), Gibbs introduced the modern operator

$$\nabla = \sum_{i=1}^{3} e_i \frac{\partial}{\partial x_i}. \tag{12}$$

Heaviside, in a series of papers on electricity in 1883 (compare §9.11), defined what he wrote as div **R** and curl **R** for the modern divergence and curl of a vector function **R**, where

$$\operatorname{div} \mathbf{R} = \nabla \cdot \mathbf{R} \quad \text{and} \quad \operatorname{curl} \mathbf{R} = \nabla \times \mathbf{R}. \tag{13}$$

This latter notation is not used by Heaviside. His sentence defining the curl probably reflects the origin of his idea, and at the same time shows how challenging his writing probably was to most readers of the journal, *The Electrician*, in which it appeared (quoted from Crowe *1967*: 164):

> When one vector or directed quantity, *B*, is related to another vector, *C*, so that the line-integral of *B round* any closed curve equals the integral of *C through* the curve, the vector *C* is called the curl of the vector *B*.

The first readers of Grassmann, Hamilton and Gibbs undoubtedly experienced the same kind of challange.

BIBLIOGRAPHY

Cartan, E. *1908*, 'Nombres complèxes', in *Encyclopédie des sciences mathématiques pures et appliquées*, Tome 1, Vol. 1, 329–468 (article I 5). [Repr. in *Oeuvres complètes*, Part 2, Vol. 1, 1953, Paris: Gauthier-Villars, 107–246. French adaptation of Study *1898* with many additional historical notes.]

Crowe, M. J. *1967*, *A History of Vector Analysis: The Evolution of the Idea of a Vectorial System*, Notre Dame, IN: University of Notre Dame Press. [Repr. with corrections and new bibliographical preface, 1985, New York: Dover.]

Glushkov, S. *1977*, 'An interpretation of Viète's "Calculus of triangles" as a precursor of the algebra of complex numbers', *Historia mathematica*, 4, 127–36.

Grassmann, H. G. *1862*, *Die Ausdehnungslehre: vollständig und in strenger Form bearbeitet*, Berlin: Enslin.

Hankel, H., *1867*, *Vorlesungen über die complexen Zahlen und ihre Functionen. I. Theil: Theorie der complexen Zahlensysteme* [...], Leipzig: Voss.

Hankins, T. L. *1980*, *Sir William Rowan Hamilton*, Baltimore, MD, and London: Johns Hopkins University Press.

Hawkins, T. *1972*, 'Hypercomplex numbers, Lie groups, and the creation of group representation theory', *Archive for History of Exact Sciences*, **8**, 243–87.

Lewis, A. C. *1977*, 'H. Grassmann's 1844 *Ausdehnungslehre* and Schleiermacher's *Dialektik*', *Annals of Science*, 34, 103–62.

Pycior, H. *1979*, 'Benjamin Peirce's *Linear Associative Algebra*', *Isis*, **70**, 537–51.

Study, E. *1898*, 'Theorie der gemeinenen und höheren complexen Grössen', in *Encyklopädie der mathematischen Wissenschaften*, Vol. 1, Part 1, 147–83 (article 1A4). [Includes historical notes. The basis of Cartan *1908*.]

6.3

Continued fractions

D. H. FOWLER

1 A SKETCH OF THE THEORY

The so-called 'Euclidean algorithm' says: given two non-negative numbers, if neither is zero, subtract the smaller from the larger; if one is zero, terminate; now repeat. For example, taking $\sqrt{2}$ and 1, we get

$$(1.414\ldots, 1) \rightarrow (0.414\ldots, 1) \rightarrow (0.414\ldots, 0.586\ldots)$$
$$\rightarrow (0.414\ldots, 0.172\ldots) \rightarrow \cdots, \quad (1)$$

a pattern of one subtraction (of the second term from the first), two subtractions (of the first term from the second), two subtractions, two subtractions, and so on. The topic of continued fractions explores and exploits the relationship between the two original numbers and these resulting patterns of subtractions. In this section are outlined the main features of a subject which is often not well known.

1.1 Procedures

For initial numbers a_0 and a_1, we can write the process as follows:

$$a_0 - n_0 a_1 = a_2 \quad \text{with} \quad 0 \leqslant a_2 < a_1. \quad \text{If } a_2 \neq 0, \text{ then}$$
$$a_1 - n_1 a_2 = a_3 \quad \text{with} \quad 0 \leqslant a_3 < a_2. \quad \text{If } a_3 \neq 0, \text{ then}$$
$$a_2 - n_2 a_3 = a_4 \quad \ldots.$$

Hence

$$\frac{a_0}{a_1} = n_0 + \frac{a_2}{a_1} = n_0 + \cfrac{1}{n_1 + \cfrac{a_3}{a_2}} = n_0 + \cfrac{1}{n_1 + \cfrac{1}{n_2 + \cfrac{a_4}{a_3}}} = n_0 + \cfrac{1}{n_1 + \cfrac{1}{n_2 + \cfrac{1}{n_3 + \cdots}}},$$

$$(2)$$

and we now see the reason for the name 'continued fractions'. These

730

expressions will be abbreviated as

$$n_0 + \frac{1}{n_1 +} \frac{1}{n_2 +} \frac{1}{n_3 +} \cdots, \tag{3}$$

or as $[n_0, n_1, n_2, \ldots]$; many other notations have been employed.

This is a 'simple (or regular) continued fraction': all the terms n_i are non-negative integers and $n_i > 1$ if $i \geqslant 1$. It 'terminates' if it contains only a finite number of terms; otherwise it can be shown that any such expression will converge rapidly in a characteristic way (see below). There are also 'general continued fractions'

$$\frac{m_0}{n_0} + \frac{m_1}{n_1 +} \frac{m_2}{n_2 +} \frac{m_3}{n_3 +} \cdots, \tag{4}$$

in which the m_i and n_i need not be restricted to non-negative integers. Although many remarkable particular examples of such general continued fractions are known (see below), they do not share some very special properties of simple continued fractions.

If we write $a_0/a_1 = x_0$, we get the 'continued fraction algorithm'. Write

$$x_0 = n_0 + f_0, \quad \text{where} \quad 0 \leqslant f_0 < 1.$$

(Here n_0 is the 'integer part' of x_0, the greatest integer less than or equal to x_0, and f_0 is the fractional part of x_0.) If $f_0 \neq 0$, then write

$$1/f_0 = x_1 = n_1 + f_1, \quad \text{where} \quad 0 \leqslant f_1 < 1. \quad \text{If } f_1 \neq 0, \text{ then write}$$
$$1/f_1 = x_2 = n_2 + f_2, \quad \ldots.$$

In this form, the algorithm is very easy to implement on the simplest of programmable calculating machines: enter x, display and record its integer part, reciprocate its fractional part, and repeat with this new number. Performing the process of $\sqrt{2}$ will yield $1, 2, 2, 2, 2, 2, \ldots$, where the 2's seem to repeat until the limit of the approximation of the machine is exhausted; and this apparent behaviour can be established by observing that

$$\sqrt{2} = 1 + (\sqrt{2} - 1),$$
$$1/(\sqrt{2} - 1) = (\sqrt{2} + 1) = 2 + (\sqrt{2} - 1), \tag{5}$$

and hereafter this step will repeat. Hence $\sqrt{2} = [1, 2, 2, 2, \ldots]$. This sequence of integers $1, 2, 2, 2, \ldots$ serves to describe $\sqrt{2}$ via continued fractions in a way different from but analogous to the way the sequence of integers $1, 4, 1, 4, 2, 1, \ldots$ serves to describe it via its decimal expansion. But the continued-fraction expansion involves no arbitrary choice (as in the base 10 of a decimal number), and it employs the simplest description of the underlying subtraction procedure. (To illustrate this, write out a subtraction algorithm that will generate the digits of the decimal or binary expansion of x.)

The most basic result about the Euclidean algorithm, to be found in Euclid's *Elements* (Book VII, Propositions 1–2 and Book X, Propositions 2–3), is that it will terminate if and only if $a_0/a_1 = p/q$, a rational number (i.e. p and q are integers), and the last non-zero remainder will then be the highest common factor. To see this, consider the Euclidean algorithm applied to (p, q): at each stage the remainder will decrease, and since it is also a non-negative integer, it must eventually become zero; hence the algorithm must terminate. Conversely, if the algorithm terminates, the expression $[n_0, n_1, \ldots, n_k]$ will simplify to a rational number.

1.2 Convergence

For non-terminating expressions $[n_0, n_1, n_2, \ldots]$, we must consider their convergence. The standard way of handling infinite processes like this applies here (§3.3 and §4.3): we chop off successively longer and longer parts of the fraction and investigate their behaviour. Formally, if $x_0 = [n_0, n_1, n_2, \ldots]$, then we define the 'convergents' of the continued fraction to be $p_k/q_k = [n_0, n_1, \ldots, n_k]$, where the p_k/q_k are always written in their lowest terms. Thus the successive convergents of $\sqrt{2}$ are

$$\frac{p_0}{q_0} = 1, \quad \frac{p_1}{q_1} = 1 + \frac{1}{2} = \frac{3}{2}, \quad \frac{p_2}{q_2} = 1 + \frac{1}{2 + \frac{1}{2}} = \frac{7}{5}, \quad \frac{p_3}{q_3} = \frac{17}{12}, \ldots, \tag{6}$$

and they oscillate around and converge towards $\sqrt{2}$ in a characteristic way:

$$1 < \frac{7}{5} < \frac{44}{29} < \cdots < \sqrt{2} < \cdots < \frac{99}{70} < \frac{17}{12} < \frac{3}{2}. \tag{7}$$

The convergents can be quickly evaluated by the recurrence relations

$$p_k = n_k p_{k-1} + p_{k-2}, \quad q_k = n_k q_{k-1} + q_{k-2},$$
$$\text{where } p_{-1} = q_{-2} = 1 \quad \text{and} \quad p_{-2} = q_{-1} = 0, \tag{8}$$

which we set out as a convenient tableau:

n_k			1	2	2	2	2	2
p_k	0	1	1	3	7	17	$2 \times 17 + 7 = 41$	etc.
q_k	1	0	1	2	5	12	$2 \times 12 + 5 = 29$	

This kind of behaviour is general: if $x = [n_0, n_1, n_2, \ldots]$, then

$$\frac{p_0}{q_0} < \frac{p_2}{q_2} < \frac{p_4}{q_4} < \cdots < x < \cdots < \frac{p_5}{q_5} < \frac{p_3}{q_3} < \frac{p_1}{q_1}, \tag{9}$$

732

where the p_k/q_k converge to, and also provide the best possible approximations to, x by fractions whose denominators (and numerators) are of a given size. That is,

$$\text{if} \quad \left| x - \frac{p'}{q'} \right| < \left| x - \frac{p_k}{q_k} \right|, \quad \text{then} \quad q' > q_k. \tag{10}$$

In fact, much more can be said. For example, there are the *a priori* estimates that

$$\frac{1}{(n_{k+1} + 2)q_k^2} < \left| x - \frac{p_k}{q_k} \right| < \frac{1}{n_k q_k^2}, \tag{11}$$

which describe precisely what we should expect: that large terms in the expansion give rise to especially good approximations by the preceding convergent. For example, the expansion $\pi = [3, 7, 15, 1, 292, 1, \ldots]$ contains coded within it the information that $p_1/q_1 = [3, 7] = 22/7$ is an overestimate to π, with

$$1.20 \times 10^{-3} < (22/7 - \pi) < 1.36 \times 10^{-3}, \tag{12}$$

and that $p_3/q_3 = [3, 7, 15, 1] = 355/113$ is also an overestimate, with

$$2.664 \times 10^{-7} < (355/113 - \pi) < 2.682 \times 10^{-7}. \tag{13}$$

Moreover, any sufficiently good approximation to x, however derived, must necessarily be the convergent of a continued fraction: if

$$\left| x - \frac{p}{q} \right| < \frac{1}{2q^2},$$

then p/q must be a convergent of x. In this way, continued fractions are always lurking in the wings of any sufficiently efficient approximation procedure (see §5.10).

For more details of these results and their history (which follows below), see Brezinski *1991*; Davenport *1952*: Chap. 4; Fowler *1987*: Chap. 9.

2 FROM BOMBELLI TO COTES

The first explicit hint of continued fractions is found in Rafael Bombelli's *Algebra*, published in Bologna in 1572, in which he gave, in effect, the examples of the continued fractions

$$\sqrt{13} = 3 + \frac{4}{6+} \frac{4}{6+} \cdots$$

and its convergents

$$3, \; 3\tfrac{2}{3}, \; 3\tfrac{3}{5}, \; 3\tfrac{20}{33}, \; 3\tfrac{66}{109}, \; 3\tfrac{109}{180},$$

and

$$\sqrt{8} = 2 - \frac{1}{6-} \; \frac{1}{6-} \cdots$$

and its convergents. Then we find very similar expressions in Pietro Cataldi's *Trattato del modo brevissimo di trouare la Radici quadra delli numeri* [...] (1613, also published in Bologna but without any reference to Bombelli); for example

$$\sqrt{18} = 4 + \frac{2}{8+} \; \frac{2}{8+} \; \frac{2}{8+} \cdots$$

(with, this time, a notation to set out this fraction; see Figure 1) and the evaluation of its first fifteen convergents. All these examples are particular cases of the expansion

$$(n^2 + r)^{1/2} = n + \frac{r}{2n+} \; \frac{r}{2n+} \; \frac{r}{2n+} \cdots, \qquad (14)$$

70

rotto totale $\frac{1}{10} \cdot \frac{1}{4} \cdot \frac{1}{0}$. via $\dfrac{1}{1040.\,\text{via}\,17}$. in che egli fupera il rotto aggiunto . Pafsiamo hora auanti nella confideratione di quefto modo in andar trouando altre radici , con andar giongendo di mano in mano al denominat. del rotto , che nella ℞ antecedente, è vltimo vn rotto eguale al primiero ; ma per maggior cômodità fupponiamo vn numero facile da pigliarne la ℞, & poniamo la prima parte d'effa numero intiero. Hor fia 18. il numero propofto , & fi dica la prima ℞ effere 4. & $\tfrac{2}{8}$. cioè $4\tfrac{1}{4}$. che farà eccedente in $\frac{1}{16}$. quadrato del rotto $\tfrac{1}{4}$.

Di 18. la ℞ fia 4. & $\dfrac{2}{8}$. & $\dfrac{2}{8}$.

cioè λ. & $\dfrac{2}{8}$. & $\dfrac{8}{33}$

che è $4\dfrac{33}{136}$ Quadrifi

$17\dfrac{16}{17}$

Il ■ è $18\dfrac{1}{18496}$. però eccede in $\dfrac{1}{18496}$.

rottó totale	aggiunto
33	8
136	33
1089	1088

$\dfrac{1}{136.\,\text{via}\,33}$.via $\dfrac{33.1}{136}$

fà $\dfrac{1}{18496}$.che è l'ecceffo.

La feconda ℞ trouandola nel modo fopradetto farà 4. & $\tfrac{2}{8}$.& $\tfrac{2}{8}$. cioè 4.& $\tfrac{2}{8}$.& $\tfrac{1}{4}$. che è 4.& $\frac{8}{34}$. quale farà fcarfa in $\frac{2}{1088}$. che nafce à moltiplicare il rotto totale $\frac{8}{34}$. via $\frac{1}{1088}$.in che effo rotto totale è minore di $\tfrac{1}{4}$. rotto aggiunto.

¶ Notifi, che nó fi potendo cômodamête nella ftâpa formare i rotti,& rotti di rotti come andariano, cioè così 4.& $\dfrac{2}{8.\,\&\,\dfrac{2}{8}}$ come ci fiamo sforzati di fare in quefto, noi da qui inâzi gli formaremo tutti à qfta fimilitudine 4.& $\tfrac{2}{8}$.& $\tfrac{2}{8}$.& $\tfrac{2}{8}$. facendo vn punto all'8. denominatore di ciafcun rotto, à fignificare, che il feguente rotto è rotto d'effo denominatore.

734

Figure 1 A passage from Cataldi *1613*: 70

The left-hand portion of the text inset on the left reads:

Of 18. let the root be

4. $\dfrac{2}{8.}$ & $\dfrac{2}{8.}$ & $\dfrac{2}{8}$

that is 4. & $\dfrac{2}{8.}$ & $\dfrac{8}{33}$

which is $4\dfrac{33}{136}$. Squaring

$17\dfrac{16}{17}$ [and] $\dfrac{1089}{18496}$

[Now $18496 = 17 \times$] 1088

[Hence:] The $\square$ is $18\dfrac{1}{18496}$, yet too large by $\dfrac{1}{18496}$.

¶Note that it is not possible to show fractions conveniently in print, nor fractions of fractions, as they should appear, that is thus: 4. $\dfrac{2}{8.}$ & $\dfrac{2}{8.}$ & $\dfrac{2}{8}$ as we have made ourselves do in this example; from now on we shall show them all this like: 4. & $\frac{2}{8}$. & $\frac{2}{8}$. & $\frac{2}{8}$., letting a point by the 8 in the denominator of each fraction indicate that the following fraction is a fraction of this denominator.

The numbers interleaved and to the right of this inset text, under the heading 'total added fraction', are connected with these evaluations.

The text on the right, starting at the end of the first line, reads:

Let us now proceed to the consideration of the manner of finding other roots, by adding step by step to the denominator of the fraction which is the last in the preceding root (R), a fraction equal to the first one; but, for greater convenience, let us take a number whose root may be easily obtained and let its first part be an integer. Now let 18. be the proposed number, and let the first root be 4. & $\frac{2}{8}$, that is $4\frac{1}{4}$; this will be too large by $\frac{1}{16}$, the square of the fraction $\frac{1}{4}$.

The second root found by the above-mentioned method will be 4. & $\frac{2}{8}$. & $\frac{2}{8}$., that is 4. & $\frac{2}{8}$. & $\frac{1}{4}$, which is 4. & $\frac{8}{33}$, which will be too small by $\frac{2}{1089}$, which happens through multiplying the whole fraction $\frac{8}{33}$ by $\frac{1}{132}$, [the amount] by which the whole fraction is less than the $\frac{1}{4}$, the fraction that was added.

The third root found by the above-mentioned method will be . . .

and Cataldi goes on like this to consider the first fifteen steps of the process.

which can be derived from the identity

$$[(n+r)^{1/2}-n]\,[(n+r)^{1/2}+n]=r. \tag{15}$$

(These are not simple continued fractions, and they do not necessarily generate particular good approximations.) Both Bombelli's and Cataldi's interest was the extraction of square roots, and Bombelli introduces his method with an engaging opinion:

> It may happen that today I teach a rule which could be more acceptable than those given in the past, but if another should be discovered later and if one of them should be found to be more vague and if another should be found to be more easy, this [latter] would then be accepted at once and mine would be rejected; for, as the saying goes, experience is our master and the workman is measured by his work.

(For extracts for Bombelli and Cataldi, see Smith *1929*: 80–84.)

The use of convergents as approximations seems common knowledge since the seventeenth century. We find it in elementary schoolbooks published by Daniel Schwenter in 1618 and 1636 (see Figure 2), where he evaluated the convergents 1/1, 3/4, 19/17, and 79/104 of 177/233. It is referred to in letters between John Wallis and John Collins in 1676; used by Christiaan Huygens in the 1680s to design suitable gearing for a planetarium; and described, in connection with the number e, by Roger Cotes in 1714 (reprinted in his posthumous book *Harmonia mensurarum*, 1722).

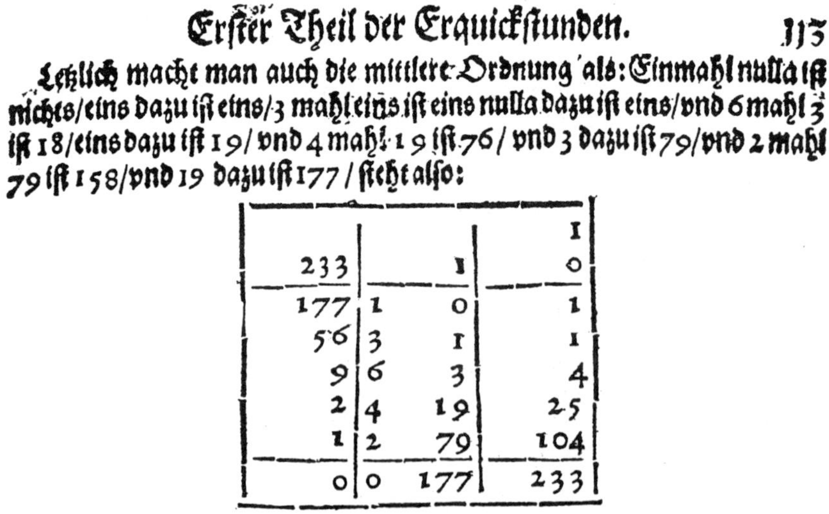
Erster Theil der Erquickstunden. 113

Letzlich macht man auch die mittlere Ordnung / als: Einmahl nulla ist nichts / eins dazu ist eins / 3 mahl eins ist eins nulla dazu ist eins / vnd 6 mahl 3 ist 18 / eins dazu ist 19 / vnd 4 mahl 19 ist 76 / vnd 3 dazu ist 79 / vnd 2 mahl 79 ist 158 / vnd 19 dazu ist 177 / steht also:

			1
233		1	0
177	1	0	1
56	3	1	1
9	6	3	4
2	4	19	25
1	2	79	104
0	0	177	233

Figure 2 An illustration of the Euclidean algorithm, with the evaluation of the convergents (from Schwenter *1636*: 113)

3 PERIODIC CONTINUED FRACTIONS AND PELL'S EQUATION

The periodic behaviour of $\sqrt{2}$, which we observed and proved at (5), is a manifestation of a remarkable general phenomenon with a rich history. If n is a non-square integer (or, more generally, a non-square rational number), then

$$\sqrt{n} = [n_0, n_1, n_2, \ldots, n_2, n_1, 2n_0, n_1, n_2, \ldots, n_2, n_1, 2n_0, \ldots], \qquad (16)$$

where the terms $n_1, n_2, \ldots, n_2, n_1, 2n_0$ will repeat indefinitely. Moreover, the final term $2n_0$ which terminates the palindromic block will be the largest term in the expansion (this result being true only when n is an integer). Hence, recalling what was said earlier about a large term giving rise to especially good approximations by its preceding convergent, we might expect that

$$p_k/q_k = [n_0, n_1, n_2, \ldots, n_2, n_1] \qquad (17)$$

and the further convergents preceding the subsequent occurrences of the term $2n_0$ all to be good approximations to $\sqrt{n}$. (Also note that if p/q is a good approximation to $\sqrt{n}$, then p^2/q^2 will be a good approximation to n.) In fact, it transpires that, for these terms,

$$\frac{p_k^2}{q_k^2} = n \pm \frac{1}{q_k^2}; \qquad \text{and so} \quad p_k^2 - nq_k^2 = \pm 1. \qquad (18)$$

The minus sign will occur when $p_k/q_k < \sqrt{n}$; hence k is an even index (recall the way the convergents alternate around their limit); hence the period contains an odd number of terms; and thus the subsequent end-of-palindrome terms will occur alternately at odd and even indices and so be associated with alternating signs. Conversely, if the first term occurs with a plus sign, the period will contain an even number of terms, and all subsequent end-of-palindrome terms will then also have plus signs.

These remarks relate to the problem of solving the equation $x^2 - ny^2 = 1$ in integers (a so-called Diophantine problem; see equation (9) in §6.10). This was proposed as a challenge by Pierre de Fermat in 1657, in an attempt to arouse interest in number theory:

> Given any number whatever which is not a square, there are also given an infinite number of squares such that, if the square is multiplied into the given number and unity is added to the product, the result is a square.

Example. Let 3, which is not a square, be the given number; when it is multiplied into the square 1, and 1 is added to the product, the result is 4, being a square.

The same 3 multiplied by the square 16 gives a product which, if increased by 1, becomes 49, a square.

And an infinite number of squares besides 1 and 16 can be found which have the same property.

But I ask for a general rule of solution when any number not a square is given. E.g. let it be required to find a square such that, if the product of the square and the number 149, or 109, or 433 etc. be increased by 1 the result is a square. . . . We await these solutions which, if England or Belgic or Celtic Gaul do not produce, Narbonese Gaul will.

Solutions were found by Bernard Frénicle de Bessy, Lord Brouncker and Wallis, and the resulting correspondence over this and other problems was then edited and published by Wallis as a book entitled *Commercium epistolicum de quaestionibus quibusdam mathematicis nuper habitum* (1658). Thus far, no connection with the nascent theory of continued fractions had been observed, and this possibility was to lie dormant for a century, until it was discovered by Leonhard Euler and published by him in his article 'De usu novi algorithmi in problemate Pelliano solvendo' (written 1759, published 1765). It was also Euler who named the problem and equation after John Pell, a minor English mathematician who had nothing to do with it; but the name 'Pell's equation' has stuck. In fact, all integral solutions of $x^2 - ny^2 = 1$ are generated by the continued-fraction expansion of $\sqrt{n}$ in the way described above, and the continued-fraction algorithm, illustrated above for $\sqrt{2}$, will quickly generate the terms of the expansion for any n.

These and other applications of continued fractions spread over Leonhard Euler's very influential writings. Joseph Louis Lagrange gave rigorous proofs of the crucial results and several other applications; and Adrien Marie Legendre wrote a short book based on continued-fraction techniques, *Essai sur la théorie des nombres* (1798), which over the next thirty years would expand into a two-volume treatise. But then, in 1801, Carl Friedrich Gauss published his *Disquisitiones arithmeticae*, an enormously influential treatise on number theory (§6.10), in which the new techniques of binary quadratic forms were used to treat problems that had hitherto been treated with continued fractions. As the influence of Gauss's book grew, so the role of continued fractions in number theory declined, and so their general importance declined also.

4 THE ANALYTIC THEORY: A VERY BRIEF SKETCH

In his *Arithmetica infinitorum* (1665), Wallis described Brouncker's expression

$$\frac{4}{\pi} = 1 + \frac{1}{2+} \frac{9}{2+} \frac{25}{2+} \frac{81}{2+} \cdots \tag{19}$$

which, he writes, has a '*denominator continuae fractus*' (a 'successively fractioned demonimator', or, in his own English words, in a brief description of this in his *Treatise of Algebra* (1685), 'the denominator of each Fraction is the number 2, with a Fraction still fracted continually'). Then, in 1714, Cotes discovered the expression

$$e = [2, 1, 2, 1, 1, 4, 1, 1, 6, 1, 1, 8, 1, \ldots], \tag{20}$$

which implies that e is irrational, although Euler was the first to prove this result (in 1737). His first reference to continued fractions is found in a letter to Christian Goldbach of 1731 concerning a particular differential equation, the Riccati equation (§3.14). Euler, Lagrange, Johann Heinrich Lambert and others developed this analytic theory; for example, Lambert, from 1761, derived and proved expansions like

$$\tan x = \frac{1}{x^{-1} -} \frac{1}{3x^{-1} -} \frac{1}{5x^{-1} -} \cdots \tag{21}$$

and used them to prove that π was not an algebraic number. Then Gauss, in a monumental paper of 1813 on the hypergeometric series (again, the name is due to Wallis, again in his *Arithmetica infinitorum*), showed how to bring some of these results into a general theory. Further developments of the analytic theory are too technical to be described here; see Brezinski *1991*.

5 THE PREHISTORY OF CONTINUED FRACTIONS

Some of the topics described here have remarkable prehistories. For example, the Euclidean algorithm was known in Chinese mathematics, where it was used to simplify fractions by finding common multiples, and to solve linear Diophantine equations. Pell's equation was known in India, in the seventh century AD by Brahmagupta, who gave rules for a partial solution, and then more completely in the eleventh century, by Jayadeva, and in the twelfth, by Bhāskara; and the most remarkable early instance of Pell's equation is in one interpretation of Archimedes' 'cattle problem'. Other hints can be found in Greek mathematics: an allusion by Aristotle at *Topics* (158b29) points to a connection between the Euclidean algorithm and early treatments of ideas of ratio or proportion, a possibility elaborated by some later Arab commentators. Among modern scholars, this connection was first described in Becker *1933*, and it has recently been developed into a proposal that the mathematicians around Plato had discovered and proved results equivalent to those, described above, concerning the expansions of square roots, and that they evaluated convergents and used them as good approximations (Fowler *1987*).

6 CONCLUDING COMMENT

It seems a characteristic of continued-fraction methods that they are repeatedly being discovered, promoted, developed, discarded, and then subsequently rediscovered, the cycle beginning anew. At present the subject seems to be in a state of modest revival after half a century of neglect. However, this article has recalled manifestations of ways in which continued fractions have permeated analysis – ways that are still comparatively little explored, are probably only partially understood, and are now neither generally well known nor part of the standard mathematical curriculum.

BIBLIOGRAPHY

Becker, O. *1933*, 'Eudoxos-Studien I. Eine voreudoxische Proportionenlehre und ihre Spuren bei Aristoteles und Euklid', *Quellen und Studien zur Geschichte der Mathematik, Astronomie und Physik*, Abteilung B, Studien, 2, 311–33.

Brezinski, C. *1991*, *History of Continued Fractions and Padé Approximants*, Berlin and New York: Springer.

Cataldi, P. *1613*, *Trattato del modo brevissimo di trovare la radice quadra delli numeri* [...], Bologna: Rossi.

Davenport, H. *1952*, *The Higher Arithmetic*, Cambridge: Cambridge University Press. [Repr. 1982.]

Fowler, D. H. *1987*, *The Mathematics of Plato's Academy*, Oxford: Clarendon Press.

Schwenter, D. *1636*, *Deliciae physico-mathematicae* [...], Nurenberg: Dümler.

Smith, D. E. *1929*, *A Source Book in Mathematics*, New York and London: McGraw-Hill.

6.4

Fundamental concepts of abstract algebra

W. PURKERT AND H. WUSSING

When David Hilbert delivered his famous talk 'Mathematical problems' in 1900 at the International Congress of Mathematicians in Paris, he put forward the question: 'is there in store for mathematics what other sciences have long experienced, namely being split into individual sub-sciences whose representatives barely understand even each other and whose cohesiveness thus becomes looser and looser?' (Hilbert *1900*: 329). He believed that this could be answered in the negative with an appeal to the affinity provided by the way ideas are formed in the whole of mathematics and to the numerous analogies between its various branches. In fact, structural thinking in modern mathematics, the construction of such integrative concepts as group, ring, field, algebra, vector space, topological space, and others, brilliantly confirms Hilbert's foresight. This article treats the basic algebraic structures that were worked out by the end of the nineteenth century as the first axiomatically defined mathematical structures. Its three sections deal in turn with groups, fields, and some other algebraic structures.

1 GROUPS

Before the word 'group' emerged in the mathematical literature, there had been a longer period of development in which mathematicians applied group-theoretical results without the concept of group being explicitly defined. This 'prehistory' is characterized as involving implicit use of the group concept. It has its historical roots in the theory of algebraic equations, number theory and geometry (including crystallography, on which see §9.17).

1.1 Groups in the theory of algebraic equations

The concept of group, in the sense of permutation group, developed from the theory of algebraic equations, and it is in this area that the word 'group' was first applied. Leonhard Euler and Etienne Bézout derived a resolvent of degree $n!$ by considering all permutations of the roots $x_1, \ldots, x_n$ of an nth-degree polynomial:

$$F(z) = \prod_{\sigma \in S_n} [z - h(x_{\sigma(1)}, x_{\sigma(2)}, \ldots, x_{\sigma(n)})], \qquad (1)$$

where

$$h(x_1, x_2, \ldots, x_n) = x_1 + \omega x_2 + \cdots + \omega^{n-1} x_n \qquad (2)$$

is a function of the roots, with ω a primitive nth root of unity; $F(z)$ was later called the 'Lagrange resolvent'. Joseph Louis Lagrange *1772–3* took up this idea of Euler and Bézout and used it to demonstrate why an algebraic solution of general equations of second, third and fourth degree is possible. In these cases there are rational functions of the roots whose coefficients are rational in the coefficients of the given polynomial, and these functions take on, for all permutations of the roots, fewer values than the degree of the equation. He attempted to extend these results to the general case. His principal results can be summarized in present-day terms by saying that he developed the Galois theory for the field $Q(x_1, \ldots, x_n)$ (Q being the field of rational numbers, and $x_1, \ldots, x_n$ indeterminates) over $Q(a_1, \ldots, a_n)$, where $a_1, \ldots, a_n$ are the elementary symmetric functions of $x_1, \ldots, x_n$. For example, he found that a rational function $g(x_1, \ldots, x_n)$ of the roots with coefficients from $Q(a_1, \ldots, a_n)$, which takes on h distinct values under all permutations of the roots, satisfies an equation of degree h over $Q(a_1, \ldots, a_n)$, where h is a divisor of $n!$. The degree is thereby reduced directly to the index of the symmetry group (subgroup) of $g(x_1, \ldots, x_n)$ in S_n. Lagrange used the group concept only implicitly; furthermore, he nowhere mentioned closure with respect to multiplication for S_n or its subgroups. His work, however, had an extraordinarily great influence on the further development of the theory of algebraic equations, in particular on Paolo Ruffini, Niels Abel and Evariste Galois (§6.1).

Lagrange had given a hint of extending his result to equations of fifth and higher degree, but Ruffini *1799* was the first to formulate and attempt a proof of the theorem that the general equation of fifth degree is insoluble by radicals. His proof relies on the determination of how many different values a rational function of the five roots can assume under all permutations of the roots. It turns out that the values 8, 4 and 3 cannot occur. This follows from the finding that, of the possible subgroups of S_5, there are none of order 15, 30 or 40. Ruffini worked expressly with systems of permutations closed with respect to multiplication for which he used the term *permutazione* (the permutations themselves were called *sostituzione*). He distinguished between transitive and non-transitive permutation groups, and among the transitive permutation groups he distinguished between primitive and non-primitive (Burkhardt *1892*). The gaps in his proof occur mainly in the field-theoretic aspect of the problem.

After Ruffini's negative result, the discovery presented by Carl Friedrich Gauss *1801* of a special class of solvable equations, the cyclotomic equations, was of great importance for the theory of equation solving. His proof of the solvability of the cyclotomic equations of prime degree p was based on successively extending the field by the adjunction of generating elements (Gaussian periods), arriving finally at the roots themselves. Gauss implicitly used the fact that the group of automorphisms of the splitting field (that is, the smallest field in which all roots lie) is isomorphic to the group of prime residue classes modulo p (i.e. it represents a cyclic group of order $p - 1$). Each of Gauss's steps consists of an exhaustive working out of the periods and their elementary symmetric functions; the group-theoretic background described here must have been something he possessed intuitively.

Abel's first work on the theory of equations was concerned with Ruffini's unsolvability theorem. Abel *1829* provided a correct proof, thanks to a more precise handling of the field-theoretic aspect of the problem; his group-theoretic reasonings were the same as Ruffini's. His study of the multiplication formulas of elliptic functions led him to a further class of solvable equations. He characterized it independently of its concrete form by two conditions. Each root x_i can be expressed as a rational function of one of the roots, say x_0: $x_i = \theta_i(x_0)$; and $\theta_i(\theta_j(x_0)) = \theta_j(\theta_i(x_0))$ for arbitrary i and j (§6.1). Later, Leopold Kronecker named such equations, of which cyclotomic equations are a special case, 'Abelian equations'. Abel's proof of the solvability of such equations relied on the structure of finite Abelian groups.

Galois, with his work on the theory of the solvability of algebraic equations (especially in 1831), stood at the beginning of a new methodological orientation in algebra, as he was well aware (Galois *1962*). It was not through voluminous calculations that the solvability question would be decided, but rather through analysis of the structure of a permutation group. Like Lagrange, Galois worked with the field $Q(a_1, \ldots, a_n)$ as the basic field, but for the coefficients of the original equation, $a_1, \ldots, a_n$, he allowed algebraic relations. For Galois the roots of the equation could be expressed (in modern terms) as rational functions of a primitive element V of the splitting field: $x_i = C_i(V)$. If V is replaced by all its conjugate elements, the result is a group of permutations of the roots (the Galois group of the equation). For the permutations as a whole, Galois stated and applied the property of being closed under multiplication, though nowhere did he prove it. He also introduced the concept 'group'. If the ground field is extended by the adjunction of all roots of an irreducible auxiliary equation, then the Galois group is reduced to one of its normal divisors. With the help of this fundamental concept, Galois could now formulate his solvability criterion. If it is assumed that p is the smallest prime divisor of the order of the Galois group G, then the original equation can, through adjunction of radicals, be reduced to an equation with a smaller group precisely when G has a normal divisor of index p. An equation is solvable in radicals exactly when the Galois group can be reduced through iterated application of such steps to the trivial group.

In parallel with the development of the theory of algebraic equations, the theory of permutation groups was developed as an independent line of research, principally by Augustin Louis Cauchy in 1815, and in numerous works from 1844–6. Cauchy introduced such concepts as the identity permutation, the order of a permutation, the generation of a group, subgroup (*système de substitutions conjuguée*) and its order, as well as the cyclic representation of permutations. There were also hints of representation theory.

In 1846, Joseph Liouville published Galois's principal works. Only a short time afterwards Victor Puiseux and Charles Hermite discovered a connection with the theory of functions. In 1850 Puiseux introduced the monodromy group of an algebraic function defined by $f(z, w) = 0$, and Hermite showed that it was the Galois group of $f(z, w)$ over the field $C(z)$ of complex numbers. Joseph Serret incorporated several of Galois's results in his influential textbook *Cours d'algèbre supérieure* (1854), and introduced the concept of conjugate subgroups. In Italy, Enrico Betti presented Galois theory in several works beginning in 1851. He developed the group-theoretic foundations in particular, and proved that the Galois group was in fact a group in the mathematical sense (i.e. it is multiplicatively closed). In 1856–8, Richard Dedekind lectured at Göttingen on higher algebra and Galois theory. Even in his lectures he presented a highly worked-out theory of permutations and of Galois theory (Purkert *1976*, Scharlau *1981*). Its historical influence was, to be sure, only slight, since only four people attended the lectures and they were not published. Dedekind later anticipated the representation of Galois theory in the language of vector spaces that Emil Artin carried out in the twentieth century (Kiernan *1971*).

Camille Jordan, in his treatise *1870* on substitutions, brought the theory of permutation groups, which had grown out of the algebraic theory of equations, to a central position in mathematics, with numerous applications in a variety of branches. This in turn prepared the way for the abstract notion of group. After presenting the known theory of permutation groups and his own results on k-fold transitive groups, Jordan turned to the solvability problem. He introduced the composition series of a group and showed that the quotients of the orders of the successive subgroups are uniquely determined up to their order by the given group. Then, in 1889, O. Hölder introduced factor groups and completed Jordan's theorem with the discovery that the factor groups of the composition series are uniquely determined up to their order. Jordan regarded a group as solvable when the above quotients are all prime numbers. He also showed the importance of isomorphic and homomorphic mappings of groups. Of particular importance was his study of the classical groups over $GF(p)$: linear substitutions $\mathbf{x}' = \mathbf{A}\mathbf{x}$, where the components of the vectors and matrices are taken from $GF(p)$, the Galois field (a finite field) with p elements. He applied this theory in a variety of ways, to the construction of solvable groups, to algebraic curves and surfaces, to division equations for the periods of Abelian functions and finally to the study of multiply transitive groups. Jordan's works from the 1860s were also precursors of the development of transformation groups in geometry, to which we now turn.

1.2 Groups in geometry; semigroups

The first half of the nineteenth century saw diverse developments in geometry that fundamentally altered its conception since the time of Euclid two thousand years before: projective geometry, affine geometry, non-Euclidean geometry and differential geometry, among others. Questions were posed above the inner unity of geometry, and the organizational principles for classifying its branches. Around 1870 these principles finally entered into group theory, though broad anticipations of it can be found in August Möbius's earlier studies of geometric relations. These relations involved groups of transformations that form a certain hierarchy, such as that Felix Klein later considered in his *Erlanger Programm* (§7.6). However, Möbius's ideas had little historical influence; their importance would come to be appreciated only in retrospect.

The algebraic invariant theory of the English school (§6.8) was a major starting point for the study of linear transformation groups, represented as groups of matrices. In particular, Arthur Cayley's ten 'Memoirs upon quantics' (starting in 1854) provided a good part of the foundations of classical invariant theory, and the 'Cayley measure' led directly to the ideas of the *Erlanger Programm*.

In Euclidean geometry, many different approaches had been made to the study of motions conceived of as composed of rotations and translations

(Euler in 1758, Michel Chasles 1830, O. Rodrigues 1840, Louis Poinsot 1851). But it was in crystallography, around 1840, that the study of symmetry systems led to an implicit characterization of all finite subgroups of $O(3, \mathbb{R})$ (i.e. the three-dimensional orthogonal group over the real numbers $\mathbb{R}$) as well as of 71 essentially discontinuous subgroups of the orientation-preserving isometries in three-dimensional Euclidean space (E^3). In fact, the work of A. Bravais around 1850 on the symmetry of spatial graphs was the decisive factor for Jordan's explicit introduction *1869* of the group concept in geometry (Scholz *1989*). Jordan, who set out to classify the groups of Euclidean motions, defined a group of motions as a collection of motions which is closed under successive applications of the motions. He silently assumed, however, the existence of the identity and of the inverses of each element. He regarded the motions as produced through finitely many motions, finite or infinitesimal, and had in mind a strict continuous discrete dichotomy. His classification contained 174 types of groups of motion in E^3, but the deepest part of his work is the analysis involved in extending the group of translations by those subgroups of the group of rotations $SO(3, \mathbb{R})$ that have a non-trivial factor system. ($SO(3, \mathbb{R})$ is the three-dimensional special orthogonal group, i.e. with determinant $+1$, over the real numbers.)

Klein and Sophus Lie, influenced by Jordan's work, applied transformation groups to geometry in two joint papers in 1870 and 1871. This extended concept of group served in Klein's classification of the various streams of geometry in his *Erlanger Programm* (Klein *1872*). The most general group is that of the projective mappings (collineations), and properties that are invariant under this group form the subject of projective geometry. Elliptic, hyperbolic, Euclidean and, later, affine geometries were characterized as subgroups of the projective group that leave certain second-order forms invariant (§7.6). In 1875 Klein succeeded in determining all finite groups of fractional linear transformations $z' = (az + b)/(cz + d)$ (discrete groups). These are the finite cyclic groups, the dihedral groups and the groups of congruence mappings of the Platonic solids, all of which play an important role in the theory of automorphic functions that Klein helped to construct. The work of Jordan and Klein also led to the discovery by E. S. Fedorov and A. Schönflies in 1890 of all three-dimensional crystallographic groups (§9.17).

Lie applied himself to the study of continuous groups, with the dream of doing for differential equations what Galois theory had done for algebraic equations. He recognized that for infinite groups the existence of an identity and of inverses must be postulated rather than derived from the other axioms. More frequently, he deliberately worked on a more general level with semi-groups of transformations (Hofmann *1991*). Lie's principal work was the three-volume *Theorie der Transformationsgruppen* (1888–93). His most important tool for the study of continuous groups was the infinitesimal transformations that can be conceived as linear operators making up a Lie algebra (on which see §6.5).

1.3 Groups in number theory

Finite Abelian groups, or finitely generated Abelian groups, made their appearance in an implicit form in number theory under the most varied guises. In 1761 Euler investigated the possible residues of a^k modulo a prime number p, where $(a, p) = 1$. He found that, if k is the smallest number such that $a^k \equiv 1 \pmod{p}$, then the residues of $1, a, a^2, \ldots, a^{k-1}$ are all distinct and k is a divisor of $p - 1$. He had thus shown that the order of an element in the group of prime residue classes modulo p is a divisor of the order of the group (although with no explicit concept of group, of course).

In his theory of cyclotomic equations, Gauss made substantial use of the fact that, in modern terms, the group of prime residue classes modulo a prime number p is cyclic. His theory of quadratic forms (Gauss *1801*) has had a very great influence on the development of number theory. He used equivalence relations defined by linear whole-number transformations of determinant 1 in order to divide the forms into classes, and found that there were only finitely many classes corresponding to a given discriminant. He recognized the composition of the forms as a binary, commutative and associative operation between the form classes (i.e. that form classes constitute a finite Abelian group). The unit element is the class represented by $x^2 - Dy^2$. While Gauss's theory deals with the special algebraic number field $Q(\sqrt{D})$, Dirichlet *1846* was the first to solve an important problem for arbitrary algebraic number fields, namely the structure of the unity group; he proved that it was finitely generated. Ernst Kummer *1847* introduced the ideal numbers and used them in creating the arithmetic of cyclotomic fields (i.e. the splitting field of cyclotomic equations). Finite Abelian groups arose while doing this, since he made use of the classes of ideal numbers. In 1869, E. Schering returned to Gauss's theory of equivalence classes of forms, and proved that the principal theorem for finite Abelian groups held in this concrete case. Kronecker *1870* has the distinction of being the one who clearly recognized the common abstract structure in the various number-theoretic results and gave an axiomatic treatment of finite Abelian groups. He did not, however, use the word 'group', though it was well known to him from the theory of equations.

1.4 The development of the abstract group concept

Cayley *1854* made the first advance towards the abstract conception of a finite group. He started with arbitrary symbols $1, \alpha, \beta, \ldots$, together with a multiplication defined between them that always resulted in one of the given symbols, and characterized the structure of the group by means of a group table. He mentioned as examples, besides permutation groups, groups of matrices and quaternions. At first no one else took up this abstract approach. Dedekind stated in the lectures mentioned in Section 1.1 above (and published in complete form in Scharlau *1981*) that the group-theoretic theorems he derived remained valid if permutations were replaced by any object for which a multiplication satisfying the group properties was defined.

Cayley *1878* returned to abstract finite groups and showed that each such group could be conceived of as a permutation group. This second advance towards the concept of abstract group soon found reinforcements. In a first step, both the permutation groups and the groups known from number theory were subsumed under one abstract-group concept, though, of course, encompassing only finite groups. Some examples are in the work of Georg Frobenius and L. Stickelberger (1879), E. Netto (1882), Heinrich Weber (1882), Frobenius (1887) and Hölder (1889). In particular, several theorems that had been proved earlier for permutation groups were recognized as general theorems holding for finite groups. Thus L. Sylow, in 1872, was one of those who proved the following theorems for permutation groups. If p^a is the highest power of the prime number p contained in the order of the group G, then G has a subgroup H of order p^a. Again, consider the normalizer of H (i.e. the set of its elements h such that $hg_1 = g_2h$ for some elements g_1, g_2 of G); if it is of order $p^\alpha m$, then there exists a non-negative integer r such that G is of order $p^\alpha m(pr + 1)$, and G contains exactly $pr + 1$ subgroups of order p^α. Frobenius's proofs of these and further Sylow theorems in 1887 were based on the abstract-group concept; they were soon followed by fundamental work by Hölder on finite groups.

W. von Dyck's 'Gruppentheoretische Studien' (*1882–3*) were abstractly conceived and went beyond the concept of finite group; they constituted an important step in the direction of a general concept of group. Working with free groups generated by finitely many elements, Dyck studied the factor groups whose kernels were determined by specified relations between the generators. He also pointed out the relationship between this general concept and the groups of fractional linear transformations.

The first complete axiom system for groups occurs in Weber *1893* and is

748

headed by four postulates:

1 well-defined composition;
2 associativity;
3 cancellation rules; and
4 unique solvability of $ax = b$ and $ya = b$, for given a and b.

He explicitly states that 4 is an independent axiom only for infinite groups; for finite groups 4 follows from 1–3. By incorporating this axiomatic characterization of group in his famous algebra text of 1895–6, it circulated rapidly and received general acceptance. The first monographs devoted to abstract group theory were J.-A. de Séguier's *Théorie des groupes finies* (1904) and O. Y. Schmidt's *Abstraktnaya teoriya grup* (1916, in Russian). Group theory has remained an interesting area of mathematical research; the recent discovery of all finite simple groups is viewed, for example, as one of the greatest mathematical achievements of the twentieth century.

2 FIELDS, RINGS AND IDEALS

The field concept was implicitly formed in two areas that were initially independent of each other – the theory of algebraic equations and number theory. The explicit concept of field, that appeared first as a number field and then as a function field, came from researchers who brought these areas together: Dedekind and Kronecker (Purkert *1973*).

2.1 Fields in the theory of algebraic equations

Lagrange, in the work mentioned in Section 1.1 above (*1772–3*), called a quantity 'rationally known' if it was a rational function over Q of the coefficients of the given equation $a_1, \ldots, a_n$ (taken as algebraically independent of one another). He was thus working implicitly with $Q(a_1, \ldots, a_n)$ as the ground field, where $a_1, \ldots, a_n$ are indeterminates. To characterize an intermediate field between $Q(a_1, \ldots, a_n)$ and the root field $Q(x_1, \ldots, x_n)$, Lagrange utilized the notion of similar functions (*fonctions semblables*): two functions of the roots $x_1, \ldots, x_n$ with rationally determined coefficients are similar when they remain unchanged for the same permutations of the roots. He was able to show that similar functions satisfied equations of the same degree over $Q(a_1, \ldots, a_n)$, and that they are rationally expressible in terms of each other over $Q(a_1, \ldots, a_n)$. Similar functions are therefore primitive elements of the same intermediate field. If a function of the roots allows all permutations, then it is rationally known.

Abel's work on the theory of algebraic equations was distinguished by the careful (although implicit) construction of the resulting sequence (now often called 'towers') of fields, each included within its successor. His starting-point was the field $Q(x', x'', \ldots, x^{(n)})$ of the rational functions of the indeterminates $x^{(i)}$. The towers of fields belonging to a solvable equation came from successive adjoining of expressions for the roots (although a technical term for this process was not introduced). Abel described the extension field by concretely presenting the form of the elements. He proved that all quantities occurring in the successive extension fields are rational functions of the roots of the original equation, thereby filling a gap in Ruffini's work. In his investigations of the division equations of the elliptic functions Abel regarded $Q(x', x'', \ldots, x^{(n)})$ as the basic field, in which algebraic dependences between the $x^{(i)}$ are now allowed. This is quite analogous to the way in which Galois did his field-theoretic work. By introducing the technical term 'adjoin', Galois was able to underline the significance of the splitting field as the ascension up from a ground field more clearly than Abel. Galois had his successors in Italy and France, such as Betti, Serret and Jordan, but they did not go beyond him as far as the field-theoretic side of the solvability problem was concerned.

With Kronecker, beginning in 1853, came a new arithmetical view of the role of the field concept in algebra. He called attention to the field concept, even if only in implicit form, as central, and made it the foundation of all his deliberations. At every point he disclosed exactly to which fields the relevant quantities belonged, recognizing the close connection between the deep questions about the classification of irrationalities and the choice of the ground field. Abelian equations over the rational number field provided completely different sorts of irrationalities from the Abelian equations over the imaginary quadratic fields (the *Jugendtraumtheorem* of Kronecker; the 'theorem of Kronecker–Weber' covers the case of the rationals). With these ideas Kronecker marked the beginning of class field theory.

Dedekind *1857* also made the field concept, along with that of group, fundamental for the formulation of Galois theory. For 'field' he used the term 'rational domain'.

2.2 Fields, rings and ideals in number theory

The extension of divisibility theory from the whole rational numbers to the whole algebraic numbers makes sense only if one is restricted to the whole numbers of an algebraic extension field of finite degree. In the field of all algebraic numbers, for example, the whole algebraic number α would have infinitely many divisors, $\sqrt{\alpha}, \sqrt[3]{\alpha}, \sqrt[4]{\alpha}, \ldots$. Accordingly, researchers who carried out this extension were necessarily led to look at the range of validity of the whole quantities of such a field or of the field itself.

After the preparatory work by Euler and Lagrange on the representation of whole rational numbers by quadratic forms, Gauss *1832* produced a work in algebraic number theory that was pioneering in two respects. First in the course of his investigations on biquadratic reciprocity laws he developed arithmetic in the range of validity of the whole quantities of the field $Q(i)$; and second, his theory of quadratic forms in the *Disquisitiones arithmeticae* (Gauss *1801*) only implicitly contained the theory of quadratic number fields. Further progress was made by J. P. G. Lejeune Dirichlet *1846* (the characterization of the unity groups of an algebraic number field), Gotthöld Eisenstein in 1850 (reciprocity laws), and Kummer *1847* (the admission of arithmetic into cyclotomic fields by the introduction of ideal divisors). They all worked with whole numbers of an algebraic number field, as their explicit presentations of the form of these numbers revealed. For example, Kummer made the following definition: if α is an imaginary root of the equation $a^\lambda = 1$, with λ a prime number, and $a, a_1, a_2, \ldots$ are whole numbers, then $f(\alpha) = a + a_1\alpha + \cdots + a_{\lambda-1}\alpha^{\lambda-1}$ is a complex whole number (*1847*: 319).

Dedekind and Kronecker constructed general arithmetic theories for arbitrary algebraic number fields of finite degree. Dedekind published his in the Supplements X and XI to Dirichlet's lectures on number theory (*1871, 1879*), where he defined a field as a system of numbers closed under the four basic operations. This concept, Dedekind wrote, is 'well suited to serve as a foundation for the higher algebra and the parts of number theory related to it' (*1871*: 224). To construct a divisibility theory for the whole numbers of an algebraic extension field of Q of finite degree, Dedekind introduced the ideal: a module, made up of whole numbers from the field, and closed with respect to multiplication by arbitrary whole numbers from the field. These ideals are appropriate for replacing the ideal numbers that were non-extensionally defined by Kummer. A divisibility theory for ideals can be constructed, and from this comes the unique decomposition of each ideal into prime ideals. An appropriate equivalence relation leads to ideal classes, which Gauss had used for quadratic fields (his theory of quadratic forms) and Kummer had used for cyclotomic fields. Dedekind gave the first convincing demonstration of the usefulness of this new conceptual apparatus in his insightful presentation of Gauss's complicated theory within the framework of his treatment of quadratic number fields. Dedekind and Weber *1882* applied Dedekind's theory to function fields.

Kronecker published his version of the arithmetic of algebraic number and function fields as *1882*, a contribution to a *Festschrift* marking the 50th anniversary of Kummer's doctorate. In keeping with his stance on the foundations of mathematics (§5.6), Kronecker defined the field in a constructive fashion. The field Q of rational numbers or the field $Q(x_1, \ldots, x_n)$, where $x_1, \ldots, x_n$ are indeterminates, form the natural domain of rationality. A simple algebraic extension of such a domain is called a 'generic domain'. Kronecker's field concept (or 'domain of rationality', as he termed it) covers only natural domains of rationality and generic domains. Fields such as the field of all algebraic numbers or the field of all real numbers, that appear freely in Dedekind's system, are not domains of rationality in Kronecker's sense since they cannot be constructed from Q by finitely many adjunctions. Kronecker's arithmetic theory rests on the concept of form. Let K be an algebraic extension of Q. Then a form (of domain of rationality K) is a whole rational function F of finitely many indeterminates $u, v, \ldots$ with coefficients from K. If the conjoined forms are constructed and multiplied together, the result is a whole rational function of indeterminates $u, v, \ldots$ with coefficients from Q. This norm, $Nm(F)$, can be split into the greatest common divisor of all coefficients and the primitive form. Two forms F and G are said to be 'absolutely equivalent' (written $F \sim G$) if two primitive forms ϕ and ψ exist such that $\phi F = \psi G$. A form H is 'divisible by' F if there is a form G such that $H \sim FG$. A form is called a 'prime form' if it is divisible in the above sense only by itself or 1.

The principal theorems of the theory are these:

1 if F is a given form, there exists also a form G such that FG is absolutely equivalent to a whole quantity of the field K;
2 if the product of two forms is divisible by a prime form P, then at least one of the two factors is divisible by P;
3 each form is decomposable into a product of prime forms which is unique up to absolute equivalence.

The relationship between this form theory and Dedekind's ideal theory of K can be sketched as follows. Let

$$F(u_i, \ldots, u_k) = \sum \alpha_{i_1 \cdots i_k} u_1^{i_1} \cdots u_k^{i_k} \tag{3}$$

be a form over K. Then the $a_{i_1 \cdots i_k}$ generate an ideal. It can be shown that absolutely equivalent forms generate the same ideal. Accordingly, to each equivalence class of absolutely equivalent forms over K there corresponds an ideal of K, and vice versa. The theorem follows: to the product of two forms there corresponds the product of the corresponding ideals. It further follows that the prime forms correspond exactly to the prime ideals, and vice versa. It can then be seen immediately that theorem 3 above is the form-theoretic version of the principal theorem of ideal theory in K.

In his report, Hilbert *1897* amalgamated the differently formed theories of Dedekind and Kronecker into a unified whole. The general concept of ring in algebra and number theory also goes back to Hilbert (§6.10, Section 3).

2.3 Finite fields

In its implicit phase, the theory of finite fields was identical to the theory of congruences between functions $\bmod p$. It had no influence on the development of the field concept, meaning number or function fields. Its incorporation into general field theory did not come about until Weber *1893*, a decisive paper in the history of abstract field theory.

The first moves towards a systematic theory of functional congruences $\bmod p$ are to be found in Gauss's *Nachlass*, which contains substantial parts of the theory of Galois fields. But it had no more direct influence on further progress since these parts of the *Nachlass* were not published until 1876. Dedekind, however, obtained a much earlier look in his capacity as an editor of the *Nachlass*, and explicitly referred to Gauss in his *1857*.

Galois founded the theory of finite fields in 1830 (Galois *1962*). He began with the irreducible congruence of nth degree, $F(x) \equiv 0 \pmod p$ with $F(x) \in Z[x]$ and p a prime number (Z represents the ring of whole numbers). While only integer roots of this congruence had been admitted previously, he considered arbitrary roots that were to be regarded as symbolic qualities, like $\sqrt{-1}$ in the usual analysis. Taking j as such a 'symbol', he formed all expressions of the form $\sum_{i=0}^{n-1} a_i j^i$ (where $a_i \in Z/pZ$) and proved that there are precisely p^n distinct expressions of this sort that satisfy the equation $\alpha^{p^n} = \alpha$. Finally, he showed that the multiplicative group of $GF(p^n)$ is cyclic. In this work all the essential elements of the field extension 'from below' occur in implicit form: that is, with no recognition of a superfield in which the roots lie. Galois wanted to make use of insights into the structure of solutions of $F(x) \equiv 0 \pmod p$ to draw conclusions for the theory of solutions of primitive algebraic equations (Wussing *1969*: Part 2, Chap. 3).

Independently of Gauss and Galois, T. Schönemann, a *Gymnasium* teacher from Bradenburg in Germany, developed the theory of Galois fields through his implicit study of the residue-class ring of $Z[x]$ $(\bmod p, F(x))$ (Schönemann *1846*). But in doing so he made recourse to the fundamental theorem of algebra, which Dedekind *1857* was to avoid.

2.4 The abstract concept of field

Weber, besides the abstract group concept, also introduced the abstract field concept. 'A group becomes a field', he wrote, 'when two types of composition are possible, the first being addition and the second multiplication.

Still we need to restrict somewhat this general characterization' (*1893*: 526). The restrictions are that both group operations are commutative, that the multiplicative group does not contain the null element, and that both operations are related to each other by the distributive property. Weber particularly emphasized that there were objects other than number and function fields that came under this abstract concept of field, objects that had not previously been regarded as fields, namely the finite fields such as Z/pZ. Among his examples Weber gave the field of residue classes $K[x]/(p(x))$, where $p(x)$ is irreducible over K, and called them 'congruence fields'. Kronecker introduced them in an implicit form in 1887 as a replacement for $K(\alpha)$, where α is a root of $p(x)$, with the object of banishing irrationals from algebra, in keeping with his philosophy.

Ernst Steinitz *1910* brought the theory of commutative fields to a stage of relative completion. Influenced by work such as K. Hensel's (1899–1908) on the theory of p-adic number fields, his goal was to obtain a comprehensive overview of all possible fields, starting from field axioms. We can trace back to Steinitz such important concepts as adjunction of a set, prime field and characteristic, equivalent extensions, algebraic and transcendent extensions, finite extensions, normal extensions, separable and inseparable extensions, and complete and incomplete fields.

Steinitz recognized that it was precisely the separable finite normal extensions for which the fundamental theorems of classical Galois theory hold. One of his principal theorems was that for each field there is an algebraically closed algebraic extension field that is unique up to equivalence extensions. For the proof he required Ernst Zermelo's well-ordering axiom, or the equivalent axiom of choice. Further important results included the theory of degree of transcendence (including Steinitz's exchange theorem) as well as the theorem that each transcendental extension can be decomposed into one that is purely transcendental, and another associated one that is algebraic. One outcome was Steinitz's comprehensive overview of all possible fields, which he expressed as follows: 'Starting with an arbitrary prime field, by taking an arbitrary, purely transcendental extension along with an arbitrary algebraic extension we have a method of arriving at any field' (*1910*: 9).

Steinitz's work marks a methodological turning-point in algebra leading to what has been termed, since van der Waerden's well-known book *1930–31*, 'modern' or abstract algebra. The school of Emmy Noether, which propagated this new concept of algebra, was based on Dedekind and, above all, on Steinitz (van der Waerden *1985*).

3 SOME OTHER ALGEBRAIC STRUCTURES

3.1 Structural algebras

The idea of carrying out calculations with formal symbols that satisfy prescribed rules arose in the first half of the nineteenth century in the English algebraic schools (Charles Babbage, R. Woodhouse, George Peacock, Augustus De Morgan and George Boole). It was expected that the usual arithmetic of the various number domains should be contained as a special case in such a general algebra, in the same way as the principles of logic (Novy *1973*, Schlote *1987*). Soon after this general formulation, the first algebras over the complex numbers were discovered. These nineteenth-century hypercomplex systems included: quaternions, discovered by William Rowan Hamilton in 1843; octaves, discovered by J. Graves in 1844 and, independently, by Cayley in 1845; and biquaternions, discovered by Hamilton in 1853 (§6.2). Cayley *1858* is an important starting-point for the algebra of matrices; in Cayley *1854*, along with the abstract group concept, we find the concept of the group algebra of a finite group (van der Waerden *1985*: 189–201.).

On the Continent, Hermann Grassmann developed the notion of n-dimensional vectors and their alternating tensor product in his *Die lineale Ausdehnungslehre* (1844) (§6.2). His new ideas were little understood because of his awkward style of presentation, and it was not until Hermann Hankel's *Theorie der complexen Zahlensysteme* (1867) that they found general acceptance. Building on Grassmann's work, William Clifford in 1878 discovered two classes of algebras (later named after him) that contain the complex numbers and the quaternions as special cases.

The structural theory of algebras began with Benjamin Peirce's *Linear Associative Algebra* (1870), in which he considered algebras in which the product of two basis elements is a linear combination of the basis elements, and introduced the concept of 'nilpotent element' and 'idempotent element'. He was able to decompose a linear associative algebra into the direct sum of two right ideals R and R', with $iR = R$ and $iR' = 0$, where i is an idempotent element (the Peirce decomposition). Peirce gave a listing of 163 algebras up to dimension 6; this was, however, incomplete, and was filled in by Eduard Study in 1889 (Shaw *1907*).

In 1884, Karl Weierstrass showed that a commutative algebra over the real numbers always has a null divisor provided its dimension is greater than 2. Frobenius, in 1878, and A. Hurwitz, in 1898, recounted important results concerning division algebras over the real numbers. Frobenius showed that an associative division algebra over $\mathbb{R}$ could only be $\mathbb{R}$ itself, the complex numbers, or the quaternions. Hurwitz showed that if associativity is dropped, octaves were the only additional object.

3.2 Lattices

The structures known today as lattices came via Ernst Schröder and Dedekind in the nineteenth century out of quite diverse beginnings (Mehrtens 1979). Schröder based his work particularly on the logical calculus of Boole, De Morgan, H. and R. Grassmann, and Charles Sanders Peirce. In the first volume of his *Vorlesungen über die Algebra der Logik* (1890), he formulated an axiom system for the logical calculus that was based on a partial ordering (§5.1). The handling of the distributive principle is a significant point. While Peirce had viewed it as provable, Schröder recognized its independence, and used this to distinguish between the general lattice structure that was realized in his 'logical calculus with groups', and the more specialized Boolean algebra (which he termed 'identity calculus'). Schröder obtained a model for a non-distributive lattice from a set of algorithms he constructed for the purpose (Mehrtens *1979*). There are close relationships between Schröder's investigations of lattices and the fundamental ideas of universal algebra that also go back essentially to Schröder (1887). Alfred North Whitehead (1898) and E. V. Huntington (1904) picked up this latter topic (see §5.8, on model theory).

Dedekind came to the concept of lattice from algebraic number theory. In the 1880s and 1890s he was studying what were actually lattices of modules when he performed the operations of taking the largest common divisor and the smallest common multiple. The axiomatic characterization of lattices ('dual groups', as he called them) followed in the two treatises *1897* and *1900*. In the former, he gave a series of examples that followed from the axiom system, examples from logic and the theories of modules, ideals, groups and fields. He investigated lattices with the distributive property and with the modular property – 'dual groups of ideal type' and 'dual groups of modular type', as he called them. The modular type followed in a natural way from the ideal type. Dedekind proved the independence of distributivity and modularity by presenting dual groups of two types, non-modular and modular, both of which were not of ideal type. This must be one of the earliest axiomatic independence arguments to use models. In *1900*, Dedekind took up the study of the free modular lattice with three generators. Its 28 elements were represented in terms of the generators, and closure was proved.

At first, the work on lattices by Schröder and Dedekind found no response. Not until thirty years later, when the investigation of abstract structures on the basis of axiom systems was practised in mathematics, did lattice theory become the object of research. K. Menger *1928* approached it geometrically, and it was developed from a logical point of view by Fritz Klein, who in 1935 introduced the German terminology for the subject, and from 1929 devoted several years to axiomatic treatments of lattices.

The motivation for G. Birkhoff and O. Ore came from algebra. Birkhoff initiated a series of works in 1933, summarized in the comprehensive monograph *Lattice Theory* (first edition 1940), in which he paid particular attention to the multitude of concrete realizations of which the lattice concept was capable. It marks the beginning of lattice theory as a self-standing subdiscipline even if, from a present-day point of view, the hopes held out for it in the 1930s could not be fulfilled. Ore aimed at an abstract treatment of algebraic decomposition theorems by means of his theory of 'structures' (1935).

3.3 Categories

The concept of category generalized those common structural elements that occur when considering mathematical objects and their characteristic mappings (morphisms). Among these are, for example, the category of sets with mappings between sets as the morphisms, the category of topological spaces with continuous mappings, and the category of groups with homomorphisms. Category theory permits a uniform handling and formulation of problem statements that occur universally across the most varied fields of mathematics. It works directly with the mathematical objects (sets, groups, topological spaces, K-modules, and so forth) without resort to the level of their elements.

The most important starting-point for the formulation of category theory was the idea of denoting mappings by arrows and considering commutative diagrams, a notion that goes back to the lectures by W. Hurewicz *1941*. Categories, functors and natural transformations were introduced and used by S. Eilenberg and S. MacLane in 1942, when the first commutative diagram appeared in print. They studied categories in their own right as a 'general theory of natural equivalences' in 1945. MacLane *1972* provides an introduction, with numerous historical notes, for 'the working mathematician'.

BIBLIOGRAPHY

Abel, N. H. *1829*, 'Mémoire sur une classe particulière d'équations résolubles algébriquement', *Journal für die reine und angewandte Mathematik*, **4**, 131–56. [Also in *Oeuvres complètes*, 2nd edn, 1881, Vol. 1, 478–507.]

Burkhardt, H. *1892*, 'Die Anfänge der Gruppentheorie und Paolo Ruffini', *Abhandlungen zur Geschichte der Mathematik*, **6**, 119–59.

Cayley, A. *1854*, 'On the theory of groups, as depending on the symbolic equation $\theta^n = 1$', *Philosophical Magazine*, Series 4, **7**, 40–47, 408. [Also in *Collected Mathematical Papers*, Vol. 2, 123–32.]

—— *1858*, 'A memoir on the theory of matrices', *Philosophical Transactions of the Royal Society of London*, **148**, 17–38. [Also in *Collected Mathematical Papers*, Vol. 2, 475–96.]

—— *1878*, 'The theory of groups', *American Journal of Mathematics*, **1**, 50–52. [Also in *Collected Mathematical Papers*, Vol. 10, 401–3.]

Chandler, B. and Magnus, W. *1982*, *The History of Combinatorial Group Theory* [...], Berlin: Springer.

Dedekind, R. *1857*, 'Abriss einer Theorie der höheren Congruenzen in Bezug auf einen reelen Primzahl-Modulus', *Journal für die reine und angewandte Mathematik*, **54**, 1–26. [Also in *Gesammelte mathematische Werke*, Vol. 1, 40–67.]

—— *1871*, 'Über die Komposition der binären quadratischen Formen', Supplement X of 2nd edn of J. P. G. Dirichlet, *Vorlesungen über Zahlentheorie*, Braunschweig: Vieweg. [Articles 159–63 also in *Gesammelte mathematische Werke*, Vol. 3, 223–61 (cited here).]

—— *1879*, 'Über die Theorie der ganzen algebraischen Zahlen', Supplement XI of 3rd edn of J. P. G. Dirichlet, *Vorlesungen über Zahlentheorie*, Braunschweig: Vieweg. [Articles 170–73 also in *Gesammelte mathematische Werke*, Vol. 3, 297–313.]

—— *1897*, 'Über Zerlegungen von Zahlen durch ihre grössten gemeinsamen Teiler', in *Festschrift der Technische Hochschule*, Braunschweig: Vieweg. [Also in *Gesammelte mathematische Werke*, Vol. 2, 103–47.]

—— *1900*, 'Über die von drei Moduln erzeugte Dualgruppe', *Mathematische Annalen*, **53**, 371–403. [Also in *Gesammelte mathematische Werke*, Vol. 2, 236–71.]

Dedekind, R. and Weber, H. *1882*, 'Theorie der algebraischen Funktionen einer Veränderlichen', *Journal für die reine und angewandte Mathematik*, **92**, 181–290. [Also in R. Dedekind, *Gesammelte mathematische Werke*, Vol. 1, 238–350.]

Dirichlet, J. P. G. *1846*, 'Zur Theorie der complexen Einheiten', *Bericht über die Verhandlungen der Preussischen Akademie der Wissenschaften*, 103–7. [Also in *Werke*, Vol. 1, 639–44.]

Dyck, W. von *1882–3*, 'Gruppentheoretische Studien. I, II', *Mathematische Annalen*, **20**, 1–44; **22**, 70–108.

Edwards, H. M. *1984*, *Galois Theory*, Berlin and New York: Springer.

Galois, E. *1962*, *Ecrits et mémoires mathématiques*, Paris: Gauthier-Villars.

Gauss, C. F. *1801*, *Disquisitiones arithmeticae*, Leipzig: Fleischer. [Also *Werke*, Vol. 1.]

—— *1832*, 'Theoria residuorum biquadraticorum. Commentatio secunda', *Mitteilungen der Königlichen Gesellschaft der Wissenschaften zu Göttingen*, **7**, 89–148. [Also in *Werke*, Vol. 2, 93–148.]

Hawkins, T. *1972*, 'Hypercomplex numbers, Lie groups, and the creation of group representation theory', *Archive for History of Exact Sciences*, **8**, 243–87.

Hilbert, D. *1897*. 'Die Theorie der algebraischen Zahlkörper', *Jahresbericht der Deutschen Mathematiker-Vereinigung*, **4**, 175–546. [Also in *Gesammelte Abhandlungen*, Vol. 1, 63–363.]

—— *1900*, 'Mathematische Probleme', in *Gesammelte Abhandlungen*, Vol. 3, 290–329.

Hofmann, K. H. *1991*, 'Einige Ideen Sophus Lies – hundert Jahre danach', *Jahrbuch Überblicke Mathematik*, 93–125.

Hurewicz, W. *1941*, 'On duality theorems', *Bulletin of the American Mathematical Society*, **47**, 562–3.

Jordan, C. *1869*, 'Mémoire sur les groupes de mouvements', *Annali di matematica*, Series 1, **2**, 167–215, 322–45. [Also in *Oeuvres*, Vol. 4, 231–302.]

—— *1870*, *Traité des substitutions et des équations algébriques*, Paris: Gauthier-Villars.

Kiernan, B. M. *1971*, 'The development of Galois theory from Lagrange to Artin', *Archive for History of Exact Sciences*, **8**, 40–154.

Klein, F. *1872*, *Vergleichende Betrachtungen über neuere geometrische Forschungen*, Erlangen: Dietrich. [Also in *Gesammelte mathematische Abhandlungen*, Vol. 1, 460–97.]

Kronecker, L. *1853*, 'Über die algebraisch auflösbaren Gleichungen', *Monatsberichte der Akademie der Wissenschaften zu Berlin*, 365–74. [Also in *Werke*, Vol. 4, 1–11.]

—— *1870*, 'Auseinandersetzung einiger Eigenschaften der Klassenanzahl idealer complexer Zahlen', *Monatsberichte der Akademie der Wissenschaften zu Berlin*, 881–9. [Also in *Werke*, Vol. 1, 271–82.]

—— *1882*, 'Grundzüge einer arithmetischen Theorie der algebraischen Grössen', *Journal für die reine und angewandte Mathematik*, **92**, 1–122. [Also in *Werke*, Vol. 2, 237–87.]

Kummer, E. E. *1847*, 'Über die Zerlegung der aus Wurzeln der Einheit gebildeten complexen Zahlen in ihre Primfactoren', *Journal für die reine und angewandte Mathematik*, **35**, 327–67. [Also in *Gesammelte Werke*, Vol. 1, 311–51 (cited here).]

Lagrange, J. L. *1772-3*, 'Réflexions sur la théorie algébrique des équations', *Nouvelles Mémoires de l'Académie des Sciences de Berlin*, (1770–71), 134–215; (1771–2), 138–253. [Also in *Oeuvres*, Vol. 3, 203–421.]

MacLane, S. *1972*, *Categories for the Working Mathematician*, Berlin: Springer.

Menger, K. *1928*, 'Bemerkungen zu Grundlagenfragen IV', *Jahresbericht der Deutschen Mathematiker-Vereinigung*, **37**, 305–25.

Mehrtens, H. *1979*, *Die Entstehung der Verbandstheorie*, Hildesheim: Olms.

Novy, L. *1973*, *Origins of Modern Algebra*, Prague: Academy of Sciences.

Purkert, W. *1973*, 'Zur Genesis des abstrakten Körperbegriffs', *NTM – Schriftenreihe für Geschichte der Naturwissenschaften, Technik und Medizin*, **10**, (1), 23–37; (2), 8–20.

—— *1976*, 'Ein Manuskript Dedekinds über Galois-Theorie', *NTM – Schriftenreihe für Geschichte der Naturwissenschaften, Technik und Medizin*, **13**, (2), 1–16.

Ruffini, P. *1799, Teoria generale delle equazioni, in cui si dimostra impossibile la soluzione algebraica delle equazioni generali di grado superiore al quarto*, Bologna: Tommaso d'Aquino. [Also in *Opere matematiche*, Vol. 1, 1–324.]

Scharlau, W. (ed.) *1981, Richard Dedekind 1831–1981*, Braunschweig and Wiesbaden: Vieweg.

Schlote, K.-H. *1987*, 'Die Entwicklung der Algebrentheorie bis zu ihrer Formulierung als abstrakte algebraische Theorie', Dissertation B, University of Leipzig. [To appear in the series 'Science Networks', Basel: Birkhäuser.]

Scholz, E. *1989, Symmetrie, Gruppe, Dualität*, Basel: Birkhäuser.

—— (ed.) *1990, Geschichte der Algebra*, Mannheim, Vienna and Zurich: Bibliographisches Institut.

Schönemann, T. *1846*, 'Grundzüge einer allgemeinen Theorie der höheren Congruenzen, deren Modul eine reelle Primzahl ist', *Journal für die reine und angewandte Mathematik*, **31**, 269–325.

Shaw, J. B. *1907, Synopsis of Linear Associative Algebra*, Washington, DC: Carnegie Institution.

Steinitz, E. *1910*, 'Algebraische Theorie der Körper', *Journal für die reine und angewandte Mathematik*, **137**, 167–309. [Book-form repr.: H. Hasse and R. Baer (eds), 1930, Berlin: Springer; repr. no date given, New York: Chelsea.]

van der Waerden, B. L. *1930–31, Moderne Algebra*, 2 vols, Berlin: Springer.

—— *1966*, 'Die Algebra seit Galois', *Jahresbericht der Deutschen Mathematiker-Vereinigung*, **68**, 155–65.

—— *1985, A History of Algebra*, Berlin: Springer.

Weber, H. *1893*, 'Die allgemeinen Grundlagen der Galois'schen Gleichungstheorie', *Mathematische Annalen*, **43**, 521–49.

—— *1895–6, Lehrbuch der Algebra*, 2 vols, Braunschweig: Vieweg.

Wussing, H. *1969, Die Genesis des abstrakten Gruppenbegriffes*, Berlin: Deutscher Verlag der Wissenschaften. [English transl.: *The Genesis of the Abstract Group Concept*, 1984, Cambridge, MA: MIT Press.]

6.5

Lie groups

IAN STEWART

1 DEFINITION AND EXAMPLES

The theory of Lie groups was created by the Norwegian Sophus Lie, in a flurry of activity beginning in the autumn of 1873. The concept of a Lie group has evolved considerably since Lie's early work. In modern terms, a Lie group is a structure having both algebraic and topological properties, the two being related. Specifically, it is a group (a set with an operation of composition that satisfies various algebraic identities, most notably the associative law) and a topological manifold (a space that locally resembles Euclidean space of some fixed dimension, but which may be curved or otherwise distorted on the global level), such that the law of composition is continuous (small changes in the elements being composed produce small changes in the result). Lie's concept was more concrete: a group of continuous transformations in many variables. He was led to study such transformation groups while seeking a theory of the solubility or insolubility of differential equations, analogous to that of Evariste Galois for algebraic equations (§6.1); but today they arise in an enormous variety of mathematical contexts, and Lie's original motivation is not the most important application.

Perhaps the simplest example of a Lie group is the set of all rotations of a circle. Each rotation is uniquely determined by an angle between $0°$ and $360°$. The set is a group because the composition of two rotations is itself a rotation – through the sum of the corresponding angles. It is a manifold of dimension 1, because angles correspond one-to-one with points on a circle, and small arcs of a circle are just slightly bent line segments, a line being a Euclidean space of dimension 1. Finally, the composition law is continuous because small changes in the angles being added produce small changes in their sum.

A more challenging example is the group of all rotations of 3-dimensional space that preserve a chosen origin. Each rotation is determined by an axis – a line through the origin in an arbitrary direction – and an angle of rotation about that axis. It takes two variables to determine an axis (say the latitude and longitude of the point in which it meets a reference sphere centred on the origin), and a third to determine the angle of rotation; therefore this group has dimension 3. Unlike the group of rotations of a circle, it is non-commutative – the result of composing two transformations depends upon the order in which they are performed.

2 THE PREHISTORY

The prehistory of Lie groups explains why Lie might have been led to such a theory. The theory of permutation groups, transformations on a finite set, was well developed by 1860 (§6.4, Section 1). The theory of invariants, algebraic expressions that do not change when certain changes of variable are performed (§6.8), had drawn attention to various infinite sets of transformations, such as the projective group – all projections of space. In 1868 Camille Jordan studied groups of motions in 3-dimensional space, and the two strands began to merge. In 1869 Lie became friendly with Felix Klein, with whom he shared an interest in line geometry (Hawkins *1989*), an offshoot of projective geometry introduced by Julius Plücker (§7.6). Lie had a highly original idea: that the solution of differential equations by classical methods of integration was possible only because the equations remained unchanged under a continuous family of transformations. Lie and Klein worked on variations of this idea in the period 1869–70; their research culminated in 1872 in Klein's famous characterization of geometry as the invariants of a transformation group, laid down in his *Erlanger Programm* (§7.4; Fano *1907*).

For a few years Lie turned his attention to partial differential equations. In 1873 he suddenly returned to transformation groups, investigating properties of 'infinitesimal' transformations, and making the key observation that their structure depends upon their Taylor expansion taken to second order. He showed that infinitesimal transformations derived from a given continuous group are not closed under composition; instead, they are closed under an operation known as the 'bracket', written $[x, y]$. The resulting algebraic structure is now known as a Lie algebra. Until about 1930 the terms 'Lie group' and 'Lie algebra' were not used: instead, these concepts were referred to as 'continuous group' and 'infinitesimal group' (see e.g. Maurer and Burkhardt *1900*). The bracket operation satisfies two basic identities:

$$[y, x] = - [x, y], \tag{1}$$

$$[x, [y, z]] + [y, [z, x]] + [z, [x, y]] = 0. \tag{2}$$

762

The second of these, known as the Jacobi identity, also appears in Lie's work on partial differential equations, which is therefore more closely related to his study of transformation groups than he himself realized. As he wrote in 1874, 'My earlier works were, as it were, already there, waiting to found the new theory of transformation groups.'

There are strong interconnections between the structure of a Lie group and that of its Lie algebra, which Lie expounded in a three-volume work *Theorie der Transformationsgruppen*, written jointly with Friedrich Engel (*1888–93*). They discussed in detail four 'classical families' of groups, two of which are the rotation groups of n-dimensional space for odd or even n. (The two cases are rather different, which is why they are distinguished. For example, in odd dimensions a rotation always possesses a fixed axis; in even dimensions it does not.) One of their major results is that every algebraic system satisfying the basic identities of a Lie algebra can indeed be obtained from a transformation group.

3 THE CONTRIBUTIONS OF KILLING AND CARTAN

The next substantial developments are due to Wilhelm Killing and Elie Cartan. In 1888, Killing laid the foundations of a structure theory for Lie algebras; in particular he classified all the simple Lie algebras (which may be thought of as the basic building blocks of which all other Lie algebras are composed) (Hawkins *1980*). His method was to associate with each simple Lie algebra a geometric structure known as a root system. He used methods of linear algebra, in particular the characteristic equation of a linear transformation, to study and classify root systems, and then derived the structure of the corresponding Lie algebra from that of the root system.

Cartan took up Killing's ideas, and his doctoral thesis of 1894 fills in a number of gaps in Killing's arguments, and extends them to the theory of representations of simple Lie algebras – the different ways of producing the same algebra as an algebra of matrices. The myth arose that Killing's work was riddled with errors, even though Cartan's thesis makes it clear that this is not the case; the myth grew in the telling, and very few mathematicians actually seem to have reading Killing's papers. As a result, many key concepts that are actually due to Killing bear the names of later mathematicians. These include 'Cartan subalgebra', a special subset of the Lie algebra that plays a central role in the structure theory; Cartan matrix', an integer matrix that summarizes the geometry of a root system in a combinatorial manner; 'Weyl group', the symmetry group of the root system; and 'Coxeter element', an important special element of the Weyl group. To add insult to injury, the one concept now named after him, the 'Killing form', was introduced by Cartan!

Killing and Cartan between them showed that the simple Lie algebras fall into four families, known as A_n, B_n, C_n and D_n, together with five exceptional algebras, G_2, F_4, E_6, E_7 and E_8. The subscript refers to the rank, defined to be the dimension (maximal number of linearly independent elements) of a Cartan subalgebra. Killing erroneously thought that there was a second exceptional algebra of rank 4, but Cartan observed that this was just F_4 in disguise.

Because of the close connections between a Lie group and its Lie algebra, the classification of simple Lie algebras also led to a classification of the simple Lie groups. In particular, the four families A_n, B_n, C_n and D_n are the Lie algebras of the four classical families of transformation groups known to Lie. These are, respectively, the group of all linear transformations in $(n + 1)$-dimensional space; the rotation group in $(2n + 1)$-dimensional space; the symplectic group in $2n$ dimensions, which is important in classical and quantum mechanics and in optics; and the rotation group in $2n$-dimensional space. A few finishing touches to this story were added later; notably the introduction by H. S. M. Coxeter and E. B. Dynkin of a graphical approach to the combinatorial analysis of root systems, now known as Coxeter or Dynkin diagrams.

4 LATER AND MODERN WORK

Up to this point the theory was mainly local – that is, it dealt only with transformations that move points through small distances. The global structure of Lie groups took a major step forward with the work of Hermann Weyl, who was inspired by two separate theories: Cartan's theory of representations of Lie algebras, and Georg Frobenius's theory of representations of finite groups. The fruits of Weyl's work – in particular his proof of the complete reducibility of representations, implying that all representations are easily deducible from a special class, the *irreducible* representations – are expounded in his celebrated monograph *The Classical Groups* (1939).

Lie groups are important in modern mathematics for many reasons. In mechanics, for example, many systems have symmetries, and those symmetries make it possible to find solutions of the dynamical equations (§12.7); the symmetries generally form a Lie group. In mathematical physics, the study of elementary particles relies heavily upon the apparatus of Lie groups, again because of certain symmetry principles. Killing's exceptional group E_8 plays a prominent role in superstring theory, which has been an important approach to the unification of quantum mechanics and general relativity. The topology of Lie groups has been investigated extensively: Simon Donaldson's epic discovery of 1983 that four-dimensional Euclidean space possesses non-standard differentiable structures rests, fundamentally, on an unusual feature of the Lie group of all rotations in 4-dimensional space. Finite analogues of Lie groups, introduced by Claude Chevalley in 1957, are important in finite group theory, where they form the backbone of the classification of the finite simple groups. The theory of Lie groups is without doubt one of the most central in the whole of mathematics.

BIBLIOGRAPHY

Fano, G. *1907*, 'Kontinuierliche geometrische Gruppen', in *Encyklopädie der mathematischen Wissenschaften*, Vol. 3, Part 1, 289–388 (article III AB 4b).

Hawkins, T. W. *1972*, 'Hypercomplex numbers, Lie groups [...]', *Archive for History of Exact Sciences*, **8**, 243–87.

—— *1980*, 'Non-Euclidean geometry and Weierstrassian mathematics: The background to Killing's work on Lie algebras', *Historia mathematica*, **7**, 289–342.

—— *1989*, 'Line geometry, differential equations, and the birth of Lie's theory of groups', in D. Rowe and J. McCleary (eds), *The History of Modern Mathematics*, Vol. 2, Boston: Academic Press, 275–327.

Lie, S. and Engel, F. *1888–93*, *Theorie der Transformationsgruppen*, 3 vols, Leipzig: Teubner.

Liebmann, H. *1914*, 'Berührungstransformationen', in *Encyklopädie der mathematischen Wissenschaften*, Vol. 3, Part 3, 442–539 (article III D 7).

Maurer, L. and Burkhardt, H. *1900*, 'Kontinuerliche Transformationsgruppen', in *Encyclopädie der mathematischen Wissenschaften*, Vol. 2, Part 1, 410–36 (article II A 6). [Important early survey article.]

van der Waerden, B. L. *1985*, *A History of Algebra*, Berlin: Springer, Chap. 9.

6.6

Determinants

EBERHARD KNOBLOCH

There are at least five different roots of determinant theory: number theory, the theory of linear equations, elimination theory, permutation theory and geometry. Today, determinants are mainly used in the theory of latent roots. Originally it was a question of homogeneous polynomials of n variables; only later was a suitable notation invented. The polynomials turned out to be the values of special homomorphisms called 'determinants'. Muir *1890–1923* provides a general history.

1 THE FAR EAST

The most extensive of all ancient Chinese mathematical treatises is the 'Computational Prescriptions in Nine Chapters' (*Jiuzhang suanshu*) written during the Han Dynasty in about 100 BC (§1.9). Its eighth chapter deals with the methods of quadratic arrangements (*fancheng* methods) of solving systems of linear equations. The equations were arranged in tabular form so that one column consisted of the coefficients and constant of one linear equation; the solution was deduced by multiplying and subtracting columns until the matrix of the system was reduced to a triangular form. The values of the unknowns were determined by successive substitutions. In substance, the *fangcheng* methods were identical to Carl Friedrich Gauss's elimination method (Martzloff *1988*).

Although the Chinese method resembles the simplification of determinants, this notion did not exist in ancient China. Only with Japanese scholars of the seventeenth century did the idea of determinants emerge in more explicit form. The old Japanese school, which began with Takakazu Seki and which was undoubtedly stimulated by the Chinese, applied determinant-like methods to the elimination of the common variable from higher-degree equations, but not to the solution of systems of linear equations. In 1683 Seki wrote his 'Method of solving the dissimulated problems' (*Kaifukudai no hō*). There was no term equivalent to our 'determinant'; despite this apparent disadvantage, Seki and his pupils were able to determine the terms of a determinant by the mechanical procedures known as *kōshiki* and *shajō*, and their signs by designating the products of the elements as *sei* or *koku* (i.e. plus or minus). The explication of the calculations was based on examples, but the applied method was general. Like the Chinese, Seki and his followers wrote the coefficients of the equations in columns, thus producing matrix-like schemes. The laws according to which the *sei* and *koku* were produced were indicated by means of diagrams for the cases of systems of two, three, four and five equations. From a modern point of view, these diagrams included the so-called rule of P. F. Sarrus (Mikami *1913*).

2 DETERMINANT-LIKE METHODS IN WESTERN EUROPE BEFORE LEIBNIZ

A 'determinant-like' method is one in which determinants do not actually appear, but there are similar, clear rules of reckoning which spare a troublesome calculation. In this weak sense, it was Girolamo Cardano who used the oldest determinant formulation. In 1545 he solved a system of two linear equations, arrived at from commercial problems, by means of the rule called by him *regula de modo* (or 'mother of rules') in his *Ars magna*. In order to solve the equation system (in modern notation)

$$a_{11}x_1 + a_{12}x_2 = b_1,$$
$$a_{21}x_2 + a_{22}x_2 = b_2,$$

(1)

he gave program-like rules of calculation that amount to forming successively

$$\frac{a_{22}(b_1/a_{12}) - b_2}{a_{22}(a_{11}/a_{12}) - a_{21}}.$$

(2)

From the last term one can easily produce the value of x_1 in the form of Cramer's rule (Section 3). Cardano did not make this step; he did not construct determinants, but rather he showed (from today's viewpoint) how one could successively arrive at one (Tropfke *1980*).

3 LEIBNIZ

Gottfried Wilhelm Leibniz was the first mathematician to elaborate a determinant theory. His contributions included coining the term 'resultant' (*resultans sc. aequatio*) to denote certain combinatorial sums of the terms in the determinant; inventing a symbol for this resultant; formulating (though not proving) some general theorems about resultants; and deducing important results in the theory of systems of linear equations and in elimination theory, formulated by means of determinants. His numerous relevant manuscripts, dating from 1678 to 1713, remained completely unpublished until recently (Knobloch *1980*).

Leibniz revealed his method of using 'fictive' numbers as coefficients with double or multiple lower index only in two publications of 1700 and 1710. His polynomial $10x^2 + 11x + 12$ and his linear equation $10 + 11x + 12y = 0$ are equivalent to the modern expressions

$$a_{10}x^2 + a_{11}x + a_{12} \quad \text{and} \quad a_{10} + a_{11}x + a_{12}y = 0. \tag{3}$$

(While the modern index notation deploys letters such as 'a', he used the index numbers themselves as coefficients.) His other ways (more than fifty) of writing fictive-number coefficients remained unknown. He was indeed particularly interested in suitable mathematical notations because in his opinion they could support the art of inventing new results. His d-notation for the differential calculus is by far the most famous example of such endeavours (§3.2).

In January 1684 Leibniz found the so-called 'Cramer's rule' for solving systems of linear equations. His expression '$\overline{1.2.3.4}$' designated the same coefficient system as the modern expression

$$\begin{vmatrix} a_{11} & \cdots & a_{14} \\ \vdots & & \vdots \\ a_{41} & \cdots & a_{44} \end{vmatrix} ; \tag{4}$$

that is, he designated a determinant by the product of the terms of its main diagonal. He stated that the whole formula $\overline{1.2.3.4}$ consisting of $4! = 24$ terms could be written down by means of one term permuting the right indices in every possible way. Those terms which originate from 11.22.33.44 by an even number of transpositions have the same sign; the other takes the opposite sign.

Leibniz knew in substance the modern combinatorial definition of a determinant. The only difference was that his sign rule was based on the number of transpositions instead of the number of inversions. Among his general theorems on determinants was the ith-column expansion (the so-called 'Laplace expansion') of a determinant.

Apart from systems of linear equations, Leibniz used determinants for the solution of the elimination problem, as Seki had done. Some time between 1679 and 1681 he anticipated J. J. Sylvester's 'dialytic method', and around 1683–4 he elaborated a method of calculating the resultant by means of auxiliary polynomials which reduced the problem to the solution of a system of linear equations. This method was rediscovered by Leonhard Euler in 1748 and Etienne Bézout in 1764. About 1692 or 1693 Leibniz knew the most important dimension and homogeneity properties of the resultant, which were again only published by Euler in 1750 (Knobloch *1990*).

4 THE EIGHTEENTH CENTURY: THE FIRST PUBLICATIONS ON DETERMINANTS

In 1748 Colin Maclaurin's *Treatise of algebra* appeared posthumously. As far as we know, it was the first publication to contain some rudimentary information on determinants. He calculated with two and three linear equations in, respectively, two and three unknowns; he gave only hints in the case of four equations. He defined the numerator determinant, then the denominator determinant, but without explaining the decisive connection between the two. His recursive rule of signs stated that opposite signs are prefixed to those products that involve two opposite coefficients (i.e. coefficients of two distinct unknowns of two distinct equations).

To a certain extent, this was an anticipation of Cramer's rule, but it remained unnoticed. The Swiss mathematician Gabriel Cramer published his *Introduction à l'analyse des lignes courbes algébriques* in 1750. He tried to determine a curve of the nth degree which goes through $\frac{1}{2}n^2 + \frac{3}{2}n$ arbitrarily given points. This geometrical problem led him to a system of linear equations. He showed how to solve such systems in an appendix, where he considered a system of n inhomogeneous linear equations. He gave without proof an explanation of the rule named after him, according to which the determinant in the denominator is to be constructed from the n^2 coefficients, while the sum of $n!$ terms with n members is constructed with the help of a sign rule based on the number of inversions of the permuted indices. In contrast to Leibniz, Cramer had no symbol for the homogeneous polynomials in n^2 variables, nor did he discuss them for their own sake.

The same is true of Bézout's publications of 1764 and 1779; he reduced the elimination problem to the solution of systems of linear equations. He noted that up to then only successive elimination methods were known.

In spite of these efforts, Pierre Simon Laplace declared in 1772 that the general rules of Cramer and Bézout were impractical and provable only by induction. Laplace wished to give a new method of integrating differential equations by approximations in his paper 'Recherches sur le calcul intégral et sur le système du monde'. His stated goal was to apply these methods to the movement of the inner planets. This led him to a system of linear equations. Without actually calculating their solution, he discussed in general the solution of such systems by means of determinants which he called resultants, thus using the same name as Leibniz, but of course without having any idea of the relevant work of his famous predecessor.

Laplace invented an index notation and deduced the expansion of determinants named after him. He did not cite Alexandre Vandermonde's 'Mémoire sur l'élimination', which appeared in the same issue of the journal of the Paris Academy of Sciences, but which had been communicated by Vandermonde to that institution one year before. He introduced lower and upper indices $\frac{\alpha}{a}$ as Leibniz had done in another, more suitable, way. He, too, emphasized that one was still far from a general unambiguous elimination formula, and that it was his goal to show the possibility of obtaining such a formula. He formulated some general theorems on expressions like

$$\frac{\alpha \ \mid \ \beta}{a \ \mid \ b}$$

which he defined by $\frac{\alpha}{a} \cdot \frac{\beta}{b} - \frac{\alpha}{b} \cdot \frac{\beta}{a}$.

In contrast to Cramer, Euler and Bézout, whom he mentioned, Joseph Louis Lagrange took an entirely different path in his elimination study of 1769. In three articles from 1773 on dynamics and on triangular pyramids, he deduced algebraic or analytic identities that agree in content with theorems of three-rowed determinants, especially with functional determinants. But this statement depends on a modern interpretation of his results; he himself saw no connection with the corresponding studies by Laplace or Vandermonde. His articles made only a very indirect contribution to determinant theory.

All these efforts had only a rather limited effect. In 1779 Bézout complained of the unsatisfactory condition of algebraic analysis or the analysis of finite quantities, that had been completely neglected in favour of infinitesimal analysis. In 1903, Leopold Kronecker gave the same reason for the long time that determinant theory had taken to reach its full development (Günther *1875*, Mellberg *1876*).

5 THE NINETEENTH CENTURY: THE BREAKTHROUGH OF DETERMINANT THEORY

5.1 Gauss: the number-theoretic approach

In 1851, J. J. Sylvester mentioned that his friend Charles Hermite had pointed out to him that some faint indications of determinant theory may be found in Gauss's *Disquisitiones arithmeticae* (1801). Gauss had investigated the binary and ternary forms

$$ax^2 + 2bx + cy^2, \tag{5}$$

$$ax^2 + a'x'^2 + a''x''^2 + 2bx'x'' + 2b'xx'' + 2b''xx'^2, \tag{6}$$

and named their 'discriminants' (Sylvester's term)

$$b^2 - ac \quad \text{and} \quad ab^2 + a'b'^2 + a''b''^2 - aa'a'' - 2bb'b'' \tag{7}$$

the 'determinants' of (5) and (6), respectively, because they determine the properties of the forms. To this extent it is correct to say that Gauss coined the term 'determinant', but not in the modern sense of the word.

Gauss also extended to ternary form a theorem which Lagrange had proved for binary forms in his 'Recherches d'arithmétique' of 1773–5: if a form arises from another by linear transformation, the discriminants D and E of the two forms differ only by the square of the modulus k of the linear transformations; that is,

$$E = k^2 D. \tag{8}$$

The modulus is the determinant in the modern sense of the coefficient matrix of the linear transformations.

Finally, Gauss showed that two linear transformations which were carried out one after the other may be replaced by a single linear transformation: its coefficient scheme can be obtained by the modern matrix multiplication. We apply the same procedure when we multiply two determinants. These are the 'faint indications' of determinant theory which Hermite mentioned to Sylvester. But neither notion, nor name nor methods were made explicit or used in the modern senses; they remained hidden behind calculating devices.

5.2 Cauchy: the function-theoretic approach

On 30 November 1812, J. P. M. Binet and Augustin Louis Cauchy each read a paper at the Institut de France. Both authors proved for the first time the multiplication theorem for determinants. Binet's model was Lagrange, his main aim to give a generalization of certain geometrical relations given by Lagrange. Cauchy wanted to generalize Gauss's formulas. Binet's paper was overshadowed by Cauchy's for at least three reasons: Binet's proof of the multiplication theorem was unsatisfactory, for it did not take the analytical approach favoured by his friend and rival Cauchy; and Cauchy laid the foundations for the systematic development of the new theory of determinants.

Apparently following Gauss, Cauchy arranged the m^2 terms in the form of a square matrix, and introduced determinants as special alternating symmetric functions. They were defined by means of difference products: multiply the product of $2^{n(n-1)/2}$ differences

$$(a_2 - a_1)(a_3 - a_1) \cdots (a_n - a_{n-1}), \quad a_i \neq a_j \text{ if } i \neq j, \tag{9}$$

by the product $a_1 a_2 \cdots a_n$, and replace the exponents by identical second indices. The expression

$$S(\pm a_{1,1} a_{2,2} \cdots a_{n,n}) \tag{10}$$

was called a 'determinant'; the operation symbol 'S' applied to the left indices. The sign rule was based on the number of cycles into which the permutation of the left indices of a special term could be decomposed.

Although Cauchy's paper overshadowed Binet's, it remained nearly unnoticed for the next thirty years. One of the possible reasons for this limited effectiveness was his practice of returning to the same subject several times later on, but using other means of dealing with it and other terminology. In 1821 he dropped the name 'determinant' in favour of 'resultant', which became a special alternating sum in 1841. He made it difficult for his readers to find lasting results (Studnička *1875*).

5.3 Jacobi: the algorithmic approach

In 1841 Carl Jacobi published three decisive treatises: 'On the formation and the properties of determinants', 'On functional determinants' and 'On alternating functions and their division by the product of the differences of the elements'. He was obviously stimulated by Cauchy's paper of 1812, and he in turn stimulated Cauchy to write again on this subject. His aim was to improve his predecessor's roundabout and unclear presentation of the algorithms used in the solution of systems of linear equations, while pursuing the same ideas.

It was Jacobi's achievement that determinant theory became well known among the mathematical community. The algorithmization of determinant theory is also due to him. He took, quite generally, $(n + 1)^2$ quantities $a_k^{(i)}$, $i, k = 0, \ldots, n$, constructed the $(n + 1)!$ products

$$aa_1' a_2'' \cdots a_n^{(n)} \tag{11}$$

by permuting the lower and upper indices in every possible way, and prescribed $+$ or $-$ according to Cauchy's sign rule for even or odd permutations. The aggregate

$$R = \sum \pm aa_1' a_2'' \cdots a_n^{(n)} \tag{12}$$

he called a 'determinant'. His approach implied a decisive level of abstraction. His 'set-theoretical' definition left entirely open what kind of things the elements of the determinant were, whether numerical coefficients or partial differential quotients.

5.4 Cayley: the array approach

Around the middle of the nineteenth century, a thorough knowledge of determinant theory was limited largely to Jacobi's pupils. But in 1841 Arthur Cayley published the first English contribution to the new theory. He was the first to delimit the array of elements by two upright lines, and he launched it into mathematical work as a new entity. A few years later he helped to found matrix and invariant theory (§6.7–6.8). He called invariants 'hyperdeterminants', because the determinants turned out to be the first and simplest invariants.

The first textbook on determinants appeared in 1851, from William Spottiswoode; numerous others followed. Determinant theory had at last become a fruitful science.

5.5 Weierstrass and Kronecker: the axiomatic approach

We know from Georg Frobenius that from 1864 (at the latest) Karl Weierstrass used a new, quasi-axiomatic approach to determinants in his mathematical seminars and lectures. A note entitled 'On determinant theory', published posthumously in 1903, showed that he introduced determinants as functions with three characteristics: namely, as normed, linear homogeneous functions. The modern definition of a determinant therefore goes back to Weierstrass.

Kronecker was especially interested in the exploration of algorithms and internally consistent operations. As a consequence, his lectures on determinant theory, published posthumously also in 1903, were one of his three standard courses. He did not use the three characteristics for a definition of determinants, but presented them as a main consequence of his theory of linear equations. This approach gained influence in the growing establishment of determinant theory in mathematics.

BIBLIOGRAPHY

Günther, S. *1875, Lehrbuch der Determinanten-Theorie für Studirende*, Erlangen: Besold, Chap. 1.

Knobloch, E. (ed.) *1980, Der Beginn der Determinantentheorie. Leibnizens nachgelassene Studien zum Determinatntenkalkül*, Hildesheim: Gerstenbery.

—— *1990*, 'Erste europäische Determinantentheorie', in G. W. Leibniz, *Das Wirken des grossen Philosophen und Universalgelehrten als Mathematiker, Physiker, Techniker* (ed. E. Stein and A. Heinekamp), Hannover: G. W. Leibniz Gesellschaft, 32–41.

Martzloff, J. C. *1988, Histoire des mathématiques chinoises*, Paris: Masson.

Mellberg, E. J. *1876, Teorin för Determinant-kalkylen*, Helsingfors: Frenckell.

Mikami, Y. *1913*, *The Development of Mathematics in China and Japan*, Leipzig: Teubner. [Repr. 1974, New York: Chelsea.]

Muir, J. *1890–1923*, *The Theory of Determinants in the Historical Order of Development*, 4 vols, London: Macmillan. [Repr. 1960, New York: Dover.]

Studnička, F. J. *1876*, 'A. L. Cauchy als formaler Begründer der Determinanten-Theorie. Eine literarisch–historische Studie', *Abhandlungen der Königlichen Böhmischen Gesellschaft der Wissenschaften*, Series 6, **8** (1875–6), Series B, *Abhandlungen der mathematischen-naturwissenschaftlichen Classe*, 46pp.

Tropfke, J. *1980*, *Geschichte der Elementarmathematik*, 4th edn, Vol. 1 (ed. K. Vogel, K. Reich and H. Gericke), Berlin and New York: de Gruyter.

Vogt, H. *1906*, 'Analyse combinatoire et théorie des déterminants', in *Encyclopédie des sciences mathématiques*, Tome 1, Vol. 1, 63–132. [Much expanded version of the article by E. Netto in the *Encyklopädie der mathematischen Wissenschaften*, Vol. 1, 22–46 (1898: article I A 2).]

6.7

Matrix theory

I. GRATTAN-GUINNESS AND

W. LEDERMANN

---

1 INTRODUCTION

Matrix theory is today one of the staples of higher-level mathematics education; so it is surprising to find that its history is fragmentary, and that only fairly recently did it acquire its present status. This fact is indicated in the final section: before that its origins and development are sketched in four sections, covering bilinear and quadratic forms, the spectral theory (where latent roots and vectors take centre stage), the conceptual approach and matrix representation. These areas did not develop completely independently, of course, although the divisions of interest are quite striking and some order of chronology is evident. Ledermann is largely responsible for Sections 4–6, Grattan-Guinness for most of the rest.

The history of this subject has been sadly neglected, and also mispresented in that excessive importance has been placed upon the conceptual approach. Luckily, an important recent series of articles by T. W. Hawkins, listed in the bibliography, repairs many of the omissions. The neighbouring articles on determinants (§6.6) and invariant theory (§6.8) should also be consulted.

2 BILINEAR AND QUADRATIC FORMS

A bilinear form in two sets of n variables, $\{x_i\}$ and $\{y_j\}$, is

$$\sum a_{ij} x_i y_j, \quad \text{or} \quad \mathbf{x}^T \mathbf{A} \mathbf{y} \tag{1}$$

in the modern vectorial and matricial notation, where $\{a_{ij}\}$ are scalars; if $x_i = y_i$ for each i (i.e. if $\mathbf{x} = \mathbf{y}$), then the form is said to be quadratic in the $\{x_i\}$. These forms occur in many branches of mathematics, some of which have stimulated the development of matrix theory, in particular differential equations, aspects of mechanics such as moments of inertia, and number theory.

Quadratic forms were studied by Gottfried Wilhelm Leibniz from the late seventeenth century on, and they appear in the works of some later writers. From the mid-eighteenth century onwards, nascent matrix theory is evident especially in the efforts of Jean d'Alembert and Joseph Louis Lagrange to solve systems of (usually) ordinary differential equations. The most striking case, which was taken up by Pierre Simon Laplace, was Lagrange's attempt of the 1780s to prove that the Solar System was stable by showing that the eccentricities and inclinations to the ecliptic of the planets' orbits were always bounded. Taking Newton's second law of motion in a certain coordinate system, they found simple-state solutions to the differential equations which, in modern terms, required knowledge of the latent roots and vectors of the characteristic equation

$$|\mathbf{C}^2 - \lambda\mathbf{I}| = 0, \tag{2}$$

where $\mathbf{C}$ is the matrix of known coefficients of the equations (see Section 4). But matrix theory as such was not present; Lagrange and Laplace worked with the corresponding quadratic form, and made assertions about its boundedness in value, which were especially hopeful for the case where a latent root might be repeated. The work was very brilliant, however; and in extending it from the astronomical point of view in the early 1800s Lagrange and the young Siméon-Denis Poisson came across the important 'bracket' solutions to Newton's equations which still bear their name (Grattan-Guinness *1990*: 324–9, 371–84; §8.1).

Two other forays of this time are worth noting. First, Carl Friedrich Gauss's *Disquisitiones arithmeticae* (1801), his masterpiece in number theory, contains a superb passage on the treatment of quadratic forms in which the coefficients are laid out as a rectangular array, and matrix inversion and multiplication and reduction to special forms are described; but nothing came of it. Second, Joseph Fourier's first model of heat diffusion (§9.4), of around 1802, featured a set of ordinary differential equations, which he solved by a long-hand equivalent of matrix inversion; and when he came to his series, one method of finding the coefficients was a wild affair which marked the origin of infinite matrix theory (Grattan-Guinness and Ravetz *1972*: Chaps 3 and 9). This topic did not really develop until the late nineteenth century, when it was more closely driven by functional analysis and integral equations (§3.9–3.10) than by mainstream matrix theory itself (Bernkopf *1968*). In this context the name 'spectral theory' has its origins.

3 THE ORIGINS OF SPECTRAL THEORY

A potentially major advance occurred at the Paris Académie des Sciences one day in July 1829, when Augustin Louis Cauchy and J. C. F. Sturm (the latter then a young emigré from Switzerland) read pertinent papers (Hawkins *1975a*). Cauchy was then strongly interested in quadratic forms in the contexts mentioned in the previous section, and he produced a paper on properties of a form in n variables. He explicitly thought of the coefficients as an array, which is crucial if any advance is to be made beyond quadratic forms themselves: he gave the matrix the name 'tableau', and considered both its latent roots (see Section 5.4) and those of the leading submatrices (to use modern terms), and its conversion to a sum of squares. Sturm took a system of ordinary differential equations (following Lagrange and Laplace) and ended up with the generalized latent-root problem, seeking the roots of

$$|\mathbf{A} - \lambda\mathbf{B}| = 0 \tag{3}$$

for matrices $\mathbf{A}$ and $\mathbf{B}$ and their properties.

In neither case did the author realise the full potential of his ideas; nor did Carl Jacobi, in his papers from the 1830s onwards on linear transformations, which however brought him to important conceptual clarifications concerning determinants. The next major advances did not occur until the 1850s and 1860s, when Leopold Kronecker examined linear transformations (introducing the Kronecker delta in the process), and Karl Weierstrass published important papers on the 'elementary divisors' of a pair of bilinear or quadratic forms $\mathbf{x}^T\mathbf{A}\mathbf{y}$ and $\mathbf{x}^T\mathbf{B}\mathbf{y}$ (i.e. the separate factors, including multiplicity, of the associated determinant $|\mathbf{A} - \lambda\mathbf{B}|$). One of Weierstrass's main contributions was his discovery of sufficient conditions for the simultaneous reduction of two quadratic forms (where one of them was positive- or negative-definite; see Section 5.2) to sums of squares; they included the case when repeated latent roots of equation (3) were involved and where he realized that the results found by Cauchy were corrigible. For the later stages, see Section 5.

The growth of knowledge about determinants and invariants also helped in these developments, and links between these theories and matrices themselves became clearer. The original questions were transformed and generalized: for example, the conversion of a quadratic form into a sum of squares became the reduction of a matrix to diagonal form. Frequently the presence of multiple latent roots in a matrix has been a source of error in the planning of theorems, and investigation of such counter-examples a fruitful source of correction. Complementary to these advances was the explicit individuation of a matrix as an array, an examination of its basic properties, and classification of matrices into their most important types (commutative, symmetric, singular, and so on). For the history of these aspects we turn, especially, to English mathematics.

4 THE CONCEPTUAL THEORY AND THE CAYLEY–HAMILTON THEOREM

Matrix algebra has two sources: bilinear forms, which were introduced in Section 2, and linear substitutions

$$y_i = \sum_{i=1}^{n} a_{ij}x_j, \quad i = 1, 2, \ldots, n. \tag{4}$$

Each of these two mathematical objects is specified by an $n \times n$ array or matrix of coefficients a_{ij}, which is conveniently denoted by a single letter as $\mathbf{A} = (a_{ij})$. But much more is true: the set of bilinear forms possesses a structure which involves multiplying a form by a scalar λ, adding two forms, and composing two forms so as to yield a third form; likewise, substitutions possess a structure involving multiplication by a scalar, addition and composition. In both cases the results are described, respectively, by the matrices $\lambda\mathbf{A}$, $\mathbf{A} + \mathbf{B}$ and $\mathbf{AB}$. Thus matrices mirror precisely the structures both of bilinear forms and of linear substitutions – a fact that must be regarded as one the most pregnant coincidences in mathematics.

As we just saw, in the first half of the nineteenth century some mathematicians, including Gauss, Gotthöld Eisenstein and Cauchy, were aware of the advantage of virtually identifying a bilinear form or a linear substitution with its array of coefficients (or system, or tableau, as they called it). Eisenstein (1844) denoted linear substitutions by single letters and explained how they can be added and multiplied (composed) much like ordinary numbers, except for the failure of the commutative law.

One more step towards abstraction had to be taken: to study the laws of matrix algebra in its own right, irrespective of whether matrices represent bilinear forms or linear substitutions or any other mathematical object. This point of view was elaborated in a 'Memoir on the theory of matrices' published by Arthur Cayley *1858*. (The term 'matrix' was introduced in 1850 by J. J. Sylvester to denote a regular array of numbers.) Guided by the structure of linear substitutions, Cayley established the formal laws of matrix algebra, including multiplication by a scalar, addition and multiplication, and the condition for the existence of the inverse $\mathbf{A}^{-1}$. Throughout his principal theme was that 'matrices comport themselves as single quantities'. Cayley was especially interested in the analogy between matrix algebra and the theory of functions, in particular, polynomials in one variable. Thus, let

$$f(t) = a_0 + a_1 t + a_2 t^2 + \cdots + a_m t^m, \tag{5}$$

where $a_0, a_1, \ldots, a_m$ are real or complex numbers. Then the matrix polynomial $f(\mathbf{A})$ is defined by

$$f(\mathbf{A}) = a_0 \mathbf{I} + a_1 \mathbf{A} + a_2 \mathbf{A}^2 + \cdots + a_m \mathbf{A}^m, \tag{6}$$

where $\mathbf{I}$ is the unit matrix. In this context, Cayley formulated his most celebrated result, now known as the Cayley–Hamilton theorem. Let

$$|t\mathbf{I} - \mathbf{A}| = t^n + c_1 t^{n-1} + \cdots + c_{n-1} t + c_n \tag{7}$$

be the characteristic function of $\mathbf{A}$. Then

$$\mathbf{A}^n + c_1 \mathbf{A}^{n-1} + \cdots + c_{n-1} \mathbf{A} + c_n \mathbf{I} = \mathbf{0}, \tag{8}$$

that is, every (square) matrix satisfies its own characteristic equation.

In his paper, Cayley proved the result only for $n = 2$, and he reported having checked it for $n = 3$. But then he went on to say: 'I have not thought it necessary to undertake the labour of a formal proof of the theorem in the general case of a matrix of any degree.' It is strange that a mathematician of Cayley's power and inventiveness should have resigned in the face of such a beautiful result. Independently of Cayley, William Rowan Hamilton proved the theorem for $n = 4$ in the course of his work on quaternions (§6.2).

Cayley also mentioned irrational functions of $\mathbf{A}$, notably $\sqrt{\mathbf{A}}$, and the problem of finding all matrices which commute with a given $\mathbf{A}$. But his answers to these questions are inaccurate. Towards the end of his paper, he introduced rectangular matrices and gave rules for their composition, though not the formula for transposition, $(\mathbf{AB})^{\mathrm{T}} = \mathbf{B}^{\mathrm{T}} \mathbf{A}^{\mathrm{T}}$.

Cayley's memoir was ignored for many years. It may be that, with a few notable exceptions, mathematicians of his generation, or even one or two subsequent generations, were reluctant to accept mathematical objects that could neither be seen (geometry) nor computed (algebra and analysis). At best, most mathematicians paid lip-service to vectors and matrices by regarding them as convenient abbreviations for certain sets of numbers which must be brought into play if concrete results are to be achieved.

The next landmark in the theory of matrices is a profound and extensive memoir by Georg Frobenius, published in 1878. In its title, 'On linear substitutions and bilinear forms' (in translation), Frobenius acknowledged the two roots of matrix algebra. In contrast to Cayley, with whose memoir on matrices he seems to have been unfamiliar in 1878, he took bilinear forms as his starting-point, and throughout his paper referred to $\mathbf{A} = (a_{ij})$ as a form. It was not until 1896 that Frobenius cited Cayley *1858*, and began to use the term 'matrix'. Having presented an ingenious proof of the fundamental theorem $f(\mathbf{A}) = 0$ in his paper of 1878, Frobenius then gave Cayley credit for discovering the theorem, albeit without a general proof.

5 CANONICAL MATRICES

Matrix algebra becomes particularly effective when it is used to classify quadratic forms or linear substitutions according to certain principles. In these circumstances both the objects and the operations performed upon them are expressed as matrices; or, as Frobenius put it: 'the distinction between operandus and operator becomes blurred'.

5.1 Definition

Let $\mathcal{M}$ be a set of matrices, for example the set of all $n \times n$ matrices over the complex field, or the set of all real symmetric matrices. Suppose that, in accordance with some equivalence relation, the set $\mathcal{M}$ has been split into equivalent classes

$$\mathcal{K}^\alpha, \mathcal{K}^\beta, \mathcal{K}^\gamma, \ldots . \tag{9}$$

From each class we choose one representative which has a simple form and which shows the characteristics of its class in an easily recognizable manner. Of course, the choice is, in general, not unique; but it is surprising how close one can get to an ideal solution. The selected representatives, say

$$\mathbf{C}^\alpha, \mathbf{C}^\beta, \mathbf{C}^\gamma, \ldots, \tag{10}$$

are called the canonical matrices for $\mathcal{M}$ under this equivalence relation. Hence every matrix of $\mathcal{M}$ is equivalent to one and only one canonical matrix, and no two distinct canonical matrices are equivalent to each other.

In his paper of 1878, Frobenius discussed the general notion of equivalence for a set of matrices. He attributed the idea to Kronecker (1874) and Weierstrass (1868), who had studied important special cases. We shall illustrate the theory of canonical matrices by examining a few typical examples.

5.2 Classification of real quadratic forms

A real quadratic form can be expressed as $q = \mathbf{x}^T \mathbf{A} \mathbf{x}$, where $\mathbf{x}^T = (x_1 x_2 \ldots x_n)$. We assume that $\mathbf{A}$ ranges over all real symmetric matrices (i.e. ones for which $\mathbf{A} = \mathbf{A}^T$). Suppose we make the substitution $\mathbf{x} = \mathbf{P}\mathbf{y}$, where **P** is an invertible matrix. Then $q = \mathbf{y}^{\mathrm{T}}\mathbf{B}\mathbf{y}$, where

$$\mathbf{B} = \mathbf{P}^{\mathrm{T}}\mathbf{A}\mathbf{P}, \quad |\mathbf{P}| \neq 0. \tag{11}$$

If equation (11) holds, we say that **A** and **B** are congruent. This is an equivalence relation for the set of all symmetric matrices, and the equivalence class of **A** is given by

$$\mathscr{K}(\mathbf{A}) = \{\mathbf{P}^{\mathrm{T}}\mathbf{A}\mathbf{P} \,|\, |\mathbf{P}| \neq 0\}. \tag{12}$$

The question of a suitable canonical feature was neatly answered by Sylvester's 'law of inertia' (1852), which states that every real symmetric $n \times n$ matrix is congruent to a unique diagonal matrix of the form

$$\mathrm{diag}(1, 1, \ldots, 1, -1, -1, \ldots, -1, 0, 0, \ldots, 0), \tag{13}$$

having p entries equal to 1, q equal to -1 and h equal to 0, where

$$p \geqslant 0, \quad q \geqslant 0, \quad h \geqslant 0; \quad p + q + h = n. \tag{14}$$

An especially important case is that in which $n = p$, so that $q = h = 0$. Such a matrix, or the quadratic form associated with it, is said to be 'positive definite' and is characterized by the property that $\mathbf{x}^{\mathrm{T}}\mathbf{A}\mathbf{x} > 0$ for all non-zero x.

5.3 Similar matrices

Let $\mathscr{M}$ be the set of all $n \times n$ matrices with complex coefficients. The matrices **A** and **B** are said to be similar if there exists an invertible matrix **P** such that

$$\mathbf{B} = \mathbf{P}^{-1}\mathbf{A}\mathbf{P}. \tag{15}$$

This is an equivalence relation on $\mathscr{M}$. The problem is to identify a suitable canonical feature in the equivalence class $\mathscr{S}(\mathbf{A}) = \{\mathbf{P}^{-1}\mathbf{A}\mathbf{P} \,|\, |\mathbf{P}| \neq 0\}$ of **A**. Similar matrices have the same characteristic function, and hence the same latent roots $\alpha_1, \alpha_2, \ldots, \alpha_n$. When the latent roots are distinct, then **A** is similar to the diagonal matrix $\mathrm{diag}(\alpha_1, \alpha_2, \ldots, \alpha_n)$. But when multiple latent roots are involved, it may be impossible to 'diagonalize' **A**. Suppose the characteristic function of **A** is

$$|t\mathbf{I} - \mathbf{A}| = (t - \alpha)^p(t - \beta)^q(t - \gamma)^r \cdots, \tag{16}$$

where $\alpha, \beta, \gamma, \ldots$ are the distinct latent roots of **A**, and $p, q, r, \ldots$ are their multiplicities. Then the various equivalence classes $\mathscr{S}(\mathbf{A})$ are in 1–1 correspondence with the set of partitions

$$p = p_1 + p_2 + \cdots, \qquad q = q_1 + q_2 + \cdots, \qquad r = r_1 + r_2 + \cdots. \quad (17)$$

In accordance with these partitions of $p, q, r, \ldots$, the characteristic polynomial of $\mathbf{A}$ is further factorized thus:

$$|t\mathbf{I} - \mathbf{A}| = (t - \alpha)^{p_1}(t - \alpha)^{p_2} \cdots (t - \beta)^{q_1}(t - \beta)^{q_2} \cdots (t - \gamma)^{r_1}(t - \gamma)^{r_2} \cdots. \quad (18)$$

The polynomials

$$(t - \alpha)^{p_1}, (t - \alpha)^{p_2}, \ldots, (t - \beta)^{q_1}, (t - \beta)^{q_2}, \ldots, (t - \gamma)^{r_1}, (t - \gamma)^{r_2}, \ldots \quad (19)$$

are called the elementary divisors of $\mathbf{A}$. The fundamental result is this:

The complex matrices $\mathbf{A}$ and $\mathbf{B}$ are similar if and only if they have the same elementary divisors. A canonical element in a typical equivalence class is furnished by the 'Jordan canonical form'

$$\mathbf{J}(\alpha : p_1) \otimes \mathbf{J}(\alpha; p_2) \otimes \cdots \otimes \mathbf{J}(\beta; q_1) \otimes \mathbf{J}(\beta; q_2) \otimes \cdots$$
$$\otimes \mathbf{J}(\gamma; r_1) \otimes \mathbf{J}(\gamma; r_2) \otimes \cdots, \quad (20)$$

being the direct sum of Jordan blocks; these are defined by the $s \times s$ Jordan matrix

$$\mathbf{J}(\alpha; s) = \begin{pmatrix} \alpha & 1 & 0 & \ldots & 0 & 0 \\ 0 & \alpha & 1 & \ldots & 0 & 0 \\ \vdots & \vdots & \vdots & & \vdots & \vdots \\ 0 & 0 & 0 & \ldots & \alpha & 1 \\ 0 & 0 & 0 & \ldots & 0 & \alpha \end{pmatrix}. \quad (21)$$

The concept of elementary divisor is due to several mathematicians. The history is confused (Hawkins *1977b*), and replete with overlooked publications and belated acknowledgements. It seems that priority should be accorded to Sylvester (1851), and more especially to H. J. S. Smith (1861), who used elementary divisors to classify matrices with integral coefficients. In a difficult paper, Weierstrass (1868) discussed the canonical form of a 'matrix pencil' $t\mathbf{A} + s\mathbf{B}$, where t and s are indeterminates. This reduces to the above similarity problem when $\mathbf{A} = \mathbf{I}$ and $s = -1$. In his classic *Traité des substitutions* (1870), Camille Jordan established the canonical form (19) for the case in which the coefficients a_{ij} are taken from a finite field of prime order; but the ideas are readily transferred to the field of complex numbers. Jordan did not mention Smith or Weierstrass. The full story was told in the erudite memoir by Frobenius (1879) on linear forms with integral coefficients. Frobenius gave credit to Jordan, and mentioned in a footnote that the latter came across Smith's papers only after the completion of his own memoir.

5.4 The principal-axes theorem

The earliest and most famous case of a canonical reduction, this theorem was discovered by Cauchy and Sturm in 1829 (Section 3). In matrix language, the result is as follows: let $\mathbf{A}$ be a real symmetric matrix; Cauchy proved that its latent roots are real, say $\alpha_1 \geqslant \alpha_2 \geqslant \cdots \geqslant \alpha_n$, and that there exists a real orthogonal matrix $\mathbf{R}$ (i.e. one for which $\mathbf{R}^T = \mathbf{R}^{-1}$) such that

$$\mathbf{R}^T \mathbf{A} \mathbf{R} = \operatorname{diag}(\alpha_1, \alpha_2, \ldots, \alpha_n). \tag{22}$$

Interest in this problem originated partly from geometry, for the equation

$$\mathbf{x}^T \mathbf{A} \mathbf{x} = 1 \tag{23}$$

represents a quadratic surface in n-dimensional space. Under the transformation $\mathbf{x} = \mathbf{R}\mathbf{y}$ (rotation of axes), the equation of the surface becomes

$$\alpha_1 y_1^2 + \alpha_2 y_2^2 + \cdots + \alpha_n y_n^2 = 1. \tag{24}$$

The directions of the new coordinate axes now coincide with the principal axes of the surface.

6 RELATED THEORIES

6.1 Grassmann algebra

In a remarkable treatise entitled (in translation) 'Theory of Extension' (1844 and 1862), Hermann Grassmann introduced an ingenious algebraic tool which serves to study the subspaces of an n-dimensional vector space. In this so-called Grassmann algebra, vectors are composed by 'outer multiplication', usually denoted by $\wedge$; its properties include the rules $\mathbf{x} \wedge \mathbf{x} = 0$ and $\mathbf{x} \wedge \mathbf{y} = -(\mathbf{y} \wedge \mathbf{x})$ for all vectors $\mathbf{x}$ and $\mathbf{y}$ (§6.2). This algebra had some influence upon the founding of matrix theory.

6.2 Group characters

The invertible matrices of degree n over a field form a multiplicative group. A powerful method of studying abstract groups consists in making a homomorphic image of them within a matrix group. Let $G: 1, x, y, \ldots$ be an abstract group, and suppose that an invertible complex matrix $\mathbf{A}(x)$ of degree n is associated with x in such a way that $\mathbf{A}(xy) = \mathbf{A}(x)\mathbf{A}(y)$, where x and y are arbitrary elements of G. Then we say that $\mathbf{A}(x)$ forms a representation of G of degree n. The trace of the matrix $\mathbf{A}(x)$, that is, the function

$$\chi(x) = \sum_{i=1}^{n} a_{ii}(x), \quad x \subset G, \tag{25}$$

is called the character of the representation. It truly characterizes most of the relevant features of the representation, such as reducibility and faithfulness, and it helps to reveal important properties of the underlying group. The theory of group characters was created by Frobenius in 1896 in response to a letter from Richard Dedekind about the evaluation of a special kind of determinant associated with a finite group (Hawkins *1971*).

7 CONSOLIDATION

A major conceptual block to the advance of matrix theory may have been the difficulty of envisaging the possible ways in which multiplication and division of (something like) a matrix could be defined; addition and subtraction would be easy to specify, but not worth the trouble on their own. The same view can be said about the late arrival of vector algebra (§6.2), for vectorial *thinking* was quite evident from the late eighteenth century onwards.

Whatever the reasons, only by the early twentieth century were all the elements described above becoming sufficiently prominent for a 'subject' to be recognizable, and textbooks to be written. Of these, the *Introduction to Higher Algebra* by the Harvard mathematician Maxime Bôcher (*1907*; note his title) was an important pioneering work, with a German translation soon appearing. After preliminaries on polynomials, he treated, in order, determinants; linear equations and matrices; invariants; bilinear and quadratic forms; polynomials and elementary divisors; and 'the equivalence and classification' of pairs of forms, where a surprisingly small amount of spectral theory was presented. Another noteworthy author was the Cambridge-trained mathematician C. E. Cullis, who disgorged his lectures at the University of Calcutta into three large volumes on *Matrices and Determinoids* (*1913–25*). Much non-standard terminology reduced its influence, but it was notable in several ways: in particular, he stressed rectangular matrices from the start (indeed, a 'determinoid' was his extension of a determinant to such matrices, using the Laplace expansion).

Even then, though, diffusion and education of matrix theory was slow; for example, when, in the mid-1920s, the creators of quantum mechanics were looking for techniques, matrix theory was still not widely known (§9.15). The early 1930s saw several new textbooks, of which Turnbull and Aitken *1932* was popular.

The rise of matrix theory to staple diet has occurred only since the 1950s. One regrettable but seemingly unavoidable consequence was the flood of boring and repetitive textbooks which turned exciting mathematical ideas into dreary and unmotivated (though obviously useful) techniques, ideal only for question-fodder; Mirsky *1955* was one of the better books in English. Another consequence (already in Bôcher, to some extent) was that the historical order of development was reversed: the textbook story begins with the matrix (and determinant) as such, the spectral theory comes in later chapters, and bilinear and quadratic forms are treated last.

A further 'benefit' of this trade has been the popularization of the appalling non-words 'eigenvalue' and 'eigenvector', created out of (absurdly partial) translations of the German words *Eigenwert* and *Eigenvektor*. The properly English phrases 'latent root' and 'latent vector' have been employed in this article. The former was introduced in Sylvester *1883*, in a charming phrase: '*Latent roots* of a matrix − latent in a somewhat similar sense as a vapour may be said to be latent in water or smoke in a tobacco leaf.' The adjective 'characteristic' was and is also deployed, to refer to the root, vector, matrix $A - \lambda I$, 'function' $|A - \lambda I|$ and equation $|A - \lambda I| = 0$.

Meanwhile, research of course has been proceeding, in established and in new directions. In particular, the arrival of computers (§5.12) has led to a rise in numerical linear algebra, where the properties (especially spectral ones) of very large matrices excite much interest.

BIBLIOGRAPHY

Bernkopf, M. *1968*, 'A history of infinite matrices [. . .]', *Archive for History of Exact Sciences*, **4**, 308–58.

Bôcher, M. *1907*, *Introduction to Higher Algebra*, New York: Macmillan. [German transl. 1910, Leipzig and Berlin: Teubner. Repr. 1964, New York: Dover. 2nd edn, 1922.]

Cayley, A. *1858*, 'Memoir on the theory of matrices', *Philosophical Transactions of the Royal Society of London*, **148**, 17–37. [Also in *Collected Mathematical Papers*, Vol. 2, 475–96.]

Cullis, C. E. *1913–25*, *Matrices and Determinoids*, 3 vols, Cambridge: Cambridge University Press.

Grattan-Guinness, I. *1990*, *Convolutions in French Mathematics, 1800–1840* [. . .], 3 vols, Basel: Birkhäuser, and Berlin: Deutscher Verlag der Wissenschaften.

Grattan-Guinness, I. and Ravetz, J. R. *1972*, *Joseph Fourier 1768–1830* [. . .], Cambridge, MA: MIT Press.

Hawkins, T. W. *1971*, 'The origins of the theory of group characters', *Archive for History of Exact Sciences*, **7**, 142–70.

—— *1972*, 'Hypercomplex numbers, Lie groups, and the creation of group representation theory', *Archive for History of Exact Sciences*, **8**, 243–87.

—— 1974, 'New light on Frobenius' creation of the theory of group characters', *Archive for History of Exact Sciences*, **12**, 217–43.

—— 1975a, 'Cauchy and the spectral theory of matrices', *Historia mathematica*, **2**, 1–29.

—— 1975b, 'The theory of matrices in the 19th century', in *Proceedings of the International Congress of Mathematicians*, Vancouver, 1974, Canadian Mathematical Congress, Vol. 2, 561–70.

—— 1977a, 'Another look at Cayley and the theory of matrices', *Archives Internationales d'Histoire des Sciences*, **26**, 82–112.

—— 1977b, 'Weierstrass and the theory of matrices', *Archive for History of Exact Sciences*, **17**, 119–63.

Mirsky, L. A. 1955, *An Introduction to Linear Algebra*, Oxford: Clarendon Press.

Muir, T. 1890–1923, *The Theory of Determinants in the Historical Order of Development*, 4 vols, London and New York: Macmillan. [Repr. 1960, New York: Dover. Information on matrices here and there.]

Schneider, H. 1977, 'Olga Taussky-Todd's influence on matrix theory and matrix theorists [...]', *Linear and Multilinear Algebra*, **5**, 197–224. [Includes interesting remarks on the earlier history, as well as noting a major recent worker.]

Seneta, E. 1973, *Non-negative Matrices*, London: Allen & Unwin. [Includes historical survey.]

Sylvester, J. J. 1883, 'On the equation to the secular inequalities in the planetary theory', *Philosophical Magazine*, Series 5, **16**, 267–9. [Also in *Collected Papers*, Vol. 4, 410–11.]

Turnbull, H. W. and Aitken, A. C. 1932, *An Introduction to the Theory of Canonical Matrices*, London and Glasgow: Blackie. [Repr. 1961, New York: Dover.]

6.8

Invariant theory

TONY CRILLY

1 INVARIANCE AND INVARIANTS

The general idea of invariance pervades science and mathematics. In the classical conservation laws of science, for example, not only time and space but also mass and energy remain constant. In the theory of relativity, the form of the basic equations remains unchanged under Lorentz transformations. In mathematics the idea of invariance is ever present. On a macroscopic level, topology is the study of those properties which remain invariant under 'elastic' transformations. On a microscopic level, even an elementary theorem in Euclidean geometry can be interpreted in terms of invariance, for example the theorem which states that a point P on the circumference of a circle subtends the same angle APB above a chord AB, whatever the position of the point P (see Figure 1).

But to mathematicians, 'invariant theory' has a precise and definite meaning – it refers to a branch of algebra which has developed over the previous century and a half. In the folklore of mathematics it summons up a vision of the nineteenth-century mathematical pioneers of Europe and America engaged in a heroic quest to seek out and catalogue invariant algebraic forms. At that time, invariant theory occupied a central position in mathematics for it held implications for branches of mathematics which

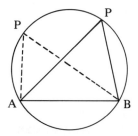

Figure 1 Invariance in elementary geometry: the angle APB is independent of the position of P

were important then but no longer attract as much attention. The subject is still studied by mathematicians, although its objectives are now expressed in terms of modern 'structural' algebra (Section 4).

2 THE CLASSICAL PERIOD: ENGLISH CONTRIBUTIONS

In the early 1840s George Boole led the way with papers in the newly founded *Cambridge Mathematical Journal*. These papers captured the imagination of the young Arthur Cayley, then only 20 years old. In the period immediately following his graduation from Cambridge University he contributed two foundational papers on invariant theory to the same journal. According to Cayley, the object of the theory was to 'find all the derivatives [invariants] of any number of functions [algebraic forms], which have the property of preserving their form unaltered after any linear transformation of the variables'. He possessed a keen appreciation of the scope and potential of the subject, and his early papers were also published in leading Continental journals where other mathematicians, notably Ferdinand Gotthold Eisenstein and Otto Hesse, contributed to the theory. Earlier than this, Carl Friedrich Gauss and Joseph Louis Lagrange had been aware of the algebraic properties of invariants; but Cayley seized upon the idea and was instrumental in developing it into a body of knowledge which became the 'modern algebra' of the nineteenth century. Boole, on the other hand, played little part in its further development.

Invariant theory is concerned with algebraic expressions or forms which Cayley called 'quantics'. The simplest type is the binary form (two variables, x and y) of which the most elementary example is the binary quadratic form

$$ax^2 + 2bxy + cy^2. \tag{1}$$

Next is the binary cubic form

$$ax^3 + 3bx^2y + 3cxy^2 + dy^3, \tag{2}$$

and in general the binary quantic of order n:

$$a_0x^n + a_1\binom{n}{1}x^{n-1}y + a_2\binom{n}{2}x^{n-2}y^2 + \cdots + a_ny^n. \tag{3}$$

Forms of any number of variables were also treated, but in the nineteenth century the binary form was the focal point of research.

Invariants are algebraic forms which are left unchanged (except for a multiplicative factor) when the variables x and y of the quantic are transformed linearly. If $ax^2 + 2bxy + cy^2$ is transformed to $Ax^2 + 2Bxy + Cy^2$, then $AC - B^2$ is a multiple of $ac - b^2$. The invariant $ac - b^2$ is the only one for the quadratic form, but higher-order binary quantics possess many invariants.

Other algebraic forms which have the invariantive property but which also involve the variables x and y are called 'covariants' (other types such as contravariants and mixed concomitants were also introduced later). The Hessian

$$\begin{vmatrix} \dfrac{\partial^2 f}{\partial x^2} & \dfrac{\partial^2 f}{\partial x \partial y} \\[2ex] \dfrac{\partial^2 f}{\partial y \partial x} & \dfrac{\partial^2 f}{\partial y^2} \end{vmatrix} \tag{4}$$

of the binary cubic is

$$(ac - b^2)x^2 + (ad - bc)xy + (bd - c^2)y^2 \tag{5}$$

and is a covariant.

Invariants and covariants are intimately connected with many branches of mathematics: the study of polynomial equations (for example, $ac - b^2$ is the discriminant of the quadratic equation); number theory; the study of determinants (but not initially with matrix algebra, which was developed later); and algebraic geometry, which centred around the study of curves and surfaces described by algebraic forms of three and four variables (if an invariant vanishes in one coordinate system it vanishes in *any* coordinate system — the vanishing of the Hessian of the cubic represents the necessary and sufficient condition for the three points defined by the binary cubic form to coincide). Cayley was concerned with all these branches of mathematics, and they provided added motivation for his ambitious invariant-theory research programme. It is curious that Cayley's early work on group theory had no influence on the early development of invariant theory; the links between these mathematical areas belong to the modern period.

It was during the 1850s and early 1860s that Cayley made his signal contribution and produced the first seven of the definitive ten memoirs on quantics, while he was a barrister-at-law at Lincoln's Inn in London. These papers established his working definition of an invariant as an algebraic form which is annihilated (to use J. J. Sylvester's terminology) by the partial differential operators $\square$ and $\dot{\square}$, defined by

$$\dot{\square} = a_0 \frac{\partial}{\partial a_1} + 2a_1 \frac{\partial}{\partial a_2} + \cdots + na_{n-1} \frac{\partial}{\partial a_n} \tag{6}$$

and

$$\dot{\Box} = na_1 \frac{\partial}{\partial a_0} + (n-1)a_2 \frac{\partial}{\partial a_1} + \cdots + a_n \frac{\partial}{\partial a_{n-1}}. \tag{7}$$

He was joined by other mathematicians, most notably J. J. Sylvester, who for a time worked in London as an actuary. Other early contributors were Francesco Brioschi, George Salmon and Charles Hermite (Crilly *1986*).

3 THE CLASSICAL PERIOD: GERMAN CONTRIBUTIONS

In the 1860s the German mathematicians Paul Gordan, Siegfried Aronhold and Alfred Clebsch developed a succinct symbolic method for invariant theory, in contrast to the English custom of expressing invariants and covariants in terms of coordinates. As part of the symbolic method, the German mathematicians defined a symbolic determinant,

$$(\alpha\beta) = \alpha_1\beta_2 - \alpha_2\beta_1. \tag{8}$$

With the quadratic form, for example, if $ax^2 + 2bxy + cy^2$ is compared with the expansion of $(\alpha_1 x + \alpha_2 y)^2$ (or equivalently, with the expansion of $(\beta_1 x + \beta_2 y)^2$) and it is possible to represent ac by $\alpha_1^2\beta_2^2$ (or equivalently, by $\alpha_2^2\beta_1^2$) and b^2 by $\alpha_1\alpha_2\beta_1\beta_2$, then

$$(\alpha\beta)^2 = \alpha_1^2\beta_2^2 - 2\alpha_1\alpha_2\beta_1\beta_2 + \alpha_2^2\beta_1^2, \tag{9}$$

which shows that $(\alpha\beta)^2$ can be used to represent the invariant which Cayley would have preferred to write as $ac - b^2$. They also defined an operation (the 'transvection operation') on algebraic forms which was a special case of the hyperdeterminant derivative introduced by Cayley in 1846, but subsequently abandoned by him as a basis for the theory (Grace and Young *1903*, Osgood *1892*).

Gordan established a high point in invariant theory in 1868 by showing that, for binary forms, all invariants and covariants could be expressed in terms of a finite number of irreducible invariants and covariants. This greatly surprised the English school, as Cayley had erroneously stated that 'finiteness' held only for binary forms of order 4 or less. A gentle rivalry developed between the two schools, the method of the English school being rather more pragmatic in character and less formal than the German method. In fact the two approaches were complementary, as the method of the English school would not necessarily find all irreducible invariants and covariants, whereas the Germans were liable to include reducible invariants and covariants in their list. If the two lists agreed, it was probable that the correct result had been gained.

It is significant that Cayley and the English school did not adopt the powerful German method. Cayley's famous remark about quaternions applies equally to invariants when expressed in symbolic form. He compared the succinct notation of quaternions 'to a pocket-map – a capital thing to put in one's pocket, but which must be unfolded: the formula, to be understood, must be translated into coordinates'. In the 1870s, spurred on by Sylvester's enthusiastic programme in the USA, there was perhaps an overemphasis on calculation, and the decade found mathematicians taking on vast calculatory feats with binary forms of quite low order. Human computers were found necessary as some invariants and covariants, when expressed in coordinates, consisted of thousands of terms (Crilly *1988*).

4 THE MODERN PERIOD

Just as Cayley in his youth had given the first impulse to the infant theory, it was the young David Hilbert, on the threshold of his mathematical career, who placed an entirely different emphasis on the subject and channelled the course of algebraic research into an abstract mode. Beginning in 1888, Hilbert provided several proofs of Gordan's theorem. The first, concerned with binary forms, used a completely novel method. Hilbert swept aside the emphasis on calculation and proceeded to demonstrate the existence of a finite basis by an existential argument. This caused Gordan to proclaim, some say ambiguously, that the new approach was 'not mathematics but theology'.

Hilbert divided the development of the theory into three stages: the *naive* (of Cayley and Sylvester), the *formal* (of Gordan and Clebsch) and his own period, the *critical*, in which the subject was revolutionized. Georg Frobenius, the leader of the Berlin school and creator of the theory of group representations, regarded the classical theory of invariants as 'hack work', but had great respect for Hilbert's abstract solution of the finiteness problem. Frobenius believed that Hilbert had simultaneously started the theory and, by solving the main problem, killed it off, thus exemplifying a traditional view held by mathematicians that once a theory has been crystallized and its main problems solved, it is time to move on to new problems.

In his famous address at the International Congress of Mathematicians in 1900, Hilbert endorsed the idea of cultivating a theory based on a broader idea of an invariant suggested by Felix Klein and Sophus Lie. Traditionally, invariants were obtained by considering all linear transformations of $GL(n, \mathbb{C})$, the projective linear group. This was implicit in the work of the pioneers, though Boole had touched upon invariants under more restricted transformations. In the case of the binary quadratic form, for example, when only transformations from the orthogonal subgroup $O(2, \mathbb{C})$ are allowed, the form $a + c$ is an invariant, though it is not an invariant under all the transformations of $GL(2, \mathbb{C})$. In dealing with a subgroup, the totality of invariants will *include* the full invariants of the whole group $GL(n, \mathbb{C})$. In Hilbert's 1900 list of 23 problems for the twentieth century, the 14th proposed considering invariants which arise when only the transformations of a subgroup are allowed: is it true that the invariants are finitely generated? In 1958 Masayoshi Nagata showed by counter-example that it is not true, contrary to Hilbert's initial belief.

Many eminent mathematicians have contributed to invariant theory. Emmy Noether's doctoral dissertation was on the theory of invariants, but she abandoned the field soon afterwards and concentrated on the abstract theory of ideals and rings, in which Hilbert's basis theorem is more easily deduced. One of Hermann Weyl's goals in writing his book *The Classical Groups* (*1939*) was to provide representation-theoretic foundations for invariant theory. A linear transformation T of the variables of a quantic induces a linear transformation of the coefficients. For the binary quantic, for example, a transformation T of x and y induces a linear transformation $U(T)$ on C^{n+1} and by this means a representation of subgroups of $GL(2, \mathbb{C})$ can be obtained.

At present, invariant theory is enjoying a period of renewed activity. It is a highly technical area of mathematics within abstract algebra. The general goal of the subject has not altered in its essentials since Cayley's day, though it is now expressed in the modern idiom and the scope has been widened. Popov *1986* outlines the modern framework of the subject: let k be an algebraically closed field (the complex numbers, for example), V a finite-dimensional vector space over k, $k[V]$ the algebra of polynomials on V, and G an algebraic subgroup of the general linear group $GL(V)$. G acts on $k[V]$. The main problem of the classical theory is to give 'the explicit description' of the algebra of invariants $k[V]^G$.

BIBLIOGRAPHY

Bell, E. T. *1945*, *The Development of Mathematics*, 2nd edn, New York: McGraw-Hill.

Cajori, F. *1919*, *A History of Mathematics*, 2nd edn, New York: Macmillan. [Useful on invariant theory.]

Crilly, T. *1986*, 'The rise of Cayley's invariant theory (1841–1862)', *Historia mathematica*, **13**, 241–54.

—— *1988*, 'The decline of Cayley's invariant theory (1863–1895)', *Historia mathematica*, **15**, 332–47.

Dieudonné, J. and Carrell, J. B. *1971*, *Invariant Theory, Old and New*, New York: Academic Press.

Elliott, E. B. *1913, An Introduction to the Algebra of Quantics*, Oxford: Clarendon Press. [Repr. 1964, New York: Chelsea. A good introduction to the work of the English school of invariant theorists.]

Fisher, C. S. *1966*, 'The death of a mathematical theory: A study in the sociology of knowledge', *Archive for History of Exact Sciences*, **3**, 137–59.

—— *1967*, 'The last invariant theorists', *Archives Européennes de Sociologie*, **8**, 216–44.

Grace, J. H. and Young, A. *1903, The Algebra of Invariants*, Cambridge: Cambridge University Press. [Repr. New York: Hafner. An introduction to the German symbolic method.]

Hawkins, T. *1986*, 'Cayley's counting problem and the representation of Lie algebras', in *Proceedings of the International Congress of Mathematicians – Berkeley, California, 1986*, Vol. 2, Providence, RI: American Mathematical Society, 1642–56.

McKinsey, J. C. C. and Suppes, P. *1955*, 'On the notion of invariance in classical mechanics', *British Journal for the Philosophy of Science*, **5**, 290–302.

MacMahon, P. A. *1910*, 'Algebraic forms', in *Encyclopaedia Britannica*, 11th edn, Vol. 1, 620–41. Cambridge: Cambridge University Press. [Broad survey of nineteenth-century invariant theory by one of its practitioners.]

Meyer, W. F. *1890–91*, 'Bericht über den gegenwärtigen Stand der Invariantentheorie', *Jahresbericht der Deutschen Mathematiker-Vereinigung*, **1**, 81–288. [Extensive survey with detailed references.]

—— *1898*, 'Invariantentheorie', in *Encyklopädie der mathematischen Wissenschaften*, Vol. 1, Part 1, 320–403 (article 1 B 2). [Incomplete rev. transl. by J. Drach in *Encyclopédie des sciences mathématiques*, Tome 1, Vol. 2, 1911, 386–520 (article I 11).]

Osgood, W. F. *1892*, 'The symbolic notation of Aronhold and Clebsch', *American Journal of Mathematics*, **12**, 251–61.

Parshall, K. H. *1989*, 'Towards a history of nineteenth-century invariant theory', in D. Rowe and J. McCleary (eds), *The History of Modern Mathematics*, Vol. 1, Boston: Academic Press, 157–208.

Popov, V. L. *1986*, 'Modern developments in invariant theory', in *Proceedings of the International Congress of Mathematicians – Berkeley, California, 1986*, Vol. 1, Providence, RI: American Mathematical Society, 394–406.

Springer, T. A. *1977, Invariant Theory*, Berlin: Springer. [Exposition of twentieth-century invariant theory and useful historical notes.]

Todd, J. A. *1947, Projective and Analytical Geometry*, London: Pitman. [Rev. edn, 1954.]

Turnbull, H. W. *1926*, 'Recent developments in invariant theory', *Mathematical Gazette*, **13**, 217–21.

Weyl, H. *1939, Classical Groups, their Invariants and Representations*, Princeton, NJ: Princeton University Press.

—— *1944*, 'David Hilbert and his mathematical work', *Bulletin of the American Mathematical Society*, **50**, 612–54.

6.9

The philosophy of algebra

HELENA M. PYCIOR

1 INTRODUCTION

Early modern algebra, which concentrated on the solution of equations by essentially arithmetical algorithms, tantalized Western mathematicians. According to Girolamo Cardano, the subject tested the limits of the human mind (Cardano *1968*: 8). Whereas geometry seemed to concern clear ideas with some sort of physical backing (at the least, imperfect representations of such concepts as points and lines), algebra was less easily justified. Indeed, the whole numbers were seen as 'natural', but algebra was never simply about whole numbers. Fractions and irrational numbers had appeared early, and even simple equations generated negative and imaginary roots (which had, however, been ignored for a long time).

In their quest for generalization, the early modern algebraists not only started to recognize such roots but also constructed a new algebraic language, which seemed to facilitate a generality and inventiveness quite beyond that supported by the prose words in which algebra had originally been expressed (§2.3–2.4). But the introduction of 'imaginary' numbers and the new symbolic language were not strictly mathematical developments. These developments impinged on philosophy (see §5.1, on algebraic logic), and through the nineteenth century mathematical and philosophical thinkers (as the current historical literature suggests, especially British thinkers, who therefore form the core of the present discussion) wrestled with the philosophy or, as it was sometimes called, the 'metaphysics' of algebra.

2 CARDANO'S *ARS MAGNA*

Together with François Viète, Girolamo Cardano significantly reshaped algebra in the sixteenth century. In his *Ars magna* ('The Great Art', 1545), Cardano announced the principles for solving cubic and quartic equations. His solutions (whereby roots were expressed in radicals, as for the quadratic equation, whose solution had been known since ancient times) were not completely original (§6.1); neither were they, in a modern sense, completely general, since he wrote and reasoned in prose and allowed only positive coefficients. Thus he divided the quadratic equation into five basic kinds, including 'the square is equal to the first power and constant' (in modern terms, $x^2 = ax + b$) and 'the number is equal to the square and first power' ($b = x^2 + ax$). For each of the five specific cases, he gave a specific algorithm. Similarly, he saw 13 kinds of cubic equation and solved each one separately. Although thus prolix and awkward from a modern perspective, Cardano's theory of equations dazzled his mathematical contemporaries, most of whom learned only from him of the possibility of solving cubic and quartic equations.

Besides extending the power of mathematics over hitherto unsolved equations, Cardano instigated a fundamental change in algebra by expanding the universe of objects accepted as legitimate algebraic entities. Through the fifteenth century, algebra was primarily the study of equations with whole-number or fractional solutions. But eventually mathematicians had found that some equations present solutions lying outside this restricted universe. In his manuscripts, the fifteenth-century Nicolas Chuquet, for example, had not only admitted negative roots but even employed negative coefficients (see G. Flegg in Hay *1988*: 63). In his *Ars magna*, Cardano publicly endorsed the negative numbers. He divided numbers into two kinds, positive and negative, and regularly gave the negative roots of equations. Perhaps pushed into consideration of imaginary numbers by the solution of the irreducible case of the cubic equation – which involves imaginary numbers that cancel out before the final solutions, which are real numbers – he very tentatively explored the consequences of working with imaginary roots (§6.1).

Even as he extended the algebraic universe, however, Cardano fell into a pattern of ambivalence towards the negative – and, more certainly, the imaginary – numbers, which was to afflict algebraists of the next three centuries. Like him, most mathematicians would admit the need of negative and (eventually) imaginary roots in general solutions of equations, and many would want to manipulate the imaginaries as if they were real numbers. Indeed, as early as 1572, Rafael Bombelli fully discussed the irreducible case of the cubic equation and, in the process, published rules of calculation for imaginary numbers. On the other hand, basically seeing algebra as a generalization of arithmetic, Cardano and his algebraic successors were hard put to explain the negative and, even more so, the imaginary numbers. As they reiterated, there really was no such number as $\sqrt{-1}$ since both positive and negative numbers, when squared, give positive numbers (§6.2). Cardano, then, referred to a positive root as a 'true solution' and a negative root as a 'fictitious one (for such we call that which is *debitum* or negative)'; when he discussed an equation involving complex roots, he called the imaginary part 'a sophistic negative' (Cardano *1968*: 10–11, 221).

3 VIÈTE'S ANALYTIC ART AND SYMBOLIC LOGISTIC

In many respects, with his *In artem analyticem isagoge* ('Introduction to the Analytic Art', 1591) and subsequent analytic pieces, Viète completed the early modern reshaping of algebra (§2.3). While Cardano excelled in new results and objects, Viète gave algebra a new language and method. The new algebraic symbolism came after centuries of sporadic experimentation with the language of algebra. The earliest algebra, which historians have described as 'rhetorical', was simply algebra expressed in complete prose statements. With the work of Diophantus of Alexandria in the third century AD, however, algebra had begun the passage from the rhetorical to the 'syncopated' stage, where abbreviations were used for common algebraic quantities and operations.

Although Cardano and other early modern algebraists interspersed their prose with algebraic abbreviations, Viète took a much bolder step in introducing 'symbolic' algebra, which employed symbols with little or no connection with the entities which they represented. Thus, in place of Diophantus's abbreviation for a single unknown (which came from the Greek word for 'number'), Viète used the vowels A, E, I, O and U for multiple unknowns. An innovation of sweeping and enduring significance, the 'symbolic logistic' (Viète's term) not only changed the language of algebra but also facilitated its generality, especially since, in addition to denoting unknowns by vowels, Viète used consonants (e.g. Z) to denote known quantities or parameters. He thus made it possible to solve algebraic problems in a general manner. Instead of solving specific problems, such as finding two numbers with a difference of 40 and a sum of 100, algebraists would henceforward solve general problems such as finding any two numbers with a difference of B and a sum of D.

In addition, Viète sketched a new approach to the justification of algebra. In place of algebra, he wrote of the 'analytic art', whose method he described as analysis and defined (in the traditional way) as 'assuming that which is sought as if it were admitted [and working] through the consequences [of that assumption] to what is admittedly true' (Viète *1983*: 11). He then moved beyond this classical definition of analysis, situating algebra not merely as a part of the 'analytic art' (which now subsumed arithmetic, geometry and trigonometry as well as algebra) but as the very core of it. (On the new relationship between algebra and analysis, see Mahoney *1973*:

33–4.) Viète thus offered mathematicians the rudiments of a new foundational approach to algebra, which suggested the possibility of an algebra on equal footing with, if not superior to, Euclidean geometry (which employed the method of synthesis).

Unlike Cardano, then, Viète did not generally rely on geometrical diagrams or proofs to confirm algebraic results. There was one respect, however, in which geometry formed a background influence. He distinguished the dimensions of both unknowns ('scalar quantities') and constants ('comparative quantities'), and in their names he took note of the three-dimensionality of space: respectively, 'latus' (A), 'quadratum' (A^2), 'cubus', 'quadrato-quadratum', ..., 'cubo-cubo-cubus' (A^9); and 'longitudo latitudove' (Z), 'planum' (Z^2), 'solidum', 'plano-planum', ..., 'solido-solido-solidum' (Z^9). Only (simple or compound) quantities of the same dimension could be added or subtracted; for example, $AAA - ZA$ was legitimate only if Z were planum. Multiplication and division respectively raised and lowered dimensions appropriately.

Viète's analytic approach was so influential that the terms 'algebra' and 'analysis' were used interchangeably into the nineteenth century. Indeed, by the second half of the eighteenth century the method of analysis came to enjoy a privileged status in Viète's own country, as the Abbé Condillac and other *philosophes* of the Enlightenment advocated analysis as the method of invention for all arts and sciences. Condillac himself 'set out to facilitate analytical thought in metaphysics by studying algebra' (Rider *1990*: 115, 117).

4 RECEPTION OF THE SYMBOLIC STYLE

The symbolic style caught on quickly. In England, it was enthusiastically embraced by William Oughtred in his *Clavis mathematicae* (1631) and by Thomas Harriot in his *Artis analyticae praxis* of the same year. Also adopting the style in *La Géométrie* of 1637 – a work which helped to introduce analytic geometry – René Descartes brought it to new heights of simplicity. The symbolic notation for powers of the unknown, in particular, underwent steady evolution from Viète through to Descartes. Using a sort of shorthand that evidenced vestiges of the traditional geometrical underpinnings of algebra, Viète had denoted the powers as A, A *quadratum* (alternatively, A *quad* or Aq), A *cubus*, and so on; Harriot had used the more abstract symbols a, aa, aaa, and so on. It was Descartes, however, who introduced the modern notation: x, x^2 (which he sometimes wrote as xx), x^3, and so on (§7.1).

Although the majority of mathematicians of the seventeenth and eighteenth centuries worried more about specific algebraic notation than about the legitimacy of the symbolic style, some major Western thinkers struggled to make philosophical sense of reasoning on arbitrary symbols. Indeed, in the 1650s Thomas Hobbes argued that, whereas algebraic symbols could provide a shorthand to record the mathematician's research, mathematical demonstration required translation from the symbols into the things they represented. Other philosophical thinkers, however, proved more receptive to the symbolic style and to algebra in general. At the beginning of the eighteenth century, the anti-abstractionist George Berkeley not only defended the early modern symbolic algebra but also held it up as a prime example of sound reasoning, devoid of consideration of particular or general ideas (Pycior *1987b*: 273, 277–83). Moreover, some of the *philosophes* extolled the symbolic style. Thus Condillac, who wanted to perfect language, turned to algebra for lessons about the influence of signs on reasoning (Rider *1990*: 120).

5 NEGATIVE AND IMAGINARY NUMBERS THROUGH THE EIGHTEENTH CENTURY

Legitimization of the negative and imaginary numbers proved more elusive in the seventeenth and eighteenth centuries than a coming to terms with the symbolic style. Like Cardano, Descartes recognized negative roots of equations, but even so referred to them as 'false' and 'less than nothing'. At the beginning of his *Arithmetica universalis* (1707), Isaac Newton described the negatives as 'quantities . . . less than nothing' and, in their defence, offered examples of quantity considered as affirmative and negative – including a possession as opposed to a debt, and a line drawn in one direction compared with a line drawn in the opposite direction. Although critics soon asked how a quantity could be less than nothing, in his *Vollständige Anleitung zur Algebra* (1770) Leonhard Euler essentially repeated Newton's definition and defence of the negatives.

Other mathematicians explored alternative approaches to the negatives. Colin Maclaurin, for example, in his posthumous *Treatise of algebra* (1748), defined the negatives essentially as quantities to be subtracted, and then gradually generalized the concept to include isolated negatives. (His initial definition covered $-b$ as a term in the expression $+a-b$, but his generalized version of a negative was that of $-b$, where $-b$ stood alone.) In the same work, Maclaurin also implied that the negatives might be justified by an appeal to the relation of contrariety or opposition. In his mid-eighteenth-century article on the negatives for the *Encyclopédie*, Jean d'Alembert admitted that there was no 'isolated negative quantity', but he nevertheless accepted negative numbers and roots, noting that a negative root exposed an error in the framing of an equation.

798

Complex numbers were even more problematic. In the seventeenth century and for most of the eighteenth, mathematicians could neither define such numbers nor explain away their impossibility nor represent them geometrically. (Although there was a long-standing tradition of geometrical representation of the negatives, geometrical representation of the complex numbers was a product of the very late eighteenth century and the early nineteenth; see §6.2.) Still, the general theory of equations that steadily evolved from Cardano's *Ars magna* demanded complex as well as negative roots. In *L'Invention nouvelle en l'algèbre* (1629), Albert Girard called complex roots *solutions impossibles*, but nevertheless formulated the fundamental theorem of algebra, which states roughly that a polynomial of degree *n* has *n* roots. Although given without proof, the theorem quickly became a mainstay of equation theory, and thus helped ensure the complex and negative numbers permanent places in the algebraic universe.

Mathematicians, however, recognized that the usefulness of the complex numbers was no complete argument for their legitimacy. Revealingly, Descartes called quantities of the form $\sqrt{-1}$ 'imaginary'; Newton wrote alternately of 'imaginary' and 'impossible' roots; Gottfried Wilhelm Leibniz saw imaginary numbers as amphibians between being and nonbeing. Over a half-century later, Euler tried to explain the appropriateness of calling such numbers 'imaginary' ('because they exist merely in the imagination'), and through the eighteenth century a host of other mathematicians attempted to explain and legitimize the complex numbers.

6 THE BRITISH PROBLEM OF THE NEGATIVE NUMBERS

Dissatisfied with all attempts to justify the negative and imaginary numbers, a few late-eighteenth-century British mathematical thinkers took the bold step of calling for their total rejection. Francis Maseres and William Frend, who had been educated at Cambridge and had come under the influence of John Locke's philosophy, demanded a traditional definition that addressed the essence of an isolated negative number. Finding no such definition, they rejected the negative (and hence imaginary) numbers. In his *Dissertation on the Use of the Negative Sign in Algebra* (1758) and a subsequent paper published in the *Philosophical Transactions* of 1778, for example, Maseres argued that mathematicians could form no idea of a negative (or imaginary) number; the subtraction of a greater quantity from a lesser (for example, $0 - a$) was impossible; and, consequently, the fundamental theorem of algebra was wrong.

For various reasons, including the British emphasis on the pedagogical value of mathematics as a deductive science based on clear ideas and self-evident axioms, as well as the related British concern for mathematical foundations (as witnessed by the history of the calculus), the theme of Maseres and Frend struck a responsive chord in the British mathematical community of the late eighteenth and early nineteenth centuries. Articles on the negative and imaginary numbers, which appeared in many of the major British scientific journals of the period, addressed the larger issues of general terms and sound reasoning as well as more narrowly mathematical concerns. Thus, in an article in the *Philosophical Transactions* of 1778, the Scotsman John Playfair focused on what he saw as a 'paradox': that reasoning on unintelligible symbols (the imaginaries) was useful. What did this say about reasoning and truth in general, he asked? In an article in the *Philosophical Transactions* of 1801 and a private letter of the same year, Robert Woodhouse, the Lucasian Professor at Cambridge, voiced both philosophical and pedagogical concerns about the negatives, while arguing that the useful negative and imaginary numbers ought to remain within the algebraic universe (Nagel *1935*; Pycior *1981*: 27–31; Pycior *1987a*: 153–6).

7 BRITISH SYMBOLICAL ALGEBRA

In Britain, the problem of the negative numbers mushroomed into the problem of the definition of algebra, and led ultimately to a new approach to the subject, called 'symbolical algebra'. Symbolical algebra was the creation of a handful of English mathematicians, including Woodhouse, Charles Babbage, John Herschel and George Peacock (Becher *1980*; Dubbey *1978*: 93–130). These mathematicians may have been led towards the new algebra by their work on the calculus of operations – a type of general calculus in which the symbols of operation were manipulated separately from the symbols operated on (Koppelman *1971*; see also §4.7, §5.1). There were also philosophical influences behind the creation of symbolical algebra. Woodhouse and Babbage, for example, had been exposed to nominalism in the writings of Berkeley and Dugald Stewart (Pycior *1984*); Peacock's algebraic system may have been shaped by Locke's views on the formation and structure of language (Durand *1990*).

Although the origins of symbolical algebra thus remain a fertile topic of ongoing research, it is generally agreed that the subject found full formulation only in Peacock's *Treatise of Algebra* (1830). Here Peacock divided algebra into two types: arithmetical algebra, or universal arithmetic in the strictest sense; and symbolical algebra, or 'the science which treats of the combinations of arbitrary signs and symbols by means of defined though arbitrary laws' (Peacock *1830*: 71). According to Peacock, the signs and symbols of symbolical algebra stood for no particular quantities or operations on quantities, although they could eventually take on many different interpretations. That interpretation would follow, not precede, manipulation was his algebraic maxim. Thus, although the negative numbers made no sense in arithmetical algebra (where subtraction was restricted to the taking of a lesser from a greater), the negatives and imaginaries entered the new symbolical algebra as (legitimate) uninterpreted symbols that could be manipulated according to defined laws. In short, Peacock and the other symbolical algebraists took algebra to a more abstract stage, in which laws of combination would be more important than meaning or traditional definition.

In his phrase 'defined though arbitrary laws', Peacock was recognizing the principle of algebraic freedom – that the algebraist somewhat arbitrarily assigns the laws of combination. 'We', he stated, 'may *assume* any laws for the combination and incorporation of such symbols, so long as our assumptions are independent, and therefore not inconsistent with each other.' Peacock was, however, bolder in theory than in practice, since he actually adopted the laws of arithmetic as the laws of symbolical algebra. To justify this action he formulated 'the principle of the permanence of equivalent forms' (Peacock *1830*: 71, 104–5; Pycior *1981*: 36–40).

8 THE QUATERNIONS AND MATHEMATICAL FREEDOM

Whereas Peacock stopped short of exercising algebraic freedom, the Irish mathematician William Rowan Hamilton liberated algebra from its traditional (arithmetically based) laws with his discovery of the quaternions in 1843. Hamilton's path to the quaternions began with his work on the complex numbers, which was inspired at least in part by his study of the philosophy of Immanuel Kant. In the late eighteenth and early nineteenth centuries, Caspar Wessel, Carl Friedrich Gauss, Jean Argand, John Warren and C. V. Mourey had published works on the geometrical representation of the complex numbers (§6.2). In the eyes of some mathematicians, geometrical representation provided the long-sought legitimization of the complex numbers (Crowe *1985*: 1–16).

For his part, the young Hamilton found in neither the geometrical representation nor Peacock's symbolical approach adequate justification for the complex numbers. In a famous paper of 1837, published in the *Transactions of the Royal Irish Academy*, Hamilton claimed to be able to construct algebra – including the negative and complex numbers – from the (Kantian) intuition of pure time. In particular, he developed the complex numbers as pairs of real numbers, which were supposed to come from couples of moments of time. Significantly, Hamilton then reasoned that, following such a process, it would be possible also to construct number-triplets (which obeyed the same rules as the complex numbers) from triplets of moments of time (Crowe *1985*: 23–7; Hankins *1980*: 258–75).

Hamilton never constructed the desired triplets since such a field does not exist. Rather, his work on the triplets led to the quaternions (Crowe *1985*: 27–33; Hankins *1980*: 283–301; van der Waerden *1980*: 179–83), the discovery of which was a mathematical event comparable to that of non-Euclidean geometry. The quaternions are elements of the form $a + bi + cj + dk$, where a, b, c and d are real numbers and i, j and k are 'imaginaries' whose multiplication is non-commutative:

$$i^2 = j^2 = k^2 = -1, \quad ij = -ji = k, \quad jk = -kj = i \text{ and } ki = -ik = j. \quad (1)$$

Before Hamilton's discovery, the commutative property of multiplication ($a \times b = b \times a$) was among those mathematical laws believed to be inviolable. Thus the non-commutative multiplication of the quaternions was, from the traditional perspective, a violation of a sacrosanct rule of arithmetic and algebra and, from the modern perspective, the first clear-cut algebraic exercise of the freedom that was to become one of the major characteristics of the modern discipline of mathematics. Some of Hamilton's contemporaries praised him for his boldness, and John Graves, Augustus De Morgan, Arthur Cayley and, later, the American Benjamin Peirce followed him in inventing linear algebras in which at least one of the standard laws of arithmetic was abandoned (Pycior *1979*).

9 MODERN ABSTRACT ALGEBRA COMES OF AGE

Although two key ingredients of modern abstract algebra – its formalism and its freedom – were thus coming into place by the 1840s, it took perhaps a further half-century for abstract algebra to come of age. Of British mathematical thinkers, Frend, William Whewell, Philip Kelland and even, for a while, Hamilton and De Morgan expressed concerns about the meaninglessness and arbitrariness of symbolical algebra and challenged its pedagogical value. 'At first sight', De Morgan wrote, Peacock's *Treatise on Algebra* 'appeared to us something like symbols bewitched, and running about the world in search of a meaning.' In addition, Hamilton's quaternions raised the question of consistency. George Biddell Airy and other critics suggested that the non-commutative quaternions would lead to 'paradoxical' or 'false' results. Hamilton's defence was extensive application of the quaternions, in the course of which he argued that the quaternions were useful, and that their application led to no inconsistencies, and, where relevant, to the same results as did traditional methods (Pycior *1987a*: 158–63).

This early criticism of the symbolical approach and algebraic freedom notwithstanding, mathematicians gravitated more towards an abstract view of their subject as the nineteenth century progressed. A year after the discovery of the quaternions, Hermann Grassmann published his *Ausdehnungslehre* (§6.2), which developed an *n*-dimensional algebra that included two kinds of multiplication, one of which (the outer product) was non-commutative (Crowe *1985*: 54–96). Grassmann, moreover, described pure mathematics as 'the theory of forms' (Lewis *1977*: 122); and in *Theorie der complexen Zahlensysteme* (1867), a work that discussed Hamilton's as well as Grassmann's new algebras, Hermann Hankel elaborated the view that mathematics was 'purely intellectual, a pure theory of forms' rather than the study of numbers or quantities. Finally, about a half-century later, David Hilbert enunciated his formalist philosophy of mathematics, which took all mathematical symbols as meaningless (Kline *1972*: 1028–31, 1203–4). Unfortunately, despite similarities between British symbolical algebra of the early nineteenth century and Hilbert's formalism, there appears to have been no detailed study of causal links between the two (compare §5.5).

As views of mathematics emphasizing formalism and creativity were evolving, algebraists introduced such key notions as group, ring, ideal and field (§6.4). Up to the later decades of the nineteenth century, however, their work concentrated mainly on specific algebraic systems, such as Hamilton's quaternions and other linear algebras, which are (in abstract terms) rings. Then, around the turn of the century, modern abstract algebra came of age. Algebraists no longer focused on specific algebraic systems, but rather probed the very structures of groups, rings, ideals, fields and the like (Kline *1972*: 1136–57). For example, the study of quaternions and other linear algebras was subsumed under the theory of abstract rings. Thus, although inspired partly by the work of Hamilton and Grassmann, and also personally by Hilbert, Emmy Noether wrote in 1933 not of specific non-commutative rings (or algebras) but of non-commutative algebra in general. 'All relations between numbers, functions, and operations become clear, generalizable, and truly fruitful only when they are separated from their particular objects and reduced to general concepts', was Noether's algebraic maxim (recorded by B. L. van der Waerden in Brewer and Smith *1981*: 93) – a maxim that celebrates the culmination of an almost four-hundred-year quest for the legitimization of a truly general, symbolic algebra.

BIBLIOGRAPHY

Becher, H. W. *1980*, 'Woodhouse, Babbage, Peacock, and modern algebra', *Historia mathematica*, **7**, 389–400.

Brewer, J. W. and Smith, M. K. (eds) *1981*, *Emmy Noether: A Tribute to Her Life and Work*, New York: Marcel Dekker.

Cardano, G. *1968*, *The Great Art or the Rules of Algebra* (transl. and ed. T. R. Witmer), Cambridge, MA: MIT Press. [Original text of *Ars magna* published in 1545.]

Crowe, M. J. *1985*, *A History of Vector Analysis: The Evolution of the Idea of a Vectorial System*, New York: Dover. [Unabridged republication of 1967 edition, with a new preface.]

Dubbey, J. M. *1978*, *The Mathematical Work of Charles Babbage*, Cambridge: Cambridge University Press.

Durand, M.-J. *1990*, 'Genèse de l'algèbre symbolique en Angleterre: Une influence possible de J. Locke', *Revue d'Histoire des Sciences et de leurs Applications*, **43**, 129–80.

Hankins, T. L. *1980*, *Sir William Rowan Hamilton*, Baltimore, MD: Johns Hopkins University Press.

Hay, C. (ed.) *1988*, *Mathematics from Manuscript to Print: 1300–1600*, Oxford: Clarendon Press.

Klein, J. *1968*, *Greek Mathematical Thought and the Origin of Algebra* (transl. E. Brann), Cambridge, MA: MIT Press.

Kline, M. *1972*, *Mathematical Thought from Ancient to Modern Times*, New York: Oxford University Press.

Koppelman, E. *1971*, 'The calculus of operations and the rise of abstract algebra', *Archive for History of Exact Sciences*, **8**, 155–242.

Lewis, A. C. *1977*, 'H. Grassmann's 1844 *Ausdehnungslehre* and Schleiermacher's *Dialektik*', *Annals of Science*, **34**, 103–62.

Mahoney, M. S. *1973*, *The Mathematical Career of Pierre de Fermat (1601–1665)*, Princeton, NJ: Princeton University Press.

Nagel, E. *1935*, ' "Impossible numbers": A chapter in the history of modern logic', *Studies in the History of Ideas*, **3**, 429–74.

Nový, L. *1973*, *Origins of Modern Algebra* (transl. J. Tauer), Prague: Academia.

Peacock, G. *1830*, *A Treatise on Algebra*, Cambridge: Deighton.

Pycior, H. M. *1979*, 'Benjamin Peirce's *Linear Associative Algebra*', *Isis*, **70**, 537–51.

—— *1981*, 'George Peacock and the British origins of symbolical algebra', *Historia mathematica*, **8**, 23–45.

—— *1984*, 'Internalism, externalism, and beyond: 19th-century British algebra', *Historia mathematica*, **11**, 424–41.

—— *1987a*, 'British abstract algebra: Development and early reception', in I. Grattan-Guinness (ed.), *History in Mathematics Education*, Paris: Belin, 152–68.

—— *1987b*, 'Mathematics and philosophy: Wallis, Hobbes, Barrow, and Berkeley', *Journal of the History of Ideas*, **48**, 265–86.

Richards, J. L. *1980*, 'The art and the science of British algebra: A study in the perception of mathematical truth', *Historia mathematica*, 7, 343–65.

Rider, R. E. *1990*, 'Measure of ideas, rule of language: Mathematics and language in the 18th century', in T. Frängsmyr, J. L. Heilbron and R. E. Rider (eds), *The Quantifying Spirit in the 18th Century*, Berkeley, CA: University of California Press, 113–40.

van der Waerden, B. L. *1980*, *A History of Algebra: From al-Khwārizmī to Emmy Noether*, Berlin: Springer.

Viète, F. *1983*, *The Analytic Art* (transl. T. R. Witmer), Kent, OH: Kent State University Press. [Original texts published between 1591 and 1631.]

6.10

Number theory

GÜNTHER FREI

1 INTRODUCTION

Number theory (arithmetic) deals with properties of numbers. Its dominant theme throughout history has been the study of solutions of so-called Diophantine equations: polynomial equations $f(x, y, z, \ldots) = n$ with integral (or rational) coefficients and integral numbers n, for which one is seeking integral (or rational) solutions $x, y, z, \ldots$. They are named after Diophantus of Alexandria (*circa* AD 250) who in his book *Arithmetica* treated various Diophantine equations of the first, second, third and even the fourth degree with two, three and sometimes more unknowns.

A typical example, probably motivated by Pythagoras's theorem, is:

THEOREM 1 *An integer n of the form $n = 4m + 3$ cannot be the sum of two squares of integers x and y, $n \neq x^2 + y^2$.*

There is no proof of this theorem in Diophantus's book, but a proof can easily be supplied by the theory of even and odd numbers (congruences modulo 4) developed by the school of Phythagoras in the sixth century BC.

The arithmetic of the Pythagoreans, dealing with for example the figurative and perfect numbers, is also present in the 13 books of Euclid's *Elements* (*circa* 300 BC). Books VII, VIII and IX of the *Elements* deal with the theory of numbers. Euclid starts with the so-called 'Euclidean algorithm' for determining the greatest common divisor of two integers, and then goes on to prove in two distinct steps the 'fundamental theorem of arithmetic':

THEOREM 2 *Every integer can be written in a unique way as a product of prime numbers (up to the sign and the order of the prime factors).*

In Book IX (Proposition 20), Euclid gives the well-known proof of:

THEOREM 3 *There are infinitely many prime numbers.*

The Euclidean algorithm is the basis for the theory of divisibility. It gives the means for the complete solution of Diophantine equations of the simplest kind, namely those of first degree,

$$ax + by + cz + \cdots = n, \tag{1}$$

where $a, b, c, \ldots, n$ are integers, as was shown by the Indians Brahmagupta (seventh century AD) and Bhāskara II (twelfth century).

Properties of irrational numbers were also studied by the Pythagoreans, who discovered them in connection with geometrical problems.

THEOREM 4 *The ratio of the diagonal to the side of a square, i.e. $\sqrt{2}$, is irrational (incommensurable),*

is stated in Book X (Proposition 117) of Euclid's *Elements* and is proved solely by means of the Pythagorean theory of even and odd numbers (§1.3).

Irrational numbers fall under two heads: the algebraic and transcendental numbers. A real or complex number α is called an 'algebraic number' if it is the root of an integral algebraic equation of the form

$$a_n x^n + a_{n-1} x^{n-1} + \cdots + a_1 x + a_0 = 0, \tag{2}$$

where the coefficients $a_n, a_{n-1}, \ldots, a_1, a_0$ are integers. Otherwise, α is called a 'transcendental number'.

The arithmetical study of algebraic numbers grew out of the study of binary quadratic forms and the generalization of the quadratic reciprocity law to higher reciprocity laws. It was initiated by Carl Friedrich Gauss at the beginning of the nineteenth century and systematically developed by Ernst Kummer, Leopold Kronecker and Richard Dedekind in the second half of the nineteenth century.

The study of transcendental numbers was begun by Joseph Liouville in the middle of the nineteenth century, but except for some noteworthy isolated results by Charles Hermite and Ferdinand Lindemann, a systematic study of transcendental numbers was launched only in the second quarter of the twentieth century.

Analytic number theory was created in the first half of the eighteenth century, when Leonhard Euler applied analytic methods (i.e. methods from differential and integral calculus) to number theory. Today, the most profound theorems in number theory are theorems linking algebraic with analytic, geometric and algebraic–geometric number theory.

2 ELEMENTARY NUMBER THEORY

2.1 Fermat

Modern number theory begins with Claude Bachet's edition of Diophantus's *Arithmetica* in 1621. This edition was carefully studied around 1638 by Pierre de Fermat, the founder of modern number theory. By reading this book Fermat discovered a wealth of new arithmetical theorems, mostly related to Diophantine problems of the second and third degrees, by taking into account not only divisibility properties but also the much finer properties of the remainders after division by a fixed (positive) integer.

An important example (of degree 2), where the geometrical background (Pythagoras's theorem) is still evident, is:

THEOREM 5 *A prime number p different from 2 is the sum of two squares, $p = x^2 + y^2$ for integers x and y, if and only if p is of the form $p = 4m + 1$.*

This observation gave rise to similar descriptions in terms of remainders with respect to the integer $4N$ for primes p which are representable as

$$p = x^2 + Ny^2, \quad N = -2, -1, 2, 3, 5. \tag{3}$$

In connection with the study of prime numbers p of the form $p = 2^n - 1$, which appear in relation to perfect numbers in Euclid's *Elements* (and probably go back to the Pythagoreans), Fermat discovered a property which is fundamental for the theory of Diophantine equations of higher degree, namely Fermat's theorem:

THEOREM 6 *For any prime number p and any integer 'a', a^p and 'a' leave the same remainder after division by p; in symbols:*

$$a^p \equiv a \ (\mathrm{mod}\, p), \tag{4}$$

or equivalently, if a is not divisible by p,

$$a^{p-1} \equiv 1 \ (\mathrm{mod}\, p). \tag{5}$$

In a letter to P. de Carcavi, Fermat gave some indications of how to prove his observation:

THEOREM 7 *The Diophantine equation of degree 4*

$$x^4 + y^4 = z^4 \tag{6}$$

has no solutions in integers x, y and z with $xyz \neq 0$.

The study of this Diophantine equation was certainly motivated by Euclid's theorem (*Elements*: Book X, Proposition 28, Lemma 1):

THEOREM 8 *The Diophantine equation*

$$x^2 + y^2 = z^2 \tag{7}$$

has infinitely many (primitive) integral solutions x, y, z, and these solutions can all be represented by two parameters, u and v.

This theorem was already known to the Babylonians in the seventeenth century BC (§1.1).

In view of this observation (Theorem 7), Fermat conjectured that, in general:

FERMAT'S CONJECTURE   *The Diophantine equation*

$$x^n + y^n = z^n \tag{8}$$

does not have any integral solutions x, y, z with xyz ≠ 0, if n is greater than 2.

This conjecture, also known as Fermat's last theorem, has still not been proved completely.

2.2 Euler

Fermat gave hardly any proofs of his discoveries. They were mostly provided by Euler, who also enriched the theory by a great many discoveries of his own. One of his most important findings (1772) is the 'quadratic reciprocity law' (this term was coined later by Adrien Marie Legendre in 1798), which must be considered as the fundamental theorem for Diophantine equations of second degree. It is, in the form given by Euler:

THEOREM 9   *Let p, q be two odd distinct prime numbers. If p is of the form 4m + 1, then $x^2 - qy^2 = ps$ is solvable with non-zero integers x, y, s if and only if $x^2 - py^2 = qt$ is solvable with non-zero integers x, y, t. If p is of the form 4m + 3, then $x^2 - qy^2 = ps$ is solvable with non-zero integers x, y, s if and only if $x^2 + py^2 = qt$ is solvable with non-zero integers x, y, t.*

The first complete proof of this theorem was given by Gauss in 1801.

A particular but important Diophantine equation of second degree is what Euler, for some unknown reason, called 'Pell's equation', although John Pell had nothing to do with it. It appears in fact in Diophantus's *Arithmetica* (Book V, Problems 9 and 11). Properties of this equation,

$$x^2 - Ny^2 = 1, \tag{9}$$

where N is a positive integer, were discovered before Euler by Brahmagupta and Bhāskara II (§1.12) and after Euler by Lagrange. The principal discovery made by Euler is the connection of this equation with the algorithm of continued fractions, which is a generalization of Euclid's division algorithm (§6.3). The main properties of Pell's equation can be stated as follows:

THEOREM 10   *Pell's equation*

$$x^2 - Ny^2 = 1 \tag{10}$$

has infinitely many integral solutions (x, y). These can be determined by the algorithm of continued fractions, and they can all be obtained from a 'smallest' solution by a composition law.

In modern terms, the integral points on the hyperbola $x^2 - Ny^2 = 1$ form a (cyclic) group generated by one particular so-called fundamental solution (x, y). The (rational) composition law was already found by Brahmagupta and Bhāskara and corresponds to the product law for numbers in the algebraic quadratic number field $\mathbb{Q}(\sqrt{N})$.

As to higher Diophantine equations, in connection with the study of prime numbers p of the form

$$x^2 + Ny^2 = pz \tag{11}$$

for integers x, y, z and a given integer N, Euler discovered an important generalization of Fermat's theorem, from a prime number p to any positive integer n, called Euler's theorem:

THEOREM 11   *Any integer 'a' relatively prime to a given positive integer n has the property that*

$$a^{\phi(n)} \equiv 1 \pmod{n}, \tag{12}$$

where $\phi(n)$ is the so-called Euler function, which counts the number of integers between 0 and n, relatively prime to n.

This result was obtained by Euler by a careful study of the structure of the remainders with respect to a given positive integer n. From this he deduced the following fact (expressed in modern terms), which is fundamental to all higher Diophantine equations:

THEOREM 12   *The remainders with respect to a prime number p form a field.*

This means that the four elementary operations can be performed without restriction (except for division by zero). Hence the remainders (or more precisely the remainder classes with respect to p) have the same algebraic properties as have the rational numbers. In addition, the $p - 1$ non-zero remainders form a cyclic group under multiplication generated by a so-called primitive root modulo p. From this last discovery Euler could easily derive 'Wilson's theorem':

THEOREM 13 *Any odd prime number* p *has the property that*

$$1 \cdot 2 \cdot 3 \cdot \ \cdots \ \cdot (p-1) \equiv -1 \ (\mathrm{mod}\, p). \tag{13}$$

Following up his investigations on extending Fermat's theorem, Euler discovered a criterion which is basic for all Diophantine equations of degree two and higher, now called Euler's power-residue criterion:

THEOREM 14 *If* p *is a prime number of the form* $p = mn + 1$, *and* 'a' *is any integer coprime to* p *and* n *a positive exponent, then* $a \equiv x^n$ (mod p) *is solvable with an integer* x *if and only if* $a^m \equiv 1$ (mod p);

or equivalently, that $a - x^n = py$ is solvable with integers x and y if and only if $a^m - 1 = pz$ is solvable for an integer z.

Another contribution of Euler, preparing the way for a study of the arithmetic of algebraic numbers, is the proof of Fermat's conjecture for the exponent $n = 3$:

THEOREM 15 *The Diophantine equation*

$$x^3 + y^3 = z^3 \tag{14}$$

has no integral solutions x, y, z *with* $xyz \neq 0$.

Euler's proof makes implicit use of the arithmetic of the cyclotomic field of the cube roots of unity, $\mathbb{Q}(\sqrt{-3})$. Other important discoveries by Euler in number theory concern the theory of partitions, magic squares, Fermat, Euler and Bernoulli numbers, and the ζ-function (Section 4.1).

2.3 Legendre and Lagrange

Euler's investigations were pursued by Joseph Louis Lagrange, partly aided by Euler. An important contribution by Lagrange (1770) is the proof of an old conjecture going back to Diophantus, the four-square theorem:

THEOREM 16 *Any positive integer n is the sum of at most four squares (of integers):*

$$n = x^2 + y^2 + z^2 + t^2. \tag{15}$$

Euler had succeeded in proving this theorem only for rational integers x, y, z, t.

Most important is Lagrange's theory of the general Diophantine equation of degree two,

$$f(x, y) = ax^2 + bxy + cy^2 = n, \tag{16}$$

with given integers a, b, c, n, which builds on Euler's study of the more special equation $ax^2 + cy^2 = n$. Lagrange's fundamental discovery is that an integral linear transformation of the variables x and y,

$$x = \alpha x' + \beta y', \qquad y = \gamma x' + \delta y' \qquad (17)$$

of the equation

$$f(x, y) = ax^2 + bxy + cy^2 = n, \qquad (18)$$

generates a new equation

$$f'(x', y') = a'x'^2 + b'x'y' + c'y'^2 = n \qquad (19)$$

of the same type, with integers a', b', c'. Hence one can at the same time obtain information on solutions for a whole class of Diophantine equations of the same type. This gave rise to a systematic study of the so-called binary quadratic forms $ax^2 + bxy + cy^2$. The important invariants of these forms are the discriminant d, defined by

$$d = b^2 - 4ac, \qquad (20)$$

and the class of quadratic forms equivalent under these invertible integral linear transformations given by $(\alpha, \beta, \gamma, \delta)$ with $\alpha\gamma - \beta\delta = \pm 1$. A fundamental property in Lagrange's theory is:

THEOREM 17 *All binary quadratic forms in the same class have the same discriminant, and to any given integer d there are only finitely many classes having this integer d as their discriminant.*

A more precise form of the four-square theorem was obtained by Legendre, who in 1785 showed that:

THEOREM 18 *A positive integer n is the sum of at most three squares,*

$$n = x^2 + y^2 + z^2, \qquad (21)$$

with integers x, y, z, if and only if $n \neq 4^r(8m + 7)$ for positive integers r and m.

Later, in 1823, Legendre was also able to prove Fermat's conjecture for the exponent $n = 5$:

THEOREM 19 *The Diophantine equation*

$$x^5 + y^5 = z^5 \qquad (22)$$

has no integral solutions x, y, z with $xyz \neq 0$.

We owe to Legendre the first book (Legendre *1798*) entirely dedicated to number theory. The status of number theory at that time was still that of an experimental science. It could not compete with the well-established theories of geometry or differential and integral calculus, and hence did not attract much attention, in spite of the fact that eminent mathematicians such as Fermat, Euler and Lagrange had devoted a considerable amount of their research to arithmetical problems. This only changed with the appearance of Gauss's fundamental treatise *Disquisitiones arithmeticae* (*1801*), which organized number theory in the axiomatic framework which Euclid had given to geometry.

In his book, Legendre made a first attempt to prove what he called the quadratic reciprocity law. However, he succeeded fully in only two out of eight cases. In the other cases he made use of the conjecture, proved by J. P. G. Lejeune Dirichlet *1837*, that every arithmetic progression contains infinitely many prime numbers.

2.4 Gauss

Gauss gave the first complete proof of the quadratic reciprocity law in his *Disquisitiones arithmeticae*, which was published when he was only 24 years old. In this book Gauss started out by giving a precise treatment of the structure and properties of remainders with respect to a positive number m by introducing the notion of congruence, denoted by $\equiv$, of two integers a and b with respect to m. He defined:

$$a \equiv b \pmod{m} \quad \text{if and only if } m \text{ divides } b - a. \tag{23}$$

Then he went on to show that the remainders, together with the relation $\equiv$, satisfy essentially the same properties with respect to addition and multiplication as does the relation $=$ for the integers (i.e. the remainders); or more precisely, the equivalence classes with respect to the relation $\equiv$ form a (commutative) ring (with unity). This result had essentially already been obtained by Euler in his 'Tractatus de numerorum doctrina' (1750; published 1849), but in a less concise form.

Gauss gave two entirely different proofs of the quadratic reciprocity law in his *Disquisitiones*. The second is based on his theory of the genus of binary quadratic forms, which must be viewed as a first example of class-field theory. The 'genus' of a binary quadratic form

$$f(x, y) = ax^2 + bxy + cy^2 \tag{24}$$

contains whole classes of forms, and it has the property that if

$$f(x, y) = ax^2 + bxy + cy^2 \quad \text{and} \quad f'(x', y') = a'x'^2 + b'x'y' + c'y'^2 \tag{25}$$

are two forms with the same discriminant

$$d = b^2 - 4ac = b'^2 - 4a'c', \tag{26}$$

such that

$$f(x,y) = n \quad \text{and} \quad f'(x',y') = n \tag{27}$$

have both solutions in integers x, y, x', y' for a given integer n, then they belong to the same genus.

In the last chapter of his treatise, Gauss developed systematically the algebraic properties; that is, the Galois theory of the so-called 'pth cyclotomic field' generated by the pth roots of unity, where p is a prime number. These are the roots of the algebraic equation $x^p = 1$.

Euler had already shown in 1772 that these roots are complex numbers, and that they can be represented by the exponential function; that is, if ζ is such a root of $x^p = 1$, then

$$\zeta = \exp(2\pi i r/p), \quad r = 1, 2, \ldots, p, \tag{28}$$

and it was this insight that allowed him to deduce that the $p - 1$ congruence classes modulo p relatively prime to p form a cyclic group under multiplication. These two properties were applied by Gauss to form certain Lagrange resolvents for these pth roots of unity, today called 'Gaussian sums', and to solve the equation $x^p = 1$ by radicals. In modern parlance, Gauss determined all subfields of the cyclotomic field $\mathbb{Q}(\zeta)$, where $\zeta = \exp(2\pi i/p)$.

3 ALGEBRAIC NUMBER THEORY

3.1 Gauss

In seeking a generalization of the quadratic reciprocity law for higher Diophantine equations, Gauss found six more proofs of this fundamental law. He was the first to state the biquadratic (1832) and to hint at the cubic reciprocity law, which represent the fundamental theorems for Diophantine equations of degrees four and three, respectively. He realized that these higher reciprocity laws (of degree n) can be formulated in full generality only within the ring of integers of the cyclotomic field (generated by the nth roots of unity).

In studying the biquadratic reciprocity law, Gauss developed (in 1832) the arithmetic of the so-called 'Gaussian integers',

$$\mathbb{Z}[i] = \{a + ib \mid a \text{ and } b \text{ integers}\}. \tag{29}$$

He examined the properties of divisibility and congruence, the notion of a

Gaussian prime and the decomposition of an ordinary prime number into Gaussian primes, Fermat's theorem (Theorem 6), the units and the fundamental theorem of arithmetic. In this way Gauss opened up the arithmetical study of algebraic numbers, defined in equation (2) as the roots of an integral algebraic equation

$$a_n x^n + a_{n-1} x^{n-1} + \cdots + a_1 x + a_0 = 0. \tag{30}$$

An algebraic number α is called an algebraic integer if $a_n = \pm 1$, and an algebraic unit if in addition $a_0 = \pm 1$.

3.2 Kummer

The proofs for the cubic and biquadratic reciprocity laws were found independently by Carl Jacobi (1837) and Ferdinand Gotthöld Eisenstein (1844). They both also made some contributions to the reciprocity laws for certain higher degrees. A systematic study of the pth reciprocity law for a prime number p was undertaken by Ernst Kummer. Following in the footsteps of Gauss, Kummer had first to develop the arithmetic of the algebraic integers of the pth cyclotomic field. He noticed that the fundamental theorem of arithmetic does not hold in general for these numbers. In order to re-enforce the fundamental theorem (i.e. the property of unique decomposition of cyclotomic integers into cyclotomic prime numbers), he introduced the notion of an 'ideal prime number' (1844), which is defined by means of Gauss's theory of the cyclotomic field and its subfields and by complicated congruence conditions. Once the fundamental theorem in terms of ideal numbers was established, Kummer was able to prove the pth-degree reciprocity law within the ring of integers of the pth cyclotomic field and, as a further consequence, Fermat's conjecture for certain prime exponents, called 'regular primes':

THEOREM 20 *If p is a regular prime number greater than* 2, *then the Diophantine equation*

$$x^p + y^p = z^p \tag{31}$$

has no integral solutions x, y, z *with* $xyz \neq 0$.

A prime number p is called a regular prime number if the number of ideal classes in the pth cylotomic field is not divisible by p. The number of ideal classes measures in some sense the proportion of the ideal numbers one has to introduce in order to get the fundamental theorem of arithmetic in the pth cyclotomic field with respect to the ordinary algebraic integers in that field. By analytic means, Kummer (and before him Dirichlet) was able to give an explicit formula for the number of ideal classes in the pth cyclotomic field. This number is always finite.

3.3 Dedekind and Kronecker: ideal theory, number fields and function fields

The complicated notion of an ideal number was made precise by Dedekind *1871*. He replaced the notion of an ideal number by the concept of an ideal. An 'ideal' J in the ring of integers R of a field of algebraic numbers F is a set of algebraic integers in that ring satisfying two conditions: (a) the sum of two algebraic numbers in J is again in J; and (b) the product of an algebraic number in J with an algebraic integer in R is always in J. Dedekind showed that the notions of addition, multiplication, divisibility, greatest common divisor, and so on can be defined for ideals, and that the usual rules for these operations remain valid. In addition, he was able to give the precise structure of the (group of) units U in any field of algebraic numbers F, as well as a description of how an ordinary prime number p from the ring of ordinary integers $\mathbb{Z}$ decomposes into prime ideals in the ring of algebraic integers R of a field F of algebraic numbers (1878).

A different interpretation of Kummer's ideal numbers was given by Leopold Kronecker in 1882 in terms of what he called 'divisors'. He represented the ideal numbers by polynomials, and thus made algebraic number theory part of algebraic geometry. This point of view was also supported by Dedekind and Heinrich Weber's theory of algebraic function fields (Dedekind and Weber *1882*), which brought to the fore a striking analogy between the arithmetic of algebraic function fields (i.e. of algebraic extensions E of the field of rational functions $\mathbb{C}(x)$ in one variable x with complex coefficients) and the arithmetic of algebraic number fields (i.e. of algebraic extensions F of the field of rational numbers $\mathbb{Q}$). Primes in number fields F correspond to points on the Riemann surface of a function field E. The particular role played by the primes which divide the discriminant of an algebraic number field F had already been clarified by Dedekind himself, who also showed that they correspond to points on the Riemann surface of an algebraic function field E where the Riemann surface is ramified. For that reason these prime numbers are said to be ramified in the field F.

A detailed and unifying report on the arithmetic theory of algebraic numbers as developed by Kummer, Kronecker, Dedekind and David Hilbert himself was given by Hilbert in *1897*. The guiding idea of this report is again the analogy between function fields and number fields. It was Hilbert's objective to discover the notions for algebraic number fields that correspond to notions already known for function fields, such as the monodromy group, differentials, Abelian integrals, the residue theorem of Cauchy (§3.12) and the Riemann–Roch theorem (§4.6). His goal was to describe all invariants of a number field, whenever possible, by means of the Galois group and the discriminant (i.e. the ramified primes) alone. He was thus following Bernhard Riemann's idea of constructing the Riemann surface for an algebraic function field solely from a given monodromy group and a given finite set of ramified points (places).

3.4 Hensel and Hasse: p-adic numbers

It was also this analogy between number fields and function fields that led Kronecker's pupil Kurt Hensel to introduce in 1897 the 'p-adic numbers', in connection with the determination of the precise prime power which divides the discriminant of an algebraic number field. A p-adic number α is a formal power series in the prime number p:

$$\alpha = \sum_{n=-N}^{\infty} a_n p^n, \quad a_i \in \{0, 1, \ldots, p-1\}. \tag{32}$$

A p-adic number α is called a 'p-adic integer' if $N = 0$, and a 'p-adic unit' if in addition $a_0 \neq 0$.

The full power of the p-adic numbers came to the fore only in 1923, when Helmut Hasse formulated the so-called 'local–global principle', or 'Hasse principle', which states that a polynomial equation (with rational coefficients) $f(x, y, z, \ldots) = n$ has a rational solution in rational numbers $(x, y, z, \ldots)$ if and only if it has a solution $(x, y, z, \ldots)$ in p-adic numbers for each prime number p as well as a solution $(x, y, z, \ldots)$ in real numbers. This principle was first observed by Hasse for quadratic forms

$$\sum_{i,j=1}^{n} a_{ij} x_i x_j = a_{11} x_1^2 + a_{12} x_1 x_2 + \cdots + a_{nn} x_n^2 \tag{33}$$

over the rational numbers, but he discovered that it holds also for quadratic forms over algebraic number fields, and for certain forms arising as norm forms from division algebras. The principle is far from being true in general, but the deviation (obstruction) to the principle is an important notion which is linked to other objects in number theory.

Two concepts which connect the local objects (e.g. the p-adic numbers) with the global objects (e.g. the rational numbers) are the (multiplicative) *idèles*, introduced by Claude Chevalley in 1936, and the (additive) *adèles* (valuation vectors) introduced by Emil Artin and George Whaples in 1945. These are infinite vectors, where one component is a real number and the other components are p-adic numbers (almost always p-adic integers), one for each prime number p. The local–global problem, among many others, indicates that further progress in the theory of algebraic numbers is made possible only when analytic methods are brought to bear on the theory.

4 ANALYTIC NUMBER THEORY

4.1 Euler's ζ-function

It was Euler who in 1737 was the first to apply analysis to number theory, by introducing the ζ-function $\zeta(s)$, defined as an infinite sum (§3.12):

$$\zeta(s) = 1 + \frac{1}{2^s} + \frac{1}{3^s} + \frac{1}{4^s} + \cdots = \sum_{n=1}^{\infty} \frac{1}{n^s}, \quad s \in \mathbb{R}. \tag{34}$$

The fruitfulness of this function for number theory stems from the fact that $\zeta(s)$ can also be represented as an infinite product over all prime numbers, thanks to the fundamental theorem of arithmetic:

THEOREM 21

$$\zeta(s) = \sum_{n=1}^{\infty} \frac{1}{n^s} = \prod_p \frac{1}{1 - p^{-s}} \quad \text{(over all primes } p\text{)}, \quad s > 1. \tag{35}$$

From an old theorem of Nicole Oresme (*circa* 1360),

THEOREM 22 *The harmonic series*

$$1 + \frac{1}{2} + \frac{1}{3} + \frac{1}{4} + \cdots = \sum_{n=1}^{\infty} \frac{1}{n} = \zeta(1) \tag{36}$$

is divergent,

Euler obtained an entirely different proof of Euclid's Theorem 3, that there are infinitely many prime numbers.

4.2 Dirichlet's L-series

Euler's proof was generalized in 1837 by Dirichlet, who introduced for that purpose the so-called 'Dirichlet L-series' (modulo m) for each character χ modulo m (Dirichlet *1837*):

$$L(s, \chi) = \sum_{n=1}^{\infty} \frac{\chi(n)}{n^s}, \quad s \in \mathbb{R} \quad \text{and} \quad s > 1, \tag{37}$$

in order to show:

THEOREM 23 *Every arithmetic progression whose first term and common difference are relatively prime contains infinitely many prime numbers.*

818

In other words, if m is a positive integer and a is an integer which is relatively prime to m, $(a, m) = 1$, then there are infinitely many prime numbers p with $p \equiv a \pmod{m}$.

A 'Dirichlet character' χ modulo m is a function which associates to each integer a, relatively prime to m, an mth root of unity, such that

$$\chi(a + m) = \chi(a), \tag{38}$$

$$\chi(ab) = \chi(a)\chi(b), \quad \text{for all integers } b, \tag{39}$$

$$\chi(b) = 0, \quad \text{if } b \text{ is an integer not relatively prime to } m. \tag{40}$$

The critical step in Dirichlet's proof was to show that $L(1, \chi)$ is different from zero, if χ is not the so-called 'principal character' χ_0; χ_0 is defined as the character which associates with each integer a, relatively prime to m, the value 1. This step was obtained by linking the value $L(1, \chi)$ with the class number of a certain field, later to be called by Weber the 'class field' of χ (or of the corresponding congruence group modulo m). The theorem now follows from the fact that these Dirichlet L-series satisfy a product relation similar to the one in Theorem 21, and that

$$L(s, \chi_0) = c(m) \zeta(s) \tag{41}$$

(the constant $c(m)$ depending only on m) is divergent for $s = 1$ (Theorem 22).

4.3 The prime-number theorem

Another important arithmetical theorem which was proved by analytic means is 'the prime-number theorem':

THEOREM 24

$$\lim_{x \to \infty} \frac{\pi(x)}{x/\ln x} = 1, \tag{42}$$

where $\pi(x)$ is the number of prime numbers which do not exceed the real number x.

This theorem was first conjectured by Gauss around 1793, and then proved independently by Jacques Hadamard and Charles Jean De La Vallée-Poussin in 1896. It can be deduced from (and is even equivalent to) the fact that $\zeta(s)$ is different from zero for all complex numbers $s = 1 + ib$, where b is any real number. In fact, a famous conjecture of Riemann claims a much stronger property (Riemann *1859*; §3.12), namely:

RIEMANN'S CONJECTURE *If* $\zeta(s) = 0$ *for a non-real complex number* $s = a + ib$, *then* $a = \frac{1}{2}$. (43)

An elementary (i.e. non-analytic) proof of the prime-number theorem was found by Paul Erdös and by Atle Selberg in 1948.

4.4 Elliptic functions in number theory

Kronecker, and before him Niels Abel and Jacobi, successfully applied the theory of elliptic functions (§4.5) to number theory. This was motivated by a remark by Gauss in his *Disquisitiones arithmeticae* (*1801*) that his theory of the cyclotomic field (i.e. of the division of the circle, which can be parametrized by the trigonometric functions and is thus related to the integral $\int (1 - x^2)^{-1/2} dx$) has a counterpart in the theory of the division of the lemniscate, related to the integral

$$\int (1 - x^4)^{-1/2} dx.$$ (44)

In his book *Fundamenta nova* (*1829*), Jacobi was able to connect Gauss's theory of quadratic forms with the theory of elliptic functions and to make the four-square theorem (Theorem 16) of Lagrange more precise by giving an explicit formula for the number of representations of a positive integer n by a sum of four squares:

$$n = x^2 + y^2 + z^2 + t^2.$$ (45)

This was made possible by means of Jacobi's θ-functions, related to the elliptic functions. Jacobi also succeeded in giving such a formula for the number of representations of a given positive integer n by a sum of six squares, and for certain positive integers n also by a sum of eight squares.

Another connection between quadratic forms and elliptic functions was discovered by Kronecker in 1857, when he noticed that each class $\mathscr{C}_i$ out of the h classes $\mathscr{C}_1, \mathscr{C}_2, \ldots, \mathscr{C}_h$ of binary quadratic forms of negative discriminant $-d$ is characterized by a unique value $j(\mathscr{C}_i)$ of the (elliptic) modular function j, termed a 'singular modulus' by Kronecker. He then proved:

THEOREM 25 *The h numbers $j(\mathscr{C}_i)$ are algebraic integers and the roots of a (monic) polynomial $H(x)$ of degree h whose coefficients are integers in the quadratic number field $F = \mathbb{Q}(\sqrt{-d})$. $H(x)$ is irreducible over F; hence the algebraic extension $E = F(j(\mathscr{C}_i))$, obtained by adjoining the algebraic integer $j(\mathscr{C}_i)$ to the field F, is independent of the class. In* *addition, E is an Abelian extension over F whose Galois group is isomorphic to the group of equivalence classes of quadratic forms of discriminant − d or (by a theorem of Dedekind's) to the group of ideal classes in F. Furthermore, the relative discriminant of E over F is 1, i.e. E has no ramified prime ideals with respect to F.*

Because of this connection between the field E and the ideal classes in F, Weber called E a class field for F.

4.5 Class-field theory

Generalizations of Dirichlet's analytic method by Dedekind, who in 1871 introduced the ζ-function for any algebraic number field F; and by Weber, who in 1897 defined the corresponding L-series for algebraic number fields, led to the notion and the theory of class fields. In order to prove a theorem on primes in arithmetic progressions for algebraic number fields F analogous to Dirichlet's theorem, Weber needed the existence of an extension field E over F satisfying certain properties on how the prime ideals in F factor in E. Since these properties depend on certain congruence and ideal classes in F, in which the prime ideals are, Weber called these fields 'class fields'. He knew that, if F is the field of rational numbers $\mathbb{Q}$ or an imaginary quadratic number field $\mathbb{Q}(\sqrt{-d})$, then such class fields E exist, namely the cyclotomic fields and the fields constructed by Kronecker by means of the modular function.

The properties of class fields were systematically studied and developed by Weber himself, by Hilbert, Teiji Takagi, Philipp Furtwängler, Artin and Hasse (Hasse *1926*, *1927*). Two fundamental theorems of this theory are the theorem of Takagi (1922):

THEOREM 26 *Every Abelian extension E of an algebraic number field F is a class field for that field F (corresponding to a congruence class group in F), and, vice versa, every class field E of F is an Abelian extension of F;*

and the reciprocity law of Artin (1927):

THEOREM 27 *There is an explicit isomorphism between the (Abelian) Galois group of E over F and the congruence class group in F corresponding to the class field E over F, given by the so-called 'Artin reciprocity map'.*

This reciprocity law must be viewed as the most general known form of Euler's reciprocity law, and the theory of class fields should be seen as the general theory of Diophantine equations arising as norm forms of relatively Abelian extensions.

A generalization of Artin's reciprocity law to non-Abelian extensions E over F is postulated by Robert Langlands's programme (1970). The aim of this programme is to establish and clarify the links between the so-called Artin L-series of irreducible representations of (Galois) groups and automorphic functions and representations.

4.6 Elliptic curves

Another class of Diophantine equations that has been studied extensively and for which analytic methods play a crucial role consists of the 'elliptic curves' (Ireland and Rosen *1982*):

$$ay^2 = bx^3 + cx^2 + dx + e, \qquad (46)$$

where a, b, c, d, e are integers. Viewed as an equation over the complex numbers (i.e. with solutions (x, y) allowed to be complex numbers), these complex solutions represent a (complex) curve in the (4-dimensional) complex plane, which can be parametrized by elliptic functions. Hasse has shown that there is a ζ-function, called the Hasse ζ-function, attached to each elliptic curve (in fact to each non-singular, complete algebraic variety over a finite algebraic number field), which contains much information on the arithmetical properties of the curve.

Two fundamental results are the theorem put forward by C. L. Siegel (1929):

THEOREM 28   *The Diophantine equation* (46) *has only finitely many solutions* (x, y) *in integers* x *and* y,

and a conjecture by Henri Poincaré of 1901, proved by Paul Mordell in 1922 (and generalized to algebraic number fields by A. Weil in 1928):

THEOREM 29   *The Diophantine equation* (46) *has infinitely many solutions* (x, y) *in rational numbers* x *and* y. *These solutions form a (commutative) group under a certain composition, and are generated by a solution* (x_0, y_0) *of finite order f and by r linearly independent solutions* (x_1, y_1), (x_2, y_2), ..., (x_r, y_r) *of infinite order.*

The natural number r is called the rank of equation (46).

A theorem by B. Mazur of 1978 asserts that:

THEOREM 30   *The order f cannot be greater than* 16.

As to the rank, a conjecture by Birch and Swinnerton-Dyer (1963) claims that the rank r is equal to the order of the zero of the ζ-function of equation (46) at the value $s = 1$.

4.7 Higher arithmetic curves; arithmetic algebraic geometry

Some theorems given in the preceding sections can be generalized to higher arithmetic curves or even to algebraic varieties, and in many cases also from the field of rational numbers to fields of algebraic numbers. One example is Siegel's theorem (Theorem 28), which in a more general (but still not the most general) form reads:

THEOREM 31   *Any curve of genus greater than or equal to 1 has at most a finite number of integral points.*

Generalizations of the arithmetical theory of elliptic curves (i.e. curves of genus 1) to curves of higher genus, to Abelian varieties and to still broader classes of Diophantine equations (i.e. algebraic varieties) are due to A. Weil, A. Grothendieck and P. Deligne. Many of the important results on these objects stem from the study of the corresponding ζ-function, which can be defined, following Hasse, for any (non-singular complete) algebraic variety over a (finite-dimensional) field of algebraic numbers.

One of the most remarkable results (Ireland and Rosen *1982*) on such higher arithmetic curves is the proof of the Conjecture of Mordell (1922) by Gerd Faltings in 1983:

THEOREM 32   *An algebraic curve (i.e. an algebraic variety of dimension 1) of genus greater than 1 has only finitely many rational points.*

In particular, a plane algebraic curve $f(x, y) = m$, with integer coefficients and of genus greater than 1, has only finitely many rational solutions (x, y). This implies that the Fermat equation

$$x^n + y^n = z^n, \tag{47}$$

which can be written as

$$f(x, y) = \left(\frac{x}{z}\right)^n + \left(\frac{y}{z}\right)^n = 1, \tag{48}$$

has only finitely many integer solutions (x, y, z) if n is greater than 2. Then in June 1993 A. Wiles announced a proof of a conjecture about elliptic curves (46) from which Fermat's last theorem follows.

4.8 Transcendental number theory

A complex number which is not an algebraic number is called a 'transcendental number' (Euler, 1744). In spite of a transcendence criterion given in

1844 by Liouville, no transcendental number was known before 1873, when Hermite proved:

THEOREM 33   *The number*

$$e = \lim_{n \to \infty} \left(1 + \frac{1}{n}\right)^n \tag{49}$$

is transcendental.

The following year Georg Cantor was able to give a procedure to show that the transcendental numbers are 'much more abundant' than the algebraic numbers (§3.6). In 1882 Lindemann applied Hermite's method to prove:

THEOREM 34   *If $\alpha_1, \alpha_2, \ldots, \alpha_r$ are distinct algebraic numbers, and $\beta_1, \beta_2, \ldots, \beta_r$ are non-zero algebraic numbers, then*

$$\beta_1 e^{\alpha_1} + \beta_2 e^{\alpha_2} + \cdots + \beta_r e^{\alpha_r} \neq 0. \tag{50}$$

From this criteria Lindemann could easily deduce that:

THEOREM 35   *The number π is transcendental.*

This last result implies the solution of a long-standing classical geometrical problem going back to the Greeks (§1.3):

THEOREM 36   *The quadrature of the circle is impossible with ruler and compass.*

A far-reaching generalization of Lindemann's theorem was conjectured by A. O. Gel'fand in 1929 and proved by A. Baker in 1966:

THEOREM 37   *If $\alpha_1, \alpha_2, \ldots, \alpha_r, \beta_1, \beta_2, \ldots, \beta_r$ are non-zero algebraic numbers, such that $\ln \alpha_1, \ln \alpha_2, \ldots, \ln \alpha_r$ are linearly independent over the rational numbers, then*

$$\beta_1 \ln \alpha_1 + \beta_2 \ln \alpha_2 + \cdots + \beta_r \ln \alpha_r \neq 0. \tag{51}$$

As a special case one obtains the solution of one of Hilbert's famous problems (posed by Hilbert in 1900 in an address given to the International Congress of Mathematicians in Paris), found independently by Gel'fand and T. Schneider in 1934:

THEOREM 38   *If α is an algebraic number different from 0 and 1, and β is an irrational algebraic number, then α^β is a transcendental number.*

4.9 Additive number theory

Additive number theory studies particular types of Diophantine equation, namely ones in which a given integer n is represented as a sum of integer numbers of a given kind, such as prime numbers or squares of integers. Classical examples of this sort are the theorems of Fermat and Euler (Theorem 5), Lagrange (Theorem 16) and Legendre (Theorem 18) on the sum of two, four and three squares.

A famous problem raised by Christian Goldbach in a letter to Euler in 1742 is still unsolved:

GOLDBACH'S CONJECTURE   *Every even positive integer n greater than 2 is the sum of two prime numbers,*

$$n = 2m = p_1 + p_2, \tag{52}$$

and every odd number n greater than 1 is the sum of at most three prime numbers.

Another conjecture, however, inspired by Lagrange's four-square theorem and stated in 1770 by Edward Waring on the sum of kth powers, was solved by Hilbert in 1909:

THEOREM 39   *For every positive power k there is a (smallest) positive number $r = r(k)$ (depending on k), such that every positive integer n is the sum of at most r k-th powers.*

The smallest such number for $k = 2$ is $r(2) = 4$, by the theorems of Lagrange and Legendre on sums of four and three squares. Waring conjectured that $r(3) = 9$ and $r(4) = 19$. This last conjecture was proved in 1986 by Deshouillers, Dress and Balasubramanian. Systematic methods for additive problems were introduced only after Hilbert, by Srinivasa Ramanujan (1918), G. H. Hardy and J. E. Littlewood (1920–28) and Ivan Vinogradov (1924) (circle methods); and by V. Brun (1920), A. Selberg (1947) and E. Bombieri (1969) (sieve methods).

5 CONCLUDING REMARKS

Thanks to its power and breadth, number theory has today become an intensive field of research that embraces almost all fields of mathematics. Very recent fields of research on which number theory has had a great impact are coding theory (used for instance for the transmission of information by satellites), cryptography and computer sciences. The availability of powerful computers has given a new impetus to the computational aspects of number theory.

ACKNOWLEDGEMENTS

This article is dedicated to the memory of my dear friend Hans-Joachim Stender. I would like to thank Professor Peter Hilton for a linguistic improvement of the text and for several useful suggestions.

BIBLIOGRAPHY

The secondary literature is confined to the principal general sources. Many further references are given there.

Dedekind, R. and Weber, H. *1882*, 'Theorie der algebraischen Funktionen einer Veränderlichen', *Journal für die reine und angewandte Mathematik*, **92**, 181–290. [Also in R. Dedekind, *Gesammelte mathematische Werke*, Vol. 1, 238–350.]

Dickson, L. E. *1919–23*, *History of the Theory of Numbers*, 3 vols, New York: Carnegie Institution. [Repr. 1934, New York: Stechert; 1952, New York: Chelsea.]

Dirichlet, J. P. G. *1837*, 'Beweis des Satzes, dass jede unbegrenzte arithmetische Progression, deren erstes Glied und Differenz ganze Zahlen ohne gemeinschaftlichen Factor sind, unendlich viele Primzahlen enthält', *Abhandlungen der Preussischen Akademie der Wissenschaften*, 45–81. [Also in *Mathematische Werke*, Vol. 1, 313–42.]

Dirichlet, J. P. G. and Dedekind, R. *1871*, *Vorlesungen über Zahlentheorie*, Braunschweig: Vieweg. [1st edn 1863, 2nd edn 1871, 3rd edn 1879, 4th edn 1894.]

Ellison, W. and Ellison, F. *1978*, 'Théorie des nombres', in J. Dieudonné (ed.), *Abrégé d'histoire des mathématiques 1700–1900*, Vol. 1, Paris: Hermann, 165–334. [The most complete historical exposition of number theory.]

Gauss, C. F. *1801*, *Disquisitiones arithmeticae*, Leipzig: Fleischer. [Also *Werke*, Vol. 1.]

Hasse, H. *1926–7*, 'Bericht über neuere Untersuchungen und Probleme aus der Theorie der algebraischen Zahlkörper', *Jahresbericht der Deutschen Mathematiker-Vereinigung*, **36**, 1–55; **37**, 233–311. [Repr. as suppl. vol. 6, 1930, Leipzig: Teubner. An important report on class field theory.]

Hilbert, D. *1897*, 'Die Theorie der algebraischen Zahlkörper', *Jahresbericht der Deutschen Mathematiker-Vereinigung*, **4**, 175–546. [Also in *Gesammelte Abhandlungen*, Vol. 1, 63–363. Influential report on algebraic number theory.]

Ireland, K. and Rosen, M. *1982*, *A Classical Introduction to Modern Number Theory*, Heidelberg: Springer. [2nd edn, 1990. An excellent modern textbook containing many historical comments.]

Jacobi, C. G. *1829*, *Fundamenta nova theoriae functionum ellipticarum*, Königsberg: Bornträger. [Also in *Gesammelte Werke*, Vol. 1, 49–239.]

Klein, F. *1926*, *Vorlesungen über die Entwicklung der Mathematik im 19. Jahrhundert*, Vol. 1, Berlin: Springer. [Repr. 1967, New York: Chelsea.]

Legendre, A. M. *1798*, *Essai sur la théorie des nombres*, Paris: Duprat. [Repr. 1955, Paris: Blanchard.]

Ore, O. *1948*, *Number Theory and its History*, New York: McGraw Hill. [Covers only elementary number theory.]

Riemann, B. *1859*, 'Über die Anzahl der Primzahlen unter einer gegebenen Grösse', *Monatsberichte der Berliner Akademie der Wissenschaften*, 671–80. [Also in *Gesammelte mathematische Werke*, 2nd edn, 1892, 245–53.]

Smith, H. J. S. *1860–66*, 'Report on the theory of numbers', *Report of the British Association for the Advancement of Science*, (1859–1865). [In six parts (nothing for 1864). Also in *Collected Mathematical Papers*, Vol. 1, 38–364. Repr. as *Number theory*, 1965, New York: Chelsea. On early algebraic number theory.]

Weber, H. *1898*, *Lehrbuch der Algebra*, 2nd edn, Vol. 1, Braunschweig: Vieweg. [An influential classic.]

Weil, A. *1984*, *Number Theory. An Approach through History. From Hammurapi to Legendre*, Basel: Birkhäuser. [An excellent history of elementary number theory.]

6.11

Linear optimization

SONJA BRENTJES

This article describes the growth of linear optimization up to the 1950s, to the point where it had achieved the status of an independent mathematical discipline. The emphasis is on the mathematical aspects; other sources and applications are mentioned only as a part of the appropriate historical context. An exception is Section 3, devoted to a short treatment of the influence of economics, which has had a special significance.

The prehistory of linear optimization can be traced back along a profusion of isolated lines into quite different scientific areas. Thus neither completeness nor a comprehensive history of the concept of optimality can be the goal of this article.

1 EARLY TRACES OF MATHEMATICAL OPTIMALITY, 1749–1840

The earliest known optimality concepts originated as central questions in geodesy and cartography (Leonhard Euler 1749, Roger Boscovich and C. Maire 1770, Pierre Simon Laplace 1799), in astronomy (Euler 1778, Laplace 1799) and in analytical mechanics (Joseph Fourier between 1798 and 1831, Claude Navier 1825, Antoine Augustin Cournot 1826–32 and Mikhail Ostrogradsky 1838). These made fruitful demands on error analysis (§10.5), estimation (§10.4) and the calculus of variations (§3.5).

By considering the problem broadly in the general form for n equations or inequalities and m variables, Augustin Louis Cauchy (between 1811 and 1831) and Fourier (posthumously in 1831, for example) had already gone far beyond the subject's origins. Up to the middle of the nineteenth century, optimization problems can be found also in other areas; for example, in context of hydraulics (Gaspard de Prony, 1804). The object studied was an over-determined system of linear equations or inequalities under the restriction of minimizing the finite sum of several variables, or of determining the minimum value of the maximum of these variables. Boscovich, de Prony and Fourier solved these geometrically; Laplace, Cauchy, Ostrogradsky and

Navier developed analytic methods. A hint of the distinction between basis and non-basis variables appears to have been made by Carl Friedrich Gauss. Fourier, who probably had the deepest insight into the structure for a general linear optimization problem and its possible applications, and to a more limited extent Cauchy, anticipated the vertex theorem of linear optimization (Farebrother *1987*, Grattan-Guinness *1993*, Sheynin *1972*).

2 THE KEY THEOREMS OF LINEAR OPTIMIZATION: FROM GORDAN TO WEYL

Between 1840 and the 1930s these topics were carried further by various authors (e.g. the Englishman Francis Edgeworth in 1888, the Hungarian J. Farkas between 1895 and 1826, the Belgian Charles Jean De La Vallée Poussin in 1911) or extended into other areas (e.g. George Boole, in probability theory, in 1854). The most important of these contributions are those of Farkas, continuing the investigations on the principle of virtual work by Fourier and Ostrogradsky (Brentjes *1985*, Prékopa *1980*). He proved among other things the lemma named after him on homogeneous linear systems of inequalities, which stands as one of the key theorems in optimization. Farkas, like Fourier, freed himself from the original problems and studied linear systems of inequalities in their own right.

Furthermore, after 1840 elements of linear optimization cropped up repeatedly within mathematics itself: in the theory of systems of linear equations and inequalities (§6.7), the theory of graphs and matrices (§7.13), geometrical and analytical number theory (§6.10), game theory, and the theory of convex sets. In 1873 Paul Gordan presented a transposition theorem in the course of investigating linear systems of equations, which could replace the Farkas lemma (Focke and Göpfert *1973*). Other theorems of this sort occur also in the geometrical number theory of Hermann Minkowski in 1896, in the dissertations on linear systems of equations by E. Stiemke in 1915 and T. Motzkin in 1936, in Hermann Weyl's treatise *1935* on convex polyhedra, in works on game theory and probability theory by the French mathematician J. Ville in 1938, and in the text by Oskar Morgenstern and John von Neumann *1944* on game theory and economics (see Section 3). These theorems were, as a rule, formulated for homogeneous systems. The Hungarian A. Haar in 1924 carried over the results of Farkas and Minkowski to inhomogeneous systems. Graph- and matrix-theoretic theorems by the Hungarian mathematicians Julius König in 1916 and E. Egerváry in 1931 formed the basis of the so-called 'Hungarian method', which H. W. Kuhn developed in 1955 to solve the assignment problem.

Farkas, Minkowski, Motzkin and Weyl gave theorems of different degrees of generality concerning the finite dimension of the solution cone of systems of inequalities, which leads to the concept of a feasible solution basis. In 1908 the Russian mathematician G. F. Voronoy analysed quadratic forms with a view to finding the dimension of such solution cones. Von Neumann's 1928 work on game theory prepared the foundations for the duality theory of optimization.

The first works on convex sets arose towards the end of the nineteenth century, in part in direct connection with linear systems of inequalities (e.g. H. Brunn in 1887 and 1889, Minkowski from 1896, C. Carathéodory in 1911, L. Féjer in 1911, and E. Schmidt in 1913). The use of properties of convex figures in the proof of a deep number-theoretic theorem was characterized by David Hilbert in 1911 as 'a pearl of the Minkowski art of discovery' (Minkowski *1911*: Vol. 2, p. x). Minkowski introduced such concepts as support plane, extreme support plane, extreme half-space, and extreme point or vertex (*1911*: 106, 136, 147, 157, 163). Other important results were the so-called 'separation theorems' (Minkowski, posthumously in 1911; P. Kirchberger in 1903; E. Steinitz in 1913).

3 ECONOMICS AND LINEAR OPTIMIZATION

Qualitative optimality considerations have permeated most economic theories since A. Smith and D. Ricardo in the late eighteenth century. A concern for quantifying optimality, however, remained the exception for a long time. Up to at least the 1830s there was hardly any systematic cooperation between economics and mathematics (§10.18). Changes in this relationship began with the discussions of price-theoretic models of German, Danish and Austrian economies which came from the mathematical school of marginal-utility theory of L. Walras and V. Pareto in the last third of the nineteenth century and took place later, for example, in the Mathematical Colloquium of the University of Vienna. An outcome was the introduction of inequalities and non-negative conditions for the variables. The modified system was solved by A. von Wald between 1934 and 1936 and characterized by Kuhn in 1956, using deep methods (the fixed-point theorem of Kakutani), as the solution of a dual linear-optimization problem.

Morgenstern and von Neumann's *Theory of Games and Economic Behavior* (*1944*) is the first prominent product of this new orientation. The rapid development of linear optimization in the USA was made possible by the fundamentally changed relationship between economists and mathematicians represented by this work.

A number of developments directly influenced the growth of linear optimization in the USA. They included: the model of an expanding economy by von Neumann from 1932; the input–output model for the

American economy of the 1930s by W. Leontief, whose ideas go back to the 'physiocratic' theories of the eighteenth century (with the 'economic tableaux' from François Quesnay), and to the attempts of the Central Statistical Office of the USSR of 1925 to produce an equilibrium model; the efforts of A. Bergson in 1938 and O. Lange in 1942 to express the so-called 'social welfare optimization of welfare economics' by means of a function which was connected through the Pareto optimality criterion with the mathematical school of marginal utility theory; and finally G. Stigler's model of a minimal-cost food table with certain nutritional conditions, and T. C. Koopmans's model for the minimization of travel distances of empty transport ships in the Second World War so as to keep the danger of being torpedoed by German U-boats as small as possible. Independently of these developments, L. Hitchcock solved the classical transportation optimization problem in 1941 (Koopmans *1951*: 1–14; Dantzig *1963*).

In the USSR the mathematization of economics took place in the 1920s (Kantorovich *1985*). After 1939, L. V. Kantorovich worked closely with the economist V. V. Novozhilov, whose principal interests lay in investment theory and planning, and after the Second World War with V. S. Nemchinov. This cooperation made it possible to overcome the discrimination against the application of mathematical methods in political economics which arose in the 1930s. This cooperation reached its maturity in 1959 with an important conference on the topic. At the end of the 1920s and into the 1930s treatises appeared, mainly by A. N. Tolstoy, on transportation networks in connection with the planning of railway traffic. The rational plan was regarded as the one with the minimal total length of track; graphical as well as algorithmic methods were proposed, and the mathematics was first worked out by L. V. Kantorovich and M. K. Gavurin (Brentjes *1985*: 315–20).

Locational optimization is a sub-area of optimization, or, in general, of non-linear optimization (Franksen and Grattan-Guinness *1989*). In the first third of the nineteenth century, the Mecklenburg landowner J. H. von Thünen attempted to solve the location problem. Later, in 1882, W. Launhardt sought the most favourable location of a factory whose raw materials were obtained from two places and its products sold at a third. Then Weber *1909* formulated a theory which identified the locations with the least transportation costs. Both Launhardt and Weber had recourse to solutions to isoperimetric problems (§3.5) by P. Fermat and J. Steiner.

4 SOVIET CONTRIBUTIONS TO THE DEVELOPMENT OF LINEAR OPTIMIZATION, 1938–1951

The starting-point for Soviet work on linear optimization was assignment problems posed by the Leningrad (St Petersburg) timber trust at the beginning of 1938; it sought the help of the Mathematical–Mechanical Faculty of the University of Leningrad. The 27-year-old Kantorovich, who undertook the contract, was the most important representative of this development. In 1938 and 1939 he studied further concrete technical–economic problems and uncovered their common mathematical structure. Both his methods of solution and theoretical foundations benefited greatly from discussions in G. M. Fichtenholz's research seminar on functional analysis and from his own works in that field. Thanks to his assiduous lecture activity in 1938–9, Kantorovich raised the interest of Leningrad economists and engineers in the new class of tasks. He described his experiences in the 1939 monograph 'On Mathematical Organization and Planning'.

An example of his methods is the following use of resolving multipliers in a distribution problem (*1939*: 35*f*): Numbers h_{ik} are sought such that

$$h_{ik} \geqslant 0, \quad i = 1, \ldots, n, \quad k = 1, \ldots, m \quad \text{and} \quad \sum_{k=1}^{m} h_{ik} = 1, \tag{1}$$

where $z_k := \Sigma_i a_{ik} h_{ik}$ satisfy

$$z' := \min(z_1, \ldots, z_m) \rightarrow \max. \tag{2}$$

The resolving multipliers, which are essentially solutions of the dual problems, are numbers $\lambda_j \geqslant 0$, $j = 1, \ldots, m$. Considering $\lambda_1 a_{i1}, \ldots, \lambda_m a_{im}$ for each i, one puts h_{ik} equal to zero when $\lambda_k a_{ik} < \max_k \lambda_k a_{ik}$, and obtains the remaining h_{ik} from the above conditions. In this connection Kantorovich included a separation theorem as a theoretical foundation for the existence proof of the λ_j, providing details of solution conditions, the relation between the variables and the resolving multipliers, and considerations which apply to the vertex and basis theorems.

In several works between 1939 and 1959, Kantorovich investigated additional concrete problems from different branches of industry, and, with students such as M. K. Gavurin, V. A. Zalgaller, A. L. Lurie and G. S. Rubinstein, developed procedures and descriptions of such methods as discrete and continuous transportation optimization and vectorial optimization, and developed new methods for the solution of general linear-optimization problems. In 1959 a large monograph brought together the research done up to that date, including material from the 1940s which could not be published at the time (Brentjes *1985*: 314–17).

In the USA the translation of two works by Kantorovich (*1939, 1942*) at the beginning of the 1960s sparked a strenuous priority debate concerning the development of linear optimization which has flared up again in the 1980s (Koopmans *1960*, Charnes and Cooper *1962*, Dorfman *1984*, Gass *1989*, Schwartz *1989*). Disregarding the differences of opinion between the authors, it is clear that, generally speaking, they were acquainted with only a few of the Soviet works.

5 THE DEVELOPMENT OF LINEAR OPTIMIZATION IN THE UNITED STATES OF AMERICA, 1945–1949

In 1946 M. K. Wood proposed that the US Air Force mechanize their planning; G. B. Dantzig could be enlisted for this since he was at the time employed in the air corps, and had just obtained his doctorate at Berkeley. The decisive years for the development of linear optimization were 1947–9. In 1947 a generalized Leontief schema was put forward as the first model, and various solution methods were fashioned from the cooperative work of, for example, Hurwicz, Koopmans and von Neumann. The concept of a linear goal-function was introduced in spite of substantial initial difficulties. Further, in collaboration with von Neumann, Dantzig established the connection with game theory and put forward the first duality theorems; in 1948 he gave the first public lecture on surface optimization, and A. W. Tucker conducted seminars on game theory.

Progress thereafter was extremely rapid. Various private and governmental research organizations conducted scientific colloquia, symposia and seminars on linear optimization and on game theory. These included the RAND Corporation (founded in 1946, mainly at the instigation of the Air Force), the National Bureau of Standards (NBS), and the US Department of Agriculture.

On 27 December 1948 Wood and Dantzig gave a lecture before the Econometric Society of the USA on the ground-breaking results of SCOOP (Scientific Programming of Optimum Programs) of the Planning Research Division (founded in 1946–7) of the Directorate of Program Standards and Cost Control of the US Air Force. The lecture, published in 1949 in reworked form, was the first thorough presentation of linear optimization. A practical problem for the generalized Leontief model was the organization of an airlift from the USA to the blockaded Western sector of Berlin. The goal-function minimized the costs of all activities over all time units. The variables were to be non-negative, and four equations were to be satisfied as side conditions. In his lecture on the mathematical model, Dantzig introduced such important concepts as admissible program, linear program, goal-function and optimal admissible program. In the published version he referred to the equivalence with the two-person zero-sum game, to the possibility of degenerate solutions, and to the putting together of the solution vector from as large a number as possible of null components. The solution methods used were not published. It was not until 1949 that the first large-scale transport problem was solved through the simplex method which

Dantzig had developed in 1947, but had rejected as ineffective. The method had been tested beforehand on electronic calculators at the NBS in 1948, it being applied in particular to Stigler's problem of diets (Section 3).

In 1947 the US Air Force assigned $400,000 to the NBS for the development of computers which formed the technical support for the rapid development of linear optimization and the success of the simplex method over other algorithms and approaches. There are two further factors in the speedy development of linear optimization: the broad scale of research establishments – both inside and outside universities – which were given financial and institutional backing by governmental bodies, and the wide-ranging interdisciplinary network, established especially by Koopmans, Dantzig, von Neumann and Tucker, which brought together economists, mathematicians and professional planners.

The high point of 1949, and the conclusion of the development phase of linear optimization in the USA, was the Chicago conference in June, organized by Koopmans, on 'Activity analysis of production and allocation' (Koopmans *1951*). After a historical introduction by Koopmans, the conference proceedings contains 26 works under the broad themes: theory of programming and allocation, applications of allocation models, mathematical properties of convex sets and problems of computation. In the first group there are contributions which, in effect, generalize the Leontiev and von Neumann models. The second group consists of examples of applications; the third contains articles on convex conics and on the relationship between linear optimization and game theory. The fourth group presents methods of solution for linear optimization problems: the simplex method applied to the transportation problem and to a game, an iterative method from game theory, and a cross-sectional method. After the methods, the most important mathematical results include the existence and duality theorems in the Gale, Kuhn and Tucker paper and in Dantzig's paper, and Dantzig's basis and vertex theorem. Beyond this, a second clear direction was indicated by the conference: the development of non-linear optimization (Kuhn and Tucker, 1951 – see Grattan-Guinness *1993*: Section 6).

6 ASPECTS OF THE DEVELOPMENT OF LINEAR OPTIMIZATION IN THE 1950s

The greatest part of 1950s research took place in the USA, and a substantial part of that was carried out by the RAND Corporation. In 1950, the US Navy began to give considerable funding to research in optimization and game theory. Significant contracts include, for example, the arrangement with the Carnegie Institute of Technology in the area of planning ($35,000 per year, 1952–63), a project which had been conducted by the Air Force for three years since 1950 (with financing of $90,000). W. W. Cooper directed the project and had among his co-workers C. Charnes, A. Henderson and B. Mellon. Cooper, Charnes and Henderson produced the first textbook on linear optimization in 1953.

Further results appeared in 1952, including the application of linear optimization methods to the problems of the Gulf Oil Company, the development of dual simplex methods in 1953 by C. E. Lemke, and the stepping-stone methods for transportation problems in 1954 by Charnes and Cooper. Further conferences on linear optimization, game theory, and linear systems of inequalities took place, now with some participation from Western Europe. There were investigations of stochastic, dynamical, parametric and integral problems (Dantzig *1963*). A significant breakthrough was the cooperation between optimization, political economics and the theory of domestic economies. In his 1951 dissertation, Dorfman applied linear-optimization methods to management theory. In 1956 a volume appeared from the RAND Corporation, written by Dorfman, Samuelson and Solow, on linear optimization and economics.

In the second half of the 1950s, optimization reached Western Europe and several Eastern European countries. At the beginning of the 1960s, after the Soviet conference of 1959 (Nemchinov *1959*), the new discipline was taken up with an intense interest. With the encouragement of the United Nations, economists began to introduce linear optimization methods into the economic planning for developing countries such as India.

In 1975 the importance of the field was acclaimed on an international scale. Koopmans and Kantorovich were jointly awarded a Nobel prize in economics for their contributions to the relationship between economics and linear optimization.

BIBLIOGRAPHY

Brentjes, S. *1985*, 'Zur Herausbildung der linearen Optimierung', in W. Lassmann and H. Schilar (eds), *Ökonomie und Optimierung*, Berlin: Akademie Verlag, 298–330.

Charnes, A. and Cooper, W. W. *1962*, 'On some works of Kantorovich, Koopmans and others', *Management Science*, **8**, 246–63.

Dantzig, G. B. *1963*, *Linear Programming and Extensions*, Princeton, NJ: Princeton University Press. [German transl.: 1966, Berlin and New York: Springer.]

Dorfman, R. *1984*, 'The discovery of linear programming', *Annals of the History of Computing*, **6**, 283–95.

Farebrother, R. W. *1987*, 'The historical development of the L_1 and L_∞ estimation procedures, 1793–1930', in *Statistical Data Analysis Bond on the L_1 Norm*, Amsterdam: North-Holland, 37–63.

Focke, J. and Göpfert, A. *1973*, '100 Jahre Gordanscher Alternativsatz für lineare Ungleichungen', *Mathematische Operationsforschung und Statistik*, **6**, 673–80.

Franksen, O. I. and Grattan-Guinness, I. *1989*, 'The earliest contribution to location theory? Spatio-economic equilibrium with Lamé and Clapeyron, 1829', *Mathematics and Computers in Simulation*, **31**, 195–220.

Gass, S. I. *1989*, 'Comments on the history of linear programming', *Annals of the History of Computing*, **11**, 147–51.

Grattan-Guinness, I. *1993*, ' "A new type of question": On the prehistory of linear and non-linear programming, 1770–1940', in E. Knobloch and D. Rowe (eds), *History of Modern Mathematics*, Vol. 3, New York: Academic Press, to appear.

Kantorovich, L. V. *1939*, *Mathematicheskie metody organizachii i planirovaniya proizvodstva*, Leningrad: University. [English transl.: *Management Science*, 1960, **6**, 366–422. German transl.: in transl. of Nemchinov *1959*: 245–385.]

—— *1942*, 'O peremeshchenii mass', *Doklady Akademii Nauk SSSR*, **37**, 227–9. [English transl.: *Management Science*, 1958, **5**, 1–4.]

—— *1985*, 'Die Entwicklung von Optimierungsmethoden in der UdSSR', in W. Lassmann and H. Schilar (eds), *Ökonomie und Optimierung*, Berlin: Akademie Verlag, 11–96.

Koopmans, T. C. (ed.) *1951*, *Activity Analysis of Production and Allocation*, New York: Wiley.

—— *1960*, 'A note about Kantorovich's paper "Mathematical methods of organizing and planning production" ', *Management Science*, **6**, 363–5.

Minkowski, H. *1911*, *Gesammelte Abhandlungen* (ed. D. Hilbert), 2 vols, Leipzig and Berlin: Teubner.

Morgenstern, O. and von Neumann, J. *1944*, *Theory of Games and Economic Behavior*, 1st edn, Princeton, NJ: Princeton University Press. [German transl.: 1961, Würzburg: Physica.]

Nemchinov, V. S. *1959*, *Primenenie matematiki v ekonomicheskich issledovaniyakh*, Moscow: Izdatel'stvo Social'no-ekonomicheskoi Literaturi. [German transl.: 1963, Leipzig: Die Wirtschaft. English transl.: 1964, Edinburgh: Oliver & Boyd; 1965, Cambridge, MA: MIT Press.]

Prékopa, A. *1980*, 'On the development of optimization theory', *American Mathematical Monthly*, **87**, 527–42.

Schwartz, B. L. *1989*, 'The invention of linear programming', *Annals of the History of Computing*, **11**, 145–7.

Sheynin, O. B. *1972*, 'On the mathematical treatment of observations by L. Euler', *Archive for History of Exact Sciences*, **9**, 45–56.

Tucker, A. W. (ed.) *1955*, *Game Theory and Programming*, Stillwater, OK: Agricultural and Mechanical College.

Weber, A. *1909*, *Über den Standort der Industrien*, Tübingen: Mohr. [English transl.: 1929, Chicago, IL: University of Chicago Press.]

Weyl, H. *1935*, 'Elementare Theorie der konvexen Polyeder', *Commentarii mathematicae Helveticae*, **7**, 290–306. [English transl.: in *Contribution to the Theory of Games*, 1950, **1**, 3–19.]

Wood, M. K. *1949*, 'Scientific techniques for program planning', *Air University Quarterly Review*, **3**, 49–65.

6.12

Operational research

ROBIN E. RIDER

The term 'operational research' (or 'operations research', in its American version) was coined early in the Second World War to refer to 'the application of the scientific method to finding the most economical and timely means for getting the maximum military effect from the available or potentially available resources in materiél [sic] or personnel' (General Research Office *1948*: 8, quoting a Canadian definition). Operations research (OR), as distinct from research at the laboratory bench, was closely connected with actual military operations; its aim was to provide those in authority with a quantitative basis for their decision-making (Kittel *1947*: 150). By the construction of mathematical, economic and statistical models for situations characterized by complexity and uncertainty, operations research seeks to analyse probable future consequences of decision choices, in order to guide the making of those choices.

1 WARTIME

The story of OR begins in Britain before the onset of the Second World War. Civilian scientists working closely with the Coastal Command had begun to identify weaknesses in the new radar network, and to recommend changes in operator techniques and system design. After war was declared, the work of this group was supplemented by the analyses of an OR group based at the Anti-Aircraft Command and headed by Manchester physicist Patrick Blackett.

The success of OR convinced other military leaders that it might prove useful elsewhere in the armed surfaces, and Blackett and his colleagues, who hailed from the physical and life sciences and other disciplines, found themselves much in demand. 'Blackett's circus' and other British OR sections were instrumental, for example, in protecting supply routes and securing defences in the Battle of Britain. Two reports Blackett drafted in 1941 concerning the organization and methodology of OR ('Scientists at the operational level' and 'A note on certain aspects of the methodology of operational research', both reprinted in Blackett *1948* and *1962*) enjoyed broad circulation – far beyond Blackett's original intent – both in Britain and elsewhere. Wartime memoranda now in the US National Archives, for example, refer to Blackett's accounts; whole sections reappear, without attribution, in American notes and reports on this new weapon in the scientific arsenal.

In the memoranda Blackett argued for 'numerical thinking on operational matters' as a way to 'avoid running the war by gusts of emotion' (*1962*: 173). He stressed the importance of Poisson distributions in setting up and solving problems, tried to decide in quantified terms the proper balance between acquiring new weapons and using existing weapons to best effect, and explained the need for a close relationship between OR scientists and the command structure (see the analysis of the memoranda in Lovell *1975*). One critic commented, upon hearing one of Blackett's reports, 'C'est magnifique, mais ce n'est pas la guerre' (quoted in Lovell *1975*: 63).

Blackett distinguished between two general methods of optimizing the performance of personnel and material under combat conditions: the *a priori* method, in which 'certain important variables ... particularly suitable for quantitative treatment', are selected, differential equations formed and solutions derived; and the variational method, which consists of finding, 'both by experimental and by analytical methods, how a real operation would be altered if certain of the variables, e.g. the tactics employed or properties of the weapons used, were varied' (Blackett *1962*: 179, 180). In more formal terms, if the yield Y of an operation of war is a function F of many operational parameters or variables, $X_1, \ldots, X_n$, the task is to 'investigate the shape of the multi-dimensional surface

$$Y = F(X_1, \ldots, X_n) \tag{1}$$

surrounding the point corresponding to a past operation, and to use this knowledge to predict the properties at a neighbouring point corresponding to a future operation' (Blackett *1962*: 182). Finding the derivatives requires the application of probability and statistics, a thorough appreciation of operational realities, a willingness to manipulate very rough data, and imagination.

The penchant for quantification allowed OR teams to buttress their claims for the importance of OR, and each success helped to secure the place of OR in a given command, while lending credence to the call for new OR sections elsewhere. Pre-existing connections between British, Canadian and American scientific personnel brought reports of British OR successes to North American attention, and prompted the establishment of American OR groups in the various services.

As OR began to prove itself in the US military, American OR groups soon found themselves plagued by the same personnel shortages that had led to earlier British appeals for Americans to learn OR at the source (and thereby fill out understaffed British OR teams). MIT physicist Philip Morse, for example, had to raid other wartime science and engineering projects to put together an OR team under a US National Defense Research Committee (NDRC) contract; the group trained pilots in the effective use of radar equipment, devised efficient geometric search strategies for finding surfaced submarines, and proposed new settings to improve the accuracy of depth charges.

2 POST-WAR APPLICATIONS

A meritorious war record, combined with the increasing complexity of modern warfare and weapons development, helped to ensure a continuing mandate for military OR after demobilization. After the war OR was also seen as a valuable tool for the reconstruction of Britain and the planning of its economy, and OR units proliferated in the British government, industrial research associations, and basic industries and utilities.

In the USA, OR began to take root during the 1940s in academe as well as in military contexts. Units like the Navy's Operations Evaluation Group at MIT and the Army's Operations Research Office at Johns Hopkins University operated for the benefit of the military services, with organizational structures that owed much to the wartime precedent of government contracts with universities and with veterans of wartime OR units on their staffs.

3 INFRASTRUCTURE

In 1948 British advocates and practitioners of OR founded an informal Operational Research Club and launched a Club publication, the *Operational Research Quarterly*, which combined abstracts of relevant articles from a wide range of journals with short non-technical articles and news of interest to the OR community.

Word of OR spread as those with wartime OR experience addressed colleagues in professional societies in various scientific fields on its virtues. The US National Research Council convened a committee on OR in 1949; the committee assembled a list of 700 persons with a professional interest in OR and issued a report in praise of OR work. The nagging problem of manpower for OR was addressed in part by establishing special courses and eventually degree courses in OR at American universities. (There were those like Blackett, however, who opposed the creation of a new academic discipline of OR.) A formal professional organization, the Operations

Research Society of America was established in 1952, with Morse as first president. Its objectives were to establish and maintain professional standards of competence, to improve methods and techniques of OR, and to encourage students. Its *Journal* began to appear in 1952; its annual meetings attracted hundreds of participants; by May 1953 there were 560 names on the ORSA membership rolls. At the same time bibliographies and compilations of abstracts addressed the problem of assembling a growing body of literature scattered among numerous journals with different readerships. A bibliography of OR compiled by James H. Batchelor and first published by 1952 occupied 95 pages; a second edition (Saint Louis University Press, 1959), covering published literature to the end of 1957, contained 4195 items and 865 pages; a subsequent volume covering 1958–60 added several thousand more entries.

A survey of several influential OR texts (written over the course of some two decades) illustrates changing contours of the field. Morse and Kimball's *Methods of Operations Research* was first published in classified form shortly after the end of the Second World War; after declassification (much delayed) it was published by MIT Press and Wiley in 1951 and had reached its eighth printing in 1962. The book, which served as Morse and Kimball's final report of wartime OR activities, was intended in part to encourage the use of OR in industry and non-military government contexts as well as the military. The work emphasized examples drawn from wartime experience (and sanitized as necessary). The examples were organized into sections on 'measures of effectiveness' (sweep rates, ratios of enemy loss to friendly loss, comparative effectiveness, evaluation of equipment performance); strategic kinematics, including mathematical theory for applying the constants of warfare (Lanchester equations, reaction-rate problems); tactical analysis, including statistical solutions and evaluation of measure and countermeasure; evaluation of weapons and their use; operational experiments of equipment and tactics; and organizational and procedural matters.

Churchman, Ackoff and Arnoff's *Introduction to Operations Research* (*1957*) was more obviously intended as a textbook, and indeed incorporated material presented in the short course on OR offered annually at Case Institute of Technology in Cleveland from 1952. The text opened with comments on the nature of OR. It offered guidance on stating a problem and setting up a model, and discussed a variety of models – inventory, allocation (including linear programming), queuing and sequencing, replacement and competitive (i.e. game theory). It concluded with OR practicalities about data, control, implementation, and administration. The latter part of the text was announced as requiring elementary calculus, and inequalities and matrices were invoked elsewhere; the presentation involved less mathematical notation and manipulation than did Morse and Kimball's book.

A decade later textbook authors viewed OR as 'now stabilizing sufficiently to permit a comprehensive survey of its basic methodology and techniques' (Hillier and Lieberman *1967*: vii). Their *Introduction to Operations Research* thus placed more emphasis on models and techniques than on applications (which are mainly relegated to problems for each chapter), which had a considerable mathematical content. After the requisite general introduction to the field and the fundamentals of probability theory, statistical inference and decision theory, the text divided techniques into mathematical programming (linear programming, program-evaluation-and-review technique (PERT), network analysis, dynamical programming, game theory), probabilistic models (queuing theory, inventory theory, Markov chains, simulation), and advanced topics in linear and non-linear programming and integer programming.

4 CONNECTIONS

Operational research has many precursors and allied fields, including Taylorism (after Frederick W. Taylor), scientific management and management science, industrial engineering and systems analysis. As one early textbook explained, the roots of OR 'are as old as science and the management function. Its name dates back only to 1940' (Churchman *et al. 1957*: 3). Certainly its practitioners have expended much energy and ink in search of an acceptable definition of OR. Morse tried unsuccessfully to halt the debate by declaring OR to be 'the activity carried on by members of the Operations Research Society' (Morse *1953*: 159). But his colleagues were not so easily dissuaded from debate. Much of the concern with definition focused on the sometimes elusive distinctions between OR and neighbouring fields; the attempt to define, or redefine, OR was also born of the desire to allow the subject to evolve beyond the orthodoxy of wartime experience. Crucial considerations included the balance between model and application, and the complexity of the mathematics invoked.

Many of the mathematical models, tools, and techniques used in OR have long and complex histories. Early works on the use of linear programming outlined its prehistory (Dantzig *1963*: Chap. 2; Dantzig *1975*; §6.11); precursors of other OR models can likewise be sought in earlier work on waiting-line theory, inventory and assignment problems, probability theory and network problems.

The application of OR to large and complex problems was made more feasible by the advent of high-speed computers like the IBM 701 and 704, and the Remington-Rand UNIVAC (§5.12). Practical use of linear programming in the American oil industry in the mid- to late 1950s, for example, was demonstrably dependent on the commercial availability of mainframe computers (Dantzig *1963*: 26; Garvin *et al. 1957*). Increases in computer power and speed and decreases in size have facilitated applications of OR, if not always a full appreciation of the theory underlying the algorithms.

BIBLIOGRAPHY

Blackett, P. M. S. *1948*, 'Operational research', *Advancement of Science*, **5**, (17), 26–38.

—— *1962*, *Studies of War. Nuclear and Conventional*, New York: Hill & Wang.

Churchman, C. W., Ackoff, R. L. and Arnoff, E. L. *1957*, *Introduction to Operations Research*, New York: Wiley.

Dantzig, G. B. *1957*, 'Concepts, origins, and use of linear programming', in *Proceedings of the First International Conference on Operational Research*, Oxford: Pergamon, 100–108.

—— *1963*, *Linear Programming and Extensions*, Princeton, NJ: Princeton University Press.

Garvin, W. W. *et al. 1957*, 'Applications of linear programming in the oil industry', *Management Science*, **3**, 407–30.

General Research Office, Johns Hopkins University *1948*, *Quarterly Report*, **1**, (1).

Hillier, F. S. and Lieberman, G. J. *1967*, *Introduction to Operations Research*, San Francisco, CA: Holden-Day. [3rd edn 1980.]

Kittel, C. *1947*, 'The nature and development of operations research', *Science*, **105**, 150–53.

Lovell, B. *1975*, 'Patrick Maynard Stuart Blackett, Baron Blackett, of Chelsea', *Biographical Memoirs of Fellows of the Royal Society of London*, **21**, 1–115.

McArthur, C. W. *1990*, *Operations Analysis in the U.S. Army Eighth Air Force in World War II*, Providence, RI: American Mathematical Society.

Morse, P. M. *1953*, 'Trends in operations research', *Journal of the Operations Research Society of America*, **1**, (4), 159–65.

Morse, P. M. and Kimball, G. E. *1951*, *Methods of Operations Research*, 1st edn (rev.), Cambridge, MA: MIT Press, and New York: Wiley.